APPLIED PHYSICS

APPLIED PHYSICS

Second Edition

DALE EWEN
Parkland Community College

RONALD J. NELSON

NEILL SCHURTER

Prentice Hall
Upper Saddle River, New Jersey Columbus, Ohio

Library of Congress Cataloging-in-Publication Data

Ewen, Dale
Applied physics / Dale Ewen, Ronald J. Nelson, Neill Schurter. —
2nd ed.
p. cm.
Includes index.
ISBN 0-13-096213-9
1. Physics. I. Nelson, Ronald J. II. Schurter, Neill.
III. Title.
QC23.E88 1999
530—dc21 98-18857
CIP

Cover photo: © Ken Whitmore/Tony Stone Images
Editor: Stephen Helba
Production Supervision: York Production Services
Design Coordinator: Karrie M. Converse
Cover Designer: Ceri Fitzgerald
Production Manager: Deidra M. Schwartz
Illustrations: York Production Services/Scientific Illustrators
Marketing Manager: Frank Mortimer, Jr.

This book was set in Times Roman by York Graphic Services, Inc. and was printed and bound by Courier/Kendallville, Inc. The cover was printed by Phoenix Color Corp.

Photo credits: Anne Vega, pp. 1, 45, 145, 371; Anne Vega/Prentice Hall, p. 290; © International Speedway Corp. All rights reserved, p. 61; Bryan Barr/USAir, p. 114; COSI, Ohio's Center of Science & Industry, p. 165; Joshua Sheldon, pp. 436, 448, 516; Cedar Point Photo by Dan Feicht, p. 218; Todd Yarrington/Prentice Hall, p. 250; The Timken Company, p. 263; Fluke Corporation, p. 400; American Electric Power, p. 466; Riverside Methodist Hospitals, a U.S. Health Affiliate, p. 539; © 1995 Nick B. Wood, Fig. 9.12; U.S. Department of Energy, Fig. 12.4; AP/WORLD WIDE PHOTOS, Fig. 13.14; © John Maher/Stock, Boston, Fig. 13.15; © Tony Freeman/PhotoEdit, Fig. 13.20; Courtesy of Bethlehem Steel Corporation, Fig. 13.24; Courtesy of Burnham Corporation, Fig. 13.26; p. 391; © (Richard Megna) Fundamental Photographs, NYC, p. 502; Ron Thomas/FPG International LLC., Figs. 11.15, 11.18, 22.17; © (Richard Megna) Fundamental Photographs, NYC, Fig. 13.19; © Leonard Lessin/Peter Arnold, Inc., Fig. 15.9; Grant Heilman Photography, Inc., Fig. 16.11; David Baumhefner, National Center for Atmospheric Research/University Corp. for Atmospheric Research/National Science Foundation; The MIT Museum, Fig. 16.8; © Tom Pantages, Figs. 16.10, 17.13; Reproduced with Permission, Fluke Corporation, Fig. 17.46; Rayovac Corporation, Fig. 18.3; © Stacey Pick/Stock, Boston, Fig. 21.1; © Berenice Abbot/Photo Researchers, Inc., Fig. 22.24.

Printed in the United States of America

10 9 8 7 6 5 4 3 2

ISBN: 0-13-096213-9

Prentice-Hall International (UK) Limited, *London*
Prentice-Hall of Australia Pty. Limited, *Sydney*
Prentice-Hall of Canada, Inc., *Toronto*
Prentice-Hall Hispanoamericana, S. A., *Mexico*
Prentice-Hall of India Private Limited, *New Delhi*
Prentice-Hall of Japan, Inc., *Tokyo*
Simon & Schuster Asia Pte. Ltd., *Singapore*
Editora Prentice-Hall do Brasil, Ltda., *Rio de Janeiro*

Contents

PREFACE

Applied Physics, Second Edition, provides comprehensive and practical coverage of physics for students considering a vocational-technical career. It emphasizes physical concepts as applied to industrial-technical fields and uses applications to improve the physics and mathematics competence of the student. This second edition has been carefully reviewed, and special efforts have been taken to emphasize clarity and accuracy of presentation.

This text is divided into five major areas: mechanics, matter and heat, wave motion and sound, electricity and magnetism, and light and modern physics.

Key Features

- Unique, highly successful problem-solving method
- Chapter introductions
- Chapter objectives
- Chapter glossaries
- Formulas summarized at the end of each chapter
- Chapter review questions
- Chapter review problems
- Problem sets that follow each problem-related section
- Important formulas and principles that are highlighted
- Abundant illustrations
- Numerous examples consistently displayed using the problem-solving format
- Two-color format that effectively highlights and illustrates important principles
- Approximately 3200 problems and questions
- Comprehensive development and consistent use of measurement and significant digits
- Algebra review in the appendices
- Basic instruction on using a scientific calculator in the appendices

Ancillaries

- Companion Laboratory Manual
- Instructor's Resource Manual with Complete Solutions, Transparency Masters, and Test Item File
- Computerized Test Item File for Windows

Illustration of Some Key Features

Problem-Solving Method Figure P.1 shows an example illustrating how the problem-solving method is used in the text. See page 53 for the detailed presentation of the problem-solving method.

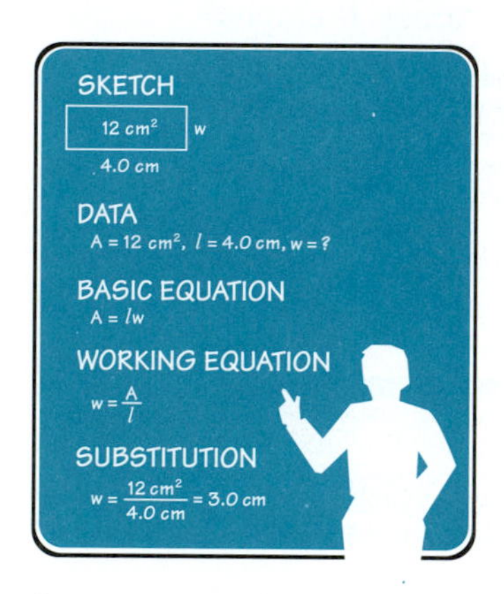

Problem-Solving Method

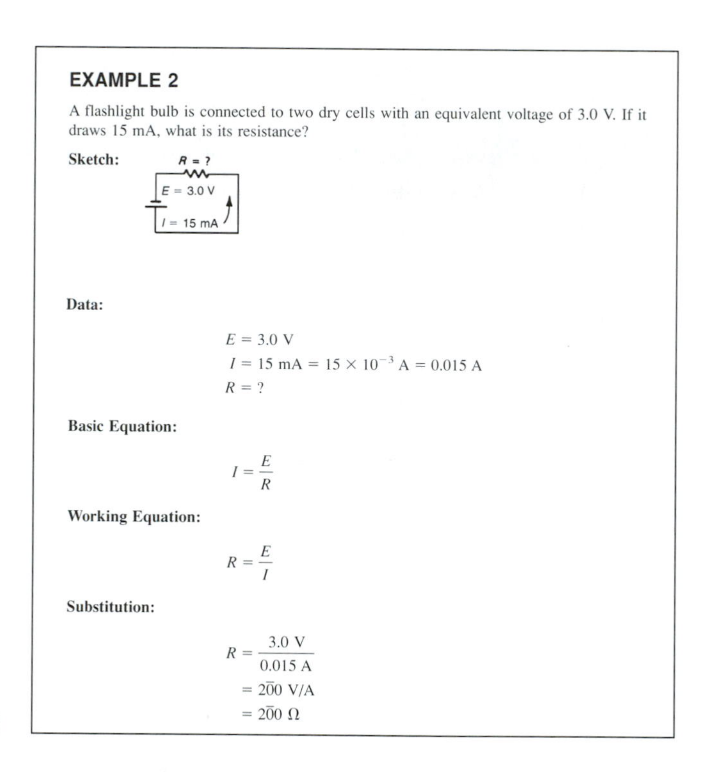

EXAMPLE 2

A flashlight bulb is connected to two dry cells with an equivalent voltage of 3.0 V. If it draws 15 mA, what is its resistance?

Sketch:

$R = ?$, $E = 3.0$ V, $I = 15$ mA

Data:

$E = 3.0 \text{ V}$

$I = 15 \text{ mA} = 15 \times 10^{-3} \text{ A} = 0.015 \text{ A}$

$R = ?$

Basic Equation:

$$I = \frac{E}{R}$$

Working Equation:

$$R = \frac{E}{I}$$

Substitution:

$$R = \frac{3.0 \text{ V}}{0.015 \text{ A}} = 2\bar{0}0 \text{ V/A} = 2\bar{0}0\ \Omega$$

Figure P.1

Chapter Objectives Figure P.2 shows the objectives for Chapter 9, "Rotational Motion."

Examples Worked examples are consistently displayed in the problem-solving format and used to illustrate and clarify basic concepts and problems. Since many students learn by example, a large number of examples are provided. The example in Figure P.3 shows how conversion factors are displayed and used.

Problem Sets Problem sets follow each section that is problem related. The problem-solving icon (shown at left) is located in the margin of problem sets as a reminder to students of the use and importance of the problem-solving method. This method is easily remembered and provides a valuable skill that can be used and applied daily in other technical and science courses and on the job.

OBJECTIVES

The major goals of this chapter are to enable you to:

1. Distinguish between rectilinear, curvilinear, and rotational motion.
2. Apply the torque equation to rotational problems.
3. Calculate the centripetal force of moving objects.
4. Find power in rotational systems.
5. Analyze how gears and gear trains are used to transfer rotational motion.

9.1 MEASUREMENT OF ROTATIONAL MOTION

Until now we have considered only motion in a straight line, called **rectilinear motion.** Technicians are often faced with many problems with motion along a curved path or objects that are rotating about an axis. Although these kinds of motion are similar, we must distinguish between them.

Motion along a curved path is called **curvilinear motion.** A satellite in orbit around the earth is an example of curvilinear motion [Fig. 9.1(a)].

(a) Curvilinear motion of an orbiting satellite.

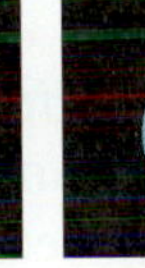

(b) Rotational motion of earth spinning on its axis.

(c) Rotational motion of a wheel spinning on its axle.

Figure 9.1

Rotational motion occurs when the body itself is spinning. Examples of rotational motion are the earth spinning on its axis, a turning wheel, a turning driveshaft, and the turning shaft of an electric motor [Fig. 9.1(b) and (c)].

We can see a wheel turn, but to gather useful information about its motion, we need a system of measurement. There are three basic systems of defining angle measurement. One unit of measurement in rotational motion is the number of rotations—how many times the object goes around. We need to know not only the number of rotations but also the time for each rotation. The unit of rotation (most often used in industry) is the **revolution** (rev). A second system of measurement divides the circle of rotation into 360 degrees ($360° = 1$ rev).

The **radian** (rad), which is approximately 57.3° or exactly $\left(\frac{360}{2\pi}\right)^{\circ}$, is a third

Sec. 9.1 Measurement of Rotational Motion

Figure P.2

EXAMPLE 2

A car accelerates from 45 km/h to 8$\bar{0}$ km/h in 3.00 s. Find its acceleration (in m/s^2).

Sketch: None needed

Data:

$$\Delta v = 8\bar{0} \text{ km/h} - 45 \text{ km/h} = 35 \text{ km/h}$$
$$t = 3.00 \text{ s}$$
$$a = ?$$

Basic Equation:

$$\Delta v = at$$

Working Equation:

$$a = \frac{\Delta v}{t}$$

Substitution:

$$a = \frac{35 \text{ km/h}}{3.00 \text{ s}} \times \frac{1000 \text{ m}}{1 \text{ km}} \times \frac{1 \text{ h}}{3600 \text{ s}}$$
$$= 3.2 \text{ m/s}^2$$

Note the use of the conversion factors to change the units km/h/s to m/s^2.

Figure P.3

Chapter End Matter A chapter glossary, a summary of chapter formulas, chapter review questions to review concept understanding, and chapter review problems help students to review for quizzes and examinations.

To the Faculty

This text is written at a language level and at a mathematics level that is cognizant of and beneficial to *most* students in technical programs that do not require a high level of mathematics rigor. The authors have assumed that the student has successfully completed one year of high school algebra or its equivalent. Simple equations and formulas are reviewed, and any mathematics beyond this level is developed in the text. The manner in which the mathematics is used in the text displays the need for mathematics in technology. For the better prepared student, the mathematics sections may be omitted with no loss in continuity.

Sections are short, and each deals with only one concept. The need for the investigation of a physical principle is developed before undertaking its study, and many diagrams are used to aid students in visualizing the concept. A large number of examples and problems allow students to develop and check their mastery of one concept before moving to another.

This text is designed to be used in a vocational-technical program in a community college, a technical institute, or a high school for students who plan to pursue a technical career. The topics were chosen with the assistance of technicians and management from several industries, as well as faculty technical consultants. Suggestions from users and reviewers of the previous edition were used extensively in this edition.

The chapter on measurement introduces students to basic units and some mathematical skills. For those students who lack a metric background or who need a review, a significant development of the metric system is found in Chapter 1, where measurements are presented as approximate numbers and then used consistently throughout the text. Chapter 2 introduces students to the problem-solving method that is used in the rest of the text. The need for vectors is developed in the chapter on motion, where graphical addition of vectors is used. One-dimensional dynamics is then discussed, and the need for trigonometry in more complex problems is developed. The chapter on trigonometry includes right triangle trigonometry and the component method for adding vectors. A more thorough treatment of dynamics and other standard topics follows.

The treatment of matter includes a discussion of the three states of matter, density, fluids, pressure, and Pascal's principle. The treatment of heat includes temperature, specific heat, thermal expansion, change of state, and gas laws.

The chapter on wave motion and sound deals with basic wave characteristics, the nature and speed of sound, the Doppler effect, and resonance.

The section on electricity and magnetism begins with a brief discussion of static electricity, followed by an extensive treatment of dc circuits and sources, Ohm's law, and series and parallel circuits. The chapter on magnetism, generators, and motors is largely descriptive, but it allows for a more in-depth study if desired. Then ac circuits and transformers are treated extensively.

The chapter on light briefly discusses the wave and particle nature of light, but deals primarily with illumination. The chapter on reflection and refraction develops the images formed by mirrors and lenses. The chapter on modern physics provides an introduction to the structure and properties of the atomic nucleus, radioactive decay, nuclear reactions, and radioactivity.

A companion laboratory manual that has been classroom tested in community colleges and high schools is available. Also included is an Instructor's Resource Manual that includes Complete Solutions, Transparency Masters and a Test Item File.

To the Student: Why Study Physics?

Physics is useful. Architects, mechanics, builders, carpenters, electricians, plumbers, and engineers are only some of the people who use physics every day in their jobs or professions. In fact, every living person uses physics principles every hour of the day. The movement of an arm can be described using principles of the lever. All building trades, as well as the entire electronics industry, also use physics.

Physics is often defined as the study of matter, energy, and their transformations. The physicist uses scientific methods to observe, measure, and predict physical events and behaviors. However, gathered data left in someone's notebook in a laboratory are of little use to society.

The basics of physics is really universal communication in the language of mathematics. The physicist describes physical phenomena in an orderly form in mathematical terms understood worldwide. Mechanics is the base on which almost all other areas of physics are built. Motion, forces, work, electricity, and light are topics confronted daily in industry and technology. The basic laws of conservation of energy are needed to understand heat, sound, wave motion, electricity, and electromagnetic radiation.

Physics is always changing as new frontiers are established in the study of the nature of matter and physics today. The topics studied in this course, however, will probably not be greatly changed with new research and will remain a classical foundation for vocational work in many, many fields. We then begin our study with the rules of the game—measurement, followed by a systematic problem-solving method. The end result will hopefully be a firm base on which to build a career study in almost any technical or vocational field.

Acknowledgments

The authors especially thank the many faculty and students who have used the previous editions and those who have offered suggestions. If anyone wishes to correspond with us regarding suggestions, criticisms, questions, or errors, please contact Dale Ewen directly at Parkland Community College, 2400 W. Bradley, Champaign, IL 61821, or through Prentice Hall.

We thank the following reviewers: Jorge Cossio, Miami-Dade Community College; Dr. Clement Y. Lam, North Harris College; Charles Oster, Sauk Valley Community College; and David E. Schiebel, Butler County Community College.

We extend our sincere and special thanks to our Prentice Hall editor Stephen Helba, as well as to photo researcher Lori Hilfinger, production editor Louise Sette, and Kirsten Kauffman at York Production Services.

Finally, we are especially grateful to Joyce Ewen for her excellent proofreading assistance and to our families for their encouragement.

Dale Ewen
Ronald J. Nelson
Neill Schurter

APPLIED PHYSICS

Chapter 1

MEASUREMENT AND THE METRIC SYSTEM

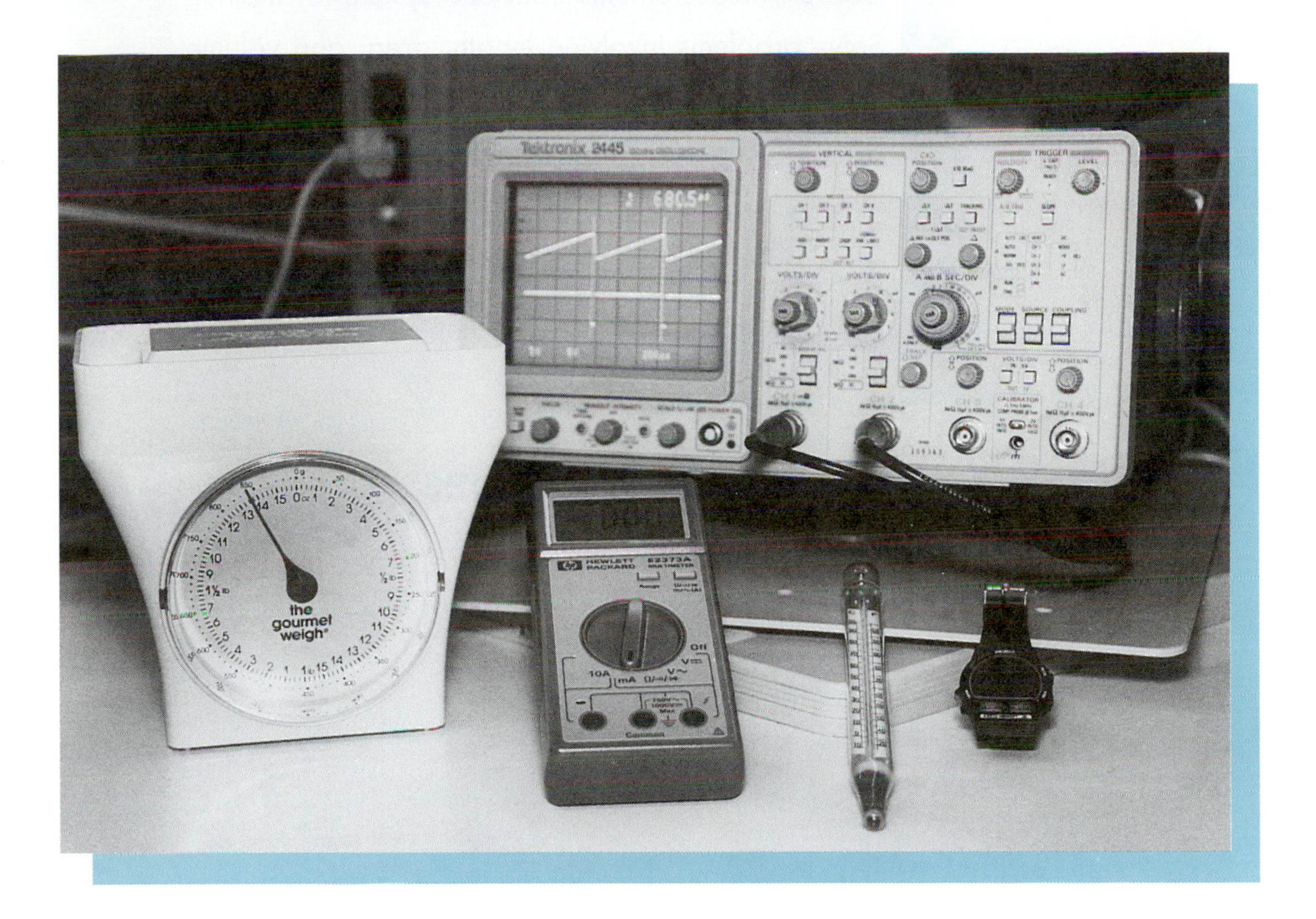

Throughout history there have been many standards by which measurements have been made. The almost universal system now in use is the metric system. Based on powers of 10, it is easy to work with very large or very small numbers that can be expressed in scientific notation.

Until now you have probably treated all numbers and measurements as exact numbers. Nearly all data of a technical nature, however, involve approximate numbers that are the result of measurements. Accuracy, precision, and significant digits are important concepts in calculations with measurements.

OBJECTIVES

The major goals of this chapter are to enable you to:

1. Understand the need for standardization of measurement.
2. Use the metric system of measurement.
3. Use scientific notation in problem solving.
4. Convert measurements from one system to another.
5. Solve problems involving length, area, and volume.
6. Distinguish between mass and weight.
7. Use significant digits to determine the accuracy of measurements.
8. Distinguish between accuracy and precision of measurements.
9. Do calculations with measurements and consistently express the results.

1.1 STANDARDS OF MEASURE

When two people work together on the same job, they should both use the same standards of measure. If not, the result can be disastrous (Fig. 1.1).

Figure 1.1
The trouble with inconsistent systems of measurement.

Standards of measure is a set of units of measurement for length, weight, and other quantities defined in a way that is useful to a large number of people. Throughout history, there have been many standards by which measurements have been made:

1. ***Chain:*** a measuring instrument of 100 links used in surveying. One chain has a length of 66 feet.
2. ***Rod:*** a length determined by having 16 men put one foot behind the foot of another man in a straight line (Fig. 1.2). The rod is now standardized as $16\frac{1}{2}$ feet.

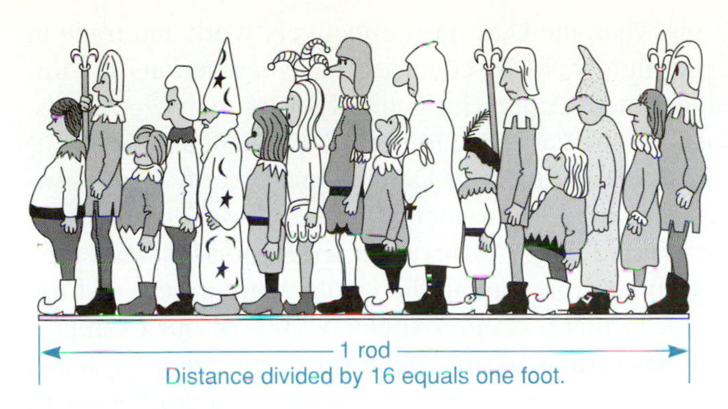

Figure 1.2
A rod used to be 16 "people feet."

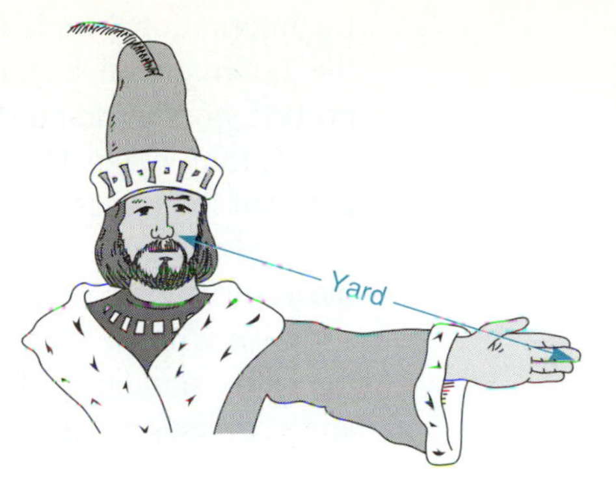

Figure 1.3
The "old" yard.

3. ***Yard:*** the distance from the tip of the king's nose to the fingertips of his outstretched hand (Fig. 1.3).
4. ***Foot:*** the rod divided by 16 was a legal foot. It was also common to use the length of one's own foot as the unit foot.
5. ***Inch:*** the length of three barley corns, round and dry, taken from the center of the ear, and laid end to end (Fig. 1.4).

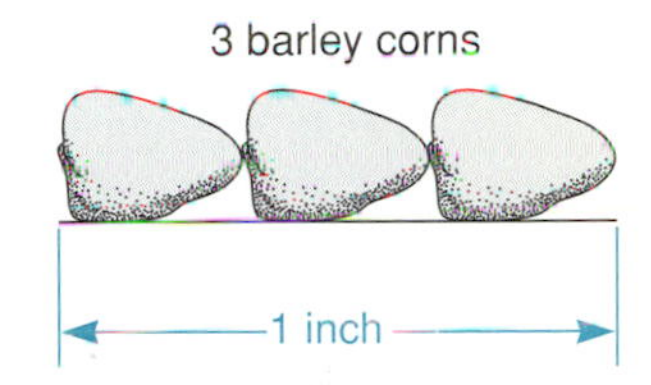

Figure 1.4
At one time, three barley corns were used to define one inch.

The U.S. units of measure are a combination of makeshift units of Anglo-Saxon, Roman, and French-Norman weights and measures.

After the standards based on parts of the human body and on other gimmicks, basic standards were accepted by world governments. They also agreed to construct and distribute accurate standard copies of all the standard units. During the 1790s, a decimal system based on our number system, the metric system, was being developed in France. Its acceptance was gained mostly because it was easy to use and easy to remember. Many nations began adopting it as their official system of measurement. By 1900, most of Europe and South America were metric. In 1866, metric measurements for official use were legalized in the United States. In 1893, the Secretary of the Treasury, by administrative order, declared the new metric standards to be the nation's "fundamental standards" of mass and length. Thus, indirectly, the United States *officially* became a metric nation. Even today, the English units are officially defined in terms of the standard metric units.

Throughout U.S. history, several attempts were made to convert the nation to the metric system. In the 1970s, the United States found itself to be the only nonmetric industrialized country left in the world. However, its government actually did little to implement the system. Industry and business, however, found their foreign markets drying up because metric products were preferred. By the mid-1980s, the United States was one of only two nonmetric nations left. Many segments of American industry and business have independently gone metric.

World trade is increasingly geared toward the metric system of measurement. Industry in the United States was often at a competitive disadvantage when dealing in international markets because of its nonstandard measurement system. It was sometimes excluded because it was unable to deliver goods measured in metric terms. The inherent simplicity of the metric system of measurement and standardization of weights and measures has led to major cost savings in industries that have converted to the metric system.

It is economically imperative that the United States complete the conversion to the metric system as soon as possible so that it can compete more favorably for international trade at a time when the U.S. must effectively work and trade in the international business community. Most countries have restrictions on importing nonmetric products, and metric countries naturally want metric products.

Many major U.S. industries, such as the automotive, aviation, and farm implement industries, as well as the Department of Defense and other federal agencies have already effectively converted to the metric system. Most others are in the process of converting or making major plans to do so. Hopefully, sometime soon we will see the benefits gained from having the whole world accept and use the same standards of measure. In this text approximately 70% of our examples and exercises are in metric in the sections where both English and metric systems are still commonly used. In some industries, you—the student and worker—will need to know both systems.

1.2 INTRODUCTION TO THE METRIC SYSTEM

The modern metric system is identified in all languages by the abbreviation **SI** (for Système International d'Unités—the international system of units of measurement written in French). The SI metric system has seven basic units [Table 1.1(a)]. All other SI units are called *derived units;* that is, they can be defined in terms of these seven basic units (Fig. 1.5). For example, the newton (N) is defined as 1 kg m/s^2 (kilogram metre per second per second). Many of the units in Fig. 1.5 will be presented and discussed in this text. This figure is used only to show how the derived units are related to the base units; do not try to memorize it. Other commonly used derived SI units are given in Table 1.1(b).

Table 1.1 SI Units of Measure

(a) Basic Unit	*SI Abbreviation*	*Used for Measuring*
metre*	m	length
kilogram	kg	mass
second	s	time
ampere	A	electric current
kelvin	K	temperature
candela	cd	light intensity
mole	mol	molecular substance
(b) Derived Unit	***SI Abbreviation***	***Used for Measuring***
litre*	L or ℓ	volume
cubic metre	m^3	volume
square metre	m^2	area
newton	N	force
metre per second	m/s	speed
joule	J	energy
watt	W	power

*At present, there is some difference of opinion in the United States on the spelling of metre and litre. We have chosen the "re" spellings for two reasons. First, this is the internationally accepted spelling for all English-speaking countries. Second, the word "meter" already has many different meanings—parking meter, electric meter, odometer, and so on. Many feel that the metric unit of length should be distinctive and readily recognizable—thus the spelling "metre" and "litre."

Because the metric system is a decimal or base 10 system, it is very similar to our decimal number system and our decimal money system. It is an easy system to use because calculations are based on the number 10 and its multiples. Special prefixes are used to name these multiples and submultiples, which may be used with most all SI units. Because the same prefixes are used repeatedly, the

RELATIONSHIPS OF SI UNITS WITH NAMES

The chart below shows graphically how the 17 SI-derived units with special names are derived in a coherent manner from the base and supplementary units. it was provided by the National Institute of Standards and Technology.

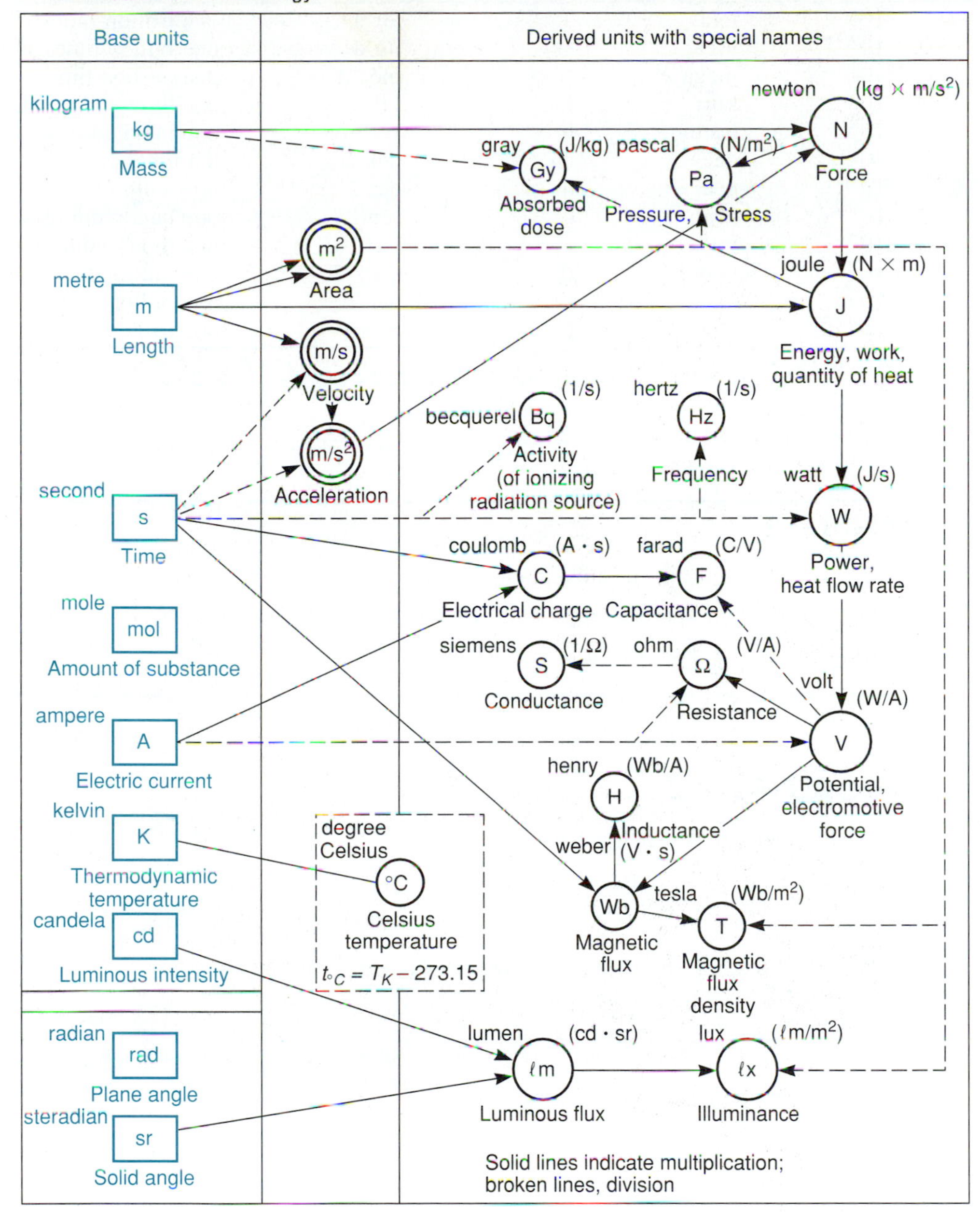

Figure 1.5

memorization of many conversions has been significantly reduced. Table 1.2 shows these prefixes and the corresponding symbols.

EXAMPLE 1

Write the SI abbreviation for 36 centimetres.

The symbol for the prefix *centi* is c.
The symbol for the unit *metre* is m.
Then, the SI abbreviation for 36 centimetres is 36 cm.

Table 1.2 Prefixes for SI Units

Multiple or Submultiple[a] *Decimal Form*	*Power of 10*	*Prefix*[b]	*Prefix Symbol*	*Pronunciation*	*Meaning*
1,000,000,000,000	10^{12}	tera	T	tĕr′ă	one trillion times
1,000,000,000	10^{9}	giga	G	jĭg′ă	one billion times
1,000,000	10^{6}	mega	M	mĕg′ă	one million times
1,000	10^{3}	kilo	k	kĭl′ō	one thousand times
100	10^{2}	hecto	h	hĕk′tō	one hundred times
10	10^{1}	deka	da	dĕk′ă	ten times
0.1	10^{-1}	deci	d	dĕs′ĭ	one tenth of
0.01	10^{-2}	centi	c	sĕnt′ĭ	one hundredth of
0.001	10^{-3}	milli	m	mĭl′ĭ	one thousandth of
0.000001	10^{-6}	micro	μ	mī′krō	one millionth of
0.000000001	10^{-9}	nano	n	năn′ō	one billionth of
0.000000000001	10^{-12}	pico	p	pē′kō	one trillionth of

[a]Factor by which the unit is multiplied.

[b]The same prefixes are used with all SI metric units.

EXAMPLE 2

Write the SI metric unit for the abbreviation 45 kg.

The prefix for k is *kilo;* the unit for g is *gram.*
Then, the SI metric unit for 45 kg is 45 kilograms.

■ PROBLEMS 1.2

Give the metric prefix for each value.

1. 1000
2. 0.01
3. 100
4. 0.1
5. 0.001
6. 10
7. 1,000,000
8. 0.000001

Give the metric symbol, or abbreviation, for each prefix.

9. hecto
10. kilo
11. milli
12. deci
13. mega
14. deka
15. centi
16. micro

Write the abbreviation for each quantity.

17. 135 millimetres
18. 83 dekagrams
19. 28 kilolitres
20. 52 centimetres
21. 49 centigrams
22. 85 milligrams
23. 75 hectometres
24. 15 decilitres

Write the SI unit for each abbreviation.

25. 24 m
26. 185 L
27. 59 g
28. 125 kg
29. 27 mm
30. 25 dL
31. 45 dam
32. 27 mg
33. 26 Mm
34. 275 μg
35. The basic metric unit of length is ______.
36. The basic unit of mass is ______.
37. Two common metric units of volume are ______ and ______.
38. The basic unit for electric current is ______.
39. The basic metric unit for time is ______.
40. The common metric unit for power is ______. ■

1.3 SCIENTIFIC NOTATION

Scientists and technicians often use very large or very small numbers that cannot be conveniently written as fractions or decimal fractions. For example, the thickness of an oil film on water is about 0.0000001 m. A more useful method of expressing such very small (or very large) numbers is known as **scientific notation.** Expressed this way, the thickness of the film is 1×10^{-7} m or 10^{-7} m. For example:

$$0.1 = 1 \times 10^{-1} \quad \text{or} \quad 10^{-1}$$

$$10{,}000 = 1 \times 10^{4} \quad \text{or} \quad 10^{4}$$

$$0.001 = 1 \times 10^{-3} \quad \text{or} \quad 10^{-3}$$

A number in scientific notation is written as a product of a number between 1 and 10 and a power of 10. General form: $M \times 10^{n}$, where

M = a number between 1 and 10

n = the exponent or power of 10

EXAMPLE 1

The following numbers are written in scientific notation:

(a) $325 = 3.25 \times 10^{2}$

(b) $100{,}000 = 1 \times 10^{5}$ or 10^{5}

To write any decimal number in scientific notation:

1. Place a decimal point after the first nonzero digit reading from left to right.
2. Place a caret (∧) at the position of the original decimal point.
3. If the decimal point is to the *left* of the caret, the exponent of 10 is the number of places from the caret to the decimal point.
   Example: $83{,}662 = 8.3662_{\wedge} \times 10^{\underline{4}}$ (4 places)
4. If the decimal point is to the *right* of the caret, the exponent of 10 is the negative of the number of places from the caret to the decimal point.
   Example: $0.00683 = {}_{\wedge}006.83 \times 10^{-\underline{3}}$ (3 places)
5. If the decimal point and the caret coincide, the exponent of 10 is zero.
   Example: $5.12 = 5\underset{\wedge}{.}12 \times 10^{0}$

A number greater than 10 is expressed in scientific notation as a product of a decimal between 1 and 10 and a *positive* power of 10.

EXAMPLE 2

Write each number greater than 10 in scientific notation.

(a) $2580 = 2.58 \times 10^3$
(b) $54{,}600 = 5.46 \times 10^4$
(c) $42{,}000{,}000 = 4.2 \times 10^7$
(d) $715.8 = 7.158 \times 10^2$
(e) $34.775 = 3.4775 \times 10^1$

A number between 0 and 1 is expressed in scientific notation as a product of a decimal between 1 and 10 and a *negative* power of 10.

EXAMPLE 3

Write each positive number less than 1 in scientific notation.

(a) $0.0815 = 8.15 \times 10^{-2}$
(b) $0.00065 = 6.5 \times 10^{-4}$
(c) $0.73 = 7.3 \times 10^{-1}$
(d) $0.0000008 = 8 \times 10^{-7}$

A number between 1 and 10 is expressed in scientific notation as a product of a decimal between 1 and 10 and the *zero* power of 10.

EXAMPLE 4

Write each number between 1 and 10 in scientific notation.

(a) $7.33 = 7.33 \times 10^0$
(b) $1.06 = 1.06 \times 10^0$

To change a number from scientific notation to decimal form:

1. Multiply the decimal part by the power of 10 by moving the decimal point *to the right* the same number of decimal places as indicated by the power of 10 if it is *positive*.
2. Multiply the decimal part by the power of 10 by moving the decimal point *to the left* the same number of decimal places as indicated by the power of 10 if it is *negative*.
3. Supply zeros as needed.

EXAMPLE 5

Write 7.62×10^2 in decimal form.

$7.62 \times 10^2 = 762$ Move the decimal point two places to the right.

EXAMPLE 6

Write 6.15×10^{-4} in decimal form.

$6.15 \times 10^{-4} = 0.000615$ Move the decimal point four places to the left and insert three zeros.

EXAMPLE 7

Write each number in decimal form.

(a) $3.75 \times 10^2 = 375$
(b) $1.09 \times 10^5 = 109{,}000$
(c) $2.88 \times 10^{-2} = 0.0288$
(d) $9.4 \times 10^{-6} = 0.0000094$
(e) $6.7 \times 10^0 = 6.7$

Since all calculators used in science and technology accept numbers entered in scientific notation and give some results in scientific notation, it is essential that you fully understand this topic before going to the next section. It may be necessary now to consult Appendix A, Section A.1 on signed numbers and Section A.2 on the powers of 10. See Appendix B, Section B.2, for using a calculator with numbers in scientific notation.

■ PROBLEMS 1.3

Write each number in scientific notation.

1. 326
2. 798
3. 2650
4. 14,500
5. 826.4
6. 24.97
7. 0.00413
8. 0.00053
9. 6.43
10. 482,300
11. 0.000065
12. 0.00224
13. 540,000
14. 1,400,000
15. 0.0000075
16. 0.0000009
17. 0.00000005
18. 3,500,000,000
19. 732,000,000,000,000,000
20. 0.00000000000000618

Write each number in decimal form.

21. 8.62×10^4
22. 8.67×10^2
23. 6.31×10^{-4}
24. 5.41×10^3
25. 7.68×10^{-1}
26. 9.94×10^1
27. 7.77×10^8
28. 4.19×10^{-6}
29. 6.93×10^1
30. 3.78×10^{-2}
31. 9.61×10^4
32. 7.33×10^3
33. 1.4×10^0
34. 9.6×10^{-5}
35. 8.4×10^{-6}
36. 9×10^8
37. 7×10^{11}
38. 4.05×10^0
39. 7.2×10^{-7}
40. 8×10^{-9}
41. 4.5×10^{12}
42. 1.5×10^{11}
43. 5.5×10^{-11}
44. 8.72×10^{-10} ■

1.4 LENGTH

In most sections that introduce units of measure, we present the units in subsections as follows: metric units, English units, and conversions between metric and English units.

Metric Length

The basic SI unit of length is the **metre** (m) (Fig. 1.6). The first standard metre was chosen in the 1790s to be one ten-millionth of the distance from the earth's equator to either pole. Modern measurements of the earth's circumference show that the first length is off by about 0.02%. The new definition is based on the speed of light in a vacuum and reads: "The metre is the length of path traveled by light in a vacuum during a time interval of 1/299,792,458 of a second." Long distances are measured in kilometres (km) (Fig. 1.7). We use the centimetre (cm) to measure short distances, such as the length of this book or the width of a board (Fig. 1.8). The millimetre (mm) is used to measure very small lengths, such as the thickness of this book or the depth of a tire tread (Fig. 1.9). A metric ruler is shown in Fig. 1.10.

To change from one unit or set of units to another, we use what is commonly called a **conversion factor.** We know that we can multiply any number or quantity by 1 without changing the value of the original quantity. We also know that any fraction equals 1 when its numerator and denominator are equal. For example,

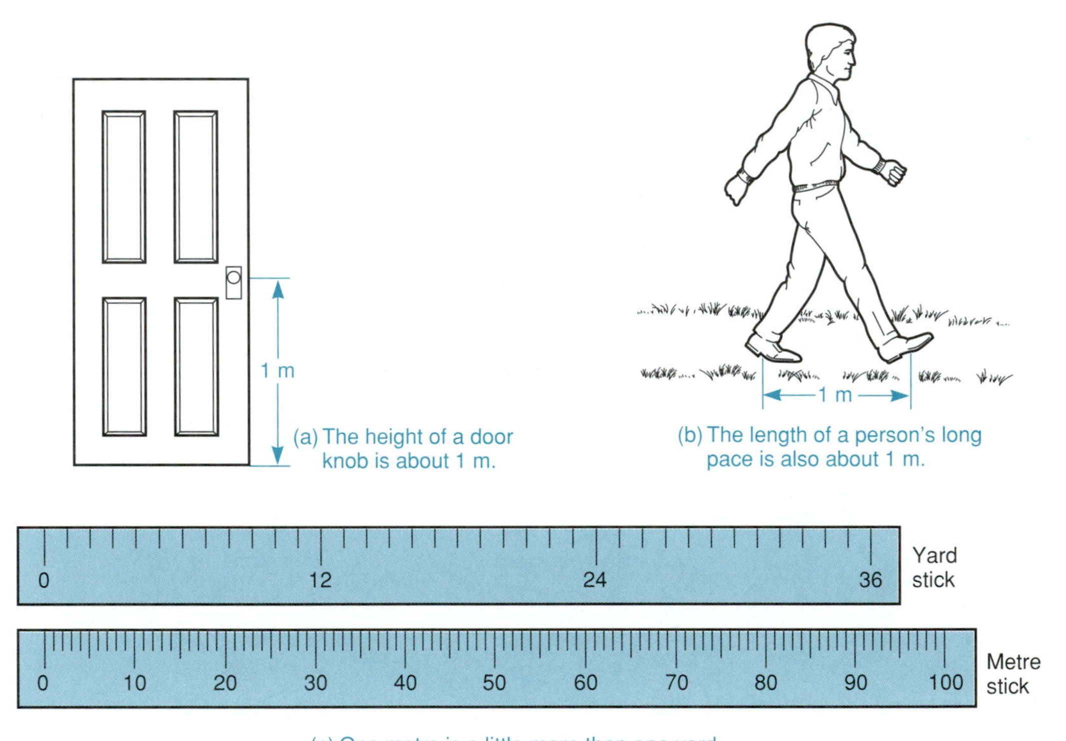

Figure 1.6
One metre

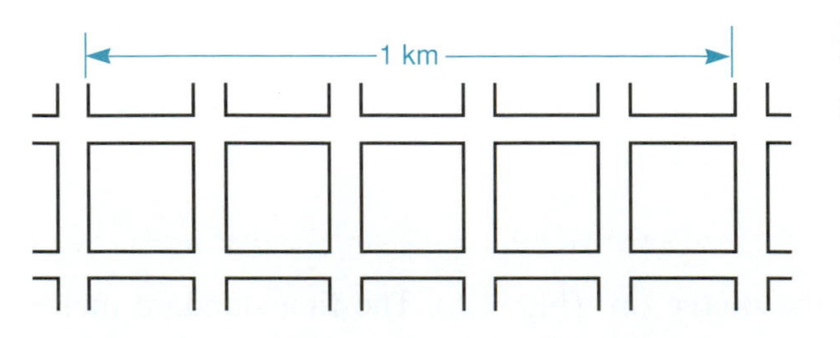

Figure 1.7
The length of five city blocks is about 1 km.

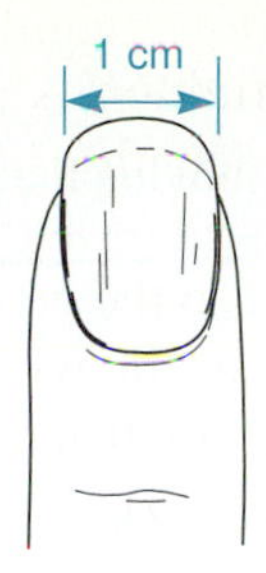

Figure 1.8
The width of your small fingernail is about 1 cm.

Figure 1.9
The thickness of a dime is about 1 mm.

Metric ruler

Figure 1.10
The large numbered divisions are centimetres.
Each cm is divided into 10 equal parts, called millimetres.

$\dfrac{5}{5} = 1$, $\dfrac{12\text{ m}}{12\text{ m}} = 1$, and $\dfrac{6.5\text{ kg}}{6.5\text{ kg}} = 1$. In addition, since 1 m = 100 cm, $\dfrac{1\text{ m}}{100\text{ cm}} = 1$. Similarly, $\dfrac{100\text{ cm}}{1\text{ m}} = 1$, because the numerator equals the denominator. We call such names for 1 *conversion factors*. The information necessary for forming a conversion factor is usually found in tables. As in the case 1 m = 100 cm, there are two conversion factors for each set of data:

$$\frac{1\text{ m}}{100\text{ cm}} \quad \text{and} \quad \frac{100\text{ cm}}{1\text{ m}}$$

CONVERSION FACTORS

Choose a conversion factor in which the old units are in the numerator of the original expression and in the denominator of the conversion factor, or in the denominator of the original expression and in the numerator of the conversion factor. That is, we want the old units to cancel each other.

EXAMPLE 1

Change 215 cm to metres.

As we saw before, the two possible conversion factors are

$$\frac{1\text{ m}}{100\text{ cm}} \quad \text{and} \quad \frac{100\text{ cm}}{1\text{ m}}$$

We choose the conversion factor with centimetres in the *denominator* so that the cm units cancel each other.

$$215 \cancel{\text{cm}} \times \frac{1 \text{ m}}{100 \cancel{\text{cm}}} = 2.15 \text{ m}$$

Note: Conversions *within* the metric system involve only moving the decimal point.

EXAMPLE 2

Change 4 m to centimetres.

$$4 \cancel{\text{m}} \times \frac{100 \text{ cm}}{1 \cancel{\text{m}}} = 400 \text{ cm}$$

EXAMPLE 3

Change 39.5 mm to centimetres.

Choose the conversion factor with millimetres in the denominator so that the mm units cancel each other.

$$39.5 \cancel{\text{mm}} \times \frac{1 \text{ cm}}{10 \cancel{\text{mm}}} = 3.95 \text{ cm}$$

EXAMPLE 4

Change 0.05 km to centimetres.

First, change to metres and then to centimetres.

$$0.05 \cancel{\text{km}} \times \frac{1000 \text{ m}}{1 \cancel{\text{km}}} = 50 \text{ m}$$

$$50 \cancel{\text{m}} \times \frac{100 \text{ cm}}{1 \cancel{\text{m}}} = 5000 \text{ cm}$$

Or,

$$0.05 \cancel{\text{km}} \times \frac{1000 \cancel{\text{m}}}{1 \cancel{\text{km}}} \times \frac{100 \text{ cm}}{1 \cancel{\text{m}}} = 5000 \text{ cm}$$

English Length

The English system has been historically used in the English-speaking countries, and its basic units are the foot, pound, and second. The foot is the basic unit of length in the English system, and it may be divided into 12 equal parts or *inches*.

Common English length conversions include:

1 foot (ft) = 12 inches (in.)

1 yard (yd) = 3 ft

1 mile (mi) = 5280 ft

See Table 1 of Appendix C for English weights and measures.

We also use a conversion factor to change from one English length unit to another.

EXAMPLE 5

Change 84 in. to feet.

Choose the conversion factor with inches in the denominator and feet in the numerator.

$$84\ \text{in.} \times \frac{1\ \text{ft}}{12\ \text{in.}} = 7\ \text{ft}$$

Metric–English Conversions

To change from an English unit to a metric unit or from a metric unit to an English unit, again use a conversion factor, such as 1 in. = 2.54 cm (Fig. 1.11).

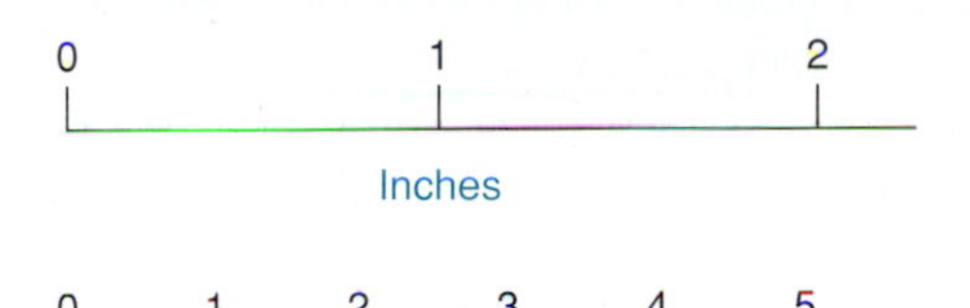

Figure 1.11
One inch equals 2.54 cm.

EXAMPLE 6

Express 10 inches in centimetres.

$$1\ \text{in.} = 2.54\ \text{cm} \quad \text{so} \quad 10\ \text{in.} \times \frac{2.54\ \text{cm}}{1\ \text{in.}} = 25.4\ \text{cm}$$

The conversion factors you will need are given in Appendix C. The following examples show you how to use these tables.

EXAMPLE 7

Change 15 miles to kilometres.

From Table 2 in Appendix C, we find 1 mile listed in the left-hand column. Moving over to the fourth column, under the heading "km," we see that 1 mile (mi) = 1.61 km. Then we have

$$15\ \text{mi} \times \frac{1.61\ \text{km}}{1\ \text{mi}} = 24.15\ \text{km}$$

EXAMPLE 8

Change 220 centimetres to inches.

Find 1 centimetre in the left-hand column and move to the fifth column under the heading "in." We find that 1 centimetre = 0.394 in. Then

$$220\ \text{cm} \times \frac{0.394\ \text{in.}}{1\ \text{cm}} = 86.68\ \text{in.}$$

EXAMPLE 9

Change 3 yards to centimetres.

Since there is no direct conversion from yards to centimetres in the tables, we must first change yards to inches and then inches to centimetres:

$$3\ \text{yd} \times \frac{36\ \text{in.}}{1\ \text{yd}} \times \frac{2.54\ \text{cm}}{1\ \text{in.}} = 274.32\ \text{cm}$$

■ PROBLEMS 1.4

Which unit is longer?

1. 1 metre or 1 centimetre
2. 1 metre or 1 millimetre
3. 1 metre or 1 kilometre
4. 1 centimetre or 1 millimetre
5. 1 centimetre or 1 kilometre
6. 1 millimetre or 1 kilometre

Which metric unit (km, m, cm, or mm) would you use to measure the following?

7. Length of a wrench
8. Thickness of a saw blade
9. Height of a barn
10. Width of a table
11. Thickness of a hypodermic needle
12. Distance around an automobile racing track
13. Distance between New York and Miami
14. Length of a hurdle race
15. Thread size on a pipe
16. Width of a house lot

Fill in each blank with the most reasonable metric unit (km, m, cm, or mm).

17. Your car is about 6 _____ long.
18. Your pencil is about 20 _____ long.
19. The distance between New York and San Francisco is about 4200 _____.
20. Your pencil is about 7 _____ thick.
21. The ceiling in my bedroom is about 240 _____ high.
22. The length of a football field is about 90 _____.
23. A jet plane usually cruises at about 9 _____ high.
24. A standard film size for cameras is 35 _____.
25. The diameter of my car tire is about 60 _____.
26. The zipper on my jacket is about 70 _____ long.
27. Juan drives 9 _____ to school each day.
28. Jacob, our basketball center, is 203 _____ tall.
29. The width of your hand is about 80 _____.
30. A handsaw is about 70 _____ long.
31. A newborn baby is usually about 45 _____ long.
32. The standard metric piece of plywood is 120 _____ wide and 240 _____ long.

Fill in each blank.

33. 1 km = _____ m
34. 1 mm = _____ m
35. 1 m = _____ cm
36. 1 m = _____ hm
37. 1 dm = _____ m
38. 1 dam = _____ m
39. 1 m = _____ mm
40. 1 m = _____ dm
41. 1 hm = _____ m
42. 1 cm = _____ m
43. 1 m = _____ km
44. 1 m = _____ dam
45. 1 cm = _____ mm
46. Change 250 m to cm.
47. Change 250 m to km.
48. Change 546 mm to cm.
49. Change 178 km to m.
50. Change 35 dm to dam.
51. Change 830 cm to m.
52. Change 75 hm to km.
53. Change 375 cm to mm.
54. Change 7.5 mm to μm.
55. Change 4 m to μm.
56. State your height in centimetres.
57. State your height in metres.
58. State your height in millimetres.
59. The wheelbase of a certain automobile is 108 in. long. Find its length
    (a) in feet.
    (b) in yards.

60. Change 43,296 ft
   (a) to miles. (b) to yards.
61. Change 6.25 mi
   (a) to yards. (b) to feet.
62. The length of a connecting rod is 7 in. What is its length in centimetres?
63. The distance between two cities is 256 mi. Find this distance in kilometres.
64. Change 5.94 m to feet.
65. Change 7.1 cm to inches.
66. Change 1.2 in. to centimetres.
67. The turning radius of an auto is 20 ft. What is this in metres?
68. Would a wrench with an opening of 25 mm be larger or smaller than a 1-in. wrench?
69. How many reamers each 20 cm long can be cut from a bar 6 ft long, allowing 3 mm for each saw cut?
70. If 214 pieces each 47 cm long are ordered to be turned from $\frac{1}{4}$-in. round steel stock and $\frac{1}{8}$ in. of waste is allowed on each piece, what length (in metres) of stock is required? ■

1.5 AREA AND VOLUME

Area

To measure a surface area of an object, you must first decide on a standard unit of **area.** Standard units of area are based on the square and are called *square inches, square centimetres, square miles,* or some other square unit of measure. An area of 1 square centimetre (cm^2) is the amount of area found within a square 1 cm on each side. An area of 1 square inch (in^2) is the amount of area found within a square of 1 in. on each side (Fig. 1.12). The area of a figure is the number of square units that it contains.

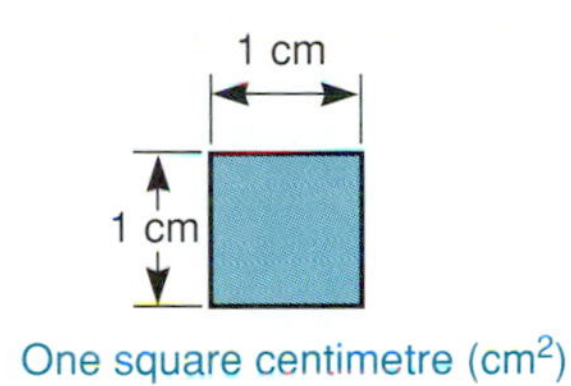

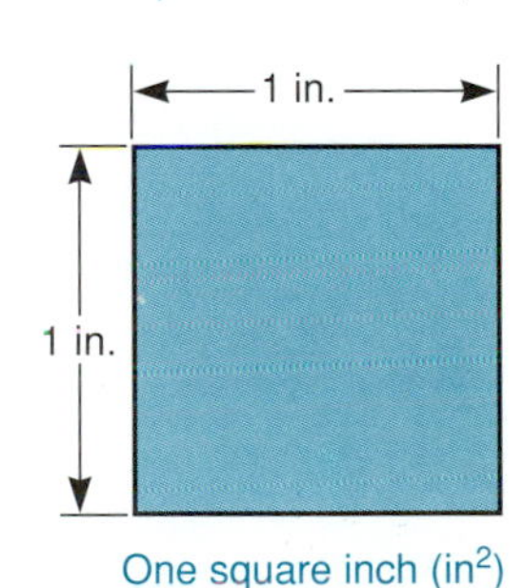

Figure 1.12

In general, when multiplying measurements of like units, multiply the numbers, and then multiply the units as follows:

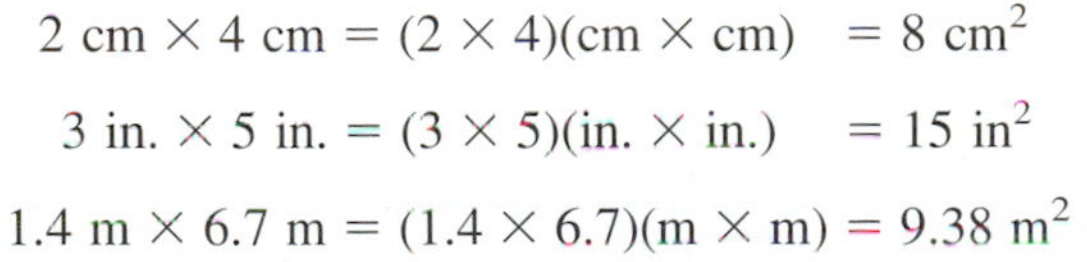

$$2 \text{ cm} \times 4 \text{ cm} = (2 \times 4)(\text{cm} \times \text{cm}) = 8 \text{ cm}^2$$

$$3 \text{ in.} \times 5 \text{ in.} = (3 \times 5)(\text{in.} \times \text{in.}) = 15 \text{ in}^2$$

$$1.4 \text{ m} \times 6.7 \text{ m} = (1.4 \times 6.7)(\text{m} \times \text{m}) = 9.38 \text{ m}^2$$

Metric Area. The basic unit of area in the metric system is the *square metre* (m^2), the area in a square whose sides are 1 m long (Fig. 1.13). The square centimetre (cm^2) and the square millimetre (mm^2) are smaller units of area. Larger units of area are the square kilometre (km^2) and the hectare (ha).

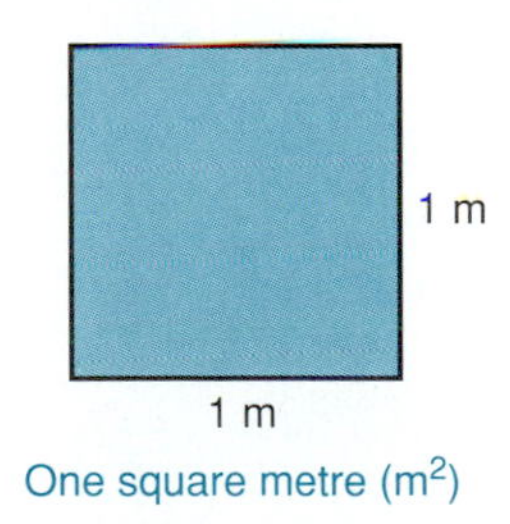

Figure 1.13

EXAMPLE 1

Find the area of a rectangle 5 m long and 3 m wide.

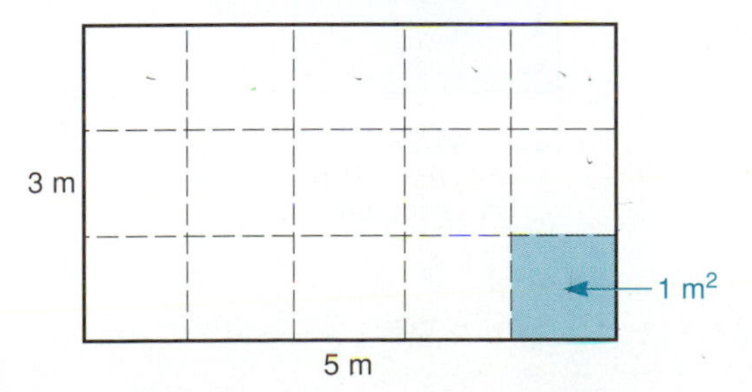

Figure 1.14

Each square in Fig. 1.14 represents 1 m^2. By simply counting the number of squares (square metres), we find that the area of the rectangle is 15 m^2. We can also find the area of the rectangle by using the formula

$$A = lw = (5 \text{ m})(3 \text{ m}) = 15 \text{ m}^2$$

Note: $\text{m} \times \text{m} = \text{m}^2$.

EXAMPLE 2

Find the area of the metal plate shown in Fig. 1.15.

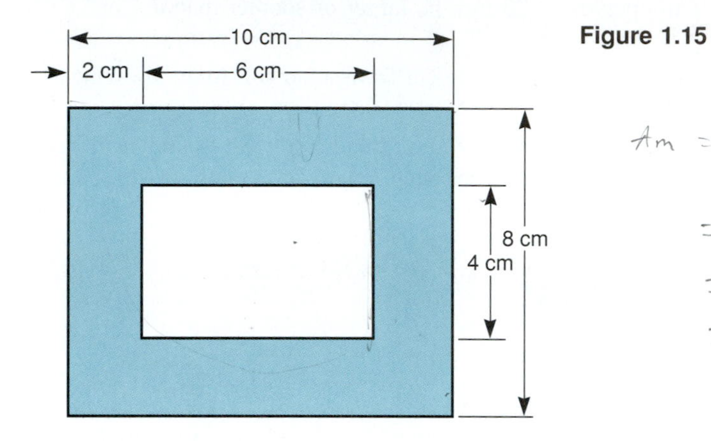

Figure 1.15

To find the area of the metal plate, find the area of each of the two rectangles and then find the difference of their areas. The large rectangle is 10 cm long and 8 cm wide. The small rectangle is 6 cm long and 4 cm wide.

Area of large rectangle: $A = lw = (10 \text{ cm})(8 \text{ cm})$	$= 80 \text{ cm}^2$
Area of small rectangle: $A = lw = (6 \text{ cm})(4 \text{ cm})$	$= 24 \text{ cm}^2$
Area of metal plate:	56 cm^2

The surface that would be seen by cutting a geometric solid with a thin plate parallel to one side of the solid represents the cross-sectional area of the solid.

EXAMPLE 3

Find the smallest cross-sectional area of the box shown in Fig. 1.16(a).

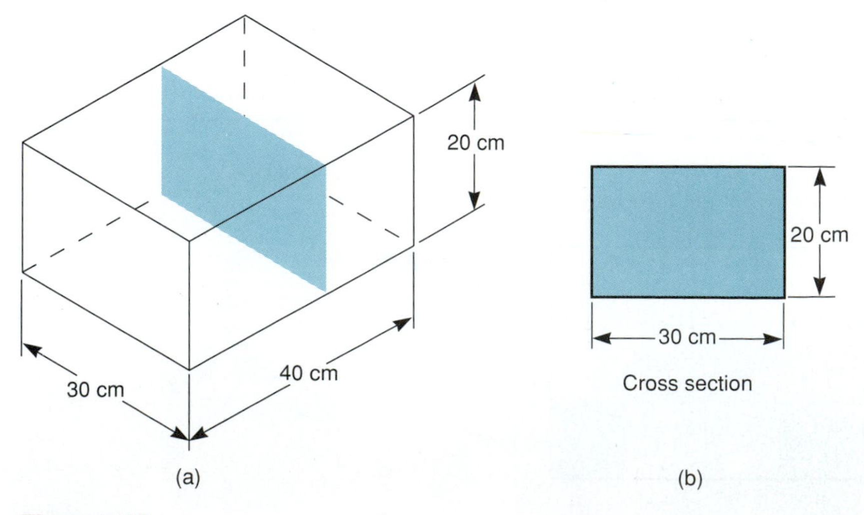

Figure 1.16

The indicated cross section of this box is a rectangle 30 cm long and 20 cm wide [Fig. 1.16(b)].

$$A = lw = (30 \text{ cm})(20 \text{ cm}) = 600 \text{ cm}^2$$

The area of this rectangle is 600 cm^2, which represents the cross-sectional area of the box.

The formulas for finding the areas of other plane figures can be found on the inside back cover.

The *hectare* is the fundamental SI unit for land area. An area of 1 hectare equals the area of a square 100 m on a side (Fig. 1.17). The hectare is used because it is more convenient to say and use than square hectometre. The metric prefixes are *not* used with the hectare unit. That is, instead of saying "2 kilohectares," we say "2000 hectares."

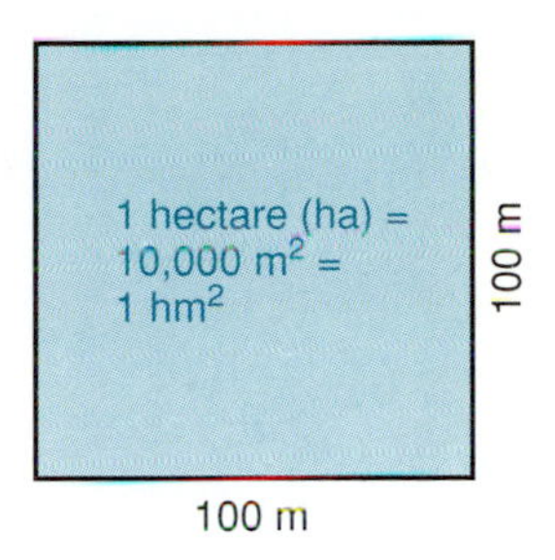

Figure 1.17
One hectare.

To convert area or square units, use a conversion factor. That is, the correct conversion factor will be in fractional form and equal to 1, with the numerator expressed in the units you wish to convert to and the denominator expressed in the units given. The conversion table for area is provided as Table 3 of Appendix C.

The conversion of area units will be shown using a method of squaring the linear or length conversion factor which you are most likely to remember. An alternative method emphasizing direct use of the conversion tables will also be shown.

EXAMPLE 4

Change 258 cm^2 to m^2.

$$258 \text{ cm}^2 \times \left(\frac{1 \text{ m}}{100 \text{ cm}}\right)^2 = 258 \cancel{\text{cm}^2} \times \frac{1^2 \text{ m}^2}{100^2 \cancel{\text{cm}^2}} = 0.0258 \text{ m}^2$$

Note: The intermediate step is usually not shown.

Alternative Method:

$$258 \cancel{\text{cm}^2} \times \frac{1 \text{ m}^2}{10{,}000 \cancel{\text{cm}^2}} = 0.0258 \text{ m}^2$$

English Area

EXAMPLE 5

Find the area of a rectangle that is 6 in. long and 4 in. wide (Fig. 1.18).

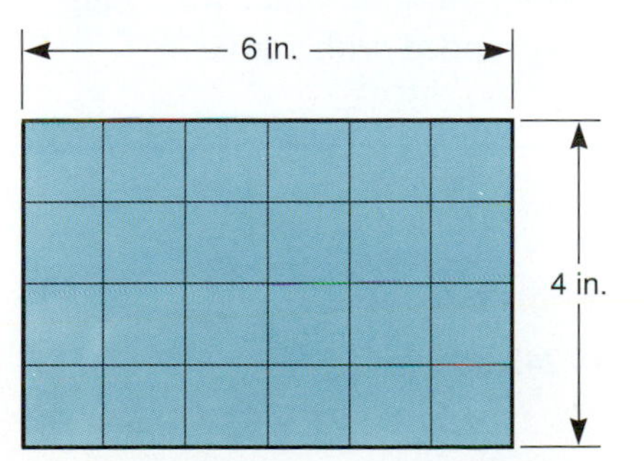

Figure 1.18

Each square is 1 in^2. To find the area of the rectangle, simply count the number of squares in the rectangle. Therefore, you find that the area = 24 in^2, or, by using the formula,

$$A = lw = (6 \text{ in.})(4 \text{ in.}) = 24 \text{ in}^2$$

EXAMPLE 6

Change 324 in^2 to yd^2.

$$324 \text{ in}^2 \times \left(\frac{1 \text{ yd}}{36 \text{ in.}}\right)^2 = 0.25 \text{ yd}^2$$

Alternative Method:

$$324 \cancel{\text{in}^2} \times \frac{1 \text{ yd}^2}{1296 \cancel{\text{in}^2}} = \frac{324}{1296} \text{ yd}^2 = 0.25 \text{ yd}^2$$

Metric–English Area Conversions

EXAMPLE 7

Change 25 cm^2 to in^2.

$$25 \text{ cm}^2 \times \left(\frac{1 \text{ in.}}{2.54 \text{ cm}}\right)^2 = 3.875 \text{ in}^2$$

Alternative Method:

$$25 \cancel{\text{cm}^2} \times \frac{0.155 \text{ in}^2}{1 \cancel{\text{cm}^2}} = 3.875 \text{ in}^2$$

Conversion factors found in tables are usually rounded. There are many rounding procedures in general use. We will use one of the simplest methods, stated as follows:

ROUNDING NUMBERS

To round a number to a particular place value:

1. If the digit in the next place to the right is less than 5, drop that digit and all other following digits. Replace any whole number places dropped with zeros.
2. If the digit in the next place to the right is 5 or greater, add 1 to the digit in the place to which you are rounding. Drop all other following digits. Replace any whole number places dropped with zeros.

EXAMPLE 8

Change 28.5 m^2 to in^2.

$$28.5 \text{ m}^2 \times \left(\frac{39.4 \text{ in.}}{1 \text{ m}}\right)^2 = 44{,}242.26 \text{ in}^2$$

Alternative Method:

$$28.5 \cancel{m^2} \times \frac{1550 \text{ in}^2}{1 \cancel{m^2}} = 44{,}175 \text{ in}^2$$

Note: The choice of rounded conversion factors will often lead to results that differ slightly. When checking your answers, you must allow for such rounding differences.

To convert between metric and English land area units, use the relationship

$$1 \text{ hectare} = 2.47 \text{ acres}$$

Volume

Standard units of **volume** are based on the cube and are called *cubic centimetres, cubic inches, cubic yards,* or some other cubic unit of measure. A volume of 1 cubic centimetre (cm^3) is the same as the amount of volume contained in a cube 1 cm on each side. One cubic inch (in^3) is the volume contained in a cube 1 in. on each side (Fig. 1.19). The volume of a figure is the number of cubic units that it contains.

Note: When multiplying measurements of like units, multiply the numbers and then multiply the units as follows:

$$3 \text{ in.} \times 5 \text{ in.} \times 4 \text{ in.} = (3 \times 5 \times 4)(\text{in.} \times \text{in.} \times \text{in.}) = 60 \text{ in}^3$$

$$2 \text{ cm} \times 4 \text{ cm} \times 1 \text{ cm} = (2 \times 4 \times 1)(\text{cm} \times \text{cm} \times \text{cm}) = 8 \text{ cm}^3$$

$$1.5 \text{ ft} \times 8.7 \text{ ft} \times 6 \text{ ft} = (1.5 \times 8.7 \times 6)(\text{ft} \times \text{ft} \times \text{ft}) = 78.3 \text{ ft}^3$$

Figure 1.19

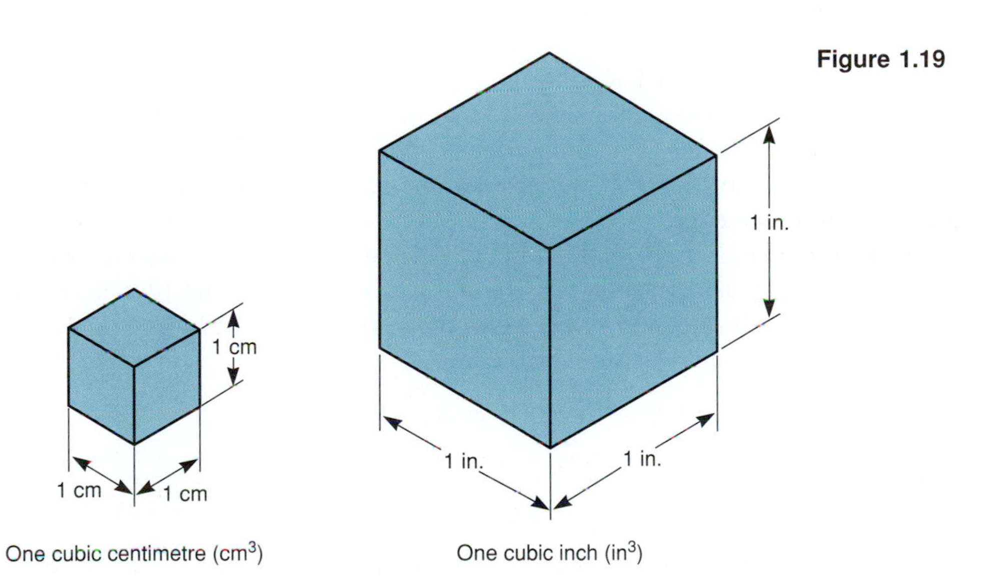

Metric Volume

EXAMPLE 9

Find the volume of a rectangular prism 6 cm long, 4 cm wide, and 5 cm high.

Each cube shown in Fig. 1.20 is 1 cm^3. To find the volume of the rectangular solid, count the number of cubes in the bottom layer of the rectangular solid and then multiply that number by the number of layers that the solid can hold. Therefore, there are 5 layers of 24 cubes, which is 120 cubes or 120 cubic centimetres.

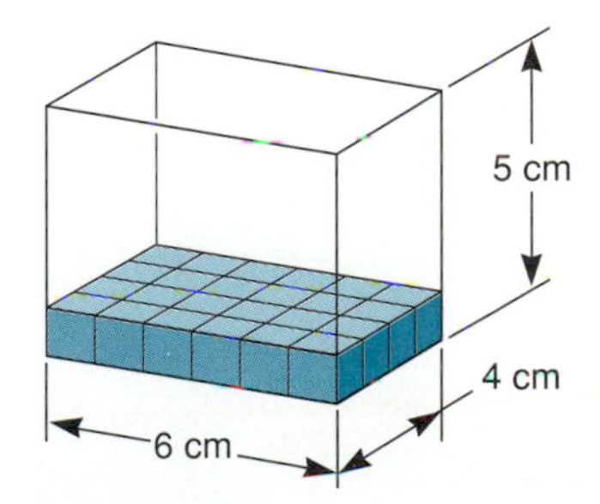

Figure 1.20

Or, by formula, $V = Bh$, where B is the area of the base and h is the height. However, the area of the base is found by lw, where l is the length and w is the width of the rectangle. Therefore, the volume of a rectangular solid can be found by the formula

$$V = lwh = (6 \text{ cm})(4 \text{ cm})(5 \text{ cm}) = 120 \text{ cm}^3$$

Note: $\text{cm} \times \text{cm} \times \text{cm} = \text{cm}^3$.

A common unit of volume in the metric system is the *litre* (L) (Fig. 1.21). The litre is commonly used for liquid volumes.

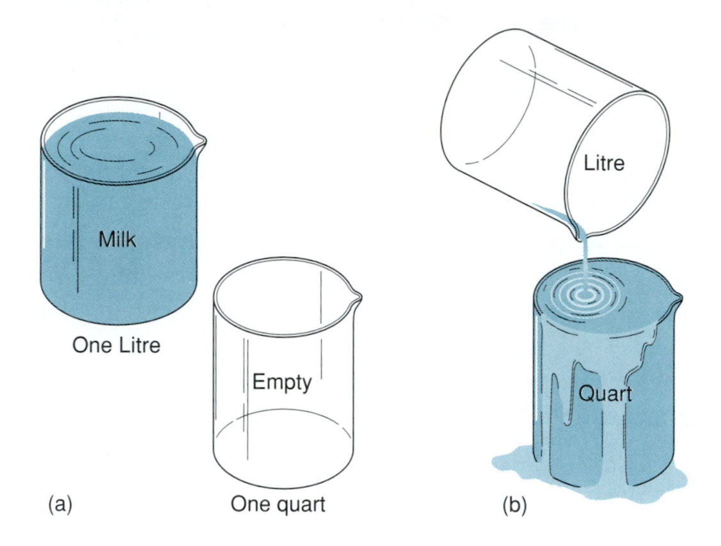

Figure 1.21
One litre of milk is a little more than one quart of milk.

The cubic metre (m^3) and the cubic centimetre (cm^3) are also used to measure volume. The cubic metre is the volume in a cube 1 m on an edge. It is used to measure large volumes. For comparison purposes, the usual teacher's desk could be boxed into 2 cubic metres side by side. The cubic centimetre is the volume in a cube 1 cm on an edge.

The relationship between the litre and the cubic centimetre deserves special mention. The litre is defined as the volume in 1 cubic decimetre (dm^3). That is, 1 litre of liquid fills a cube 1 dm (10 cm) on an edge (Fig. 1.22). The volume of this cube can be found by using the formula

$$V = lwh = (10 \text{ cm})(10 \text{ cm})(10 \text{ cm}) = 1000 \text{ cm}^3$$

That is,

$$1 \text{ L} = 1000 \text{ cm}^3$$

Then

$$\frac{1}{1000} \text{ L} = 1 \text{ cm}^3$$

But

$$\frac{1}{1000} \text{ L} = 1 \text{ mL}$$

Therefore,

$$1 \text{ mL} = 1 \text{ cm}^3$$

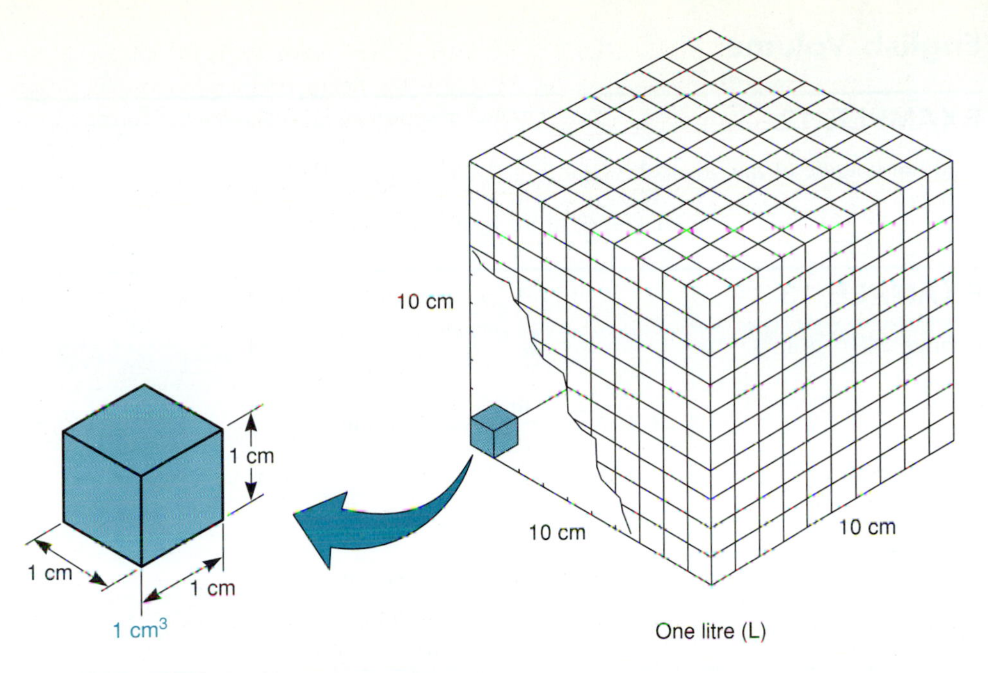

Figure 1.22
One litre contains 1000 cm³.

Milk, soda, and gasoline are usually sold by the litre in countries using the metric system. Liquid medicine, vanilla extract, and lighter fluid are usually sold by the millilitre. Many metric cooking recipes are given in millilitres. Very large quantities of oil are sold by the kilolitre (1000 L).

EXAMPLE 10

Change 0.75 L to millilitres.

$$0.75\ \cancel{L} \times \frac{1000\text{ mL}}{1\ \cancel{L}} = 750\text{ mL}$$

Similarly, the conversion of volume cubic units will be shown using a method of cubing the linear or length conversion factor that you are most likely to remember. An alternative method emphasizing direct use of the conversion tables will also be shown.

EXAMPLE 11

Change 0.65 cm^3 to cubic millimetres.

$$0.65\text{ cm}^3 \times \left(\frac{10\text{ mm}}{1\text{ cm}}\right)^3 = 0.65\ \cancel{\text{cm}^3} \times \frac{10^3\text{ mm}^3}{1^3\ \cancel{\text{cm}^3}} = 650\text{ mm}^3$$

Note: The intermediate step is usually not shown.

Alternative Method:

$$0.65\ \cancel{\text{cm}^3} \times \frac{1000\text{ mm}^3}{1\ \cancel{\text{cm}^3}} = 650\text{ mm}^3$$

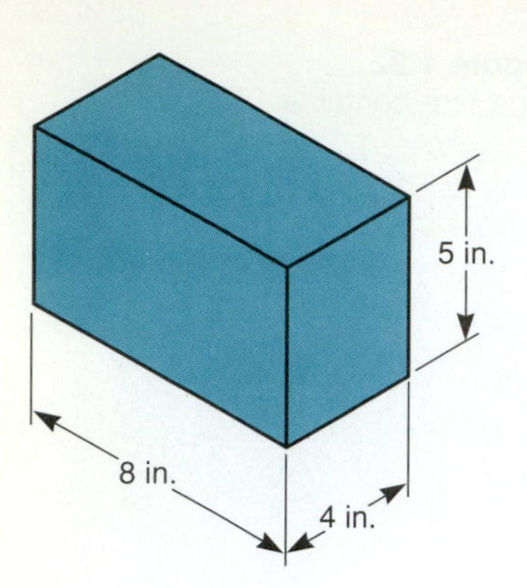

Figure 1.23

English Volume

EXAMPLE 12

Find the volume of the prism shown in Fig. 1.23.

$$V = lwh = (8 \text{ in.})(4 \text{ in.})(5 \text{ in.}) = 160 \text{ in}^3$$

EXAMPLE 13

Change 24 ft^3 to in^3.

$$24 \text{ ft}^3 \times \left(\frac{12 \text{ in.}}{1 \text{ ft}}\right)^3 = 41{,}472 \text{ in}^3$$

Alternative Method:

$$24 \cancel{\text{ft}^3} \times \frac{1728 \text{ in}^3}{1 \cancel{\text{ft}^3}} = 41{,}472 \text{ in}^3$$

Metric–English Volume Conversions

EXAMPLE 14

Change 56 in^3 to cm^3.

$$56 \text{ in}^3 \times \left(\frac{2.54 \text{ cm}}{1 \text{ in.}}\right)^3 = 917.68 \text{ cm}^3$$

Alternative Method:

$$56 \cancel{\text{in}^3} \times \frac{16.4 \text{ cm}^3}{1 \cancel{\text{in}^3}} = 918.4 \text{ cm}^3$$

EXAMPLE 15

Change 28 m^3 to ft^3.

$$28 \text{ m}^3 \times \left(\frac{3.28 \text{ ft}}{1 \text{ m}}\right)^3 = 988.1 \text{ ft}^3$$

Alternative Method:

$$28 \cancel{\text{m}^3} \times \frac{35.3 \text{ ft}^3}{1 \cancel{\text{m}^3}} = 988.4 \text{ ft}^3$$

Surface Area

The lateral (side) **surface area** of any geometric solid is the area of all the lateral faces. The total surface area of any geometric solid is the lateral surface area plus the area of the bases.

EXAMPLE 16

Find the lateral surface area of the prism shown in Fig. 1.24.

$$\text{area of lateral face } 1 = (6 \text{ in.})(5 \text{ in.}) = 30 \text{ in}^2$$
$$\text{area of lateral face } 2 = (5 \text{ in.})(4 \text{ in.}) = 20 \text{ in}^2$$
$$\text{area of lateral face } 3 = (6 \text{ in.})(5 \text{ in.}) = 30 \text{ in}^2$$
$$\text{area of lateral face } 4 = (5 \text{ in.})(4 \text{ in.}) = \underline{20 \text{ in}^2}$$
$$\text{lateral surface area} = 100 \text{ in}^2$$

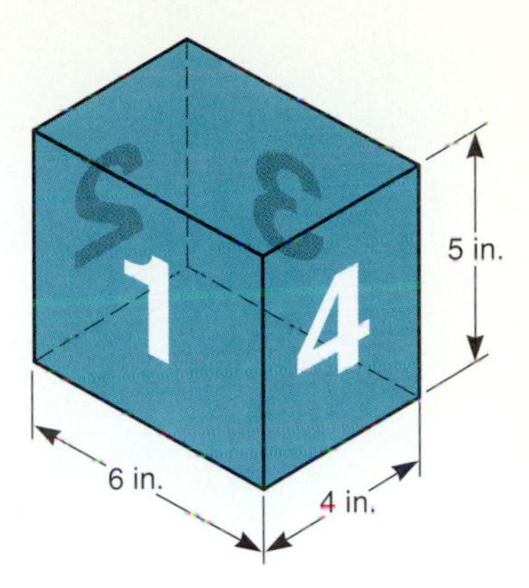

Figure 1.24

EXAMPLE 17

Find the total surface area of the prism shown in Fig. 1.24.

$$\text{total surface area} = \text{lateral surface area} + \text{area of the bases}$$
$$\text{area of base} = (6 \text{ in.})(4 \text{ in.}) = 24 \text{ in}^2$$
$$\text{area of both bases} = 2(24 \text{ in}^2) = 48 \text{ in}^2$$
$$\text{total surface area} = 100 \text{ in}^2 + 48 \text{ in}^2 = 148 \text{ in}^2$$

Area formulas, volume formulas, and lateral surface area formulas are provided on the inside back cover.

■ PROBLEMS 1.5

Find the area of each figure.

1.

5 cm

8 cm

2.

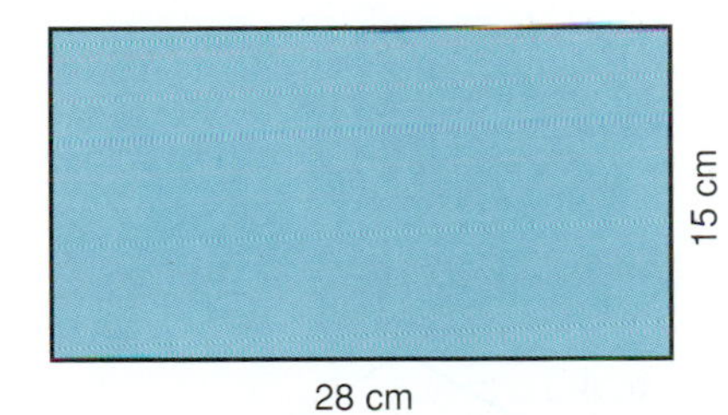

3.

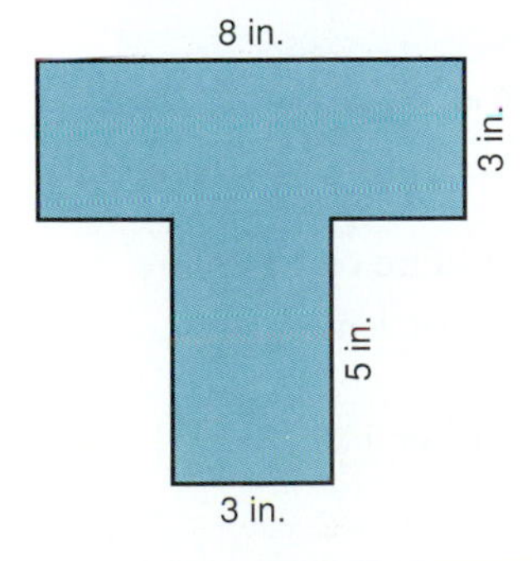

4.

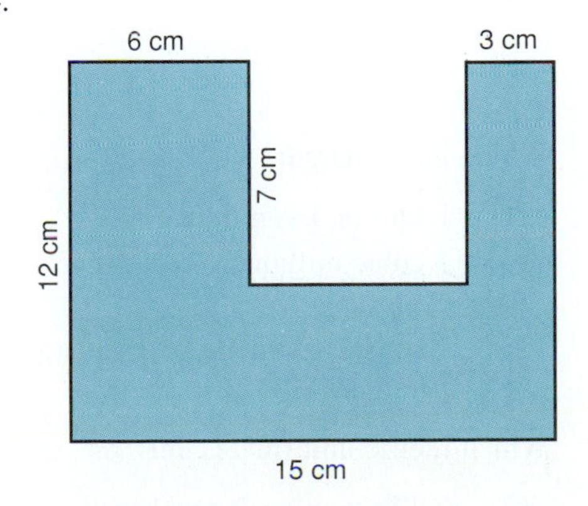

5. Find the cross-sectional area of the I-beam.

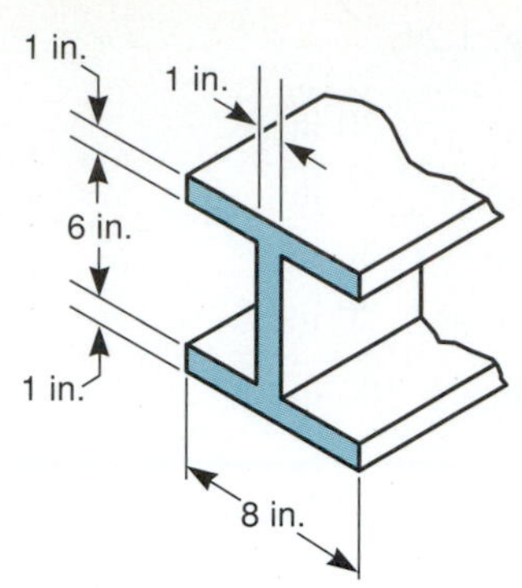

6. Find the largest cross-sectional area of the figure.

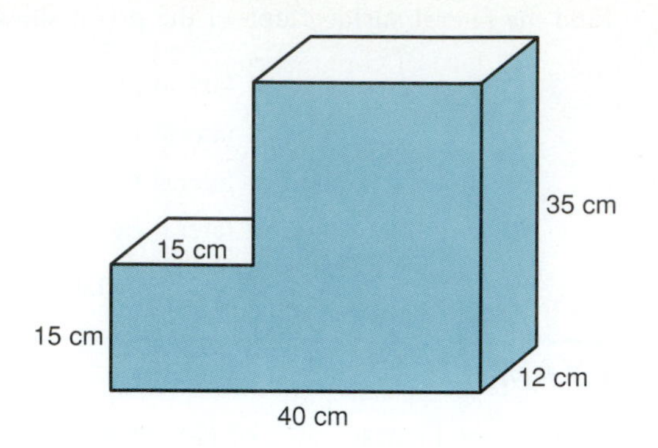

Find the volume in each figure.

7.

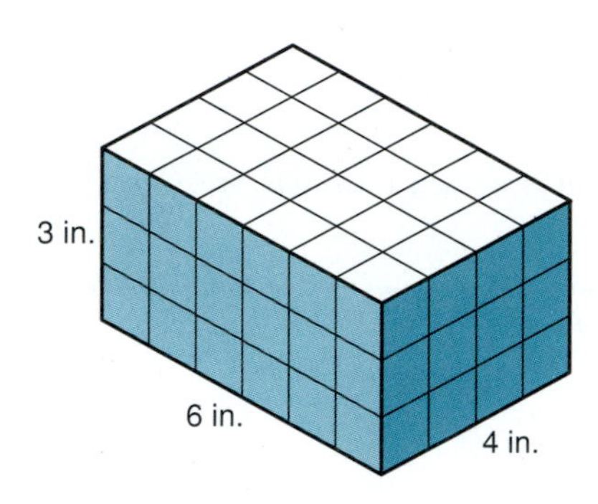

8.

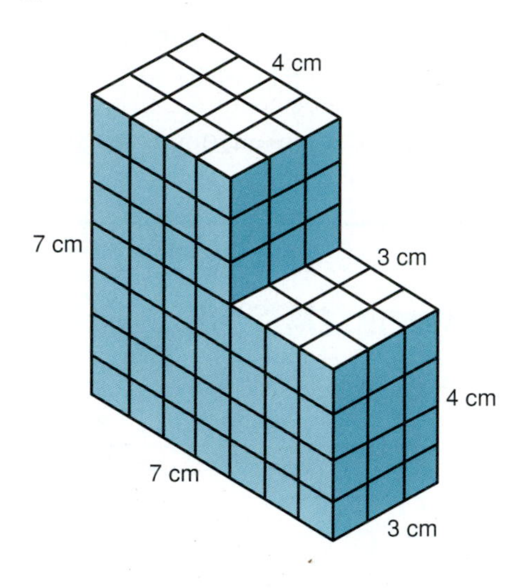

9.

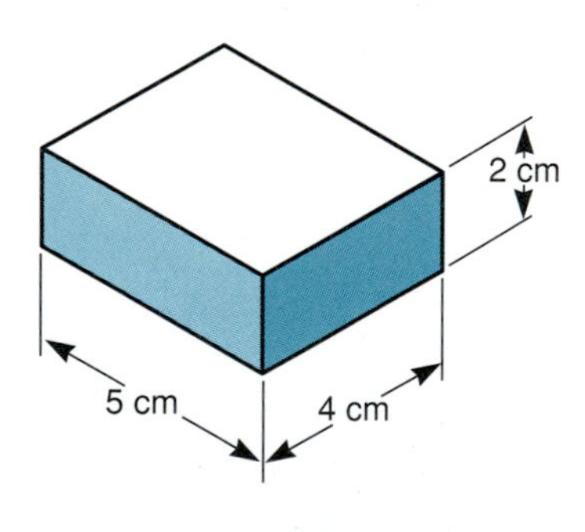

10.

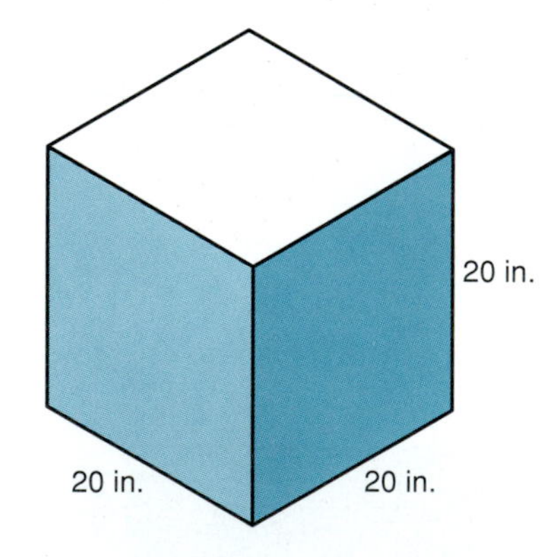

Which unit is larger?

11. 1 litre or 1 centilitre

12. 1 millilitre or 1 kilolitre

13. 1 cubic millimetre or 1 cubic centimetre

14. 1 cm^3 or 1 m^3

15. 1 square kilometre or 1 square hectometre

16. 1 mm^2 or 1 dm^2

Which metric unit (m^3, L, mL, m^2, cm^2, ha) would you use to measure the following?

17. Oil in your car's crankcase

18. Water in a bathtub

19. Floor space in a house
20. Cross section of a piston
21. Storage space in a miniwarehouse
22. Coffee in an office coffeepot
23. Size of a field of corn
24. Page size of a newspaper
25. A dose of cough syrup
26. Size of a cattle ranch
27. Cargo space in a truck
28. Gasoline in your car's gas tank
29. Piston displacement of an engine
30. Paint needed to paint a house
31. Drops to put into your eyes
32. Size of a plot of timber

Fill in the blank with the most reasonable metric unit (m^3, L, mL, m^2, cm^2, ha).

33. Go to the store and buy 4 _____ of root beer for the party.
34. I drank 200 _____ of orange juice for breakfast.
35. Craig bought a 30-_____ tarpaulin for his truck.
36. The cross section of a log is 3200 _____.
37. A farmer's gasoline storage tank holds 4000 _____.
38. Our city water tower holds 500 _____ of water.
39. Brian planted 60 _____ of soybeans this year.
40. David needs some copper tubing with a cross section of 3 _____.
41. Paula ordered 15 _____ of concrete for her new driveway.
42. Barbara heats 420 _____ of living space in her house.
43. Joyce's house has 210 _____ of floor space.
44. Kurt mows 5 _____ of grass each week.
45. Amy is told by her doctor to drink 2 _____ of water each day.
46. My favorite coffee cup holds 225 _____ of coffee.

Fill in each blank.

47. 1 L = _____ mL
48. 1 kL = _____ L
49. 1 L = _____ daL
50. 1 L = _____ kL
51. 1 L = _____ hL
52. 1 L = _____ dL
53. 1 daL = _____ L
54. 1 mL = _____ L
55. 1 mL = _____ cm^3
56. 1 L = _____ cm^3
57. 1 m^3 = _____ cm^3
58. 1 cm^3 = _____ mL
59. 1 cm^3 = _____ L
60. 1 dm^3 = _____ L
61. 1 m^2 = _____ cm^2
62. 1 km^2 = _____ m^2
63. 1 cm^2 = _____ mm^2
64. 1 mm^2 = _____ m^2
65. 1 dm^2 = _____ m^2
66. 1 ha = _____ m^2
67. 1 km^2 = _____ ha
68. 1 ha = _____ km^2
69. Change 7500 mL to L.
70. Change 0.85 L to mL.
71. Change 1.6 L to mL.
72. Change 9 mL to L.
73. Change 275 cm^3 to mm^3.
74. Change 5 m^3 to cm^3.
75. Change 4 m^3 to mm^3.
76. Change 520 mm^3 to cm^3.
77. Change 275 cm^3 to mL.
78. Change 125 cm^3 to L.
79. Change 1 m^3 to L.
80. Change 150 mm^3 to L.
81. Change 7.5 L to cm^3.
82. Change 450 L to m^3.
83. Change 5000 mm^2 to cm^2.
84. Change 1.75 km^2 to m^2.
85. Change 5 m^2 to cm^2.
86. Change 250 cm^2 to mm^2.
87. Change 4×10^8 m^2 to km^2.
88. Change 5×10^7 cm^2 to m^2.
89. Change 5 yd^2 to ft^2.
90. How many m^2 are in 225 ft^2?
91. Change 15 ft^2 to cm^2.
92. How many ft^2 are in a rectangle 15 m long and 12 m wide?
93. Change 108 in^2 to ft^2.
94. How many in^2 are in 51 cm^2?
95. How many in^2 are in a square 11 yd on a side?
96. How many m^2 are in a doorway whose area is 20 ft^2?

97. Change 19 yd^3 to ft^3.
98. How many in^3 are in 29 cm^3?
99. How many yd^3 are in 23 m^3?
100. How many cm^3 are in 88 in^3?
101. Change 8 ft^3 to in^3.
102. How many in^3 are in 12 m^3?
103. The volume of a casting is 38 in^3. What is its volume in cm^3?
104. How many castings of 14 cm^3 can be made from a 12-ft^3 block of steel?
105. Find the lateral surface area of the figure in Problem 9.
106. Find the lateral surface area of the figure in Problem 10.
107. Find the total surface area of the figure in Problem 9.
108. Find the total surface area of the figure in Problem 10.
109. How many mL of water would the figure in Problem 9 hold?
110. How many mL of water would the figure in Problem 8 hold? ■

1.6 MASS AND WEIGHT

The **mass** of an object is the quantity of material making up the object. One unit of mass in the metric system is the *gram* (g) (Fig. 1.25). The gram is defined as the mass of 1 cubic centimetre (cm^3) of water at its maximum density. Since the gram is so small, the **kilogram** (kg) is the basic unit of mass in the metric system. One kilogram is defined as the mass of 1 cubic decimetre (dm^3) of water at its maximum density. The standard kilogram is a special platinum–iridium cylinder at the International Bureau of Weights and Measures near Paris, France. Since 1 dm^3 = 1 L, 1 litre of water has a mass of 1 kilogram.

A common paper clip has a mass of about 1 g.

(a)

Three aspirin have a mass of about 1 g.

(b)

Figure 1.25
One gram.

For very, very small masses, such as medicine dosages, we use the *milligram* (mg). One grain of salt has a mass of about 1 mg. The metric ton (1000 kg) is used to measure the mass of very large quantities, such as coal on a barge, a trainload of grain, or a shipload of ore.

EXAMPLE 1

Change 74 kg to grams.

Choose the conversion factor with kilograms in the denominator so that the kg units cancel each other.

$$74\ \cancel{kg} \times \frac{1000\ g}{1\ \cancel{kg}} = 74{,}000\ g$$

EXAMPLE 2

Change 600 mg to grams.

$$600\ \cancel{mg} \times \frac{1\ g}{1000\ \cancel{mg}} = 0.6\ g$$

The **weight** of an object is a measure of the gravitational force or pull acting on an object. The weight unit in the metric system is the *newton* (N). An apple weighs about one newton.

The pound (lb), a unit of force, is one of the basic English system units. It is defined as the pull of the earth on a cylinder of a platinum–iridium alloy that is stored in a vault at the U.S. Bureau of Standards.

The *ounce* (oz) is another common unit of weight in the English system. The relationship between pounds and ounces is

$$1 \text{ lb} = 16 \text{ oz}$$

The following relationships can be used for conversion between systems of units:

$$1 \text{ N} = 0.225 \text{ lb} \qquad \text{or} \qquad 1 \text{ lb} = 4.45 \text{ N}$$

The mass of an object remains constant, but its weight changes according to its distance from earth or another planet. Mass and weight and their units of measure are discussed later in more detail.

A common method of measuring weights is based on the **spring balance** (Fig. 1.26). The basis of this method is the fact that the distance a spring stretches when a body is supported by it is proportional to the weight of the body. A pointer can be attached to the spring and a calibrated scale added so that the device will read directly in pounds or newtons. The common bathroom scale uses this principle to measure weights.

The other common device for measuring mass is the **platform balance** (Fig. 1.27). Two platforms are connected by a horizontal rod that balances on a knife edge. This device compares the pull of gravity on objects that are on the two platforms. The platforms are at the same height only when the unknown mass of the object on the left is equal to the known mass placed on the right. It is also possible to use one platform and a mass that slides along a calibrated scale. Variations of this basic design are used in scales such as those found in some meat markets and truck scales.

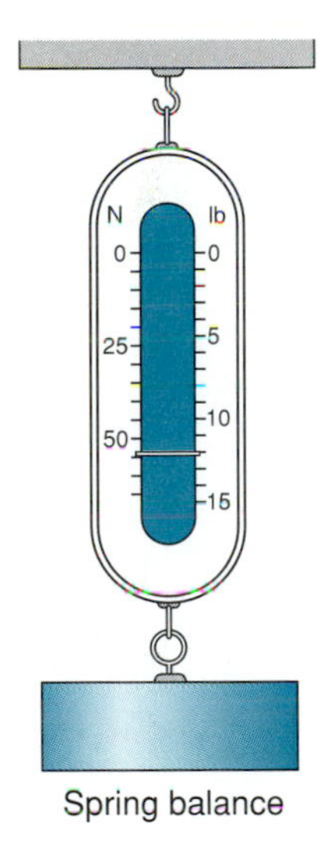

Figure 1.26

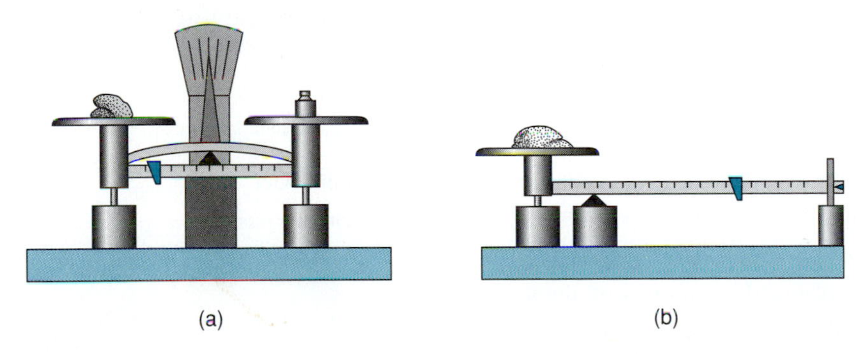

Figure 1.27
Platform balances.

EXAMPLE 3

The weight of the intake valve of an auto engine is 0.18 lb. What is its weight in ounces and in newtons?

To find the weight in ounces, we simply use a conversion factor as follows:

$$0.18 \not{lb} \times \frac{16 \text{ oz}}{1 \not{lb}} = 2.88 \text{ oz}$$

To find weight in newtons, we again use a conversion factor:

$$0.18 \not{lb} \times \frac{4.45 \text{ N}}{1 \not{lb}} = 0.801 \text{ N}$$

■ PROBLEMS 1.6

Which unit is larger?

1. 1 gram or 1 centigram
2. 1 gram or 1 milligram
3. 1 gram or 1 kilogram
4. 1 centigram or 1 milligram
5. 1 centigram or 1 kilogram
6. 1 milligram or 1 kilogram

Which metric unit (kg, g, mg, or metric ton) would you use to measure the following?

7. Your mass
8. An aspirin
9. A bag of lawn fertilizer
10. A bar of hand soap
11. A trainload of grain
12. A sewing needle
13. A small can of corn
14. A channel catfish
15. A vitamin capsule
16. A car

Fill in each blank with the most reasonable metric unit (kg, g, mg, or metric ton).

17. A newborn's mass is about 3 _____.
18. An elevator in a local department store has a load limit of 2000 _____.
19. Margie's diet calls for 250 _____ of meat.
20. A 200-car train carries 11,000 _____ of soybeans.
21. A truckload shipment of copper pipe has a mass of 900 _____.
22. A carrot has a mass of 75 _____.
23. A candy recipe calls for 175 _____ of chocolate.
24. My father has a mass of 70 _____.
25. A pencil has a mass of 10 _____.
26. Postage rates for letters would be based on the _____.
27. A heavyweight boxing champion has a mass of 93 _____.
28. A nickel has a mass of 5 _____.
29. My favorite spaghetti recipe calls for 1 _____ of ground beef.
30. My favorite spaghetti recipe calls for 150 _____ of tomato paste.
31. Our local grain elevator shipped 10,000 _____ of wheat last year.
32. A slice of bread has a mass of about 25 _____.
33. I bought a 5-_____ bag of potatoes at the store today.
34. My grandmother takes 250-_____ capsules for her arthritis.

Fill in each blank.

35. 1 kg = _____ g
36. 1 mg = _____ g
37. 1 g = _____ cg
38. 1 g = _____ hg
39. 1 dg = _____ g
40. 1 dag = _____ g
41. 1 g = _____ mg
42. 1 g = _____ dg
43. 1 hg = _____ g
44. 1 cg = _____ g
45. 1 g = _____ kg
46. 1 g = _____ dag
47. 1 g = _____ μg
48. 1 mg = _____ μg
49. Change 575 g to mg.
50. Change 575 g to kg.

51. Change 650 mg to g.
52. Change 375 kg to g.
53. Change 50 dg to g.
54. Change 485 dag to dg.
55. Change 30 kg to mg.
56. Change 4 metric tons to kg.
57. Change 25 hg to kg.
58. Change 58 μg to g.
59. Change 400 μg to mg.
60. Change 30,000 kg to metric tons.
61. What is the mass of 750 mL of water?
62. What is the mass of 1 m^3 of water?
63. The weight of a car is 3500 lb. Find its weight in newtons.
64. A certain bridge is designed to support 150,000 lb. Find the maximum weight that it will support in newtons.
65. Jose weighs 200 lb. What is his weight in newtons?
66. Change 80 lb to newtons.
67. Change 2000 N to pounds.
68. Change 2000 lb to newtons.
69. Change 120 oz to pounds.
70. Change 3.5 lb to ounces.
71. Change 10 N to ounces.
72. Change 25 oz to newtons.
73. Find the metric weight of a 94-lb bag of cement.
74. What is the weight in newtons of 500 blocks if each weighs 3 lb? ■

1.7 OTHER UNITS

Time

Airlines and other transportation systems run on time schedules that would be meaningless if we did not have a common unit for time measurement. All the common units for time measurement are the same in both the metric and English systems. These units are based on the motion of the earth and the moon (Fig. 1.28). The year is approximately the time required for one complete revolution of the earth about the sun. The month is approximately the time for one complete revolution of the moon about the earth. The day is the time for one rotation of the earth about its axis.

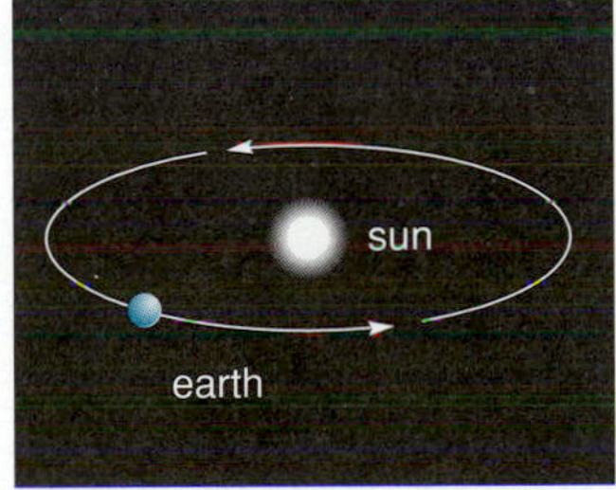

Revolution of earth about the sun

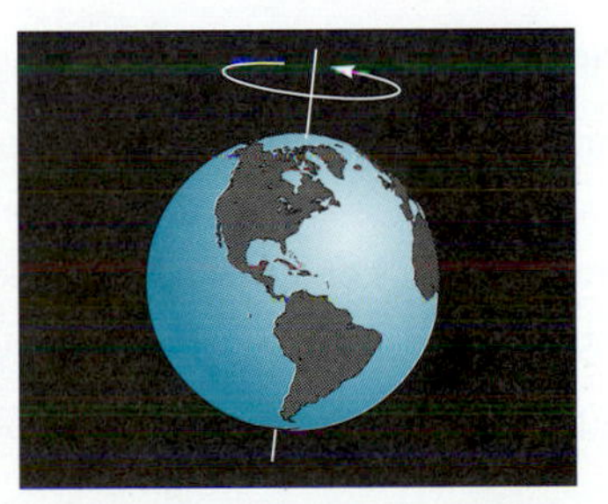

Rotation of earth about its axis

Figure 1.28

The basic time unit is the **second** (s). For many years, the second was defined as 1/86,400 of a mean solar day. Now the standard second is defined more precisely in terms of the frequency of radiation emitted by cesium atoms when they pass between two particular states; that is, the time required for 9,192,631,770 periods of this radiation. The second is not always convenient to use, so other units are necessary. The *minute* (min) is 60 seconds, the *hour* (h) is 60 minutes, and the *day* is 24 hours. The *year* is 365 days in length except for every fourth year, when it is 366 days long. This difference is necessary to keep the seaons at the same time each year, since one revolution of the earth about the sun takes

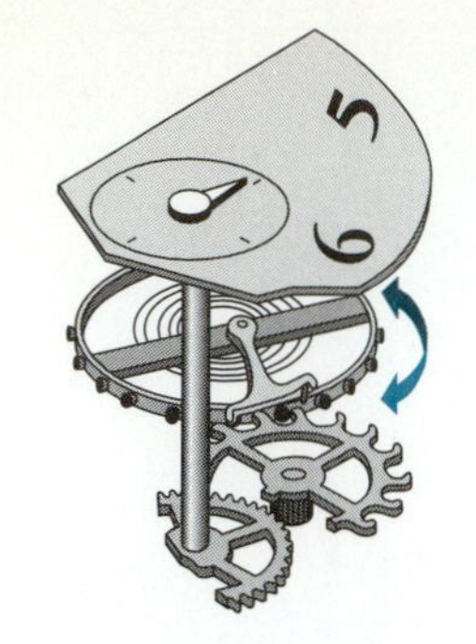

Oscillation of balance wheel

Figure 1.29

$365\frac{1}{4}$ days. Other small variations are compensated for each year by adding or subtracting seconds from the last day of the year. This is done by international agreement.

Common devices for time measurement are the electric clock, the mechanical watch, and the quartz crystal watch. The accuracy of an electric clock depends on how accurately the 60-Hz (hertz = cycles per second) line voltage is controlled. In the United States this is controlled very accurately. Most mechanical watches have a balance wheel that oscillates near a given frequency, usually 18,000 to 36,000 vibrations per hour, and drives the hands of the watch (Fig. 1.29). The quartz crystal in a watch is excited by a small power cell and vibrates 32,768 times per second. The accuracy of the watch depends on how well the frequency of oscillation is controlled.

EXAMPLE 1

Change 2 h 15 min to seconds.

First,

$$2 \cancel{h} \times \frac{60 \text{ min}}{1 \cancel{h}} = 120 \text{ min}$$

Then

$$2 \text{ h } 15 \text{ min} = 120 \text{ min} + 15 \text{ min} = 135 \text{ min}$$

and

$$135 \cancel{\text{min}} \times \frac{60 \text{ s}}{1 \cancel{\text{min}}} = 8100 \text{ s}$$

Very short periods of time are measured in parts of a second, given with the appropriate metric prefix. Such units are commonly used in electronics.

EXAMPLE 2

What is the meaning of each unit?

(a) 1 ms = 1 millisecond = 10^{-3} s and means one-thousandth of a second.

(b) 1 μs = 1 microsecond = 10^{-6} s and means one one-millionth of a second.

(c) 1 ns = 1 nanosecond = 10^{-9} s and means one one-billionth of a second.

(d) 1 ps = 1 picosecond = 10^{-12} s and means one one-trillionth of a second.

Note: The Greek letter μ is pronounced "mu." However, 1 μs is stated or read as "one microsecond."

EXAMPLE 3

Change 45 ms to seconds.

Since 1 ms = 10^{-3} s,

$$45 \cancel{\text{ms}} \times \frac{10^{-3} \text{ s}}{1 \cancel{\text{ms}}} = 45 \times 10^{-3} \text{ s} = 0.045 \text{ s}$$

EXAMPLE 4

Change 0.000000025 s to nanoseconds.

Since 1 ns = 10^{-9} s,

$$0.000000025 \not{s} \times \frac{1 \text{ ns}}{10^{-9} \not{s}} = 25 \text{ ns}$$

Electrical Units

Later in this book we will study electricity and magnetism. **Electricity** is the flow of energy by charge transported through wires. The importance of electricity to our industrialized society cannot be overestimated. Every household and every industry depends on energy to run appliances and large machinery.

For this study of electricity we need to define several units. The **ampere** (A) is a basic unit and is a measure of the amount of electric current. The **coulomb** (C) is a measure of the amount of electrical charge. The **volt** (V) is a measure of electric potential. The **watt** (W) is a measure of power. The **kilowatt-hour** (kWh) is a measure of work or electrical energy used.

The basic unit of electric current is the ampere, or amp (A), which is the same as in the English system. The ampere is a fairly large amount of current, so smaller currents are measured in parts of an ampere and are used with the appropriate metric prefix.

EXAMPLE 5

What is the meaning of each unit?

(a) 1 mA = 1 milliampere = 10^{-3} A and means one one-thousandth of an ampere.
(b) 1 μA = 1 microampere = 10^{-6} A and means one one-millionth of an ampere.

EXAMPLE 6

Change 450 μA to amperes.
Since 1 μA = 10^{-6} A,

$$450 \not{\mu A} \times \frac{10^{-6}}{1 \not{\mu A}} = 0.00045 \text{ A}$$

EXAMPLE 7

Change 0.0065 A to milliamperes.
Since 1 mA = 10^{-3} A,

$$0.0065 \not{A} \times \frac{1 \text{ mA}}{10^{-3} \not{A}} = 6.5 \text{ mA}$$

The common metric unit for both mechanical and electrical power is the *watt* (W).

EXAMPLE 8

What is the meaning of each unit?

(a) 1 MW = 1 megawatt = 10^6 W and means 1 million watts.
(b) 1 kW = 1 kilowatt = 10^3 W and means 1000 watts.
(c) 1 mW = 1 milliwatt = 10^{-3} W and means one one-thousandth of a watt.

EXAMPLE 9

Change 8 MW to watts.

Since 1 MW = 10^6 W,

$$8 \cancel{\text{MW}} \times \frac{10^6 \text{ W}}{1 \cancel{\text{MW}}} = 8 \times 10^6 \text{ W} = 8{,}000{,}000 \text{ W}$$

EXAMPLE 10

Change 0.375 W to milliwatts.

Since 1 mW = 10^{-3} W,

$$0.375 \cancel{\text{W}} \times \frac{1 \text{ mW}}{10^{-3} \cancel{\text{W}}} = 375 \text{ mW}$$

■ PROBLEMS 1.7

Fill in each blank.

1. The basic metric unit of time is _____. Its abbreviation is _____.
2. The basic metric unit of electric current is _____. Its abbreviation is _____.
3. The common metric unit of power is _____. Its abbreviation is _____.

Which is larger?

4. 1 watt or 1 milliwatt
5. 1 milliampere or 1 microampere
6. 1 millisecond or 1 nanosecond
7. 1 MW or 1 mW
8. 1 ps or 1 μs
9. 1 mA or 1 A
10. 1 mW or 1 μW

Write the abbreviation for each unit.

11. 4.7 microamperes
12. 8.6 microseconds
13. 7.5 kilowatts
14. 250 milliamperes
15. 45 nanoseconds
16. 15 amperes
17. 125 megawatts
18. 75 picoseconds

Fill in each blank.

19. 1 kW = _____ W
20. 1 A = _____ μA
21. 1 s = _____ ns
22. 1 mA = _____ A
23. 1 MW = _____ W
24. 1 μs = _____ s
25. 1 mA = _____ μA
26. 1 ns = _____ ps
27. 1 MW = _____ μW
28. 1 μs = _____ ns
29. Change 4 A to mA.
30. Change 7500 W to kW.
31. Change 52 mW to μW.
32. Change 345 μs to s.
33. Change 6800 mA to A.
34. Change 1 h 25 min to min.
35. Change 4 h 25 min 15 s to s.
36. Change 7×10^6 s to h.
37. Change 4 s to ns.
38. Change 8×10^{10} μW to W.
39. Change 3×10^{10} μW to MW.
40. Change 1 h to ps. ■

1.8 MEASUREMENT: SIGNIFICANT DIGITS AND ACCURACY

Up to this time in your studies, probably all numbers and all measurements have been treated as exact numbers. An **exact number** is a number that has been determined as a result of counting, such as 24 students are enrolled in this class, or by some definition, such as 1 h = 60 min or 1 in. = 2.54 cm, a conversion definition agreed to by the world governments' bureaus of standards. Generally, the treatment of the addition, subtraction, multiplication, and division of exact numbers is the emphasis or main content of elementary mathematics.

However, nearly all data of a technical nature involve **approximate numbers;** that is, they have been determined as a result of some measurement process—some direct, as with a ruler, and some indirect, as with a surveying transit or reading an electric meter. First, realize that no measurement can be found exactly. The length of the cover of this book can be found using many instruments. The better the measuring device used, the better the measurement.

A measurement may be expressed in terms of its accuracy or its precision. The **accuracy** of a measurement refers to the number of digits, called **significant digits,** which indicates the number of units that we are reasonably sure of having counted when making a measurement. The greater the number of significant digits given in a measurement, the better the accuracy and vice versa.

EXAMPLE 1

The average distance between the moon and the earth is 385,000 km. This measurement indicates measuring 385 thousands of kilometres; its accuracy is indicated by three significant digits.

EXAMPLE 2

A measurement of 0.025 cm indicates measuring 25 thousandths of a centimetre; its accuracy is indicated by two significant digits.

EXAMPLE 3

A measurement of 0.0500 A indicates measuring $50\bar{0}$ ten-thousandths of an ampere; its accuracy is indicated by three significant digits.

Notice that sometimes a zero is significant and sometimes it is not. To clarify this, we use the following rules for significant digits:

SIGNIFICANT DIGITS

1. All nonzero digits are significant: 156.4 m has four significant digits (this measurement indicates 1564 tenths of metres).
2. All zeros between significant digits are significant: 306.02 km has five significant digits (this measurement indicates 30,602 hundredths of kilometres).
3. In a number greater than 1, a zero that is specially tagged, such as by a bar above it, is significant: $23\bar{0},000$ km has three significant digits (this measurement indicates $23\bar{0}$ thousands of kilometres).
4. All zeros to the right of a significant digit *and* a decimal point are significant: 86.10 cm has four significant digits (this measurement indicates $861\bar{0}$ hundredths of centimetres).

5. In whole-number measurements, zeros at the right that are not tagged are *not* significant: 2500 m has two significant digits (25 hundreds of metres).

6. In measurements of less than 1, zeros at the left are *not* significant: 0.00752 m has three significant digits (752 hundred-thousandths of a metre).

When a number is written in scientific notation, the decimal part indicates the number of significant digits. For example, $20\bar{0},000$ m would be written in scientific notation as 2.00×10^5 m.

In summary:

To find the number of significant digits:

1. All nonzero digits are significant.

2. Zeros are significant when they
  (a) Are between significant digits;
  (b) Follow the decimal point and a significant digit; or
  (c) Are in a whole number and a bar is placed over the zero.

EXAMPLE 4

Determine the accuracy (the number of significant digits) of each measurement.

Measurement	*Accuracy (significant digits)*
(a) 2642 ft	4
(b) 2005 m	4 (Both zeros are significant.)
(c) 2050 m	3 (Only the first zero is significant.)
(d) 2500 m	2 (No zero is significant.)
(e) $25\bar{0}0$ m	3 (Only the first zero is significant.)
(f) $250\bar{0}$ m	4 (Both zeros are significant.)
(g) 34,000 mi	2 (No zeros are significant.)
(h) 15,670,000 lb	4 (No zeros are significant.)
(i) 203.05 km	5 (Both zeros are significant.)
(j) 0.000345 kg	3 (No zeros are significant.)
(k) 75 V	2
(l) 2.3 A	2
(m) 0.02700 g	4 (Only the right two zeros are significant.)
(n) 2.40 cm	3 (The zero is significant.)
(o) 4.050 μA	4 (All zeros are significant.)
(p) 100.050 km	6 (All zeros are significant.)
(q) 0.004 s	1 (No zeros are significant.)
(r) 2.03×10^4 m^2	3 (The zero is significant.)
(s) 1.0×10^{-3} A	2 (The zero is significant.)
(t) 5×10^6 kg	1
(u) 3.060×10^8 m^3	4 (Both zeros are significant.)

■ PROBLEMS 1.8

Determine the accuracy (the number of significant digits) of each measurement.

1. 536 V
2. 307.3 mi
3. 5007 m
4. 5.00 cm
5. 0.0070 in.
6. 6.010 cm

7. $84\bar{0}0$ km
8. $30\bar{0}0$ ft
9. 187.40 m
10. $5\bar{0}0$ g
11. 0.00700 in.
12. 10.30 cm
13. 376.52 m
14. 3.05 mi
15. 4087 kg
16. 35.00 mm
17. 0.0160 in.
18. $37\bar{0}$ lb
19. $4\bar{0}00$ Ω
20. 5010 ft^3
21. 7 A
22. 32,000 tons
23. 70.00 m^2
24. 0.007 m
25. 2.4×10^3 kg
26. 1.20×10^{-5} ms
27. 3.00×10^{-4} A
28. 4.0×10^6 ft
29. 5.106×10^7 V
30. 1×10^{-9} m ■

1.9 MEASUREMENT: PRECISION

The **precision** of a measurement refers to the smallest unit with which a measurement is made, that is, the position of the last significant digit.

EXAMPLE 1

The **precision** of the measurement 385,000 km is 1000 km. (The position of the last significant digit is in the thousands place.)

EXAMPLE 2

The precision of the measurement 0.025 cm is 0.001 cm. (The position of the last significant digit is in the thousandths place.)

EXAMPLE 3

The precision of the measurement 0.0500 A is 0.0001 A. (The position of the last significant digit is in the ten-thousandths place.)

Unfortunately, the terms *accuracy* and *precision* have several different common meanings. Here we will use each term consistently as we have defined them. A measurement of 0.0004 cm has good precision and poor accuracy when compared with the measurement 378.0 cm.

Measurement	*Precision*	*Accuracy*
0.0004 cm	0.0001 cm	1 significant digit
378.0 cm	0.1 cm	4 significant digits

EXAMPLE 4

Determine the precision of each measurement given in Example 4 of Section 1.8.

Measurement	*Precision*	*Accuracy (significant digits)*
(a) 2642 ft	1 ft	4
(b) 2005 m	1 m	4
(c) 2050 m	10 m	3
(d) 2500 m	100 m	2
(e) $25\bar{0}0$ m	10 m	3
(f) $250\bar{0}$ m	1 m	4
(g) 34,000 mi	1000 mi	2
(h) 15,670,000 lb	10,000 lb	4
(i) 203.05 km	0.01 km	5

Measurement	Precision	Accuracy (significant digits)
(j) 0.000345 kg	0.000001 kg	3
(k) 75 V	1 V	2
(l) 2.3 A	0.1 A	2
(m) 0.02700 g	0.00001 g	4
(n) 2.40 cm	0.01 cm	3
(o) 4.050 μA	0.001 μA	4
(p) 100.050 km	0.001 km	6
(q) 0.004 s	0.001 s	1
(r) 2.03×10^4 m^2	0.01×10^4 m^2 or 100 m^2	3
(s) 1.0×10^{-3} A	0.1×10^{-3} A or 0.0001 A	2
(t) 5×10^6 kg	1×10^6 kg or 1,000,000 kg	1
(u) 3.060×10^8 m^3	0.001×10^8 m^3 or 1×10^5 m^3 or 100,000 m^3	4

■ PROBLEMS 1.9

Determine the precision of each measurement.

1. 536 V
2. 307.3 mi
3. 5007 m
4. 5.00 cm
5. 0.0070 in.
6. 6.010 cm
7. $84\bar{0}0$ km
8. $30\bar{0}0$ ft
9. 187.40 m
10. $5\bar{0}0$ g
11. 0.00700 in.
12. 10.30 cm
13. 376.52 m
14. 3.05 mi
15. 4087 kg
16. 35.00 mm
17. 0.0160 in.
18. $37\bar{0}$ lb
19. $4\bar{0}00$ Ω
20. 5010 ft^3
21. 7 A
22. 32,000 tons
23. 70.00 m^2
24. 0.007 m
25. 2.4×10^3 kg
26. 1.20×10^{-5} ms
27. 3.00×10^{-4} A
28. 4.0×10^6 ft
29. 5.106×10^7 V
30. 1×10^{-9} m

In each set of measurements, find the measurement that is (a) the most accurate, and (b) the most precise.

31. 15.7 in.; 0.018 in.; 0.07 in.
32. 368 ft; 600 ft; 180 ft
33. 0.734 cm; 0.65 cm; 16.01 cm
34. 3.85 m; 8.90 m; 7.00 m
35. 0.0350 A; 0.025 A; 0.00040 A; 0.051 A
36. 125.00 g; 8.50 g; 9.000 g; 0.05 g
37. $27{,}0\bar{0}0$ L; 350 L; 27.6 L; 4.75 L
38. 8.4 m; 15 m; 180 m; 0.40 m
39. 500 Ω; 10,000 Ω; 500,000 Ω; 50 Ω
40. 7.5 mA; 14.2 mA; 10.5 mA; 120.0 mA

In each set of measurements, find the measurement that is (a) the least accurate, and (b) the least precise.

41. 16.4 in.; 0.075 in.; 0.05 in.
42. 475 ft; 300 ft; 360 ft
43. 27.5 m; 0.65 m; 12.02 m
44. 5.7 kg; 120 kg; 0.025 kg
45. 0.0250 g; 0.015 g; 0.00005 g; 0.75 g
46. 185.0 m; 6.75 m; 5.000 m; 0.09 m
47. 45,000 V; 250 V; 16.8 V; 0.25 V; 3 V
48. 2.50 kg; 42.0 kg; $15\bar{0}$ kg; 0.500 kg
49. $20\bar{0}0$ Ω; $10{,}\bar{0}00$ Ω; $40\bar{0}{,}000$ Ω; 20 Ω
50. 80 V; 250 V; 12,550 V; $26\bar{0}0$ V ■

1.10 CALCULATIONS WITH MEASUREMENTS

If one person measured the length of one of two parts of a shaft with a micrometer calibrated in 0.01 mm as 42.28 mm and another person measured the second part with a ruler calibrated in mm as 54 mm, would the total length be 96.28 mm? Note that the sum 96.28 mm indicates a precision of 0.01 mm. The precision of the ruler is 1 mm, which means that the measurement 54 mm with the ruler could actually be anywhere between 53.50 mm and 54.50 mm using the micrometer (which has a precision of 0.01 mm). That is, using the ruler, any measurement between 53.50 mm and 54.50 mm can only be read as 54 mm. Of course, this means that the tenths and hundredths digits in the sum 96.28 mm are really meaningless. In other words, *the sum or difference of measurements can be no more precise than the least precise measurement.* That is,

To add or subtract measurements:

1. Make certain that all the measurements are expressed in the same units. If they are not, convert them all to the same units.
2. Add or subtract.
3. Round the result to the same precision as the least precise measurement.

EXAMPLE 1

Add the measurements: 16.6 mi; 124 mi; 3.05 mi; 0.837 mi.

All measurements are in the same units, so add.

$$\begin{array}{r@{\,}l} 16.6 & \text{mi} \\ 124 & \text{mi} \\ 3.05 & \text{mi} \\ 0.837 & \text{mi} \\ \hline 144.487 & \text{mi} \rightarrow 144 \text{ mi} \end{array}$$

Then, round this sum to the same precision as the least precise measurement, which is 124 mi. Thus, the sum is 144 mi.

EXAMPLE 2

Add the measurements: 1370 cm; 1575 mm; 2.374 m; 8.63 m.

First, convert all measurements to the same units, say m.

$$1370 \text{ cm} = 13.7 \text{ m}$$
$$1575 \text{ mm} = 1.575 \text{ m}$$

Then add,

$$\begin{array}{r@{\,}l} 13.7 & \text{m} \\ 1.575 & \text{m} \\ 2.374 & \text{m} \\ 8.63 & \text{m} \\ \hline 26.279 & \text{m} \rightarrow 26.3 \text{ m} \end{array}$$

Then, round this sum to the same precision as the least precise measurement, which is 13.7 m. Thus, the sum is 26.3 m.

EXAMPLE 3

Subtract the measurements: 3457.8 g − 2.80 kg.

First, convert both measurements to the same unit, say g.

$$2.80 \text{ kg} = 28\bar{0}0 \text{ g}$$

Then subtract.

$$\begin{array}{r} 3457.8 \text{ g} \\ \underline{28\bar{0}0 \quad \text{g}} \\ 657.8 \text{ g} \rightarrow 660 \text{ g} \end{array}$$

Then, round this difference to the same precision as the least precise measurement, which is $28\bar{0}0$ g. Thus, the difference is 660 g.

Now suppose that you wish to find the area of the base of a rectangular building. You measure its length as 54.7 m and its width as 21.5 m. Its area is then

$$A = lw$$

$$A = (54.7 \text{ m})(21.5 \text{ m})$$

$$= 1176.05 \text{ m}^2$$

Note that the result contains six significant digits, whereas each of the original measurements contains only three significant digits. To rectify this inconsistency, we say that the product or quotient of measurements can be no more accurate than the least accurate measurement. That is,

> To multiply or divide measurements:
>
> **1.** Multiply or divide the measurements as given.
>
> **2.** Round the result to the same number of significant digits as the measurement with the least number of significant digits.

Using the preceding rules the area of the base of the rectangular building is 1180 m^2.

Note: We assume throughout that you are using a calculator to do all calculations.

EXAMPLE 4

Multiply the measurements: 124 ft × 187 ft.

$$124 \text{ ft} \times 187 \text{ ft} = 23{,}188 \text{ ft}^2$$

Round this product to three significant digits, which is the accuracy of the least accurate measurement (and also the accuracy of each measurement in the example). That is,

$$124 \text{ ft} \times 187 \text{ ft} = 23{,}200 \text{ ft}^2$$

EXAMPLE 5

Multiply the measurements: (2.75 m)(1.25 m)(0.75 m).

$$(2.75 \text{ m})(1.25 \text{ m})(0.75 \text{ m}) = 2.578125 \text{ m}^3$$

Round this product to two significant digits, which is the accuracy of the least accurate measurement (0.75 m). That is,

$$(2.75 \text{ m})(1.25 \text{ m})(0.75 \text{ m}) = 2.6 \text{ m}^3$$

EXAMPLE 6

Divide the measurements: 144,000 $\text{ft}^3 \div 108$ ft.

$$144{,}000 \text{ ft}^3 \div 108 \text{ ft} = 1333.333\ldots \text{ ft}^2$$

Round this quotient to three significant digits, which is the accuracy of the least accurate measurement (the accuracy of both measurements in this example). That is,

$$144{,}000 \text{ ft}^3 \div 108 \text{ ft} = 1330 \text{ ft}^2$$

EXAMPLE 7

Find the value of $\dfrac{68 \text{ ft} \times 10{,}\bar{0}00 \text{ lb}}{95.6 \text{ s}}$.

$$\frac{68 \text{ ft} \times 10{,}\bar{0}00 \text{ lb}}{95.6 \text{ s}} = 7112.9707\ldots \frac{\text{ft lb}}{\text{s}}$$

Round this result to two significant digits, which is the accuracy of the least accurate measurement (68 ft). That is,

$$\frac{68 \text{ ft} \times 10{,}\bar{0}00 \text{ lb}}{95.6 \text{ s}} = 7100 \text{ ft lb/s}$$

EXAMPLE 8

Find the value of $\dfrac{(58.0 \text{ kg})(2.40 \text{ m/s})^2}{5.40 \text{ m}}$.

$$\frac{(58.0 \text{ kg})(2.40 \text{ m/s})^2}{5.40 \text{ m}} = 61.8666\ldots \frac{\text{kg m}}{\text{s}^2}$$

Carefully simplify the units:

$$\frac{(\text{kg})(\text{m/s})^2}{\text{m}} = \frac{(\text{kg})(\overset{\text{m}}{\cancel{\text{m}^2}}/\text{s}^2)}{\cancel{\text{m}}} = \frac{\text{kg m}}{\text{s}^2}$$

Round this result to three significant digits, which is the accuracy of the least accurate measurement (the accuracy of all measurements in this example). That is,

$$\frac{(58.0 \text{ kg})(2.40 \text{ m/s})^2}{5.40 \text{ m}} = 61.9 \text{ kg m/s}^2$$

Note: To multiply or divide measurements, the units do not need to be the same. The units must be the same to add or subtract measurements. Also, the units are multiplied and/or divided in the same manner as the corresponding numbers.

Any power or root of a measurement should be rounded to the same accuracy as the given measurement.

COMBINATIONS OF OPERATIONS WITH MEASUREMENTS

For combinations of additions, subtractions, multiplications, divisions, and powers involving measurements, follow the usual order of operations used in mathematics as follows:

1. Perform all operations inside parentheses first.
2. Evaluate all powers.
3. Perform any multiplications or divisions, in order, from left to right; then express each product or quotient using its correct accuracy.
4. Perform any additions or subtractions, in order, from left to right; then express the final result using the correct precision.

EXAMPLE 9

Find the value of (4.00 m)(12.65 m) + $(24.6 \text{ m})^2$.

$$(4.00 \text{ m})(12.65 \text{ m}) + (24.6 \text{ m})^2 =$$
$$50.6 \text{ m}^2 + 605 \text{ m}^2 = 656 \text{ m}^2$$

Obviously, such calculations with measurements should be done with a calculator. When no calculator is available, you may round the original measurements or any intermediate results to one more digit than the required accuracy or precision as required in the final result.

If both exact numbers and approximate numbers (measurements) occur in the same calculation, only the approximate numbers are used to determine the accuracy or precision of the result.

The procedures for operations with measurements shown here are based on methods followed and presented by the American Society for Testing and Materials. There are even more sophisticated methods for dealing with the calculations of measurements. The method one uses, and indeed whether one should even follow any given procedure, depends on the number of measurements and the sophistication needed for a particular situation.

In this book, we generally follow the customary practice of expressing measurements in terms of three significant digits, which is the accuracy used in most engineering and design work.

■ PROBLEMS 1.10

Use the rules for addition of measurements to add each set of measurements.

1.	2.	3.	4.
3847 ft	8,560 m	42.8 cm	0.456 g
5800 ft	84,000 m	16.48 cm	0.93 g
4520 ft	18,476 m	1.497 cm	0.402 g
	12,500 m	12.8 cm	0.079 g
		9.69 cm	0.964 g

5. 39,000 V; 19,600 V; 8470 V; 2500 V

6. 6800 ft; 2760 ft; $4\bar{0}00$ ft; $20\bar{0}0$ ft
7. 467 m; 970 cm; $12\bar{0}0$ cm; 1352 cm; $30\bar{0}$ m
8. 36.8 m; 147.5 cm; 1.967 m; 125.0 m; 98.3 cm
9. 12 A; 1.004 A; 0.040 A; 3.9 A; 0.87 A
10. 160,000 V; 84,200 V; 4300 V; 239,000 V; 17,450 V

Use the rules for subtraction of measurements to subtract each second measurement from the first.

11. $\begin{array}{r} 2876 \text{ kg} \\ \underline{2400 \text{ kg}} \end{array}$

12. $\begin{array}{r} 14.73 \text{ m} \\ \underline{9.378 \text{ m}} \end{array}$

13. $\begin{array}{r} 45.585 \text{ g} \\ \underline{4.6 \text{ g}} \end{array}$

14. $\begin{array}{r} 34{,}500 \text{ kg} \\ \underline{9{,}5\bar{0}0 \text{ kg}} \end{array}$

15. 4200 km − 975 km
16. 64.73 g − 9.4936 g
17. 1,600,000 V − 685,000 V
18. 170 mm − 10.2 cm
19. 3.00 m − $26\bar{0}$ cm
20. 1.40 mA − 0.708 mA

Use the rules for multiplication of measurements to multiply each set of measurements.

21. 125 m × 39 m
22. 470 ft × 1200 ft
23. (1637 km)(857 km)
24. (9100 m)($6\bar{0}0$ m)
25. (18.70 m)(39.45 m)
26. (565 cm)(180 cm)
27. 14.5 cm × 18.7 cm × 20.5 cm
28. (0.046 m)(0.0317 m)(0.0437 m)
29. ($45\bar{0}$ in.)(315 in.)(205 in.)
30. (18.7 kg)(217 m)

Use the rules for division of measurements to divide.

31. $360 \text{ ft}^3 \div 12 \text{ ft}^2$
32. $125 \text{ m}^2 \div 3.0 \text{ m}$
33. $275 \text{ cm}^2 \div 90.0 \text{ cm}$
34. 185 mi ÷ 4.5 h
35. $\dfrac{347 \text{ km}}{4.6 \text{ h}}$
36. $\dfrac{2700 \text{ m}^3}{9\bar{0}0 \text{ m}^2}$
37. $\dfrac{8800 \text{ V}}{8.5 \text{ A}}$
38. $\dfrac{4960 \text{ ft}}{2.95 \text{ s}}$

Use the rules for multiplication and division of measurements to find the value of each of the following.

39. $\dfrac{(18 \text{ ft})(290 \text{ lb})}{4.6 \text{ s}}$
40. $\dfrac{18.5 \text{ kg} \times 4.65 \text{ m}}{19.5 \text{ s}}$
41. $\dfrac{4500 \text{ V}}{12.3 \text{ A}}$
42. $\dfrac{48.9 \text{ kg}}{(1.5 \text{ m})(3.25 \text{ m})}$
43. $\dfrac{(48.7 \text{ m})(68.5 \text{ m})(18.4 \text{ m})}{(35.5 \text{ m})(40.0 \text{ m})}$
44. $\frac{1}{2}(270 \text{ kg})(16.4 \text{ m/s})^2$
45. $\dfrac{(115 \text{ V})^2}{25\ \Omega}$
46. $\dfrac{(45.2 \text{ kg})(13.7 \text{ m})}{(2.65 \text{ s})^2}$
47. $\dfrac{(85.7 \text{ kg})(25.7 \text{ m/s})^2}{12.5 \text{ m}}$
48. $\dfrac{(120 \text{ V})^2}{275\ \Omega}$
49. $\frac{4}{3}\pi(13.5 \text{ m})^3$
50. $\dfrac{140 \text{ g}}{(3.4 \text{ cm})(2.8 \text{ cm})(5.6 \text{ cm})}$
51. (213 m)(65.3 m) − (175 m)(44.5 m)
52. $(4.5 \text{ ft})(7.2 \text{ ft})(12.4 \text{ ft}) + (5.42 \text{ ft})^3$
53. $\dfrac{(125 \text{ ft})(295 \text{ ft})}{44.7 \text{ ft}} + \dfrac{(215 \text{ ft})^3}{(68.8 \text{ ft})(12.4 \text{ ft})} + \dfrac{(454 \text{ ft})^3}{(75.5 \text{ ft})^2}$
54. $(12.5 \text{ m})(46.75 \text{ m}) + \dfrac{(6.76 \text{ m})^3}{4910 \text{ m}} - \dfrac{(41.5 \text{ m})(21 \text{ m})(28.8 \text{ m})}{31.7 \text{ m}}$ ■

GLOSSARY

Accuracy The number of digits, called significant digits, in a measurement that indicates the number of units that we are reasonably sure of having counted. The greater the number of significant digits, the better the accuracy. (p. 33)

Ampere The basic unit of electric current. (p. 31)

Approximate Number A number that has been determined by some measurement or estimation process. (p. 33)

Area The number of square units contained in a figure. (p. 15)

Conversion Factor An expression that is used to convert from one set of units to another. Often expressed as a fraction whose numerator and denominator are equal to each other although in different units. (p. 10)

Coulomb The basic unit of electrical charge. (p. 31)

Electricity The flow of energy by charge transported through wires. (p. 31)

Exact Number A number that has been determined as a result of counting, such as 21 students enrolled in a class, or by some definition, such as 1 h = 60 min. (p. 32)

Kilogram The basic metric unit of mass. (p. 26)

Kilowatt-hour A measure of work or electrical energy used. (p. 31)

Mass A measure of the quantity of material making up an object. (p. 26)

Metre The basic metric unit of length. (p. 9)

Platform Balance An instrument consisting of two platforms connected by a horizontal rod that balances on a knife edge. The pull of gravity on objects placed on the two platforms is compared. (p. 27)

Precision Refers to the smallest unit with which a measurement is made, that is, the position of the last significant digit. (p. 35)

Scientific Notation A form in which a number can be written as a product of a number between 1 and 10 and a power of 10. General form is $M \times 10^n$, where M is a number between 1 and 10 and n is the exponent or power of 10. (p. 7)

Second The basic metric unit of time. (p. 29)

SI (Système International d'Unités) The international system of units of measurement in the modern metric system. Seven basic units are included. (p. 4)

Significant Digits The number of digits in a number that indicates the number of units we are reasonably sure of having counted. (p. 33)

Spring Balance An instrument containing a spring, which stretches in proportion to the force applied to it, and a pointer that can be used to determine a force or weight of an object. (p. 27)

Standards of Measure A set of units of measurement for length, weight, and other quantities defined in such a way as to be useful to a large number of people. (p. 2)

Surface Area The total area of all the surfaces of a solid figure. (p. 22)

Volt The basic unit of electrical energy. (p. 31)

Volume The number of cubic units contained in a figure. (p. 19)

Watt The basic unit of power. (p. 31)

Weight A measure of the gravitational force or pull acting on an object. (p. 27)

REVIEW QUESTIONS

1. What are the basic metric units for length, mass, and time?
   (a) Foot, pound, hour (b) Newton, litre, second
   (c) Metre, kilogram, second (d) Mile, ton, day
2. When a value is multiplied or divided by 1, the value is
   (a) increased. (b) unchanged.
   (c) decreased. (d) none of the above.
3. The lateral surface area of a solid is
   (a) always equal to total surface area.

(b) never equal to total surface area.
(c) usually equal to total surface area.
(d) rarely equal to total surface area.

4. Accuracy is
(a) the same as precision.
(b) the smallest unit with which a measurement is made.
(c) the number of significant digits.
(d) all of the above.
5. When multiplying or dividing two or more measurements, the units
(a) must be the same. (b) must be different.
(c) can be different.
6. Cite three examples of problems that would arise in the construction of a home by workers using different systems of measurement.
7. Why do you think the metric system is preferred to the English system of measurement?
8. List two common very large or very small measured numbers that could be usefully written in scientific notation.
9. How are negative exponents used in scientific notation?
10. What is the meaning of zero as an exponent?
11. When using conversion factors, can units be treated in an equation like other algebraic quantities?
12. What is the meaning of cross-sectional area?
13. Can a brick have more than one cross-sectional area?
14. What is the fundamental metric unit for land area?
15. Which is larger, a litre or a quart?
16. List three things that might conveniently be measured in millilitres.
17. How do weight and mass differ?
18. What is the basic metric unit of weight?
19. A microsecond is one-_____ of a second.
20. Why must we concern ourselves with significant digits?
21. Can the sum or difference of two measurements ever be more precise than the least precise measurement?
22. When rounding the product or quotient of two measurements, is it necessary to consider significant digits?

REVIEW PROBLEMS

Give the metric prefix for each value:

1. 1000
2. 0.001

Give the metric symbol, or abbreviation, for each prefix:

3. micro
4. mega

Write the abbreviation for each quantity:

5. 45 milligrams
6. 138 centimetres

Write each number in scientific notation:

7. 214,000,000
8. 3.36
9. 0.0045

Write each number in decimal form:

10. 1.72×10^4
11. 6.6×10^{-3}
12. 9.03×10^0

Which is larger?

13. 1 L or 1 mL
14. 1 MW or 1 kW
15. 1 L or 1 m^3

Fill in each blank: (Round to three significant digits when necessary.)

16. 250 m = _____ km
17. 850 mL = _____ L
18. 5.4 kg = _____ g
19. 0.55 A = _____ μA
20. 25 MW = _____ W
21. 75 μs = _____ ns
22. 275 cm^2 = _____ mm^2
23. 350 cm^2 = _____ m^2
24. 0.15 m^3 _____ cm^3
25. 500 cm^3 = _____ mL
26. 150 lb = _____ kg
27. 36 ft = _____ m
28. 250 cm = _____ in.
29. 150 in^2 = _____ cm^2
30. 24 yd^2 = _____ ft^2
31. 6 m^3 = _____ ft^3
32. 16 lb = _____ N
33. 15,600 s = _____ h _____ min

Determine the accuracy (the number of significant digits) in each measurement:

34. 5.08 kg
35. 20,570 lb
36. 0.060 A
37. 2.00×10^{-4} W

Determine the precision of each measurement:

38. 30.6 ft
39. 0.0500 s
40. 18,000 mi
41. 4×10^5 V

For each set of measurements, find the measurement that is
(a) the most accurate.
(b) the least accurate.
(c) the most precise.
(d) the least precise.

42. 12.00 m; 0.150 m; 2600 m; 0.008 m
43. 208 L; 18,050 L; 21.5 L; 0.75 L

Use the rules of measurements to add the following measurements:

44. 0.0250 A; 0.075 A; 0.00080 A; 0.024 A
45. 2100 V; 36,800 V; 24,000 V; 14.5 V; 470 V

Use the rules for multiplication and division of measurements to find the value of each of the following:

46. (450 cm)(18.5 cm)(215 cm)
47. $\dfrac{1480 \text{ m}^3}{9.6 \text{ m}}$
48. $\dfrac{(25.0 \text{ kg})(1.20 \text{ m/s})^2}{3.70 \text{ m}}$
49. Find the area of a rectangle 4.50 m long and 2.20 m wide.
50. Find the volume of a rectangular box 9.0 cm long, 6.0 cm wide, and 13 cm high.

Chapter 2

PROBLEM SOLVING

A formula is an equation, usually expressed in letters, called *variables,* and numbers. Much technical work includes the substitution of measured data into known formulas or relationships to find solutions to problems. A systematic approach to solving problems is a valuable tool.

The problem-solving method presented will assist you in processing data, analyzing the problems present, and finding the solution in an orderly manner.

OBJECTIVES

The major goals of this chapter are to enable you to:

1. Understand the use of formulas in problem solving.
2. Develop a systematic approach to solving technical problems.
3. Analyze technical problems using a problem-solving method.

2.1 FORMULAS

A **formula** is an equation, usually expressed in letters (called *variables* in algebra) and numbers. A **variable** is a symbol, usually a letter, used to represent some unknown number or quantity.

EXAMPLE 1

The formula $s = vt$ states that the distance traveled, s, equals the product of the velocity, v, and the time, t.

EXAMPLE 2

The formula $I = \frac{Q}{t}$ states that the current, I, equals the quotient of the charge, Q, and the time, t.

To solve a formula for a given letter means to express the given letter or variable in terms of all the remaining letters. That is, by using the equation-solving principles in Section A.3 of Appendix A, rewrite the formula so that the given letter appears on one side of the equation by itself and all the other letters appear on the other side.

EXAMPLE 3

Solve $s = vt$ for v.

$$s = vt$$

$$\frac{s}{t} = \frac{vt}{t} \quad \text{Divide both sides by } t.$$

$$\frac{s}{t} = v$$

EXAMPLE 4

Solve $I = Q/t$

(a) for Q. (b) for t.

(a)

$$I = \frac{Q}{t}$$

$$(I)t = \left(\frac{Q}{t}\right)t \qquad \text{Multiply both sides by } t.$$

$$It = Q$$

(b) Starting with $It = Q$, we obtain

$$\frac{It}{I} = \frac{Q}{I} \qquad \text{Divide both sides by } I.$$

$$t = \frac{Q}{I}$$

EXAMPLE 5

Solve $V = E - Ir$ for r.

Method 1:

$$V = E - Ir$$

$$V - E = E - Ir - E \qquad \text{Subtract } E \text{ from both sides.}$$

$$V - E = -Ir$$

$$\frac{V - E}{-I} = \frac{-Ir}{-I} \qquad \text{Divide both sides by } -I.$$

$$\frac{V - E}{-I} = r$$

Method 2:

$$V = E - Ir$$

$$V + Ir = E - Ir + Ir \qquad \text{Add } Ir \text{ to both sides.}$$

$$V + Ir = E$$

$$V + Ir - V = E - V \qquad \text{Subtract } V \text{ from both sides.}$$

$$Ir = E - V$$

$$\frac{Ir}{I} = \frac{E - V}{I} \qquad \text{Divide both sides by } I.$$

$$r = \frac{E - V}{I}$$

Note that the two results are equivalent. Take the first result,

$$\frac{V - E}{-I}$$

and multiply numerator and denominator by -1. That is,

$$\left(\frac{V - E}{-I}\right)\left(\frac{-1}{-1}\right) = \frac{-V + E}{I} = \frac{E - V}{I}$$

which is the same as the second result.

It is often convenient to use the same quantity in more than one way in a formula. For example, we may wish to use a certain measurement of a quantity, such as velocity, at a given time, say at $t = 0$ s, then use the velocity at a later time, say at $t = 6$ s. To write these desired values of the velocity is rather awkward. We simplify this written statement by using *subscripts* (small letters or numbers printed a half space below the printed line and to the right of the variable) to shorten what we must write.

From the example given, v at time $t = 0$ s will be written as v_i (initial velocity); v at time $t = 6$ s will be written as v_f (final velocity). Mathematically, v_i and v_f are two different quantities, which in most cases are unequal. The sum of v_i and v_f is written as $v_i + v_f$. The product of v_i and v_f is written as $v_i v_f$. The subscript notation is used only to distinguish the general quantity, v, velocity, from the measure of that quantity at certain specified times.

EXAMPLE 6

Solve the formula $x = x_i + v_i t + \frac{1}{2}at^2$ for v_i.

Method 1:

$$x = x_i + v_i t + \tfrac{1}{2}at^2$$

$$x - v_i t = x_i + v_i t + \tfrac{1}{2}at^2 - v_i t \qquad \text{Subtract } v_i t \text{ from both sides.}$$

$$x - v_i t = x_i + \tfrac{1}{2}at^2$$

$$x - v_i t - x = x_i + \tfrac{1}{2}at^2 - x \qquad \text{Subtract } x \text{ from both sides.}$$

$$-v_i t = x_i + \tfrac{1}{2}at^2 - x$$

$$\frac{-v_i t}{-t} = \frac{x_i + \frac{1}{2}at^2 - x}{-t} \qquad \text{Divide both sides by } -t.$$

$$v_i = \frac{x_i + \frac{1}{2}at^2 - x}{-t}$$

Method 2:

$$x = x_i + v_i t + \tfrac{1}{2}at^2$$

$$x - x_i - \tfrac{1}{2}at^2 = x_i + v_i t + \tfrac{1}{2}at^2 - x_i - \tfrac{1}{2}at^2 \qquad \text{Subtract } x_i \text{ and } \tfrac{1}{2}at^2 \text{ from both sides.}$$

$$x - x_i - \tfrac{1}{2}at^2 = v_i t$$

$$\frac{x - x_i - \frac{1}{2}at^2}{t} = \frac{v_i t}{t} \qquad \text{Divide both sides by } t.$$

$$\frac{x - x_i - \frac{1}{2}at^2}{t} = v_i$$

EXAMPLE 7

Solve the formula $v_{avg} = \frac{1}{2}(v_f + v_i)$ for v_f (avg is used here as a subscript meaning average).

$$v_{avg} = \tfrac{1}{2}(v_f + v_i)$$

$$2v_{avg} = v_f + v_i \qquad \text{Multiply both sides by 2.}$$

$$2v_{avg} - v_i = v_f \qquad \text{Subtract } v_i \text{ from both sides.}$$

EXAMPLE 8

Solve $A = \frac{\pi d^2}{4}$ for d, where d is a diameter.

$$A = \frac{\pi d^2}{4}$$

$$4A = \left(\frac{\pi d^2}{4}\right)(4) \quad \text{Multiply both sides by 4.}$$

$$4A = \pi d^2$$

$$\frac{4A}{\pi} = \frac{\pi d^2}{\pi} \quad \text{Divide both sides by } \pi.$$

$$\frac{4A}{\pi} = d^2$$

$$\pm\sqrt{\frac{4A}{\pi}} = d \quad \text{Take the square root of both sides.}$$

In this case, a negative diameter has no physical meaning, so the result is

$$d = \sqrt{\frac{4A}{\pi}}$$

■ PROBLEMS 2.1

Solve each formula for the quantity given.

1. $v = \frac{s}{t}$ for s
2. $a = \frac{v}{t}$ for v
3. $w = mg$ for m
4. $F = ma$ for a
5. $E = IR$ for R
6. $V = lwh$ for w
7. $\text{PE} = mgh$ for g
8. $\text{PE} = mgh$ for h
9. $v^2 = 2gh$ for h
10. $X_L = 2\pi fL$ for f
11. $P = \frac{w}{t}$ for w
12. $p = \frac{F}{A}$ for F
13. $P = \frac{W}{t}$ for t
14. $p = \frac{F}{A}$ for A
15. $\text{KE} = \frac{1}{2}mv^2$ for m
16. $\text{KE} = \frac{1}{2}mv^2$ for v^2
17. $W = Fs$ for s
18. $v_f = v_i + at$ for a
19. $V = E - Ir$ for I
20. $v_2 = v_1 + at$ for t
21. $R = \frac{\pi}{2P}$ for P
22. $R = \frac{kL}{d^2}$ for L
23. $F = \frac{9}{5}C + 32$ for C
24. $C = \frac{5}{9}(F - 32)$ for F
25. $X_C = \frac{1}{2\pi fC}$ for f
26. $R = \frac{\rho L}{A}$ for L
27. $R_T = R_1 + R_2 + R_3 + R_4$ for R_3
28. $Q_1 = P(Q_2 - Q_1)$ for Q_2
29. $\frac{I_S}{I_P} = \frac{N_P}{N_S}$ for I_P
30. $\frac{V_P}{V_S} = \frac{N_P}{N_S}$ for N_S
31. $v_{avg} = \frac{1}{2}(v_f + v_i)$ for v_i
32. $2a(s - s_i) = v^2 - v_i^2$ for a
33. $2a(s - s_i) = v^2 - v_i^2$ for s
34. $Ft = m(V_2 - V_1)$ for V_1
35. $Q = \frac{I^2Rt}{J}$ for R
36. $x = x_i + v_it + \frac{1}{2}at^2$ for x_i

37. $A = \pi r^2$ for r, where r is a radius

38. $V = \pi r^2 h$ for r, where r is a radius

39. $R = \dfrac{kL}{d^2}$ for d, where d is a diameter

40. $V = \frac{1}{3}\pi r^2 h$ for r, where r is a radius

41. $Q = \dfrac{I^2 Rt}{J}$ for I

42. $F = \dfrac{mv^2}{r}$ for v ■

2.2 SUBSTITUTING DATA INTO FORMULAS

An important part of problem solving is being able to substitute the given data into the appropriate formula and to find the value of the unknown quantity. Basically, there are two ways of substituting data into formulas to solve for the unknown quantity:

1. Solve the formula for the unknown quantity and then make the substitution of the data.
2. Substitute the data into the formula first, and then solve for the unknown quantity.

When using a calculator, the first way is more useful. We will be using this way most of the time in this text.

EXAMPLE 1

Given the formula $A = bh$, $A = 120\text{ m}^2$, and $b = 15$ m, find h.

First, solve for h.

$$A = bh$$

$$\frac{A}{b} = \frac{bh}{b} \qquad \text{Divide both sides by } b.$$

$$\frac{A}{b} = h$$

Then substitute the data:

$$h = \frac{A}{b} = \frac{120\text{ m}^2}{15\text{ m}} = 8.0\text{ m}$$

(Remember to follow the rules of measurement discussed in Chapter 1. We use them consistently throughout.)

EXAMPLE 2

Given the formula $P = 2a + 2b$, $P = 824$ cm, and $a = 292$ cm, find b.

First, solve for b.

$$P = 2a + 2b$$

$$P - 2a = 2a + 2b - 2a \qquad \text{Subtract } 2a \text{ from both sides.}$$

$$P - 2a = 2b$$

$$\frac{P - 2a}{2} = \frac{2b}{2} \qquad \text{Divide both sides by 2.}$$

$$\frac{P - 2a}{2} = b \qquad \left(\text{or } b = \frac{P}{2} - a\right)$$

Then substitute the data:

$$h = \frac{P - 2a}{2} = \frac{824 \text{ cm} - 2(292 \text{ cm})}{2}$$

$$= \frac{824 \text{ cm} - 584 \text{ cm}}{2}$$

$$= \frac{24\bar{0} \text{ cm}}{2} = 12\bar{0} \text{ cm}$$

EXAMPLE 3

Given the formula $A = \left(\frac{a + b}{2}\right)h$, $A = 15\bar{0}$ m^2, $b = 18.0$ m, and $h = 10.0$ m, find a.

First, solve for a.

$$A = \left(\frac{a + b}{2}\right)h$$

$$2A = \left[\left(\frac{a + b}{2}\right)h\right](2) \qquad \text{Multiply both sides by 2.}$$

$$2A = (a + b)h$$

$$2A = ah + bh \qquad \text{Remove the parentheses.}$$

$$2A - bh = ah + bh - bh \qquad \text{Subtract } bh \text{ from both sides.}$$

$$2A - bh = ah$$

$$\frac{2A - bh}{h} = \frac{ah}{h} \qquad \text{Divide both sides by } h.$$

$$\frac{2A - bh}{h} = a \qquad \left(\text{or } a = \frac{2A}{h} - b\right)$$

Then substitute the data:

$$a = \frac{2A - bh}{h}$$

$$= \frac{2(15\bar{0} \text{ m}^2) - (18.0 \text{ m})(10.0 \text{ m})}{10.0 \text{ m}}$$

$$= \frac{30\bar{0} \text{ m}^2 - 18\bar{0} \text{ m}^2}{10.0 \text{ m}}$$

$$= \frac{12\bar{0} \text{ m}^2}{10.0 \text{ m}} = 12.0 \text{ m}$$

EXAMPLE 4

Given the formula $V = \frac{1}{3}\pi r^2 h$, $V = 64{,}400$ mm^3, and $h = 48.0$ mm, find r, where r is a radius.

First, solve for r.

$$V = \tfrac{1}{3}\pi r^2 h$$

$$3V = (\tfrac{1}{3}\pi r^2 h)(3) \qquad \text{Multiply both sides by 3.}$$

$$3V = \pi r^2 h$$

$$\frac{3V}{\pi h} = \frac{\pi r^2 h}{\pi h} \qquad \text{Divide both sides by } \pi h.$$

$$\frac{3V}{\pi h} = r^2$$

$$\pm\sqrt{\frac{3V}{\pi h}} = r \qquad \text{Take the square root of both sides.}$$

In this case, a negative radius has no physical meaning, so the result is

$$r = \sqrt{\frac{3V}{\pi h}}$$

Then substitute the data:

$$r = \sqrt{\frac{3(64{,}400 \text{ mm}^3)}{\pi(48.0 \text{ mm})}}$$
$$= 35.8 \text{ mm}$$

■ PROBLEMS 2.2

For each formula, (a) solve for the indicated letter and then (b) substitute the given data to find the value of the indicated letter. Follow the rules of calculations with measurements. ***Note:*** In Problems 14 and 16, r is a radius, and in Problem 15, b is the length of the side of a square.

Formula	*Data*	*Find*
1. $A = bh$	$b = 14.5$ cm, $h = 11.2$ cm	A
2. $V = lwh$	$l = 16.7$ m, $w = 10.5$ m, $h = 25.2$ m	V
3. $A = bh$	$A = 34.5 \text{ cm}^2$, $h = 4.60$ cm	b
4. $P = 4b$	$P = 42\bar{0}$ in.	b
5. $P = a + b + c$	$P = 48.5$ cm, $a = 18.2$ cm, $b = 24.3$ cm	c
6. $C = \pi d$	$C = 495$ ft	d
7. $C = 2\pi r$	$C = 68.5$ yd	r
8. $A = \frac{1}{2}bh$	$A = 468 \text{ m}^2$, $b = 36.0$ m	h
9. $P = 2(a + b)$	$P = 88.7$ km, $a = 11.2$ km	b
10. $V = \pi r^2 h$	$r = 61.0$ m, $h = 125.3$ m	V
11. $V = \pi r^2 h$	$V = 368 \text{ m}^3$, $r = 4.38$ m	h
12. $A = 2\pi rh$	$A = 51\bar{0} \text{ cm}^2$, $r = 14.0$ cm	h
13. $V = Bh$	$V = 2185 \text{ m}^3$, $h = 14.2$ m	B
14. $A = \pi r^2$	$A = 463.5 \text{ m}^2$	r
15. $A = b^2$	$A = 465 \text{ in}^2$	b
16. $V = \frac{1}{3}\pi r^2 h$	$V = 2680 \text{ m}^3$, $h = 14.7$ m	r
17. $C = 2\pi r$	$r = 19.36$ m	C
18. $V = \frac{4}{3}\pi r^3$	$r = 25.65$ m	V
19. $V = \frac{1}{3} Bh$	$V = 19{,}850 \text{ ft}^3$, $h = 486.5$ ft	B
20. $A = \left(\frac{a + b}{2}\right)h$	$A = 205.2 \text{ m}^2$, $a = 16.50$ m, $b = 19.50$ m	h

■

2.3 PROBLEM-SOLVING METHOD

Problem solving in technical fields is more than substituting numbers and units into formulas. You must develop skill in taking data, analyzing the problem, and finding the solution in an orderly manner.

Understanding the principle involved in solving a problem is more important than blindly substituting into a formula. By following an orderly procedure for problem solving, we develop an approach to problem solving that you can use in your studies and on the job.

In all problems in the remainder of the book, the following **problem-solving method** will be applied to all problems where appropriate.

1. ***Read the problem carefully.*** This might appear obvious, but it is the most important step in solving a problem. As a matter of habit, you should read the problem at least twice.
   (a) The first time you should read the problem straight through from beginning to end. Do not stop to think about setting up an equation or formula. You are only trying to get a general overview of the problem during this first reading.
   (b) Read through a second time slowly and *completely*, beginning to think ahead to the following steps.
2. ***Make a sketch.*** Some problems may not lend themselves to a sketch. However, make a sketch whenever it is possible. Many times, seeing a sketch of the problem will show if you have forgotten important parts of the problem and may suggest the solution. This is a *very important* part of problem solving and is often overlooked.
3. ***Write all given information including units.*** This is necessary to get all essential facts in mind before looking for the solution. There are some common phrases that have understood physical meanings. For example, the term *from rest* means the initial velocity equals zero or $v_i = 0$; the term *smooth surface* means assume that no friction is present.
4. ***Write the unknown or quantity asked for in the problem.*** Many students have difficulty solving problems because they don't know what they are looking for and solve for the wrong quantity.
5. ***Write the basic equation or formula that relates the known and unknown quantities.*** Find the basic formula or equation to use by studying what is given and what you are asked to find. Then look for a formula or equation that relates these quantities. Sometimes, you may need to use more than one equation or formula in working a problem.
6. ***Find a working equation by solving the basic equation or formula for the unknown quantity.***
7. ***Substitute the data in the working equation, including the appropriate units.*** It is important that you *carry the units all the way through the problem* as a check that you have solved the problem correctly. For example, if you are asked to find the weight of an object in newtons and the units of your answer work out to be metres, you need to review your solution for the error. (When the unit analysis is not obvious, we will go through it step by step in a box nearby.)
8. ***Perform the indicated operations and work out the solution.*** Although this will be your final written step in the solution, in every case you should ask yourself, "Is my answer reasonable?" Here and on the job you will be dealing with practical problems. A quick estimate will often reveal an error in your calculations.
9. ***Check your answer.*** Also ask yourself, "Did I answer the questions?"

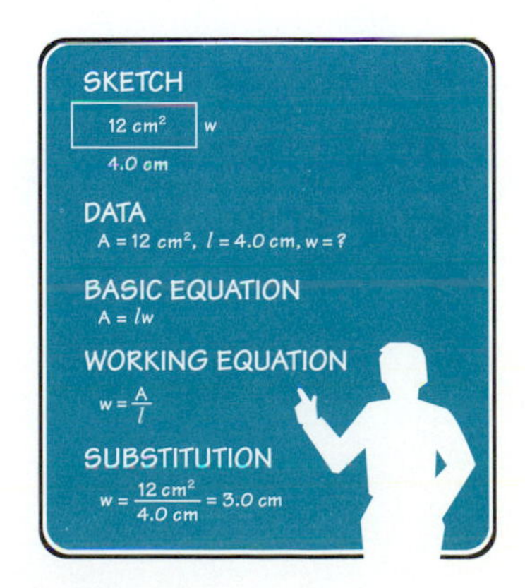

Figure 2.1

To help you recall this procedure, with almost every problem set that follows, you will find Fig. 2.1 as shown here. This figure is not meant to be complete, but is only an outline to assist you in remembering and following the procedure for solving problems. *You should follow this outline in solving all problems in this course.*

This problem-solving method will now be demonstrated in terms of relationships and formulas with which you are probably familiar. The formulas for finding area and volume can be found on the inside back cover.

EXAMPLE 1

Find the volume of concrete required to fill a rectangular bridge abutment whose dimensions are 6.00 m × 3.00 m × 15.0 m.

Sketch:

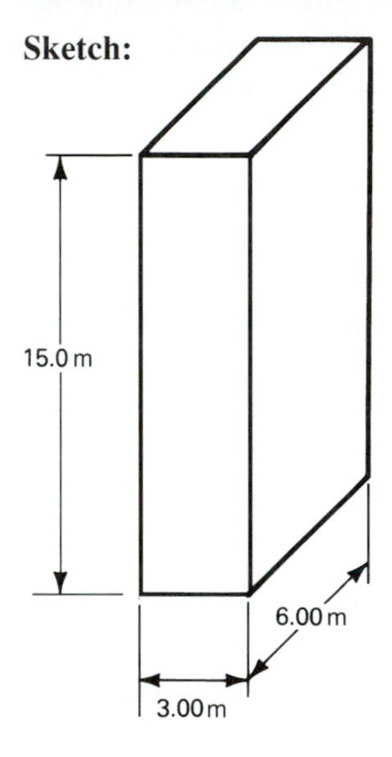

Data:

$$\left.\begin{aligned} l &= 6.00 \text{ m} \\ w &= 3.00 \text{ m} \\ h &= 15.0 \text{ m} \end{aligned}\right\} \quad \text{This is a listing of the information that is known.}$$

$$V = ? \quad \text{This identifies the unknown.}$$

Basic Equation:

$$V = lwh$$

Working Equation: Same

Substitution:

$$\begin{aligned} V &= (6.00 \text{ m})(3.00 \text{ m})(15.0 \text{ m}) \\ &= 27\overline{0} \text{ m}^3 \end{aligned}$$

Note: m × m × m = m^3

EXAMPLE 2

A rectangular holding tank 24.0 m in length and 15.0 m in width is used to store water for short periods of time in an industrial plant. If 2880 m^3 of water are pumped into the tank, what is the depth of the water?

Sketch:

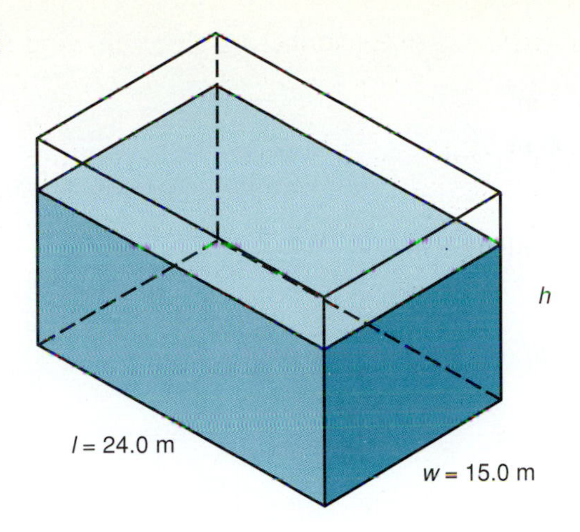

Data:

$$V = 2880 \text{ m}^3$$
$$l = 24.0 \text{ m}$$
$$w = 15.0 \text{ m}$$
$$h = ?$$

Basic Equation:

$$V = lwh$$

Working Equation:

$$h = \frac{V}{lw}$$

Substitution:

$$h = \frac{2880 \text{ m}^3}{(24.0 \text{ m})(15.0 \text{ m})}$$
$$= 8.00 \text{ m}$$

$$\frac{\text{m}^3}{\text{m} \times \text{m}} = \text{m}$$

EXAMPLE 3

A storage bin in the shape of a cylinder contains 814 m^3 of storage spacc. If its radius is 6.00 m, find its height.

Sketch:

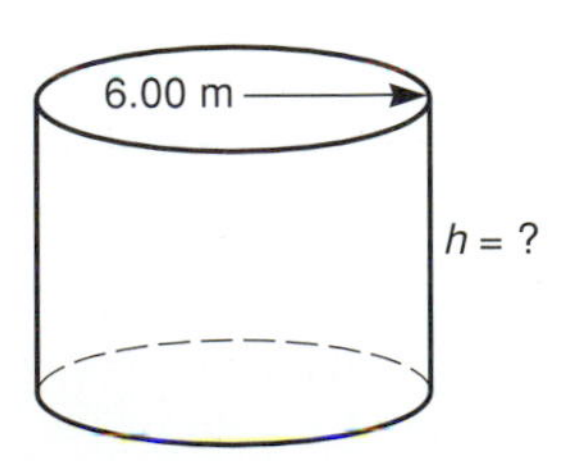

Data:

$$V = 814 \text{ m}^3$$
$$r = 6.00 \text{ m}$$
$$h = ?$$

If your calculator does not have a button for π, use 3.14.

Basic Equation:

$$V = \pi r^2 h$$

Working Equation:

$$h = \frac{V}{\pi r^2}$$

Substitution:

$$h = \frac{814 \text{ m}^3}{\pi(6.00 \text{ m})^2}$$

$$= 7.20 \text{ m} \qquad \boxed{\frac{\text{m}^3}{\text{m}^2} = \text{m}}$$

EXAMPLE 4

A rectangular piece of sheet metal measures 45.0 cm by 75.0 cm. A 10.0-cm square is then cut from each corner. The metal is then folded to form a box-like container without a top. Find the volume of the container.

Sketch:

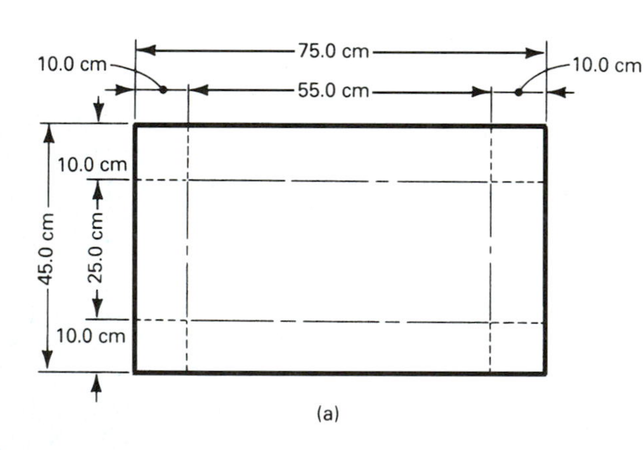

(a)

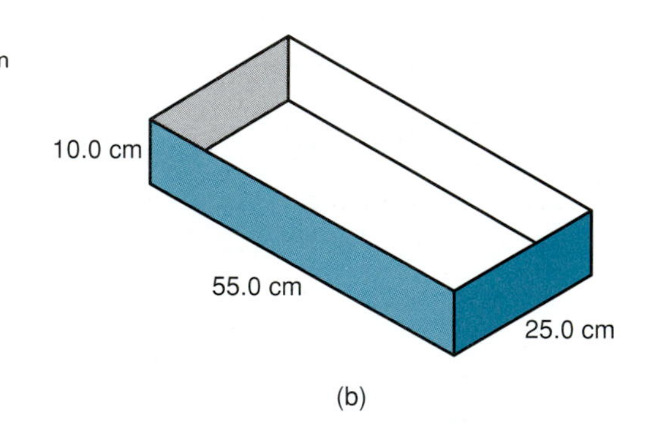

(b)

Data:

$$l = 55.0 \text{ cm}$$
$$w = 25.0 \text{ cm}$$
$$h = 10.0 \text{ cm}$$
$$V = ?$$

Basic Equation:

$$V = lwh$$

Working Equation: Same

Substitution:

$$V = (55.0 \text{ cm})(25.0 \text{ cm})(10.0 \text{ cm}) = 13{,}800 \text{ cm}^3$$

$\text{cm} \times \text{cm} \times \text{cm} = \text{cm}^3$

EXAMPLE 5

The cross-sectional area of a hole is 725 cm². Find its radius.

Sketch:

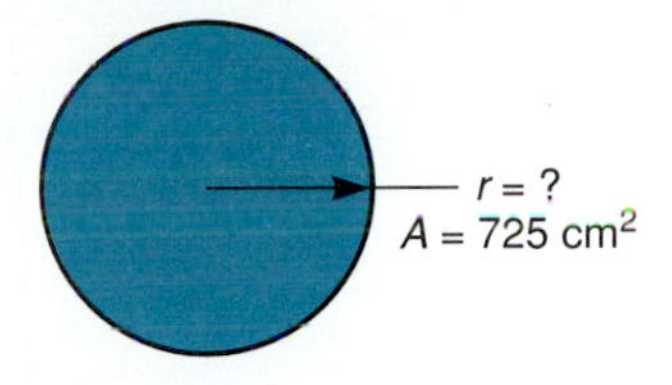

Data:

$$A = 725 \text{ cm}^2$$
$$r = ?$$

Basic Equation:

$$A = \pi r^2$$

Working Equation:

$$r = \sqrt{\frac{A}{\pi}}$$

Substitution:

$$r = \sqrt{\frac{725 \text{ cm}^2}{\pi}} = 15.2 \text{ cm}$$

$\sqrt{\text{cm}^2} = \text{cm}$

■ PROBLEMS 2.3

Use the problem-solving method to work each problem. (Here, as throughout the text, follow the rules for calculations with measurements.)

1. Find the volume of the box in Fig. 2.2.
2. Find the volume of a cylinder whose height is 7.50 in. and diameter is 4.20 in. (Fig. 2.3).
3. Find the volume of a cone whose height is 9.30 cm if the radius of the base is 5.40 cm (Fig. 2.4).

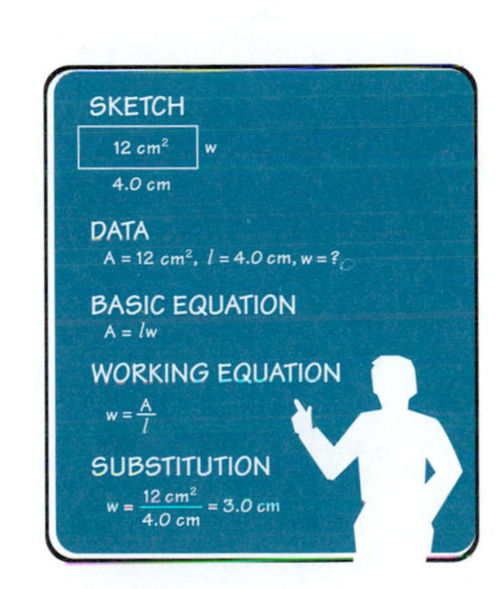

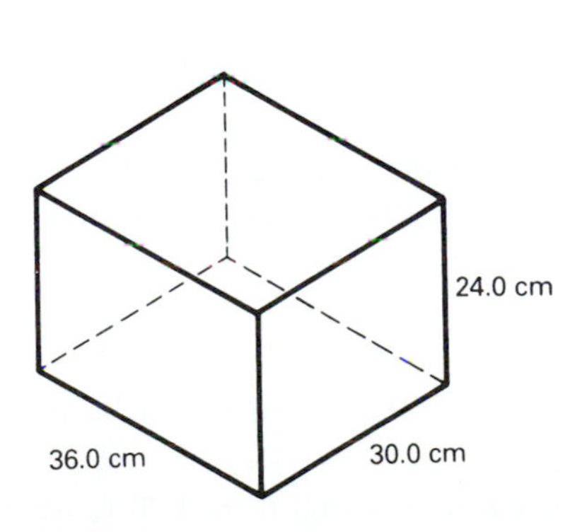

Figure 2.2

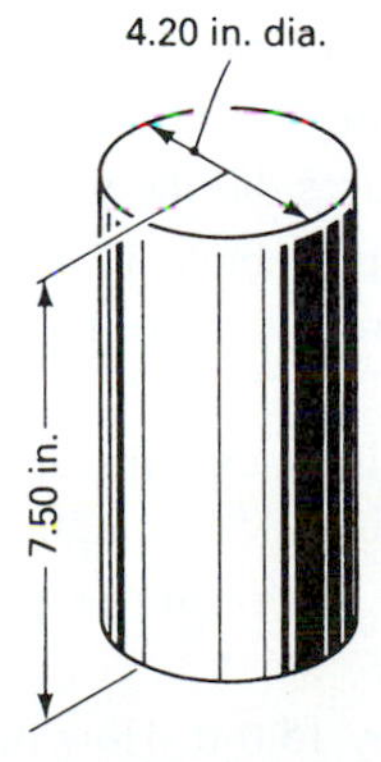

Figure 2.3

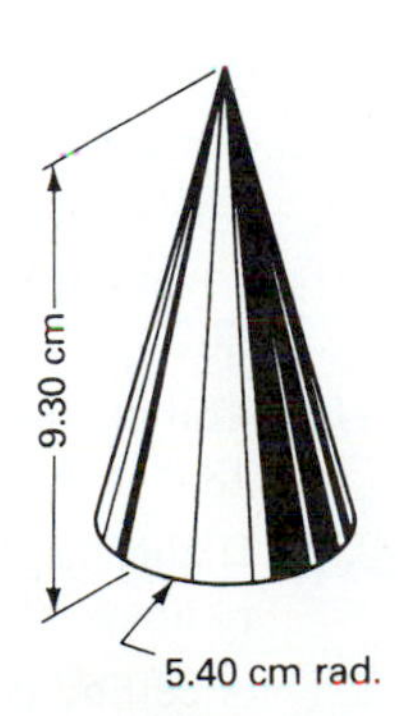

Figure 2.4

The cylinder in an engine of a road grader as shown in Fig. 2.5 is 11.40 cm in diameter and 24.00 cm high. Use Fig. 2.5 for Problems 4 through 6.

4. Find the volume of the cylinder.
5. Find the cross-sectional area of the cylinder.
6. Find the lateral surface area of the cylinder.

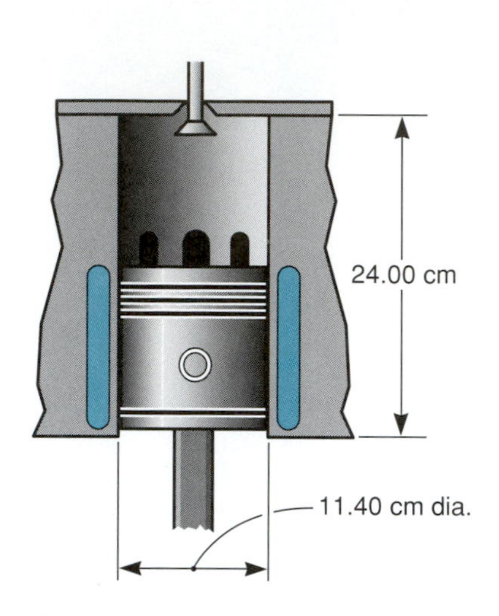

Figure 2.5

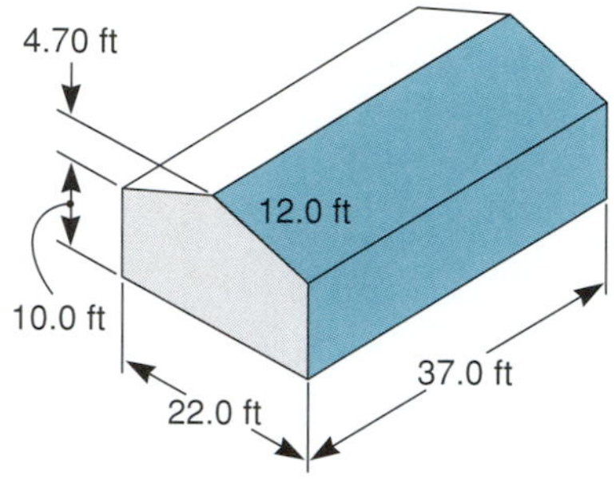

Figure 2.6

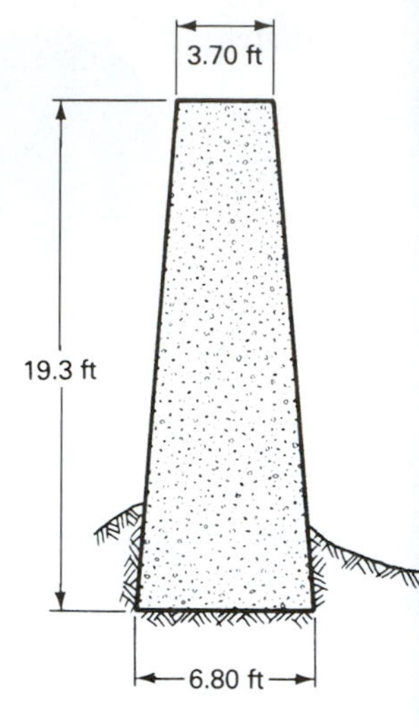

Figure 2.7

7. Find the total volume of the building shown in Fig. 2.6.
8. Find the cross-sectional area of the concrete retaining wall shown in Fig. 2.7.
9. Find the volume of a rectangular storage facility 9.00 ft by 12.0 ft by 8.00 ft.
10. Find the cross-sectional area of a piston head with a diameter of 3.25 cm.
11. Find the area of a right triangle that has legs of 4.00 cm and 6.00 cm.
12. Find the length of the hypotenuse of the right triangle in Problem 11.
13. Find the cross-sectional area of a pipe whose outer diameter is 3.50 cm and inner diameter is 3.20 cm.
14. Find the volume of a spherical water tank whose radius is 8.00 m.
15. The area of a rectangular parking lot is $90\bar{0}$ m^2. If the length is 25.0 m, what is the width?
16. The volume of a rectangular crate is 192 ft^3. If the length is 8.00 ft and the width is 4.00 ft, what is the height?
17. A cylindrical silo has a circumference of 29.5 m. Find its diameter.
18. If the silo in Problem 17 has a capacity of $100\bar{0}$ m^3, what is its height?
19. A wheel 30.0 cm in diameter moving along level ground made 145 complete rotations. How many metres did the wheel travel?
20. The side of the silo in Problem 17 and 18 needs to be painted. If each litre of paint covers 5.0 m^2, how many litres of paint will be needed? (Round up to the nearest litre.)
21. You are asked to design a cylindrical water tank that holds $50\bar{0}$,000 gal with radius 18.0 ft. Find its height. (1 ft^3 = 7.50 gal)
22. If the height of the water tank in Problem 21 were 42.0 ft, what would be its radius?
23. A ceiling is 12.0 ft by 15.0 ft. How many suspension panels 1.00 ft by 3.00 ft are needed to cover the ceiling?
24. Find the cross-sectional area of the dovetail slide shown in Fig. 2.8.

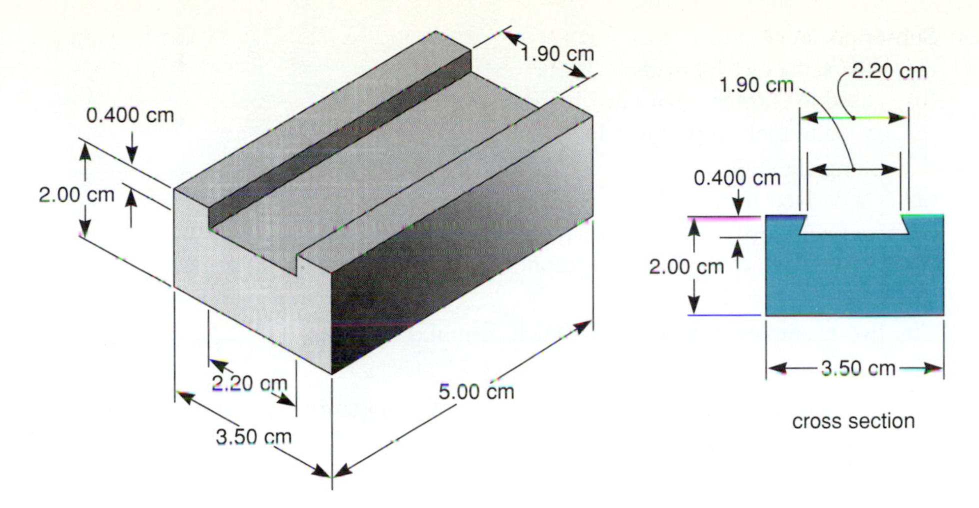

Figure 2.8

25. Find the volume of the storage bin shown in Fig. 2.9.
26. The maximum cross-sectional area of a spherical propane storage tank is 3.05 m^2. Will it fit into a 2.00-m-wide trailer?
27. How many cubic yards of concrete are needed to pour a patio 12.0 ft × 20.0 ft and 6.00 in. thick?
28. What length of sidewalk 4.00 in. thick and 4.00 ft wide could be poured with 2.00 yd^3 of concrete?

1.50 m
2.75 m
2.00 m

Figure 2.9

Find the volume of each figure.

29. 30.

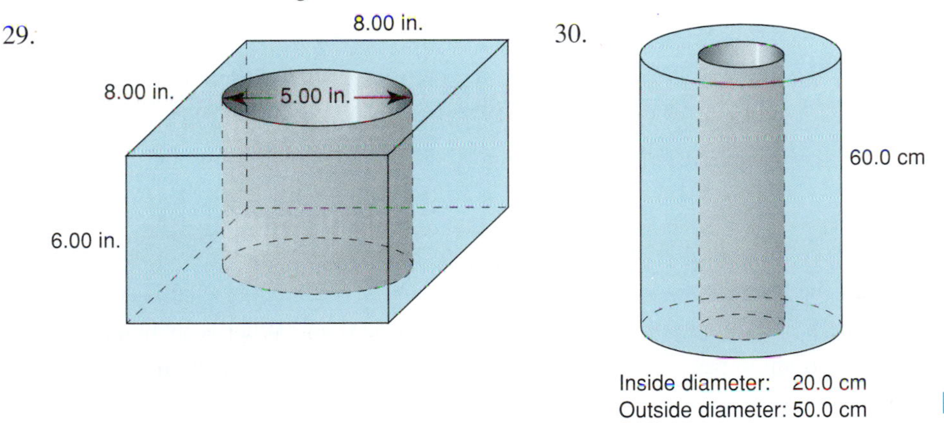

Inside diameter: 20.0 cm
Outside diameter: 50.0 cm

GLOSSARY

Formula An equation, usually expressed in letters (called *variables*) and numbers. (p. 46)
Problem-Solving Method An orderly procedure that aids in understanding and solving problems. (p. 53)
Variable A symbol, usually a letter, used to represent some unknown number or quantity. (p. 46)

REVIEW QUESTIONS

1. A formula is
   (a) the amount of each value needed.
   (b) a solution for problems.
   (c) an equation usually expressed in letters and numbers.

2. Subscripts are
   (a) the same as exponents.
   (b) used to shorten what must be written.
   (c) used to make a problem look hard.
3. A working equation
   (a) is derived from the basic equation.
   (b) is totally different from the basic equation.
   (c) comes before the basic equation in the problem.
   (d) none of the above.
4. Cite two examples in industry in which formulas are used.
5. How are subscripts used in measurement?
6. Why is reading the problem carefully the most important step in problem solving?
7. How can making a sketch help in problem solving?
8. What do we call the relationship between data that are given and what we are asked to find?
9. How is a working equation different from a basic equation?
10. How can analysis of the units in a problem assist in solving the problem?
11. How can making an estimate of your answer assist in the correct solution of problems?

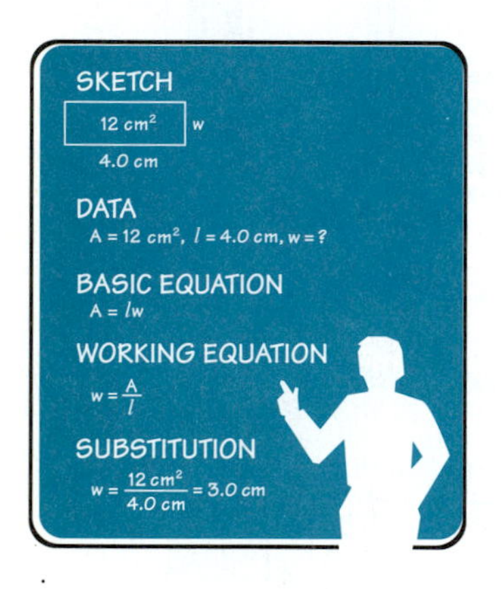

REVIEW PROBLEMS

1. Solve $F = ma$ for (a) m and (b) a.
2. Solve $v = \sqrt{2gh}$ for h.
3. Solve $s = \frac{1}{2}(v_f + v_i)\, t$ for v_f.
4. Solve $\text{KE} = \frac{1}{2}mv^2$ for v.
5. Given the formula $P = a + b + c$, with $P = 36$, $a = 12$, and $c = 6$, find b.
6. Given the formula $A = \left(\frac{a + b}{2}\right)h$, with $A = 21\bar{0}\text{ m}^2$, $b = 16.0$ m, and $h = 15.0$ m, find a.
7. Given the formula $A = \pi r^2$, if $A = 15.0\text{ m}^2$, what is r?
8. Given the formula $A = \frac{1}{2}bh$, if $b = 12.2$ cm and $h = 20.0$ cm, what is A?
9. A cone has a volume of 314 cm^3 and radius of 5.00 cm. What is its height?
10. A right triangle has a side of 41.2 mm and a side of 9.80 mm. Find the length of the hypotenuse.
11. Given a cylinder with a radius of 7.20 cm and a height of 13.4 cm, find the lateral surface area.
12. A rectangle has a perimeter of 40.0 cm. One side has a length of 14.0 cm. What is the length of an adjacent side?
13. The formula for the volume of a cylinder is $V = \pi r^2 h$. If $V = 21\bar{0}0\text{ m}^3$ and $h = 17.0$ m, find r.
14. The area of a triangle is found by using the formula $A = \frac{1}{2}bh$. If $b = 12.3$ m and $A = 88.6\text{ m}^2$, find h.
15. Find the volume of the lead sleeve with the cored hole in Fig. 2.10.
16. A rectangular plot of land measures 40.0 m by $12\bar{0}$ m with a parcel 10.0 m by 12.0 m out of one corner for an electrical transformer. What is the area of the remaining plot?

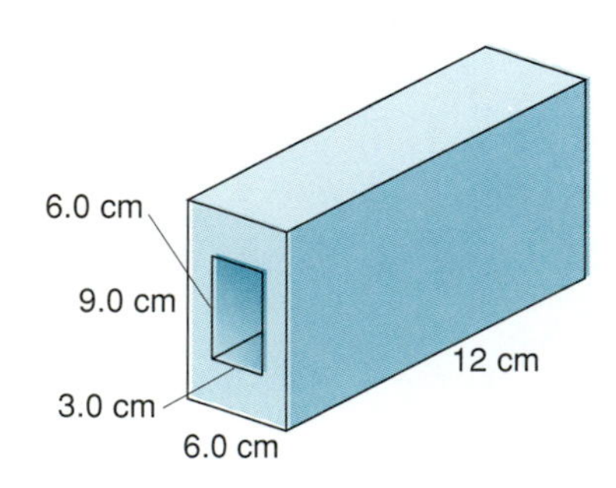

Figure 2.10

Chapter 3

MOTION

Motion is a change of position. Scalar quantities have magnitude but not direction.

Vectors have both magnitude and direction and are therefore more useful in describing motion. Velocity and acceleration describe important kinds of motion. An analysis of motion helps introduce the real nature of physics—to understand the nature and behavior of the physical world.

OBJECTIVES

The major goals of this chapter are to enable you to:

1. Express linear motion in terms of distance and direction.
2. Describe quantities using vectors and scalars.
3. Use vectors in solving velocity and acceleration problems.
4. Analyze uniformly accelerated motion.
5. Understand simple harmonic motion.
6. Use the pendulum formula.

3.1 DISPLACEMENT

The part of physics that is concerned with motion is called *mechanics*. Motion is involved in almost every area of science and technology.

Automotive technicians are concerned not only with the motion of the entire auto, but also with the motion of pistons, valves, driveshafts, and so on. Obviously, the motion of each of the internal parts has a direct and very important effect on the motion of the entire automobile.

The highway engineer must determine the correct banking angle of a curve if he or she is to design a road (Fig. 3.1). This angle is determined from several laws of motion that we will soon study.

Figure 3.1

In the next few chapters we develop the skills necessary for you to understand the basic aspects of motion.

Motion can be defined as a change of position. An airplane is in motion when it flies through the air because its position is changing as it flies from one city to another (Fig. 3.2). To describe the change of position of an object, such as an airplane, we use the term *displacement*. **Displacement** is the net distance, or direct distance, traveled as a result of motion.

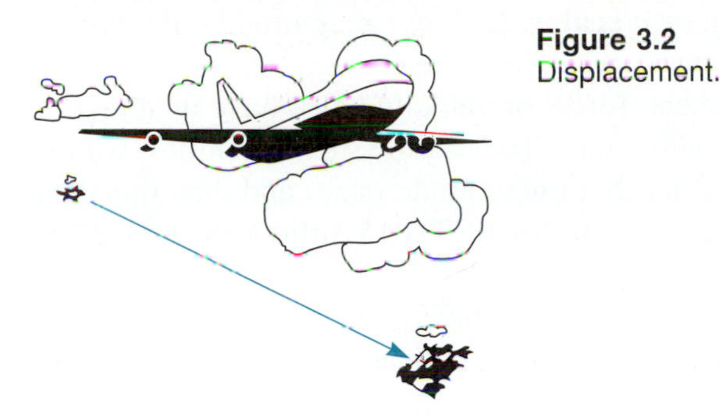

Figure 3.2
Displacement.

Suppose that a friend asks you how to reach your home from school. If you replied that he should walk four blocks, you would not have given him enough information [Fig. 3.3(a)]. Obviously, you would need to tell him which direction to go. If you had replied, "four blocks north," your friend could then find your home [Fig. 3.3(b)].

Displacement involves all the necessary information about a change in position; that is, it includes both *distance* and *direction*. It does not contain any information about the path that has been followed. *The units of displacement are length units,* such as metres, feet, or miles. If your friend decides to walk one block west, four blocks north, and then one block east, he will still arrive at your house. This resultant displacement is the same as if he had walked four blocks north [Fig. 3.3(c)].

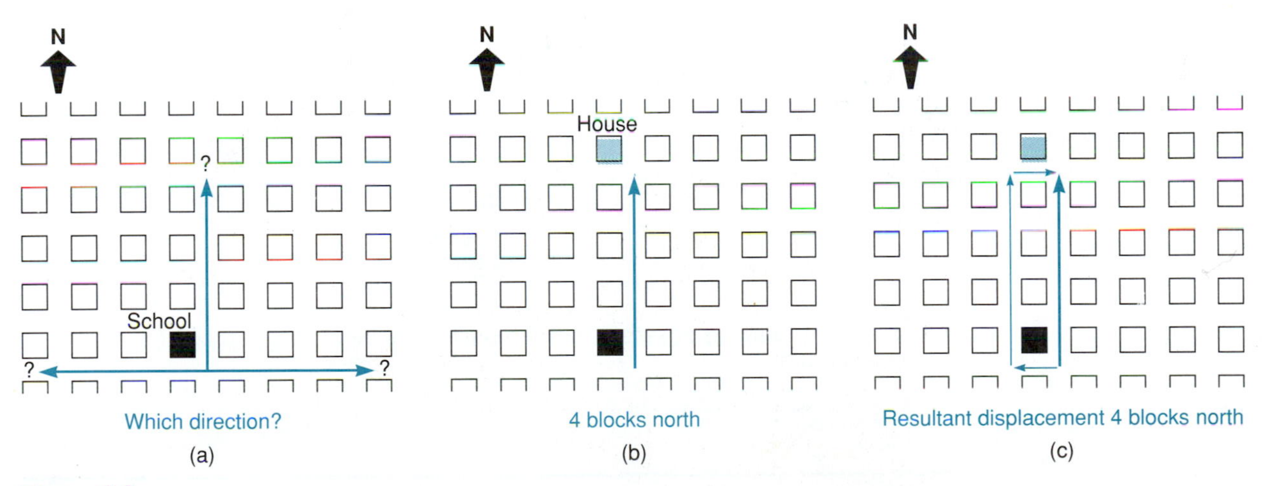

Figure 3.3
Displacement involves both a distance and a direction.

3.2 VECTORS AND SCALARS

Displacement problems are easily solved by graphical methods. To solve this type of problem we need to examine the difference between what are called scalar and vector quantities. In Chapter 1 we discussed quantities of length, time,

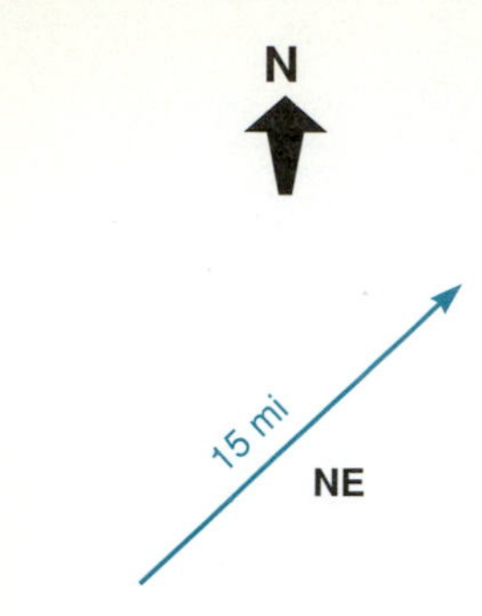

Figure 3.4

and volume. All these quantities can be expressed by a number with the appropriate units. For example, the weight of a given steel beam may be expressed as 2250 N; the temperature at 11:00 A.M. is 15°C; and the volume of a classroom is 300 m^3. *A quantity that can be completely described by a number and a unit is called a* **scalar quantity** *or a* **scalar.** It shows magnitude only, not direction.

A quantity, such as displacement, force, or velocity, must have its direction specified in addition to a number with a unit. To describe such quantities, we use vectors. A **vector** is a quantity that has both magnitude (size) and direction. The magnitude of the displacement vector "15 miles NE" is 15 miles (Fig. 3.4). *Thus, a vector has both magnitude and direction.*

To represent a vector in our diagrams, we draw an arrow that points in the correct direction. The magnitude of the vector is indicated by the length of the arrow. We usually choose a scale, such as 1.0 cm = 25 mi, for this purpose (Fig. 3.5). Thus, a displacement of $10\bar{0}$ mi west would be drawn as an arrow (pointing west) 4.0 cm long [Fig. 3.5(a)] since

$$10\bar{0} \text{ mi} \times \frac{1.0 \text{ cm}}{25 \text{ mi}} = 4.0 \text{ cm}$$

One end of the vector is called the initial point, and the other is called the terminal point, as shown in Fig. 3.5(a). Displacements of $5\bar{0}$ mi north [Fig. 3.5(b)] and $5\bar{0}$ mi east [Fig. 3.5(c)] using the same scale are also shown.

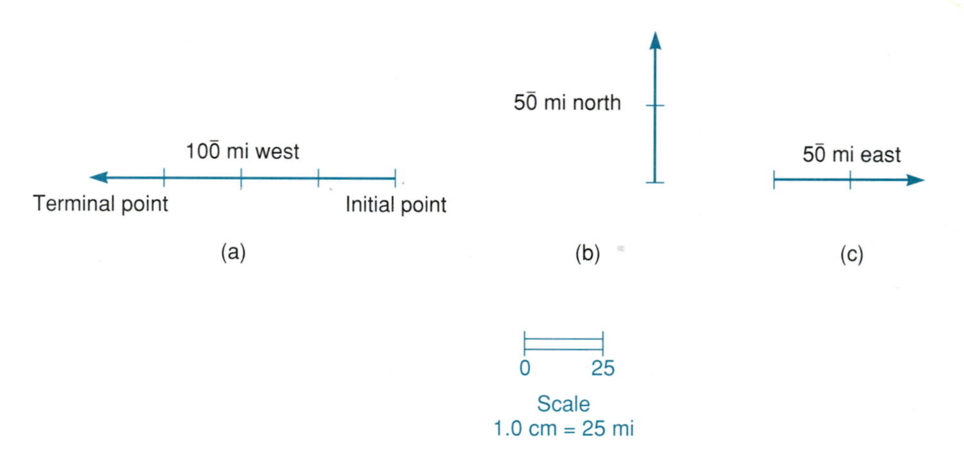

Figure 3.5
Use of a scale in drawing objects allows the drawing to be in proper proportion to the actual object drawn.

EXAMPLE 1

Using the scale 1.0 cm = $5\bar{0}$ km, draw the displacement vector 275 km at 45° north of west.

First, find the length of the vector.

$$275 \text{ km} \times \frac{1.0 \text{ cm}}{5\bar{0} \text{ km}} = 5.5 \text{ cm}$$

Then draw the vector at an angle 45° north of west (Fig. 3.6).

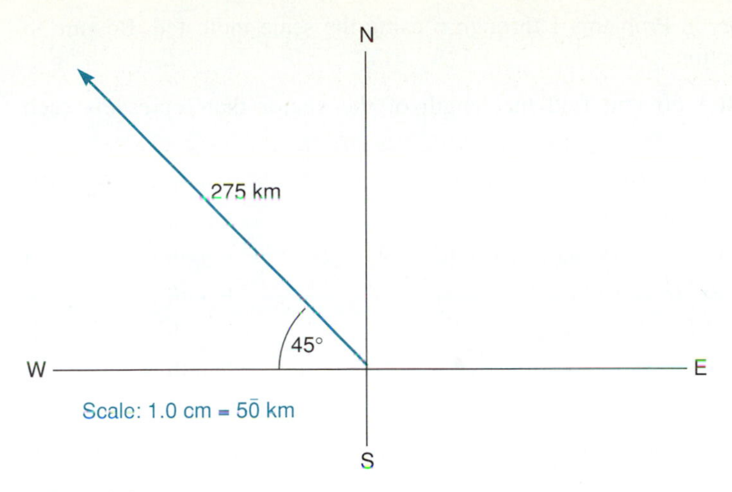

Figure 3.6

EXAMPLE 2

Using the scale $\frac{1}{4}$ in. = 2$\bar{0}$ mi, draw the displacement vector 15$\bar{0}$ mi at 22° east of south.

First, find the length of the vector.

$$15\bar{0} \text{ mi} \times \frac{\frac{1}{4} \text{ in.}}{2\bar{0} \text{ mi}} = 1\tfrac{7}{8} \text{ in.}$$

Then draw the vector at 22° east of south (Fig. 3.7).

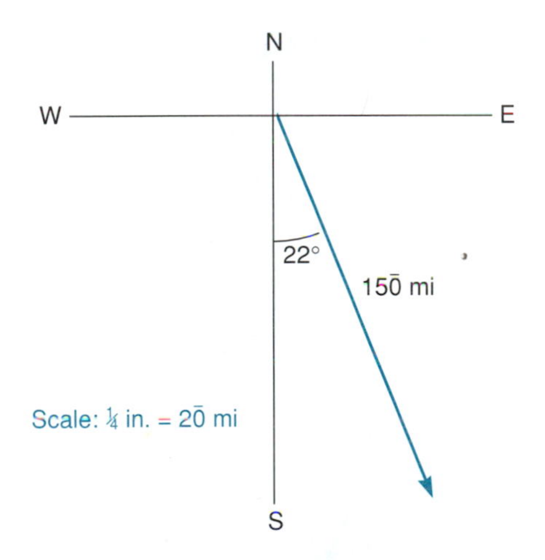

Figure 3.7

■ PROBLEMS 3.2

Using the scale of 1.0 cm = 25 mi, find the length of the vector that represents each displacement.

1. Displacement 75 mi north — length = ______ cm
2. Displacement 1$\bar{0}$0 mi west — length = ______ cm
3. Displacement 35 mi at 45° south of east — length = ______ cm
4. Displacement 160 mi at 35° west of north — length = ______ cm
5. Displacement 9$\bar{0}$ mi at 1$\bar{0}$° east of north — length = ______ cm
6. Displacement 6$\bar{0}$ mi at 55° east of south — length = ______ cm

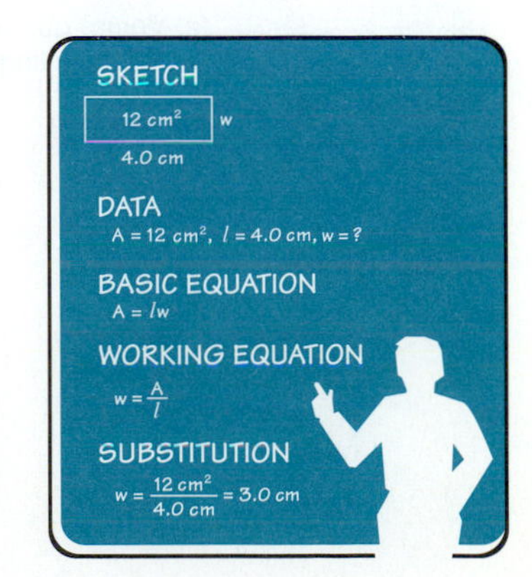

7–12. Draw the vectors in Problems 1 through 6 using the scale indicated. Be sure to include the direction.

Using the scale 1.0 cm = $5\bar{0}$ km, find the length of the vector that represents each displacement.

13. Displacement $10\bar{0}$ km east length = ____ cm
14. Displacement 125 km south length = ____ cm
15. Displacement $14\bar{0}$ km at 45° east of south length = ____ cm
16. Displacement $26\bar{0}$ km at $3\bar{0}$° south of west length = ____ cm
17. Displacement 315 km at 65° north of east length = ____ cm
18. Displacement 187 km at 17° north of west length = ____ cm

19–24. Draw the vectors in Problems 13 through 18 using the scale indicated.

Using the scale $\frac{1}{4}$ in. = $2\bar{0}$ mi, find the length of the vector that represents each displacement.

25. Displacement $10\bar{0}$ mi west length = ____ in.
26. Displacement $17\bar{0}$ mi north length = ____ in.
27. Displacement $21\bar{0}$ mi at 45° south of west length = ____ in.
28. Displacement 145 mi at $6\bar{0}$° north of east length = ____ in.
29. Displacement 75 mi at 25° west of north length = ____ in.
30. Displacement $16\bar{0}$ mi at 72° west of south length = ____ in.

31–36. Draw the vectors in Problems 25 through 30 using the scale indicated. ■

3.3 GRAPHICAL ADDITION OF VECTORS

Vectors may be denoted by a single letter with a small arrow above, such as $\vec{A}$, $\vec{v}$, or $\vec{R}$ [Fig. 3.8(a)]. This notation is especially useful when writing vectors on paper or on a chalkboard. In this book we use the traditional boldface type to denote vectors, such as **A**, **v**, or **R** [Fig. 3.8(b)]. The length of vector $\vec{A}$ is written $|\vec{A}|$; the length of vector **A** is written $|\mathbf{A}|$.

(a) Vector quantities $\vec{A}$, $\vec{v}$, and $\vec{R}$ usually have arrows when writing them on paper or on a chalkboard.

(b) Vector quantities **A**, **v**, and **R** usually are written in boldface type in textbooks.

Figure 3.8

Any given displacement can be the result of many different combinations of displacements. In Fig. 3.9, the displacement represented by the arrow, labeled **R** for resultant, is the result of either of the two paths shown. This vector is called the **resultant** of the vectors which make up either path 1 or path 2. *The resultant vector is the sum of a set of vectors.* The resultant vector, **R**, in Fig. 3.9 is the sum of the vectors **A**, **B**, **C**, and **D**. It is also the sum of vectors **E** and **F**. That is,

$$\mathbf{A} + \mathbf{B} + \mathbf{C} + \mathbf{D} = \mathbf{R} \quad \text{and} \quad \mathbf{E} + \mathbf{F} = \mathbf{R}$$

To solve a vector addition problem such as displacement:

1. Choose a suitable scale and calculate the length of each vector.
2. Draw the north–south reference line. Graph paper should be used.
3. Using a ruler and protractor, draw the first vector and then draw the other vectors so that the initial end of each vector is placed at the terminal end of the previous vector.
4. Draw the sum or resultant vector from the initial end of the first vector to the terminal end of the last vector.
5. Measure the length of the resultant and use the scale to find the magnitude of the vector. Use a protractor to measure the angle of the resultant.

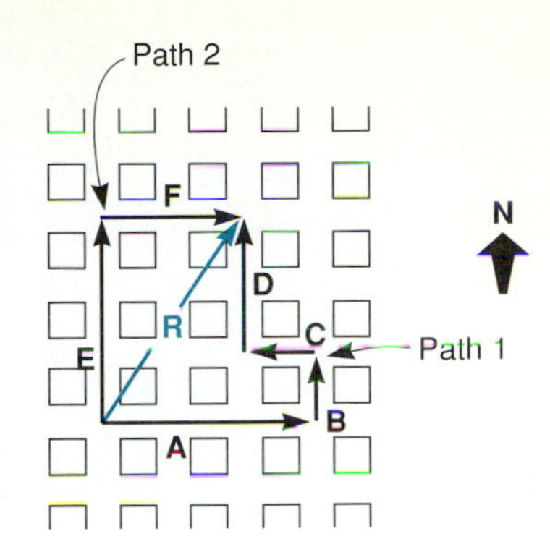

Figure 3.9
The resultant vector **R** is the graphic sum of the component set of vectors **A, B, C,** and **D** and **E** and **F.** That is, **A** + **B** + **C** + **D** = **R** and **E** + **F** = **R.**

EXAMPLE 1

Find the resultant displacement of an airplane that flies $2\bar{0}$ mi due east, then $3\bar{0}$ mi due north, and then $1\bar{0}$ mi at $6\bar{0}°$ west of south.

We choose a scale of 1.0 cm = 5.0 mi so that the vectors are large enough to be accurate and small enough to fit on the paper. (Here each block represents 0.5 cm.) The length of the first vector is

$$|\mathbf{A}| = 2\bar{0}\ \cancel{\text{mi}} \times \frac{1.0\text{ cm}}{5.0\ \cancel{\text{mi}}} = 4.0\text{ cm}$$

The length of the second vector is

$$|\mathbf{B}| = 3\bar{0}\ \cancel{\text{mi}} \times \frac{1.0\text{ cm}}{5.0\ \cancel{\text{mi}}} = 6.0\text{ cm}$$

The length of the third vector is

$$|\mathbf{C}| = 1\bar{0}\ \cancel{\text{mi}} \times \frac{1.0\text{ cm}}{5.0\ \cancel{\text{mi}}} = 2.0\text{ cm}$$

Draw the north–south reference line and draw the first vector as shown in Fig. 3.10(a). The second and third vectors are then drawn as shown in Fig. 3.10(b) and 3.10(c).

Using a ruler, the length of the resultant measures 5.5 cm [Fig. 3.10(d)]. Since 1.0 cm = 5.0 mi, this represents a displacement with magnitude:

$$|\mathbf{R}| = 5.5\ \cancel{\text{cm}} \times \frac{5.0\text{ mi}}{1.0\ \cancel{\text{cm}}} = 28\text{ mi}$$

The angle measures 24°, so the resultant is 28 mi at 24° east of north.

EXAMPLE 2

Find the resultant of the displacements: $15\bar{0}$ km due west, then $20\bar{0}$ km due east, and then 125 km due south.

Choose a scale of 1.0 cm = $5\bar{0}$ km. Follow the procedure in Fig. 3.11. The length of the resultant measures 2.6 cm. Since 1.0 cm = $5\bar{0}$ km,

$$|\mathbf{R}| = 2.6\ \cancel{\text{cm}} \times \frac{5\bar{0}\text{ km}}{1.0\ \cancel{\text{cm}}} = 130\text{ km}$$

The angle measures 22°, so the resultant is 130 km at 22° east of south.

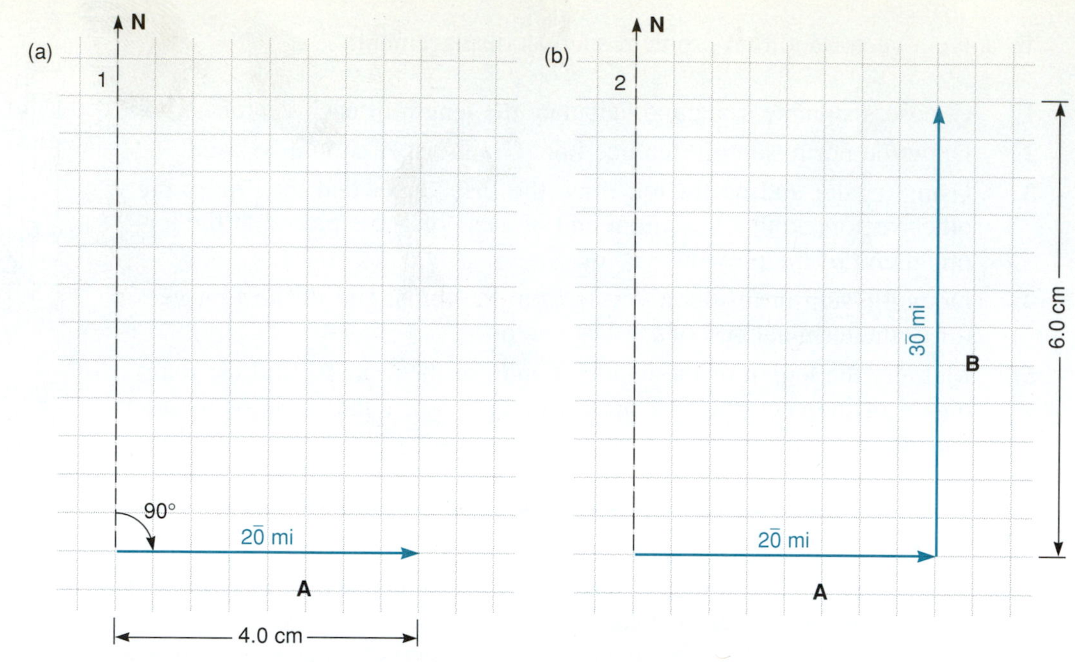

1. Draw the north-south reference line and the first vector: $2\bar{0}$ mi due east.

2. Draw the second vector: $3\bar{0}$ mi due north.

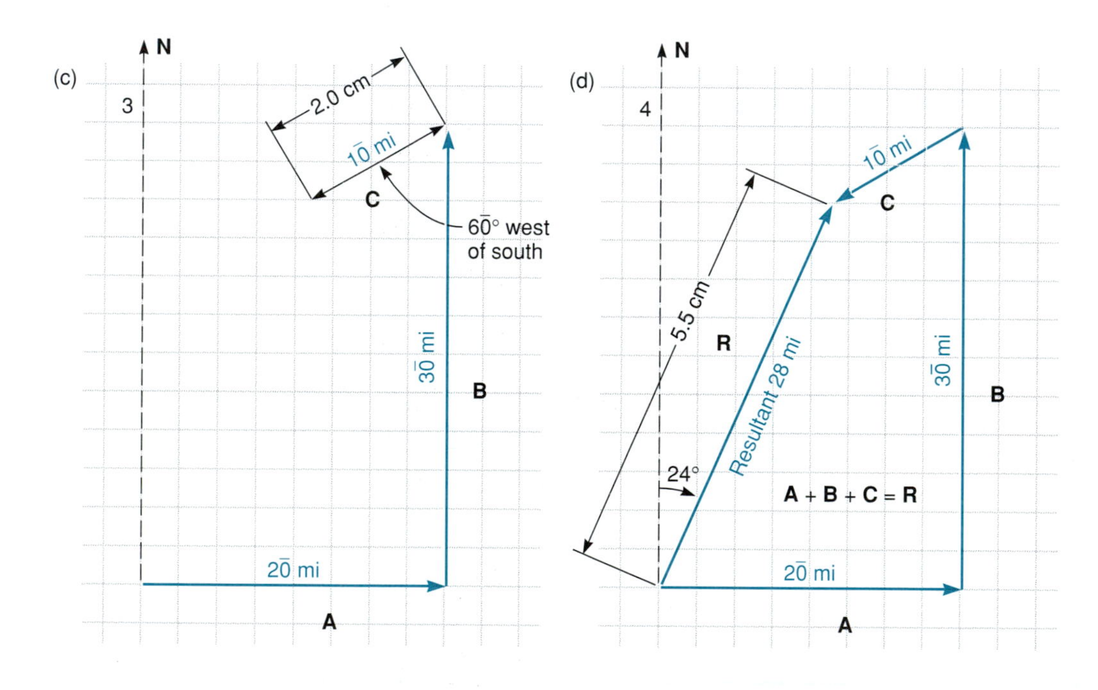

3. Draw the third vector: $1\bar{0}$ mi at $6\bar{0}$° west of south.

4. Draw the resultant vector, which is 28 mi at 24° east of north.

Scale: 1.0 cm = 5.0 mi

Figure 3.10

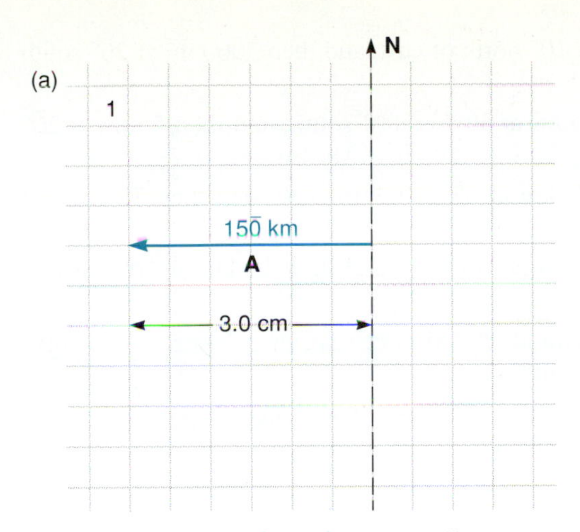

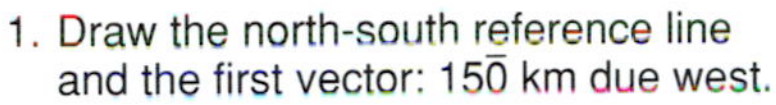

1. Draw the north-south reference line and the first vector: $15\bar{0}$ km due west.

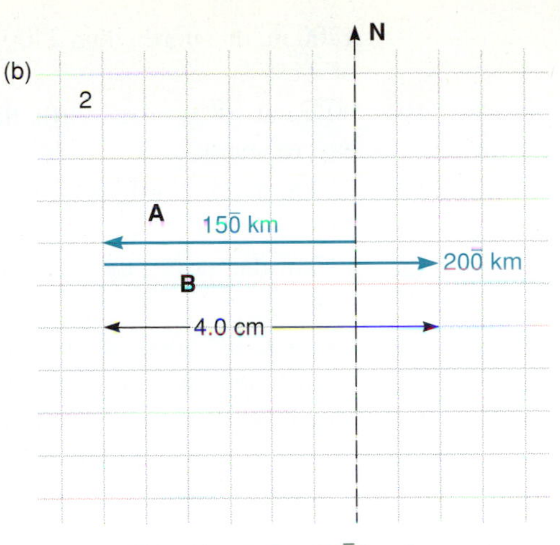

2. Draw the vector: $20\bar{0}$ km due east.

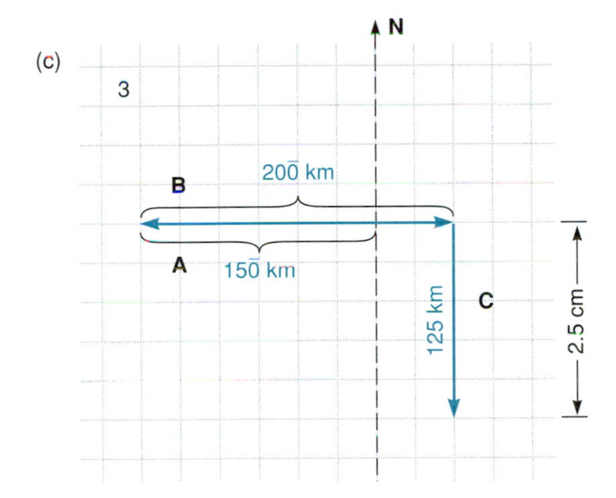

3. Draw the vector: 125 mi due south.

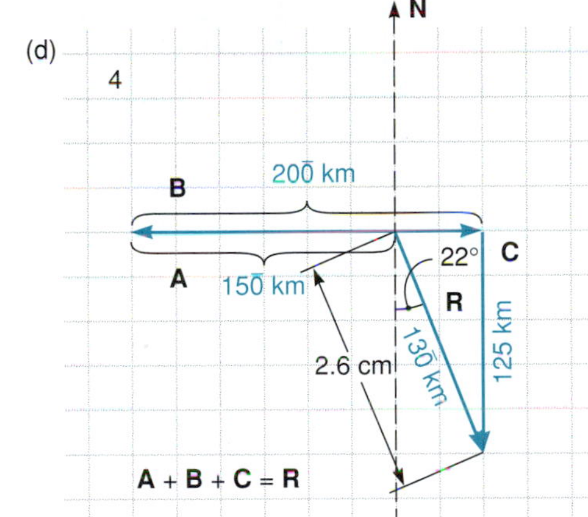

4. The length of the resultant is 2.6 cm, which represents $13\bar{0}$ km at 22° east of south.

Scale: 1.0 cm = $5\bar{0}$ km

Figure 3.11

■ PROBLEMS 3.3

Use graph paper to find the resultant of each displacement pair.

1. 35 km due east, then $5\bar{0}$ km due north.
2. $6\bar{0}$ km due west, then $9\bar{0}$ km due south
3. $5\bar{0}\bar{0}$ mi at 75° east of north, then $15\bar{0}\bar{0}$ mi at $2\bar{0}$° west of south
4. $2\bar{0}$ mi at 3° north of east, then 17 mi at 9° west of south
5. 67 km at 55° north of west, then 46 km at 25° south of east
6. 4.0 km at 25° west of south, then 2.0 km at 15° north of east

Use graph paper to find the resultant of each combination of displacements.

7. $6\bar{0}$ km due south, then $9\bar{0}$ km at 15° north of west, and then 75 km at 45° north of east
8. 110 km at $5\bar{0}$° north of east, then 170 km at $3\bar{0}$° east of south, and then 145 km at $2\bar{0}$° north of east

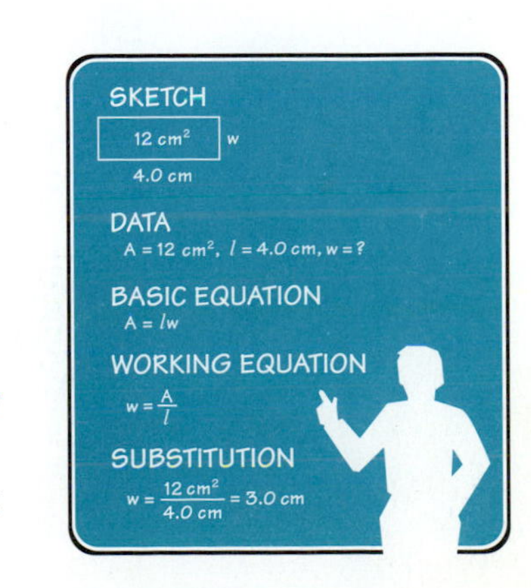

9. 1700 mi due north, then 2400 mi at 10° north of east, and then $2\bar{0}00$ mi at $2\bar{0}$° south of west
10. $9\bar{0}$ mi at $1\bar{0}$° west of north, then 75 mi at $3\bar{0}$° west of south, and then 55 mi at $2\bar{0}$° east of south
11. 75 km at 25° north of east, then 75 km at 65° south of west, and then 75 km due south
12. 17 km due north, then $1\bar{0}$ km at 7° south of east, and then 15 km at $1\bar{0}$° west of south
13. 12 mi at 58° north of east, then 16 mi at 78° north of east, then $1\bar{0}$ mi at 45° north of east, and then 14 mi at $1\bar{0}$° north of east
14. $1\bar{0}$ km at 15° west of south, then 27 km at 35° north of east, then 31 km at 5° north of east, and then 22 km at $2\bar{0}$° west of north ■

3.4 VELOCITY

When an automobile travels a certain distance, we are interested in how fast the distance was traveled. *The distance traveled per unit of time is called the* **speed.** Speed is a scalar (showing only magnitude, not direction) because it is described by a number and a unit. The unit of time is usually an hour or a second, so that the units of speed are kilometres per hour (km/h), metres per second (m/s), miles per hour (mi/h), or feet per second (ft/s).

The average speed equals the total distance traveled divided by the total time. That is,

$$\text{average speed} = \frac{\text{total distance traveled}}{\text{total time}}$$

If you drive 350 mi in 7 h, your average speed is

$$\frac{350 \text{ mi}}{7 \text{ h}} = 50 \text{ mi/h}$$

In everyday usage the terms *speed* and *velocity* often are given the same meaning. However, in physics they have separate meanings. The speed of an automobile shows how fast it is moving, such as 50 mi/h, as above. Speed tells nothing about the direction. Suppose that you start from Chicago and drive 50 mi/h for 6 h. Where did you end your trip? Obviously, with no given direction, the question cannot be answered. You may have driven 50 mi/h southwest to St. Louis, 50 mi/h northeast to Detroit, 50 mi/h southeast to Louisville, or 50 mi/h in a loop that returned you to Chicago.

The quantity that combines both the speed of an object and its direction of motion is called *velocity.* The **velocity** of an object is the time rate of change of its displacement, or its rate of motion in a particular direction. That is, velocity is a vector that gives the *direction* of travel and the *distance* traveled per unit of time. This relationship may be expressed by the equation

$$v_{\text{avg}} = \frac{s}{t}$$

or

$$s = v_{\text{avg}} t$$

where s = displacement or distance
v_{avg} = average velocity or speed
t = time

Note: The equation $v_{avg} = \frac{s}{t}$ is used to find either average speed (a scalar quantity) or the magnitude of velocity (a vector quantity).

A useful conversion is $6\bar{0}$ mi/h = 88 ft/s.

EXAMPLE 1

Find the average speed of an automobile that travels 160 km in 2.0 h.

Sketch: None needed

Data:

$$s = 160 \text{ km}$$
$$t = 2.0 \text{ h}$$
$$v_{avg} = ?$$

Basic Equation:

$$s = v_{avg}t$$

Working Equation:

$$v_{avg} = \frac{s}{t}$$

Substitution:

$$v_{avg} = \frac{160 \text{ km}}{2.0 \text{ h}} = 8\bar{0} \text{ km/h}$$

EXAMPLE 2

An airplane flies $35\bar{0}0$ mi in 5.00 h. Find its average speed.

Sketch: None needed

Data:

$$s = 35\bar{0}0 \text{ mi}$$
$$t = 5.00 \text{ h}$$
$$v_{avg} = ?$$

Basic Equation:

$$s = v_{avg}t$$

Working Equation:

$$v_{avg} = \frac{s}{t}$$

Substitution:

$$v_{avg} = \frac{350\overline{0} \text{ mi}}{5.00 \text{ h}}$$
$$= 70\overline{0} \text{ mi/h}$$

EXAMPLE 3

Find the velocity of a plane that travels $60\overline{0}$ km due north in 3 h 15 min.

Sketch: None needed

Data:

$$s = 60\overline{0} \text{ km}$$
$$t = 3 \text{ h } 15 \text{ min} = 3.25 \text{ h}$$
$$v_{avg} = ?$$

Basic Equation:

$$s = v_{avg}t$$

Working Equation:

$$v_{avg} = \frac{s}{t}$$

Substitution:

$$v_{avg} = \frac{60\overline{0} \text{ km}}{3.25 \text{ h}}$$
$$= 185 \text{ km/h}$$

The direction is north. Thus, the velocity is 185 km/h due north.

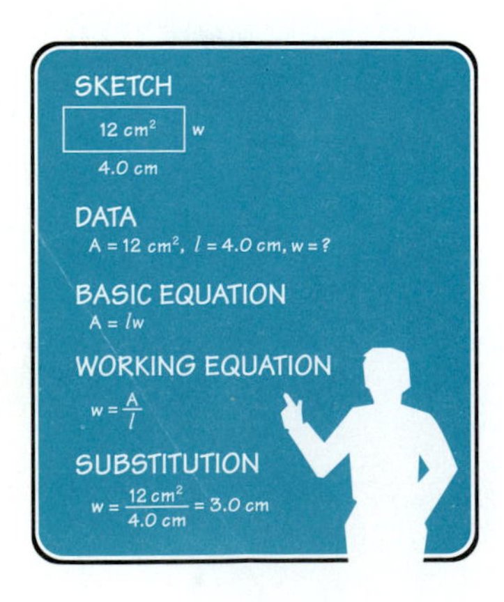

■ PROBLEMS 3.4

Find the average speed (in the given units) of an auto that travels each distance in the given time.

1. Distance of 65 km in 1.0 h (in km/h)
2. Distance of $12\overline{0}$ ft in 2.0 s (in ft/s)
3. Distance of 270 km in 3.0 h (in km/h)
4. Distance of 190 m in 8.5 s (in m/s)
5. Distance of 150 mi in 3.0 h (in mi/h)
6. Distance of 45 km in 0.50 h (in km/h)
7. Distance of 8550 m in 6 min 35 s (in m/s)
8. Distance of 785 ft in 11.5 s (in ft/s)
9. Find the average speed (in mi/h) of a racing car that turns a lap on a 1.00-mi oval track in 30.0 s.
10. While driving at $9\overline{0}$ km/h, how far can you travel in 3.5 h?
11. While driving at $9\overline{0}$ km/h, how far (in metres) do you travel in 1.0 s?

An automobile is traveling at 55 mi/h. Find its speed

12. in ft/s. 13. in m/s. 14. in km/h.

An automobile is traveling at 22.0 m/s. Find its speed

15. in km/h. 16. in mi/h. 17. in ft/s.

Find the velocity for each displacement and time.

18. $1\bar{0}0$ km north in 3.0 h
19. 160 km east in 2.0 h
20. 31.0 mi west in 0.500 h
21. $10\bar{0}0$ mi south in 8.00 h
22. 426 km at 45° north of west in 2.75 h
23. 275 km at $3\bar{0}$° south of east in 4.50 h
24. 870 mi at 68° west of north in 3.5 h
25. Milwaukee is 121 mi (air miles) due west of Grand Rapids. Maria drives 255 mi in 4.75 h from Grand Rapids to Milwaukee around Lake Michigan. Find (a) her average driving speed and (b) her average travel velocity.
26. Telluride, Colorado, is 45 air miles at 11° east of north of Durango. Chuck drove 120 mi from Durango to Telluride around a mountain in $4\frac{1}{4}$ h on a winter day including a traffic delay. Find (a) his average driving speed and (b) his average travel velocity. ■

3.5 ACCELERATION

When the dragster shown in Fig. 3.12 travels down a quarter-mile track, its velocity changes. Its velocity in the last few feet of the race is much greater than its velocity at the start. The faster the velocity of the dragster changes, the less

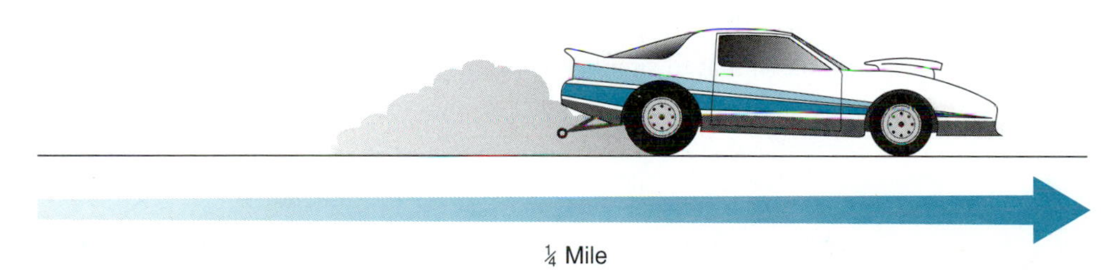

Figure 3.12 The velocity of the dragster changes in magnitude from zero at the start to its final velocity at the finish.

its travel time will be. And the faster the velocity changes, the larger the acceleration will be. **Acceleration** *is the change in velocity per unit time*. That is,

$$\text{average acceleration} = \frac{\text{change in velocity (or speed)}}{\text{elapsed time}}$$

$$= \frac{\text{final velocity} - \text{initial velocity}}{\text{time}}$$

This relationship may be expressed by the equation

$$a = \frac{\Delta v}{t} = \frac{v_f - v_i}{t}$$

or

$$\Delta v = at$$

where Δv = change in velocity (or speed)
a = acceleration
t = time

The Greek letter Δ (capital delta) is used to mean "change in."

EXAMPLE 1

A dragster starts from rest (velocity = 0 ft/s) and attains a speed of $15\overline{0}$ ft/s in 10.0 s. Find its acceleration.

Sketch: None needed

Data:

$$\Delta v = 15\overline{0} \text{ ft/s} - 0 \text{ ft/s} = 15\overline{0} \text{ ft/s}$$
$$t = 10.0 \text{ s}$$
$$a = ?$$

Basic Equation:

$$\Delta v = at$$

Working Equation:

$$a = \frac{\Delta v}{t}$$

Substitution:

$$a = \frac{15\overline{0} \text{ ft/s}}{10.0 \text{ s}}$$
$$= 15.0 \frac{\text{ft/s}}{\text{s}} \text{ or } \underline{15.0 \text{ feet per second per second}}$$

Recall from arithmetic that to simplify fractions in the form

$$\frac{\frac{a}{b}}{\frac{c}{d}}$$

we divide by the denominator; that is, invert and multiply:

$$\frac{\frac{a}{b}}{\frac{c}{d}} = \frac{a}{b} \div \frac{c}{d} = \frac{a}{b} \cdot \frac{d}{c} = \frac{ad}{bc}$$

Use this idea to simplify the units 15.0 feet per second per second from above:

$$\frac{\frac{15.0 \text{ ft}}{\text{s}}}{\frac{\text{s}}{1}} = \frac{15.0 \text{ ft}}{\text{s}} \div \frac{\text{s}}{1} = \frac{15.0 \text{ ft}}{\text{s}} \cdot \frac{1}{\text{s}} = \frac{15.0 \text{ ft}}{\text{s}^2} \text{ or } 15.0 \text{ ft/s}^2$$

The units of acceleration are usually ft/s^2 or m/s^2.

When the speed of an automobile increases from rest to 5 mi/h in the first second, to 10 mi/h in the next second, and to 15 mi/h in the third second, its acceleration is $5 \frac{\text{mi/h}}{\text{s}}$. That is, its increase in speed is 5 mi/h during each second. If an object increases in speed from 6 m/s to 9 m/s in the first second, to 12 m/s in the next second, and to 15 m/s in the third second, its acceleration is $3 \ \dfrac{\text{m/s}}{\text{s}}$, usually written 3 m/s^2. This means that the speed of the object increases 3 m/s during each second.

EXAMPLE 2

A car accelerates from 45 km/h to $8\bar{0}$ km/h in 3.00 s. Find its acceleration (in m/s^2).

Sketch: None needed

Data:

$$\begin{aligned} \Delta v &= 8\bar{0} \text{ km/h} - 45 \text{ km/h} = 35 \text{ km/h} \\ t &= 3.00 \text{ s} \\ a &= ? \end{aligned}$$

Basic Equation:

$$\Delta v = at$$

Working Equation:

$$a = \frac{\Delta v}{t}$$

Substitution:

$$\begin{aligned} a &= \frac{35 \cancel{\text{km}}/\cancel{\text{h}}}{3.00 \text{ s}} \times \frac{1000 \text{ m}}{1 \cancel{\text{km}}} \times \frac{1 \cancel{\text{h}}}{3600 \text{ s}} \\ &= 3.2 \text{ m/s}^2 \end{aligned}$$

Note the use of the conversion factors to change the units km/h/s to m/s^2.

EXAMPLE 3

A plane accelerates at 8.5 m/s^2 for 4.5 s. Find its increase in speed (in m/s).

Sketch: None needed

Data:

$$\begin{aligned} a &= 8.5 \text{ m/s}^2 \\ t &= 4.5 \text{ s} \\ \Delta v &= ? \end{aligned}$$

Basic Equation:

$$\Delta v = at$$

Working Equation: Same

Substitution:

$$\begin{aligned} \Delta v &= (8.5 \text{ m/s}^2)(4.5 \text{ s}) \\ &= 38 \text{ m/s} \end{aligned}$$

$$\boxed{\frac{\text{m}}{\text{s}^2} \times \text{s} = \frac{\text{m}}{\text{s}}}$$

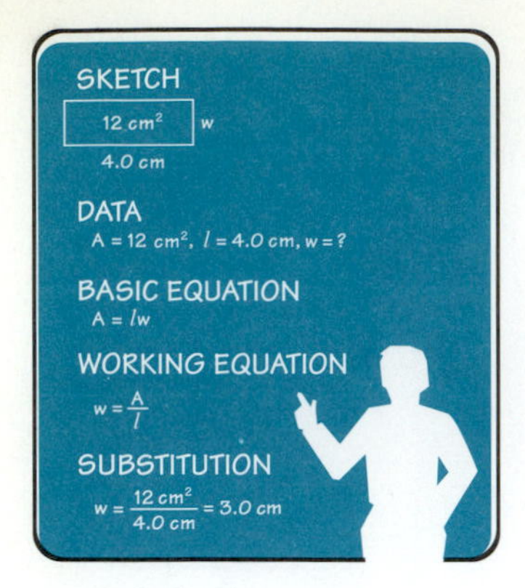

■ PROBLEMS 3.5

An automobile changes speed as shown below. Find its acceleration.

	Speed change	*Time interval*	*Find a*
1.	From 0 to 15 m/s	1.0 s	in m/s^2
2.	From 0 to 18 m/s	3.0 s	in m/s^2
3.	From $6\overline{0}$ ft/s to $7\overline{0}$ ft/s	1.0 s	in ft/s^2
4.	From 45 m/s to 65 m/s	2.0 s	in m/s^2
5.	From 25 km/h to $9\overline{0}$ km/h	5.6 s	in m/s^2
6.	From $1\overline{0}$ mi/h to $5\overline{0}$ mi/h	3.5 s	in ft/s^2

7. A dragster starts from rest and reaches a speed of 62.5 m/s in 10.0 s. Find its acceleration (in m/s^2).
8. A car accelerates from 25 mi/h to 55 mi/h in 4.5 s. Find its acceleration (in ft/s^2).
9. A train accelerates from $1\overline{0}$ km/h to $11\overline{0}$ km/h in 2 min 15 s. Find its acceleration (in m/s^2).

A plane accelerates at 30.0 ft/s^2 for 3.30 s. Find its increase in speed

10. in ft/s.
11. in mi/h.

A rocket accelerates at 10.0 m/s^2 from rest for 20.0 s. Find its increase in speed

12. in m/s.
13. in km/h.
14. How long (in seconds) does it take for a rocket sled accelerating at 15.0 m/s^2 to change its speed from 20.0 m/s to 65.0 m/s?
15. How long (in seconds) does it take for a truck accelerating at 1.50 m/s^2 to go from rest to 90.0 km/h?
16. How long (in seconds) does it take for a car accelerating at 3.50 m/s^2 to go from rest to $12\overline{0}$ km/h? ■

3.6 UNIFORMLY ACCELERATED MOTION

Every time a truck speeds up or slows down, its velocity changes. This change of velocity is called **acceleration.** Acceleration may be an increase or decrease in velocity. A negative (−) acceleration is commonly called **deceleration,** meaning the object is slowing down.

Because we lack the mathematical tools to study all kinds of motion, we must limit our study to one kind—uniformly accelerated motion. The most common example of this kind of motion is that of a freely falling body. Because of the complexity of this kind of problem, we must assume that falling bodies are unaffected by the resistance of the air. However, air resistance is, in fact, an important factor in the design of machines that must move through the atmosphere. In motion problems, we assume air resistance is negligible. For freely falling bodies the **acceleration** *(a)* **due to gravity** is $a = 9.80\ m/s^2$ (metric system) or $a = 32.2\ ft/s^2$ (English system).

What does $a = 9.80\ m/s^2$ mean? When a ball is dropped from a building, the speed of the ball increases by 9.80 m/s during each second. That is, its speed is 9.80 m/s at the end of 1 second, 19.6 m/s at the end of 2 seconds, 29.4 m/s at the end of 3 seconds, 39.2 m/s at the end of 4 seconds, and so on, until it hits the ground.

A number of formulas and equations apply to freely falling bodies and uniformly accelerated motion in general.

$$s = v_{avg}t \qquad s = v_i t + \tfrac{1}{2}a_{avg}t^2$$

$$v_{avg} = \frac{v_f + v_i}{2} \qquad v_f = v_i + a_{avg}t$$

$$a_{avg} = \frac{v_f - v_i}{t} \qquad s = \tfrac{1}{2}(v_f + v_i)t$$

$$2a_{avg}\, s = v_f^2 - v_i^2$$

where s = displacement
v_f = final velocity
v_i = initial velocity
v_{avg} = average velocity
a_{avg} = average acceleration
t = time

Now consider some problems using these equations, applying our problem-solving method.

EXAMPLE 1

The average velocity of a rolling freight car is 2.00 m/s. How long does it take for the car to roll 15.0 m?

Sketch: None needed

Data:

$$s = 15.0 \text{ m}$$
$$v_{avg} = 2.00 \text{ m/s}$$
$$t = ?$$

Basic Equation:

$$s = v_{avg}t$$

Working Equation:

$$t = \frac{s}{v_{avg}}$$

Substitution:

$$t = \frac{15.0 \text{ m}}{2.00 \text{ m/s}} = 7.50 \text{ s}$$

$$\frac{\text{m}}{\text{m/s}} = \text{m} \div \frac{\text{m}}{\text{s}} = \text{m} \cdot \frac{\text{s}}{\text{m}} = \text{s}$$

EXAMPLE 2

A dragster starting from rest reaches a final velocity of 318 km/h. Find its average velocity.

Sketch: None needed

Data:

$$v_i = 0$$
$$v_f = 318 \text{ km/h}$$
$$v_{avg} = ?$$

Basic Equation:

$$v_{avg} = \frac{v_f + v_i}{2}$$

Working Equation: Same

Substitution:

$$v_{avg} = \frac{318 \text{ km/h} + 0 \text{ km/h}}{2}$$
$$= 159 \text{ km/h}$$

EXAMPLE 3

A rock is thrown straight down from a cliff with an initial velocity of 10.0 ft/s. Its final velocity when it strikes the water below is $31\bar{0}$ ft/s. The acceleration due to gravity is 32.2 ft/s^2. How long is the rock in flight?

Sketch: None needed

Data:

$$v_i = 10.0 \text{ ft/s}$$
$$a = 32.2 \text{ ft/s}^2$$
$$v_f = 31\bar{0} \text{ ft/s}$$
$$t = ?$$

Note the importance of listing all the data as an aid to finding the basic equation.

Basic Equation:

$$v_f = v_i + a_{avg}t \qquad \text{or} \qquad a_{avg} = \frac{v_f - v_i}{t}$$

(two forms of the same equation)

Working Equation:

$$t = \frac{v_f - v_i}{a_{avg}}$$

Substitution:

$$t = \frac{31\bar{0} \text{ ft/s} - 10.0 \text{ ft/s}}{32.2 \text{ ft/s}^2}$$
$$= \frac{30\bar{0} \text{ ft/s}}{32.2 \text{ ft/s}^2}$$
$$= 9.32 \text{ s}$$

$$\frac{\text{ft/s}}{\text{ft/s}^2} = \frac{\text{ft}}{\text{s}} \div \frac{\text{ft}}{\text{s}^2} = \frac{\text{ft}}{\text{s}} \cdot \frac{\text{s}^2}{\text{ft}} = \text{s}$$

EXAMPLE 4

A train slowing to a stop has an average acceleration of $-3.00\ \text{m/s}^2$. [Note that a minus (−) acceleration is commonly called *deceleration,* meaning that the train is slowing down.] If its initial velocity is 30.0 m/s, how far does it travel in 4.00 s?

Sketch: None needed

Data:

$$a_{\text{avg}} = -3.00\ \text{m/s}^2$$
$$v_i = 30.0\ \text{m/s}$$
$$t = 4.00\ \text{s}$$
$$s = ?$$

Basic Equation:

$$s = v_i t + \tfrac{1}{2} a_{\text{avg}} t^2$$

Working Equation: Same

Substitution:

$$\begin{aligned} s &= (30.0\ \text{m/s})(4.00\ \text{s}) + \tfrac{1}{2}(-3.00\ \text{m/s}^2)(4.00\ \text{s})^2 \\ &= 12\bar{0}\ \text{m} - 24.0\ \text{m} \\ &= 96\ \text{m} \end{aligned}$$

EXAMPLE 5

An automobile accelerates from 67.0 km/h to 96.0 km/h in 7.80 s. What is its acceleration (in m/s^2)?

Sketch: None needed

Data:

$$v_f = 96.0\ \text{km/h}$$
$$t = 7.80\ \text{s}$$
$$v_i = 67.0\ \text{km/h}$$
$$a = ?$$

Basic Equation:

$$a_{\text{avg}} = \frac{v_f - v_i}{t}$$

Working Equation: Same

Substitution:

$$\begin{aligned} a_{\text{avg}} &= \frac{96.0\ \text{km/h} - 67.0\ \text{km/h}}{7.80\ \text{s}} \\ &= \frac{29.0\ \text{km/h}}{7.80\ \text{s}} \\ &= 29.0\ \frac{\frac{\cancel{\text{km}}}{\cancel{\text{h}}} \times \frac{10^3\ \text{m}}{1\ \cancel{\text{km}}} \times \frac{1\ \cancel{\text{h}}}{3600\ \text{s}}}{7.80\ \text{s}} \\ &= 1.03\ \text{m/s}^2 \end{aligned}$$

When any object is thrown or hurled vertically upward, its upward speed is uniformly decreased by the force of gravity until it stops for an instant at its peak before falling back to the ground. As it is falling to the ground, it is uniformly accelerated by gravity the same as it would have been if dropped from its peak height. If an object is thrown vertically upward and if the initial velocity is known, the previous acceleration/gravity formulas may be used to find how high the object rises, how long it is in flight, and so on.

Note: When we consider a problem involving an object being thrown upward, we will consider an upward direction to be negative and the opposing gravity in its normal downward direction to be positive.

EXAMPLE 6

A baseball is thrown vertically upward with an initial velocity of 25.0 m/s.

(a) How high does it go?

(b) How long does it take to reach its maximum height?

(c) How long is it in flight?

Sketch:

(a) Data:

$v_i = -25.0$ m/s (v_i is negative because the initial velocity is directed opposite gravity, g.)

$v_f = 0$ (At the instant of the ball's maximum height, its velocity is zero.)

$a_{avg} = g = 9.80 \text{ m/s}^2$

$s = ?$

Basic Equation:

$$2a_{avg}s = v_f^2 - v_i^2$$

Working Equation:

$$s = \frac{v_f^2 - v_i^2}{2a_{avg}}$$

Substitution:

$$s = \frac{0^2 - (-25.0 \text{ m/s})^2}{2(9.80 \text{ m/s}^2)}$$

$$\frac{(\text{m/s})^2}{\text{m/s}^2} = \frac{\text{m}^2/\text{s}^2}{\text{m/s}^2} = \frac{\text{m}^2}{\text{s}^2} \div \frac{\text{m}}{\text{s}^2} = \frac{\overset{\text{m}}{\cancel{\text{m}^2}}}{\cancel{\text{s}^2}} \times \frac{\cancel{\text{s}^2}}{\cancel{\text{m}}} = \text{m}$$

$s = -31.9$ m (*s* being negative indicates an upward displacement.)

(b) Data:

$$v_i = -25.0 \text{ m/s}$$
$$v_f = 0$$
$$a_{avg} = g = 9.80 \text{ m/s}^2$$
$$t = ?$$

Basic Equation:

$$v_f = v_i + a_{avg}t$$

$$a = \frac{v_f - v_i}{t}$$

Working Equation:

$$t = \frac{v_f - v_i}{a_{avg}}$$

Substitution:

$$t = \frac{0 - (-25.0 \text{ m/s})}{9.80 \text{ m/s}^2}$$
$$= 2.55 \text{ s}$$

$$\frac{\text{m/s}}{\text{m/s}^2} = \frac{\text{m}}{\text{s}} \div \frac{\text{m}}{\text{s}^2} = \frac{\cancel{\text{m}}}{\cancel{\text{s}}} \times \frac{\overset{\text{s}}{\cancel{\text{s}^2}}}{\cancel{\text{m}}} = \text{s}$$

(c) The ball decelerates on the way up and accelerates on the way down at the same rate because the force of gravity is constant (9.80 m/s^2). Therefore, the time for the ball to reach its peak is the same as the time for it to fall to the ground. The total time in flight is 2(2.55 s) = 5.10 s.

With what speed does the ball in Example 6 hit the ground? The answer is 25.0 m/s. Can you explain why?

Earlier in this section, we assumed no air resistance. We know that two dense and compact objects, such a bowling ball and a marble, will fall at the same rate in air. We also know that two unlike objects, such as a marble and a feather, fall at different rates because of the air's resistance.

A parachute takes advantage of the air's resistance to slow a sky diver's descent. What would happen if a parachute does not open? Does this mean that his or her velocity will increase constantly until he or she hits the ground? As the velocity increases, the air resistance also increases. Since the gravitational pull and the air resistance are directed opposite each other, they tend to oppose or equalize each other. (Here, the velocity and acceleration are both directed downward while the air resistance is directed upward.) This equalization occurs when the friction of the air's resistance equals the force of gravity. When this equalization occurs, the falling object stops accelerating and continues to fall at a constant velocity, called *terminal velocity.* The terminal velocity of a ball or a person is approximately 120 mi/h or 190 km/h.

In general, the terminal velocity of an object varies with its weight and its aerodynamic features that include the following:

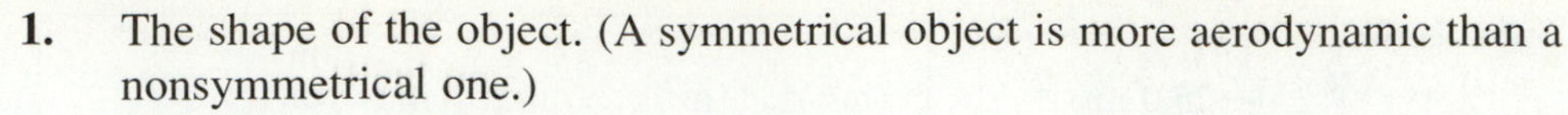

1. **1.** The shape of the object. (A symmetrical object is more aerodynamic than a nonsymmetrical one.)
2. **2.** The orientation of the object as it is traveling. (A sky diver slows the fall by spreading out his or her arms and legs and falling horizontally. The speed of the fall increases if the person falls head or feet first.)
3. **3.** The smoothness of the surface. (A body with a smooth surface provides less air resistance than a body with a rough surface and falls or flies faster as a result.)

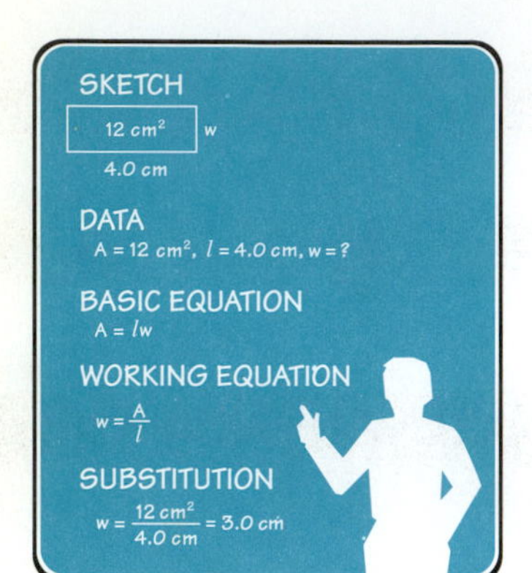

■ PROBLEMS 3.6

Substitute in the given equation and find the unknown quantity.

1. Given: $v_{avg} = \dfrac{v_f + v_i}{2}$
   $v_f = 6.20$ m/s
   $v_i = 3.90$ m/s
   $v_{avg} = ?$

2. Given: $a_{avg} = \dfrac{v_f - v_i}{t}$
   $a_{avg} = 3.07$ m/s^2
   $v_f = 16.8$ m/s
   $t = 4.10$ s
   $v_i = ?$

3. Given: $s = v_i t + \frac{1}{2} a_{avg} t^2$
   $t = 3.00$ s
   $a_{avg} = 6.40$ m/s^2
   $v_i = 33.0$ m/s
   $s = ?$

4. Given: $2a_{avg}s = v_f^2 - v_i^2$
   $a_{avg} = 8.41$ m/s^2
   $s = 4.81$ m
   $v_i = 1.24$ m/s
   $v_f = ?$

5. Given: $v_f = v_i + a_{avg}t$
   $v_f = 10.40$ ft/s
   $v_i = 4.01$ ft/s
   $t = 3.00$ s
   $a_{avg} = ?$

6. The average velocity of a mini-bike is 15.0 km/h. How long does it take for the bike to go 35.0 m?
7. A sprinter starting from rest reaches a final velocity of 18.0 mi/h. What is her average velocity?
8. A coin is dropped with no initial velocity. Its final velocity when it strikes the earth is 50.0 ft/s. The acceleration due to gravity is 32.2 ft/s^2. How long does it take to strike the earth?
9. A rocket lifting off from earth has an average acceleration of 44.0 ft/s^2. Its initial velocity is zero. How far into the atmosphere does it travel during the first 5.00 s, assuming that it goes straight up?
10. The final velocity of a truck is 74.0 ft/s. If it accelerates at a rate of 2.00 ft/s^2 from an initial velocity of 5.00 ft/s, how long will it take for it to attain its final velocity?
11. A truck accelerates from 85 km/h to 12$\overline{0}$ km/h in 9.2 s. Find its acceleration in m/s^2.
12. How long does it take a rock to drop 95.0 m from rest?
13. Find the final speed of the rock in Problem 12.
14. A ball is thrown downward from the top of a 43.0-ft building with an initial speed of 62.0 ft/s. Find its final speed as it strikes the ground.
15. A car is traveling at 7$\overline{0}$ km/h. It then uniformly decelerates to a complete stop in 12 s. Find its acceleration (in m/s^2).
16. A car is traveling at 6$\overline{0}$ km/h. It then accelerates at 3.6 m/s^2 to 9$\overline{0}$ km/h.
    (a) How long does it take to reach the new speed?
    (b) How far does it travel while accelerating?

17. A rock is dropped from a bridge to the water below. It takes 2.40 s for the rock to hit the water.
    (a) Find the speed (in m/s) of the rock as it hits the water.
    (b) How high (in metres) is the bridge above the water?
18. A bullet is fired vertically upward from a gun and reaches a height of $70\bar{0}0$ ft.
    (a) Find its initial velocity.
    (b) How long does it take to reach its maximum height?
    (c) How long is it in flight?
19. A bullet is fired vertically upward from a gun with an initial velocity of $25\bar{0}$ m/s.
    (a) How high does it go?
    (b) How long does it take to reach its maximum height?
    (c) How long is it in flight?
20. A rock is thrown down with an initial speed of 30.0 ft/s from a bridge to the water below. It takes 3.50 s for the rock to hit the water.
    (a) Find the speed (in ft/s) of the rock as it hits the water.
    (b) How high is the bridge above the water?
21. A rock is thrown straight up with an initial speed of 10.0 m/s from a deck that is 25.0 m above the ground.
    (a) How long does it take to reach its maximum height?
    (b) What maximum height above the deck does it reach?
    (c) At what speed does it hit the ground?
    (d) What total length of time is the rock in the air?
22. A rock is thrown straight up and reaches a height of 15.0 m above a deck that is 40.0 m above the ground.
    (a) What is the initial speed of the rock?
    (b) How long does it take to reach its maximum height?
    (c) At what speed does it hit the ground?
    (d) What total length of time is the rock in the air?
23. John is standing on a steel beam 255.0 ft above the ground. Linda is standing 30.0 ft directly above John.
    (a) For John to throw a hammer up to Linda, at what initial speed must John throw the hammer for it to just reach Linda?
    (b) Suppose the hammer reaches the correct height, but Linda just misses catching it. How long does someone on the ground have to move out of the way from the time the hammer reaches its maximum height?
    (c) At what speed does the hammer hit the ground?

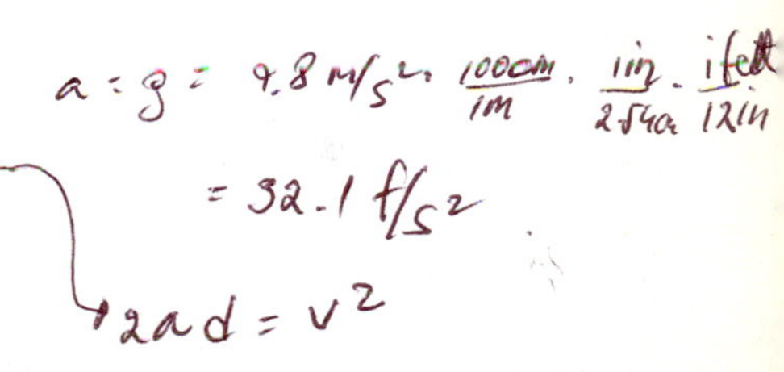

24. Kurt is standing on a steel beam 275.0 ft above the ground and throws a hammer straight up at an initial speed of 40.0 ft/s. At the instant he releases the hammer, he also drops a wrench from his pocket. Assume that neither the hammer nor the wrench hits anything while in flight.
    (a) Find the time difference between when the wrench and the hammer hit the ground.
    (b) Find the speed at which the wrench hits the ground.
    (c) Find the speed at which the hammer hits the ground.
    (d) How long does it take for the hammer to reach its maximum height?
    (e) How high above the ground is the wrench at the time the hammer reaches its maximum height?

3.7 SIMPLE HARMONIC MOTION

Periodic motion occurs when an object moves repeatedly over the same path in equal time intervals. Attach a mass m to a spring suspended from a support [Fig. 3.13(a)]. Pull the mass down (b) and release it (c). The mass moves up and down in periodic motion.

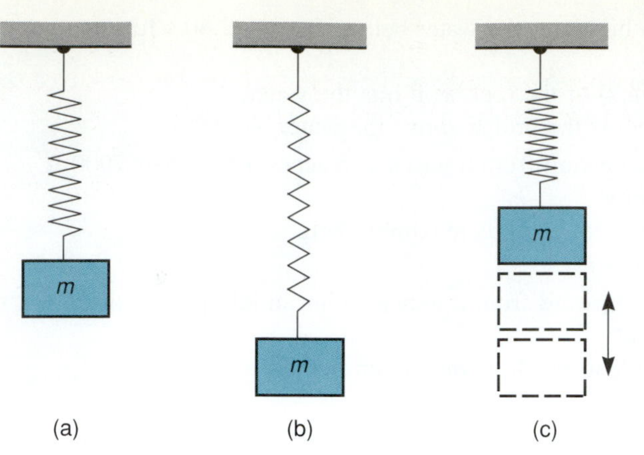

Figure 3.13
Mass suspended by spring from support in simple harmonic motion.

Simple harmonic motion is a type of linear motion in which the acceleration of an object is directly proportional to its displacement from its equilibrium position, and the motion is always directed to the equilibrium position. That is, the farther the spring is pulled down, the faster the spring moves when it is released, and the motion is always directed toward the equilibrium (rest) position. The mass on the spring in Fig. 3.13 is an example of an object in simple harmonic motion.

Next, we will compare simple harmonic motion and circular motion and discuss some of the corresponding terms. Assume that we have a Ferris wheel rotating uniformly with you in the only seat and with the sun directly overhead (Fig. 3.14). Now compare your position with the position of your shadow. When you are in position *a,* your shadow is in position *a′;* when you are in position *b,* your shadow is in position *b′;* and so on. As you complete one revolution on the Ferris wheel, your shadow makes one complete vibration (cycle) on the ground in simple harmonic motion. When your shadow is at *b′*, the **displacement** of your shadow is the distance *b′O′*, which is the distance from your shadow to the midpoint of its vibration, *O′*. The **amplitude** of the vibration is the maximum dis-

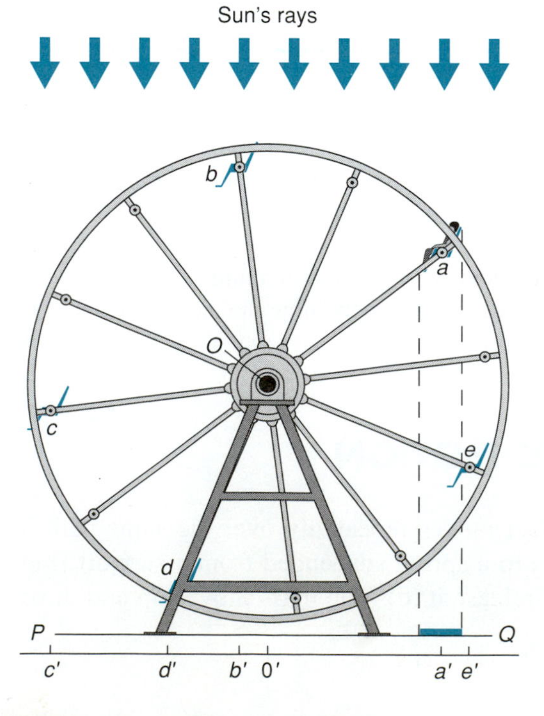

Figure 3.14
Simple harmonic motion of shadow of a person on line *PQ,* where the person is rotating at uniform speed in a circle on a Ferris wheel.

placement $O'P$ or $O'Q$, which is also the radius of the Ferris wheel. The **period** is the time required for one complete vibration—the time required for you to make one complete revolution on the Ferris wheel and the time required for your shadow to make one complete vibration on the ground. The **frequency** is the number of complete vibrations per unit of time or the number of complete revolutions that you make on the Ferris wheel per unit of time. The frequency f equals the reciprocal of the period T. That is, $f = \frac{1}{T}$. The equilibrium position of the shadow is the midpoint of its path, O'.

3.8 THE PENDULUM

A **pendulum** consists of an object suspended so that it swings freely back and forth about a pivot (Fig. 3.15). Pendulums have been commonly used in clocks for many years. The motion of a pendulum, when the displacement is small, very closely approximates simple harmonic motion. There are three basic properties of a pendulum:

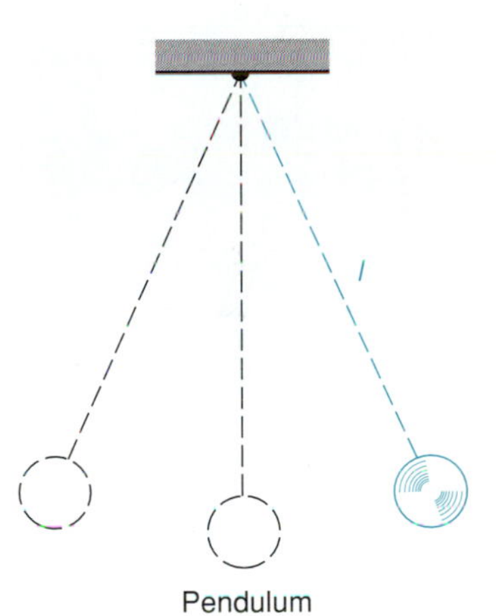

Figure 3.15
Free-swinging pendulum of length l.

1. Its period is independent of its mass. (Air resistance is more affected by the size and shape of the bob than by its mass.)
2. Its period is independent of the amplitude when the arc is small, that is, when its arc is less than 10°.
3. Subject to these conditions, its period is given by

$$T = 2\pi \sqrt{\frac{l}{g}}$$

where T = period (usually, in seconds)
l = length of pendulum (m or ft)
$g = 9.80\ \text{m/s}^2$ or $32.2\ \text{ft/s}^2$

EXAMPLE

Find the length (in cm) of a pendulum with a period of 1.50 s.

Data:

$$T = 1.50\ \text{s}$$
$$g = 9.80\ \text{m/s}^2$$
$$l = ?$$

Basic Equation:

$$T = 2\pi \sqrt{\frac{l}{g}}$$

$$T^2 = 4\pi^2 \frac{l}{g} \qquad \text{Square both sides.}$$

$$gT^2 = 4\pi^2 l \qquad \text{Multiply both sides by } g.$$

$$\frac{gT^2}{4\pi^2} = l \qquad \text{Divide both sides by } 4\pi^2.$$

Working Equation:

$$l = \frac{gT^2}{4\pi^2}$$

Substitution:

$$l = \frac{(9.80\ \text{m/s}^2)(1.50\ \text{s})^2}{4\pi^2}$$

$$= 0.559\ \text{m} = 55.9\ \text{cm}$$

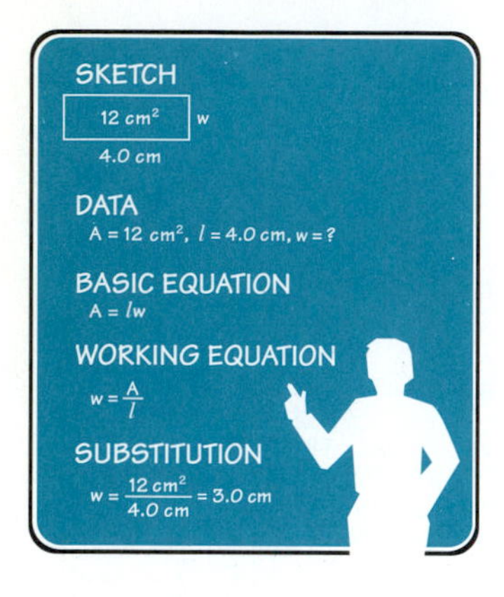

PROBLEMS 3.8

1. Find the length (in cm) of a pendulum to have a period of 1.00 s.
2. Find the length (in ft) of a pendulum to have a period of 1.00 s.
3. Find the period of a pendulum 1.25 m long.
4. Find the period of a pendulum 2.00 ft long.
5. Find the length (in in.) of a pendulum to have a period of 2.25 s.
6. Find the length (in m) of a pendulum to have a period of 0.700 s.
7. Find the period of a pendulum 18.0 in. long.
8. Find the period of a pendulum 35.0 cm long.
9. If you double the length of a pendulum, what happens to its period?
10. If you double the period of a pendulum, what happens to its length?

GLOSSARY

Acceleration Change in velocity per unit time. (p. 73)

Acceleration Due to Gravity The acceleration of a freely falling object. On the earth's surface, the acceleration due to gravity is 9.80 m/s^2 (metric) or 32.2 ft/s^2 (English). (p. 76)

Amplitude The maximum displacement of a vibration or oscillation. (p. 84)

Deceleration A negative acceleration, indicating that an object is slowing down. (p. 76)

Displacement The net distance, or direct distance, traveled as the result of motion. Measured in units of length. (p. 63) In simple harmonic motion, the distance of an object from its equilibrium, or rest, position. (p. 84)

Frequency The number of complete vibrations per unit time in simple harmonic motion. (p. 85)

Motion A change of position. (p. 62)

Pendulum An object suspended so that it swings freely back and forth about a pivot. (p. 85)

Period The time required for one complete vibration in simple harmonic motion. (p. 85)

Resultant The sum of a set of vectors. (p. 66)

Scalar A quantity that can be described completely by a number and a unit. Shows magnitude only, not direction. (p. 64)

Simple Harmonic Motion A type of linear motion of an object in which the acceleration is directly proportional to its displacement from its equilibrium position and the motion is always directed to the equilibrium position. (p. 84)

Speed The distance traveled per unit of time. A scalar because it is described by a number and a unit, not a direction. (p. 70)

Vector A quantity that has both magnitude (size) and direction. Vectors are often represented by an arrow pointing in the correct direction. (p. 64)

Velocity The rate of motion in a particular direction. The time rate of change of an object's displacement. Velocity is a vector that gives the direction of travel and the distance traveled per unit of time. (p. 70)

FORMULAS

3.4 $s = v_{avg}t$

3.5 $\Delta v = at$

3.6 $v_{avg} = \dfrac{v_f + v_i}{2}$

$a_{avg} = \dfrac{v_f - v_i}{t}$

$s = v_i t + \frac{1}{2}a_{avg}t^2$

$v_f = v_i + a_{avg}t$

$s = \frac{1}{2}(v_f + v_i)t$

$2a_{avg}s = v_f^2 - v_i^2$

3.8 $T = 2\pi\sqrt{\dfrac{l}{g}}$

REVIEW QUESTIONS

1. Displacement
   (a) can be interchanged with direction.
   (b) is a measurement of volume.
   (c) can be described only with a number.
   (d) is the net distance an object travels, showing direction and distance.
2. A vector can be described/represented
   (a) the same way as a scalar.
   (b) with both a magnitude and direction.
   (c) by an arrow in diagrams.
   (d) both (b) and (c).
3. Velocity is
   (a) the distance traveled per unit of time.
   (b) the same as speed.
   (c) direction of travel and distance traveled per unit of time.
   (d) only the direction of travel.
4. The period is
   (a) the time required for one revolution.
   (b) the same as frequency.
   (c) the number of revolutions per unit of time.
5. Is the study of mechanics in physics limited to automobiles?
6. Is motion the same as displacement?
7. Give three examples of numbers that are scalars.
8. What is a vector?
9. Can vectors and scalars be added together?
10. Is the resultant of two vectors having 3 units and 5 units of magnitude always 8 units?
11. Distinguish between velocity and speed.

12. Is velocity always constant?
13. Give three familiar examples of acceleration.
14. Distinguish between acceleration, deceleration, and average acceleration.
15. State the values of the acceleration due to gravity for freely falling bodies in both the metric and English systems.
16. Distinguish between amplitude and displacement.
17. Distinguish between period and frequency.
18. Does the period of a pendulum depend on its mass, and if so, how?

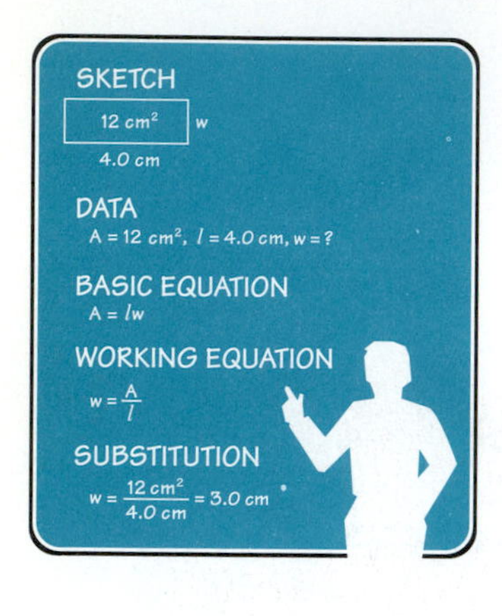

REVIEW PROBLEMS

1. Draw a vector 35° north of west that is 3.0 cm long.
2. Using a scale of 1 cm = 3 mi, find the resultant displacement of an airplane that first flies 12 mi south, then 9 mi west, then 3 mi north.
3. A boat travels at 17.0 mi/h for 1.50 h. How far does the boat travel?
4. A jet flies at $11\bar{0}0$ mi/h for $30\bar{0}0$ mi. How long is the jet flying?
5. A runner starts from rest and attains a speed of 8.00 ft/s after 2.00 s. What is the runner's acceleration?
6. A race car goes from rest to 110 km/h with an acceleration of 6.0 m/s^2. How many seconds does it take?
7. A sailboat has an initial velocity of 10.0 km/h and accelerates to 20.0 km/h. Find its average velocity.
8. A skateboarder starts from rest and accelerates at a rate of 1.30 m/s^2 for 3.00 s. What is his final velocity?
9. A plane has an average velocity of $5\bar{0}0$ km/h. How long does it take to travel 1.5×10^4 km?
10. A train has a final velocity of 110 km/h. It accelerated for 36 s at 0.50 m/s^2. What was the initial velocity?
11. A boulder is rolling down a hill at 8.00 m/s before it comes to rest 17.0 s later. What is the average velocity?
12. A truck accelerates from rest to 120 km/h in 13 s. Find its acceleration.
13. A bullet is fired vertically upward and reaches a height of 2150 m.
    (a) Find its initial velocity.
    (b) How long does it take to reach its maximum height?
    (c) How long is it in flight?
14. A rock is thrown down with an initial speed of 10.0 m/s from a bridge to the water below. It takes 2.75 s for the rock to hit the water.
    (a) Find the speed of the rock as it hits the water.
    (b) How high is the bridge above the water?
15. A pendulum has a length of 0.450 m. What is its period?
16. A pendulum has a period of 0.700 s. Find the length of the pendulum in inches.

Chapter 4

FORCES IN ONE DIMENSION

Classical physics is sometimes called Newtonian physics in honor of Sir Isaac Newton who lived from 1642 to 1727 and formulated three laws of motion that summarize much of the behavior of moving bodies.

Forces may cause motion. Inertia tends to resist the influence of an applied force. Forces, inertia, friction, and how they relate to motion are considered now.

OBJECTIVES

The major goals of this chapter are to enable you to:

1. Relate force and the law of inertia.
2. Apply the law of acceleration.
3. Understand the importance of friction.
4. Analyze forces in one dimension.
5. Distinguish between weight, mass, and gravity.
6. Understand the law of action and reaction.
7. Use momentum, impulse, and conservation of momentum in describing motion.

4.1 FORCE AND THE LAW OF INERTIA

To understand the causes of the various types of motion, we need to study forces. Many types of forces are responsible for the motion of an automobile. The force produced by a hot expanding gas on each piston causes it to move (Fig. 4.1).

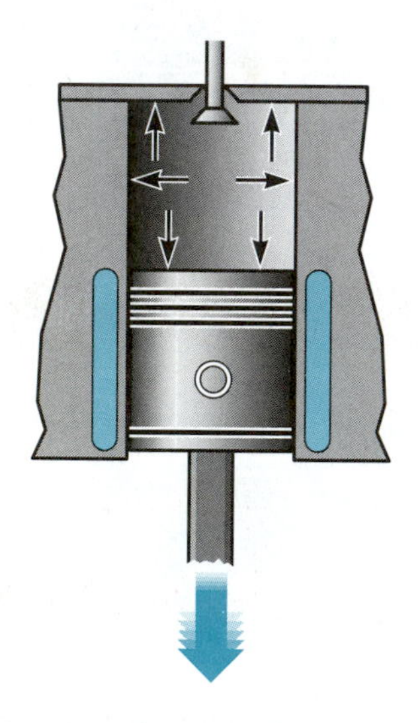

Figure 4.1
Force on piston produced by hot expanding gas causes it to move.

When a structural engineer designs the supports for a bridge, he or she must allow for the weight of the vehicles on it and also the weight of the bridge itself. These forces do not cause motion but are still very important.

A ***force*** *is a push or a pull that tends to cause motion or tends to prevent motion. Force is a vector quantity and thus has both magnitude and direction.* The force tends to produce an acceleration in the direction of its application. Some forces, such as the weight of the bridge shown in Fig. 4.2, do not cause motion because they are balanced by other forces. The downward force of the bridge's

Figure 4.2
The weight of the bridge is balanced by the supporting force.

weight is balanced by the upward force supplied by the supports. If the supports were weakened and could not supply this force, the downward force would no longer be balanced, and the bridge would move; that is, it would collapse.

The units for measuring force are the newton (N) in the metric system and the pound (lb) in the English system. The conversion factor is

$$4.45 \text{ N} = 1 \text{ lb}$$

Let's examine the relationship between forces and motion. There are three relationships or laws that were discovered by Isaac Newton during the late seventeenth century. The three laws are often called Newton's laws. The first law is:

LAW OF INERTIA: NEWTON'S FIRST LAW

A body that is in motion continues in motion with the same velocity (at constant speed and in a straight line), and a body at rest continues at rest unless an unbalanced (outside) force acts upon it.

If an automobile is stopped (at rest) on level ground, it resists being moved. That is, a person is required to exert a tremendous push to get it moving. Similarly, if an automobile is moving—even slowly—it takes a large force to stop it. This property of resisting a change in motion is called *inertia*. ***Inertia*** is the property of a body that causes it to remain at rest if it is at rest or to continue moving with a constant velocity unless an unbalanced force acts upon it. When the accelerating force of an automobile engine is no longer applied to a moving car, it will slow down. This is not a violation of the **law of inertia** because there are forces being applied to the car through air resistance, friction in the bearings, and the rolling resistance of the tires [Fig. 4.3(a)]. If these forces could be removed, the auto would continue moving with a constant velocity. Anyone who has tried to stop quickly on ice knows the effect of the law of inertia when frictional forces are small [Fig. 4.3(b)].

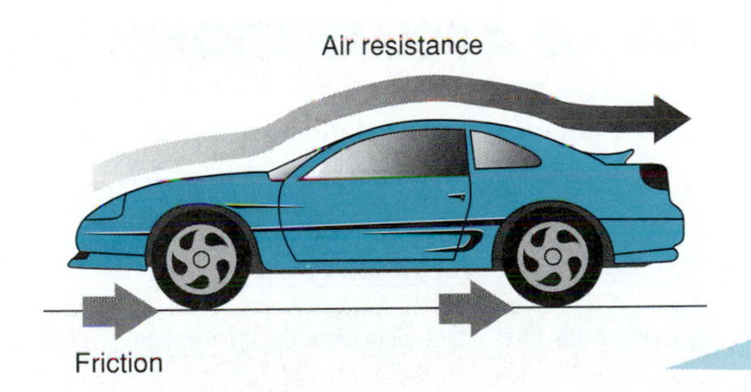

(a) Air resistance and friction slow the car.

(b) Inertia makes it hard to stop a car on ice.

Figure 4.3

Assume that you place a quarter on a card on a glass as shown in Fig. 4.4(a). If you quickly flick the card horizontally, the inertia of the coin tends to keep it at rest until gravity pulls it straight down into the glass as in Fig. 4.4(b).

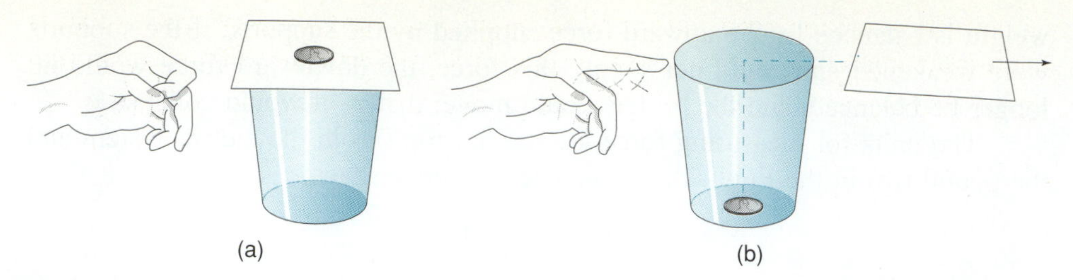

Figure 4.4
When the card is flicked, the inertia of the quarter keeps it at rest until it falls into the glass.

Some objects tend to resist changes in their motion more than others. It is much easier to push a small automobile than to push a large truck into motion (Fig. 4.5). **Mass** *is a measure of the inertia of a body;* that is, a measure of the resistance a body has to change its motion. The common units of mass are the kilogram (kg) in the metric system and the slug in the English system.

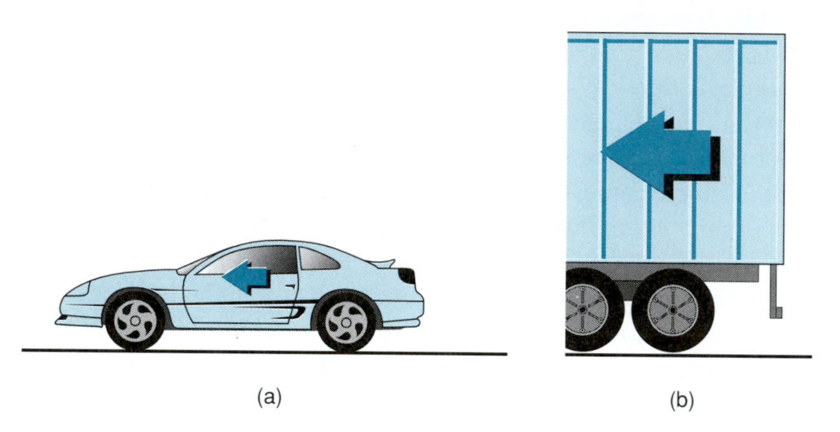

Figure 4.5
A larger body (in mass) has greater resistance to a change in its motion than does a smaller one.

4.2 FORCE AND THE LAW OF ACCELERATION

The second law of motion, called the **law of acceleration,** relates the applied force and the acceleration of an object:

> *LAW OF ACCELERATION: NEWTON'S SECOND LAW*
>
> The total force acting on a body is equal to the mass of the body times its acceleration.

In equation form this law is

$$F = ma$$

where F = total force
m = mass
a = acceleration

In SI units, the mass unit is the kilogram (kg), and the acceleration unit is metre/second/second (m/s^2). The force required to accelerate 1 kg of mass at a rate of 1 m/s^2 is

$$\begin{aligned} F &= ma \\ &= (1 \text{ kg})(1 \text{ m/s}^2) \\ &= 1 \text{ kg m/s}^2 \end{aligned}$$

The SI force unit is the newton (N), named in honor of Isaac Newton, and is defined as

$$1 \text{ N} = 1 \text{ kg m/s}^2$$

In the English system, the mass unit is the slug, and the acceleration unit is foot/second/second (ft/s^2). The force required to accelerate 1 slug of mass at the rate of 1 ft/s^2 is

$$\begin{aligned} F &= ma \\ &= (1 \text{ slug})(1 \text{ ft/s}^2) \\ &= 1 \text{ slug ft/s}^2 \end{aligned}$$

The English force unit is the pound (lb) and is defined as

$$1 \text{ lb} = 1 \text{ slug ft/s}^2$$

We should note here that the other metric unit of force is the dyne. One dyne is the force required to accelerate 1 g of mass at the rate of 1 cm/s^2. The dyne is not an SI unit, and its use is becoming less common as the world more universally accepts and converts to the SI system of metric units.

EXAMPLE 1

What force is necessary to produce an acceleration of 6.00 m/s^2 on a mass of 5.00 kg?

Data:

$$\begin{aligned} m &= 5.00 \text{ kg} \\ a &= 6.00 \text{ m/s}^2 \\ F &= ? \end{aligned}$$

Basic Equation:

$$F = ma$$

Working Equation: Same

Substitution:

$$\begin{aligned} F &= (5.00 \text{ kg})(6.00 \text{ m/s}^2) \\ &= 30.0 \text{ kg m/s}^2 \\ &= 30.0 \text{ N} \qquad (1 \text{ N} = 1 \text{ kg m/s}^2) \end{aligned}$$

EXAMPLE 2

What force is necessary to produce an acceleration of 2.00 ft/s^2 on a mass of 3.00 slugs?

Data:

$$m = 3.00 \text{ slugs}$$
$$a = 2.00 \text{ ft/s}^2$$
$$F = ?$$

Basic Equation:

$$F = ma$$

Working Equation: Same

Substitution:

$$\begin{aligned} F &= (3.00 \text{ slugs})(2.00 \text{ ft/s}^2) \\ &= 6.00 \text{ slug ft/s}^2 \\ &= 6.00 \text{ lb} \qquad (1 \text{ lb} = 1 \text{ slug ft/s}^2) \end{aligned}$$

EXAMPLE 3

Find the acceleration produced by a force of $50\overline{0}$ N applied to a mass of 20.0 kg.

Sketch: None needed

Data:

$$F = 50\overline{0} \text{ N}$$
$$m = 20.0 \text{ kg}$$
$$a = ?$$

Basic Equation:

$$F = ma$$

Working Equation:

$$a = \frac{F}{m}$$

Substitution:

$$\begin{aligned} a &= \frac{50\overline{0} \text{ N}}{20.0 \text{ kg}} \\ &= 25.0 \frac{\text{N}}{\text{kg}} \\ &= 25.0 \frac{\cancel{\text{N}}}{\cancel{\text{kg}}} \times \frac{1 \cancel{\text{kg}} \text{ m/s}^2}{1 \cancel{\text{N}}} \\ &= 25.0 \text{ m/s}^2 \end{aligned}$$

Note: We use a conversion factor to obtain acceleration units.

■ PROBLEMS 4.2

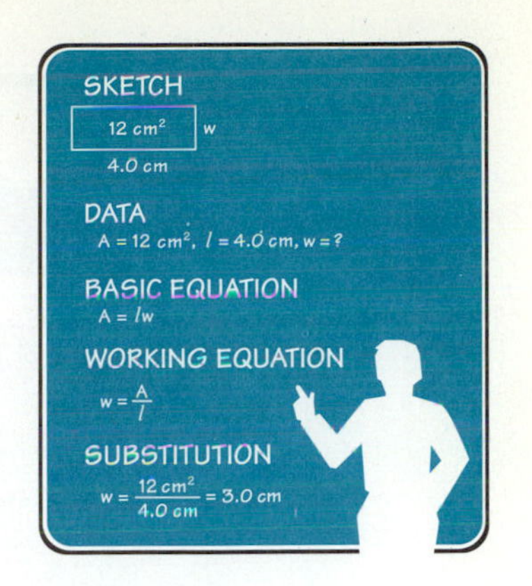

Find the total force necessary to give each mass the given acceleration.

1. $m = 15.0$ kg, $a = 2.00$ m/s^2
2. $m = 4.00$ kg, $a = 0.500$ m/s^2
3. $m = 111$ slugs, $a = 6.70$ ft/s^2
4. $m = 91.0$ kg, $a = 6.00$ m/s^2
5. $m = 28.0$ slugs, $a = 9.00$ ft/s^2
6. $m = 42.0$ kg, $a = 3.00$ m/s^2
7. $m = 59.0$ kg, $a = 3.90$ m/s^2
8. $m = 2.20$ slugs, $a = 1.53$ ft/s^2

Find the acceleration of each mass with the given total force.

9. $m = 19\overline{0}$ kg, $F = 76\overline{0}0$ N
10. $m = 7.00$ slugs, $F = 12.0$ lb
11. $m = 3.60$ kg, $F = 42.0$ N
12. $m = 0.790$ kg, $F = 13.0$ N
13. $m = 11\overline{0}$ kg, $F = 57.0$ N
14. $m = 84.0$ kg, $F = 33.0$ N
15. $m = 9.97$ slugs, $F = 13.9$ lb
16. $m = 21\overline{0}$ kg, $F = 41.0$ N
17. Find the total force necessary to give an automobile of mass 1750 kg an acceleration of 3.00 m/s^2.
18. Find the acceleration produced by a total force of 93.0 N on a mass of 6.00 kg.
19. Find the total force necessary to give an automobile of mass $12\overline{0}$ slugs an acceleration of 11.0 ft/s^2.
20. Find the total force necessary to give a rocket of mass $25,\overline{0}00$ slugs an acceleration of 28.0 ft/s^2.
21. Find the total force necessary to give a $14\overline{0}$ kg mass an acceleration of 41.0 m/s^2.
22. Find the acceleration produced by a total force of $30\overline{0}$ N on a mass of 0.750 kg.
23. Find the mass of an object that has an acceleration of 15.0 m/s^2 when an unbalanced force of 90.0 N acts on it.
24. An automobile has a mass of $10\overline{0}$ slugs. The passengers it carries have a mass of 5.00 slugs each.
    (a) Find the acceleration of the auto and one passenger if the total force acting on it is $15\overline{0}0$ lb.
    (b) Find the acceleration of the auto and six passengers if the total force is again $15\overline{0}0$ lb.
25. Find the acceleration produced by a force of 6.75×10^6 N on a rocket of mass 5.27×10^5 kg.
26. An astronaut has a mass of 80.0 kg. His space suit has a mass of 15.5 kg. Find the acceleration of the astronaut during his space walk when his backpack propulsion unit applies a force to him (and his suit) of 85.0 N. ■

4.3 FRICTION

When we attempt to slide or roll two objects across each other, a force that resists the motion is produced. This force is called **friction.** Friction can be caused by the irregularities of the two surfaces sliding or rolling across each other, which tend to catch on each other (Fig. 4.6). In general, it is found that if two rough surfaces are polished, the frictional force between them is lessened. However, there is a point beyond which this decrease in friction is not observed. If two objects are polished such that the surfaces are too smooth, then the frictional force actually increases. This is related to the fact that friction can also be caused by the adhesion of molecules of one surface to the molecules of the other surface. This adhesive force is similar to the electrical forces which hold atoms together in solids.

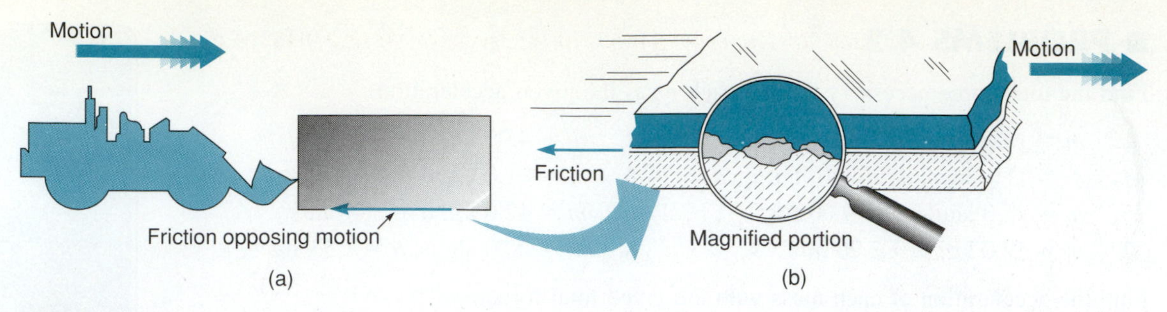

Figure 4.6
Friction resists motion of objects in contact with each other.

Friction is both a necessity and a hindrance to our everyday living. Experiments with frictional forces indicate the following general characteristics:

1. ***Friction is a force that always acts parallel to the surface in contact and opposite to the direction of motion.*** If there is no motion, friction acts in the direction opposite any force that tends to produce motion [Fig. 4.7(a)].
2. ***Starting friction is greater than moving friction.*** If you have ever pushed a car by hand, you probably noticed that it took more force to start the car moving than it did to keep it moving.
3. ***Friction increases as the force between the surfaces increases.*** It is much easier to slide a light crate across the floor than a heavy one (Fig. 4.7). *The area of contact is not relevant.*

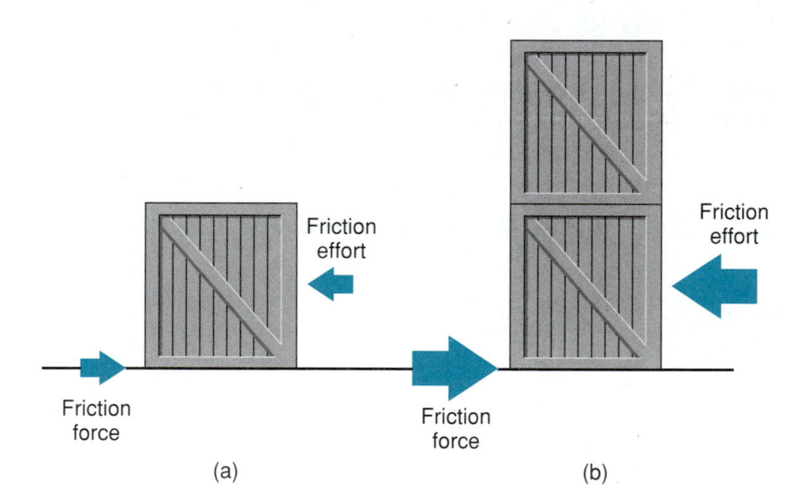

Figure 4.7
Friction increases as the force between the surfaces increases.

The characteristics of friction can be described by the following equation:

$$F_f = \mu F_N$$

where F_f = frictional force
F_N = normal force (force perpendicular to the contact surfaces)
μ = coefficient of friction

The Greek letter μ is pronounced "myu."

Representative values for the coefficients of friction for some surfaces are given in Table 4.1. Values may vary with surface conditions.

Table 4.1 Coefficients of Friction (μ)

Material	*Starting Friction*	*Sliding Friction*
Steel on steel	0.58	0.20
Steel on steel (lubricated)	0.13	0.13
Glass on glass	0.95	0.40
Hardwood on hardwood	0.40	0.25
Steel on concrete		0.30
Aluminum on aluminum	1.9	
Rubber on dry concrete	2.0	1.0
Rubber on wet concrete	1.5	0.97
Aluminum on wet snow	0.4	0.02
Steel on Teflon	0.04	0.04

In general:

To reduce sliding friction:

1. Use smoother surfaces.
2. Use lubrication to provide a thin film between surfaces.
3. Use Teflon to greatly reduce friction between surfaces when an oil lubricant is not desirable, such as in electric motors.
4. Substitute rolling friction for sliding friction. Using ball bearings and roller bearings greatly reduces friction.

EXAMPLE 1

A force of 170 N is needed to keep a 530-N wooden box sliding on a wooden floor. What is the coefficient of sliding friction?

Sketch:

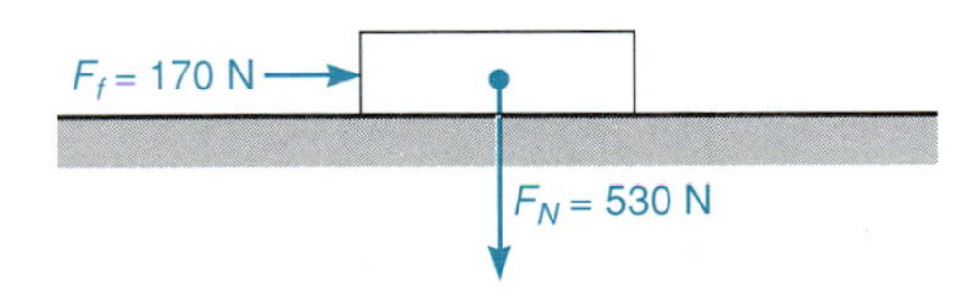

Data:

$$F_f = 170 \text{ N}$$
$$F_N = 530 \text{ N}$$
$$\mu = ?$$

Basic Equation:

$$F_f = \mu F_N$$

Working Equation:

$$\mu = \frac{F_f}{F_N}$$

Substitution:

$$\mu = \frac{170\ \cancel{N}}{530\ \cancel{N}}$$
$$= 0.32$$

Note that μ does not have a unit because the force units always cancel.

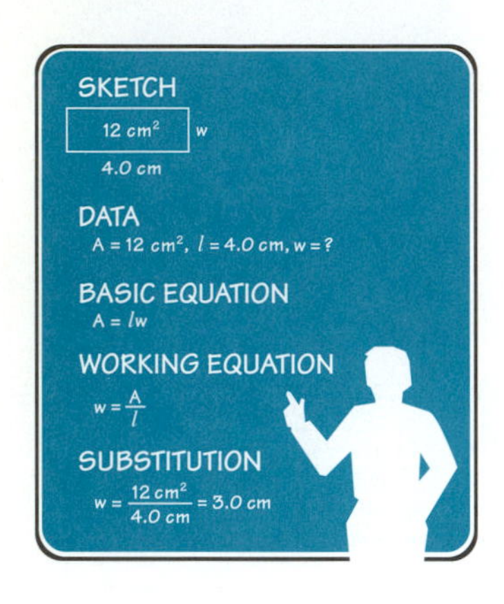

■ PROBLEMS 4.3

1. A cart on wheels weighs 2400 N. The coefficient of rolling friction between the wheels and floor is 0.16. What force is needed to keep the cart rolling uniformly?
2. A wooden crate weighs 780 lb. What force is needed to start the crate sliding on a wooden floor when the coefficient of starting friction is 0.40?
3. A piano weighs 4700 N. What force is needed to start the piano rolling across the floor when the coefficient of starting friction is 0.23?
4. A force of 850 N is needed to keep the piano in Problem 3 rolling uniformly. What is the coefficient of rolling friction?
5. A dog sled weighing 750 lb is pulled over level snow at a uniform speed by a dog team exerting a force of $6\overline{0}$ lb. Find the coefficient of friction.
6. A horizontal conveyor belt system has a coefficient of moving friction of 0.65. The motor driving the system can deliver a maximum force of 2.5×10^6 N. What maximum total weight can be placed on the conveyor system?
7. A tow truck can deliver 2500 lb of pulling force. What is the maximum-weight truck that can be pulled by the tow truck if the coefficient of rolling friction of the truck is 0.10?
8. A snowmobile is pulling a large sled across a snow-covered field. The weight of the sled is 3560 N. If the coefficient of friction of the sled is 0.12, what is the pulling force supplied by the snowmobile?
9. An automobile weighs 12,000 N and has a coefficient of starting friction of 0.13. What force is required to start the auto rolling?
10. If the coefficient of rolling friction of the auto in Problem 9 is 0.080, what force is required to keep it moving once it is in motion? ■

4.4 TOTAL FORCES IN ONE DIMENSION

In the examples used to illustrate the law of acceleration, we discussed total forces only. We need to remember that forces are vectors and have magnitude and direction. The total force acting on an object is the resultant of the separate forces. *When forces act in the same or opposite directions (in one dimension), the total force can be found by adding the forces which act in one direction and subtracting the forces which act in the opposite direction.* It is useful to draw the forces as vectors (arrows) in the sketch before working the problem.

EXAMPLE 1

Two workers push in the same direction (to the right) on a crate. The force exerted by one worker is $15\overline{0}$ lb. The force exerted by the other is 175 lb. Find the net force exerted.

Sketch:

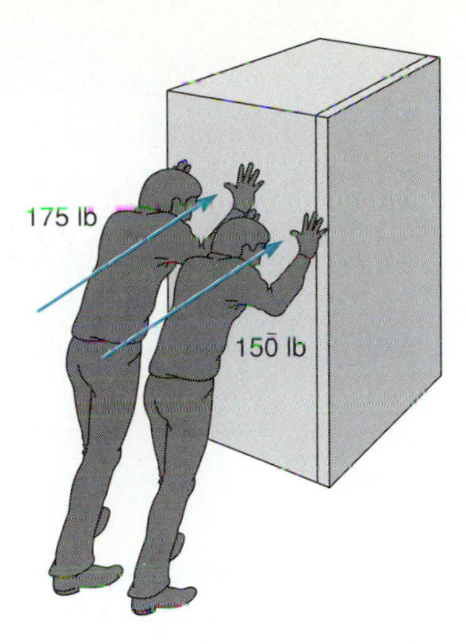

Both forces act in the same direction, so the total force is the sum of the two. Note: The Greek letter Σ (sigma) means "sum of."

$$\begin{aligned}\Sigma \mathbf{F} &= 15\bar{0} \text{ lb} + 175 \text{ lb} \\ &= 325 \text{ lb to the right}\end{aligned}$$

EXAMPLE 2

The same two workers push the crate to the right, and the motion is opposed by a frictional force of $30\bar{0}$ lb. Find the net force.

Sketch:

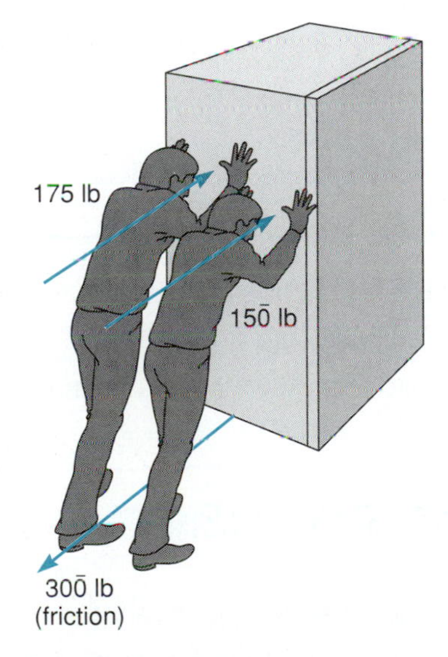

The workers push in one direction and friction pushes in the opposite direction, so we add the forces exerted by the workers and subtract the frictional force.

$$\begin{aligned}\Sigma \mathbf{F} &= 175 \text{ lb} + 15\bar{0} \text{ lb} - 30\bar{0} \text{ lb} \\ &= 25 \text{ lb to the right}\end{aligned}$$

EXAMPLE 3

The crate in Example 2 has a mass of 5.00 slugs. What is its acceleration when the workers are pushing against the frictional force?

Sketch: None needed

Data:

$$F = 25 \text{ lb (from Example 2)}$$
$$m = 5.00 \text{ slugs}$$
$$a = ?$$

Basic Equation:

$$F = ma$$

Working Equation:

$$a = \frac{F}{m}$$

Substitution:

$$a = \frac{25 \text{ lb}}{5.00 \text{ slugs}}$$
$$= 5.0 \frac{\cancel{\text{lb}}}{\cancel{\text{slugs}}} \times \frac{1 \cancel{\text{slug}} \text{ ft/s}^2}{1 \cancel{\text{lb}}}$$
$$= 5.0 \text{ ft/s}^2$$

Note: We use a conversion factor to obtain acceleration units.

EXAMPLE 4

Two workers push in the same direction on a large pallet. The force exerted by one worker is 645 N. The force exerted by the other worker is 755 N. The motion is opposed by a frictional force of 1175 N. Find the net force.

$$\Sigma \mathbf{F} = 645 \text{ N} + 755 \text{ N} - 1175 \text{ N}$$
$$= 225 \text{ N}$$

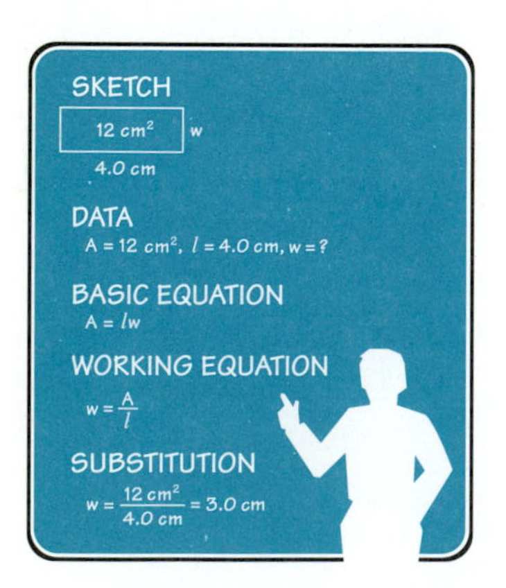

■ PROBLEMS 4.4

Find the net force including its direction when each force acts in the direction indicated.

1. 17.0 N to the left, 20.0 N to the right.
2. 265 N to the left, 85 N to the right.
3. 100.0 N to the left, 75.0 N to the right, and 10.0 N to the right.
4. $19\bar{0}$ lb to the left, 87 lb to the right, and 49 lb to the right.
5. 346 N to the right, 247 N to the left, and 103 N to the left.
6. 37 N to the right, 24 N to the left, 65 N to the right, and 85 N to the right.

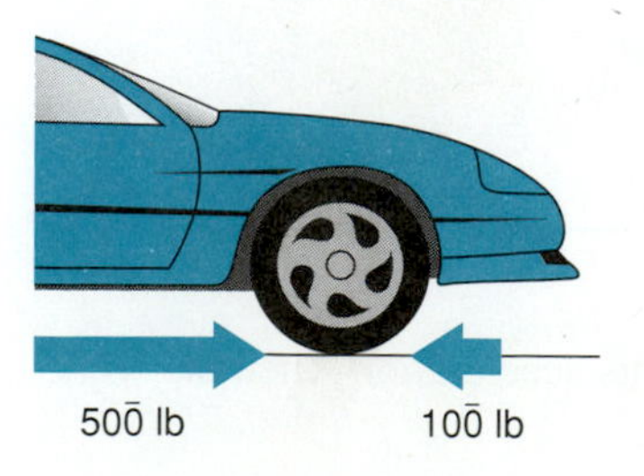

Figure 4.8

7. Find the acceleration of an automobile of mass $10\bar{0}$ slugs acted upon by a driving force of $50\bar{0}$ lb that is opposed by a frictional force of $10\bar{0}$ lb (Fig. 4.8).
8. Find the acceleration of an automobile of mass $15\bar{0}0$ kg acted upon by a driving force of $22\bar{0}0$ N that is opposed by a frictional force of 450 N.
9. A truck of mass 13,100 kg is acted upon by a driving force of $89\bar{0}0$ N. The motion is opposed by a frictional force of 2230 N. Find the acceleration.
10. A speedboat of mass 30.0 slugs has a $30\bar{0}$-lb force applied by the propellers. The friction of the water on the hull is a force of $10\bar{0}$ lb. Find the acceleration.

4.5 GRAVITY AND WEIGHT

We have said that the weight of an object is the amount of **gravitational pull** exerted on an object by the earth. If this force is not balanced by other forces, an acceleration is produced. When you hold a brick in your hand as in Fig. 4.9, you exert an upward force on the brick that balances the downward force (weight). If you remove your hand, the brick moves downward due to the unbalanced force. The velocity of the falling brick increases, but the acceleration (rate of change of the velocity) is constant.

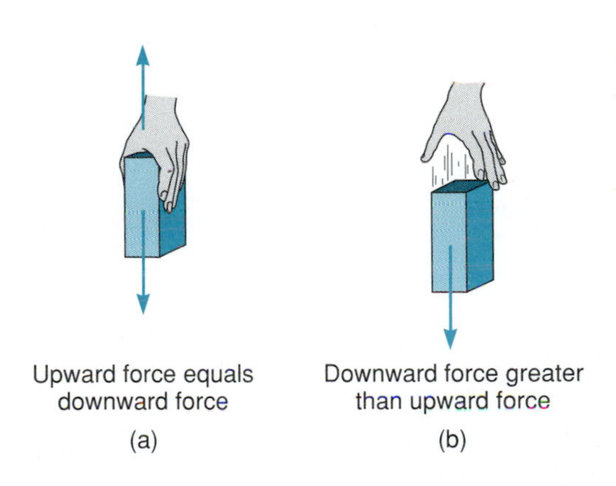

Figure 4.9
Balancing gravity in holding a brick.

The acceleration of all objects near the surface of earth is the same if air resistance is ignored. We call this acceleration due to the gravitational pull of the earth g. Its value is 9.80 m/s^2 in the metric system and 32.2 ft/s^2 in the English system.

The **weight** of an object is the force exerted by the earth or by another large body and gives the body acceleration g. This force can be found using $F = ma$, where $a = g$. If we abbreviate weight by F_w, the equation for weight is

$$F_w = mg$$

where F_w = weight
m = mass
g = acceleration due to gravity
g = 9.80 m/s^2 (metric)
g = 32.2 ft/s^2 (English)

EXAMPLE 1

Find the weight of 5.00 kg.

Data:

$$m = 5.00 \text{ kg}$$
$$a = 9.80 \text{ m/s}^2$$
$$F_w = ?$$

Basic Equation:

$$F_w = mg$$

Working Equation: Same

Substitution:

$$\begin{aligned} F_w &= (5.00 \text{ kg})(9.80 \text{ m/s}^2) \\ &= 49.0 \text{ kg m/s}^2 \\ &= 49.0 \text{ N} \qquad (1 \text{ N} = 1 \text{ kg m/s}^2) \end{aligned}$$

EXAMPLE 2

Find the weight of 12.0 slugs.

Data:

$$m = 12.0 \text{ slugs}$$
$$a = 32.2 \text{ ft/s}^2$$
$$F_w = ?$$

Basic Equation:

$$F_w = mg$$

Working Equation: Same

Substitution:

$$\begin{aligned} F_w &= (12.0 \text{ slugs})(32.2 \text{ ft/s}^2) \\ &= 386 \text{ slug ft/s}^2 \\ &= 386 \text{ lb} \qquad (1 \text{ lb} = 1 \text{ slug ft/s}^2) \end{aligned}$$

Mass versus Weight

Note that the mass of an object remains the same, but its weight varies according to the gravitational pull. For example, an astronaut of mass 75 kg has a weight of

$$F_w = mg = (75 \text{ kg})(9.80 \text{ m/s}^2) = 740 \text{ N}$$

on earth. When the spacecraft is enroute to the moon, we say the astronaut is "weightless." Actually, his or her weight is almost zero (not because there is no gravity, but because of the greater distance between bodies); the astronaut's mass remains 75 kg. The following example shows the astronaut's weight on the moon.

EXAMPLE 3

Find the 75-kg astronaut's weight on the moon, where $g = 1.63 \text{ m/s}^2$.

Sketch: None needed

Data:

$$m = 75 \text{ kg}$$
$$g = 1.63 \text{ m/s}^2$$
$$F_w = ?$$

Basic Equation:

$$F_w = mg$$

Working Equation: Same

Substitution:

$$F_w = (75 \text{ kg})(1.63 \text{ m/s}^2)$$
$$= 120 \text{ N} \qquad (1 \text{ N} = 1 \text{ kg m/s}^2)$$

Note that the astronaut's mass on the moon remains 75 kg.

■ PROBLEMS 4.5

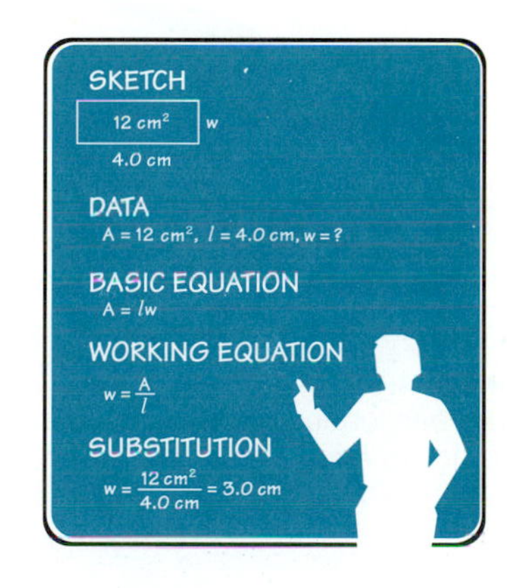

Find the weight for each mass.

1. $m = 30.0 \text{ kg}$
2. $m = 60.0 \text{ kg}$
3. $m = 10.0 \text{ slugs}$
4. $m = 9.00 \text{ kg}$

Find the mass for each weight.

5. $F_w = 17.0 \text{ N}$
6. $F_w = 21.0 \text{ lb}$
7. $F_w = 12,\bar{0}00 \text{ N}$
8. $F_w = 25,\bar{0}00 \text{ N}$
9. $F_w = 6.7 \times 10^{12} \text{ N}$
10. $F_w = 5.5 \times 10^6 \text{ lb}$
11. Find the weight of an 1150-kg automobile.
12. Find the weight of an 81.5-slug automobile.
13. Find the mass of a 2750-lb automobile.
14. Find the mass of an 11,500-N automobile.
15. Find the weight of a 1350-kg automobile (a) on earth and (b) on the moon.
16. Maria weighs 115 lb on earth. What are her (a) mass and (b) weight on the moon?
17. John's mass is 65 kg on earth. What are his (a) mass and (b) weight on the moon?
18. What is your weight in newtons and in pounds?
19. What is your mass in kilograms and in slugs?
20. What are your English mass and weight on the moon?
21. What are your metric mass and weight on the moon?
22. John's mass is 65 kg on earth. What are his English (a) mass and (b) weight $40\bar{0}0$ mi above the surface of the earth, where $g = 7.85 \text{ ft/s}^2$?
23. Maria weighs 115 lb on earth. What are her English (a) mass and (b) weight on Jupiter, where $g = 85.0 \text{ ft/s}^2$?
24. John's mass is 65 kg on earth. What are his metric (a) mass and (b) weight on Mars, where $g = 3.72 \text{ m/s}^2$?
25. What are your metric mass and weight on Jupiter, where $g = 25.9 \text{ m/s}^2$?
26. What are your metric mass and weight on Mars, where $g = 3.72 \text{ m/s}^2$? ■

4.6 LAW OF ACTION AND REACTION

When an automobile accelerates, we know that a force is being applied to it. What applies this force? You may think that the tires exert this force on the auto. This is not correct, because the tires move along with the auto and there must be a force applied to them also. The ground below the tires actually supplies the force that accelerates the car (Fig. 4.10). This force is called a *reaction* to the force exerted by the tires on the ground, which is called the *action force*. The third law of motion, the **law of action and reaction,** can be stated as follows:

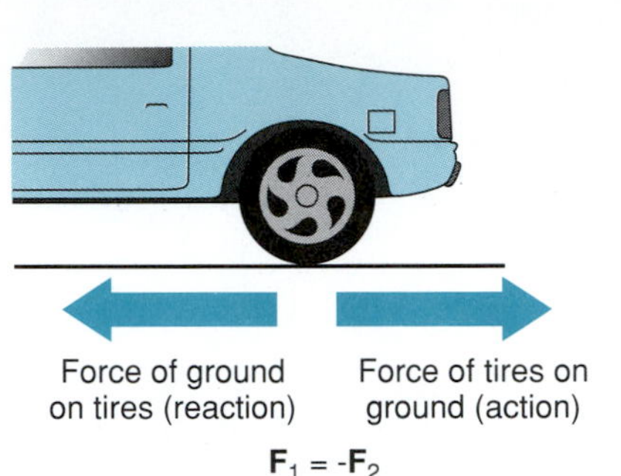

Figure 4.10
The ground supplies the force to accelerate the car.

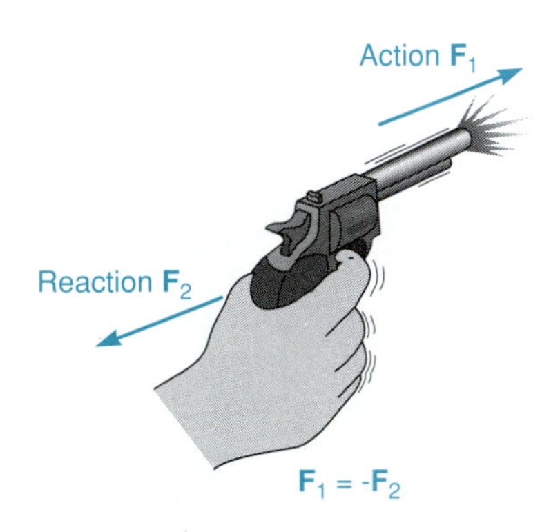

Figure 4.11
The reaction force always equals the action force.

> *LAW OF ACTION AND REACTION: NEWTON'S THIRD LAW*
>
> For every force applied by object *A* to object *B* (action), there is a force exerted by object *B* on object *A* (reaction) which has the same magnitude but is opposite in direction.

When a bullet is fired from a handgun (action), the recoil felt is the reaction. These forces are shown in Fig. 4.11. Note that the action and reaction forces *never* act on the same object.

4.7 MOMENTUM

Consider a heavy ball and a light ball, each made of the same hard material, rolling toward each other on a smooth surface with the same speed as in Fig. 4.12(a). After they collide as in Fig. 4.12(b), the path of the light ball is changed much more than the path of the heavy ball.

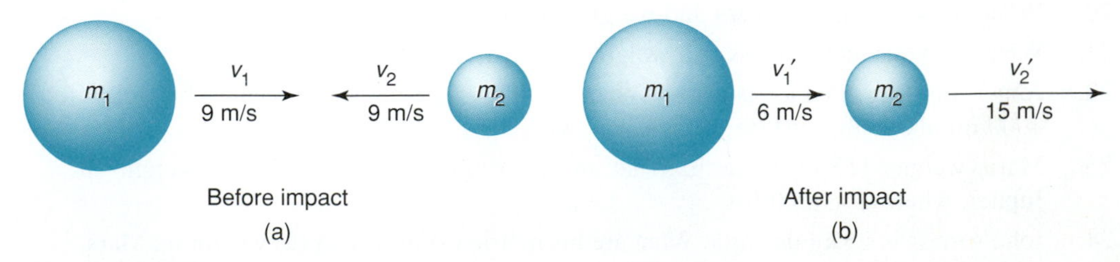

Figure 4.12
A heavy ball has more momentum, (mv), than a light ball.

The heavy ball has more momentum than the light ball. Momentum is a measure of the difficulty in bringing a moving body to rest. **Momentum** *equals the product of the mass times the velocity of an object.*

$$p = mv$$

where p = momentum
m = mass
v = velocity

The units of momentum are kg m/s in the metric system and slug ft/s in the English system. Momentum is a vector quantity whose direction is the same as the velocity.

EXAMPLE 1

Find the momentum of an auto with a mass of 105 slug and with a velocity of 60.0 mi/h.

Sketch: None needed

Data:

$$m = 105 \text{ slugs}$$
$$v = 60.0 \text{ mi/h} = 88.0 \text{ ft/s}$$
$$p = ?$$

Basic Equation:

$$p = mv$$

Working Equation: Same

Substitution:

$$p = (105 \text{ slugs})(88.0 \text{ ft/s})$$
$$= 9240 \text{ slug ft/s}$$

EXAMPLE 2

Find the momentum of an auto that has a mass of 1350 kg at a speed of 75.0 km/h.

Sketch: None needed

Data:

$$m = 1350 \text{ kg}$$
$$v = 75.0 \frac{\cancel{\text{km}}}{\cancel{\text{h}}} \times \frac{1000 \text{ m}}{1 \cancel{\text{km}}} \times \frac{1 \cancel{\text{h}}}{3600 \text{ s}} = 20.8 \text{ m/s}$$
$$p = ?$$

Basic Equation:

$$p = mv$$

Working Equation: Same

Substitution:

$$p = (1350 \text{ kg})(20.8 \text{ m/s}) \\ = 28{,}100 \text{ kg m/s}$$

EXAMPLE 3

Find the velocity of a bullet of mass 1.00×10^{-2} kg so that it has the same momentum as a lighter bullet of mass 1.80×10^{-3} kg and a velocity of 325 m/s.

Sketch:

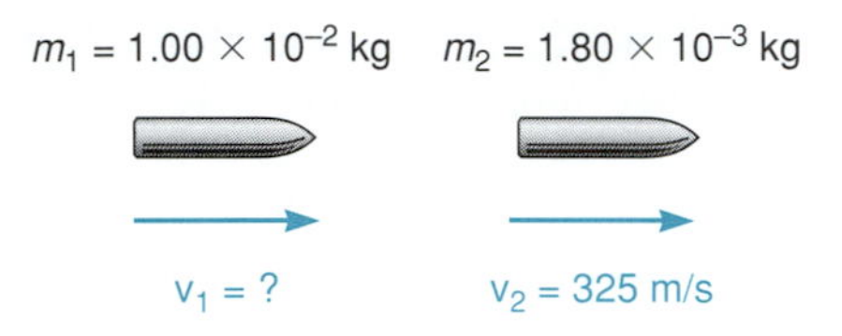

Data:

Heavier bullet	*Lighter bullet*
$m_1 = 1.00 \times 10^{-2}$ kg	$m_2 = 1.80 \times 10^{-3}$ kg
$v_1 = ?$	$v_2 = 325$ m/s
$p_1 = ?$	$p_2 = ?$

Basic Equation:

$$p_1 = m_1 v_1$$
$$p_2 = m_2 v_2$$

We want:

$$p_1 = p_2$$

or

$$m_1 v_1 = m_2 v_2$$

Working Equation:

$$v_1 = \frac{m_2 v_2}{m_1}$$

Substitution:

$$v_1 = \frac{(1.80 \times 10^{-3} \cancel{\text{kg}})(325 \text{ m/s})}{1.00 \times 10^{-2} \cancel{\text{kg}}} \\ = 58.5 \text{ m/s}$$

The **impulse** on an object is the product of the force and the time interval during which the force acts on the object. That is,

$$\text{impulse} = Ft$$

where F = force
t = time interval during which the force acts

How are impulse and momentum related? Recall that

$$a = \frac{v_f - v_i}{t}$$

If we substitute this equation into Newton's second law of motion, we have

$$F = ma$$

$$F = m\left(\frac{v_f - v_i}{t}\right)$$

$$F = \frac{mv_f - mv_i}{t} \qquad \text{Remove parentheses.}$$

$$Ft = mv_f - mv_i \qquad \text{Multiply both sides by } t.$$

Note that mv_f is the final momentum, and mv_i is the initial momentum. That is,

impulse = change in momentum

A common example that illustrates this relationship is a golf club hitting a golf ball. During the time that the club and ball are in contact, the force of the swinging club is transferring most of its momentum to the ball. The impulse given to the ball is directly related to the force with which the ball is hit and the length of time that the club and ball are in contact. After the ball leaves the club, the ball has a momentum equal to its mass times its velocity. This principle is the basis for the necessity of "follow-through" for long drives.

EXAMPLE 4

A 175-g bullet is fired at a muzzle velocity of 582 m/s from a gun with a mass of 8.00 kg and a barrel length of 75.0 cm.

(a) How long is the bullet in the barrel?

(b) What is the force of the bullet as it leaves the barrel?

(c) Find the impulse given to the bullet while it is in the barrel.

(d) Find the bullet's momentum as it leaves the barrel.

Sketch:

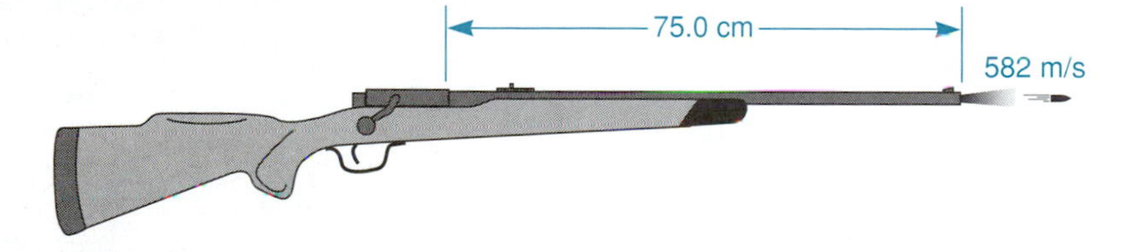

(a) Data:

$$s = 75.0 \text{ cm} = 0.750 \text{ m}$$

$$v_f = 582 \text{ m/s}$$

$$v_i = 0 \text{ m/s}$$

$$v_{\text{avg}} = \frac{v_f + v_i}{2} = \frac{582 \text{ m/s} + 0 \text{ m/s}}{2} = 291 \text{ m/s}$$

$$t = ?$$

Basic Equation:

$$s = v_{\text{avg}}t$$

Working Equation:

$$t = \frac{s}{v_{avg}}$$

Substitution:

$$t = \frac{0.750 \text{ m}}{291 \text{ m/s}}$$
$$= 0.00258 \text{ s}$$

Note: This is the length of time that the force is applied to the bullet.

(b) Data:

$$t = 0.00258 \text{ s}$$
$$m = 175 \text{ g} = 0.175 \text{ kg}$$
$$v_f = 582 \text{ m/s}$$
$$v_i = 0 \text{ m/s}$$
$$F = ?$$

Basic Equation:

$$Ft = mv_f - mv_i$$

Working Equation:

$$F = \frac{mv_f - mv_i}{t}$$

Substitution:

$$F = \frac{(0.175 \text{ kg})(582 \text{ m/s}) - (0.175 \text{ kg})(0 \text{ m/s})}{0.00258 \text{ s}}$$
$$= 39{,}500 \text{ kg m/s}^2$$
$$= 39{,}500 \text{ N} \qquad (1 \text{ N} = 1 \text{ kg m/s}^2)$$

(c) Data:

$$t = 0.00258 \text{ s}$$
$$F = 39{,}500 \text{ N}$$
$$\text{impulse} = ?$$

Basic Equation:

$$\text{impulse} = Ft$$

Working Equation: Same

Substitution:

$$\text{impulse} = (39{,}500 \text{ N})(0.00258 \text{ s})$$
$$= 102 \text{ N s}$$
$$= 102 \text{ (kg m/s}^2\text{)(s)} \qquad (1 \text{ N} = 1 \text{ kg m/s}^2)$$
$$= 102 \text{ kg m/s}$$

(d) Data:

$$m = 175 \text{ g} = 0.175 \text{ kg}$$
$$v = 582 \text{ m/s}$$
$$p = ?$$

Basic Equation:

$$p = mv$$

Working Equation: Same

Substitution:

$$p = (0.175 \text{ kg})(582 \text{ m/s})$$
$$= 102 \text{ kg m/s}$$

Note: The impulse equals the change in momentum.

One of the most important laws of physics is the following:

LAW OF CONSERVATION OF MOMENTUM

When no outside forces are acting on a system of moving objects, the total momentum of the system remains constant.

Let us look at some examples. Consider a 35-kg boy and a 75-kg man standing next to each other on ice skates on "frictionless ice" (Fig. 4.13). The man pushes on the boy, which gives the boy a velocity of 0.40 m/s. What happens to the man? Initially, the total momentum was zero because the initial velocity of each was zero. Because of the **law of conservation of momentum,** the total momentum must still be zero. That is,

$$m_{\text{boy}}v_{\text{boy}} + m_{\text{man}}v_{\text{man}} = 0$$
$$(35 \text{ kg})(0.40 \text{ m/s}) + (75 \text{ kg})v_{\text{man}} = 0$$
$$v_{\text{man}} = -0.19 \text{ m/s}$$

Note: The minus sign indicates that the man's velocity and the boy's velocity are in opposite directions.

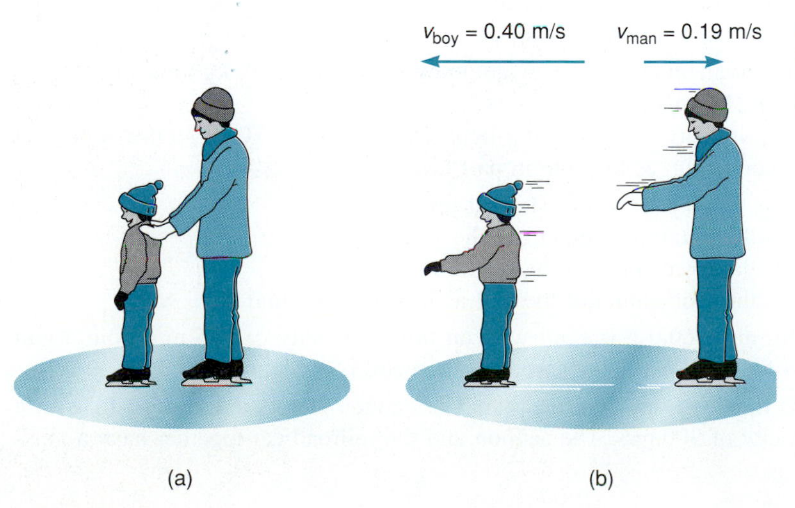

Figure 4.13
Momentum is conserved by the lighter boy moving faster than the heavier man.

Rocket propulsion is another illustration of conservation of momentum. Like the skaters, the total momentum of a rocket on the launch pad is zero. When the rocket engines are fired, hot exhaust gases (actually gas molecules) are expelled downward through the rocket nozzle at tremendous speeds. As the rocket takes off, the sum of the total momentums of the rocket and the gas particles must remain zero. The total momentum of the gas particles is the sum of the products of each mass and its corresponding velocity and is directed down. The momentum of the rocket is the product of its mass and its velocity and is directed up.

When the rocket is in space, its propulsion works in the same manner. The conservation of momentum is still valid except that when the rocket engines are fired, the total momentum is a nonzero constant. This is because the rocket has velocity.

Actually, repair work is more difficult in space than it is on earth because of the conservation of momentum and the "weightlessness" of objects in orbit. On earth, when a hammer is swung, the person is coupled to the earth by frictional forces, so that the person's mass includes that of the earth. In space orbit, because the person is weightless, there is no friction to couple him or her to the spaceship. A person in space has roughly the same problem driving a nail as a person on earth would have wearing a pair of "frictionless" roller skates.

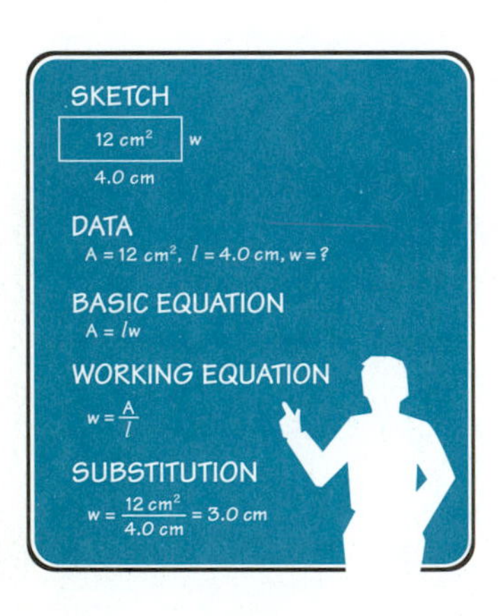

■ PROBLEMS 4.7

Find the momentum of each object.

1. $m = 2.00$ kg, $v = 40.0$ m/s
2. $m = 5.00$ kg, $v = 90.0$ m/s
3. $m = 17.0$ slugs, $v = 45.0$ ft/s
4. $m = 38.0$ kg, $v = 97.0$ m/s
5. $m = 3.8 \times 10^5$ kg, $v = 2.5 \times 10^3$ m/s
6. $m = 3.84$ kg, $v = 1.6 \times 10^5$ m/s
7. $F_w = 1.50 \times 10^5$ N, $v = 4.50 \times 10^4$ m/s
8. $F_w = 3200$ lb, $v = 6\overline{0}$ mi/h (change to ft/s)
9. (a) Find the momentum of a heavy automobile of mass $18\overline{0}$ slugs traveling with velocity 70.0 ft/s.
   (b) Find the velocity of a light auto of mass 80.0 slugs so that it has the same momentum as the auto in part (a).
   (c) Find the weight (in lb) of each auto in parts (a) and (b).
10. (a) Find the momentum of a bullet of mass 1.00×10^{-3} slug and velocity $70\overline{0}$ ft/s.
    (b) Find the velocity of a bullet of mass 5.00×10^{-4} slug so that it has the same momentum as the bullet in part (a).
11. (a) Find the momentum of a heavy automobile of mass 2630 kg traveling at a velocity of 21.0 m/s.
    (b) Find the velocity (in km/h) of a light auto of mass 1170 kg so that it has the same momentum as the auto in part (a).
12. A ball of mass 0.50 kg is thrown straight up at 6.0 m/s.
    (a) What is the initial momentum of the ball?
    (b) What is the momentum of the ball at its peak?
    (c) What is the momentum of the ball as it hits the ground?
13. A bullet with mass 60.0 g is fired with an initial velocity of 575 m/s from a gun with mass 4.50 kg. What is the speed of the recoil of the gun?
14. A cannon is mounted on a railroad car. The cannon shoots a 1.75-kg ball with a muzzle velocity of $30\overline{0}$ m/s. The cannon and the railroad car together have a mass

of 4500 kg. If the ball, cannon, and railroad car are initially at rest, what is the recoil velocity of the car and cannon?

15. A 125-kg pile driver falls from a height of 10.0 m to hit a piling.
    (a) What is its speed as it hits the piling?
    (b) With what momentum does it hit the piling?
16. A person is traveling in an automobile at 75.0 km/h and throws a bottle of mass 0.500 kg out the window.
    (a) With what momentum does it hit a roadway sign?
    (b) With what momentum does it hit an oncoming automobile traveling in the opposite direction at 85.0 km/h?
    (c) With what momentum does it hit an automobile passing and traveling in the same direction at 85.0 km/h?
17. A 75.0-g bullet is fired with a muzzle velocity of 460 m/s from a gun with mass 3.75 kg and barrel length of 66.0 cm.
    (a) How long is the bullet in the barrel?
    (b) What is the force of the bullet as it leaves the barrel?
    (c) Find the impulse given to the bullet while it is in the barrel.
    (d) Find the bullet's momentum as it leaves the barrel.
18. A 60.0-g bullet is fired at a muzzle velocity of 525 m/s from a gun with mass 4.50 kg and a barrel of 55.0 cm.
    (a) How long is the bullet in the barrel?
    (b) What is the force of the bullet as it leaves the barrel?
    (c) Find the impulse given to the bullet while it is in the barrel.
    (d) Find the bullet's momentum as it leaves the barrel.

GLOSSARY

Force A push or a pull that tends to cause motion or tends to prevent motion. Force is a vector quantity and thus has both magnitude and direction. (p. 90)

Friction A force that resists the motion of an object as a result of irregularities of two surfaces sliding or rolling across each other. (p. 95)

Gravitational Pull The attractive gravitational force exerted by one object, such as earth, on another object. (p. 101)

Impulse The product of the force exerted and the time interval during which the force acts on the object. Impulse equals the change in momentum of an object in response to the exerted force. (p. 106)

Inertia The property of a body that causes it to remain at rest if it is at rest or to continue moving with a constant velocity unless an unbalanced force acts upon it. (p. 91)

Law of Acceleration The total force acting on a body is equal to the mass of the body times its acceleration. (Newton's second law) (p. 92)

Law of Action and Reaction For every force applied by object *A* to object *B* (action), there is a force exerted by object *B* on object *A* (reaction) that has the same magnitude but is opposite in direction. (Newton's third law) (p. 104)

Law of Conservation of Momentum When no outside forces are acting on a system of moving objects, the total momentum of the system remains constant. (p. 109)

Law of Inertia A body that is in motion continues in motion with the same velocity (at constant speed and in a straight line), and a body at rest continues at rest unless an unbalanced (outside) force acts upon it. (Newton's first law) (p. 91)

Mass A measure of the inertia of a body. (p. 92)

Momentum A measure of the difficulty in bringing a moving body to rest. Momentum equals the mass times the velocity of an object. (p. 105)

Weight The force exerted on an object by the earth or by another large body. (p. 101)

FORMULAS

4.2 $F = ma$

4.3 $F_f = \mu F_N$

4.5 $F_w = mg$

where $g = 9.80 \text{ m/s}^2$ (metric)
$g = 32.2 \text{ ft/s}^2$ (English)

4.7 $p = mv$
impulse = Ft
impulse = change in momentum

REVIEW QUESTIONS

1. Force
   (a) is a vector quantity. (b) may be different from weight.
   (c) does not always cause motion. (d) all of the above.
2. The metric weight of a 10-lb bag of sugar is approximately
   (a) 4.45 N. (b) 44.5 N.
   (c) 445 N. (d) Pounds cannot be changed to newtons.
3. Mass and weight
   (a) are the same. (b) are different.
   (c) do not change wherever you are.
4. According to Newton's second law, the law of acceleration,
   (a) acceleration is equal to mass times force.
   (b) mass is equal to mass times acceleration.
   (c) force is equal to mass times acceleration.
   (d) none of the above.
5. Friction
   (a) always acts parallel to the surface of contact and opposite to the direction of motion.
   (b) acts in the direction of motion.
   (c) is smaller when starting than moving.
   (d) is an imaginary force.
6. Momentum is
   (a) equal to speed times weight. (b) equal to mass times velocity.
   (c) the same as force.
7. Cite three examples of forces acting without motion being produced.
8. Does a pound of feathers have more inertia than a pound of lead?
9. Does the pound of feathers have more mass than the pound of lead?
10. How is inertia a factor in multi-car pile-ups?
11. Would the mass of an object be different if the object were deposited on the moon?
12. Using your own words, state Newton's first law, the law of inertia.
13. Distinguish between velocity and acceleration.
14. What acceleration does a force of 11.0 lb give to a mass of 1.00 slug?
15. When the same force is applied to two masses, will they both have the same acceleration?
16. Is 3 pounds heavier than 10 newtons?

17. Would life be easier or more difficult without friction?
18. Does the pull of gravity always produce acceleration?
19. Explain how the weight of an astronaut can be different on the moon from on earth.
20. Explain the difference between action and reaction forces.
21. State Newton's third law of motion, the law of action and reaction, in your own words.
22. Can the action and reaction forces act on the same object?
23. Why is "follow-through" important in hitting a baseball?
24. Describe in your own words the law of conservation of momentum.
25. Describe conservation of momentum in terms of a rocket being fired.

REVIEW PROBLEMS

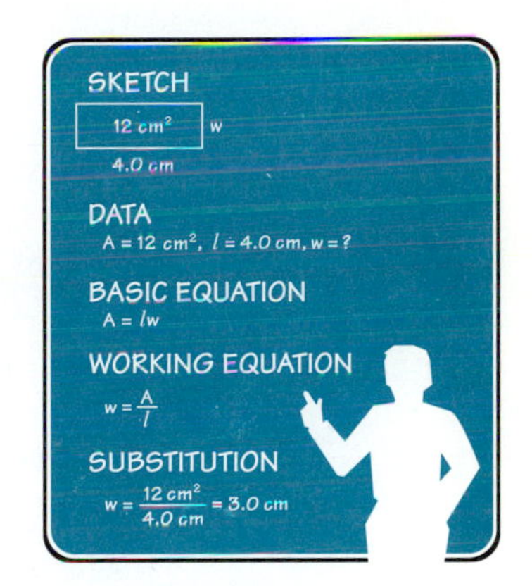

1. A boy pulls his wagon along a frictionless surface. The wagon has a mass of 15.0 kg and is pulled with an acceleration of 1.50 m/s^2. What is the force generated?
2. A crate having a mass of 6.00 kg requires a force of 18.0 N to move. What is the acceleration?
3. A 825-N force is required to pedal a bike with an acceleration of 11.0 m/s^2. What is the mass of the bike and person?
4. A block having mass 0.89 slug moves with a force of 17.0 lb. What is the block's acceleration?
5. What is the force necessary for a 2400-kg truck to accelerate at a rate of 8.0 m/s^2?
6. Two movers push a piano across a frictionless surface. One pushes with 29.0 N of force, and the other mover exerts 35.0 N. What is the total force?
7. A 340-N box has a frictional force of 57 N. What is the coefficient of sliding friction?
8. A truck pulls a trailer with a frictional force of 870 N and a coefficient of friction of 0.23. What is the trailer's normal force?
9. A steel box is slid along a steel surface. It has a normal force of 57 N. What is the frictional force?
10. A rock with a mass of 13.0 kg is dropped from a cliff. Find its weight.
11. A projectile has a mass of 0.37 slug. What is the weight of the projectile?
12. A truck travels at 57.0 mi/h and has a mass of 1475 slugs. What is the momentum of the truck?
13. A projectile is fired with a momentum of 5.50 kg m/s and has a mass of 27.0 kg. What is the velocity at which it travels?
14. A box is pushed with a force of 125 N for 2.00 min. What is the impulse?
15. What is the momentum of a bullet traveling at 250 m/s with a mass of 0.034 kg?
16. A 4.0-g bullet is fired from a 4.5-kg gun with a muzzle velocity of 625 m/s. What is the speed of the recoil of the gun?
17. A 150-kg pile driver falls from a height of 7.5 m to hit a piling.
    (a) What is its speed as it hits the piling?
    (b) With what momentum does it hit the piling?
18. A 15.0-g bullet is fired at a muzzle velocity of 3250 m/s from a high-powered rifle with a mass of 4.75 kg and barrel of 75.0 cm.
    (a) How long is the bullet in the barrel?
    (b) What is the force of the bullet as it leaves the barrel?
    (c) Find the impulse given to the bullet while it is in the barrel.
    (d) Find the bullet's momentum as it leaves the barrel.

Chapter 5

VECTORS AND TRIGONOMETRY

Trigonometry includes the study of triangles and the relationships between the sides and angles. Rarely do events and motion in the physical world occur in a straight line or at right angles. Therefore, some knowledge of trigonometry is essential.

Triangles and vectors are the focus of this chapter.

OBJECTIVES

The major goals of this chapter are to enable you to:

1. Learn the basic concepts of right-triangle trionometry.
2. Understand and use vector notation and the Pythagorean theorem.
3. Use vectors to solve motion problems.

5.1 RIGHT-TRIANGLE TRIGONOMETRY

To this point we have discussed vectors in terms of a north–south reference. To study vectors more thoroughly, trigonometry of the right triangle is needed (Fig. 5.1). A **right triangle** is a triangle with one right angle (90°), two acute angles (less than 90°), two legs, and a hypotenuse (the side opposite the right angle).

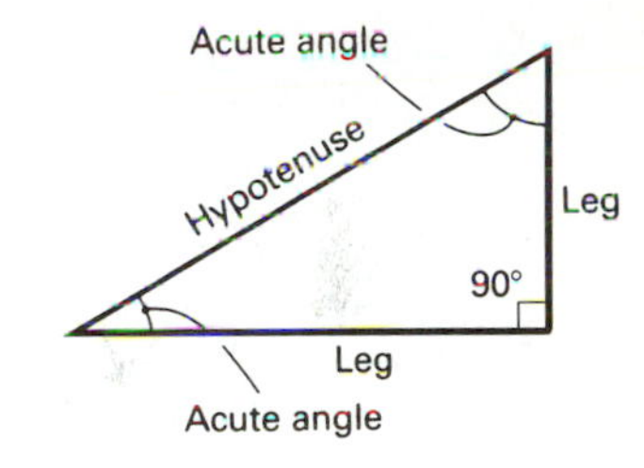

Figure 5.1
Parts of a right triangle.

When it is necessary to label a triangle, the vertices are often labeled using capital letters and the sides opposite the vertices are labeled using the corresponding lowercase letters (Fig. 5.2).

The side opposite angle A is a.
The side adjacent to angle A is b.
The side opposite angle B is b.
The side adjacent to angle B is a.
The side opposite angle C is c and is called the *hypotenuse*.

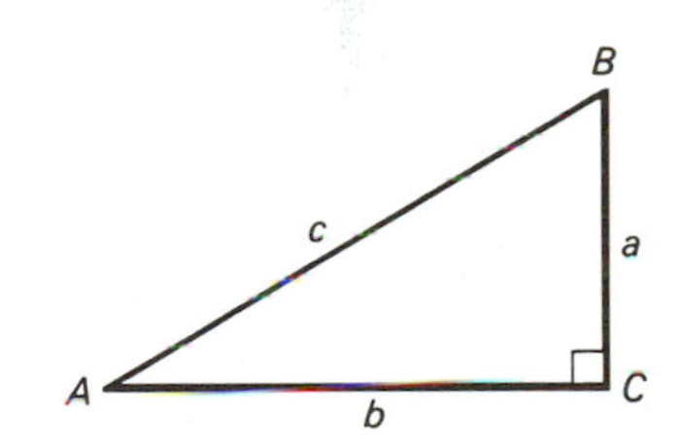

Figure 5.2
Labeling a right triangle.

If we consider a certain acute angle of a right triangle, the two legs can be identified as the side opposite or the side adjacent to the acute angle.

The side opposite angle A is the same as the side adjacent to angle B.
The side adjacent to angle A is the same as the side opposite angle B.
The side opposite angle B is the same as the side adjacent to angle A.
The side adjacent to angle B is the same as the side opposite angle A.

A **ratio** is a comparison of two quantities by division. In a right triangle, there are three ratios that are very important. Consider the right triangle shown in Fig. 5.3. These ratios are:

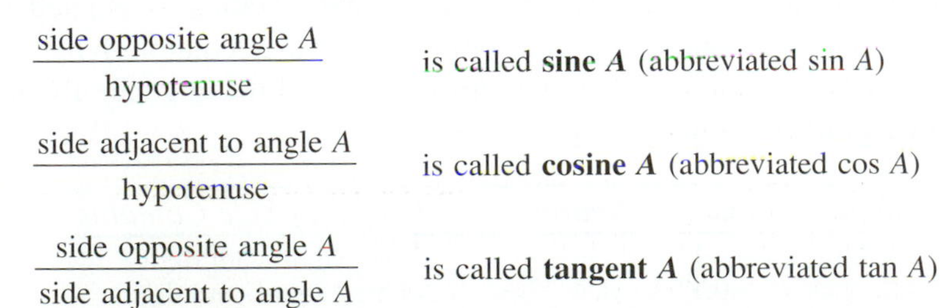

$\dfrac{\text{side opposite angle } A}{\text{hypotenuse}}$ is called **sine *A*** (abbreviated sin A)

$\dfrac{\text{side adjacent to angle } A}{\text{hypotenuse}}$ is called **cosine *A*** (abbreviated cos A)

$\dfrac{\text{side opposite angle } A}{\text{side adjacent to angle } A}$ is called **tangent *A*** (abbreviated tan A)

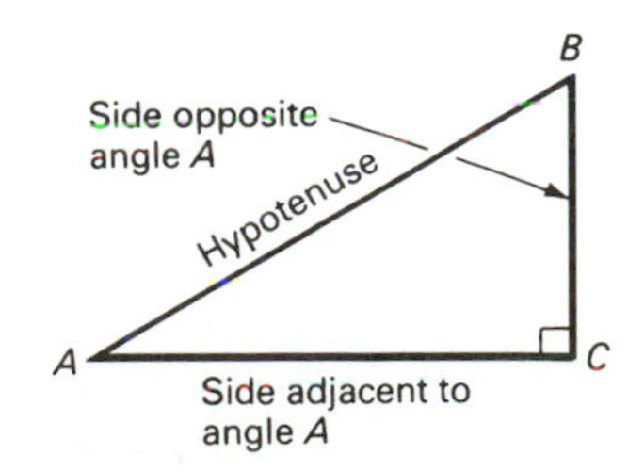

Figure 5.3

$$\sin A = \frac{\text{side opposite angle } A}{\text{hypotenuse}}$$

$$\cos A = \frac{\text{side adjacent to angle } A}{\text{hypotenuse}}$$

$$\tan A = \frac{\text{side opposite angle } A}{\text{side adjacent to angle } A}$$

The ratios are defined similarly for angle B.

$$\sin B = \frac{\text{side opposite angle } B}{\text{hypotenuse}}$$

$$\cos B = \frac{\text{side adjacent to angle } B}{\text{hypotenuse}}$$

$$\tan B = \frac{\text{side opposite angle } B}{\text{side adjacent to angle } B}$$

EXAMPLE 1

Find the three trigonometric ratios of angle A in Fig. 5.4.

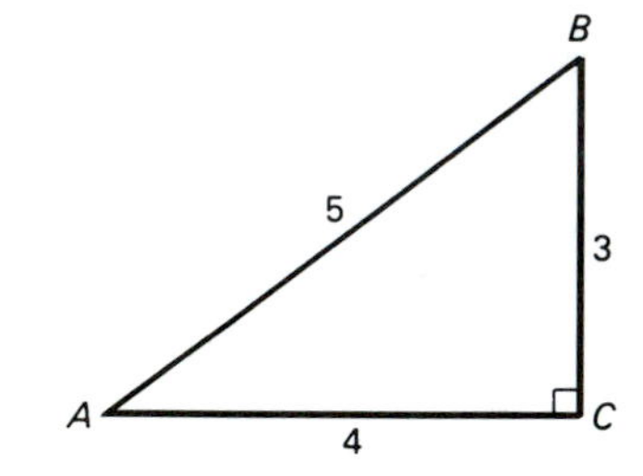

Figure 5.4

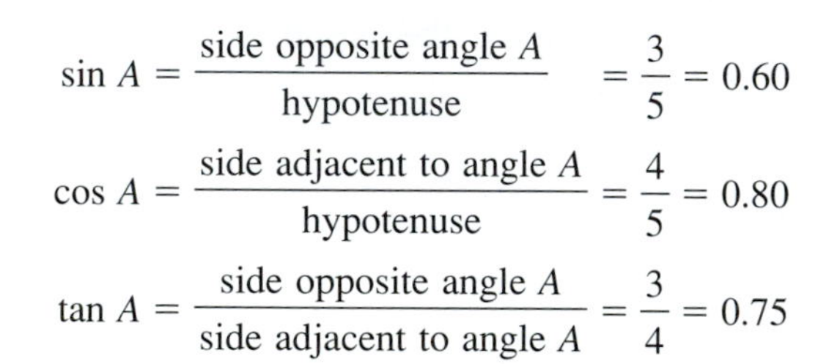

$$\sin A = \frac{\text{side opposite angle } A}{\text{hypotenuse}} = \frac{3}{5} = 0.60$$

$$\cos A = \frac{\text{side adjacent to angle } A}{\text{hypotenuse}} = \frac{4}{5} = 0.80$$

$$\tan A = \frac{\text{side opposite angle } A}{\text{side adjacent to angle } A} = \frac{3}{4} = 0.75$$

EXAMPLE 2

Find the three trigonometric ratios of angle B in Fig. 5.4.

$$\sin B = \frac{\text{side opposite angle } B}{\text{hypotenuse}} = \frac{4}{5} = 0.80$$

$$\cos B = \frac{\text{side adjacent to angle } B}{\text{hypotenuse}} = \frac{3}{5} = 0.60$$

$$\tan B = \frac{\text{side opposite angle } B}{\text{side adjacent to angle } B} = \frac{4}{3} = 1.33$$

Note: Every acute angle has three trigonometric ratios associated with it.

In this book we assume that you will be using a calculator. When calculations involve a trigonometric ratio, we will use the following generally accepted practice for significant digits:

Angle Expressed to Nearest:	*Length of Side Contains:*
1°	Two significant digits
0.1°	Three significant digits
0.01°	Four significant digits

A useful and time-saving fact about right triangles is that *the sum of the two acute angles of any right triangle is always* 90°. That is,

$$A + B = 90°$$

Why is this true? We know that the sum of the three interior angles of any triangle is 180°. A right triangle must contain a right angle, whose measure is 90°. This leaves 90° to be divided between the two acute angles. Therefore, if one acute angle is known, the other acute angle may be found by subtracting the known angle from 90°. That is,

$$A = 90° - B$$
$$B = 90° - A$$

EXAMPLE 3

Find angle B and side a in the right triangle in Fig. 5.5.

To find angle B, we use

$$B = 90° - A = 90° - 30.0° = 60.0°$$

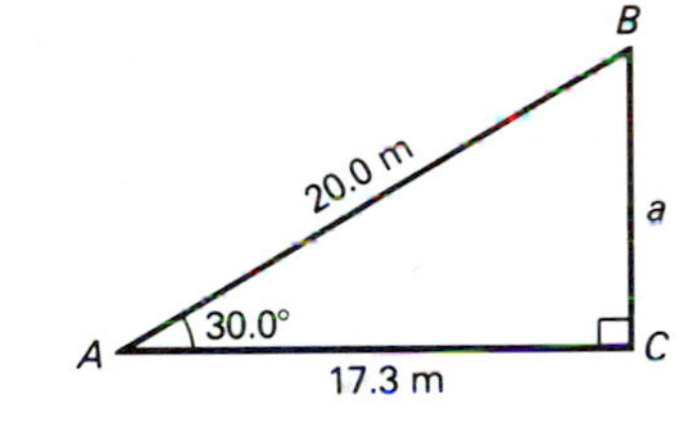

Figure 5.5

To find side a, we use a trigonometric ratio. Note that we are looking for the *side opposite* angle A and that the *hypotenuse* is given. The trigonometric ratio having these two quantities is sine.

$$\sin A = \frac{\text{side opposite angle } A}{\text{hypotenuse}}$$

$$\sin 30.0° = \frac{a}{20.0 \text{ m}}$$

$$(\sin 30.0°)(20.0 \text{ m}) = \left(\frac{a}{20.0 \text{ m}}\right)(20.0 \text{ m}) \qquad \text{Multiply both sides by 20.0 m.}$$

$$10.0 \text{ m} = a$$

EXAMPLE 4

Find angle A, angle B, and side a in the right triangle in Fig. 5.6.

First, find angle A. The *side adjacent* to angle A and the *hypotenuse* are given. Therefore, we use cos A to find angle A because cos A uses these two quantities:

$$\cos A = \frac{\text{side adjacent to } A}{\text{hypotenuse}}$$

$$\cos A = \frac{13 \text{ ft}}{19 \text{ ft}} = 0.684$$

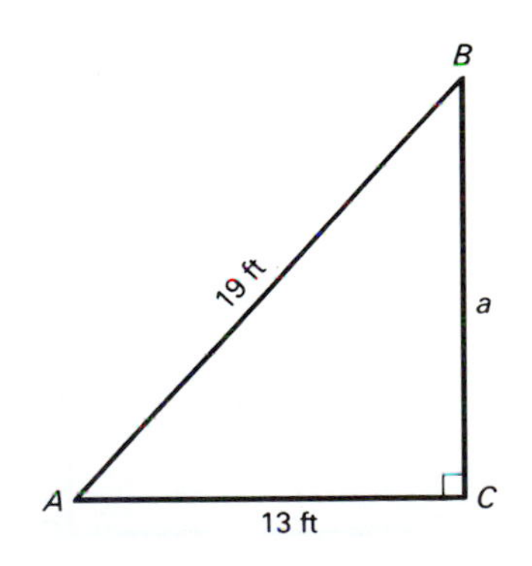

Figure 5.6

Using a calculator as in Section B.5 (in Appendix B), we find that $A = 47°$.

To find angle B, we use

$$B = 90° - A = 90° - 47° = 43°$$

To find side a, we use sin A because the *hypotenuse* is given and side a is the *side opposite* angle A.

$$\sin A = \frac{\text{side opposite angle } A}{\text{hypotenuse}}$$

$$\sin 47° = \frac{a}{19 \text{ ft}}$$

$$(\sin 47°)(19 \text{ ft}) = \left(\frac{a}{19 \text{ ft}}\right)(19 \text{ ft}) \qquad \text{Multiply both sides by 19 ft.}$$

$$14 \text{ ft} = a$$

EXAMPLE 5

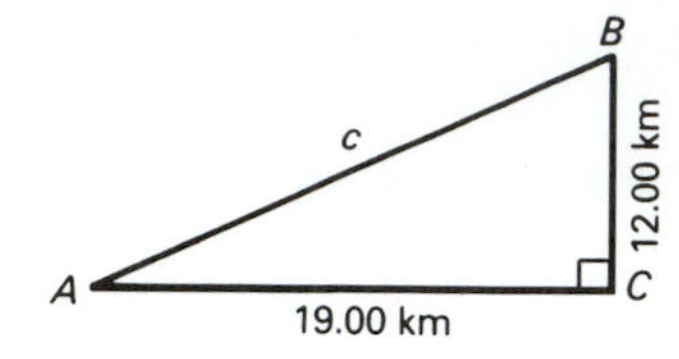

Figure 5.7

Find angle A, angle B, and the hypotenuse in the right triangle in Fig. 5.7.

To find angle A, use $\tan A$:

$$\tan A = \frac{\text{side opposite angle } A}{\text{side adjacent to angle } A}$$

$$\tan A = \frac{12.00 \text{ km}}{19.00 \text{ km}} = 0.6316$$

$$A = 32.28°$$

To find angle B,

$$B = 90° - A = 90° - 32.28° = 57.72°$$

To find the hypotenuse, use $\sin A$.

$$\sin A = \frac{\text{side opposite angle } A}{\text{hypotenuse}}$$

$$\sin 32.28° = \frac{12.00 \text{ km}}{c}$$

$$(\sin 32.28°)(c) = \left(\frac{12.00 \text{ km}}{c}\right)(c) \qquad \text{Multiply both sides by } c.$$

$$(\sin 32.28°)(c) = 12.00 \text{ km}$$

$$\frac{c(\sin 32.28°)}{\sin 32.28°} = \frac{12.00 \text{ km}}{\sin 32.28°} \qquad \text{Divide both sides by } \sin 32.28°.$$

$$c = \frac{12.00 \text{ km}}{\sin 32.28°}$$

$$= 22.47 \text{ km}$$

When the two legs of a right triangle are given, the hypotenuse can be found without using trigonometric ratios. From geometry, *the sum of the squares of the legs of a right triangle is equal to the square of the hypotenuse* (**Pythagorean theorem;** see Fig. 5.8):

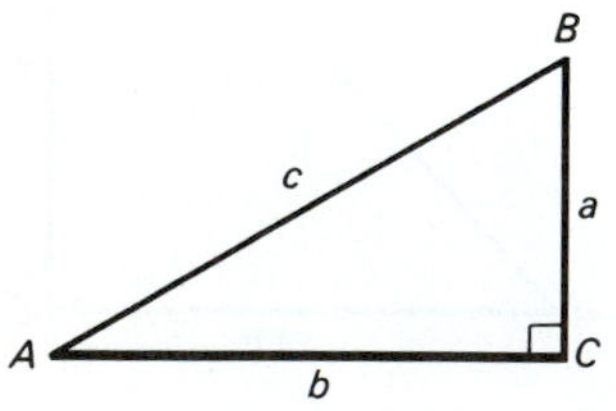

Figure 5.8

$$a^2 + b^2 = c^2$$

or, taking the square root of each side of the equation,

$$c = \sqrt{a^2 + b^2}$$

Also, if one leg and the hypotenuse are given, the other leg can be found by

$$a = \sqrt{c^2 - b^2}$$
$$b = \sqrt{c^2 - a^2}$$

EXAMPLE 6

Find the hypotenuse of the right triangle in Fig. 5.9.

$$\begin{aligned} c &= \sqrt{a^2 + b^2} \\ c &= \sqrt{(13.0 \text{ m})^2 + (11.0 \text{ m})^2} \\ &= \sqrt{169 \text{ m}^2 + 121 \text{ m}^2} \\ &= \sqrt{290 \text{ m}^2} \\ &= 17.0 \text{ m} \end{aligned}$$

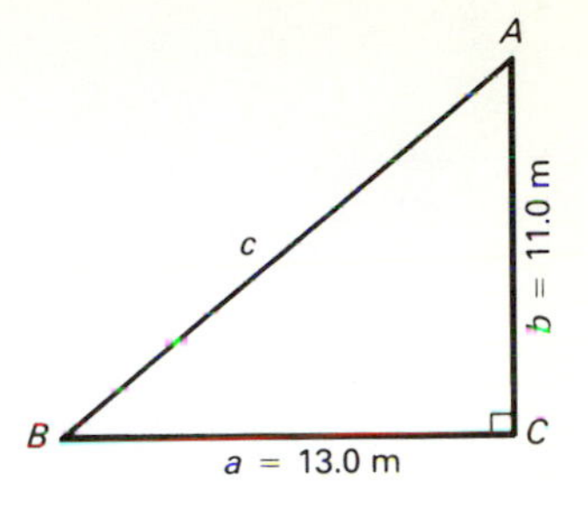

Figure 5.9

EXAMPLE 7

Find side b in the right triangle in Fig. 5.10.

$$\begin{aligned} b &= \sqrt{c^2 - a^2} \\ b &= \sqrt{(12.2 \text{ km})^2 - (7.30 \text{ km})^2} \\ &= 9.77 \text{ km} \end{aligned}$$

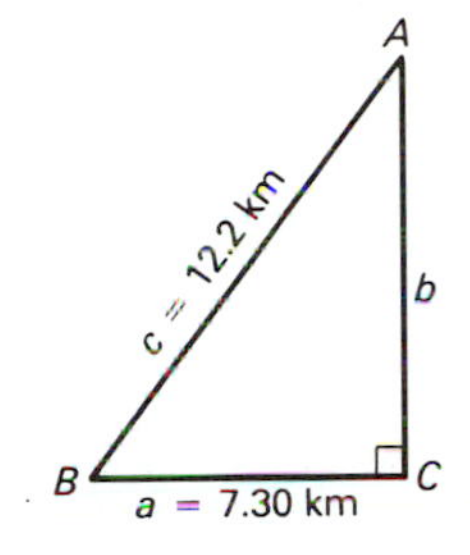

Figure 5.10

■ PROBLEMS 5.1

Use right triangle ABC in Fig. 5.11 to fill in each blank.

1. The side opposite angle A is ______.
2. The side opposite angle B is ______.
3. The hypotenuse is ______.
4. The side adjacent to angle A is ______.
5. The side adjacent to angle B is ______.
6. The angle opposite side a is ______.
7. The angle opposite side b is ______.
8. The angle opposite side c is ______.
9. The angle adjacent to side a is ______.
10. The angle adjacent to side b is ______.

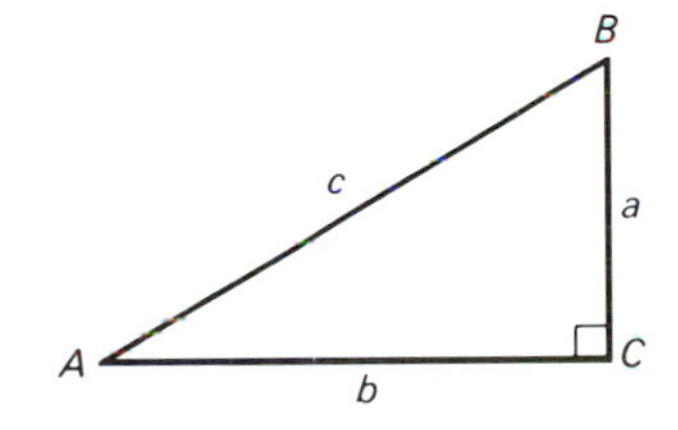

Figure 5.11

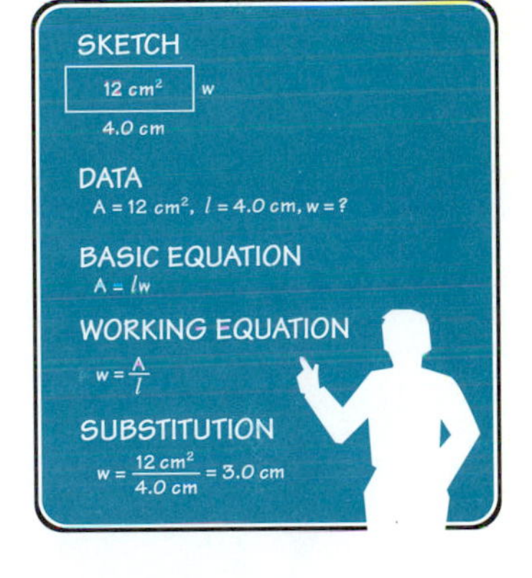

Use a calculator to find each trigonometric ratio rounded to four significant digits.

11. $\sin 71°$
12. $\cos 40°$
13. $\tan 61°$
14. $\tan 41.2°$
15. $\cos 11.5°$
16. $\sin 79.4°$
17. $\cos 49.63°$
18. $\tan 53.45°$
19. $\tan 17.04°$
20. $\cos 34°$
21. $\sin 27.5°$
22. $\cos 58.72°$

Find each angle rounded to the nearest whole degree.

23. $\sin A = 0.2678$
24. $\cos B = 0.1046$
25. $\tan A = 0.9237$
26. $\sin B = 0.9253$
27. $\cos B = 0.6742$
28. $\tan A = 1.351$

Find each angle rounded to the nearest tenth of a degree.

29. $\sin B = 0.5963$
30. $\cos A = 0.9406$
31. $\tan B = 1.053$
32. $\sin A = 0.9083$
33. $\cos A = 0.8660$
34. $\tan B = 0.9433$

Find each angle rounded to the nearest hundredth of a degree.

35. $\sin A = 0.3792$
36. $\cos B = 0.06341$
37. $\tan B = 0.3010$
38. $\sin A = 0.4540$
39. $\cos B = 0.8141$
40. $\tan A = 2.369$

Solve each triangle (find the missing angles and sides) using trigonometric ratios.

41.

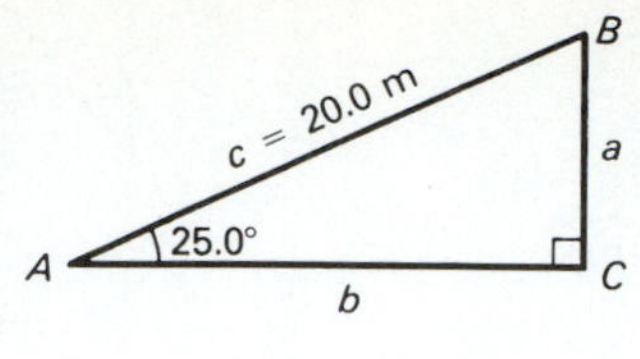

42.

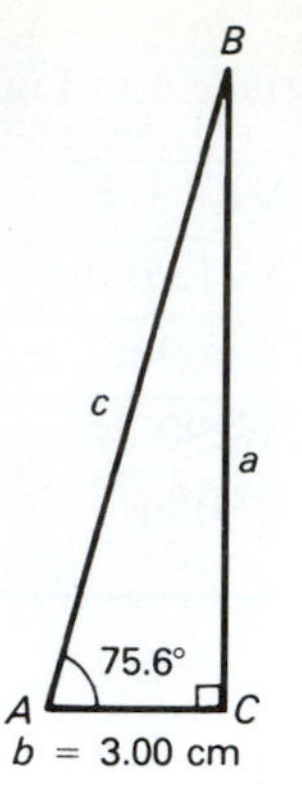

43.

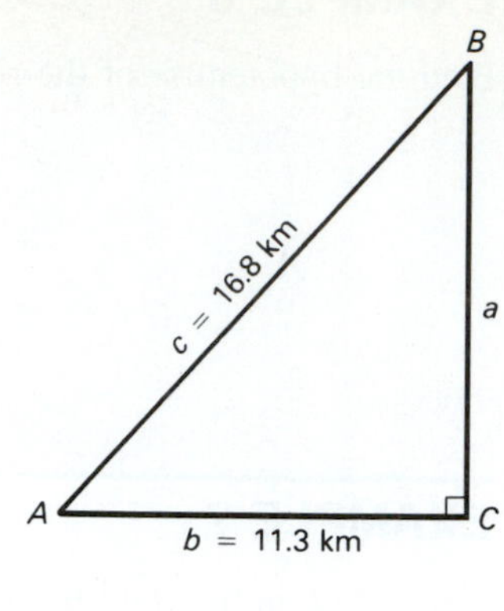

44.

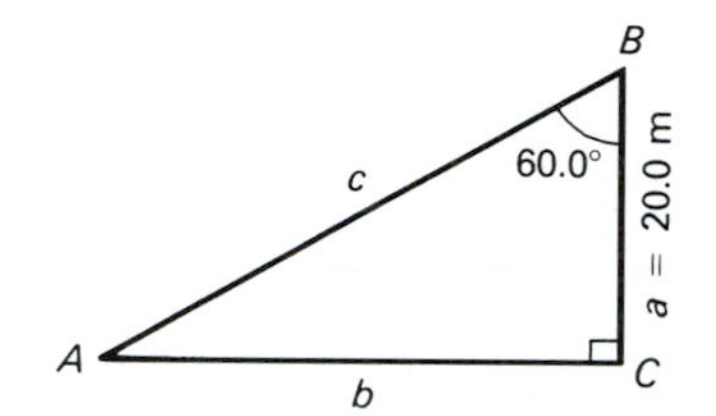

45.

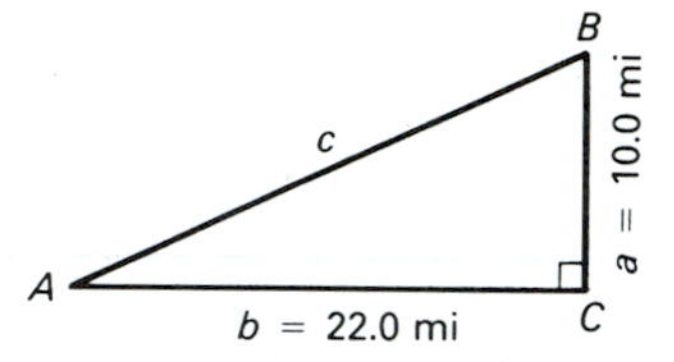

46.

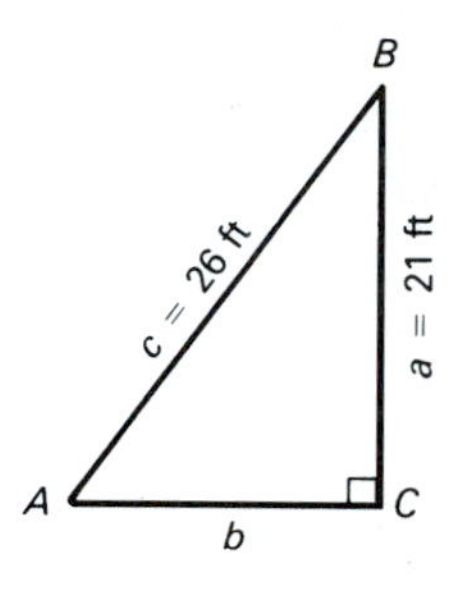

47.

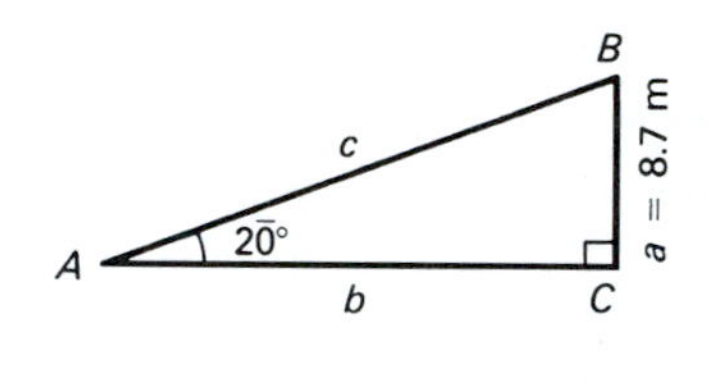

48.

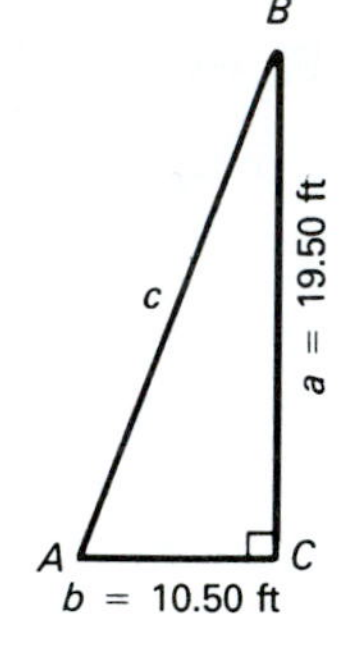

49.

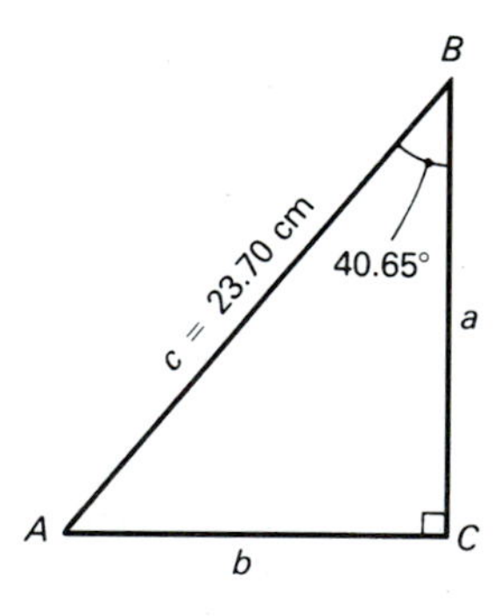

50.

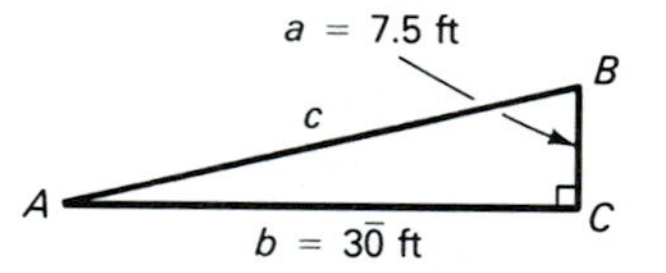

Find the missing side in each right triangle using the Pythagorean theorem.

51.

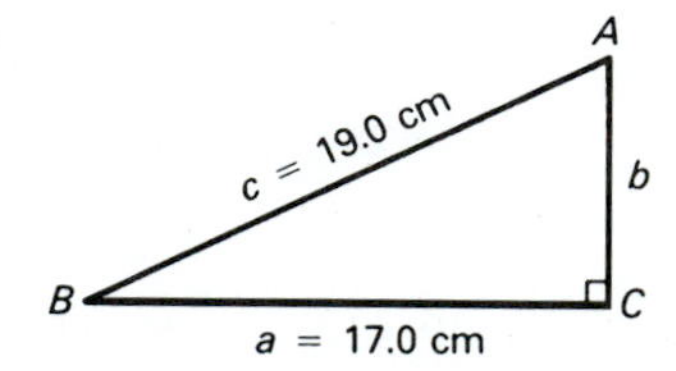

52.

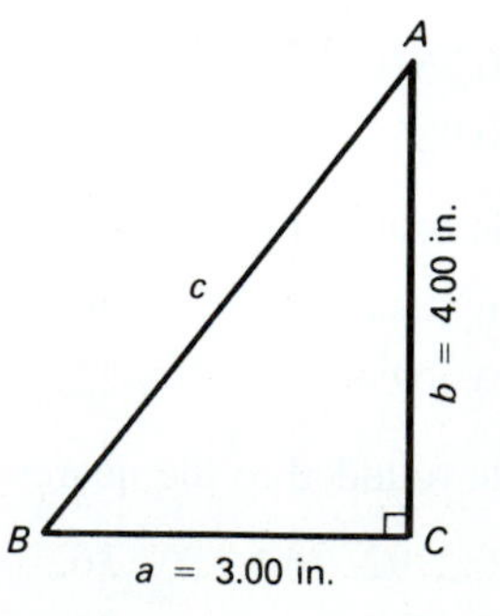

53.

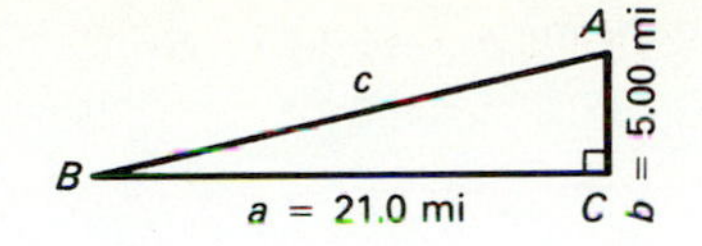

54.

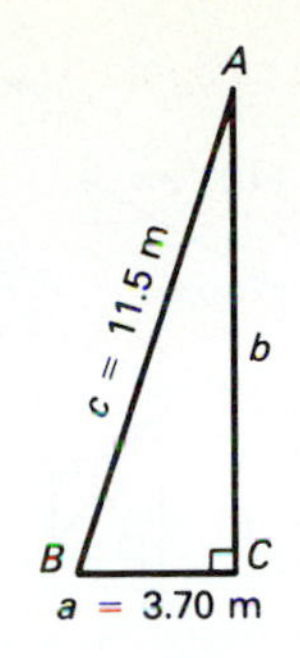

55.

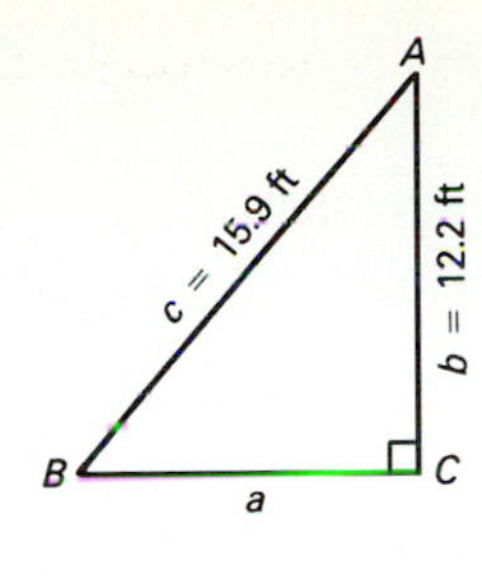

56.

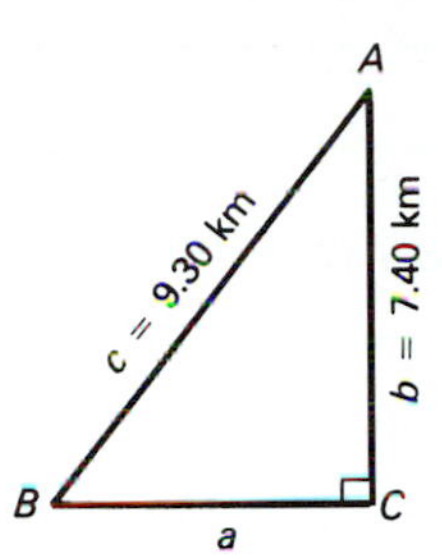

57.

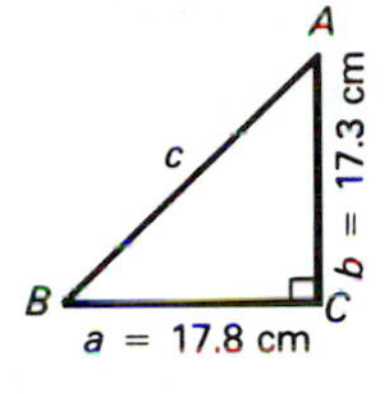

58.

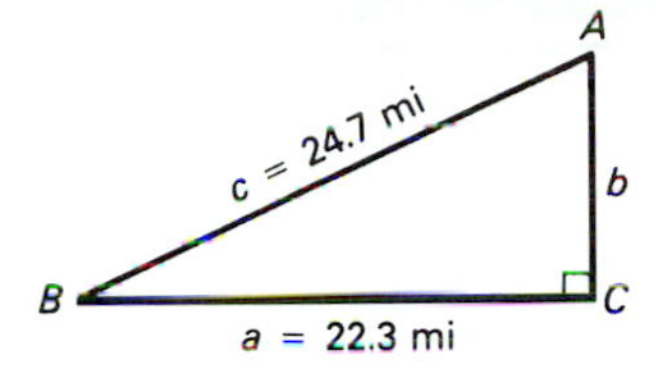

59.

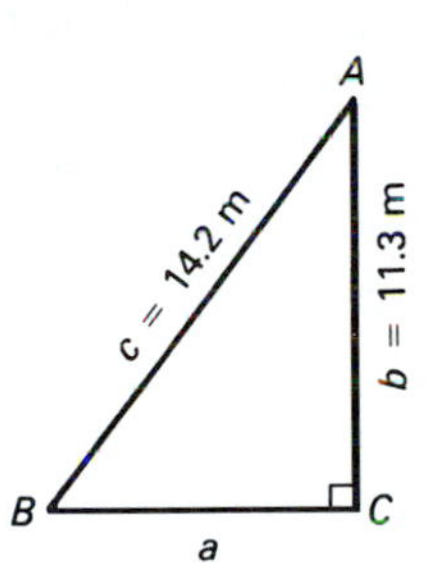

60.

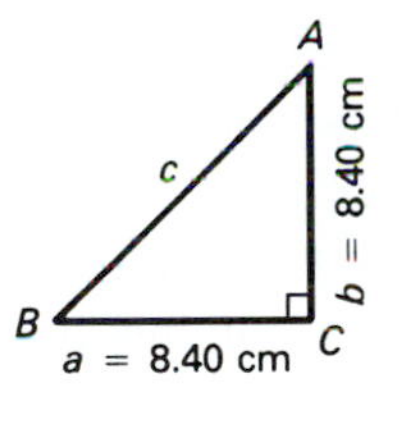

61. A round taper is shown in Fig. 5.12.
    (a) Find $\angle BAC$.
    (b) Find the length BC.
    (c) Find the diameter of end x.
62. Across the flats *(a)* a hexagonal nut is $\frac{3}{4}$ in. Find the distance across the corners *(b)* (Fig. 5.13).
63. Find distances C and D between the holes of the plate shown in Fig. 5.14.

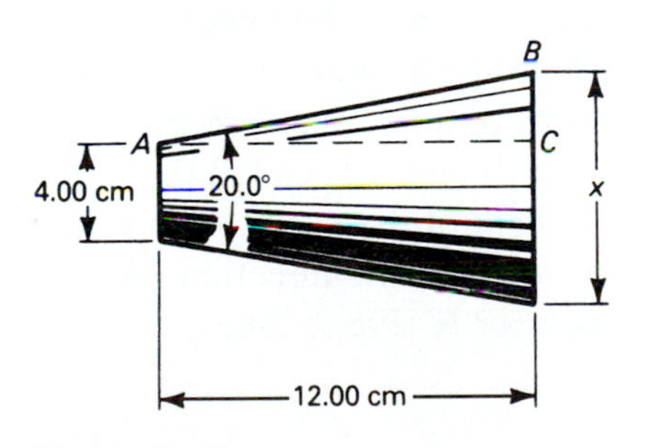

Figure 5.12

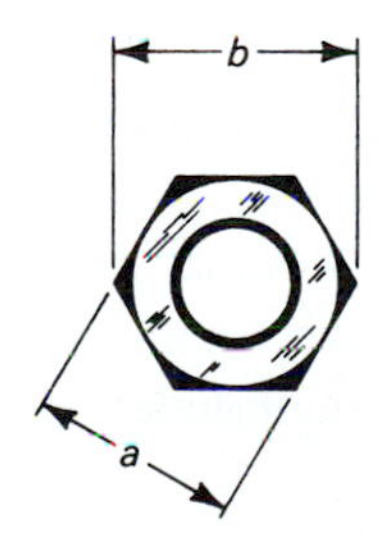

Figure 5.13

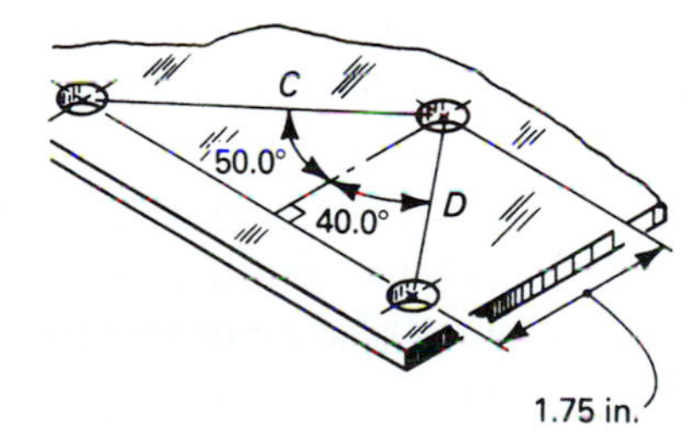

Figure 5.14

64. A piece of electrical conduit cuts across a corner of a room $24\bar{0}$ cm from the corner. It meets the adjoining wall $35\bar{0}$ cm from the corner. Find length AB of the conduit (Fig. 5.15).
65. Find the distances between holes on the plate shown in Fig. 5.16.
66. Find length x in Fig. 5.17. (*Note:* $AB = BC$.)

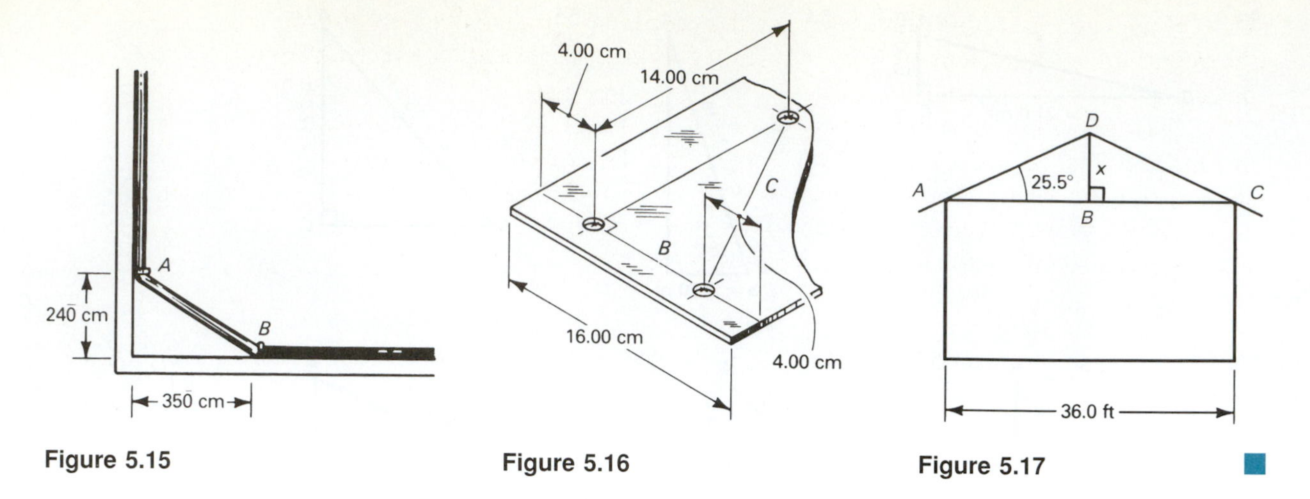

Figure 5.15

Figure 5.16

Figure 5.17

5.2 COMPONENTS OF A VECTOR

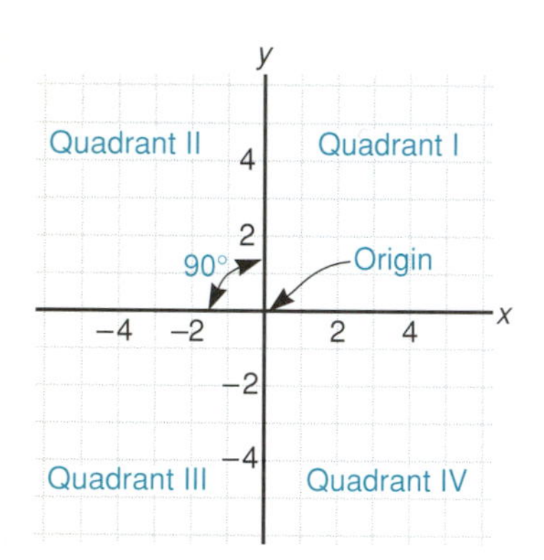

Figure 5.18
Number plane.

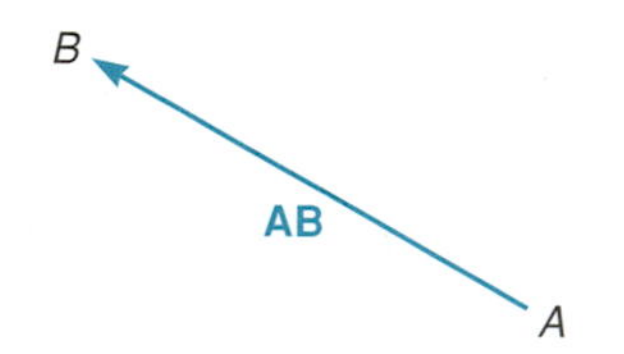

Figure 5.19
Vector from A to B.

Before further study of vectors, we need to discuss components of vectors. This requires using the number plane. The **number plane** (sometimes called the *Cartesian coordinate system*) is determined by a horizontal line called the x-axis and a vertical line called the y-axis intersecting at right angles as shown in Fig. 5.18. These two lines divide the number plane into four quadrants which we label as quadrants I, II, III, and IV. Each axis has a scale, and the intersection of the two axes is called the *origin.* The x-axis contains positive numbers to the right of the origin and negative numbers to the left of the origin. The y-axis contains positive numbers above the origin and negative numbers below the origin.

A quantity such as length, volume, time, or temperature is completely described when its magnitude (size) is given. These quantities are called **scalars.** Other physical quantities require both a magnitude and a direction to be completely described. For example, to completely describe wind velocity, we need not only the speed, such as 25 mph, but also the direction, such as from the north. A quantity that requires both a magnitude and a direction to be completely described is called a **vector.** Other examples of vector quantities include force, torque, and certain quantities in electricity.

Graphically, a vector is usually represented by a directed line segment. The length of the line segment indicates the magnitude of the quantity. An arrowhead indicates the direction. If A and B are the end points of a line segment as in Fig. 5.19, the symbol **AB** denotes the *vector from A to B*. Point A is called the *initial point.* Point B is called the *terminal point* or *end point* of the vector. The vector **BA** has the same length as the vector **AB** but has the opposite direction. Vectors may also be denoted by a single letter, such as **u, v,** or **R** [Fig. 5.20(a)].

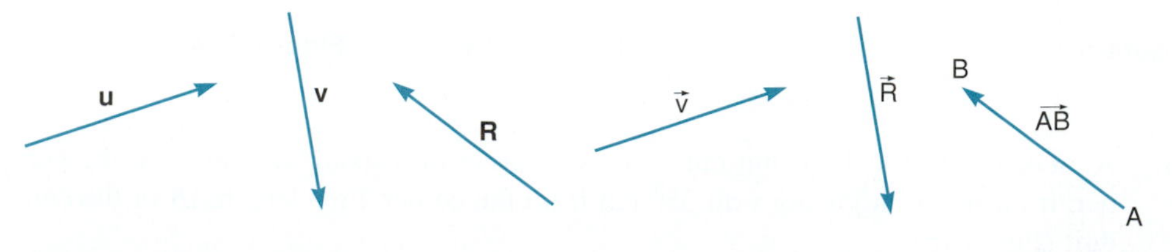

(a) Vectors in textbooks are indicated by letters in boldface type.

(b) Vectors written on paper or a chalkboard are indicated by small arrows above the letters.

Figure 5.20

When writing vectors on paper or a chalkboard, we use a small arrow above the vector quantity, such as $\vec{v}$, $\vec{R}$, or $\overrightarrow{AB}$ [Fig. 5.20(b)]. The magnitude or length of vector **v** or $\vec{w}$ is written $|\mathbf{v}|$ or $|\vec{w}|$.

The sum of two or more vectors is called the **resultant vector.** When two or more vectors are added, each of the vectors is called a **component** of the resultant, or sum, vector. The components of vector **R** in Fig. 5.21(a) are vectors **A, B,** and **C.** ***Note:*** A vector may have more than one set of component vectors. The components of vector **R** in Fig. 5.21(b) are vectors **E** and **F.**

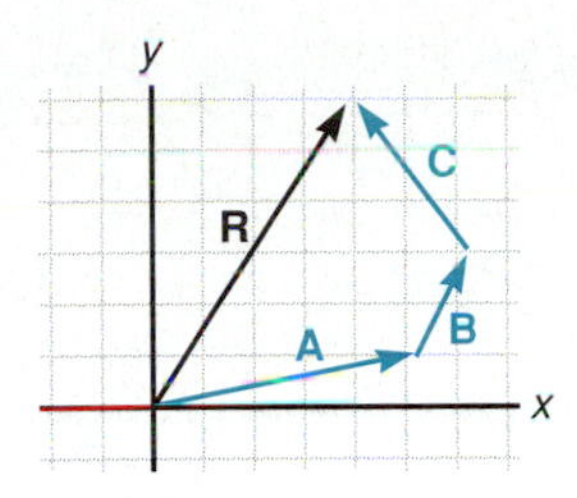

(a) Vectors **A**, **B**, and **C** are components of the resultant, or sum, vector **R**.

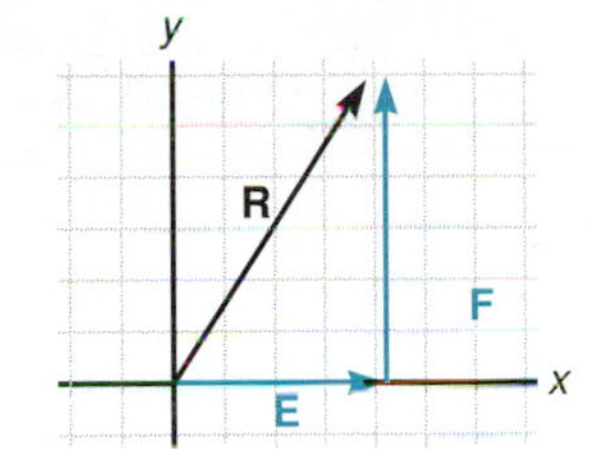

(b) Vector **E** is a horizontal component and vector **F** is a vertical component of the resultant, or sum, vector **R**.

Figure 5.21

We are most interested in the components of a vector that are perpendicular to each other and that are on or parallel to the x- and y-axes. In particular, we are interested in the type of component vectors shown in Fig. 5.21(b) (component vectors **E** and **F**). The horizontal component vector that lies on or is parallel to the x-axis is called the ***x*-component.** The vertical component vector that lies on or is parallel to the y-axis is called the ***y*-component.** Three examples are shown in Fig. 5.22.

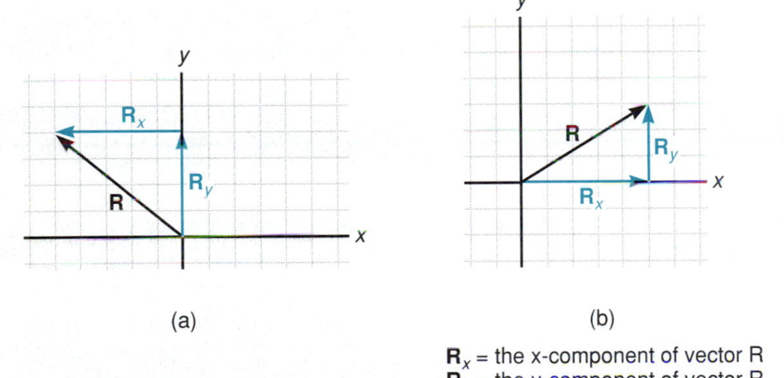

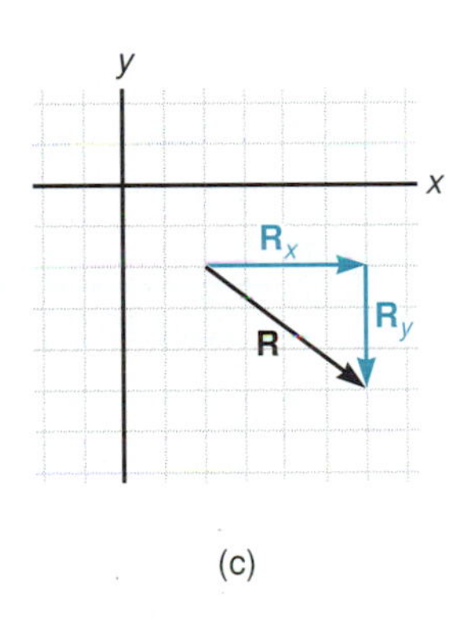

$\mathbf{R}_x$ = the x-component of vector R
$\mathbf{R}_y$ = the y-component of vector R

Figure 5.22

The x- and y-components of vectors can also be expressed as signed numbers. The absolute value of the signed number corresponds to the magnitude of the vector. The sign of the number corresponds to the direction of the component as follows:

x-component	*y-component*
+, if right	+, if up
−, if left	−, if down

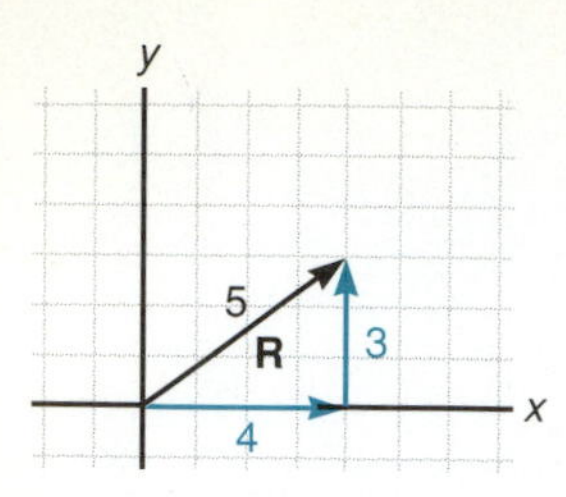

Figure 5.23

EXAMPLE 1

Find the x- and y-components of vector **R** in Fig. 5.23.

$$\mathbf{R}_x = x\text{-component of } \mathbf{R} = +4$$
$$\mathbf{R}_y = y\text{-component of } \mathbf{R} = +3$$

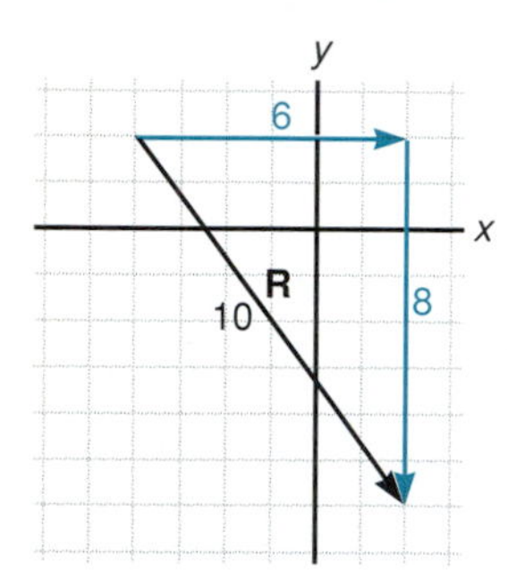

Figure 5.24

EXAMPLE 2

Find the x- and y-components of vector **R** in Fig. 5.24.

$$\mathbf{R}_x = x\text{-component of } \mathbf{R} = +6$$
$$\mathbf{R}_y = y\text{-component of } \mathbf{R} = -8$$

(y-component points in a negative direction)

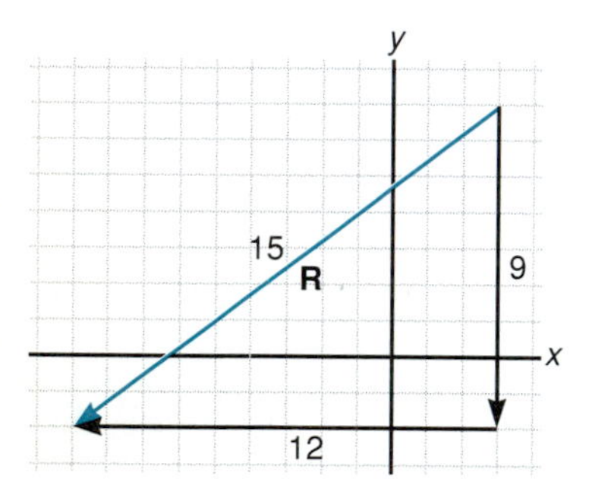

Figure 5.25

EXAMPLE 3

Find the x- and y-components of vector **R** in Fig. 5.25.

$$\mathbf{R}_x = -12$$
$$\mathbf{R}_y = -9$$

(both x- and y-components point in a negative direction)

Now that we have expressed the x- and y-components as signed numbers, we can find the resultant vector of several vectors using arithmetic and graphing. To find the resultant vector of several vectors, find the x-component of each vector and find the sum of the x-components. Then find the y-component of each vector and find the sum of the y-components. The two sums are the x- and y-components of the resultant vector. This is shown in the following examples.

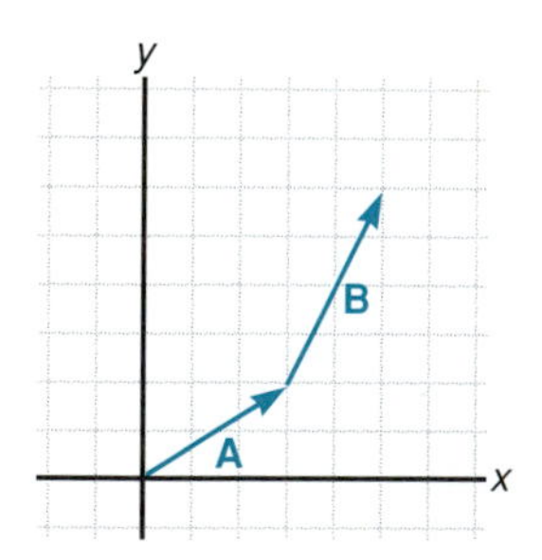

Figure 5.26

EXAMPLE 4

Given vectors **A** and **B** in Fig. 5.26, graph and find the x- and y-components of the resultant vector **R.**

The graph of the resultant vector **R** is found by connecting the initial point of vector **A** to the end point of vector **B** [Fig. 5.27(a)]. The resultant vector **R** is shown in Fig. 5.27(b).

The x-component of **R** is found by adding the x-components of **A** and **B.**

$$\begin{aligned} \mathbf{A}_x &= +3 \\ \mathbf{B}_x &= \underline{+2} \\ \mathbf{R}_x &= +5 \end{aligned}$$

The y-component of **R** is found by adding the y-components of **A** and **B.**

$$\begin{aligned} \mathbf{A}_y &= +2 \\ \mathbf{B}_y &= \underline{+4} \\ \mathbf{R}_y &= +6 \end{aligned}$$

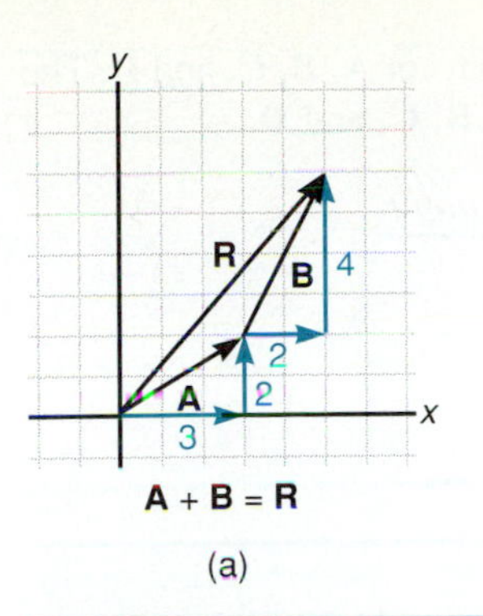

(a)

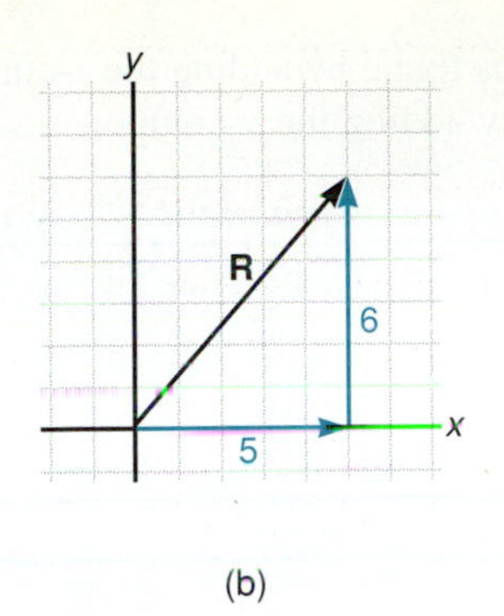

(b)

Figure 5.27

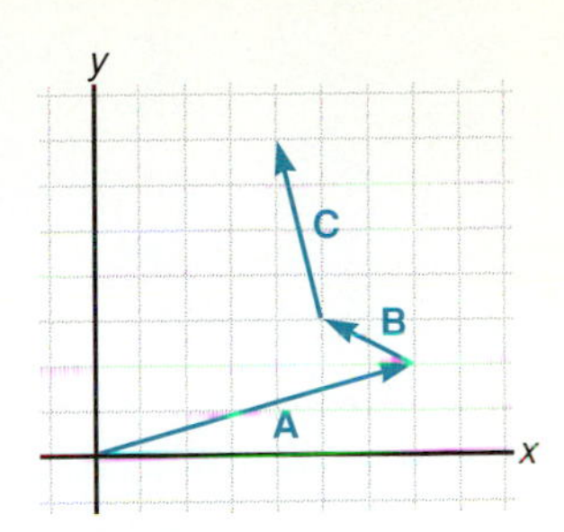

Figure 5.28

EXAMPLE 5

Given vectors **A, B,** and **C** in Fig. 5.28, graph and find the *x*- and *y*-components of the resultant vector **R.**

The graph of the resultant vector **R** is found by connecting the initial point of vector **A** to the end point of vector **C** [Fig. 5.29(a)]. The resultant vector **R** is shown in Fig. 5.29(b).

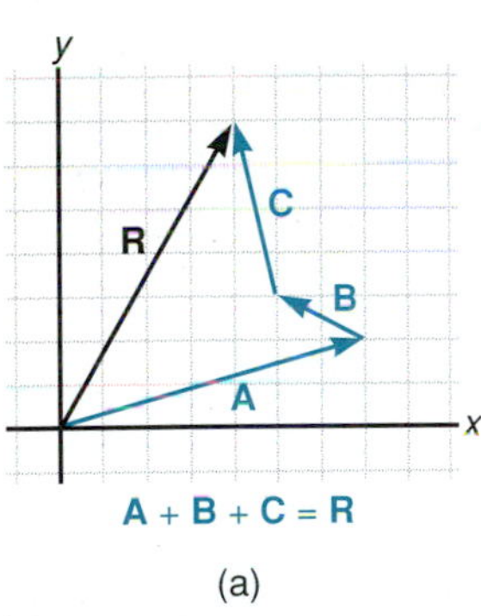

(a)

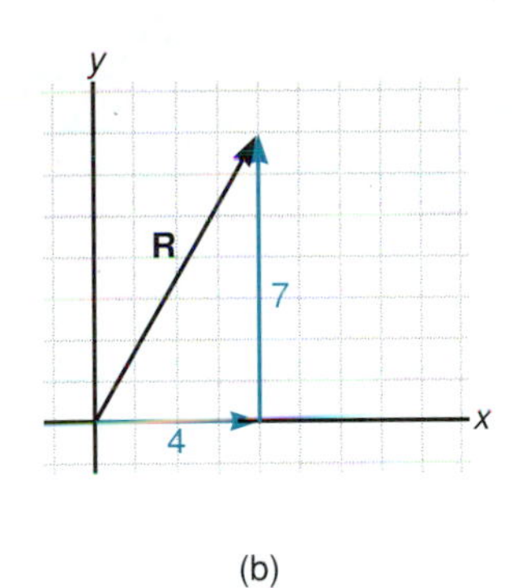

(b)

Figure 5.29

The *x*-component of **R** is found by adding the *x*-components of **A, B,** and **C.** The *y*-component of **R** is found by adding the *y*-components of **A, B,** and **C.**

Vector	*x-component*	*y-component*
A	+7	+2
B	−2	+1
C	−1	+4
R	+4	+7

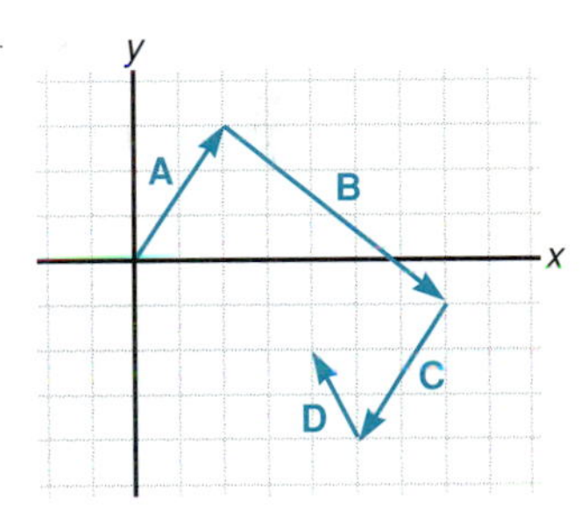

Figure 5.30

EXAMPLE 6

Given vectors **A, B, C,** and **D** in Fig. 5.30, graph and find the *x*- and *y*-components of the resultant vector **R.**

The graph of the resultant vector **R** is found by connecting the initial point of vector **A** to the end point of vector **D** [Fig. 5.31(a)]. The resultant vector **R** is shown in Fig. 5.31(b).

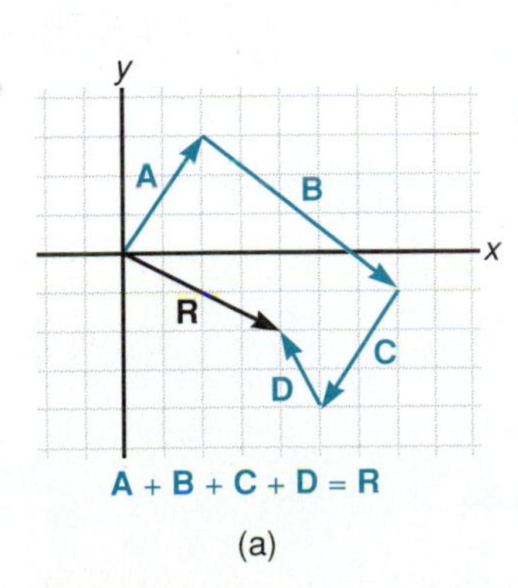

(a)

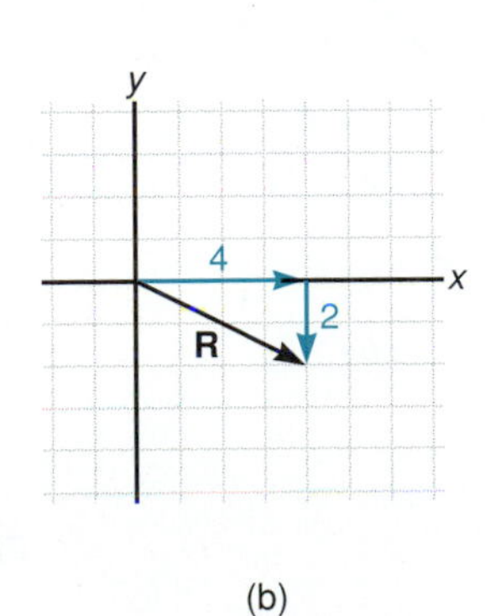

(b)

Figure 5.31

The x-component of **R** is found by adding the x-components of **A, B, C,** and **D.** The y-component of **R** is found by adding the y-components of **A, B, C,** and **D.**

Vector	*x-component*	*y-component*
A	+2	+3
B	+5	−4
C	−2	−3
D	−1	+2
R	+4	−2

Two vectors are equal when they have the same magnitude and the same direction [Fig. 5.32(a)]. Two vectors are opposites or negatives of each other when they have the same magnitude but opposite directions [Fig. 5.32(b)].

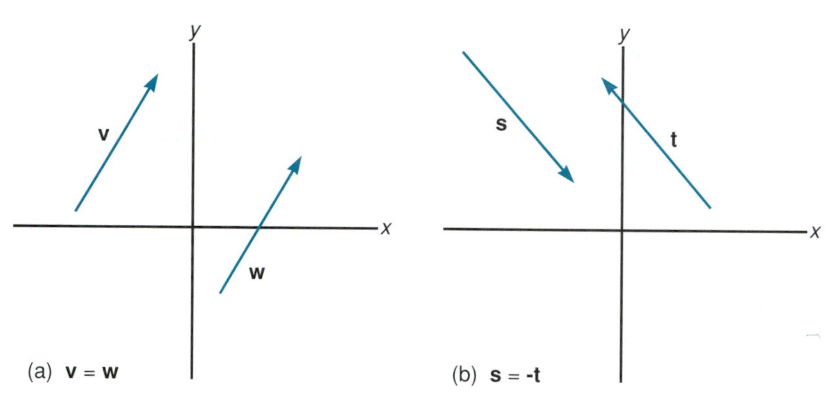

Figure 5.32

A vector may be placed in any position in the number plane as long as its magnitude and direction are not changed (Fig. 5.33). Note that the vectors in Fig. 5.34 are equal because they have the same magnitude (length) and the same direction.

To add two or more vectors graphically, construct the first vector with its initial point at the origin and parallel to its given position. Then, construct the second vector with its initial point on the end point of the first vector and parallel to its given position. Then, construct the third vector with its initial point on

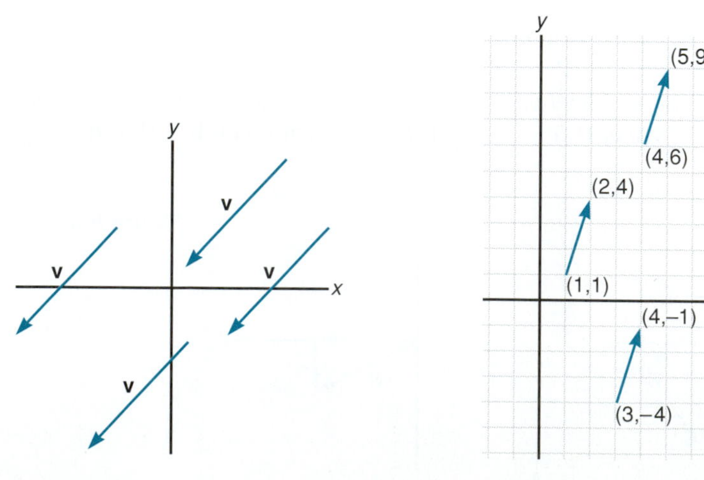

Figure 5.33

Figure 5.34

the end point of the second vector and parallel to its given position. Continue this process until all vectors have been so constructed. The sum or resultant vector is the vector joining the initial point of the first vector (origin) to the end point of the last vector. (The order of adding or constructing the given vectors does not matter.)

EXAMPLE 7

Given vectors **A, B,** and **C** in Fig. 5.35(a), graph and find the x- and y-components of the resultant vector **R.**

Construct vector **A** with its initial point at the origin and parallel to its given position as in Fig. 5.35(b). Next, construct vector **B** with its initial point on the end point of vector **A** and parallel to its given position. Then, construct vector **C** with its initial point on the end point of vector **B** and parallel to its given position. The resultant vector **R** is the vector with its initial point at the origin and its end point at the end point of vector **C.**

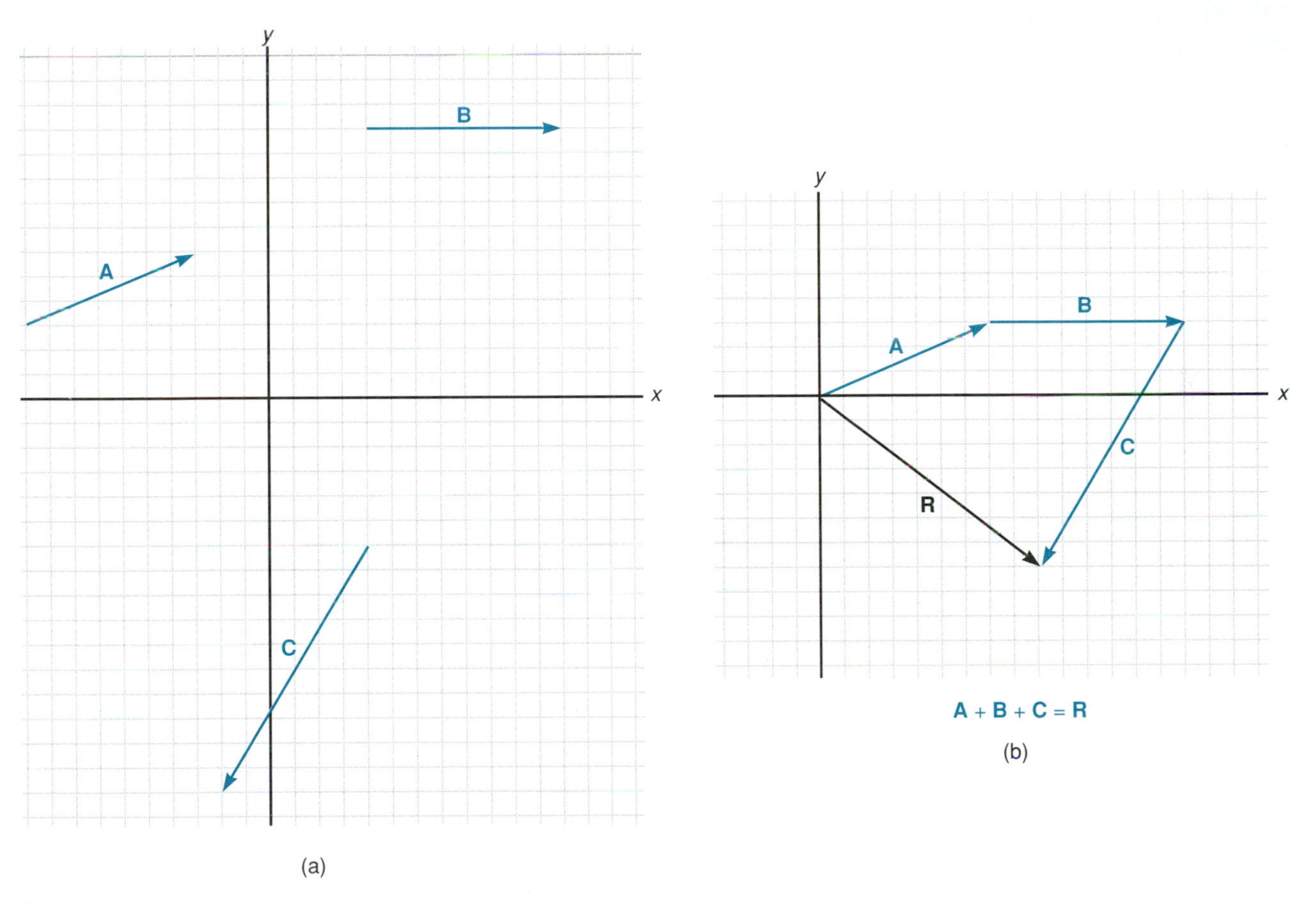

Figure 5.35

From the graph in Fig. 5.35(b), we read the x-component of **R** as 9 and the y-component of **R** as -7.

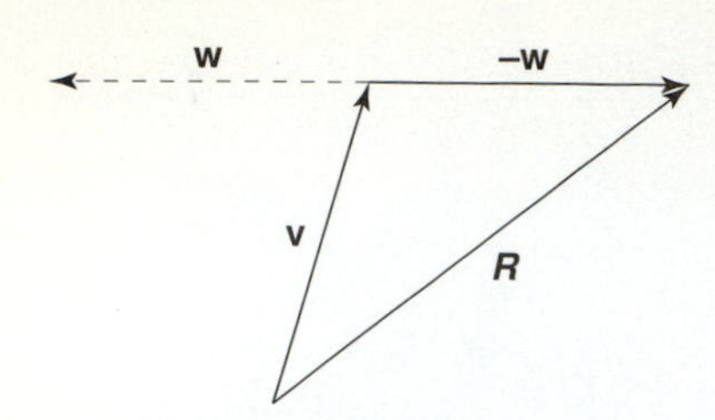

Figure 5.36
$\mathbf{R} = \mathbf{v} - \mathbf{w} = \mathbf{v} + (-\mathbf{w})$

One vector may be subtracted from a second vector by adding its negative to the first. That is, $\mathbf{v} - \mathbf{w} = \mathbf{v} + (-\mathbf{w})$. Construct **v** as usual. Then construct $-\mathbf{w}$ and find the resultant **R** as shown in Fig. 5.36.

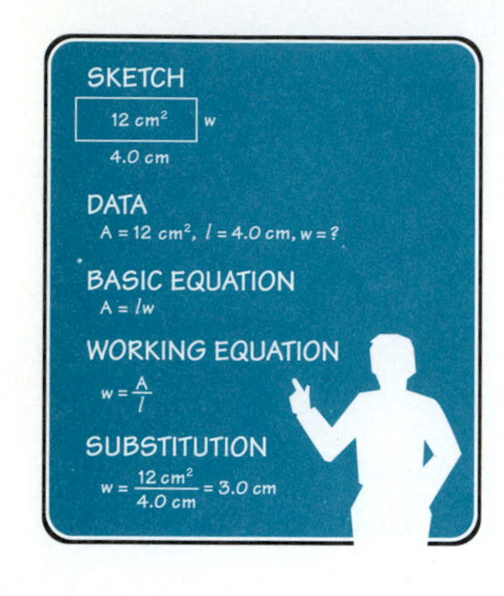

■ PROBLEMS 5.2

Find the x- and y-components of each vector in the following diagram. (Express them as signed numbers and then graph them as vectors.)

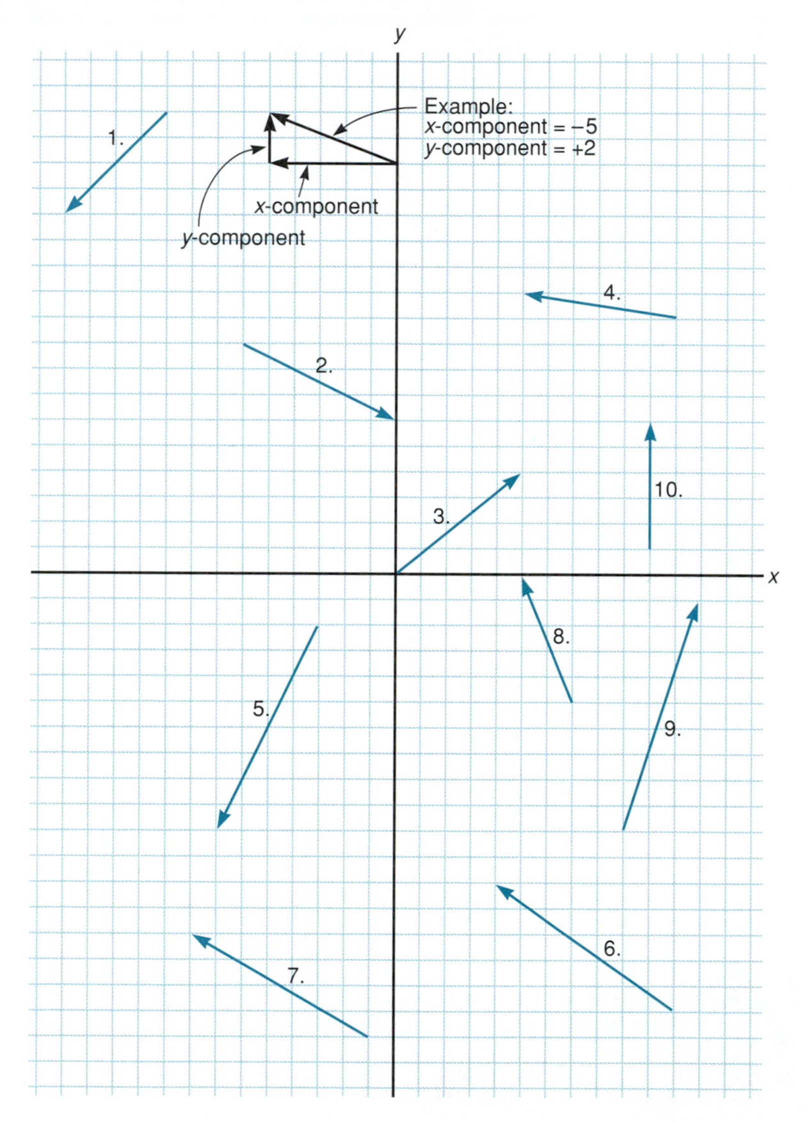

Find the x- and y-components of each resultant vector **R** and graph the resultant vector **R**.

	Vector	x-component	y-component		Vector	x-component	y-component
11.	**A**	+2	+3	12.	**A**	+9	−5
	B	+7	+2		**B**	−4	−6
	R				**R**		

	Vector	x-component	y-component
13.	**A**	−2	+13
	B	−11	+1
	C	+3	−4
	R		
15.	**A**	+17	+7
	B	−14	+11
	C	+7	+9
	D	−6	−15
	R		
17.	**A**	+1.5	−1.5
	B	−3	−2
	C	+7.5	−3
	D	+2	+2.5
	R		
19.	**A**	+1.5	+2.5
	B	−2	−3
	C	+3.5	−7.5
	D	−4	+6
	E	−5.5	+2
	R		

	Vector	x-component	y-component
14.	**A**	+10	−5
	B	−13	−9
	C	+4	+3
	R		
16.	**A**	+1	+7
	B	+9	−4
	C	−4	+13
	D	−11	−4
	R		
18.	**A**	+1	−1
	B	−4	−2
	C	+2	+4
	D	+5	−3
	E	+3	+5
	R		
20.	**A**	−7	+15
	B	+13.5	−17.5
	C	−7.5	−20
	D	+6	+13.5
	E	+2.5	+2.5
	F	−11	+11.5
	R		

For each set of vectors, graph and find the x- and y-components of the resultant vector **R**.

21.

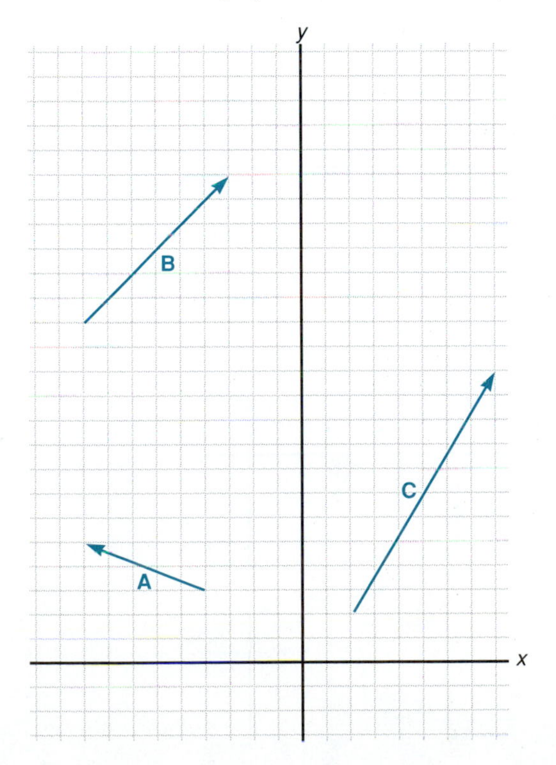

22.

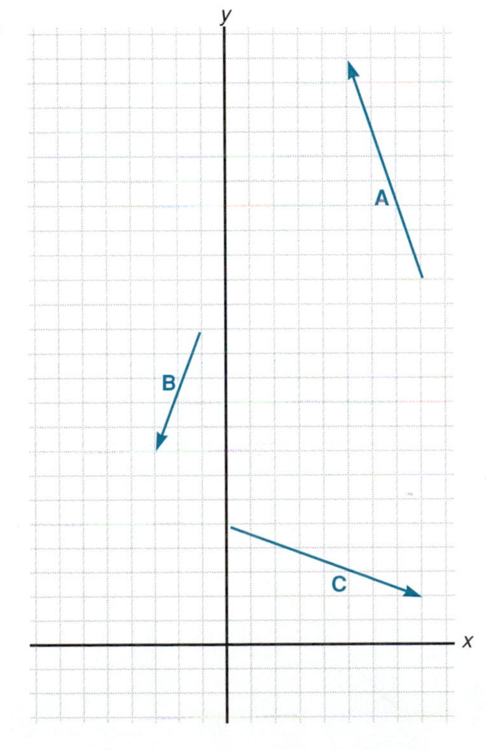

23.

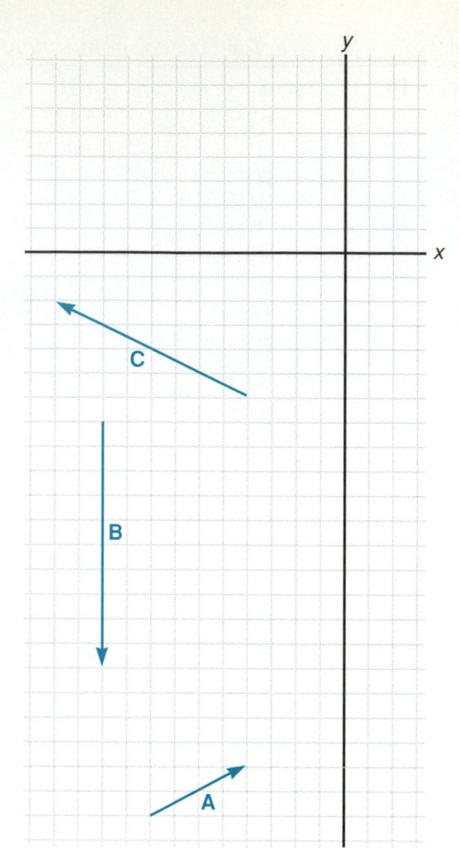

24.

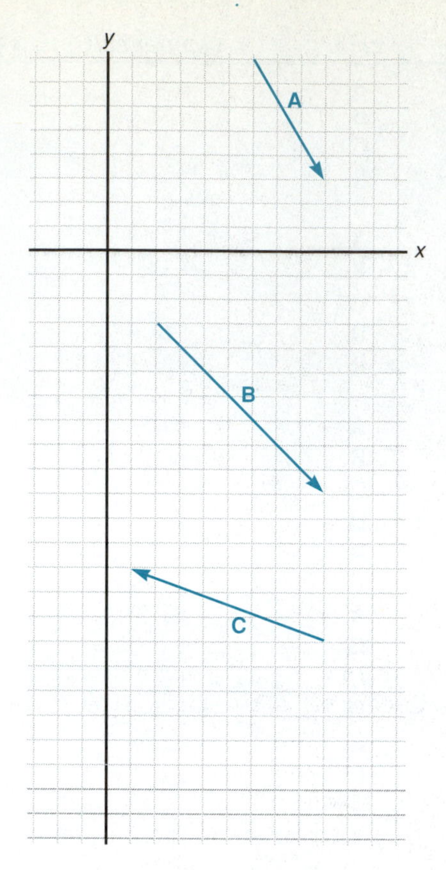

25.

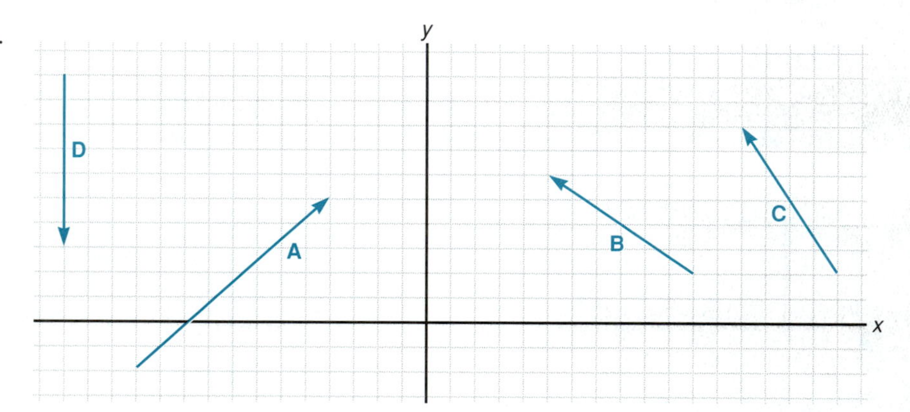

26.

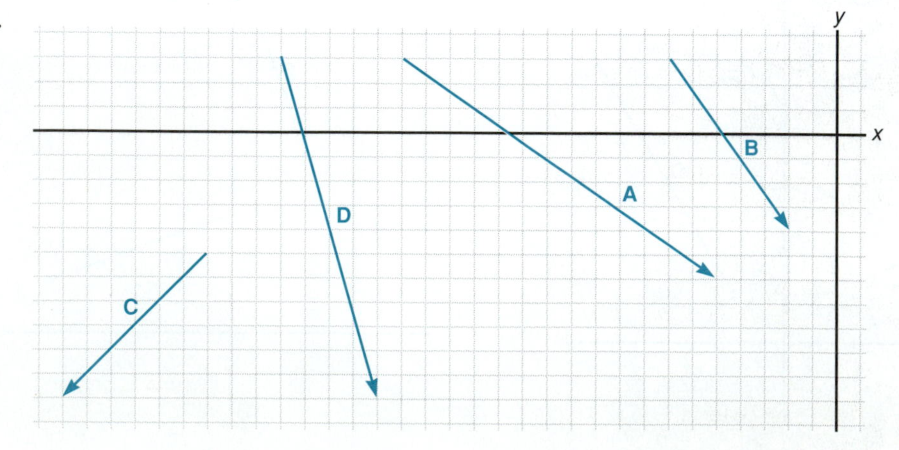

5.3 VECTORS IN STANDARD POSITION

Any vector may be placed in any position in the number plane as long as its magnitude and direction are not changed. A vector is in **standard position** when its initial point is at the origin of the number plane. A vector in standard position is expressed in terms of its length and its angle θ, where θ *is measured counterclockwise from the positive x-axis to the vector.* The vectors shown in Fig. 5.37 are in standard position.

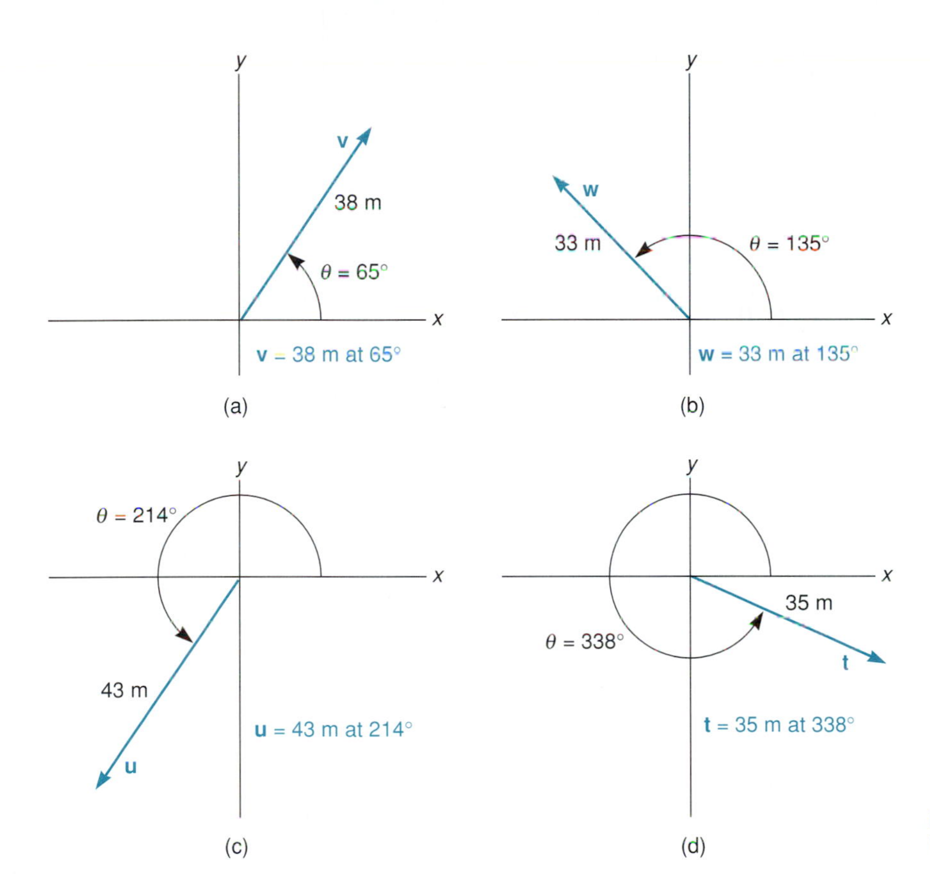

Figure 5.37
Vectors in standard position.

Finding Components of a Vector

EXAMPLE 1

Find the x- and y-components of the vector $\mathbf{v} = 10.0$ m at $60.0°$.

First, draw the vector in standard position [Fig. 5.38(a)]. Then, draw a right triangle where the legs represent the x- and y-components [Fig. 5.38(b)]. The absolute value of the x-component of the vector is the length of the side adjacent to the $60.0°$ angle. Therefore, to find the x-component

$$\cos 60.0° = \frac{\text{side adjacent to } 60.0°}{\text{hypotenuse}}$$

Since

$$\text{hypotenuse} = 10.0 \text{ m}$$
$$\text{side adjacent to } 60.0° = |\mathbf{v}_x|$$

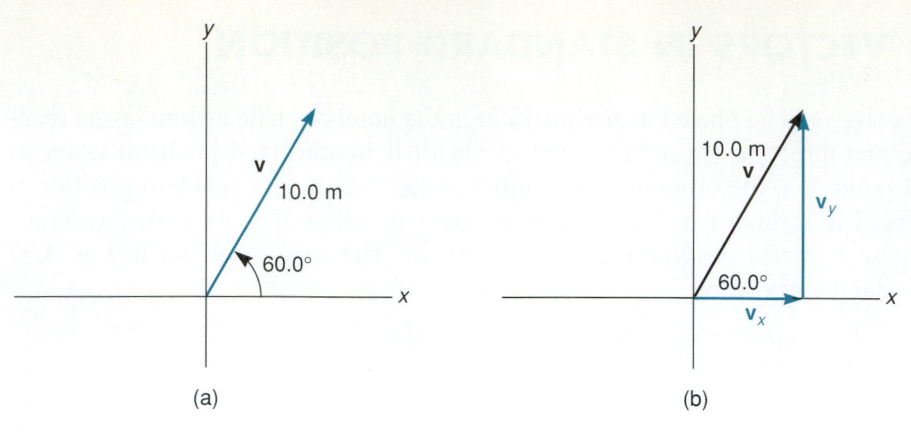

Figure 5.38

we have

$$\cos 60.0° = \frac{|\mathbf{v}_x|}{10.0 \text{ m}}$$

$$(\cos 60.0°)(10.0 \text{ m}) = \left(\frac{|\mathbf{v}_x|}{10.0 \text{ m}}\right)(10.0 \text{ m}) \qquad \text{Multiply both sides by 10.0 m.}$$

$$5.00 \text{ m} = |\mathbf{v}_x|$$

Since the x-component is pointing in the positive x-direction, $\mathbf{v}_x = +5.00$ m.

The absolute value of the y-component of the vector is the length of the side opposite the 60.0° angle. Therefore, to find the y-component,

$$\sin 60.0° = \frac{\text{side opposite } 60.0°}{\text{hypotenuse}}$$

Since

$$\text{hypotenuse} = 10.0 \text{ m}$$

$$\text{side opposite } 60.0° = |\mathbf{v}_y|$$

we have

$$\sin 60.0° = \frac{|\mathbf{v}_y|}{10.0 \text{ m}}$$

$$(\sin 60.0°)(10.0 \text{ m}) = \left(\frac{|\mathbf{v}_y|}{10.0 \text{ m}}\right)(10.0 \text{ m}) \qquad \text{Multiply both sides by 10.0 m.}$$

$$8.66 \text{ m} = |\mathbf{v}_y|$$

Since the y-component is pointing in the positive y-direction, $\mathbf{v}_y = +8.66$ m.

EXAMPLE 2

Find the x- and y-components of the vector $\mathbf{w} = 13.0$ km at 220.0°.

First, draw the vector in standard position [Fig. 5.39(a)]. Then, complete a right triangle with the x- and y-components being the two legs [Fig. 5.39(b)].

Find angle A as follows:

$$180° + A = 220.0°$$

$$A = 40.0°$$

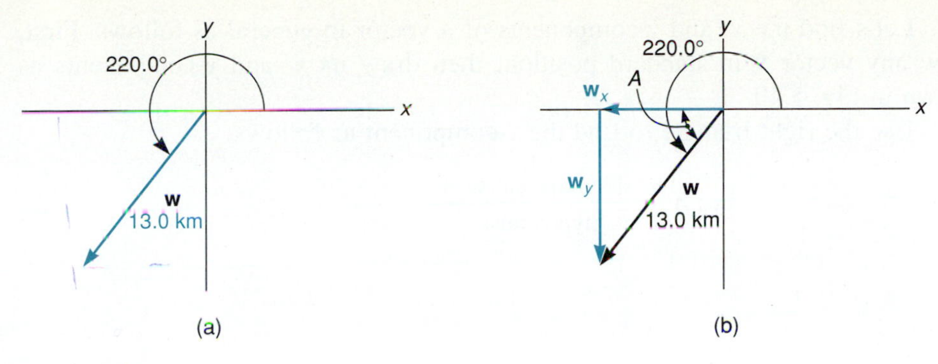

Figure 5.39

The absolute value of the x-component is the length of the side adjacent to angle A. Therefore, to find the x-component,

$$\cos A = \frac{\text{side adjacent to } A}{\text{hypotenuse}}$$

Since

$$\text{hypotenuse} = 13.0 \text{ km}$$
$$\text{side adjacent to } A = |\mathbf{w}_x|$$

we have

$$\cos 40.0° = \frac{|\mathbf{w}_x|}{13.0 \text{ km}}$$

$$(\cos 40.0°)(13.0 \text{ km}) = \left(\frac{|\mathbf{w}_x|}{13.0 \text{ km}}\right)(13.0 \text{ km}) \quad \text{Multiply both sides by 13.0 km.}$$

$$9.96 \text{ km} = |\mathbf{w}_x|$$

Since the x-component is pointing in the negative x-direction, $\mathbf{w}_x = -9.96$ km.

The absolute value of the y-component of the vector is the length of the side opposite angle A. Therefore, to find the y-component,

$$\sin A = \frac{\text{side opposite } A}{\text{hypotenuse}}$$

Since

$$\text{hypotenuse} = 13.0 \text{ km}$$
$$\text{side opposite } A = |\mathbf{w}_y|$$

we have

$$\sin 40.0° = \frac{|\mathbf{w}_y|}{13.0 \text{ km}}$$

$$(\sin 40.0°)(13.0 \text{ km}) = \left(\frac{|\mathbf{w}_y|}{13.0 \text{ km}}\right)(13.0 \text{ km}) \quad \text{Multiply both sides by 13.0 km.}$$

$$8.36 \text{ km} = |\mathbf{w}_y|$$

Since the y-component is pointing in the negative y-direction, $\mathbf{w}_y = -8.36$ km.

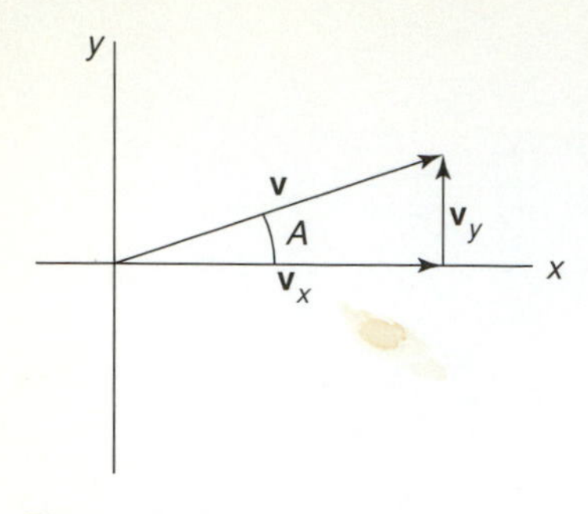

Figure 5.40
Vector **v** in standard position with its horizontal component $\mathbf{v}_x$ and its vertical component $\mathbf{v}_y$.

Let's find the x- and y-components of a vector in general as follows. First, draw any vector **v** in standard position; then draw its x- and y-components as shown in Fig. 5.40.

Use the right triangle to find the x-component as follows:

$$\cos A = \frac{\text{side adjacent to } A}{\text{hypotenuse}}$$

Or,

$$\cos A = \frac{|\mathbf{v}_x|}{|\mathbf{v}|}$$

$$(|\mathbf{v}|)(\cos A) = \left(\frac{|\mathbf{v}_x|}{|\mathbf{v}|}\right)|\mathbf{v}| \qquad \text{Multiply both sides by } |\mathbf{v}|.$$

$$(|\mathbf{v}|)(\cos A) = |\mathbf{v}_x|$$

Similarly, we use the right triangle to find the y-component as follows:

$$\sin A = \frac{\text{side opposite } A}{\text{hypotenuse}}$$

Or,

$$\sin A = \frac{|\mathbf{v}_y|}{|\mathbf{v}|}$$

$$(|\mathbf{v}|)(\sin A) = \left(\frac{|\mathbf{v}_y|}{|\mathbf{v}|}\right)|\mathbf{v}| \qquad \text{Multiply both sides by } |\mathbf{v}|.$$

$$(|\mathbf{v}|)(\sin A) = |\mathbf{v}_y|$$

The signs of the x- and y-components are determined by the quadrants in which the vector in standard position lies.

In general:

To find the x- and y-components of a vector **v** given in standard position:

1. Complete the right triangle with the legs being the x- and y-components of the vector.
2. Find the lengths of the legs of the right triangle as follows:

$$|\mathbf{v}_x| = (|\mathbf{v}|)(\cos A)$$
$$|\mathbf{v}_y| = (|\mathbf{v}|)(\sin A)$$

3. Determine the signs of the x- and y-components.

EXAMPLE 3

Find the x- and y-components of the vector $\mathbf{v} = 27.0$ ft/s at $125.0°$.

First, draw the vector in standard position [Fig. 5.41(a)]. Then, complete a right triangle, with the x- and y-components being the two legs [Fig. 5.41(b)]. Find angle A as follows:

$$A + 125.0° = 180°$$
$$A = 55.0°$$

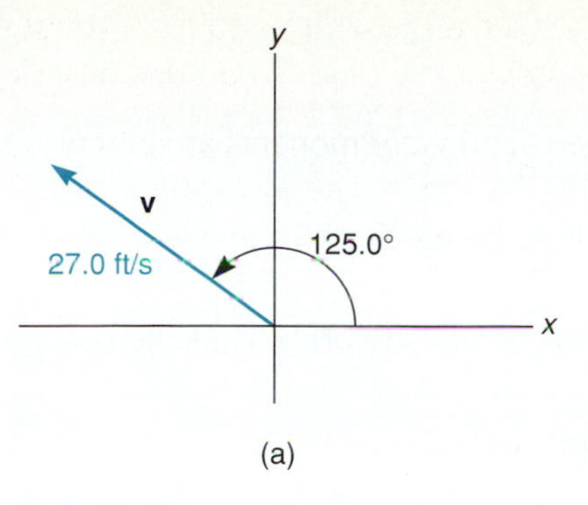

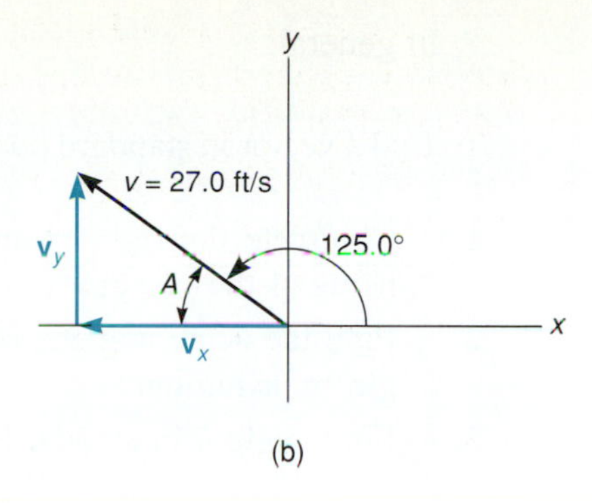

Figure 5.41

First find the x-component as follows:

$$|\mathbf{v}_x| = (|\mathbf{v}|)(\cos A)$$
$$|\mathbf{v}_x| = (27.0 \text{ ft/s})(\cos 55.0°)$$
$$= 15.5 \text{ ft/s}$$

Since the x-component is pointing in the negative x-direction,

$$\mathbf{v}_x = -15.5 \text{ ft/s}.$$

Then find the y-component as follows:

$$|\mathbf{v}_y| = (|\mathbf{v}|)(\sin A)$$
$$|\mathbf{v}_y| = (27.0 \text{ ft/s})(\sin 55.0°)$$
$$= 22.1 \text{ ft/s}$$

Since the y-component is pointing in the positive y-direction,

$$\mathbf{v}_y = +22.1 \text{ ft/s}.$$

Finding a Vector from Its Components

EXAMPLE 4

Find vector **R** in standard position with $\mathbf{R}_x = +3.00$ m and $\mathbf{R}_y = +4.00$ m.

First, graph the x- and y-components (Fig. 5.42) and complete the right triangle. The hypotenuse is the resultant vector. Find angle A as follows:

$$\tan A = \frac{\text{side opposite } A}{\text{side adjacent to } A}$$

$$\tan A = \frac{4.00 \text{ m}}{3.00 \text{ m}} = 1.333$$

$$A = 53.1° \qquad \text{(see Appendix B, Section B.5)}$$

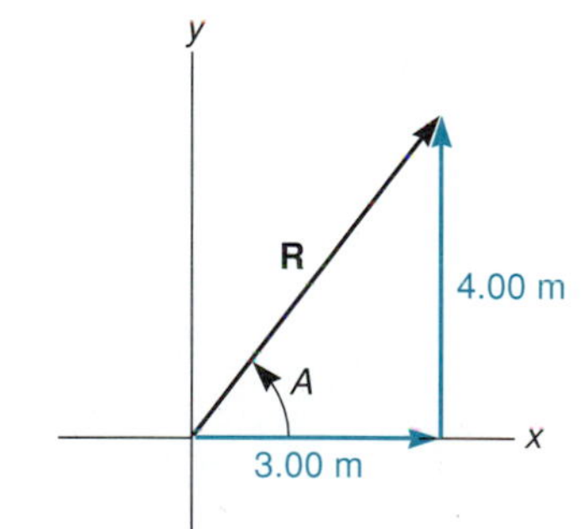

Figure 5.42

Find the magnitude of **R** using the Pythagorean theorem:

$$|\mathbf{R}| = \sqrt{|\mathbf{R}_x|^2 + |\mathbf{R}_y|^2}$$
$$|\mathbf{R}| = \sqrt{(3.00 \text{ m})^2 + (4.00 \text{ m})^2}$$
$$= 5.00 \text{ m}$$

That is, **R** = 5.00 m at 53.1°.

In general:

To find a vector in standard position when the x- and y-components are given:

1. Complete the right triangle with the legs being the x- and y-components of the vector.
2. Find the acute angle A of the right triangle whose vertex is at the origin by using $\tan A$.
3. Find angle θ in standard position as follows:

$$\theta = A \qquad (\theta \text{ in first quadrant})$$

$$\theta = 180° - A \qquad (\theta \text{ in second quadrant})$$

$$\theta = 180° + A \qquad (\theta \text{ in third quadrant})$$

$$\theta = 360° - A \qquad (\theta \text{ in fourth quadrant})$$

4. Find the magnitude of the vector using the Pythagorean theorem:

$$c = \sqrt{a^2 + b^2}$$

The Greek letter θ (theta) is often used to represent the measure of an angle.

EXAMPLE 5

Find vector **R** in standard position whose x-component is $+7.00$ mi and y-component is -5.00 mi.

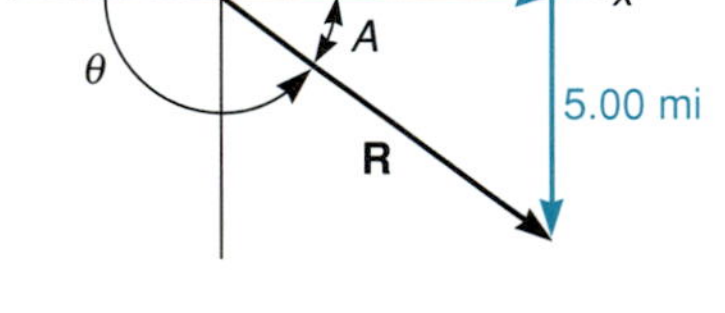

Figure 5.43

First, graph the x- and y-components (Fig. 5.43) and complete the right triangle. The hypotenuse is the resultant vector. Find angle A as follows:

$$\tan A = \frac{\text{side opposite } A}{\text{side adjacent to } A}$$

$$\tan A = \frac{5.00 \text{ mi}}{7.00 \text{ mi}} = 0.7143$$

$$A = 35.5°$$

But

$$\begin{aligned} \theta &= 360° - A \\ &= 360° - 35.5° \\ &= 324.5° \end{aligned}$$

Find the magnitude of **R** using the Pythagorean theorem:

$$|\mathbf{R}| = \sqrt{|\mathbf{R}_x|^2 + |\mathbf{R}_y|^2}$$

$$|\mathbf{R}| = \sqrt{(7.00 \text{ mi})^2 + (5.00 \text{ mi})^2}$$

$$= 8.60 \text{ mi}$$

That is, $\mathbf{R} = 8.60$ mi at $324.5°$.

EXAMPLE 6

Find vector **R** in standard position with $\mathbf{R}_x = -115$ km/h and $\mathbf{R}_y = +175$ km/h.

First, graph the x- and y-components (Fig. 5.44) and complete the right triangle. The hypotenuse is the resultant vector. Find angle A as follows:

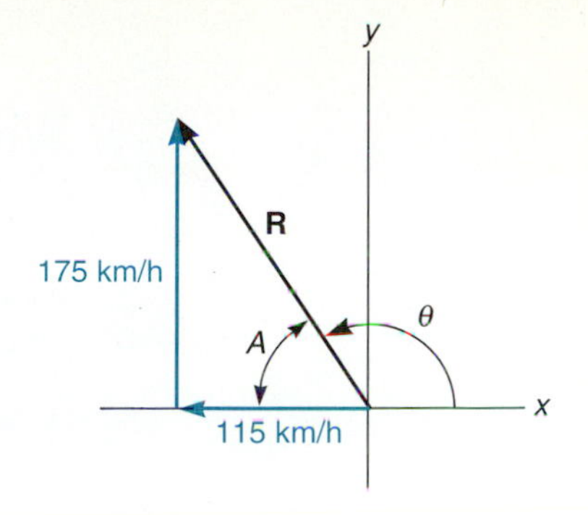

Figure 5.44

$$\tan A = \frac{\text{side oppposite } A}{\text{side adjacent to } A}$$

$$\tan A = \frac{175 \text{ km/h}}{115 \text{ km/h}} = 1.522$$

$$A = 56.7°$$

But

$$\theta = 180° - A$$
$$= 180° - 56.7°$$
$$= 123.3°$$

Find the magnitude of **R** using the Pythagorean theorem:

$$|\mathbf{R}| = \sqrt{|\mathbf{R}_x|^2 + |\mathbf{R}_y|^2}$$
$$|\mathbf{R}| = \sqrt{(115 \text{ km/h})^2 + (175 \text{ km/h})^2}$$
$$= 209 \text{ km/h}$$

That is, **R** = 209 km/h at 123.3°.

EXAMPLE 7

A plane is flying due north (at 90°) at 265 km/h. Suddenly there is a wind from the east (at 180°) at 55.0 km/h. What is the plane's velocity with respect to the ground in standard position?

First, graph the plane's velocity as the y-component and the wind velocity as the x-component (Fig. 5.45). The resultant vector is the plane's velocity with respect to the ground. Find angle A as follows:

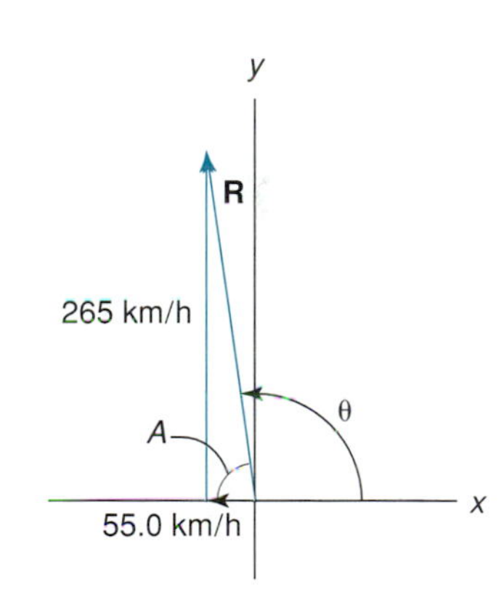

Figure 5.45

$$\tan A = \frac{\text{side opposite } A}{\text{side adjacent to } A}$$

$$\tan A = \frac{265 \text{ km/h}}{55.0 \text{ km/h}} = 4.818$$

$$A = 78.3°$$

then

$$\theta = 180° - 78.3° = 101.7°$$

Find the magnitude of the velocity (ground speed) using the Pythagorean theorem:

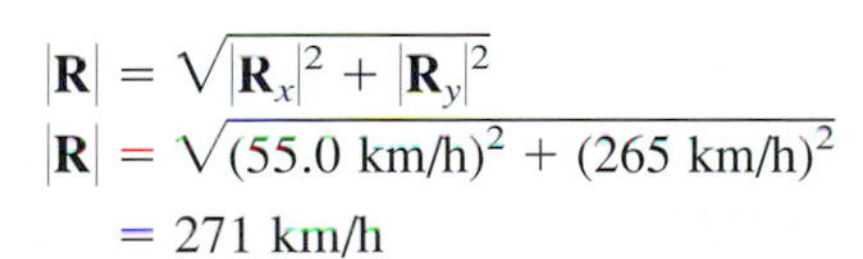

$$|\mathbf{R}| = \sqrt{|\mathbf{R}_x|^2 + |\mathbf{R}_y|^2}$$
$$|\mathbf{R}| = \sqrt{(55.0 \text{ km/h})^2 + (265 \text{ km/h})^2}$$
$$= 271 \text{ km/h}$$

That is, the velocity of the plane is 271 km/h at 101.7°.

EXAMPLE 8

A plane is flying northwest (at 135.0°) at 315 km/h. Suddenly there is a wind from 30.0° south of west (at 30.0°) at 65.0 km/h. What is the plane's velocity with respect to the ground in standard position?

First, graph the plane's velocity and the wind velocity as vectors in standard position (Fig. 5.46). The resultant vector is the plane's velocity with respect to the ground.

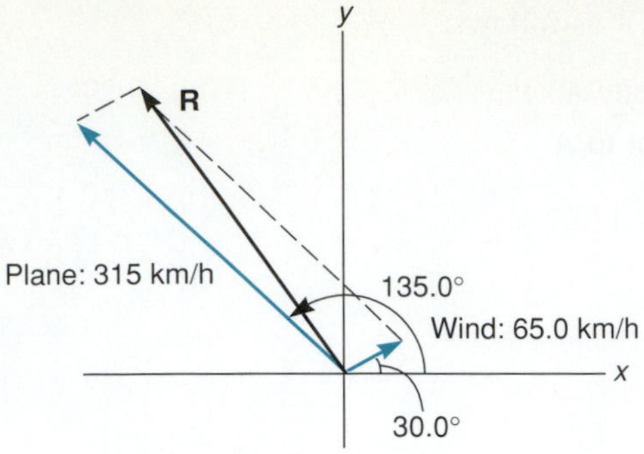

Figure 5.46

Then, find the x- and y-components of the plane's velocity and the wind velocity using Fig. 5.47.

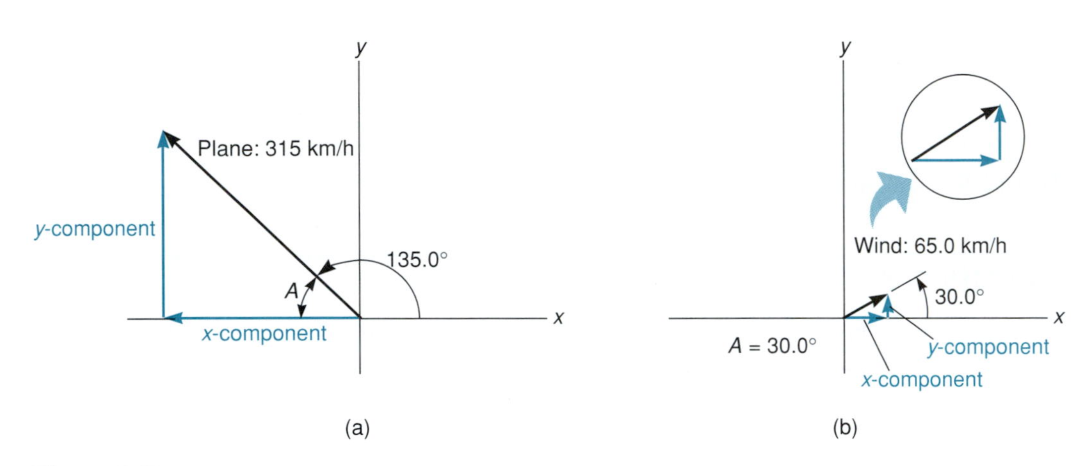

Figure 5.47

Plane: See Fig. 5.47(a).

$$A = 180° - 135.0° = 45.0°$$

To find:

x-*component*

$$\cos A = \frac{\text{side adjacent to } A}{\text{hypotenuse}}$$

$$\cos 45.0° = \frac{x\text{-component}}{315 \text{ km/h}}$$

$$(\cos 45.0°)(315 \text{ km/h}) = x\text{-component}$$

$$223 \text{ km/h} = x\text{-component}$$

$$\text{Thus, } x\text{-component} = -223 \text{ km/h}$$

y-*component*

$$\sin A = \frac{\text{side opposite } A}{\text{hypotenuse}}$$

$$\sin 45.0° = \frac{y\text{-component}}{315 \text{ km/h}}$$

$$(\sin 45.0°)(315 \text{ km/h}) = y\text{-component}$$

$$223 \text{ km/h} = y\text{-component}$$

$$y\text{-component} = +223 \text{ km/h}$$

Wind: See Fig. 5.47(b).

To find:

x-*component*

$$\cos A = \frac{\text{side adjacent to } A}{\text{hypotenuse}}$$

$$\cos 30.0° = \frac{x\text{-component}}{65.0 \text{ km/h}}$$

y-*component*

$$\sin A = \frac{\text{side opposite } A}{\text{hypotenuse}}$$

$$\sin 30.0° = \frac{y\text{-component}}{65.0 \text{ km/h}}$$

x-component	*y-component*
$(\cos 30.0°)(65.0 \text{ km/h}) = x\text{-component}$	$(\sin 30.0°)(65.0 \text{ km/h}) = y\text{-component}$
$56.3 \text{ km/h} = x\text{-component}$	$32.5 \text{ km/h} = y\text{-component}$
Thus, $x\text{-component} = +56.3 \text{ km/h}$	$y\text{-component} = +32.5 \text{ km/h}$

To find **R**:

	x-component	*y-component*	
Plane:	−223 km/h	+223 km/h	
Wind:	+56.3 km/h	+32.5 km/h	(Round each component sum
Sum:	−167 km/h	+256 km/h	to its least precise component.)

We find angle A from Fig. 5.48 as follows:

$$\tan A = \frac{\text{side opposite } A}{\text{side adjacent to } A}$$

$$\tan A = \frac{256 \text{ km/h}}{167 \text{ km/h}} = 1.533$$

$$A = 56.9°$$

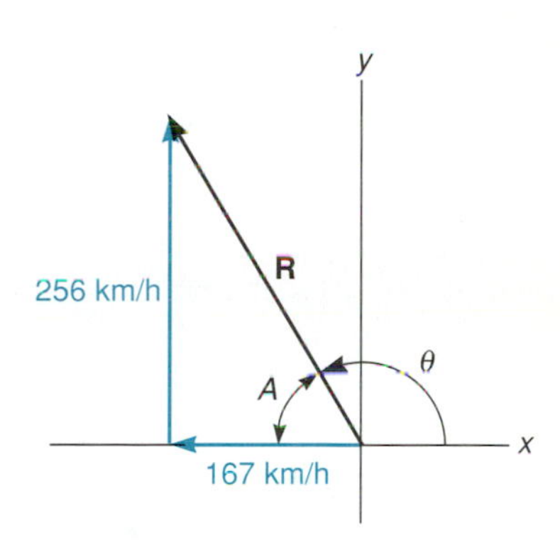

Figure 5.48

and

$$\theta = 180° - 56.9° = 123.1°$$

Find the magnitude of **R** using the Pythagorean theorem:

$$|\mathbf{R}| = \sqrt{|\mathbf{R}_x|^2 + |\mathbf{R}_y|^2}$$

$$|\mathbf{R}| = \sqrt{(167 \text{ km/h})^2 + (256 \text{ km/h})^2}$$

$$= 306 \text{ km/h}$$

That is, the velocity of the plane is 306 km/h at 123.1°.

■ PROBLEMS 5.3

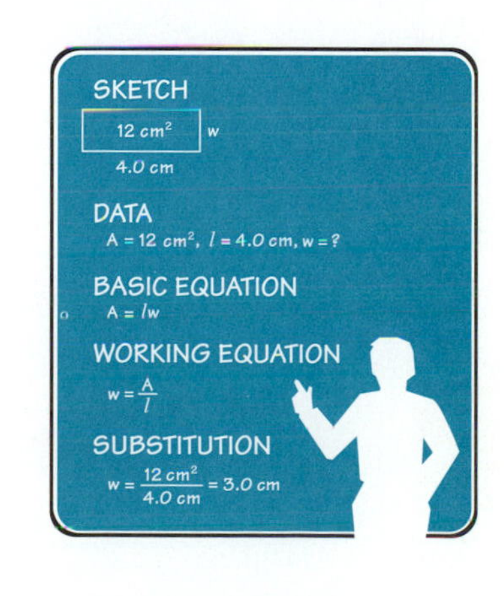

Make a sketch of each vector in standard position. Use the scale 1.0 cm = $1\overline{0}$ m.

1. **v** = $2\overline{0}$ m at 25°
2. **w** = 25 m at 125°
3. **u** = 25 m at 245°
4. **s** = $2\overline{0}$ m at 345°
5. **t** = 15 m at 105°
6. **r** = 35 m at 291°
7. **m** = $3\overline{0}$ m at 405°
8. **n** = 25 m at 525°

Find the x- and y-components of each vector.

9.

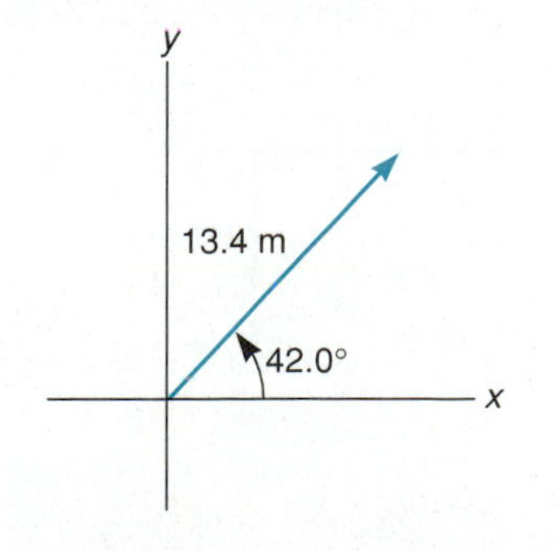

10.

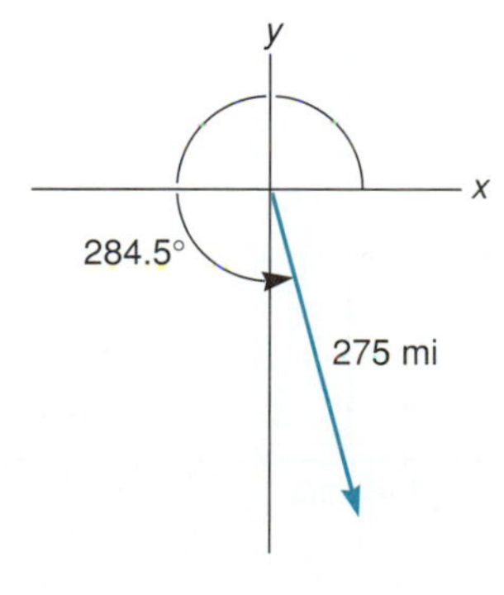

11. 

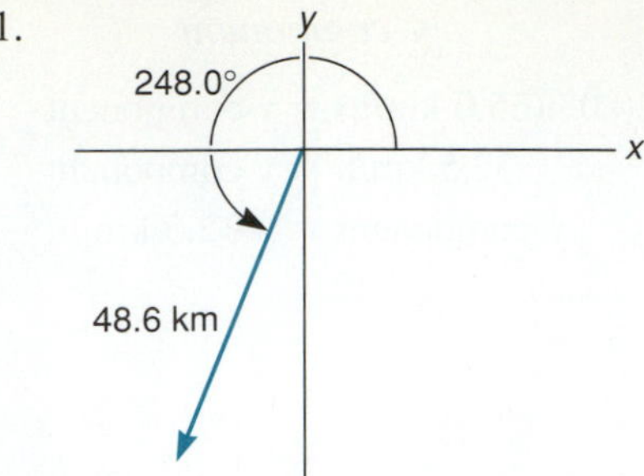

12.

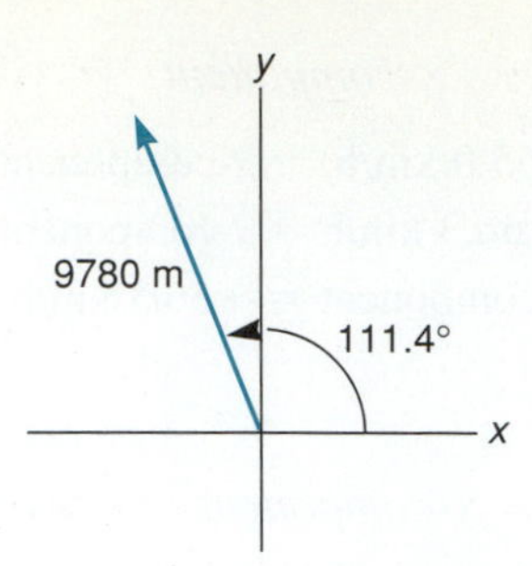

13.

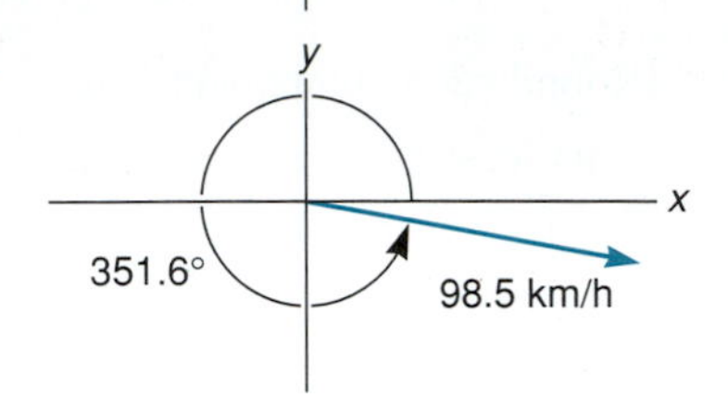

14.

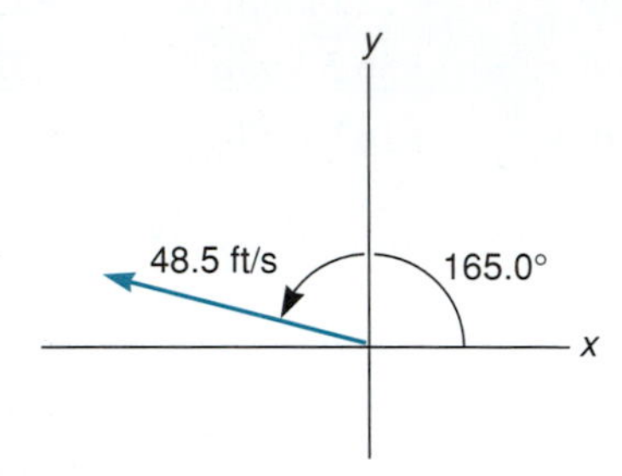

Find the x- and y-components of each vector given in standard position.

15. **v** = 38.9 m at 10.5°
16. **u** = 478 ft at 195.0°
17. **w** = 9.60 km at 310.0°
18. **s** = 5430 mi at 153.7°
19. **t** = 29.5 m/s at 101.5°
20. **m** = 154 mi/h at 273.2°

In Problems 21 through 32, find each resultant vector **R.** Give **R** in standard position.

21.

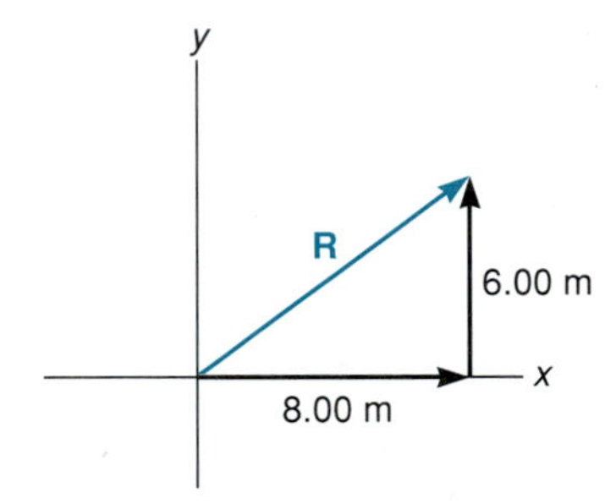

22.

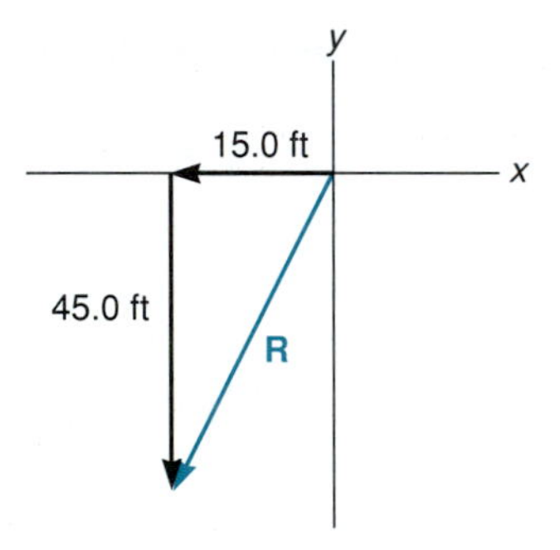

23.

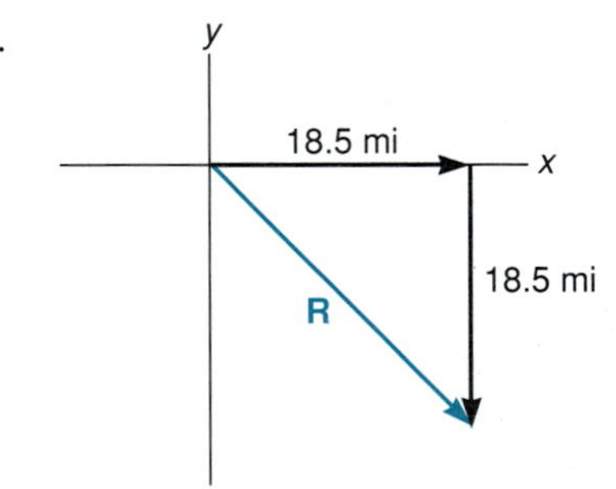

24.

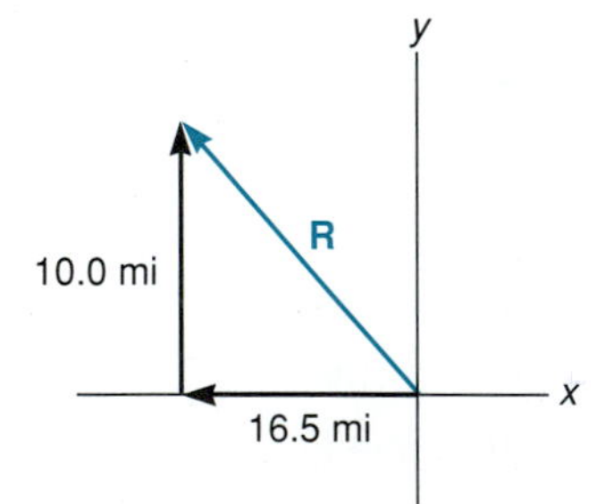

25.

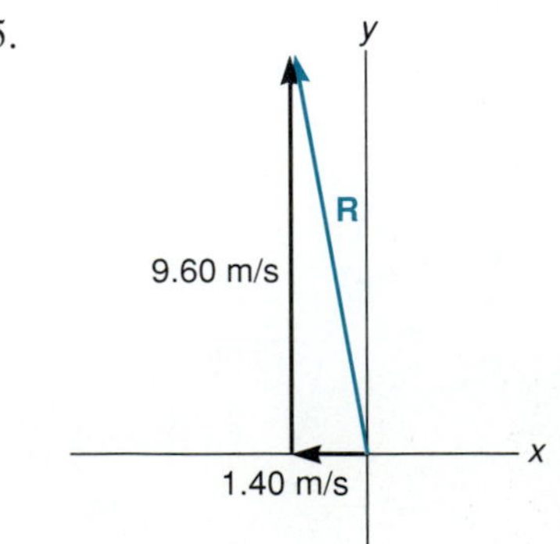

26.

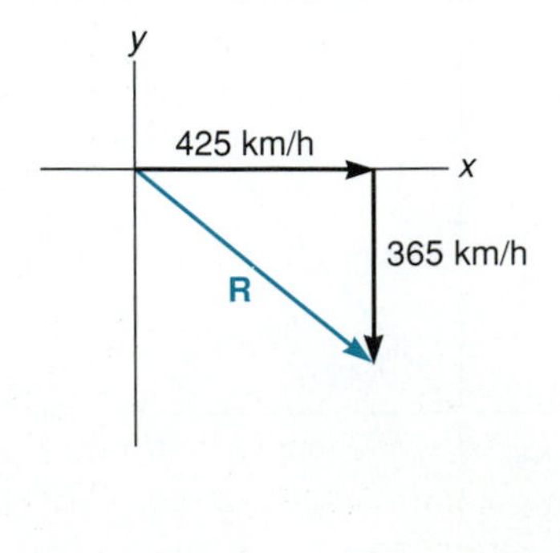

	x-*component*	y-*component*
27.	+19.5 m	−49.6 m
28.	−158 km	+236 km
29.	+14.7 mi	+16.8 mi
30.	−3240 ft	−1890 ft
31.	−9.65 m/s	+4.36 m/s
32.	+375 km/h	−408 km/h

33. A plane is flying due north at 325 km/h. Suddenly there is a wind from the south at 45 km/h. What is the plane's velocity with respect to the ground in standard position?
34. A plane is flying due west at 275 km/h. Suddenly there is a wind from the west at $8\bar{0}$ km/h. What is the plane's velocity with respect to the ground in standard position?
35. A plane is flying due west at 235 km/h. Suddenly there is a wind from the north at 45.0 km/h. What is the plane's velocity with respect to the ground in standard position?
36. A plane is flying due north at 185 mi/h. Suddenly there is a wind from the west at 35.0 mi/h. What is the plane's velocity with respect to the ground in standard position?
37. A plane is flying southwest at 155 mi/h. Suddenly there is a wind from the west at 45.0 mi/h. What is the plane's velocity with respect to the ground in standard position?
38. A plane is flying southeast at 215 km/h. Suddenly there is a wind from the north at 75.0 km/h. What is the plane's velocity with respect to the ground in standard position?
39. A plane is flying at 25.0° north of west at $19\bar{0}$ km/h. Suddenly there is a wind from 15.0° north of east at 45.0 km/h. What is the plane's velocity with respect to the ground in standard position?
40. A plane is flying at 36.0° south of west at $15\bar{0}$ mi/h. Suddenly there is a wind from 75.0° north of east at 55.0 mi/h. What is the plane's velocity with respect to the ground in standard position? ■

GLOSSARY

Component Vector When two or more vectors are added, each of the vectors is called a component of the resultant, or sum, vector. (p. 123)

Cosine *A* The ratio of the side adjacent to angle *A* to the hypotenuse of a right triangle. (p. 115)

Number Plane A plane determined by the horizontal line called the *x*-axis and a vertical line called the *y*-axis intersecting at right angles. These two lines divide the number plane into four quadrants. The *x*-axis contains positive numbers to the right of the origin and negative numbers to the left of the origin. The *y*-axis contains positive numbers above the origin and negative numbers below the origin. (p. 122)

Pythagorean Theorem A theorem from geometry that states that the sum of the squares of the legs of a right triangle is equal to the square of the hypotenuse ($c^2 = a^2 + b^2$). (p. 118)

Ratio A comparison of two quantities by division. (p. 115)

Resultant Vector The sum of two or more vectors. (p. 123)

Right Triangle A triangle with one right angle (90°), two acute angles (less than 90°), two legs, and a hypotenuse (the side opposite the right angle). (p. 115)

Scalar A quantity that is completely described when its magnitude (size) is given. (p. 122)

Sine A The ratio of the side opposite angle A to the hypotenuse of a right triangle. (p. 115)

Standard Position A vector is in standard position when its initial point is at the origin of the number plane. The vector is expressed in terms of its length and its angle θ, where θ is measured counterclockwise from the positive x-axis to the vector. (p. 131)

Tangent A The ratio of the side opposite angle A to the side adjacent to angle A in a right triangle. (p. 115)

Vector A quantity that requires both magnitude and direction to be completely described. (p. 122)

x-component The horizontal component of a vector that lies along the x-axis. (p. 123)

y-component The vertical component of a vector that lies along the y-axis. (p. 123)

FORMULAS

5.1 $\sin A = \dfrac{\text{side opposite angle } A}{\text{hypotenuse}}$

$$\cos A = \frac{\text{side adjacent to angle } A}{\text{hypotenuse}}$$

$$\tan A = \frac{\text{side opposite angle } A}{\text{side adjacent to angle } A}$$

$$c = \sqrt{a^2 + b^2}$$

$$a = \sqrt{c^2 - b^2}$$

$$b = \sqrt{c^2 - a^2}$$

5.3 To find the x- and y-components of a vector **v** given in standard position:

1. Complete the right triangle with the legs being the x- and y-components of the vector.
2. Find the lengths of the legs of the right triangle as follows:

$$|\mathbf{v}_x| = (|\mathbf{v}|)(\cos A)$$
$$|\mathbf{v}_y| = (|\mathbf{v}|)(\sin A)$$

3. Determine the signs of the x and y components.

To find vector **R** in standard position when the x- and y-components are given:

1. Complete the right triangle with the legs being the x- and y-components of the vector.
2. Find the acute angle A of the right triangle whose vertex is at the origin by using $\tan A$.
3. Find angle θ in standard position as follows:

$\theta = A$	(θ in first quadrant)
$\theta = 180° - A$	(θ in second quadrant)
$\theta = 180° + A$	(θ in third quadrant)
$\theta = 360° - A$	(θ in fourth quadrant)

4. Find the magnitude of the vector using the Pythagorean theorem:

$$\mathbf{R} = \sqrt{|\mathbf{R}_x|^2 + |\mathbf{R}_y|^2}$$

REVIEW QUESTIONS

1. A right triangle
   (a) always has two equal sides. (b) always has two equal angles.
   (c) always has two equal sides and angles. (d) has one angle of 90°.
2. A triangle has one angle of 90°. The sum of the other two angles is _____.
   (a) 45° (b) 90° (c) 180° (d) 360°
3. When adding vectors, the order in which they are added
   (a) is not important. (b) is important.
   (c) is important only in certain cases.
4. A vector is in standard position when its initial point is
   (a) at the origin. (b) along the x-axis. (c) along the y-axis.
5. How can you identify the hypotenuse in a right triangle?
6. State the names of three trigonometric ratios.
7. State the Pythagorean theorem in your own words.
8. Can the length of the third leg of a right triangle be found using the Pythagorean theorem if the lengths of any two legs are given?
9. Discuss number plane, origin, and axis in your own words.
10. Can every vector be described in terms of its components?
11. Can a vector have more than one set of component vectors?
12. Describe how to add two or more vectors graphically.
13. Describe how to find a resultant vector if given its x- and y-components.
14. Is a vector limited to a single position in the number plane?
15. Is the angle of a vector in standard position measured clockwise or counterclockwise?
16. What are the limits on the angle measure of a vector in standard position in the third quadrant?
17. Describe how to find the x- and y-components of a vector given in standard position.
18. Describe how to find a vector in standard position when the x- and y-components are given.

REVIEW PROBLEMS

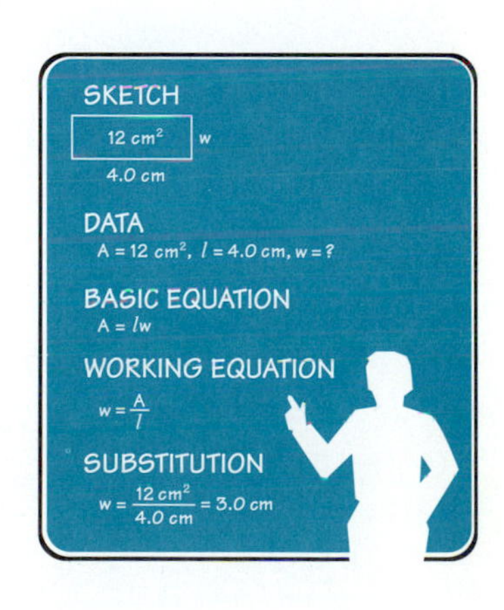

Use a calculator to find each trigonometric ratio rounded to four significant digits.

1. sin 56°
2. cos 66.65°
3. tan 75.125°

Use a calculator to find each angle rounded to the nearest tenth of a degree.

4. $\cos A = 0.3429$
5. $\tan B = 1.197$
6. $\sin B = 0.2994$

Solve each triangle (find the missing angles and sides).

7.

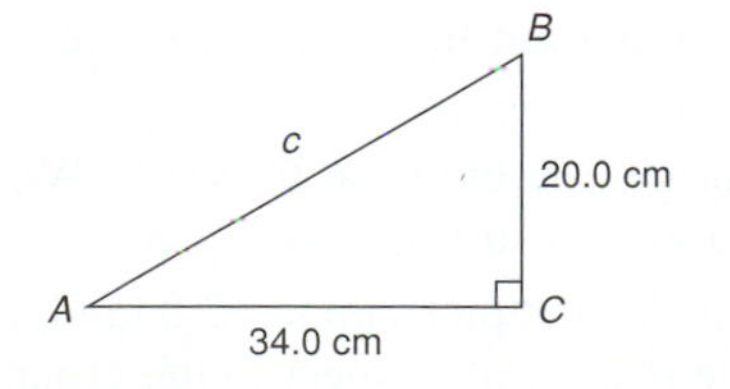

8.

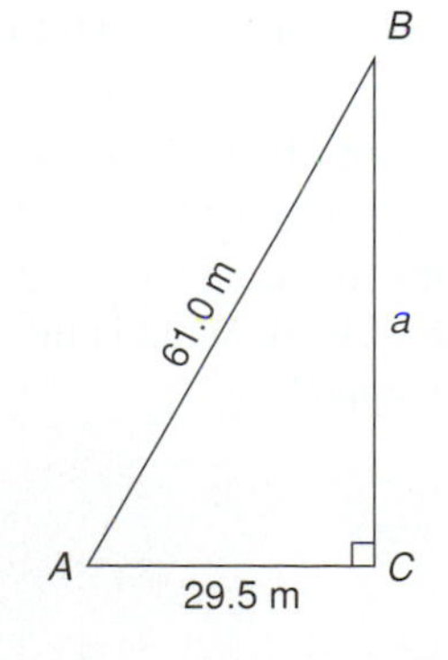

9.

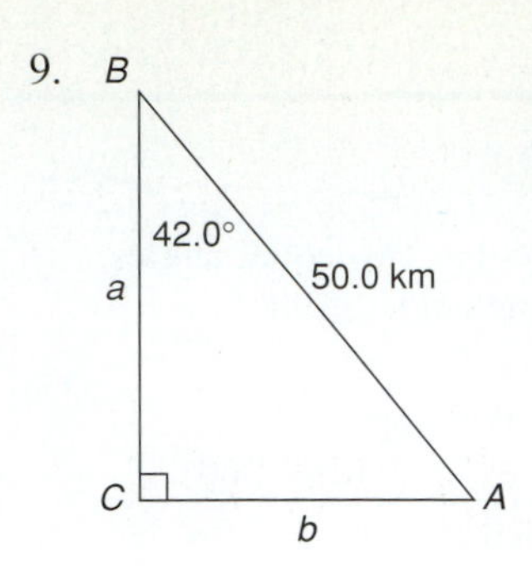

10.

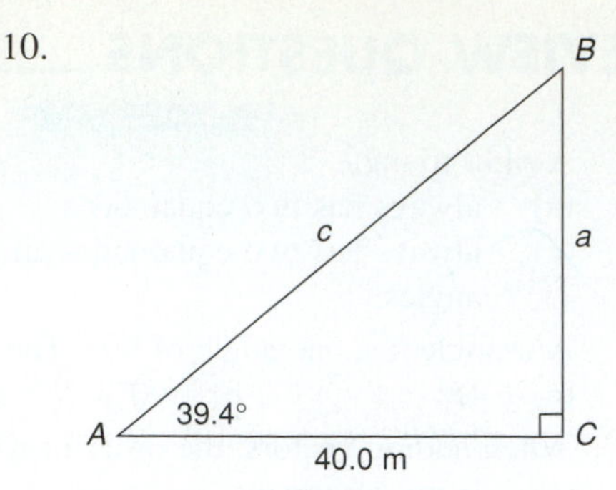

11. A right triangle has legs of 14.0 cm and 9.00 cm. What is the length of the hypotenuse?
12. A right triangle has a hypotenuse of 0.750 km and one side of 0.350 km. What is the length of the other side?
13. A right triangle, ABC, has sides $BC = 22.0$ ft and $AC = 14.0$ ft. Angle C is a right angle. Solve for all other angles and sides of the triangle.
14. Vector **R** has x-component $= +14.0$ and y-component $= +3.00$. What is the length of vector **R?**
15. Vectors **A, B,** and **C** are given. Vector **A** has x-component $= +3.00$ and y-component $= +4.00$. Vector **B** has x-component $= +5.00$ and y-component $= -7.00$. Vector **C** has x-component $= -2.00$ and y-component $= +1.00$. Find the resultant vector **R.**
16. Graph and find the x- and y-*components of the resultant vector* ***R.***

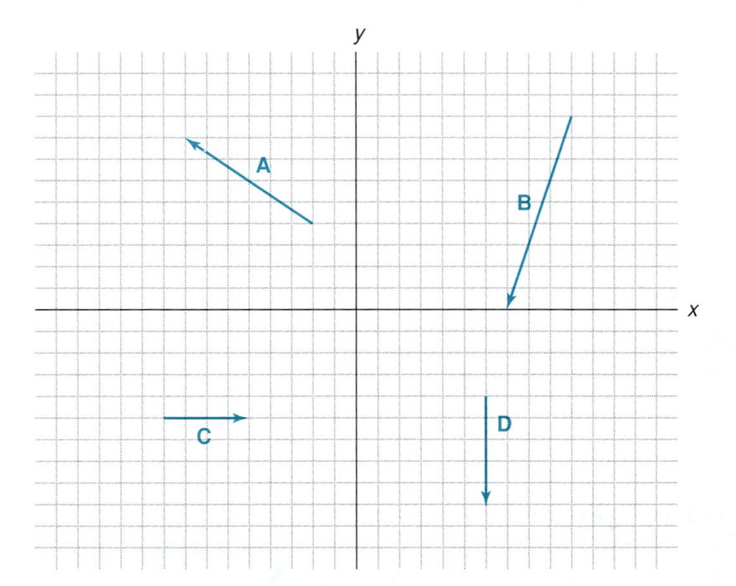

17. Find the x- and y-components of vector **R,** which has a length 13.0 mm at 30.0°.
18. Vector **R** has a length of 9.00 cm at 240.0°. What are the x- and y-components of **R?**
19. A plane flies north at 215 km/h. A wind from the east blows at 69 km/h. What is the plane's velocity with respect to the ground in standard position?
20. A glider flies southeast (at 320.0°) at 25.0 km/h. A wind picks up to 12.0 km/h from 15.0° south of west. What is the velocity of the plane with respect to the ground in standard position?

Chapter 6

CONCURRENT FORCES

Not all forces actually cause motion. A body that is static, or not moving, may be in equilibrium. Huge forces may be acting on a bridge without producing motion. We will now consider concurrent forces (forces acting on the same point) in equilibrium.

OBJECTIVES

The major goals of this chapter are to enable you to:

1. Find the vector sum of concurrent forces graphically and using the component method.
2. Understand equilibrium in one dimension.
3. Analyze concurrent force situations using free-body diagrams.
4. Distinguish compression and tension.

6.1 FORCES IN TWO DIMENSIONS

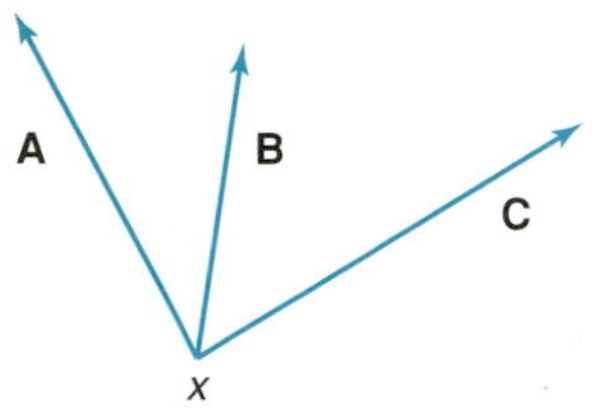

Figure 6.1
Concurrent forces are applied to or act on the same point.

Concurrent forces are those forces that are applied to, or acting on, the same point as in Fig. 6.1.

When two or more forces act at the same point, the resultant force is the sum of the forces applied at that point. The **resultant force** is the single force that has the same effect as the two or more forces acting together.

As we saw in Section 4.4, when forces act in the same or opposite directions (in one dimension), the total force can be found by adding the forces that act in one direction and subtracting the forces that act in the opposite direction. What is the result when the forces are acting in two dimensions? In Section 5.2, addition of two vectors was shown by connecting the end point of the first vector to the initial point of the second vector as shown in Fig. 6.2. This is often called the *vector triangle method*.

$\mathbf{R} = \mathbf{v} + \mathbf{w}$

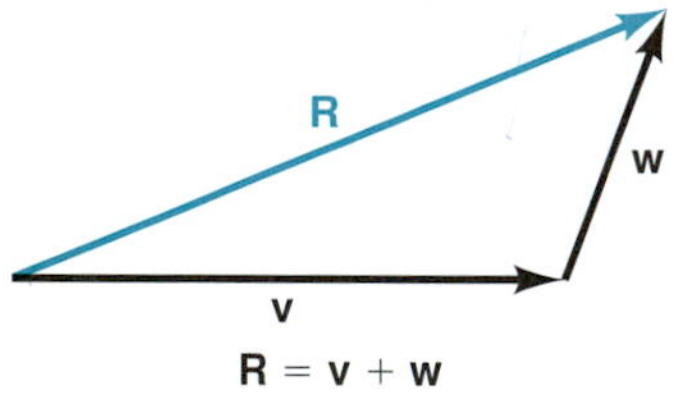

Figure 6.2
Vector triangle method of adding two vectors.

This sum may also be obtained by constructing a parallelogram using the two vectors as adjacent sides and then constructing the opposite sides parallel as shown in Fig. 6.3. The diagonal of the parallelogram is the resultant, or sum, of the two vectors. This is often called the *parallelogram method*.

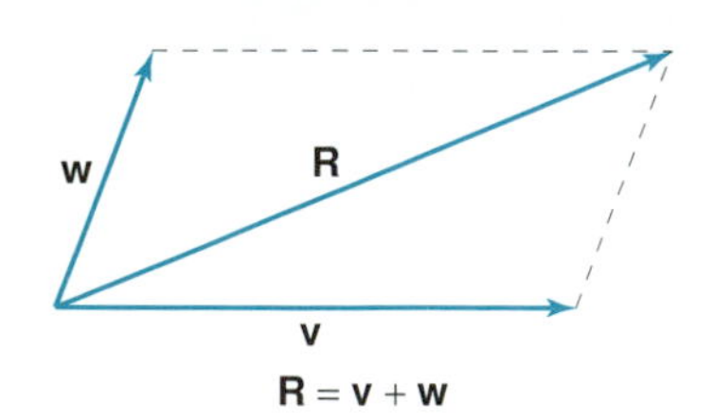

Figure 6.3
Parallelogram method of adding two vectors.

EXAMPLE 1

Suppose two workers move a large crate by applying two ropes at the same point. The first worker applies a force of 525 N while the second worker applies a force of 763 N at the same point at right angles as shown in Fig. 6.4. Find the resultant force.

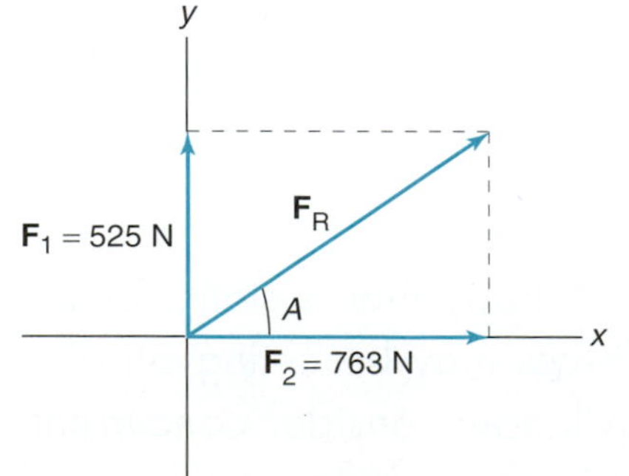

Figure 6.4

To find $\mathbf{F}_R$, find the x- and y-components of each vector and add the components as follows:

Vector	x-*component*	y-*component*
$\mathbf{F}_1$	0 N	525 N
$\mathbf{F}_2$	763 N	0 N
$\mathbf{F}_R$	763 N	525 N

Find angle A as follows:

$$\tan A = \frac{\text{side opposite } A}{\text{side adjacent to } A} = \frac{|\mathbf{F}_{Ry}|}{|\mathbf{F}_{Rx}|}$$

$$\tan A = \frac{525 \text{ N}}{763 \text{ N}} \quad \text{(In a parallelogram, opposite sides are equal.)}$$

$$= 0.6881$$

$$A = 34.5°$$

Find the magnitude of $\mathbf{F}_R$ using the Pythagorean theorem:

$$|\mathbf{F}_R| = \sqrt{|\mathbf{F}_{Rx}|^2 + |\mathbf{F}_{Ry}|^2}$$

$$|\mathbf{F}_R| = \sqrt{(763 \text{ N})^2 + (525 \text{ N})^2}$$

$$= 926 \text{ N}$$

That is, $\mathbf{F}_R = 926$ N at 34.5°.

EXAMPLE 2

Suppose two workers move a large crate by applying two ropes at the same point. The first worker applies a force of 525 N while the second worker applies a force of 763 N at the same point as shown in Fig. 6.5. Find the resultant force.

Figure 6.5

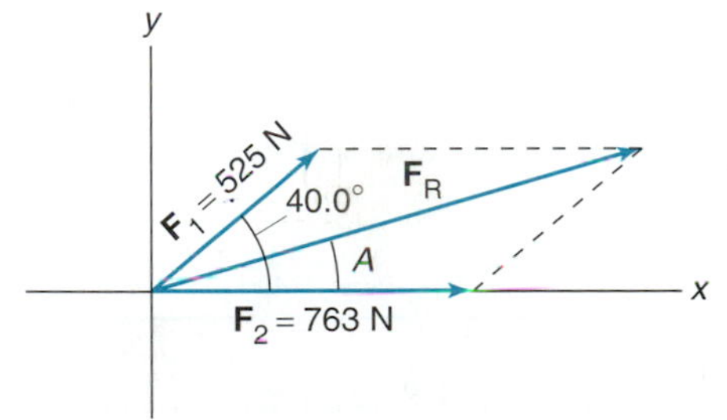

To find $\mathbf{F}_R$, find the x- and y-components of each vector and add the components as follows:

Vector	x-*component*	y-*component*
$\mathbf{F}_1$	$|\mathbf{F}_1| \cos \theta =$ (525 N) cos 40.0° = 402 N	$|\mathbf{F}_1| \sin \theta =$ (525 N) sin 40.0° = 337 N
$\mathbf{F}_2$	763 N	0 N
$\mathbf{F}_R$	1165 N	337 N

Find angle A as follows:

$$\tan A = \frac{|\mathbf{F}_{Ry}|}{|\mathbf{F}_{Rx}|}$$

$$\tan A = \frac{337 \text{ N}}{1165 \text{ N}} \quad \text{(In a parallelogram, opposite sides are equal.)}$$

$$= 0.2893$$

$$A = 16.1°$$

Find the magnitude of $\mathbf{F}_R$ using the Pythagorean theorem:

$$|\mathbf{F}_R| = \sqrt{|\mathbf{F}_{Rx}|^2 + |\mathbf{F}_{Ry}|^2}$$
$$|\mathbf{F}_R| = \sqrt{(1165 \text{ N})^2 + (337 \text{ N})^2}$$
$$= 1210 \text{ N}$$

That is, $\mathbf{F}_R = 1210$ N at 16.1°.

EXAMPLE 3

Forces of $\mathbf{F}_1 = 375$ N, $\mathbf{F}_2 = 575$ N, and $\mathbf{F}_3 = 975$ N are applied to the same point. The angle between $\mathbf{F}_1$ and $\mathbf{F}_2$ is 60.0° while the angle between $\mathbf{F}_2$ and $\mathbf{F}_3$ is 80.0°. $\mathbf{F}_2$ is between $\mathbf{F}_1$ and $\mathbf{F}_3$. Find the resultant force.

First, draw a force diagram as in Fig. 6.6. Place the point of application at the origin and one of the forces on the x-axis for ease in computing the components.

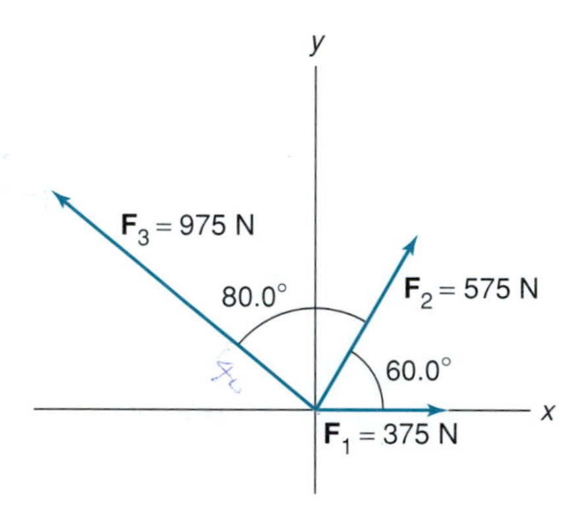

Figure 6.6

To find $\mathbf{F}_R$, find the x- and y-components of each vector and add the components as follows:

Vector	x-*component*		y-*component*	
$\mathbf{F}_1$		375 N		0 N
$\mathbf{F}_2$	$|\mathbf{F}_2| \cos A =$		$|\mathbf{F}_2| \sin A =$	
	$(575 \text{ N}) \cos 60.0° =$	288 N	$(575 \text{ N}) \sin 60.0° =$	498 N
$\mathbf{F}_3$	$|\mathbf{F}_3| \cos A =$		$|\mathbf{F}_3| \sin A =$	
	$(975 \text{ N}) \cos 40.0° =$	−747 N	$(975 \text{ N}) \sin 40.0° =$	627 N
***Note:* $A = 180° - 140.0° = 40.0°$**				
$\mathbf{F}_R$		−84 N		1125 N

Find angle A of the resultant or sum vector as follows:

$$\tan A = \frac{|\mathbf{F}_{Ry}|}{|\mathbf{F}_{Rx}|}$$
$$\tan A = \frac{1125 \text{ N}}{84 \text{ N}}$$
$$= 13.39$$
$$A = 85.7°$$

Note: The x-component of $\mathbf{F}_R$ is negative and its y-component is positive; this means that $\mathbf{F}_R$ is in the second quadrant. And, its angle in standard position is $180° - 85.7° = 94.3°$.

Find the magnitude of $\mathbf{F}_R$ using the Pythagorean theorem:

$$|\mathbf{F}_R| = \sqrt{|\mathbf{F}_{Rx}|^2 + |\mathbf{F}_{Ry}|^2}$$
$$|\mathbf{F}_R| = \sqrt{(84 \text{ N})^2 + (1125 \text{ N})^2}$$
$$= 1130 \text{ N}$$

That is, $\mathbf{F}_R = 1130$ N at 94.3°, or 94.3° from $\mathbf{F}_1$.

The resultant vector is shown in Fig. 6.7.

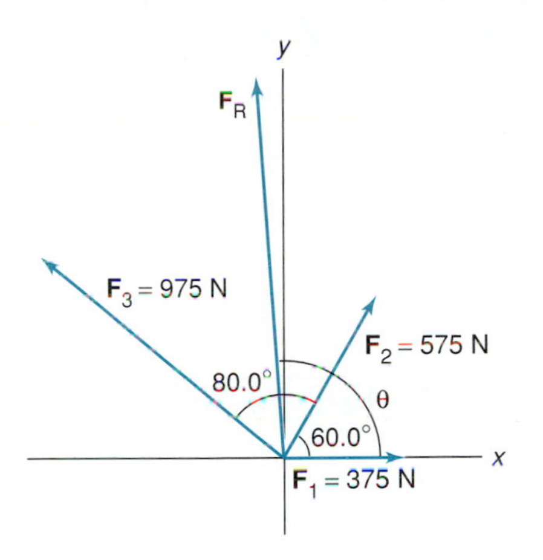

Figure 6.7

■ PROBLEMS 6.1

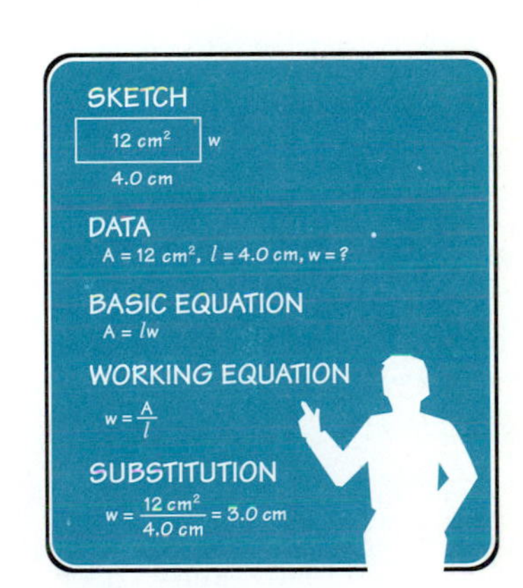

Find the sum of each set of forces acting on the same point in a straight line.

1. 355 N (right); 475 N (right); 245 N (left); 555 N (left)
2. 703 N (right); 829 N (left); 125 N (left); 484 N (left)
3. Forces of 225 N and 175 N act on the same point.
   (a) What is the magnitude of the maximum net force the two forces can exert together?
   (b) What is the magnitude of the minimum net force the two forces can exert together?
4. Three forces with magnitudes of 225 N, 175 N, and 125 N act on the same point.
   (a) What is the magnitude of the maximum net force the three forces can exert together?
   (b) What is the magnitude of the minimum net force the three forces can exert together?

Find the sum of each set of vectors. Give angles in standard position.

5.

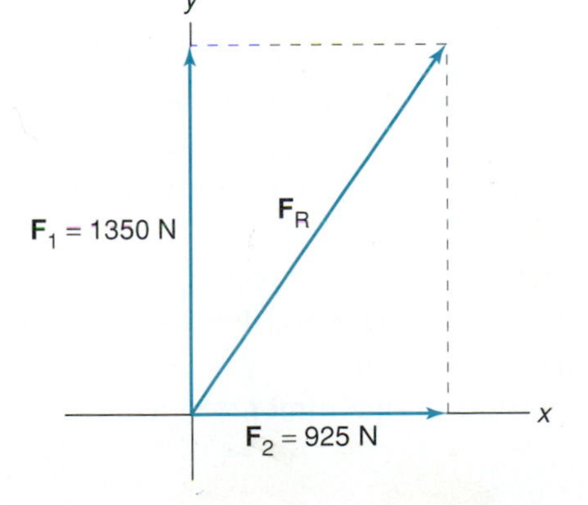

6.

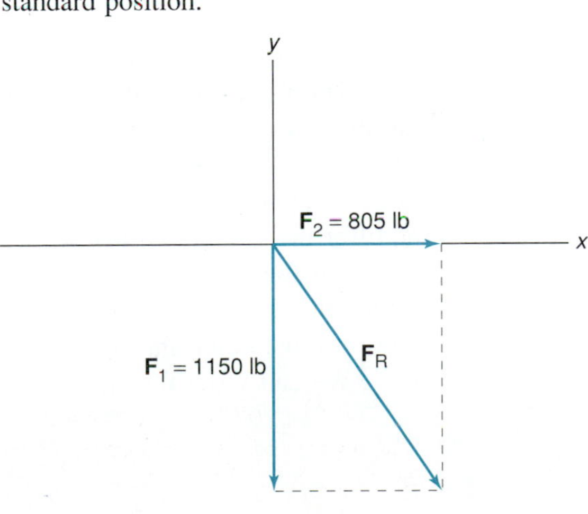

7.

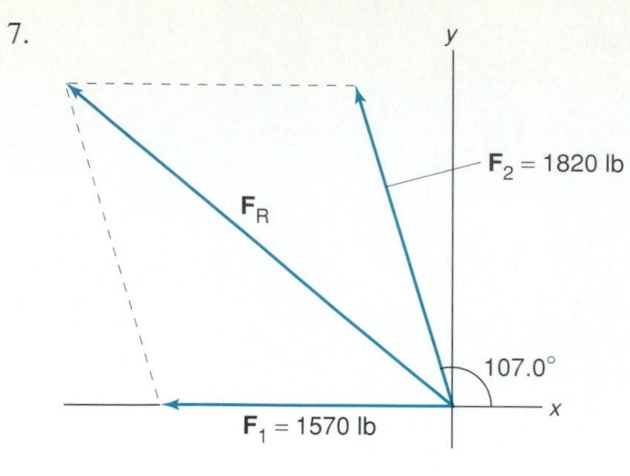

8.

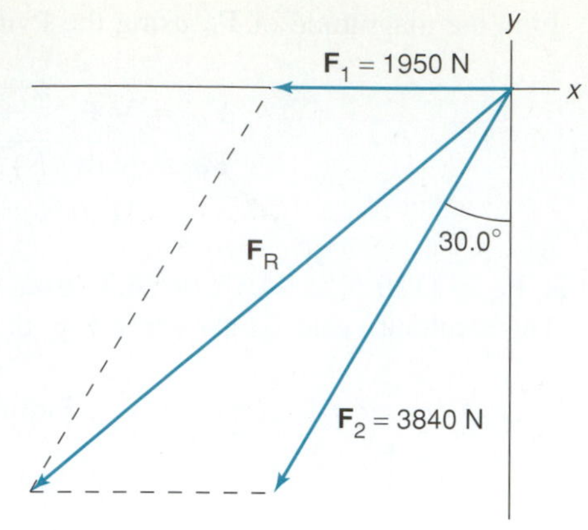

9.

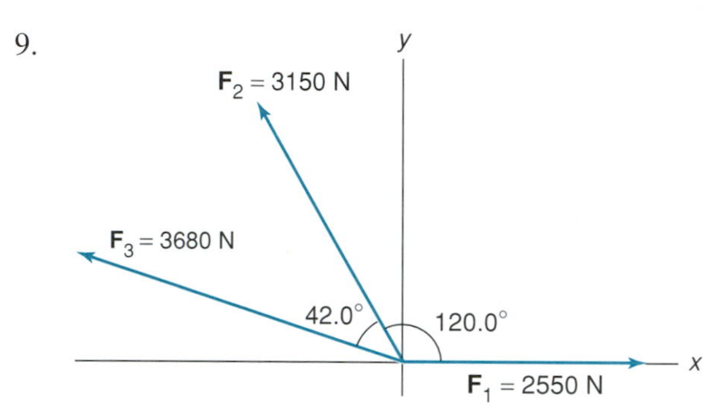

10.

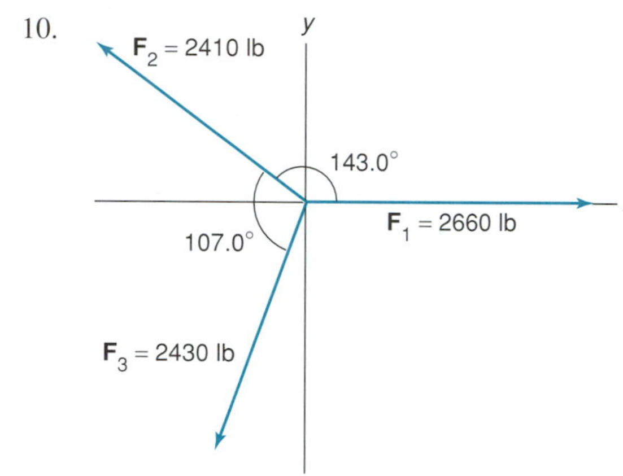

11. Forces of $\mathbf{F}_1 = 1150$ N, $\mathbf{F}_2 = 875$ N, and $\mathbf{F}_3 = 1450$ N are applied to the same point. The angle between $\mathbf{F}_1$ and $\mathbf{F}_2$ is 90.0°, and the angle between $\mathbf{F}_2$ and $\mathbf{F}_3$ is 120.0°. $\mathbf{F}_2$ is between $\mathbf{F}_1$ and $\mathbf{F}_3$. Find the resultant force.
12. Four forces, each of magnitude 2750 lb, act on the same point. The angle between adjacent forces is 30.0°. Find the resultant force. ■

6.2 CONCURRENT FORCES IN EQUILIBRIUM

Equilibrium in One Dimension

In Chapter 4 we studied forces that actually caused motion. A force may be applied to an object but not actually cause motion (Fig. 6.8). The forces acting on the bridge do not actually cause motion because they balance each other out. The weight is balanced by the supporting force. Because the forces balance each other out, the net force on the bridge is zero.

There are many important engineering situations in which the net force on an object is zero. *An object is said to be in* **equilibrium** *when the net force acting on it is zero.* A body is in equilibrium when it is not accelerating; that is, the body is either at rest or moving at a constant velocity. The study of objects in equilibrium is called **statics.**

Let the forces applied to an object act in the same direction or in opposite directions (in one dimension). For the net force to be zero, the forces in one direction must equal the forces in the opposite direction.

We can write the equation for equilibrium in one dimension as

$$\mathbf{F}_+ = \mathbf{F}_-$$

Figure 6.8
The forces shown do not cause motion, but balance each other.

where $\mathbf{F}_+$ is the sum of all forces acting in one direction (call it the positive direction) and $\mathbf{F}_-$ is the sum of all the forces acting in the opposite (negative) direction.

EXAMPLE 1

A cable supports a large crate of weight 1250 N (Fig. 6.9). What is the upward force on the crate if it is in equilibrium?

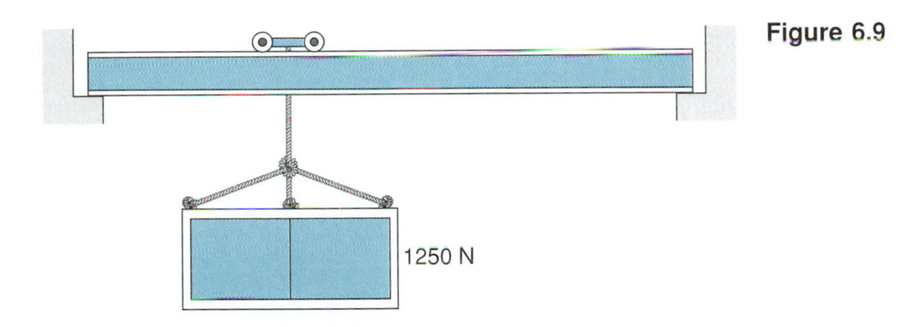

Figure 6.9

Sketch: It is helpful to draw a "free-body diagram" of the crate. This is a sketch in which we draw only the object in equilibrium and show the forces that act on it. Note that we call the upward direction positive as indicated by the arrow.

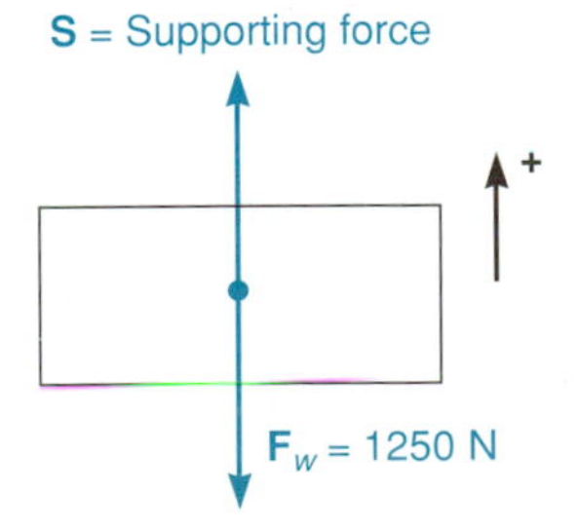

Data:

$$\mathbf{F}_w = 1250 \text{ N}$$
$$\mathbf{S} = ?$$

Basic Equation:

$$\mathbf{F}_+ = \mathbf{F}_-$$

Working Equation:

$$\mathbf{S} = \mathbf{F}_w$$

Substitution:

$$\mathbf{S} = 1250 \text{ N}$$

EXAMPLE 2

Four persons are having a tug-of-war with a rope. Harry and Mary are on the left; Bill and Jill are on the right. Mary pulls with a force of 105 lb, Harry pulls with a force of 255 lb, and Jill pulls with a force of 165 lb. With what force must Bill pull to produce equilibrium?

Sketch:

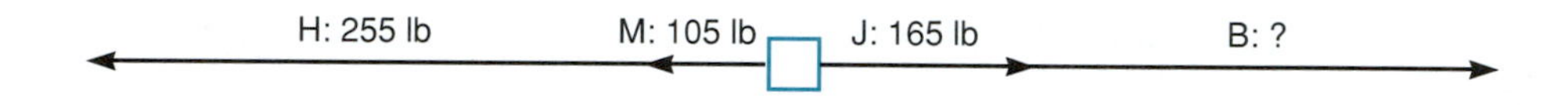

Data:

$$M = 105 \text{ lb}$$
$$H = 255 \text{ lb}$$
$$J = 165 \text{ lb}$$
$$B = ?$$

Basic Equation:

$$\mathbf{F}_+ = \mathbf{F}_- \text{ or}$$
$$M + H = J + B$$

Working Equation:

$$B = M + H - J$$

Substitution:

$$B = 105 \text{ lb} + 255 \text{ lb} - 165 \text{ lb}$$
$$= 195 \text{ lb}$$

Equilibrium in Two Dimensions

A body is in equilibrium when it is either at rest or moving at a constant speed in a straight line. Fig. 6.10(a) shows the resultant force of the sum of two forces from Example 1 in Section 6.1. When two or more forces are acting at a point, the **equilibrant force** is the force that when applied at that same point produces equilibrium. *The equilibrant force is equal in magnitude to that of the resultant force but it acts in the opposite direction.* [See Fig. 6.10(b)]. In this case, the equilibrant force is 926 N at 214.5° (180° + 34.5°).

If an object is in equilibrium in two dimensions, the net force acting on it must be zero. For the net force to be zero, the sum of the x-components must be zero, and the sum of the y-components must be zero. For forces **A, B,** and **C** with

Figure 6.10

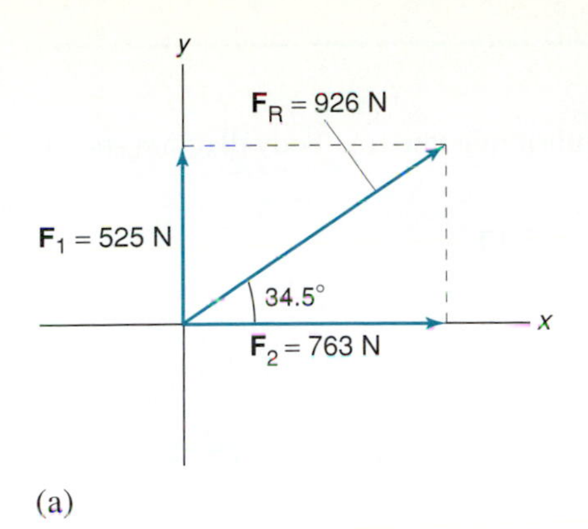

(a)

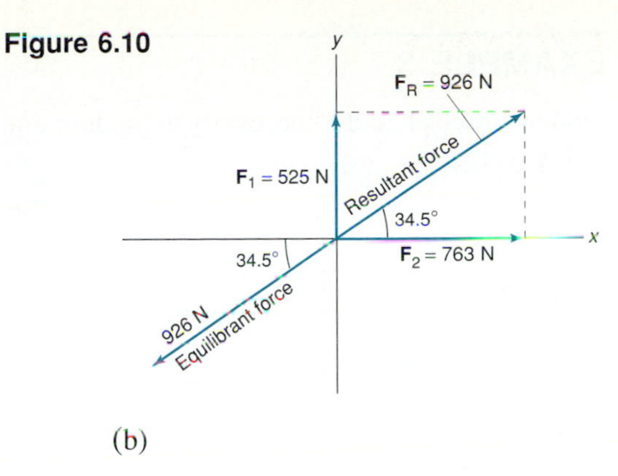

(b)

x-components $\mathbf{A}_x$, $\mathbf{B}_x$, and $\mathbf{C}_x$, respectively, and with y-components $\mathbf{A}_y$, $\mathbf{B}_y$, and $\mathbf{C}_y$, respectively, to be in equilibrium:

> *CONDITIONS FOR EQUILIBRIUM*
>
> **1.** The sum of x-components = 0; that is, $\mathbf{A}_x + \mathbf{B}_x + \mathbf{C}_x = 0$ and
>
> **2.** The sum of y-components = 0; that is, $\mathbf{A}_y + \mathbf{B}_y + \mathbf{C}_y = 0$.

In general, to solve equilibrium problems:

1. Draw a free-body diagram from the point at which the unknown forces act.

2. Find the x- and y-component of each force.

3. Substitute the components in the equations

$$\text{sum of } x\text{-components} = 0$$
$$\text{sum of } y\text{-components} = 0$$

4. Solve for the unknowns. This may involve two simultaneous equations.

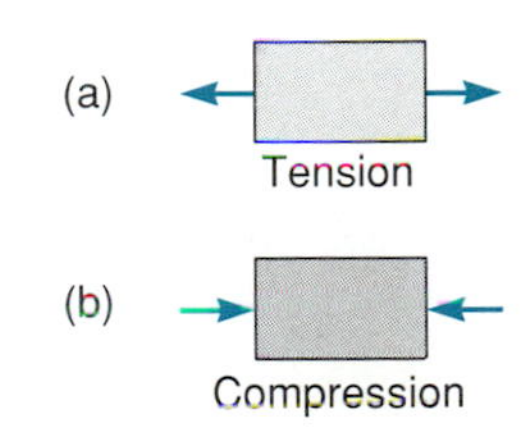

Figure 6.11
Tension and compression forces.

We may need to find the tension or compression in part of a structure, such as in a beam or a cable. **Tension** is a stretching force produced by forces pulling outward on the ends of the object [Fig. 6.11(a)]. **Compression** is a force produced by forces pushing inward on the ends of an object [Fig. 6.11(b)]. A rubber band being stretched is an example of tension [Fig. 6.12(a)]. A valve spring whose ends are pushed together is an example of compression [Fig. 6.12(b)].

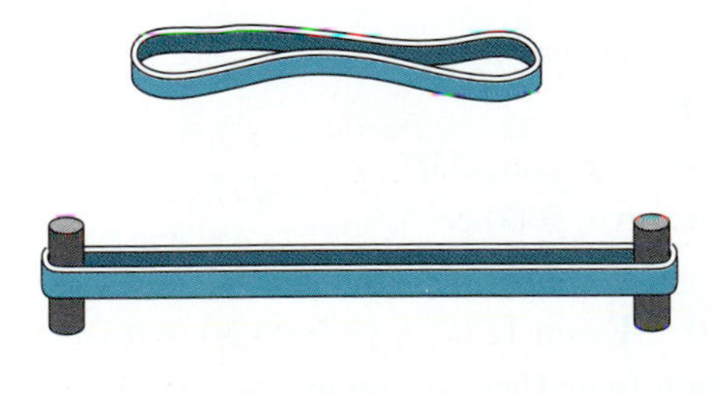

(a) Tension in a rubber band.

(b) Compression in a valve spring.

Figure 6.12

EXAMPLE 3

Find the forces **F** and **F′** necessary to produce equilibrium in the free-body diagram shown in Fig. 6.13.

1.

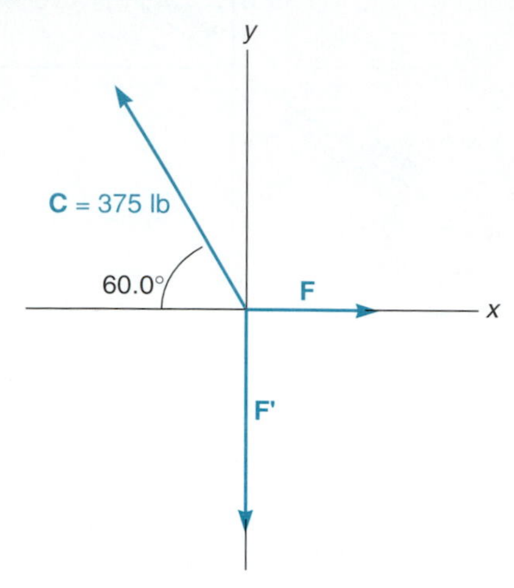

Figure 6.13

2. *x-components*

$\mathbf{F}_x = \mathbf{F}$

$\mathbf{F}'_x = 0$

$\mathbf{C}_x = -(375 \text{ lb})(\cos 60.0°) = -188 \text{ lb}$

y-components

$\mathbf{F}_y = 0$

$\mathbf{F}'_y = -\mathbf{F}'$

$\mathbf{C}_y = (375 \text{ lb})(\sin 60.0°) = 325 \text{ lb}$

3. Sum of x-components = 0

$\mathbf{F} + 0 + (-188 \text{ lb}) = 0$

Sum of y-components = 0

$0 + (-\mathbf{F}') + 325 \text{ lb} = 0$

4. $\mathbf{F} = 188 \text{ lb}$ $\qquad$ $\mathbf{F}' = 325 \text{ lb}$

EXAMPLE 4

Find the forces **F** and **F′** necessary to produce equilibrium in the free-body diagram shown in Fig. 6.14.

1.

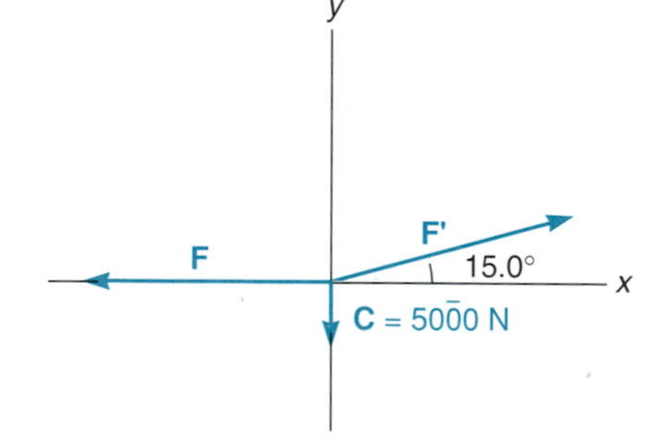

Figure 6.14

2. *x-components*

$\mathbf{F}_x = -\mathbf{F}$

$\mathbf{F}'_x = \mathbf{F}' \cos 15.0°$

$\mathbf{C}_x = 0$

y-components

$\mathbf{F}_y = 0$

$\mathbf{F}'_y = \mathbf{F}' \sin 15.0°$

$\mathbf{C}_y = -50\bar{0}0 \text{ N}$

3. Sum of x-components = 0

$(-\mathbf{F}) + \mathbf{F}' \cos 15.0° + 0 = 0$

Sum of y-components = 0

$0 + \mathbf{F}' \sin 15.0° + (-50\bar{0}0 \text{ N}) = 0$

4. ***Note:*** Solve for **F′** in the right-hand equation first. Then substitute this value in the left-hand equation to solve for **F**.

$$\mathbf{F}' = \frac{50\bar{0}0 \text{ N}}{\sin 15.0°} = 19{,}300 \text{ N}$$

$$\mathbf{F} = \mathbf{F}' \cos 15.0°$$
$$= (19{,}300 \text{ N})(\cos 15.0°)$$
$$= 18{,}600 \text{ N}$$

EXAMPLE 5

The crane shown in Fig. 6.15 is supporting a beam that weighs $60\bar{0}0$ N. Find the tension in the horizontal supporting cable and the compression in the boom.

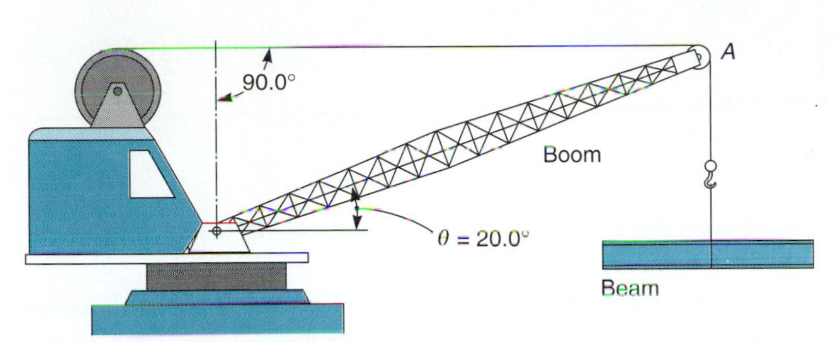

Figure 6.15

1. Draw the free-body diagram showing the forces *on* the point labeled *A*.

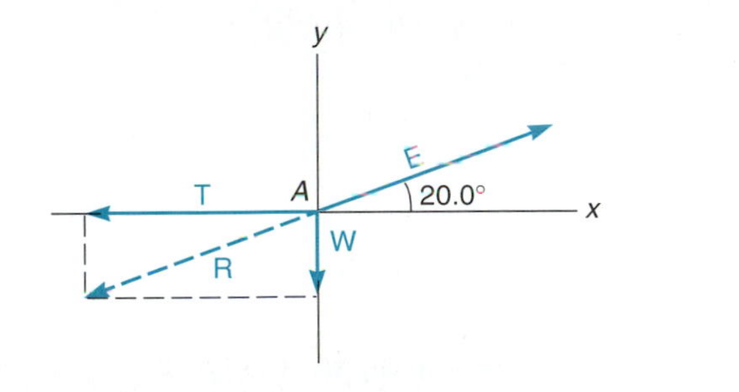

T = tension

R = compression

E = equilibrant force

T is the force exerted on *A* by the horizontal supporting cable.

E is the force exerted by the boom on *A*.

W is the force (weight of the beam) pulling straight down on *A*.

R is the sum of forces **W** and **T**, which is equal in magnitude but opposite in direction to force **E**. (**R** = −**E**.)

2. x-*components*

$\mathbf{E}_x = \mathbf{E} \cos 20.0°$

$\mathbf{T}_x = -\mathbf{T}$

$\mathbf{W}_x = 0$

y-*components*

$\mathbf{E}_y = \mathbf{E} \sin 20.0°$

$\mathbf{T}_y = 0$

$\mathbf{W}_y = -60\bar{0}0 \text{ N}$

3. Sum of *x*-components = 0

$\mathbf{E} \cos 20.0° + (-\mathbf{T}) = 0$

Sum of *y*-components = 0

$\mathbf{E} \sin 20.0° + (-60\bar{0}0 \text{ N}) = 0$

4. $\mathbf{T} = \mathbf{E} \cos 20.0°$

$$\mathbf{E} = \frac{60\bar{0}0\text{N}}{\sin 20.0°} = 17{,}500 \text{ N}$$

$$\mathbf{T} = (17{,}500 \text{ N})(\cos 20.0°)$$
$$= 16{,}400 \text{ N}$$

EXAMPLE 6

Figure 6.16

A homeowner pushes a 40.0-lb lawn mower at a constant velocity (Fig. 6.16). The frictional force on the mower is 20.0 lb. What force must the person exert on the handle, which makes an angle of 30.0° with the ground? Also, find the normal (perpendicular to ground) force.

This is an equilibrium problem because the mower is not accelerating and the net force is zero.

1. Draw the free-body diagram of the mower.

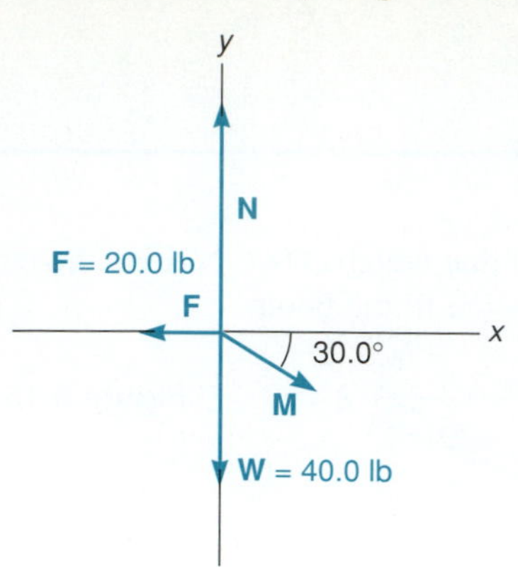

M is the force exerted on the mower by the person; this compression force is directed down along the handle.

W is the weight of the mower directed straight down.

N is the force exerted upward on the mower by the ground, which keeps the mower from falling through the ground.

F is the frictional force that opposes the motion.

2. x-*components*

$\mathbf{N}_x = 0$

$\mathbf{W}_x = 0$

$\mathbf{F}_x = -20.0 \text{ lb}$

$\mathbf{M}_x = \mathbf{M} \cos 30.0°$

y-*components*

$\mathbf{N}_y = \mathbf{N}$

$\mathbf{W}_y = -40.0 \text{ lb}$

$\mathbf{F}_y = 0$

$\mathbf{M}_y = -\mathbf{M} \sin 30.0°$

3. Sum of x-components = 0

$0 + 0 + (-20.0 \text{ lb}) + \mathbf{M} \cos 30.0° = 0$

Sum of y-components = 0

$\mathbf{N} + (-40.0 \text{ lb}) + 0 + (-\mathbf{M} \sin 30.0°) = 0$

$\mathbf{N} = \mathbf{M} \sin 30.0 + 40.0 \text{ lb}$

4. $$\mathbf{M} = \frac{20.0 \text{ lb}}{\cos 30.0°} = 23.1 \text{ lb}$$

$$\mathbf{N} = (23.1 \text{ lb})(\sin 30.0°) + 40.0 \text{ lb} = 51.6 \text{ lb}$$

EXAMPLE 7

The crane shown in Fig. 6.17 is supporting a beam that weighs $60\bar{0}0$ N. Find the tension in the supporting cable and the compression in the boom.

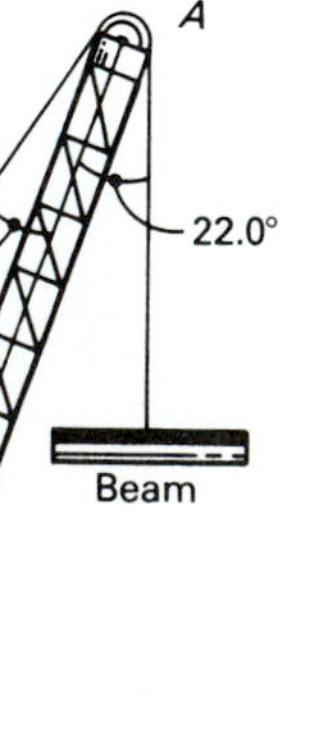

Figure 6.17

1. Draw the free-body diagram showing the forces *on* point A.

W is the weight of the beam, which pulls straight down.

T is the force exerted on *A* by the supporting cable.

E is the force exerted by the boom on *A*.

R is the sum of forces **W** and **T**, which is equal in magnitude but opposite in direction to force **E**. (**R** = −**E**.)

2. *x-components*

$\mathbf{E}_x = \mathbf{E} \cos 68.0°$

$\mathbf{T}_x = -\mathbf{T} \cos 53.0°$

$\mathbf{W}_x = 0$

y-components

$\mathbf{E}_y = \mathbf{E} \sin 68.0°$

$\mathbf{T}_y = -\mathbf{T} \sin 53.0°$

$\mathbf{W}_y = -60\bar{0}0 \text{ N}$

3. Sum of *x*-components = 0

$\mathbf{E} \cos 68.0° + (-\mathbf{T} \cos 53.0°) + 0 = 0$

Sum of *y*-components = 0

$\mathbf{E} \sin 68.0° + (-\mathbf{T} \sin 53.0°) + (-60\bar{0}0 \text{ N}) = 0$

4. ***Note:*** Solve the left equation for **E**. Then substitute this quantity in the right equation and solve for **T**.

$$\mathbf{E} = \frac{\mathbf{T} \cos 53.0°}{\cos 68.0°}$$

$$\left(\frac{\mathbf{T} \cos 53.0°}{\cos 68.0°}\right)(\sin 68.0°) - \mathbf{T} \sin 53.0° = 60\bar{0}0 \text{ N}$$

$$1.490\mathbf{T} - 0.799\mathbf{T} = 60\bar{0}0 \text{ N}$$

$$0.691\mathbf{T} = 60\bar{0}0 \text{ N}$$

$$\mathbf{T} = \frac{60\bar{0}0 \text{ N}}{0.691} = 8680 \text{ N}$$

$$\mathbf{E} = \frac{(8680 \text{ N})(\cos 53.0°)}{\cos 68.0°} = 13{,}900 \text{ N}$$

Alternate Method: You can orient a free-body diagram any way you want on the *x–y* axes. You should orient it so that as many of the vectors as possible are on an *x*- or *y*-axis. The result will be the same. Let's rework Example 7 as follows:

1. Draw the free-body diagram showing the forces *on* point *A* using the same notation as follows:

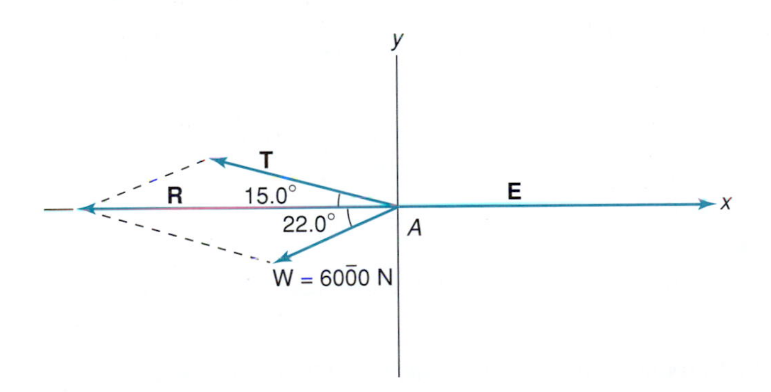

2. *x-components*

$\mathbf{E}_x = \mathbf{E}$

$\mathbf{T}_x = -\mathbf{T} \cos 15.0°$

$\mathbf{W}_x = -(60\bar{0}0 \text{ N})(\cos 22.0°)$

y-components

$\mathbf{E}_y = 0$

$\mathbf{T}_y = \mathbf{T} \sin 15.0°$

$\mathbf{W}_y = -(60\bar{0}0 \text{ N})(\sin 22.0°)$

3. Sum of *x*-components = 0

$\mathbf{E} + (-\mathbf{T} \cos 15.0°) + (-60\bar{0}0 \text{ N})(\cos 22.0°) = 0$

Sum of *y*-components = 0

$0 + \mathbf{T} \sin 15.0° + (-60\bar{0}0 \text{ N})(\sin 22.0°) = 0$

4. Solve the right equation for **T** (since it has only one variable). Then solve the left equation for **E** and substitute this quantity.

$$\mathbf{T} = \frac{(60\bar{0}0\text{ N})(\sin 22.0^\circ)}{\sin 15.0^\circ}$$
$$= 8680\text{ N}$$

$$\mathbf{E} = \mathbf{T}\cos 15.0^\circ + (60\bar{0}0\text{ N})(\cos 22.0^\circ)$$
$$\mathbf{E} = (8680\text{ N})(\cos 15.0^\circ) + (60\bar{0}0\text{ N})(\cos 22.0^\circ)$$
$$= 13{,}900\text{ N}$$

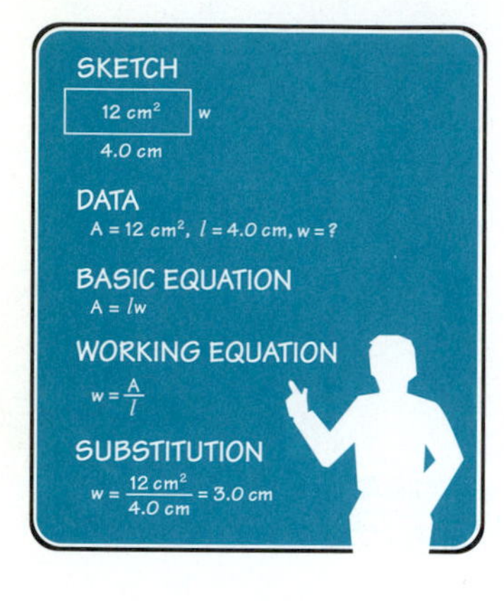

■ PROBLEMS 6.2

Find the force **F** that will produce equilibrium in each free-body diagram. Use the same procedure as in Examples 3 and 4.

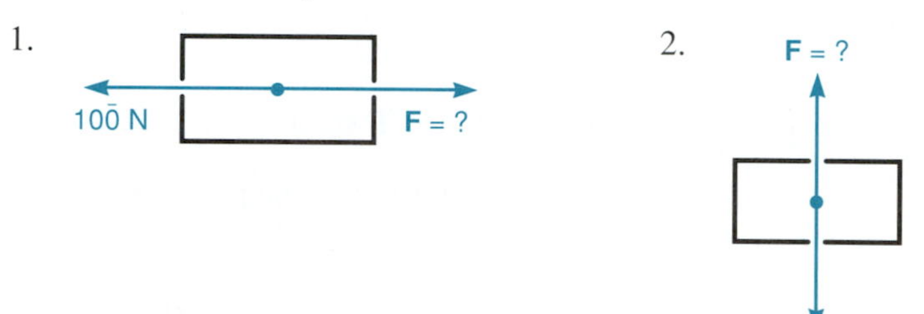

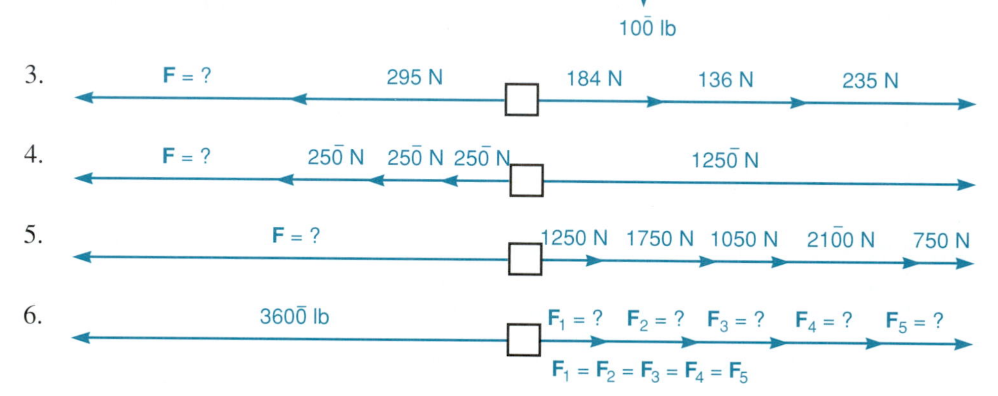

7. Five persons are having a tug-of-war. Kurt and Brian are on the left; Amy, Barbara, and Joyce are on the right. Amy pulls with a force of 225 N, Barbara pulls with a force of 495 N, Joyce pulls with a force of 455 N, and Kurt pulls with a force of 605 N. With what force must Brian pull to produce equilibrium?
8. A certain wire can support 6450 lb before it breaks. Seven 820-lb weights are suspended from the wire. Can the wire support an eighth weight of 820 lb?
9. The frictional force of a loaded pallet in a warehouse is 385 lb. Can three workers, each exerting a force of 135 lb, push it to the side?
10. A bridge has a weight limit of 7.0 tons. How heavy a load can a 2.5-ton truck carry across?

Find the forces $\mathbf{F}_1$ and $\mathbf{F}_2$ that produce equilibrium in each free-body diagram.

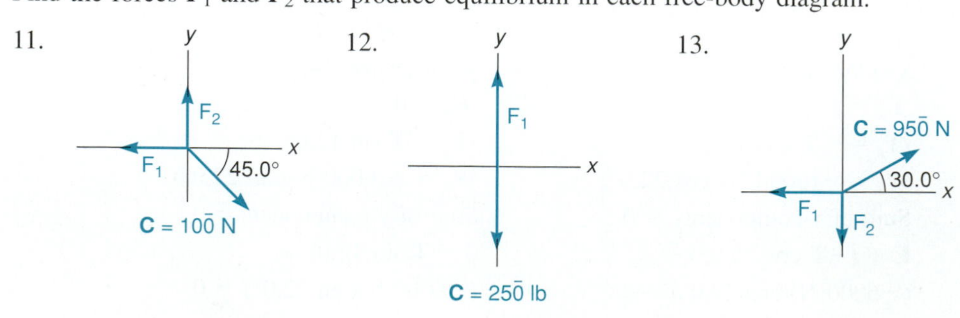

14.

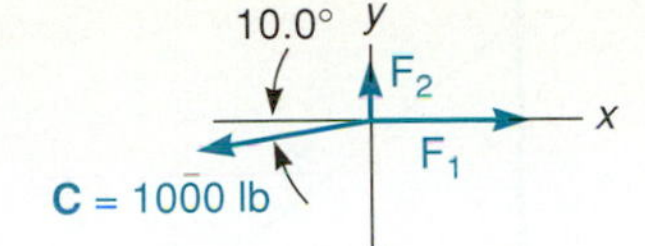

15.

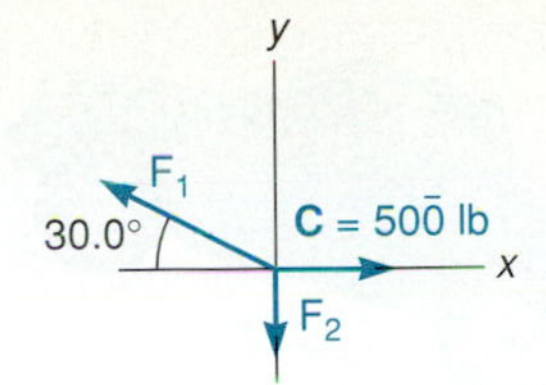

16.

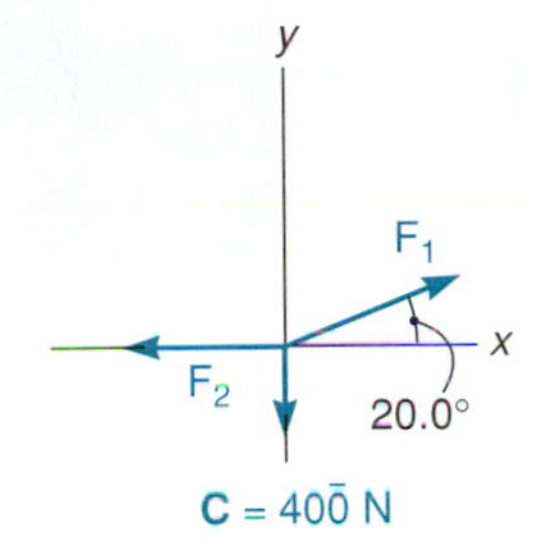

17.

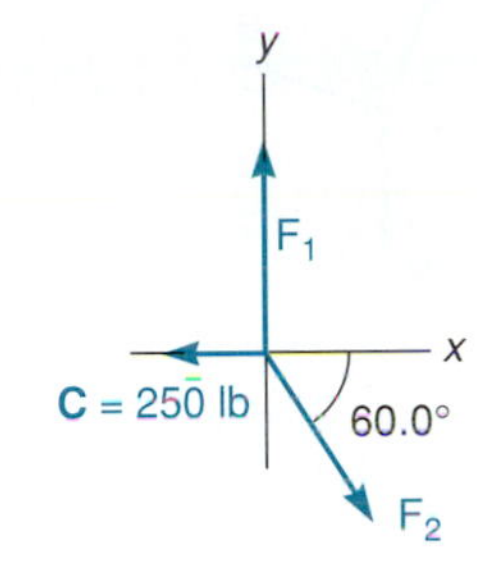

18. A rope is attached to two buildings and supports a $50\bar{0}$-lb sign (Fig. 6.18). Find the tension in the two ropes, $\mathbf{T}_1$ and $\mathbf{T}_2$. (*Hint:* Draw the free-body diagram of the forces acting *on* the point labeled *A*.)

19. If the angle between the horizontal and the ropes in Problem 18 is changed to 10.0°, what are the tensions in the two ropes, $\mathbf{T}_1$ and $\mathbf{T}_2$?

20. Find the tension in the horizontal supporting cable and the compression in the boom of the crane shown in Fig. 6.19, which supports an $89\bar{0}0$-N beam.

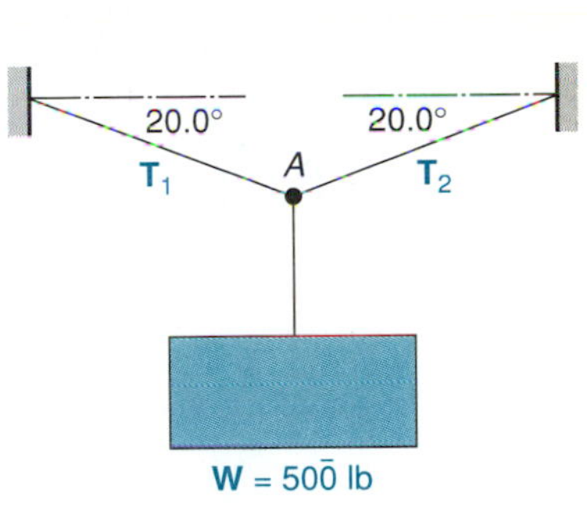

Figure 6.18

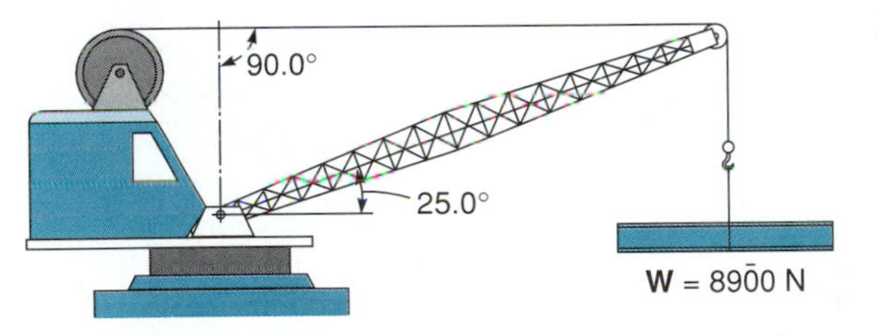

Figure 6.19

21. Find the tension in the horizontal supporting cable and the compression in the boom of the crane shown in Fig. 6.20, which supports a $15\bar{0}0$-lb beam.

22. The frictional force of the mower shown in Fig. 6.21 is $2\bar{0}$ lb. What force must the man exert along the handle to push it at a constant velocity?

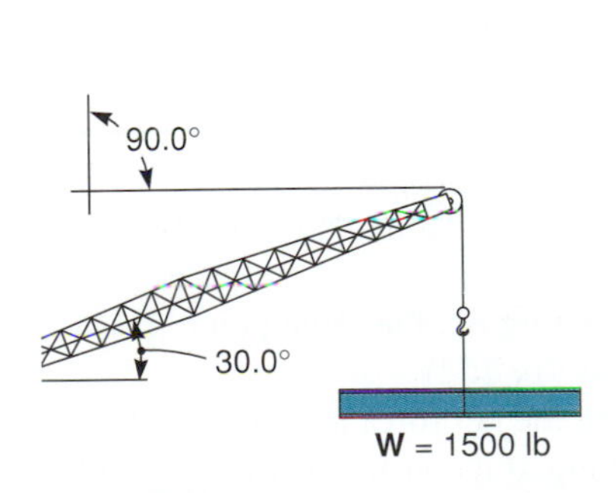

Figure 6.20

Figure 6.21

23. A vehicle that weighs 16,200 N is parked on a 20.0° hill (Fig. 6.22). What braking force is necessary to keep it from rolling? Neglect frictional forces. (*Hint:* When you draw the free-body diagram, tilt the *x*- and *y*-axes as shown. **B** is the braking force directed up the hill and along the *x*-axis.)

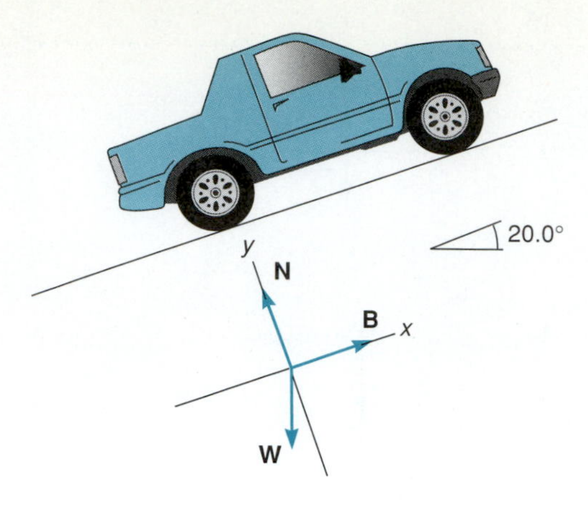

Figure 6.22

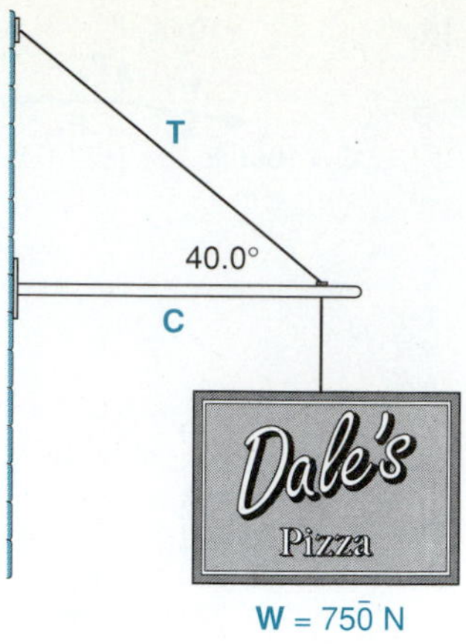

T
40.0°
C
Dale's
Pizza
W = $75\bar{0}$ N

Figure 6.23

24. Find the tension in the cable and the compression in the support of the sign shown in Fig. 6.23.
25. The crane shown in Fig. 6.24 is supporting a load of 1850 lb. Find the tension in the supporting cable and the compression in the boom.
26. The crane shown in Fig. 6.25 is supporting a load of 11,500 N. Find the tension in the supporting cable and the compression in the boom.

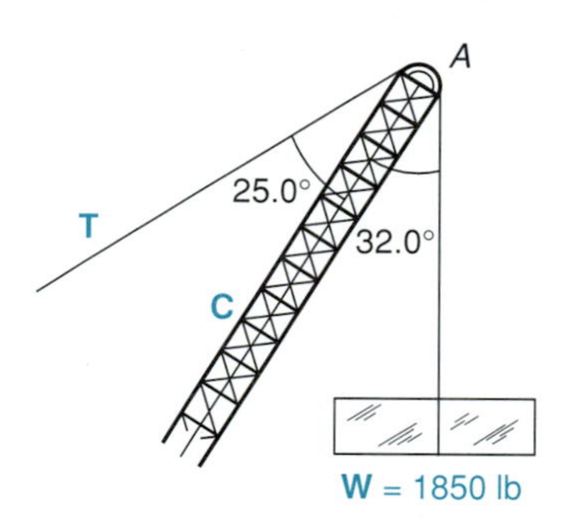

Figure 6.24

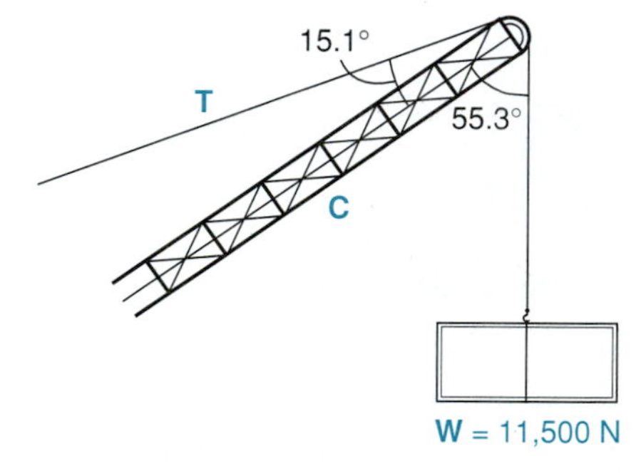

Figure 6.25

GLOSSARY

Compression A force produced by forces pushing inward on the ends of an object. (p. 153)
Concurrent Forces Two or more forces applied to, or acting on, the same point. (p. 146)
Equilibrant Force The force that produces equilibrium. (p. 152)
Equilibrium An object is said to be in equilibrium when the net force acting on it is zero. A body that is in equilibrium is either at rest or moving at a constant velocity. (p. 150)
Resultant Force The sum of the forces applied to the same point; the single force that has the same effect as the two or more forces acting together. (p. 146)
Statics The study of objects that are in equilibrium. (p. 150)
Tension A stretching force produced by forces pulling outward on the ends of an object. (p. 153)

FORMULAS

6.1 To find the resultant vector $\mathbf{F}_R$ of two or more vectors:
(a) find the x- and y-components of each vector and add the components.
(b) find angle A as follows:

$$\tan A = \frac{|\text{sum of } y\text{-components}|}{|\text{sum of } x\text{-components}|} = \frac{|\mathbf{F}_{Ry}|}{|\mathbf{F}_{Rx}|}$$

Determine the quadrant of the angle from the signs of the sum of the x- and y-components.
(c) find the magnitude of $\mathbf{F}_R$ using the Pythagorean theorem:

$$|\mathbf{F}_R| = \sqrt{|\mathbf{F}_{Rx}|^2 + |\mathbf{F}_{Ry}|^2}$$

6.2 Condition for equilibrium in one dimension:

$$\mathbf{F}_+ = \mathbf{F}_-$$

where $\mathbf{F}_+$ is the sum of the forces acting in one direction (call it the positive direction) and $\mathbf{F}_-$ is the sum of the forces acting in the opposite (negative) direction.

Conditions for equilibrium in two dimensions:

(a) The sum of x-components = 0; that is $\mathbf{A}_x + \mathbf{B}_x + \mathbf{C}_x = 0$ and
(b) The sum of y-components = 0; that is $\mathbf{A}_y + \mathbf{B}_y + \mathbf{C}_y = 0$.

To solve equilibrium problems:

1. Draw a free-body diagram from the point at which the unknown forces act.
2. Find the x- and y-component of each force.
3. Substitute the components in the equations.

$$\text{sum of } x\text{-components} = 0$$
$$\text{sum of } y\text{-components} = 0$$

4. Solve for the unknowns. This may involve two simultaneous equations.

REVIEW QUESTIONS

1. Concurrent forces act at
   (a) two or more different points. (b) the same point.
   (c) the origin.
2. The resultant force is
   (a) the last force applied.
   (b) the single force which has the same effect as the two or more forces acting together.
   (c) equal to either diagonal when using the parallelogram method to add vectors.
3. A moving object
   (a) can be in equilibrium. (b) is never in equilibrium.
   (c) has no force being applied.
4. The study of an object in equilibrium is called
   (a) dynamics. (b) astronomy.
   (c) statics. (d) biology.
5. Is motion produced every time a force is applied to an object?
6. What is the relationship between opposing forces on a body that is in equilibrium?

7. Describe the forces acting on a bridge in equilibrium.
8. Define *equilibrium*.
9. In what direction does the force due to gravity always act?
10. What may be said about concurrent forces whose sum of x-components equals zero and whose sum of y-components equals zero?
11. What name is given to forces pressing inward on the ends of an object?
12. What name is given to forces pulling outward on the ends of an object?
13. What is a free-body diagram?
14. Why must tension and compression be represented by vectors?
15. Does it matter if opposing forces of equal magnitude are *not* acting on the same point?

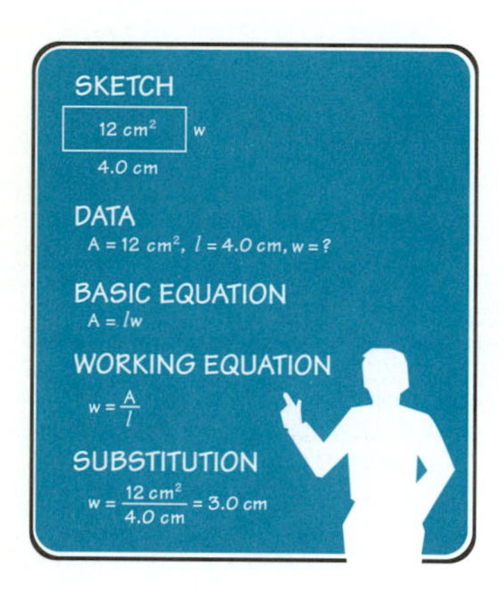

REVIEW PROBLEMS

1. Find the sum of the following forces acting on the same point in a straight line: 345 N (right); 108 N (right); 481 N (left); 238 N (left); 303 N (left).
2. Forces of 275 lb and 225 lb act on the same point.
   (a) What is the magnitude of the maximum net force the two forces can exert together?
   (b) What is the magnitude of the minimum net force the two forces can exert together?

Find the sum of each set of vectors. Give angles in standard position.

3.

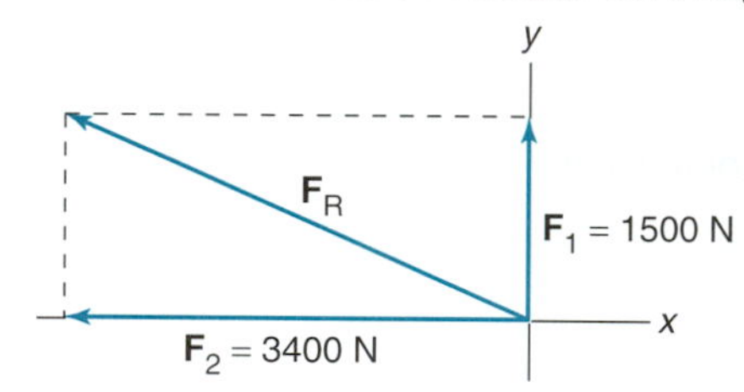

4.

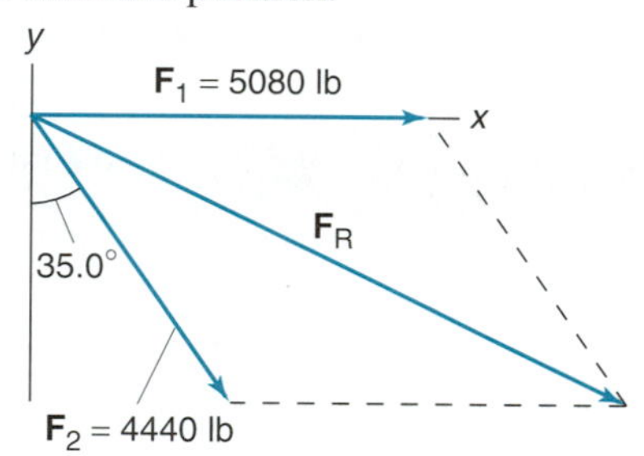

5.

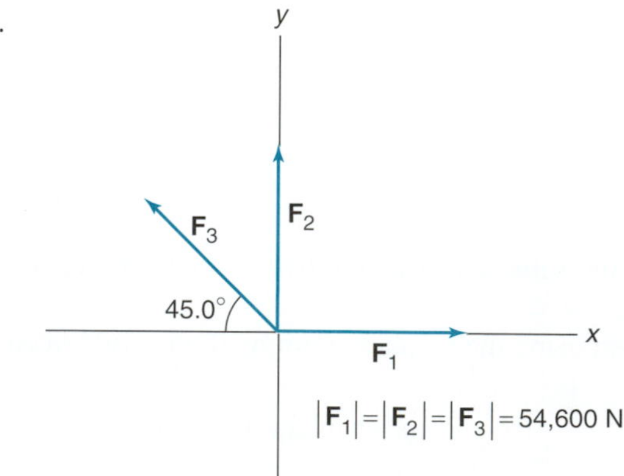

6. Forces of $\mathbf{F}_1$ = 1250 N, $\mathbf{F}_2$ = 625 N, and $\mathbf{F}_3$ = 1850 N are applied to the same point. The angle between $\mathbf{F}_1$ and $\mathbf{F}_2$ is 120.0°, and the angle between $\mathbf{F}_2$ and $\mathbf{F}_3$ is 30.0°. $\mathbf{F}_2$ is between $\mathbf{F}_1$ and $\mathbf{F}_3$. Find the resultant force.

7. A motorcycle on a bridge exerts a downward force of 2150 N. What is the upward force if it is in equilibrium?
8. Eight people are involved in a tug-of-war. The blue team members pull with forces of 220 N, 340 N, 180 N, and 560 N. Three members of the red team pull with forces of 250 N, 160 N, and 420 N. With what force must the fourth person pull to remain in equilibrium?
9. A table will support $23\bar{0}$ lb. If a 187-lb person sits on the table, how much more weight can be added?
10. A bridge has a weight limit of 14.0 tons. What is the maximum weight an 8.0-ton truck can carry across and still maintain equilibrium?
11. The x-components of three vectors are $\mathbf{F}_x$, 375 units, and 150 units. If their sum is equal to zero, what is $\mathbf{F}_x$?
12. The sum of the y-components of two vectors is zero. If one is 650 units, what is the other ($\mathbf{F}_y$)?
13. If $\mathbf{N} = \mathbf{M} \sin 30.0° + 35.0$ lb and $\mathbf{M} = 38.0$ lb, find $\mathbf{N}$.
14. If $\mathbf{W} = \mathbf{C} \cos 45.0° + 53.0 \text{ N} - 38.0 \text{ N}$ and $\mathbf{C} = 20.0$ N, find $\mathbf{W}$.
15. If $\mathbf{W}_y = 60\bar{0}$ N and $\mathbf{W}_x = 90\bar{0}$ N, what are the magnitude and direction of the resultant $\mathbf{W}$?

Find forces $\mathbf{F}_1$ and $\mathbf{F}_2$ that produce equlibrium in each free-body diagram.

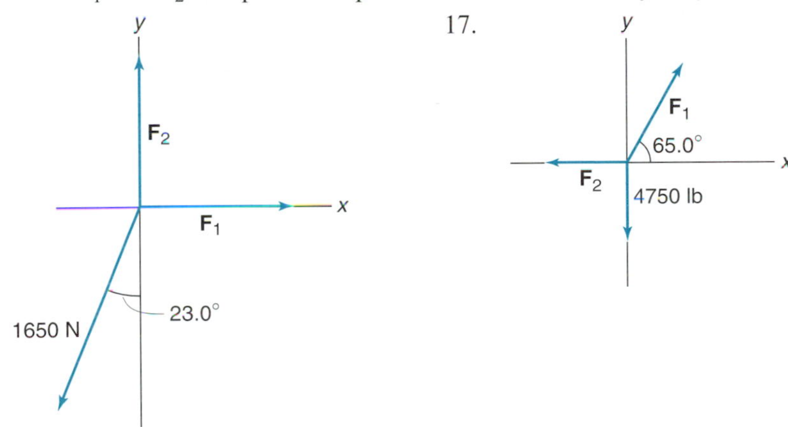

18. Find the tension in the cable and the compression in the support of the sign shown in Fig. 6.26.
19. Find the tension in each cable in Fig. 6.27.

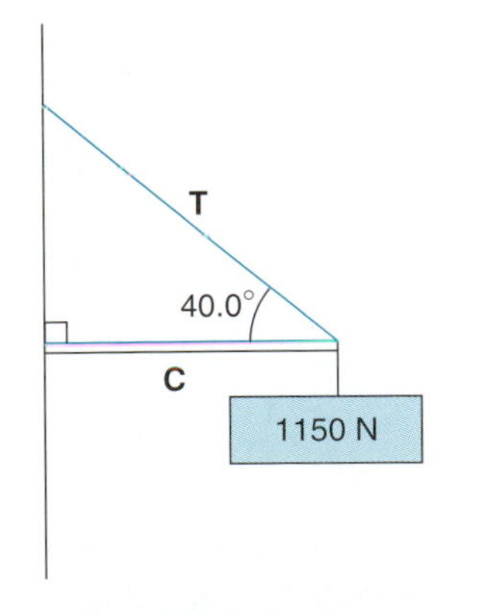

Figure 6.26

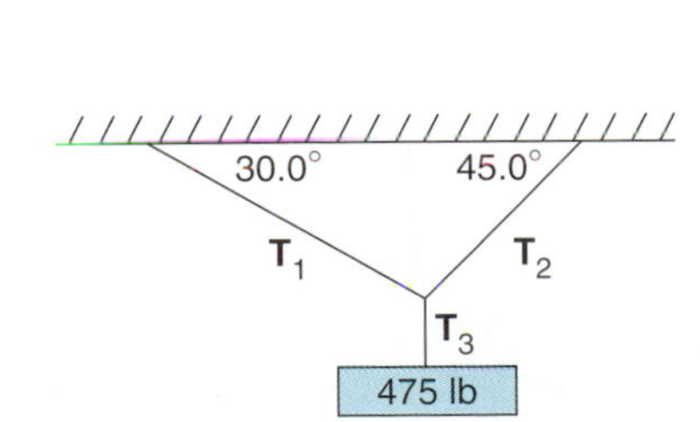

Figure 6.27

20. Find the tension in each cable in Fig. 6.28.
21. Find the tension and the compression in Fig. 6.29.

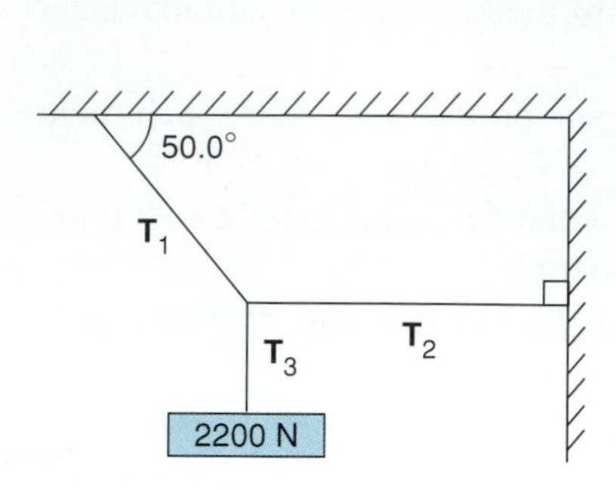

Figure 6.28

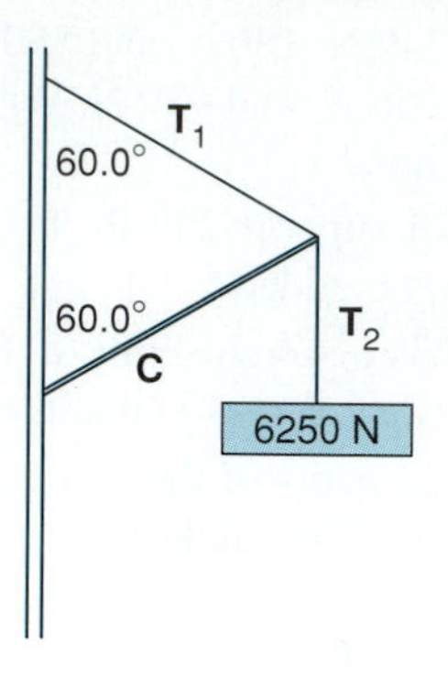

Figure 6.29

Chapter 7

WORK AND ENERGY

Work, power, and energy are common terms used to describe changes in physical activity. In science each of these terms has a limited definition. Work, for example, is accomplished only where there is movement of the object on which the applied force acts. Effort alone is insufficient.

We will now study how the scientist and engineer use work, power, and energy—all scalars; how they are related; and how they differ from their everyday meanings.

OBJECTIVES

The major goals of this chapter are to enable you to:

1. Distinguish the common and technical definitions of work.
2. Understand power and how it is described in technical terms.
3. Relate kinetic and potential energy to the law of conservation of mechanical energy.

7.1 WORK

What is work? The common idea of the definition of work is quite different from the engineering definition. We often associate work with physical or mental effort that leads to fatigue. The technical definition of work is more limited. If we try to lift a heavy crate that doesn't budge [Fig. 7.1(a)], we would probably say that we have done work because we strained our muscles and feel tired, but in a technical sense, no work was done, because the crate did not move. Work would be done *on* the crate if we were to move it across the floor [Fig. 7.1(b)]. In this case, work was done *by* us, and work was done *on* the crate.

Figure 7.1
Work is done only in the case shown in part (b).

The technical meaning of work requires that work must be done *by* one object *on* another object. When a stake is driven into the ground (Fig. 7.2), work is done *by* the moving sledgehammer, and work is done *on* the stake. When a bull-

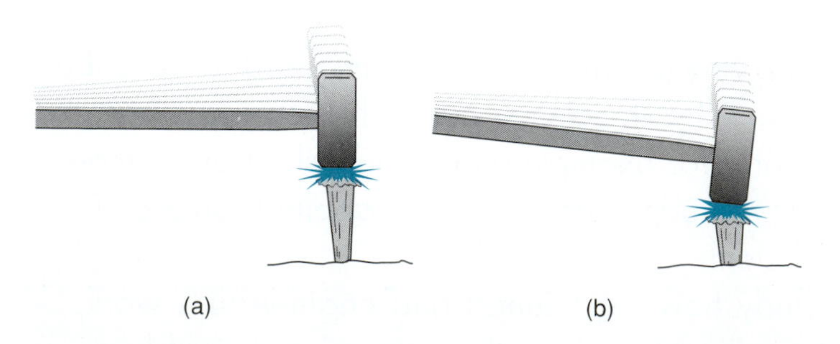

Figure 7.2
Work is done by one object on another.

Figure 7.3
Work is done *by* the bulldozer *on* the boulder.

dozer pushes a boulder (Fig. 7.3), work is done *by* the bulldozer and work is done *on* the boulder.

The previous examples show a limited meaning of work: that work is done when a force acts through a distance. The physical definition of **work** is even narrower: *Work is the product of the force in the direction of the motion and the displacement.*

$$W = Fs$$

where W = work
F = force applied
s = displacement *in the direction of the force*

Now, let us apply our technical definition of work to trying unsuccessfully to lift the crate. We applied a force by lifting on the crate. Have we done work? No work has been done. We were unable to move the crate. Therefore, the displacement was zero, and the product of the force and the displacement must also be zero. Therefore, no work was done.

By studying the equation for work, we can determine the correct units for work. In the metric system, force is expressed in newtons and the displacement in metres:

$$\text{work} = \text{force} \times \text{displacement} = \text{newton} \times \text{metre} = \text{N m}$$

This unit (N m) has a special name in honor of an English physicist, James P. Joule. It is the joule (J) [pronounced jo͞ol].

$$1 \text{ N m} = 1 \text{ joule} = 1 \text{ J}$$

In the English system, force is expressed in pounds (lb) and the displacement in feet (ft):

$$\text{work} = \text{force} \times \text{displacement} = \text{pounds} \times \text{feet} = \text{ft lb}$$

The English unit of work is called the foot-pound.

Work is not a vector quantity because it has no particular direction. It is a scalar and has only magnitude.

EXAMPLE 1

Find the amount of work done by a worker lifting 225 N of bricks to a height of 1.75 m as shown in Fig. 7.4.

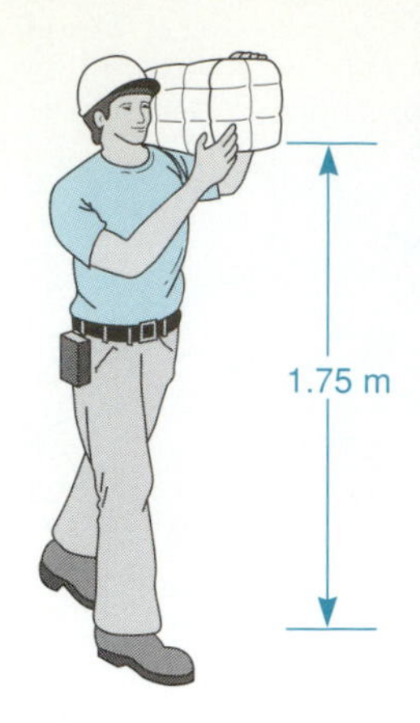

Figure 7.4

Data:

$$F = 225 \text{ N}$$
$$s = 1.75 \text{ m}$$
$$W = ?$$

Basic Equation:

$$W = Fs$$

Working Equation: Same

Substitution:

$$W = (225 \text{ N})(1.75 \text{ m})$$
$$= 394 \text{ N m} \quad \text{or} \quad 394 \text{ J}$$

EXAMPLE 2

A worker pushes a 350-lb cart a distance of $3\overline{0}$ ft by exerting a constant force of $4\overline{0}$ lb as shown in Fig. 7.5. How much work does the person do?

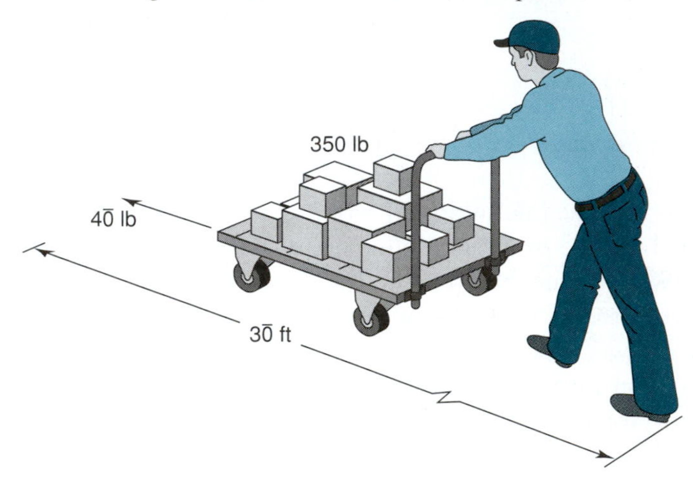

Figure 7.5

Data:

$$F = 4\overline{0} \text{ lb}$$
$$s = 3\overline{0} \text{ ft}$$
$$W = ?$$

Basic Equation:

$$W = Fs$$

Working Equation: Same

Substitution:

$$W = (3\overline{0} \text{ lb})(4\overline{0} \text{ ft})$$
$$= 1200 \text{ ft lb}$$

Note: In Example 2 the cart weighs 350 lb but $F = 4\overline{0}$ lb. (Recall that the weight of an object is the measure of its gravitational attraction to the earth and is represented by a vertical vector pointing down to the center of the earth.) There is no motion in the direction *this* force is exerted. Therefore, the weight of the box is not the force used to determine the work being done.

Work is being done by the worker pushing the pallet. He is exerting a force of $4\overline{0}$ lb, and there is a resulting displacement in the direction the force is applied. The work done is the product of this force ($4\overline{0}$ lb) and the displacement ($3\overline{0}$ ft) in the direction the force is applied.

Recall that the definition of work states that work is the product of the *force in the direction of the motion* and the displacement. To determine the work when the force is not applied in the direction of the motion, consider a block being pulled by a rope with a force **F** that makes an angle θ with level ground as shown in Fig. 7.6. First, draw the horizontal component $\mathbf{F}_x$ and complete the right triangle. Note that $\mathbf{F}_x$ is the force in the direction of the motion. From the right triangle we have

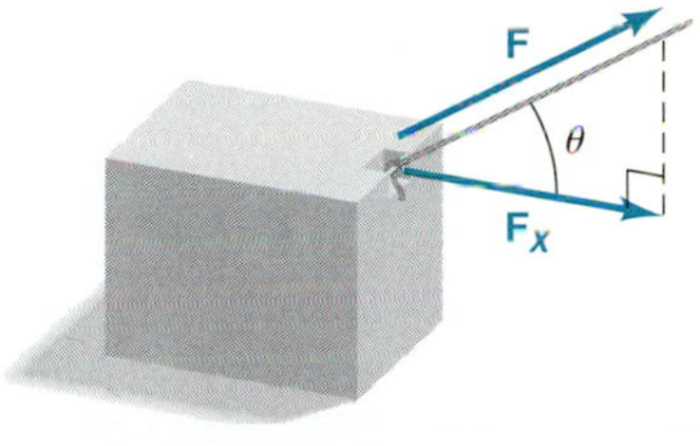

$$\cos\theta = \frac{\text{side adjacent to }\theta}{\text{hypotenuse}} = \frac{|\mathbf{F}_x|}{|\mathbf{F}|}$$

Figure 7.6

Or,

$$|\mathbf{F}_x| = |\mathbf{F}|\cos\theta$$

That is, when the applied force is not in the direction of the motion, the work done is

$$W = Fs\cos\theta$$

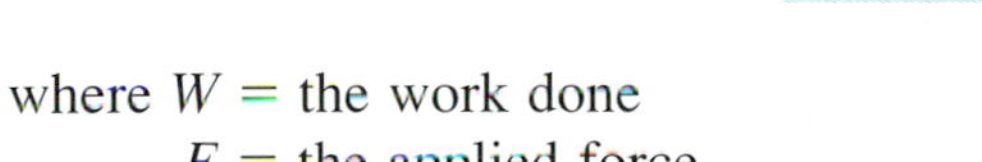

where W = the work done
F = the applied force
s = the displacement
θ = the angle between the applied force and the direction of the motion

EXAMPLE 3

A person pulls a sled along level ground a distance of 15.0 m by exerting a constant force of 215 N at an angle of 30.0° with the ground (Fig. 7.7). How much work does he do?

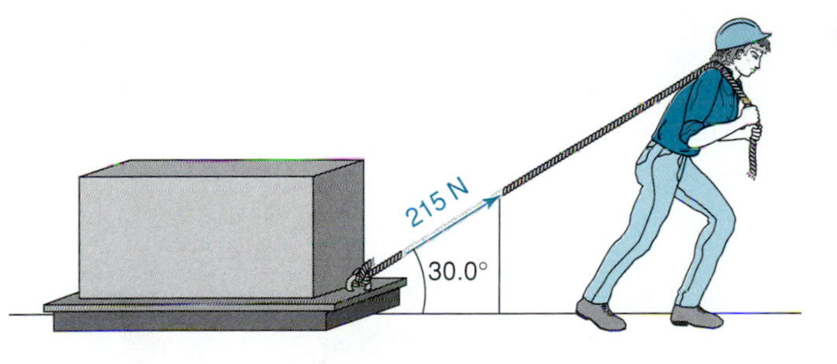

Figure 7.7

Data:

$F = 215$ N
$s = 15.0$ m
$\theta = 30.0°$
$W = ?$

Basic Equation:

$$W = Fs\cos\theta$$

Working Equation: Same

Substitution:

$$W = (215 \text{ N})(15.0 \text{ m}) \cos 30.0°$$
$$= 2790 \text{ N m}$$
$$= 2790 \text{ J} \qquad (1 \text{ N m} = 1 \text{ J})$$

EXAMPLE 4

Juan and Sonja use a push mower to mow a lawn. Juan, who is taller, pushes at a constant force of 33.1 N on the handle at an angle of 55.0° with the ground. Sonja, who is shorter, pushes at a constant force of 23.2 N on the handle at an angle of 35.0° with the ground. Assume they each push the mower $30\bar{0}0$ m. Who does more work and by how much?

Sketch:

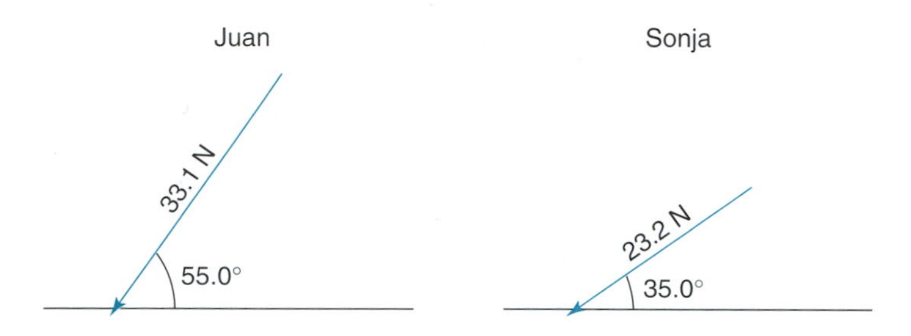

Data:

Juan	Sonja
$F = 33.1$ N	$F = 23.2$ N
$s = 30\bar{0}0$ m	$s = 30\bar{0}0$ m
$\theta = 55.0°$	$\theta = 35.0°$
$W = ?$	$W = ?$

Basic Equation:

$W = Fs \cos \theta$ $\qquad$ $W = Fs \cos \theta$

Working Equation: Same $\qquad$ Same

Substitution:

$$W = (33.1 \text{ N})(30\bar{0}0 \text{ m}) \cos 55.0° \qquad W = (23.2 \text{ N})(30\bar{0}0 \text{ m}) \cos 35.0°$$
$$= 57,\bar{0}00 \text{ N m} \qquad = 57,\bar{0}00 \text{ N m}$$
$$= 57,\bar{0}00 \text{ J} \quad (1 \text{ N m} = 1 \text{ J}) \qquad = 57,\bar{0}00 \text{ J}$$

They do the same amount of work. However, Juan must exert more energy because he pushes into the ground more than Sonja, who pushes more in the direction of the motion.

EXAMPLE 5

Find the amount of work done in vertically lifting a steel beam with mass 750 kg at uniform speed a distance of 45 m.

Here the force is the weight of the mass and the distance is the height.

Data:

$$F = mg$$
$$m = 750 \text{ kg}$$
$$g = 9.80 \text{ m/s}^2$$
$$s = 45 \text{ m}$$
$$W = ?$$

Basic Equation:

$$W = Fs = mgs$$

Working Equation: Same

Substitution:

$$\begin{aligned} W &= (750 \text{ kg})(9.80 \text{ m/s}^2)(45 \text{ m}) \\ &= 3.3 \times 10^5 \text{ (kg m/s}^2\text{)(m)} \\ &= 3.3 \times 10^5 \text{ N m} \qquad (1 \text{ N} = 1 \text{ kg m/s}^2) \\ &= 3.3 \times 10^5 \text{ J} \qquad (1 \text{ J} = 1 \text{ N m}) \end{aligned}$$

Do you see that 330 kJ would also be an acceptable answer?

■ PROBLEMS 7.1

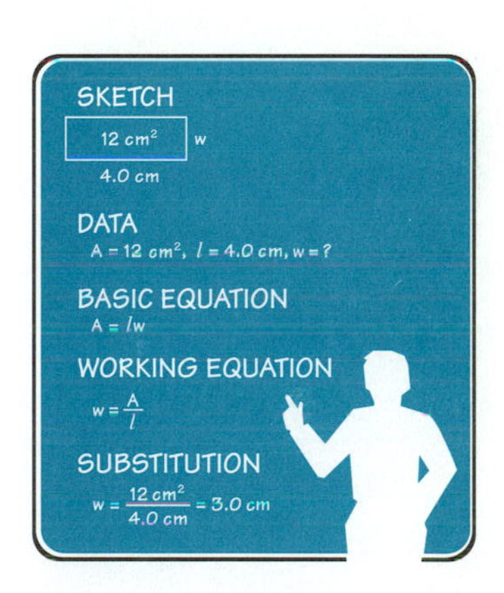

1. Given: $F = 10.0$ N
   $s = 3.43$ m
   $W = ?$
2. Given: $F = 125$ N
   $s = 4875$ m
   $W = ?$
3. Given: $F = 1850$ N
   $s = 625$ m
   $\theta = 37.5°$
   $W = ?$
4. Given: $W = 697$ ft lb
   $s = 976$ ft
   $F = ?$
5. Given: $F = 25{,}700$ N
   $s = 238$ m
   $W = 5.57 \times 10^6$ J
   $\theta = ?$
6. Given: $F = ma$
   $m = 16.0$ kg
   $a = 9.80 \text{ m/s}^2$
   $s = 13.0$ m
   $W = ?$
7. How much work is required for a mechanical hoist to lift a $70\bar{0}0$ N automobile to a height of 1.80 m for repairs?
8. A hay wagon is used to move bales from the field to the barn. The tractor pulling the wagon exerts a constant force of 350 lb. The distance from field to barn is $\frac{1}{2}$ mi. How much work (ft lb) is done in moving one load of hay to the barn?
9. A worker lifts 75 concrete blocks a distance of 1.50 m to the bed of a truck. Each block has a mass of 4.00 kg. How much work is done to lift all the blocks to the truck bed?
10. The work required to lift eleven 94.0-lb bags of cement from the ground to the back of a truck is 4340 ft lb. What is the distance from the ground to the bed of the truck?
11. How much work is done in lifting 450 lb of cement 75 ft above the ground?
12. How much work is done lifting a $20\bar{0}$-kg wrecking ball 6.50 m above the ground?
13. A gardener pushes a mower a distance of $90\bar{0}$ m in mowing a yard. The handle of the mower makes an angle of 40.0° with the ground. The gardener exerts a force of 35.0 N along the handle of the mower (Fig. 7.8). How much work does the gardener do in mowing the lawn?

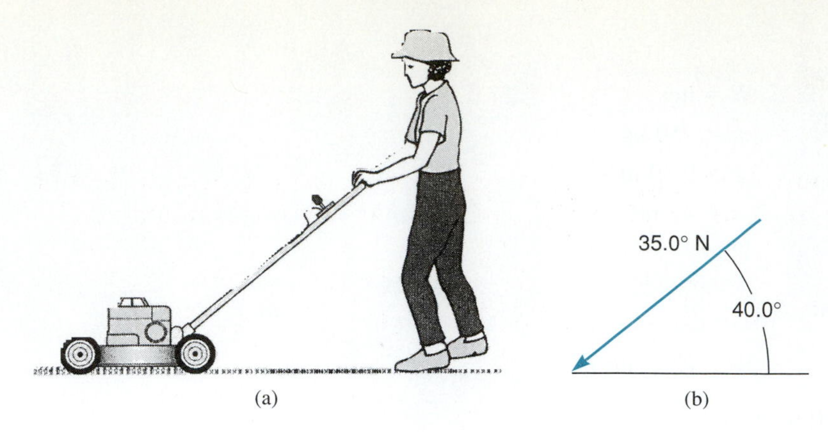

Figure 7.8

14. The handle of a vegetable wagon makes an angle of 25.0° with the horizontal (Fig. 7.9). If the peddler exerts a force of 35.0 lb along the handle, how much work does the peddler do in pulling the cart 1.00 mi?

Figure 7.9

15. A crate is pulled 675 ft across a warehouse floor by a rope that makes an angle of 50.0° with the floor. If 375 lb is exerted on the rope, how much work is done in pulling the crate across the floor?
16. A man pulls a sled a distance of 231 m. The rope attached to the sled makes an angle of 30.0° with the ground. The man exerts a force of 775 N on the rope. How much work does the man do in pulling the sled?
17. A tractor tows a barge through a canal with a towrope that makes an angle of 21° with the bank of the canal. If the tension in the rope is 12,000 N, how much work is done in moving the barge 550 m?
18. Two tractors tow a barge through a canal; each tractor uses a towrope that makes an angle of 21° with the bank of the canal. If the tension in each rope is 12,000 N, how much work is done in moving the barge 550 m?

7.2 POWER

Power is the rate of doing work; that is,

$$P = \frac{W}{t}$$

where P = power
W = work
t = time

The units of power are familiar to most of us. In the metric system the unit of power is the **watt.** Power is often expressed in kilowatts and megawatts.

$$P = \frac{W}{t} = \frac{Fs}{t} = \frac{\text{N m}}{\text{s}} = \frac{\text{J}}{\text{s}} = \text{watt}$$

and

$$1000 \text{ watts (W)} = 1 \text{ kilowatt (kW)}$$

$$1{,}000{,}000 \text{ watts} = 1 \text{ megawatt (MW)}$$

In the English system the unit of power is either ft lb/s or horsepower.

$$P = \frac{W}{t} = \frac{Fs}{t} = \frac{\text{ft lb}}{\text{s}}$$

Horsepower (hp) is a unit defined by and named after James Watt (who also designed the first practical steam engine).

1 horsepower (hp) = 550 ft lb/s = 33,000 ft lb/min

One horsepower is equivalent to moving a force of 550 lb one foot in one second or moving one pound 550 ft in one second.

Note: Since the above is a definition, treat any conversion factor as an exact number, which does not affect the number of significant digits in any calculation.

EXAMPLE 1

A freight elevator with operator weighs $50\bar{0}0$ N. If it is raised to a height of 15.0 m in 10.0 s, how much power is developed?

Data:

$$F = 50\bar{0}0 \text{ N}$$
$$s = 15.0 \text{ m}$$
$$t = 10.0 \text{ s}$$
$$P = ?$$

Basic Equation:

$$P = \frac{W}{t} \quad \text{and} \quad W = Fs$$

Working Equation:

$$P = \frac{Fs}{t}$$

Substitution:

$$P = \frac{(50\bar{0}0 \text{ N})(15.0 \text{ m})}{10.0 \text{ s}}$$
$$= 75\bar{0}0 \text{ N m/s}$$

EXAMPLE 2

The power expended in lifting a 825-lb girder to the top of a building $10\bar{0}$ ft high is 10.0 hp. How much time is required to raise the girder?

Data:

$$F = 825 \text{ lb}$$
$$s = 10\bar{0} \text{ ft}$$
$$P = 10.0 \text{ hp}$$
$$t = ?$$

Basic Equation:

$$P = \frac{W}{t} \quad \text{and} \quad W = Fs$$

Working Equation:

$$t = \frac{W}{P} = \frac{Fs}{P}$$

Substitution:

$$t = \frac{(825 \text{ lb})(10\bar{0} \text{ ft})}{10.0 \text{ hp}}$$
$$= \frac{(825 \text{ lb})(10\bar{0} \text{ ft})}{10.0 \text{ hp}} \times \frac{1 \text{ hp}}{550 \frac{\text{ft lb}}{\text{s}}}$$
$$= 15.0 \text{ s}$$

$$\frac{\text{lb ft}}{\text{hp}} \times \frac{\text{hp}}{\frac{\text{ft lb}}{\text{s}}} = \frac{\text{lb ft}}{\text{hp}} \times \left(\text{hp} \div \frac{\text{ft lb}}{\text{s}}\right) = \frac{\cancel{\text{lb}}\ \cancel{\text{ft}}}{} \times \left(\cancel{\text{hp}} \times \frac{\text{s}}{\cancel{\text{ft}}\ \cancel{\text{lb}}}\right) = \text{s}$$

Note: We use a conversion factor to obtain time units.

EXAMPLE 3

The mass of a large steel wrecking ball is $20\bar{0}0$ kg. What power is used to raise it to a height of 40.0 m if the work is done in 20.0 s?

Data:

$$m = 20\bar{0}0 \text{ kg}$$
$$s = 40.0 \text{ m}$$
$$t = 20.0 \text{ s}$$
$$P = ?$$

Basic Equation:

$$P = \frac{W}{t} \quad \text{and} \quad W = Fs$$

Working Equation:

$$P = \frac{Fs}{t}$$

Substitution: Note that we cannot directly substitute into the working equation because our data are given in terms of *mass* and we must find *force* to substitute in $P = Fs/t$. Recall that the force is the weight of the ball.

$$F = mg = (20\bar{0}0 \text{ kg})(9.80 \text{ m/s}^2) = 19{,}600 \text{ kg m/s}^2 = 19{,}600 \text{ N}$$

Then

$$\begin{aligned} P &= \frac{(19{,}600 \text{ N})(40.0 \text{ m})}{20.0 \text{ s}} \\ &= 39{,}200 \text{ N m/s} \\ &= 39{,}200 \text{ W} \quad \text{or} \quad 39.2 \text{ kW} \end{aligned}$$

EXAMPLE 4

A machine is designed to perform a given amount of work in a given amount of time. A second machine does the same amount of work in half the time. Find the power of the second machine compared with the first.

Data: (for the second machine given in terms of the first)

$$\begin{aligned} W &= W \\ t &= \tfrac{1}{2}t = \frac{t}{2} \\ P &= ? \end{aligned}$$

Basic Equation:

$$P = \frac{W}{t}$$

Working Equation: Same

Substitution:

$$P = \frac{W}{\frac{t}{2}} = W \div \frac{t}{2} = W \times \frac{2}{t} = 2\left(\frac{W}{t}\right) = 2P$$

Thus, the power is doubled when the time is halved.

EXAMPLE 5

A motor is capable of developing 10.0 kW of power. How large a mass can it lift 75.0 m in 20.0 s?

Data:

$$\begin{aligned} P &= 10.0 \text{ kW} = 10{,}\bar{0}00 \text{ W} \\ s &= 75.0 \text{ m} \\ t &= 20.0 \text{ s} \\ F &= ? \end{aligned}$$

Basic Equation:

$$P = \frac{W}{t} \quad \text{and} \quad W = Fs \quad \text{or} \quad P = \frac{Fs}{t}$$

Working Equation:

$$F = \frac{Pt}{s}$$

Substitution:

$$F = \frac{(10{,}\bar{0}00 \text{ W})(20.0 \text{ s})}{75.0 \text{ m}}$$

$$= 2670 \frac{\cancel{\text{W}}\ \cancel{\text{s}}}{\cancel{\text{m}}} \times \frac{1 \text{ N } \cancel{\text{m}}/\cancel{\text{s}}}{1 \cancel{\text{W}}} \qquad (1 \text{ W} = 1 \text{ J/s} = 1 \text{ N m/s})$$

$$= 2670 \text{ N}$$

Next, change the weight to mass as follows:

Data:

$F = 2670$ N
$g = 9.80 \text{ m/s}^2$
$m = ?$

Basic Equation:

$$F = mg$$

Working Equation:

$$m = \frac{F}{g}$$

Substitution:

$$m = \frac{2670 \cancel{\text{N}}}{9.80 \cancel{\text{m/s}^2}} \times \frac{1 \text{ kg } \cancel{\text{m/s}^2}}{1 \cancel{\text{N}}} \qquad (1 \text{ N} = 1 \text{ kg m/s}^2)$$

$$= 272 \text{ kg}$$

EXAMPLE 6

A pump is needed to lift $15\bar{0}0$ L of water per minute a distance of 45.0 m. What power, in kW, must the pump be able to deliver? (1 L of water has a mass of 1 kg.)

Data:

$m = 15\bar{0}0 \text{ L} \times \frac{1 \text{ kg}}{1 \text{ L}} = 15\bar{0}0 \text{ kg}$
$s = 45.0$ m
$t = 1 \text{ min} = 60.0$ s
$g = 9.80 \text{ m/s}^2$
$P = ?$

Basic Equation:

$$P = \frac{W}{t}, \; W = Fs, \quad \text{and} \quad F = mg \quad \text{or} \quad P = \frac{mgs}{t}$$

Working Equation:

$$P = \frac{mgs}{t}$$

Substitution:

$$P = \frac{(15\bar{0}0 \text{ kg})(9.80 \text{ m/s}^2)(45.0 \text{ m})}{60.0 \text{ s}}$$

$$= 1.10 \times 10^4 \text{ kg m}^2\text{/s} \qquad \left(1 \text{ W} = \frac{1 \text{ J}}{\text{s}} = \frac{1 \text{ N m}}{\text{s}} = \frac{1 \text{ (kg m/s}^2\text{)(m)}}{\text{s}} = 1 \text{ kg m}^2\text{/s}\right)$$

$$= 1.10 \times 10^4 \text{ W} \times \frac{1 \text{ kW}}{10^3 \text{ W}}$$

$$= 11.0 \text{ kW}$$

■ PROBLEMS 7.2

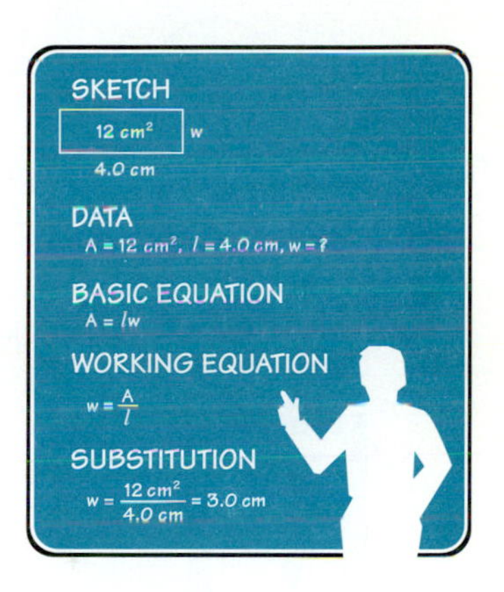

1. Given: $W = 132$ J
   $t = 7.00$ s
   $P = ?$
2. Given: $P = 231$ ft lb/s
   $t = 14.3$ s
   $W = ?$
3. Given: $P = 75.0$ W
   $W = 40.0$ J
   $t = ?$
4. Given: $W = 55.0$ J
   $t = 11.0$ s
   $P = ?$
5. The work required to lift a crate is $31\bar{0}$ J. If the crate can be lifted in 25.0 s, what power is developed?
6. A $36\bar{0}0$-lb automobile is pushed by its unhappy driver a quarter of a mile (0.250 mi) when it runs out of gas. To keep the car rolling, the driver must exert a constant force of 175 lb.
   (a) How much work does he do?
   (b) If it takes him 15.0 min, how much power does he develop?
   (c) Expressed in horsepower, how much power does he develop?
7. An electric golf cart develops 1.25 kW of power while moving at a constant speed.
   (a) Express its power in horsepower.
   (b) If the cart travels $20\bar{0}$ m in 35.0 s, what force is exerted by the cart?
8. How many seconds would it take a 7.00-hp motor to raise a 475-lb boiler to a platform 38.0 ft high?
9. How long would it take a $95\bar{0}$-W motor to raise a $36\bar{0}$-kg mass to a height of 16.0 m?
10. A $15\bar{0}0$-lb casting is raised 22.0 ft in 2.50 min. Find the required horsepower.
11. What is the rating in kW of a 2.00-hp motor?
12. A wattmeter shows that a motor is drawing $22\bar{0}0$ W. What horsepower is being delivered?
13. A 525-kg steel beam is raised 30.0 m in 25.0 s. How many kilowatts of power are needed?
14. How long would it take a 4.50-kW motor to raise a 175-kg boiler to a platform 15.0 m above the floor?
15. An escalator is needed to carry 75 passengers per minute a vertical distance of 8.0 m. Assume that the mass of each passenger is $7\bar{0}$ kg.
    (a) What is the power (in kW) of the motor needed?
    (b) Express this power in horsepower.
    (c) What is the power (in kW) of the motor needed if 35% of the power is lost to friction and other losses?
16. A pump is needed to lift $75\bar{0}$ L of water per minute a distance of 25.0 m. What power (in kW) must the pump be able to deliver? (1 L of water has a mass of 1 kg.)
17. A machine is designed to perform a given amount of work in a given amount of time. A second machine does twice the same amount of work in half the time. Find the power of the second machine compared with the first.
18. A machine is designed to perform a given amount of work in a given amount of

time. A second machine does 2.5 times the same amount of work in one-third of the time. Find the power of the second machine compared with the first.

19. A motor on an escalator is capable of developing 12 kW of power.
    (a) How many passengers of mass 75 kg each can it lift a vertical distance of 9.0 m per min, assuming no power loss?
    (b) What power, in kW, of motor is needed to move the same number of passengers at the same rate if 45% of the actual power developed by the motor is lost to friction and heat loss?
20. A pump is capable of developing 4.00 kW of power. How many litres of water per minute can be lifted a distance of 35.0 m? (1 L of water has a mass of 1 kg.) ■

7.3 ENERGY

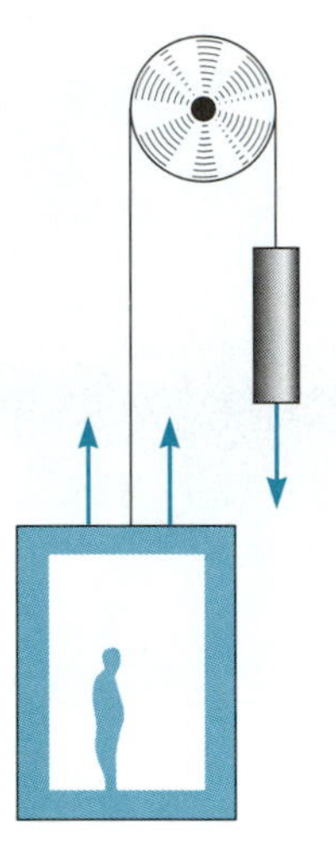

Figure 7.10
Potential energy of position.

Energy is defined as the ability to do work. There are many forms of energy, such as mechanical, electrical, thermal, fluid, chemical, atomic, and sound.

The mechanical energy of a body or a system is due to its position, its motion, or its internal structure. There are two kinds of mechanical energy: potential energy and kinetic energy. **Potential energy** is the stored energy of a body due to its internal characteristics or its position. **Kinetic energy** is the energy due to the mass and the velocity of a moving object.

Internal potential energy is determined by the nature or condition of the substance; for example, gasoline, a compressed spring, or a stretched rubber band has internal potential energy due to its internal characteristics. **Gravitational potential energy** is determined by the position of an object relative to a particular reference level; for example, a rock lying on the edge of a cliff, the raised counterweight on an elevator (Fig. 7.10), or a raised pile driver has potential energy due to its position. Each weight has the ability to do work because of the pull of gravity on it. The unit of energy is the joule (J) in the metric system and the foot-pound (ft lb) in the English system.

The formula for gravitational potential energy is:

$$\text{potential energy (PE)} = mgh$$

where m = mass
$g = 9.80 \text{ m/s}^2$ or 32.2 ft/s^2
h = height above reference level

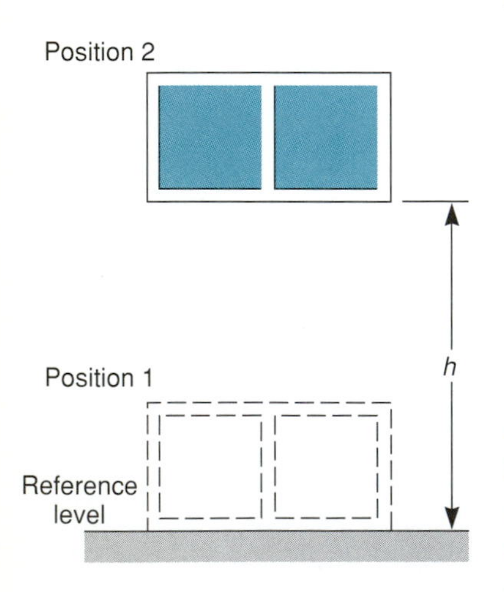

Figure 7.11
Work done in raising the crate gives it potential energy.

In position 1 (Fig. 7.11), the crate is at rest on the floor. It has no ability to do work as it is in its lowest position. To raise the crate to position 2, work must be done to lift it. In the raised position, however, it now has stored ability to do work (by falling to the floor!). Its PE (potential energy) can be calculated by multiplying the mass of the crate times acceleration of gravity *(g)* times height above reference level *(h)*. Note that we could calculate the potential energy of the crate with respect to any level we may choose. Here we have chosen the floor as the zero or lowest reference level.

EXAMPLE 1

A wrecking ball of mass $20\overline{0}$ kg is poised 4.00 m above a concrete platform whose top is 2.00 m above the ground.

(a) With respect to the platform, what is the potential energy of the ball?
(b) With respect to the ground, what is the potential energy of the ball?

Sketch:

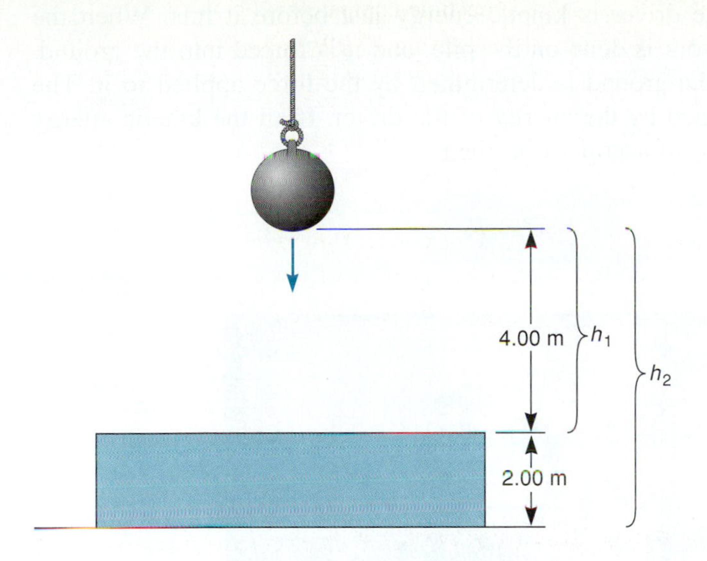

Data:

$$m = 20\bar{0}\ \text{kg}$$
$$h_1 = 4.00\ \text{m}$$
$$h_2 = 6.00\ \text{m}$$
$$\text{PE} = ?$$

Basic Equation:

$$\text{PE} = mgh$$

Working Equation: Same

(a) Substitution:

$$\text{PE} = (20\bar{0}\ \text{kg})(9.80\ \text{m/s}^2)(4.00\ \text{m})$$

$$= 7840\ \frac{\text{kg m}^2}{\text{s}^2} \times \frac{1\ \text{J}}{\text{kg m}^2/\text{s}^2} \qquad [1\ \text{J} = 1\ \text{N m} = 1\ (\text{kg m/s}^2)(\text{m}) = 1\ \text{kg m}^2/\text{s}^2]$$

$$= 7840\ \text{J}$$

(b) Substitution:

$$\text{PE} = (20\bar{0}\ \text{kg})(9.80\ \text{m/s}^2)(6.00\ \text{m})$$

$$= 11{,}800\ \frac{\text{kg m}^2}{\text{s}^2} \times \frac{1\ \text{J}}{\text{kg m}^2/\text{s}^2}$$

$$= 11{,}800\ \text{J}$$

Kinetic energy is due to the mass and the velocity of a moving object and given by the formula:

$$\text{kinetic energy (KE)} = \tfrac{1}{2}mv^2$$

where m = mass of moving object
v = velocity of moving object

A pile driver (Fig. 7.12) shows the relation of energy of motion to useful work. The energy of the driver is kinetic energy just before it hits. When the driver strikes the pile, work is done on the pile, and it is forced into the ground. The depth it goes into the ground is determined by the force applied to it. The force applied is determined by the energy of the driver. If all the kinetic energy of the driver is converted to useful work, then

$$\frac{1}{2}mv^2 = Fs.$$

Figure 7.12
Energy of motion becomes useful work in the pile driver.

EXAMPLE 2

The driver has a mass of $10,\bar{0}00$ kg and upon striking the pile has a velocity of 10.0 m/s.

(a) What is the kinetic energy of the driver as it strikes the pile?

(b) If the pile is driven 20.0 cm into the ground, what force is applied to the pile? Assume that all the kinetic energy of the driver is converted to work.

Sketch:

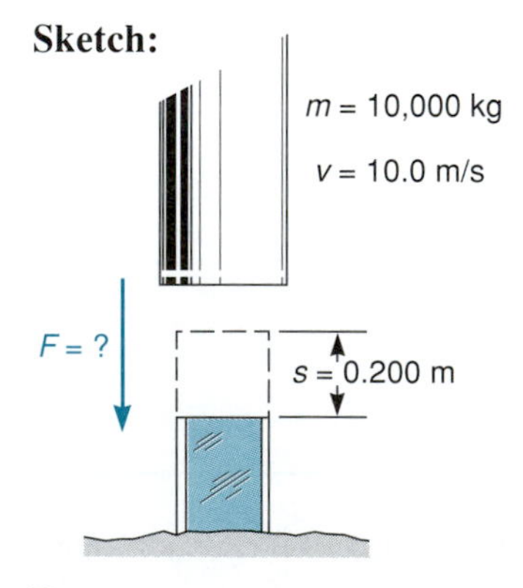

Data:

$$m = 1.00 \times 10^4 \text{ kg}$$
$$v = 10.0 \text{ m/s}$$
$$s = 20.0 \text{ cm} = 0.200 \text{ m}$$
$$F = ?$$

(a) Basic Equation:

$$KE = \tfrac{1}{2}mv^2$$

Working Equation: Same

Substitution:

$$KE = \tfrac{1}{2}(1.00 \times 10^4 \text{ kg})(10.0 \text{ m/s})^2$$

$$= 5.00 \times 10^5 \frac{\cancel{\text{kg}}\,\cancel{\text{m}^2}}{\cancel{\text{s}^2}} \times \frac{1 \text{ J}}{\cancel{\text{kg}}\,\cancel{\text{m}^2}/\cancel{\text{s}^2}} \qquad [1 \text{ J} = 1 \text{ N m} = 1 \text{ (kg m/s}^2\text{)(m)} = 1 \text{ kg m}^2/\text{s}^2]$$

$$= 5.00 \times 10^5 \text{ J} \quad \text{or} \quad 50\bar{0} \text{ kJ}$$

(b) Basic Equation:

$$KE = W = Fs$$

Working Equation:

$$F = \frac{KE}{s} \qquad \text{[Use KE from part (a).]}$$

Substitution:

$$F = \frac{5.00 \times 10^5 \cancel{\text{J}}}{0.200 \cancel{\text{m}}} \times \frac{1 \text{ N} \cancel{\text{m}}}{1 \cancel{\text{J}}} \qquad (1 \text{ J} = 1 \text{ N m})$$

$$= 2.50 \times 10^6 \text{ N}$$

EXAMPLE 3

A 60.0-g bullet is fired from a gun and has 3150 J of kinetic energy. Find its velocity.

Data:

$KE = 3150$ J

$m = 60.0 \text{ g} = 0.0600 \text{ kg}$

$v = ?$

Basic Equation:

$$KE = \tfrac{1}{2}mv^2$$

Working Equation:

$$v = \sqrt{\frac{2(KE)}{m}}$$

Substitution:

$$v = \sqrt{\frac{2(3150 \cancel{\text{J}})}{0.0600 \cancel{\text{kg}}} \times \frac{1 \cancel{\text{kg}} \text{ m}^2/\text{s}^2}{1 \cancel{\text{J}}}} \qquad [1 \text{ J} = 1 \text{ N m} = 1 \text{ (kg m/s}^2\text{)(m)} = 1 \text{ kg m}^2/\text{s}^2]$$

$$v = 324 \text{ m/s}$$

We have discussed only two types of energy—kinetic and potential. Keep in mind that energy exists in many forms—chemical, atomic, electrical, sound, and heat. These forms and the conversion of energy from one form to another will be studied later.

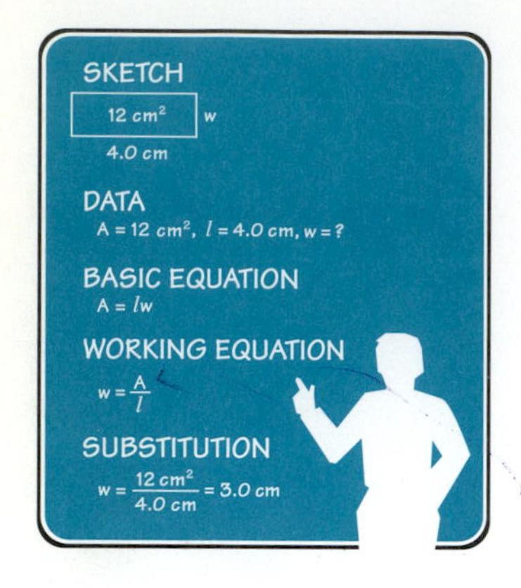

■ PROBLEMS 7.3

1. Given: $m = 11.4$ kg
   $g = 9.80$ m/s^2
   $h = 22.0$ m
   PE = ?
2. Given: $m = 3.50$ kg
   $g = 9.80$ m/s^2
   $h = 15.0$ m
   PE = ?
3. Given: $m = 4.70$ kg
   $v = 9.60$ m/s
   KE = ?
4. Given: PE = 93.6 J
   $g = 9.80$ m/s^2
   $m = 2.30$ kg
   $h = ?$
5. A truck with mass $95\overline{0}$ slugs is driven along a highway with a velocity of 55.0 mi/h.
   (a) What is its velocity in ft/s?
   (b) What is the kinetic energy of the truck?
6. A bullet with mass 12.0 g travels at 415 m/s. What is its kinetic energy? (*Hint:* Convert 12.0 g to kg.)
7. A bicycle and rider together have a mass of 7.40 slugs. If the kinetic energy is 742 ft lb, what is the velocity?
8. A crate of mass 475 kg is raised to a height of 17.0 m above the floor. What potential energy has it acquired with respect to the floor?
9. A tank of water containing $25\overline{0}0$ L of water is stored on the roof of a building.
   (a) What is its potential energy with respect to the floor, which is 12.0 m below the roof?
   (b) What is its potential energy with respect to the basement, which is 4.0 m below the first floor?
10. The potential energy possessed by a girder after being lifted to the top of a building is 5.17×10^5 ft lb. If the mass of the girder is 173 slugs, how high is the girder?
11. A 30.0-g bullet is fired from a gun and possesses 1750 J of kinetic energy. Find its velocity.
12. Hoover Dam is 726 ft high. Find the potential energy of 1.00 million ft^3 of water at the top of the dam. *Note:* 1 ft^3 of water weighs 62.4 lb.
13. A 250-kg part falls from a plane and hits the ground at 150 km/h. What is its kinetic energy?
14. A meteorite is a solid composed of stone and/or metal material from outer space that passes through the earth's atmosphere and hits the surface of the earth. Find the kinetic energy of a meteorite with mass 250 kg that collides with the earth at a speed of 25 km/s.
15. Water is pumped at the rate of $25\overline{0}$ m^3/min from a lake into a tank that is 65.0 m above the lake.
    (a) What power (in kW) must be delivered by the pump?
    (b) What horsepower rating does this pump motor have?
    (c) What is the increase in potential energy of the water each minute?
16. Oil is pumped at the rate of 25.0 m^3/min into a tank 10.0 m above the ground. (1 L of oil has a mass of 0.68 kg.)
    (a) What power, in kW, must be delivered by the pump?
    (b) What is the increase in potential energy of the oil after 10.0 min?
    (c) What would be the increase in potential energy of the oil after 10.0 min if the tank were 5.00 m above the ground?
17. If the velocity of an object is doubled, by what factor is its kinetic energy increased?
18. If the kinetic energy of an object is doubled, by what factor is its velocity increased? ■

7.4 CONSERVATION OF MECHANICAL ENERGY

Kinetic and potential energy are related by an important principle:

> *LAW OF CONSERVATION OF MECHANICAL ENERGY*
>
> The sum of the kinetic energy and the potential energy in a system is constant if no resistant forces do work.

A pile driver shows this energy conservation. When the driver is at its highest position, the potential energy is maximum and the kinetic energy is zero [Fig. 7.13(a)]. Its potential energy is

$$\text{PE} = mgh$$

and it kinetic energy is

$$\text{KE} = \tfrac{1}{2}mv^2 = \tfrac{1}{2}m(0)^2 = 0$$

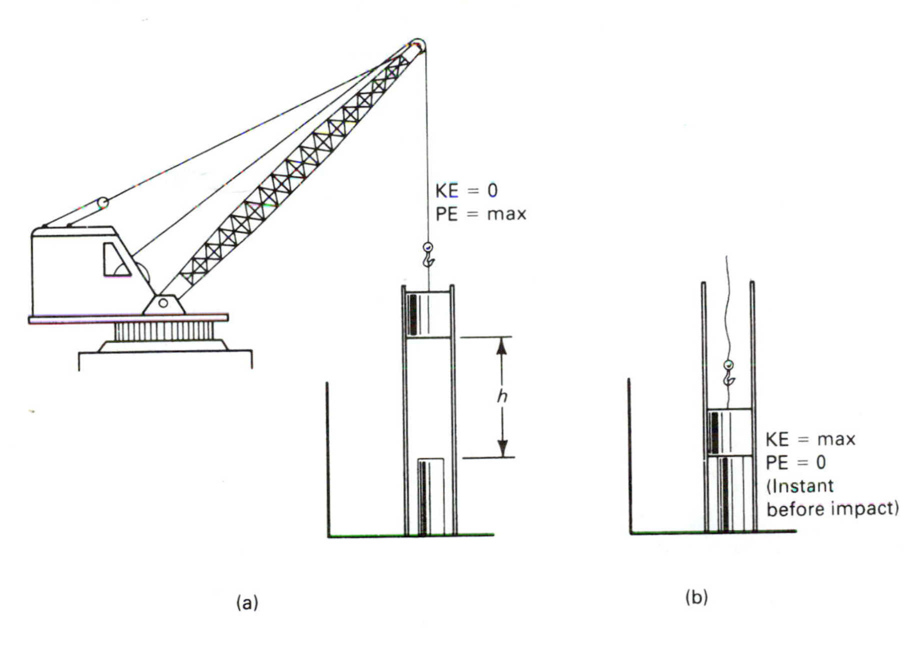

Figure 7.13

When the driver hits the top of the pile [Fig. 7.13(b)], it has its maximum kinetic energy and the potential energy is

$$\text{PE} = mgh = mg(0) = 0$$

Since the total energy in the system must remain constant, the maximum potential energy must equal the maximum kinetic energy.

$$\text{PE}_{\text{max}} = \text{KE}_{\text{max}}$$

$$mgh = \tfrac{1}{2}mv^2$$

Solving for the velocity of the driver just before it hits the pile gives

$$v = \sqrt{2gh}$$

EXAMPLE 1

A pile driver falls freely from a height of 3.50 m above a pile. What is its velocity as it hits the pile?

Data:

$$h = 3.50 \text{ m}$$
$$g = 9.80 \text{ m/s}^2$$
$$v = ?$$

Basic Equation:

$$v = \sqrt{2gh}$$

Working Equation: Same

Substitution:

$$v = \sqrt{2(9.80 \text{ m/s}^2)(3.50 \text{ m})}$$
$$= 8.28 \text{ m/s}$$

$$\boxed{\sqrt{\text{m}^2/\text{s}^2} = \text{m/s}}$$

The conservation of mechanical energy can also be illustrated by considering a swinging pendulum bob where there is no resistance involved. Pull the bob over to the right side so that the string makes an angle of 65° with the vertical [Fig. 7.14(a)]. At this point, the bob contains its maximum potential energy and its minimum kinetic energy (zero). Note that a larger maximum potential energy is possible when an initial deflection of greater than 65° is made.

An instant later, the bob has lost some of its potential or stored energy, but it has gained in kinetic energy due to its motion [Fig. 7.14(b)]. At the bottom of

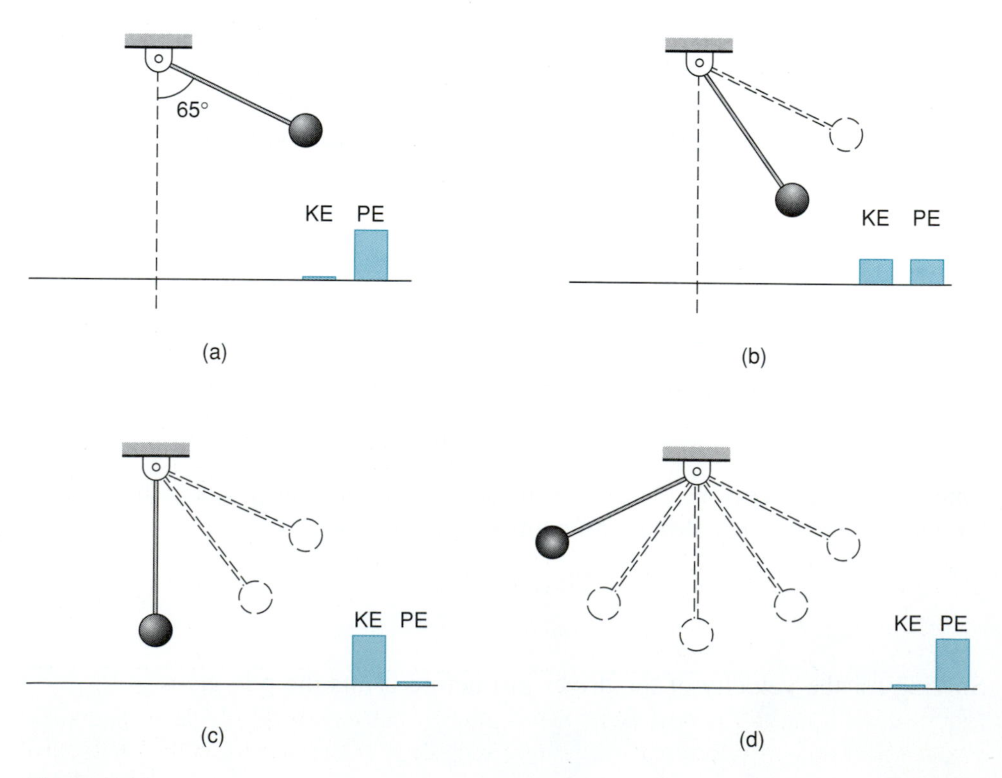

Figure 7.14
Kinetic and potential energy changes in a pendulum.

its arc of swing [Fig. 7.14(c)], its potential energy is zero and its kinetic energy is maximum (its velocity is maximum). The kinetic energy of the bob then causes the bob to swing upward to the left. As it completes its swing [Fig. 7.14(d)], its kinetic energy is decreasing and its potential energy is increasing. That is, its kinetic energy is changing to potential energy.

According to the law of conservation of mechanical energy, the sum of the kinetic energy and the potential energy of the bob at any instant is a constant. Assuming no resistant forces, such as friction or air resistance, the bob would swing uniformly "forever."

EXAMPLE 2

Drop a 5.000-kg mass from a hot air balloon 400.0 m above the ground. Find its kinetic energy, its potential energy, and the sum of the kinetic energy and the potential energy in one-second intervals until the mass hits the ground. (Assume no air resistance.)

Data:

$$m = 5.000 \text{ kg}$$
$$g = 9.80 \text{ m/s}^2$$

Fill in each column as follows:

In column (1), list the times, t, in 1.000 s increments until the mass hits the ground.

In column (2), use $s = \frac{1}{2}gt^2$ to find the distance, s, the mass has fallen from the air balloon at each time, t, rounded to the nearest 0.1 m.

In column (3), use $v = \sqrt{2gh} = \sqrt{2gs}$ to find its velocity at each time, t, rounded to the nearest 0.01 m/s.

In column (4), use $\text{KE} = \frac{1}{2}mv^2$ to find the kinetic energy at each time, t, rounded to the nearest 10 J.

In column (5), use $h = 400.0 \text{ m} - s$ to find the height above the ground at each time, t, rounded to the nearest 0.1 m.

In column (6), use $\text{PE} = mgh$ to find the potential energy at each time, t, rounded to the nearest 10 J.

In column (7), find the sum of the KE and PE columns at each time, t.

(1) t (s)	(2) s (m)	(3) v (m/s)	(4) KE (J)	(5) h (m)	(6) PE (J)	(7) Total (J)
0.000	0.0	0.00	0	400.0	19,6$\bar{0}$0	19,6$\bar{0}$0
1.000	4.9	9.80	240	395.1	19,360	19,6$\bar{0}$0
2.000	19.6	19.60	960	380.4	18,640	19,6$\bar{0}$0
3.000	44.1	29.40	2,160	355.9	17,440	19,6$\bar{0}$0
4.000	78.4	39.20	3,840	321.6	15,760	19,6$\bar{0}$0
5.000	122.5	49.00	6,0$\bar{0}$0	277.5	13,6$\bar{0}$0	19,6$\bar{0}$0
6.000	176.4	58.80	8,640	223.6	10,960	19,6$\bar{0}$0
7.000	240.1	68.60	11,760	159.9	7,840	19,6$\bar{0}$0
8.000	313.6	78.40	15,370	86.4	4,230	19,6$\bar{0}$0
9.000	396.9	88.20	19,450	3.1	150	19,6$\bar{0}$0
9.035	400.0	88.54	19,6$\bar{0}$0	0.0	0	19,6$\bar{0}$0

As you can see from the table, the sum of the kinetic energy and the potential energy at each time, t, is constant according to the **Law of Conservation of Mechanical Energy.**

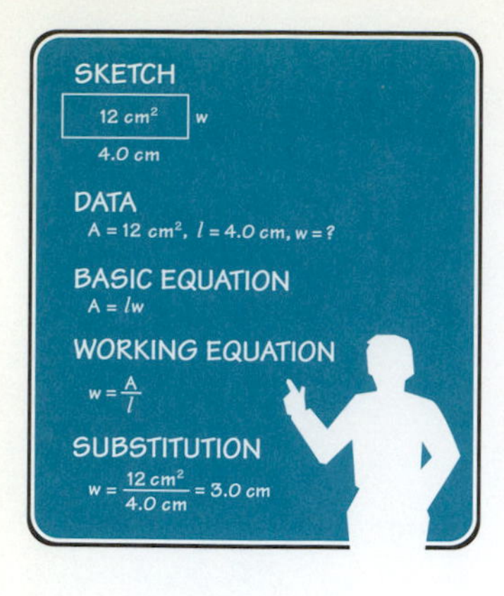

■ PROBLEMS 7.4

1. A pile driver falls through a distance of 2.50 m before hitting a pile. What is the velocity of the driver as it hits the pile?
2. A sky diver jumps out of a plane at a height of $50\bar{0}0$ ft. If her parachute does not open until she reaches $10\bar{0}0$ ft, what is her velocity at that point if air resistance is neglected?
3. A piece of shattered glass falls from the 82nd floor of a skyscraper, $27\bar{0}$ m above the ground. What is the velocity of the glass when it hits the ground, if air resistance is neglected?
4. A 10.0 kg mass is dropped from an air balloon at a height of 325 m above the ground. Find its speed at points $30\bar{0}$ m, $20\bar{0}$ m, and $10\bar{0}$ m above the ground and as it hits the ground.
5. A 0.175-lb ball is thrown upward with an initial velocity of 75.0 ft/s. What is the maximum height reached by the ball?
6. A pile driver falls through a distance of 1.75 m before hitting a pile. Find the velocity of the driver as it hits the pile.
7. A sandbag is dropped from a hot air balloon at a height of 125 m above the ground. What is the velocity of the sandbag just before it hits the ground? Ignore the effects of air resistance.
8. A window breaks near the top of a skyscraper. If the glass fragments drop 175 m before hitting the ground, what is the velocity of the fragments as they reach the ground? Ignore air resistance.
9. A ball is thrown downward from the top of a building at a speed of 75 ft/s. Find the velocity of the ball as it hits the ground 475 ft below. Ignore air resistance.
10. Find the maximum height reached by a ball thrown upward at a velocity of 95 ft/s.
11. Drop a 4.000-kg mass from an air balloon 300.0 m above the ground. Find its kinetic energy, its potential energy, and the sum of the kinetic energy and the potential energy in one-second intervals until the mass hits the ground. (Assume no air resistance.)
12. A 2.00-kg projectile is fired vertically upward with an initial velocity of 98.0 m/s. Find its kinetic energy, its potential energy, and the sum of its kinetic and potential energies at each of the following times:

(a)	the instant of its being fired	(e)	$t = 10.00$ s
(b)	$t = 1.00$ s	(f)	$t = 12.00$ s
(c)	$t = 2.00$ s	(g)	$t = 15.00$ s
(d)	$t = 5.00$ s	(h)	$t = 20.00$ s

■

GLOSSARY

Energy The ability to do work. There are many forms of energy, such as mechanical, electrical, thermal, fluid, chemical, atomic, and sound. (p. 178)

Gravitational Potential Energy The energy determined by the position of an object relative to a particular reference level; for example, a rock lying on the edge of a cliff, the raised counterweight on an elevator, or a raised pile driver has potential energy due to its position. (p. 178)

Horsepower The English system unit of power, defined as equivalent to 550 ft lb/s. (p. 173)

Internal Potential Energy The energy determined by the nature or condition of the substance; for example, gasoline, a compressed spring, or a stretched rubber band has internal potential energy due to its internal characteristics. (p. 178)

Kinetic Energy The energy due to the mass and the velocity of a moving object. (p. 178)

Law of Conservation of Mechanical Energy The sum of the kinetic energy and the potential energy in a system is constant, if no resistant forces do work. (p. 183)

Potential Energy The stored energy of a body due to its internal characteristics or its position. (p. 178)

Power The rate of doing work (work divided by time). (p. 172)

Work The product of the force in the direction of motion and the displacement. (p. 167)

FORMULAS

7.1 $W = Fs$

$W = Fs \cos \theta$

7.2 $P = \dfrac{W}{t}$

7.3 $\text{PE} = mgh$

$\text{KE} = \frac{1}{2}mv^2$

7.4 $v = \sqrt{2gh}$

REVIEW QUESTIONS

1. Work is done when
   (a) a force is applied.
   (b) a person tries unsuccessfully to move a crate.
   (c) force is applied and an object is moved.
2. Power
   (a) is work divided by time.
   (b) is measured in newtons.
   (c) is time divided by work.
   (d) none of the above.
3. A large boulder at rest possesses
   (a) potential energy. (b) kinetic energy. (c) no energy.
4. A large boulder rolling down a hill possesses
   (a) potential energy. (b) kinetic energy. (c) no energy.
   (d) both kinetic and potential energy.
5. With no air resistance and no friction, a pendulum would
   (a) not swing. (b) swing for a short time.
   (c) swing forever.
6. Can work be done by a moving object on itself?
7. Has a man done work if in swinging a sledgehammer he misses the stake at which he is swinging?
8. Develop the units associated with work from the components of the definition: work = force × displacement.
9. Is work a vector quantity?
10. Is work being done on a boulder by gravity?
11. Is work being done by a grandfather clock weight?
12. How could the power developed by a man pushing a stalled car be measured?
13. How does water above a waterfall possess potential energy?
14. What are two devices possessing gravitational potential energy?
15. Is kinetic energy dependent on time?
16. At what point is the kinetic energy of a swinging pendulum bob at a maximum?
17. At what point is the potential energy of a swinging pendulum bob at a maximum?

18. Is either kinetic or potential energy a vector quantity?
19. Can an object possess both kinetic and potential energy at the same time?
20. Why is a person more likely to be severely injured by a bolt falling from the fourth floor of a job site than by one falling from the second floor?

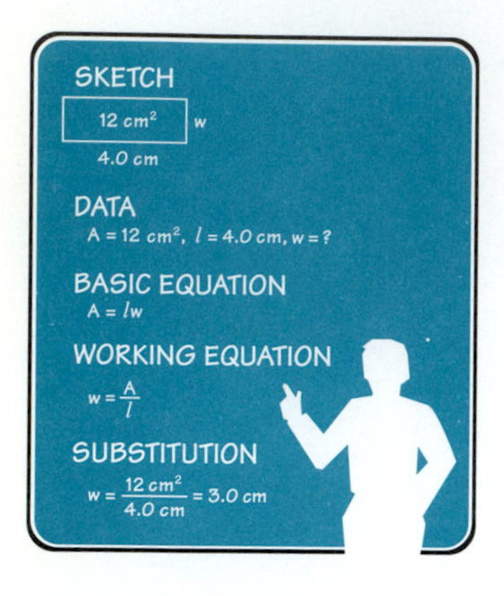

REVIEW PROBLEMS

1. How many joules are in one kilowatt-hour?
2. An endloader holds $20\bar{0}0$ kg of sand 2.00 m off the ground for 2.00 min. How much work does it do?
3. How high can a 10.0-kg mass be lifted by $10\bar{0}0$ J of work?
4. A 40.0-kg pack is carried up a $25\bar{0}0$-m-high mountain in 10.0 h. How much work is performed?
5. What is the average power output in Problem 4 in (a) watts; (b) horsepower?
6. A 10.0-kg mass has a potential energy of 10.0 J when it is at what height?
7. A 10.0-lb weight has a potential energy of 20.0 ft lb at what height?
8. At what speed does a 1.00-kg mass have a kinetic energy of 1.00 J?
9. At what speed does a 10.0-N weight have a kinetic energy of 1.00 J?
10. What is the kinetic energy of a $30\bar{0}0$-lb automobile moving at 55.0 mi/h?
11. What is the potential energy of an 80.0-kg diver standing 3.00 m above the water?
12. What is the kinetic energy of a 0.020-kg bullet having a velocity of 550 m/s?
13. What is the potential energy of an 85.0-kg high jumper clearing a 2.00-m bar?
14. A worker pushes a crate 10.0 m by exerting a force of $20\bar{0}$ N. How much work does he do?
15. How much work is done by the person pulling the crate in Problem 14 a distance of 10.0 m by exerting the same force at an angle of 20.0° with the horizontal?
16. A hammer falls from a scaffold on a building 50.0 m above the ground. What is its speed as it hits the ground?

Chapter 8

SIMPLE MACHINES

Machines may be used to transfer energy, multiply force, or multiply speed. We use them to obtain a mechanical advantage—to do something more efficiently that would be difficult or impossible without mechanical help. In this chapter we examine simple machines and their usefulness in technology.

OBJECTIVES

The major goals of this chapter are to enable you to:

1. Determine how energy is transferred using simple machines.
2. Analyze the efficiency and mechanical advantages of simple machines.
3. Distinguish the three types of levers and the mechanical advantage of each.
4. Analyze the mechanical advantage of the wheel-and-axle, the pulley, the inclined plane, and the screw.

8.1 MACHINES AND ENERGY TRANSFER

A **machine** is used to transfer energy from one place to another and allows us to do work that could not otherwise be done or could not be done as easily. By using a pulley system [Fig. 8.1(a)], one person can easily lift an engine from an automobile. Pliers [Fig. 8.1(b)] allow a person to cut a wire or turn a nut with the strength of his or her hand.

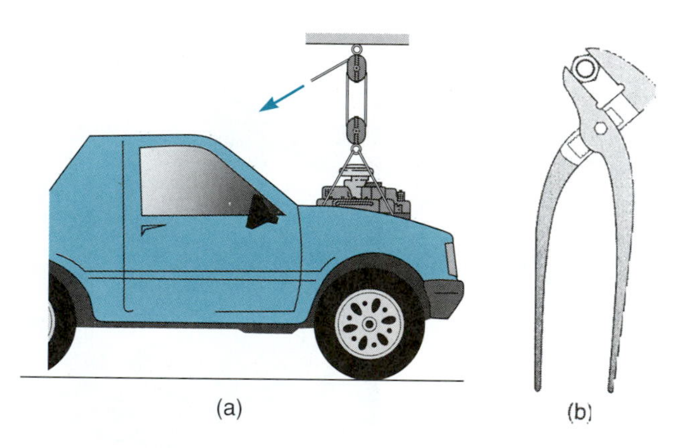

Figure 8.1
Simple machines used to transfer energy.

Machines are sometimes used to multiply force. By pushing with a small force, we can use a machine to jack up an automobile [Fig. 8.2(a)]. Machines are sometimes used to multiply speed. The gears on a bicycle [Fig. 8.2(b)] are used to multiply speed. Machines are used to change direction. When we use a single fixed pulley on a flag pole to raise a flag [Fig. 8.2(c)], the only advantage we get is the change in direction. (We pull the rope down, and the flag goes up.)

There are six basic or simple machines (Fig. 8.3). All other machines—no matter how complex—are combinations of two or more of these simple machines.

In every machine we are concerned with two forces—effort and resistance. The **effort** is the force applied *to* the machine. The **resistance** is the force over-

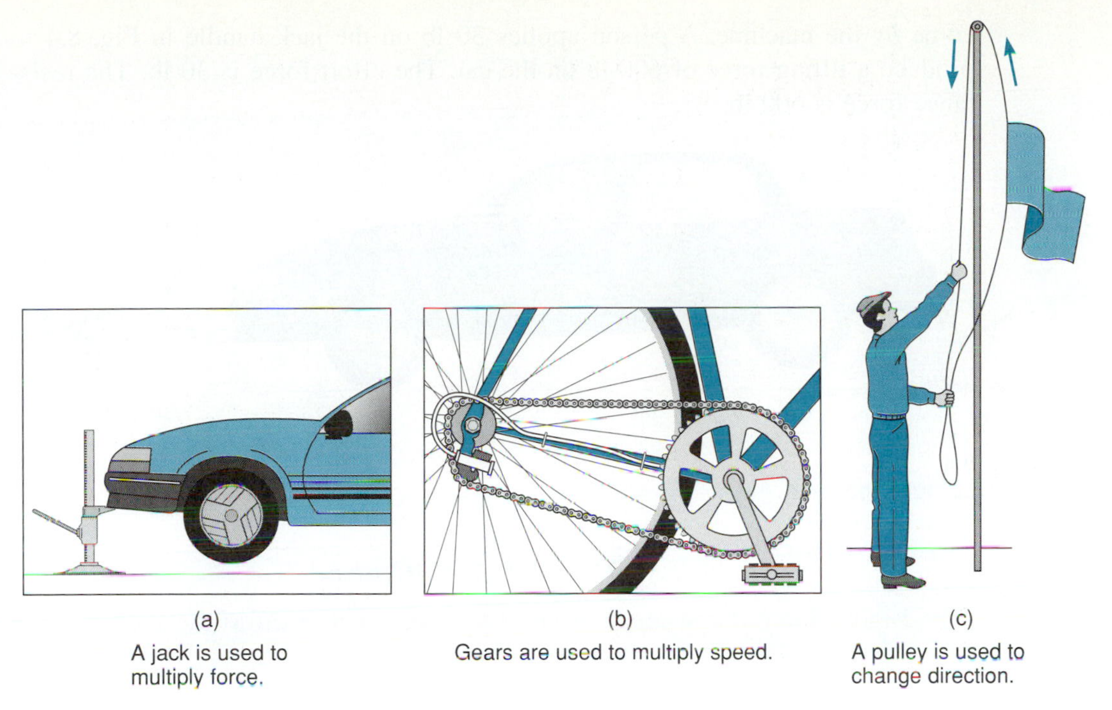

(a) A jack is used to multiply force.

(b) Gears are used to multiply speed.

(c) A pulley is used to change direction.

Figure 8.2
Simple machines may be used to multiply force or speed, or change direction.

Figure 8.3
Six basic or simple machines.

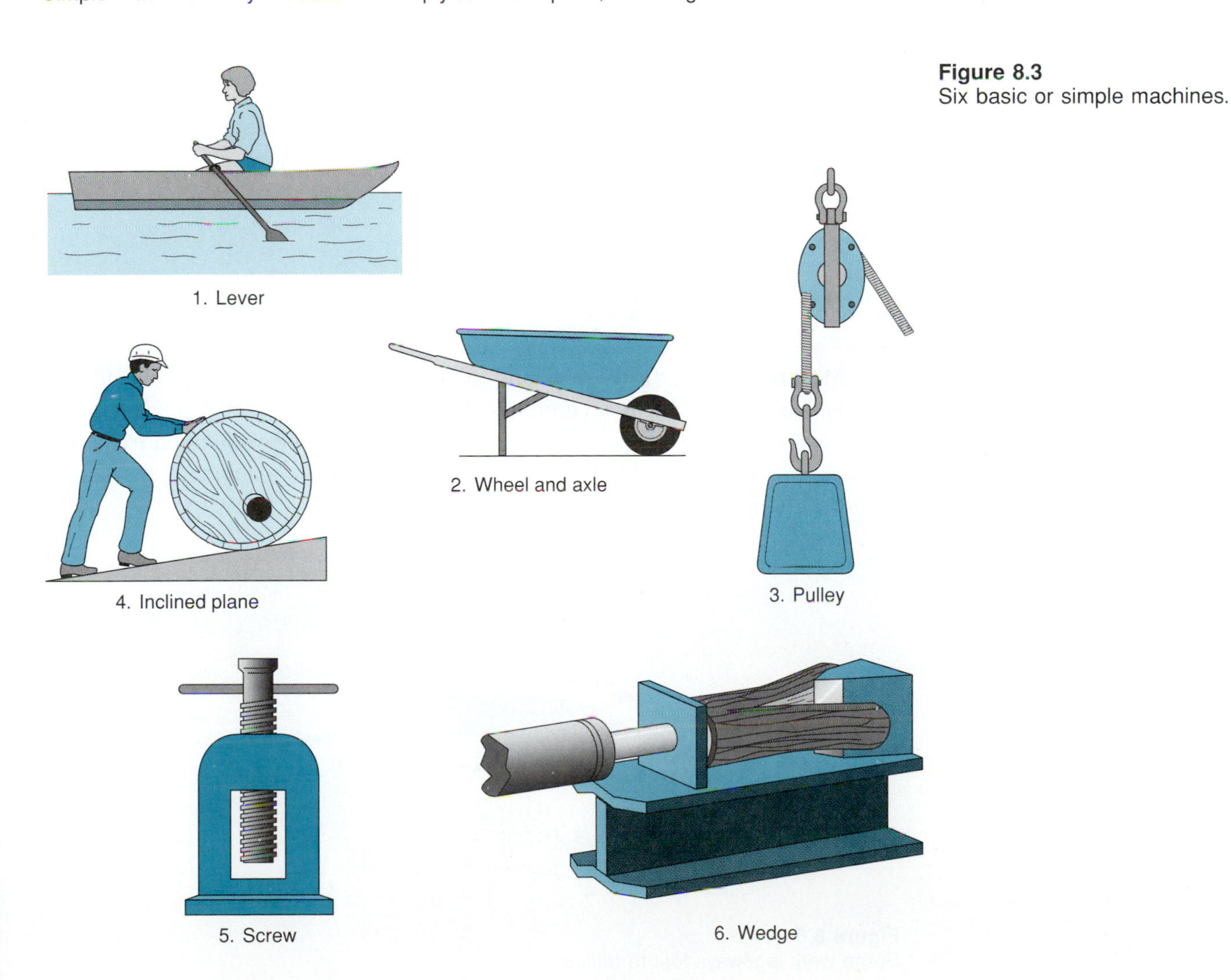

come *by* the machine. A person applies $3\bar{0}$ lb on the jack handle in Fig. 8.4 to produce a lifting force of $6\bar{0}0$ lb on the car. The effort force is $3\bar{0}$ lb. The resistance force is $6\bar{0}0$ lb.

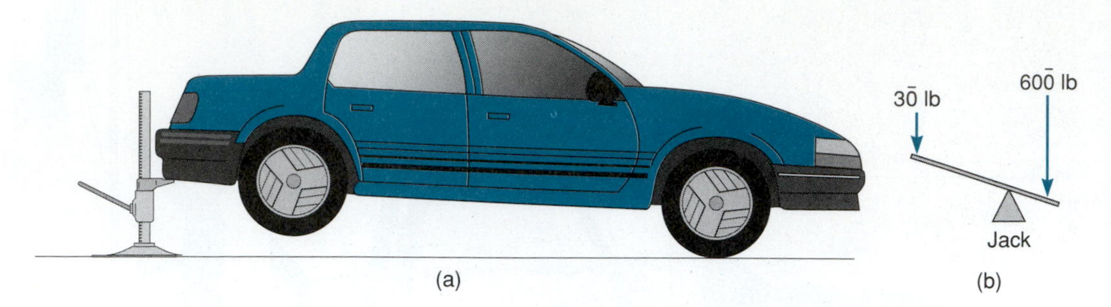

Figure 8.4
This lever multiplies force.

> *LAW OF SIMPLE MACHINES*
>
> resistance force × resistance distance = effort force × effort distance

8.2 MECHANICAL ADVANTAGE AND EFFICIENCY

The **mechanical advantage** is the ratio of the resistance force to the effort force. By formula:

$$\text{MA} = \frac{\text{resistance force}}{\text{effort force}}$$

The MA of the jack in Fig. 8.4 is found as follows:

$$\text{MA} = \frac{\text{resistance force}}{\text{effort force}} = \frac{6\bar{0}0 \text{ lb}}{3\bar{0} \text{ lb}} = \frac{2\bar{0}}{1}$$

This MA means that, for each pound applied by the person, he or she lifts $2\bar{0}$ pounds. Note that MA has no units. Why?

Each time a machine is used, part of the energy or effort applied to the machine is lost due to friction (Fig. 8.5). The **efficiency** of a machine is the ratio of the work output to the work input. By formula:

$$\text{efficiency} = \frac{\text{work output}}{\text{work input}} \times 100\% = \frac{F_{\text{output}} \times s_{\text{output}}}{F_{\text{input}} \times s_{\text{input}}} \times 100\%$$

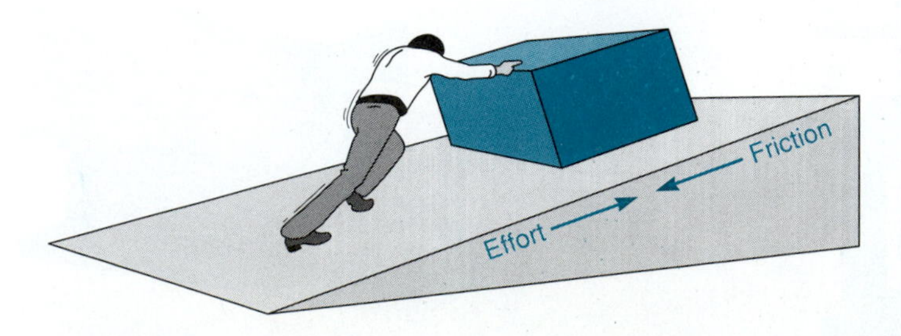

Figure 8.5
Some work is always lost to friction.

8.3 THE LEVER

A **lever** consists of a rigid bar free to turn on a pivot called a **fulcrum** (Fig. 8.6). The mechanical advantage (MA) is the ratio of the effort arm (d_E) to the resistance arm (d_R).

$$MA_{lever} = \frac{\text{effort arm}}{\text{resistance arm}} = \frac{d_E}{d_R}$$

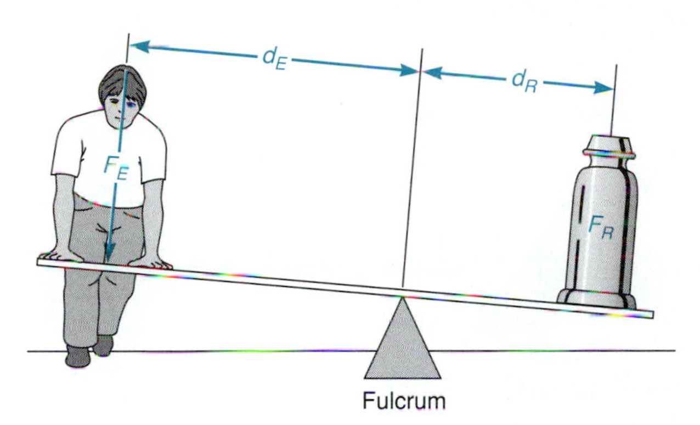

Figure 8.6
Mechanical advantage of the lever: MA $= \frac{d_E}{d_R}$.

The **effort arm** is the distance from the effort to the fulcrum. The **resistance arm** is the distance from the fulcrum to the resistance. The three types or classes of levers are shown in Fig. 8.7. The law of simple machines as applied to levers (basic equation) is

$$F_R \cdot d_R = F_E \cdot d_E$$

where F_R = resistance force
d_R = length of resistance arm
F_E = effort force
d_E = length of effort arm

EXAMPLE 1

A bar is used to raise an $18\bar{0}0$-N stone. The pivot is placed 30.0 cm from the stone. The worker pushes 2.50 m from the pivot. What is the mechanical advantage? What force is exerted?

Sketch:

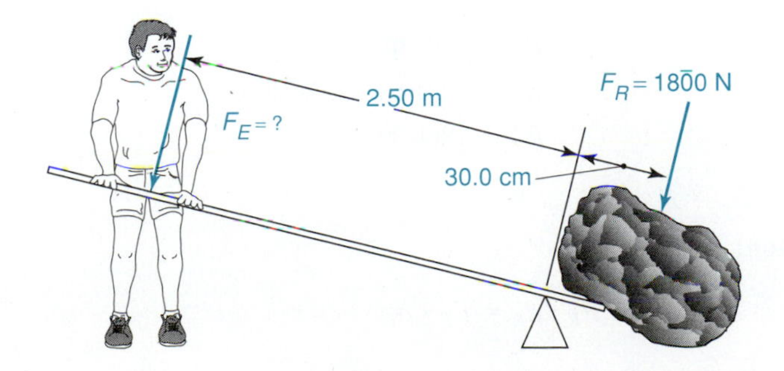

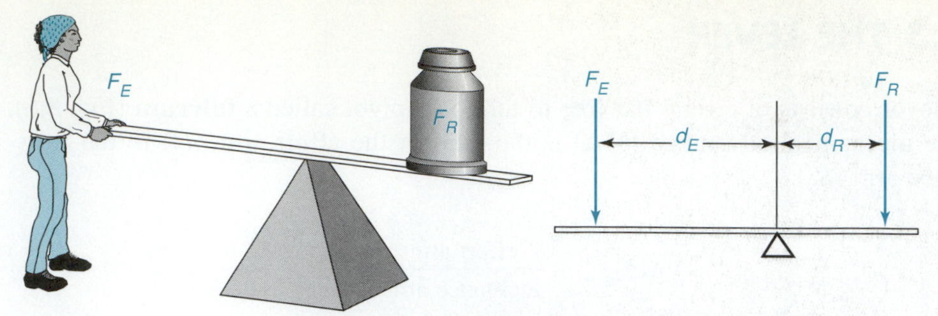

First class: The fulcrum is between the resistance force (F_R) and the effort force (F_E).

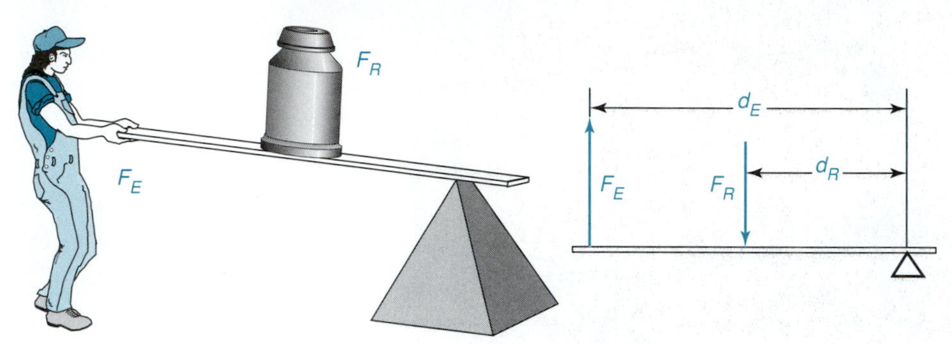

Second class: The resistance force (F_R) is between the fulcrum and the effort force (F_E).

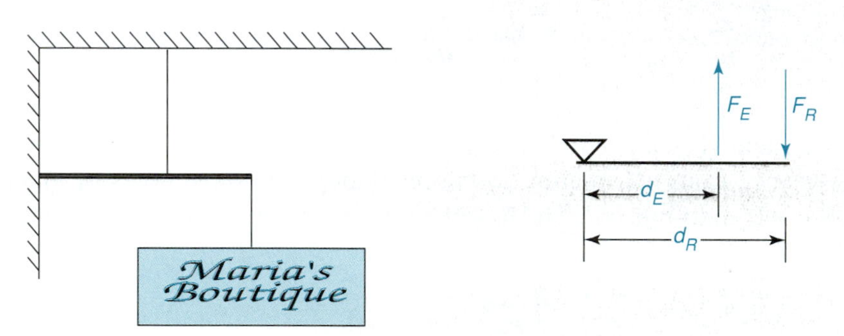

Third class: The effort force (F_E) is between the fulcrum and the resistance force (F_R).

Figure 8.7
Three classes of levers.

First, find MA:

$$\text{MA}_{\text{lever}} = \frac{d_E}{d_R} = \frac{2.50 \text{ m}}{0.300 \text{ m}} = \frac{8.33}{1}$$

To find the force:

Data:

$$d_E = 2.50 \text{ m}$$
$$d_R = 30.0 \text{ cm} = 0.300 \text{ m}$$
$$F_R = 18\bar{0}0 \text{ N}$$
$$F_E = ?$$

Basic Equation:

$$F_R \cdot d_R = F_E \cdot d_E$$

Working Equation:

$$F_E = \frac{F_R \cdot d_R}{d_E}$$

Substitution:

$$F_E = \frac{(18\bar{0}0 \text{ N})(0.300 \cancel{\text{m}})}{2.50 \cancel{\text{m}}}$$
$$= 216 \text{ N}$$

EXAMPLE 2

A wheelbarrow 2.00 m long has a $90\bar{0}$-N load 50.0 cm from the axle. What is the MA? What force is needed to lift the wheelbarrow?

Sketch:

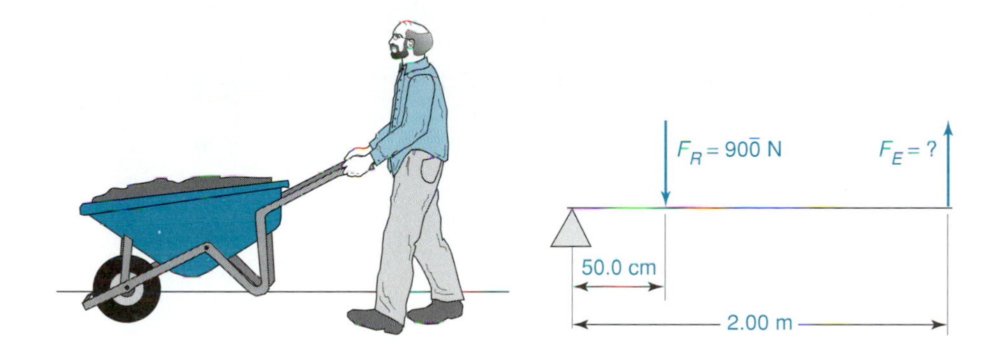

First, find MA:

$$\text{MA} = \frac{d_E}{d_R} = \frac{2.00 \cancel{\text{m}}}{0.500 \cancel{\text{m}}} = \frac{4.00}{1}$$

To find the force:

Data:

$$d_E = 2.00 \text{ m}$$
$$d_R = 0.500 \text{ m}$$
$$F_R = 90\bar{0} \text{ N}$$
$$F_E = ?$$

Basic Equation:

$$F_R \cdot d_R = F_E \cdot d_E$$

Working Equation:

$$F_E = \frac{F_R \cdot d_R}{d_E}$$

Substitution:

$$F_E = \frac{(90\bar{0} \text{ N})(0.500 \cancel{\text{m}})}{2.00 \cancel{\text{m}}}$$
$$= 225 \text{ N}$$

EXAMPLE 3

The MA of a pair of pliers is 6.0/1. A force of 8.0 lb is exerted on the handle. What force is exerted on a wire in the pliers?

MA = 6.0/1 means that for each pound of force applied on the handle, 6.0 lb is exerted on the wire. Therefore, if a force of 8.0 lb is applied on the handle, a force of (6.0)(8.0 lb) or 48 lb is exerted on the wire.

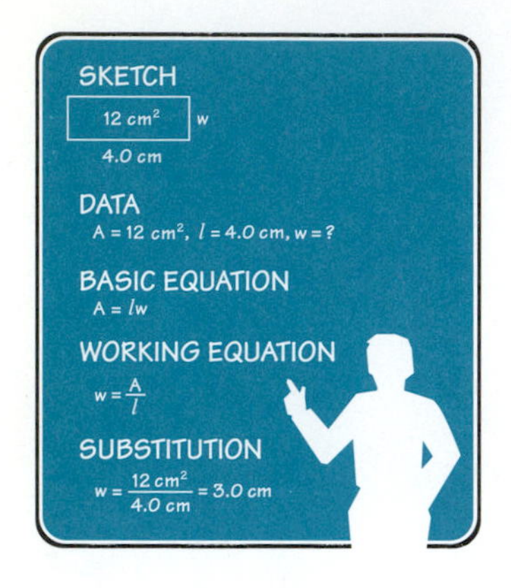

■ PROBLEMS 8.3

Given $F_R \cdot d_R = F_E \cdot d_E,$ find each missing quantity.

	F_R	F_E	d_R	d_E
1.	20.0 N	5.00 N	3.70 cm	____ cm
2.	____ N	176 N	49.2 cm	76.3 cm
3.	37.0 N	12.0 N	____ cm	112 cm
4.	23.4 lb	9.80 lb	____ in.	53.9 in.
5.	119 N	____ N	29.7 cm	67.4 cm

Given $MA_{lever} = \frac{F_R}{F_E}$, find each missing quantity.

	MA	F_R	F_E
6.	____	20.0 N	5.00 N
7.	____	23.4 lb	9.80 lb
8.	7.00	119 N	____ N
9.	4.00	____ lb	12.2 lb
10.	____	37.0 N	12.0 N

Given $MA_{lever} = \frac{d_E}{d_R}$, find each missing quantity.

	MA	d_R	d_E
11.	____	49.2 cm	76.3 cm
12.	7.00	29.7 in.	____ in.
13.	____	29.7 cm	67.4 cm
14.	4.00	____ cm	67.4 cm
15.	3.00	13.7 in.	____ in.

16. A pole is used to lift a car that fell off a jack (Fig. 8.8). The pivot is 2.00 ft from the car. Two people together exert 275 lb of force 8.00 ft from the pivot. What force is applied to the car? (Ignore the weight of the pole.)

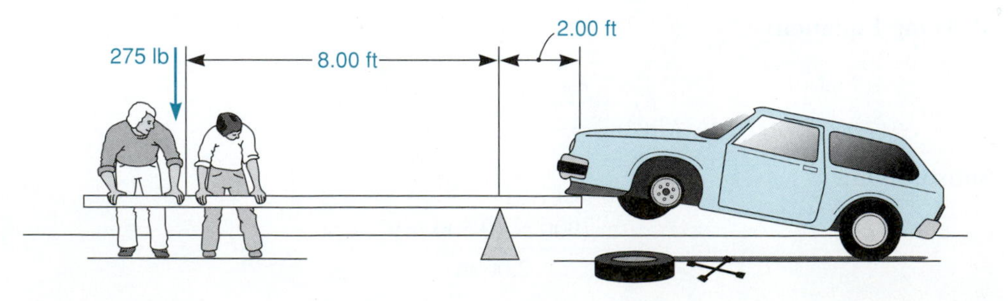

Figure 8.8

17. Calculate the MA of Problem 16.
18. A bar is used to lift a 10$\bar{0}$-kg block of concrete. The pivot is 1.00 m from the block. If the worker pushes down on the other end of the bar a distance of 2.50 m from the pivot, what force (in N) must he apply?
19. Calculate the MA of Problem 18.
20. A wheelbarrow 6.00 ft long is used to haul a 18$\bar{0}$-lb load. How far from the wheel is the load placed so that a person can lift the load with a force of 45.0 lb?
21. Calculate the MA of Problem 20.
22. Find the force, F_E, pulling up on the beam holding the sign shown in Fig. 8.9.
23. Calculate the MA of Problem 22.

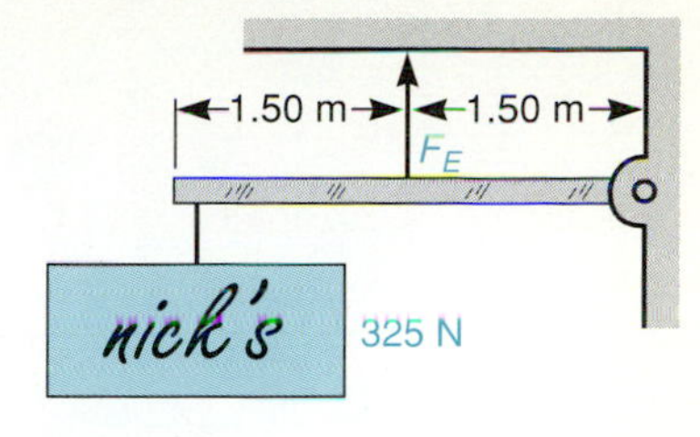

Figure 8.9

8.4 THE WHEEL-AND-AXLE

The **wheel-and-axle** consists of a large wheel attached to an axle so that both turn together (Fig. 8.10). Other examples include a doorknob, a screwdriver with a thick handle, and a wheelbarrow.

The law of simple machines as applied to the wheel-and-axle (basic equation) is

$$F_R \cdot r_R = F_E \cdot r_E$$

where F_R = resistance force
r_R = radius of resistance wheel
F_E = effort force
r_E = radius of effort wheel

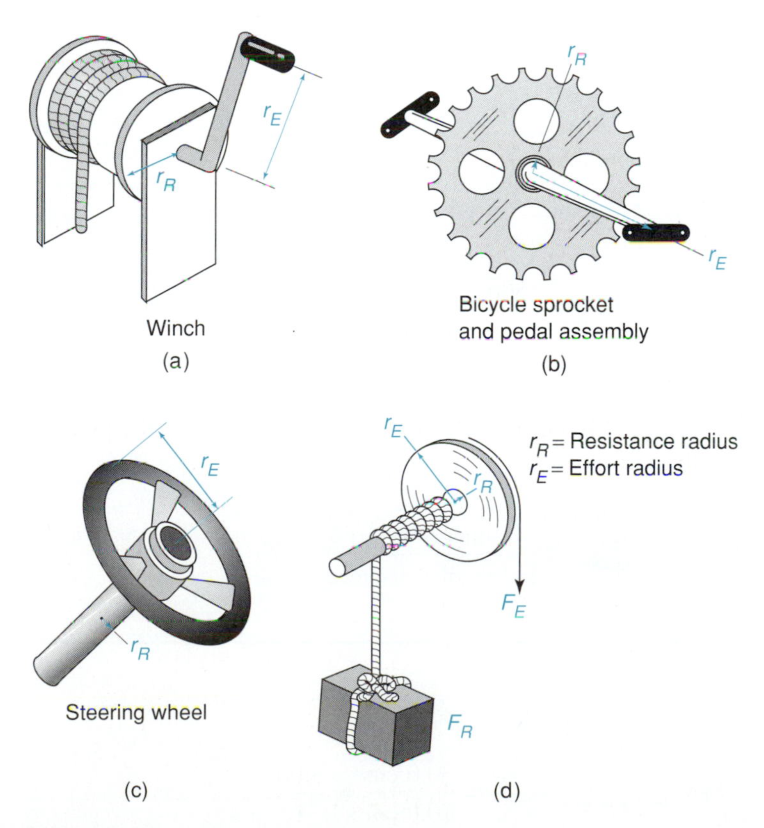

Figure 8.10
Examples of the wheel-and-axle.

EXAMPLE 1

A winch has a handle that turns in a radius of 30.0 cm. The radius of the drum or axle is 10.0 cm. Find the force required to lift a bucket weighing $50\overline{0}$ N.

Sketch:

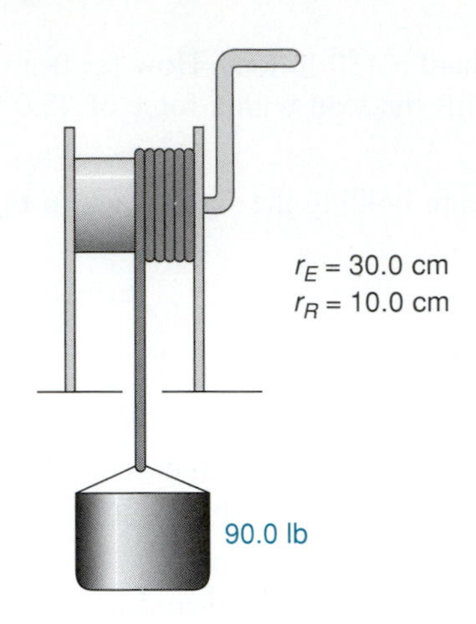

Data:

$$F_R = 50\overline{0} \text{ N}$$
$$r_E = 30.0 \text{ cm}$$
$$r_R = 10.0 \text{ cm}$$
$$F_E = ?$$

Basic Equation:

$$F_R \cdot r_R = F_E \cdot r_E$$

Working Equation:

$$F_E = \frac{F_R \cdot r_R}{r_E}$$

Substitution:

$$F_E = \frac{(50\overline{0} \text{ N})(10.0 \text{ cm})}{30.0 \text{ cm}}$$
$$= 167 \text{ N}$$

The mechanical advantage (MA) of the wheel-and-axle is the ratio of the radius of the effort to the radius of the resistance.

$$\text{MA}_{\text{wheel-and-axle}} = \frac{\text{radius of effort}}{\text{radius of resistance}} = \frac{r_E}{r_R}$$

EXAMPLE 2

Calculate the MA of the winch in Example 1.

$$\text{MA}_{\text{wheel-and-axle}} = \frac{r_E}{r_R} = \frac{30.0 \text{ cm}}{10.0 \text{ cm}} = \frac{3.00}{1}$$

■ PROBLEMS 8.4

Given $F_R \cdot r_R = F_E \cdot r_E$, find each missing quantity.

	F_R	F_E	r_R	r_E
1.	20.0 N	5.30 N	3.70 cm	____ cm
2.	37$\overline{0}$ N	12$\overline{0}$ N	____ m	1.12 m
3.	____ N	175 N	49.2 cm	76.3 cm
4.	23.4 lb	9.80 lb	____ in.	53.9 in.
5.	1190 N	____ N	29.7 cm	67.4 cm

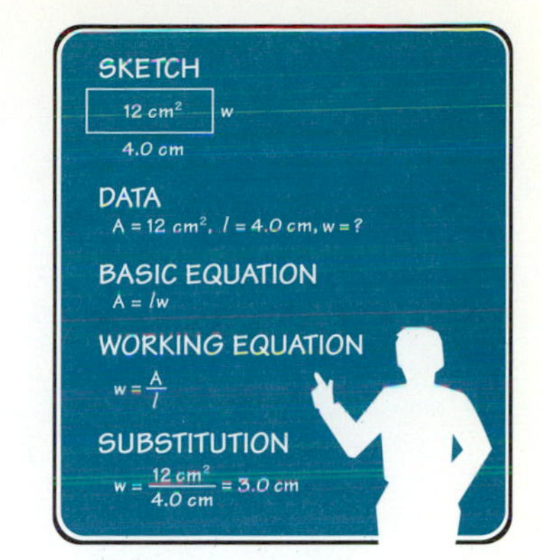

Given $\text{MA}_{\text{wheel-and-axle}} = \dfrac{r_E}{r_R}$, find each missing quantity.

	MA	r_E	r_R
6.	7.00	119 mm	____ mm
7.	4.00	____ in.	12.2 in.
8.	____	49.2 cm	31.7 cm
9.	3.00	61.3 cm	____ cm
10.	____	67.4 mm	29.7 mm

11. The radius of the axle of a winch is 3.00 in. The length of the handle (radius of wheel) is 1.50 ft. What weight will be lifted by an effort of 73.0 lb?
12. Calculate the MA of Problem 11.
13. A wheel having a radius of 70.0 cm is attached to an axle of radius 20.0 cm. What force must be applied to the rim of the wheel to raise a weight of 15$\overline{0}$0 N?
14. Calculate the MA of Problem 13.
15. What weight can be lifted in Problem 13 if a force of 575 N is applied?
16. The diameter of the wheel of a wheel-and-axle is 10.0 m. If a force of 475 N is raised by applying a force of 142 N, find the diameter of the axle.
17. Calculate the MA of Problem 16. ■

8.5 THE PULLEY

A **pulley** is a grooved wheel that turns readily on an axle and is supported in a frame. It can be fastened to a fixed object or it may be fastened to the resistance that is to be moved. If the pulley is fastened to a fixed object, it is called a **fixed pulley** [Fig. 8.11(a)]. If the pulley is fastened to the resistance to be moved, it is called a **movable pulley** [Fig. 8.11(b)]. A pulley system consists of combinations of fixed and movable pulleys [Fig. 8.11(c)–(e)].

The law of simple machines as applied to pulleys (Fig. 8.12) is

$$F_R \cdot d_R = F_E \cdot d_E$$

Here, d refers to the distance moved, not the diameter of the pulley. From the preceding equation,

$$\frac{F_R}{F_E} = \frac{d_E}{d_R} = \text{MA}_{\text{pulley}}$$

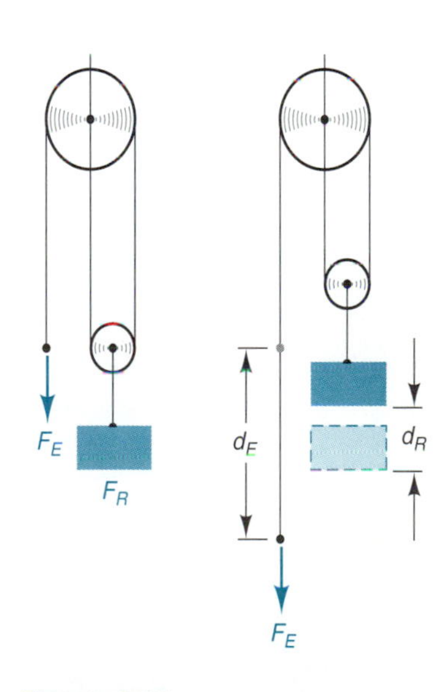

Figure 8.12
Law of simple machines applied to pulleys: $F_R \cdot d_R = F_E \cdot d_E$.

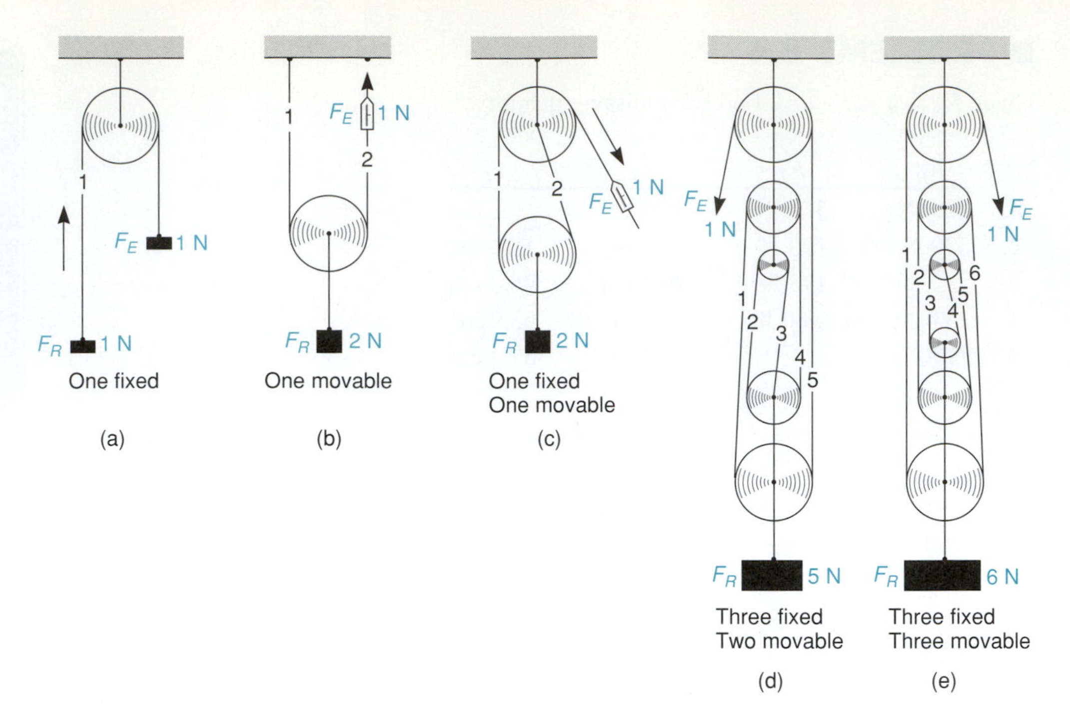

Figure 8.11
Pulleys and pulley systems.

However, when one continuous cord is used, this ratio reduces to the number of strands holding the resistance in the pulley system. Therefore,

$$\text{MA}_{\text{pulley}} = \text{number of strands holding the resistance}$$

This result may be explained as follows: When a weight is supported by two strands, each individual strand supports one half of the total weight. Thus, the MA = 2. If a weight is supported by three strands, each individual strand supports one third of the total weight. Thus, the MA = 3. If a weight is supported by four strands, each individual strand supports one fourth of the total weight. In general, when a weight is supported by n strands, each individual strand supports $\frac{1}{n}$ of the total weight. Thus, the MA = n.

Stated another way, the resistance force (F_R) is spread equally among the supporting strands. Thus, $F_R = n \times T$, where n is the number of strands holding the resistance and T is the tension in each supporting strand.

The effort force (F_E) equals the tension, T, in each supporting strand. The equation above may then be written

$$\text{MA}_{\text{pulley}} = \frac{F_R}{F_E} = \frac{nT}{T} = n$$

Note: The mechanical advantage of the pulley does not depend on the diameter of the pulley.

A number of examples are shown in Fig. 8.13.

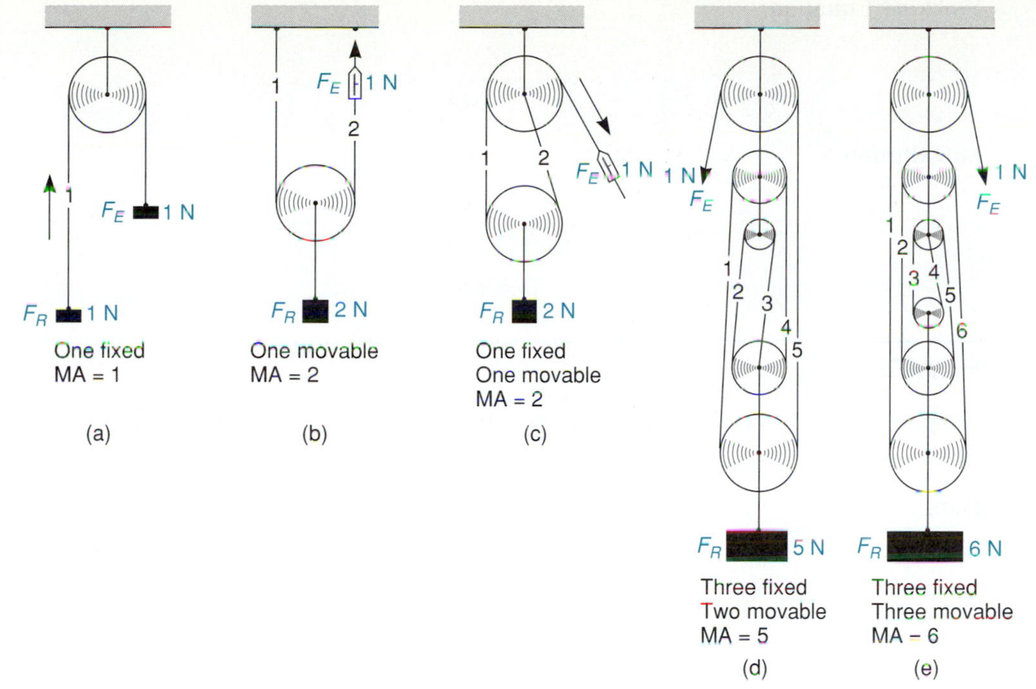

Figure 8.13
Mechanical advantage of pulleys and pulley systems.

EXAMPLE 1

Draw two different sets of pulleys, each with an MA of 4.

Sketch:

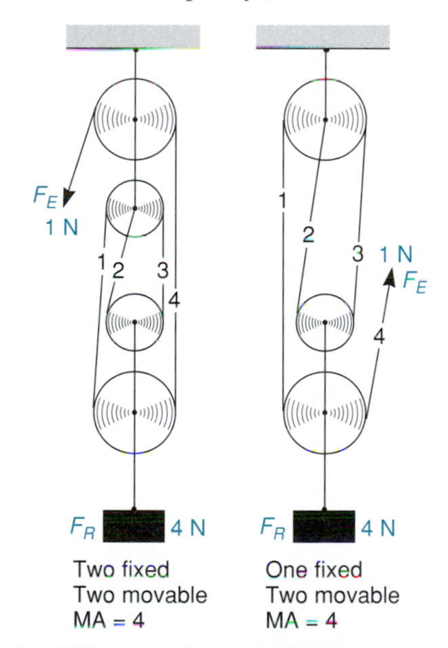

EXAMPLE 2

What effort will lift a resistance of 480 N in the pulley system in Example 1?

Data:

$$MA_{\text{pulley}} = 4$$
$$F_R = 480 \text{ N}$$
$$F_E = ?$$

Basic Equation:

$$MA_{\text{pulley}} = \frac{F_R}{F_E}$$

Working Equation:

$$F_E = \frac{F_R}{\text{MA}_{\text{pulley}}}$$

Substitution:

$$F_E = \frac{480 \text{ N}}{4}$$
$$= 120 \text{ N}$$

EXAMPLE 3

If the resistance moves 7.00 ft, what is the effort distance of the pulley system in Example 1?

Data:

$$\text{MA}_{\text{pulley}} = 4$$
$$d_R = 7.00 \text{ ft}$$
$$d_E = ?$$

Basic Equation:

$$\text{MA}_{\text{pulley}} = \frac{d_E}{d_R}$$

Working Equation:

$$d_E = d_R(\text{MA}_{\text{pulley}})$$

Substitution:

$$d_E = (7.00 \text{ ft})(4)$$
$$= 28.0 \text{ ft}$$

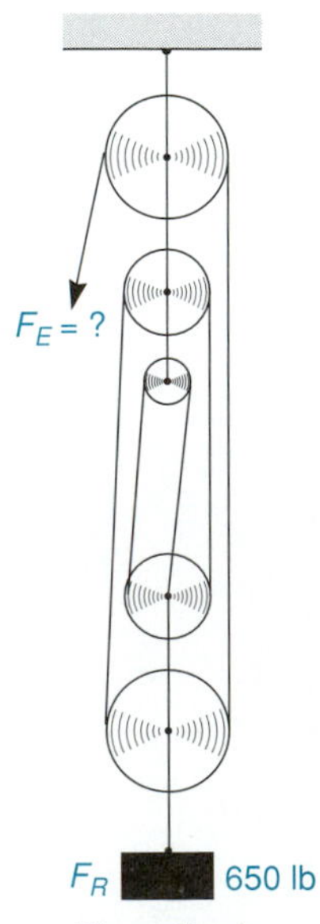

Figure 8.14

EXAMPLE 4

The pulley system in Fig. 8.14 is used to raise a 650-lb object 25 ft. What is the mechanical advantage? What force is exerted?

$$\text{MA}_{\text{pulley}} = \text{number of strands holding the resistance}$$
$$= 5$$

To find the force exerted:

Data:

$$\text{MA}_{\text{pulley}} = 5$$
$$F_R = 650 \text{ lb}$$
$$F_E = ?$$

Basic Equation:

$$\text{MA}_{\text{pulley}} = \frac{F_R}{F_E}$$

Working Equation:

$$F_E = \frac{F_R}{\text{MA}_{\text{pulley}}}$$

Substitution:

$$F_E = \frac{650 \text{ lb}}{5}$$
$$= 130 \text{ lb}$$

■ PROBLEMS 8.5

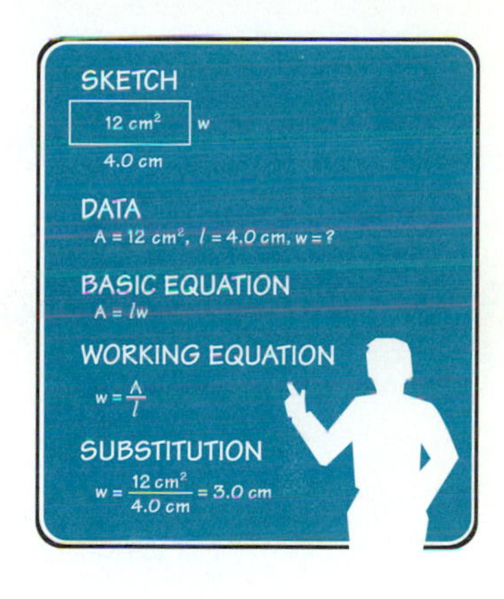

Find the mechanical advantage of each system.

1. 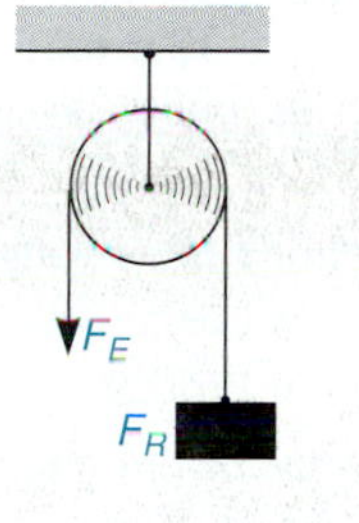

2. 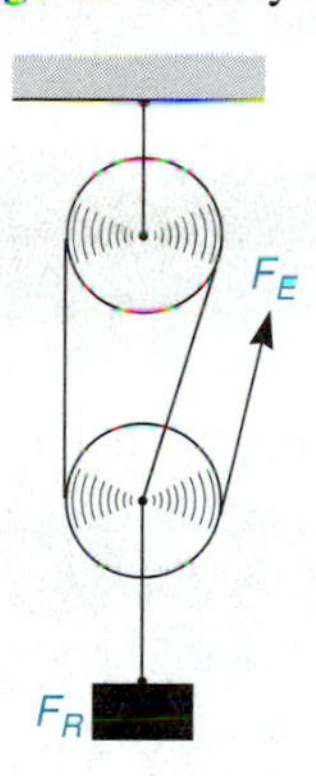

3. 4. 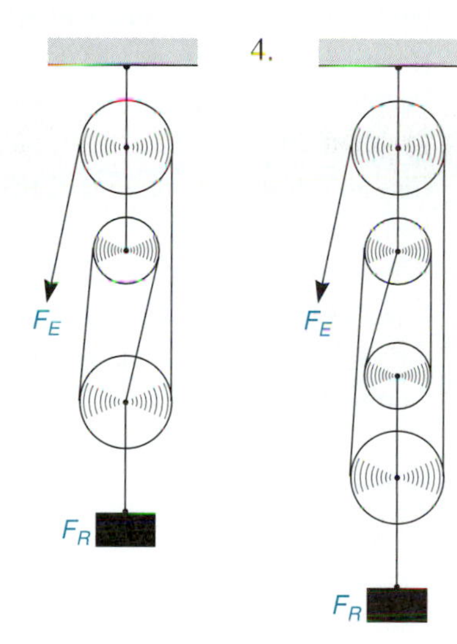

5. 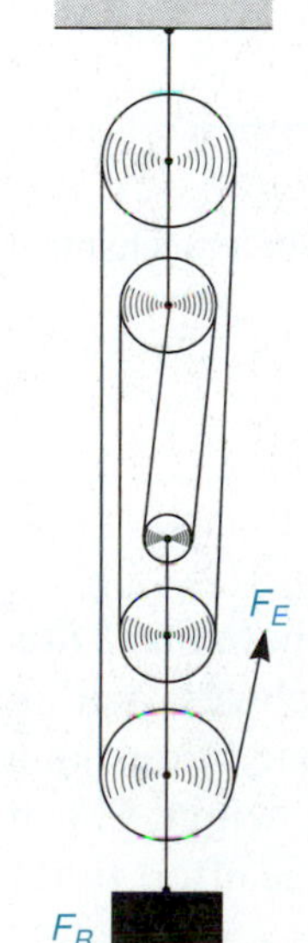

6. 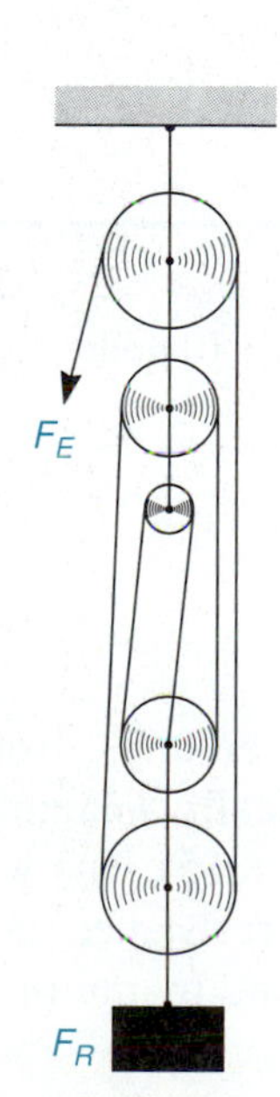

7. 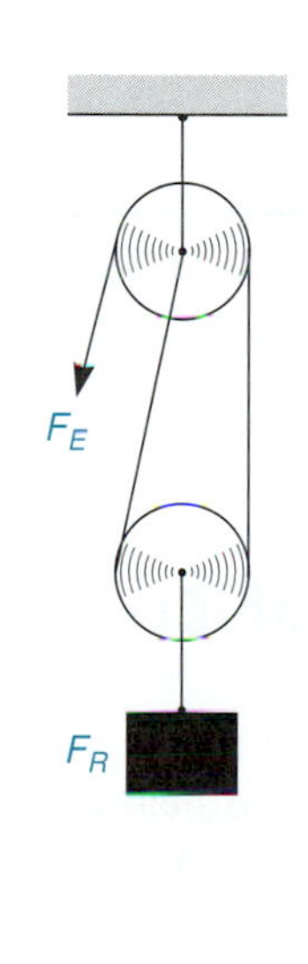

8. 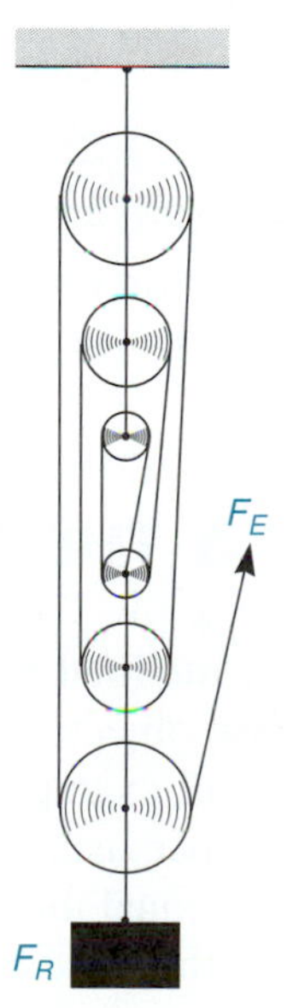

Draw each pulley system for Problems 9–14.

9. One fixed and two movable. Find the system's MA.
10. Two fixed and two movable with an MA of 5.

11. Three fixed and three movable with an MA of 6.
12. Four fixed and three movable. Find the system's MA.
13. Four fixed and four movable with an MA of 8.
14. Three fixed and four movable with an MA of 8.
15. What is the MA of a single movable pulley?
16. What effort will lift a $25\bar{0}$-lb weight by using a single movable pulley?
17. If the weight is moved 15.0 ft in Problem 16, how many feet of rope are pulled by the person exerting the effort?
18. A system consisting of two fixed pulleys and two movable pulleys has a mechanical advantage of 4. If a force of 97.0 N is exerted, what weight can be raised?
19. If the weight in Problem 18 is raised 20.5 m, what length of rope is pulled?
20. A $40\bar{0}$-lb weight is lifted 30.0 ft. Using a system of one fixed and two movable pulleys, find the effort force and effort distance.
21. If an effort force of 65.0 N is applied through an effort distance of 13.0 m, find the weight of the resistance and the distance it is moved using the pulley system in Problem 20.
22. Can an effort force of 75.0 N lift a 275-N weight using the pulley system in Problem 3?
23. What effort will lift a 1950-N weight using the pulley system in Problem 5?
24. If the weight in Problem 23 is moved 3.00 m, how much rope must be pulled through the pulley system by the person exerting the force?
25. Complete the following pulley system mechanical advantage chart, which lists two possible arrangements of fixed and movable pulleys for each given mechanical advantage.

	Mechanical Advantage (MA)							
Pulleys	1	2	3	4	5	6	7	8
Fixed	1	1	2					
Movable	0	1	1					
Fixed		0	1					
Movable		1	1					

26. Can you arrange a pulley system containing 10 pulleys and obtain a mechanical advantage of 12? Why or why not? ■

8.6 THE INCLINED PLANE

An **inclined plane** is a plane surface set at an angle from the horizontal used to raise objects that are too heavy to lift vertically. Gangplanks, chutes, and ramps are all examples of the inclined plane (Fig. 8.15). The work done in raising a resistance using the inclined plane equals the resistance times the height. This must also equal the work input, which can be found by multiplying the effort times the length of the plane.

$$F_R \cdot d_R = F_E \cdot d_E \qquad \text{(law of machines)}$$

$$F_R \cdot \text{height of plane} = F_E \cdot \text{length of plane}$$

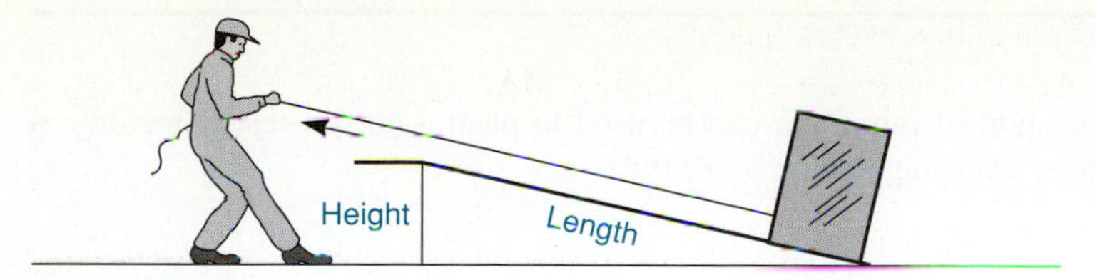

Figure 8.15
Inclined plane.

From the preceding equation:

$$\frac{F_R}{F_E} = \frac{\text{length of plane}}{\text{height of plane}} = \text{MA}_{\text{inclined plane}}$$

EXAMPLE 1

A worker is pushing a box weighing $15\bar{0}0$ N up a ramp 6.00 m long onto a platform 1.50 m above the ground. What is the mechanical advantage? What effort is applied?

Sketch:

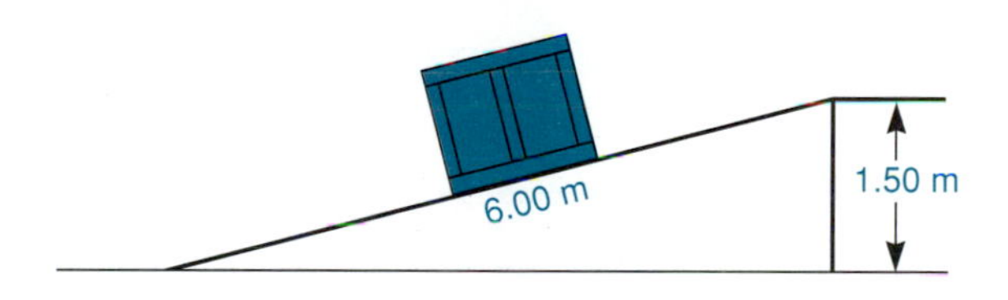

$$\text{MA}_{\text{inclined plane}} = \frac{\text{length of plane}}{\text{height of plane}} = \frac{6.00 \not{m}}{1.50 \not{m}} = 4.00$$

To find the effort force:

Data:

$$F_R = 15\bar{0}0 \text{ N}$$
$$\text{MA}_{\text{inclined plane}} = 4.00$$
$$F_E = ?$$

Basic Equation:

$$\text{MA}_{\text{inclined plane}} = \frac{F_R}{F_E}$$

Working Equation:

$$F_E = \frac{F_R}{\text{MA}_{\text{inclined plane}}}$$

Substitution:

$$F_E = \frac{15\bar{0}0 \text{ N}}{4.00}$$
$$= 375 \text{ N}$$

EXAMPLE 2

Find the length of the shortest ramp that can be used to push a $60\bar{0}$-lb resistance onto a platform 3.50 ft high by exerting a force of 72.0 lb.

Data:

$$F_R = 60\bar{0}\text{ lb}$$
$$F_E = 72.0\text{ lb}$$
$$\text{height} = 3.50\text{ ft}$$
$$\text{length} = ?$$

Basic Equation:

$$F_R \cdot \text{height} = F_E \cdot \text{length}$$

Working Equation:

$$\text{length} = \frac{F_R \cdot \text{height}}{F_E}$$

Substitution:

$$\text{length} = \frac{(60\bar{0}\text{ }\cancel{\text{lb}})(3.50\text{ ft})}{72.0\text{ }\cancel{\text{lb}}}$$
$$= 29.2\text{ ft}$$

EXAMPLE 3

An inclined plane is 13.0 m long and 5.00 m high. What is the mechanical advantage and what weight can be raised by exerting a force of 375 N?

$$\text{MA}_{\text{inclined plane}} = \frac{\text{length of plane}}{\text{height of plane}} = \frac{13.0\text{ }\cancel{\text{m}}}{5.00\text{ }\cancel{\text{m}}} = 2.60$$

To find the weight of the resistance:

Data:

$$\text{MA}_{\text{inclined plane}} = 2.60$$
$$F_E = 375\text{ N}$$
$$F_R = ?$$

Basic Equation:

$$\text{MA}_{\text{inclined plane}} = \frac{F_R}{F_E}$$

Working Equation:

$$F_R = (F_E)(\text{MA}_{\text{inclined plane}})$$

Substitution:

$$F_R = (375\text{ N})(2.60)$$
$$= 975\text{ N}$$

■ PROBLEMS 8.6

Given $F_R \cdot \text{height} = F_E \cdot \text{length}$, find each missing quantity.

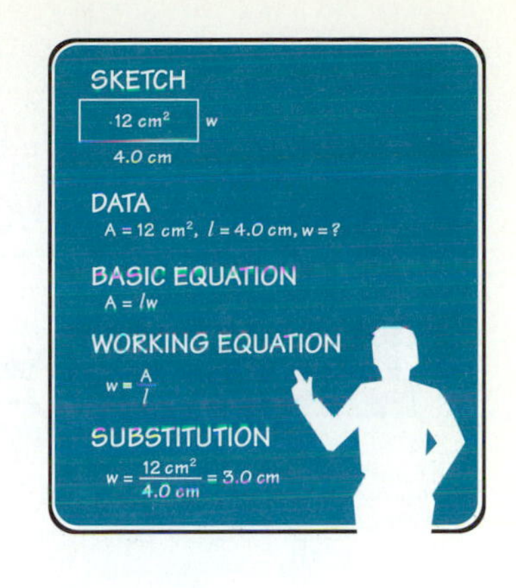

	F_R	F_E	*Height of plane*	*Length of plane*
1.	20.0 N	5.30 N	3.40 cm	___ cm
2.	$98\bar{0}0$ N	2340 N	___ m	3.79 m
3.	119 lb	___ lb	13.2 in.	74.0 in.
4.	___ N	1760 N	82.1 cm	3.79 m
5.	$37\bar{0}0$ N	$12\bar{0}0$ N	___ cm	112 cm

Given $\text{MA}_{\text{inclined plane}} = \dfrac{\text{length of plane}}{\text{height of plane}}$, find each missing quantity.

	MA	*Length of plane*	*Height of plane*
6.	9.00	3.40 ft	___ ft
7.	___	3.79 m	0.821 m
8.	1.30	___ ft	9.72 ft
9.	___	74.0 cm	13.2 cm
10.	17.4	___ in.	13.4 in.

11. An inclined plane is 10.0 m long and 2.50 m high. What is the mechanical advantage?
12. A resistance of 727 N is pushed up the plane in Problem 11. What effort is needed?
13. An effort of $2\bar{0}0$ N is applied to push an 815-N resistance up the inclined plane in Problem 11. Is the effort enough?
14. A safe is loaded onto a truck whose bed is 5.50 ft off the ground. The safe weighs 538 lb. If the effort applied is $14\bar{0}$ lb, what length of ramp is needed to raise the safe?
15. What is the MA of the inclined plane in Problem 14?
16. Another safe weighing 257 lb is loaded onto the same truck as in Problem 14. If the ramp is 21.1 ft long, what effort is needed?
17. A resistance of 325 N is raised by using a ramp 5.76 m long and by applying a force of 75.0 N. How high can it be raised?
18. What is the MA of the ramp in Problem 17? ■

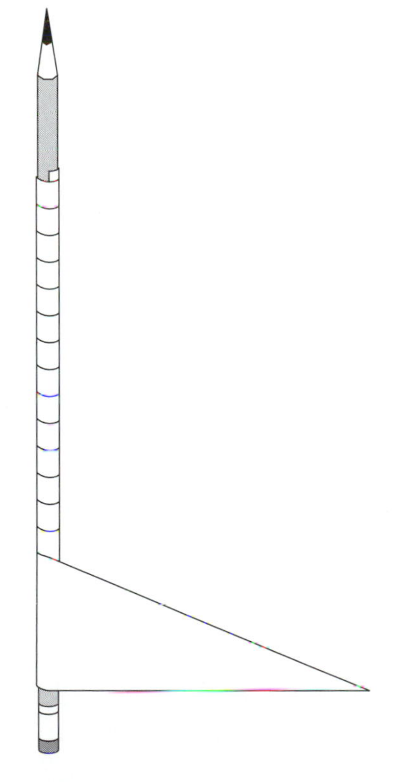

Figure 8.16 The screw is an inclined plane wound around a cylinder. The hypotenuse of the triangular section of paper corresponds to the inclined plane (threads) of a screw as it is wound around the pencil.

8.7 THE SCREW

A **screw** is an inclined plane wrapped around a cylinder. To illustrate, cut a sheet of paper in the shape of a right triangle and wind it around a pencil as shown in Fig. 8.16. The jackscrew, wood screw, and auger are examples of this simple machine (Fig. 8.17). The distance a beam rises or the distance the wood screw advances into a piece of wood in one revolution is called the **pitch** of the screw. Therefore, the pitch of a screw is actually the distance between two successive threads. From the law of machines,

$$F_R \cdot d_R = F_E \cdot d_E$$

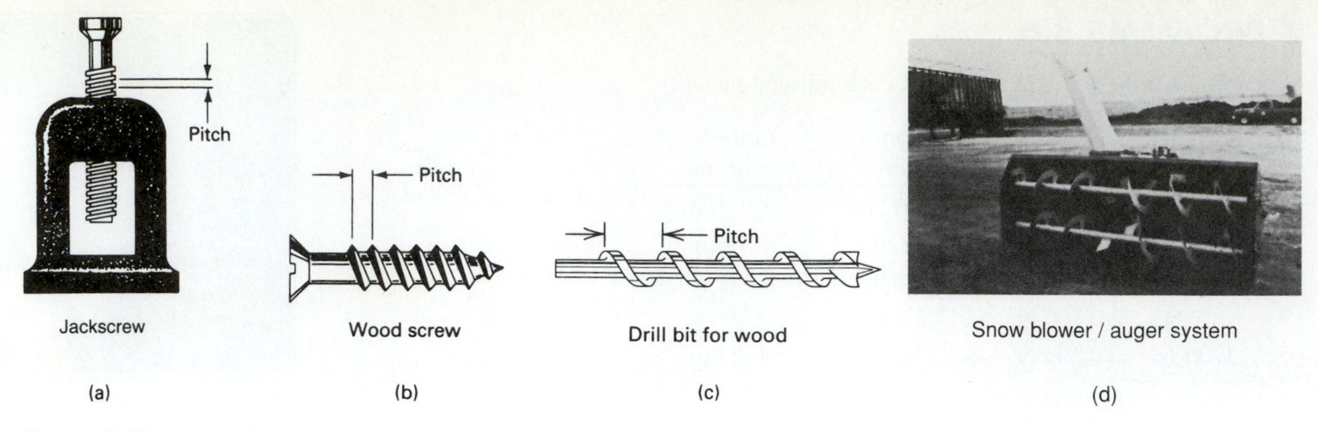

Figure 8.17
(c) is courtesy of Custom Products of Litchfield, Inc. (d) is courtesy of Farm King Allied Inc. in Morden, Manitoba, Canada.

However, for advancing a screw with a screwdriver

$$d_R = \text{pitch of screw}$$
$$d_E = \text{circumference of the handle of the screwdriver}$$

or

$$d_E = 2\pi r$$

where r is the radius of the handle of the screwdriver. Therefore,

$$F_R \cdot \text{pitch} = F_E \cdot 2\pi r$$

so

$$\frac{F_R}{F_E} = \frac{2\pi r}{\text{pitch}} = \text{MA}_{\text{screw}}$$

In the case of a jackscrew, *r is the length of the handle turning the screw and not the radius of the screw.*

EXAMPLE 1

Find the mechanical advantage of a jackscrew having a pitch of 25.0 mm and a handle radius of 35.0 cm.

Data:

$$\text{pitch} = 25.0 \text{ mm} = 2.50 \text{ cm}$$
$$r = 35.0 \text{ cm}$$
$$\text{MA}_{\text{screw}} = ?$$

Basic Equation:

$$\text{MA}_{\text{screw}} = \frac{2\pi r}{\text{pitch}}$$

Working Equation: Same

Substitution:

$$\text{MA}_{\text{screw}} = \frac{2\pi(35.0 \text{ cm})}{2.50 \text{ cm}}$$
$$= 88.0$$

EXAMPLE 2

What resistance can be lifted using the jackscrew in Example 1 if an effort of 203 N is exerted?

Data:

$$\text{MA}_{\text{screw}} = 88.0$$
$$F_E = 203 \text{ N}$$
$$F_R = ?$$

Basic Equation:

$$\text{MA}_{\text{screw}} = \frac{F_R}{F_E}$$

Working Equation:

$$F_R = (F_E)(\text{MA}_{\text{screw}})$$

Substitution:

$$F_R = (203 \text{ N})(88.0)$$
$$= 17{,}900 \text{ N}$$

EXAMPLE 3

A 19,400-N weight is raised using a jackscrew having a pitch of 5.00 mm and a handle length of 255 mm. What force must be applied?

Data:

$$\text{pitch} = 5.00 \text{ mm}$$
$$r = 255 \text{ mm}$$
$$F_R = 19{,}400 \text{ N}$$
$$F_E = ?$$

Basic Equation:

$$F_R \cdot \text{pitch} = F_E \cdot 2\pi r$$

Working Equation:

$$F_E = \frac{F_R\,(\text{pitch})}{2\pi r}$$

Substitution:

$$F_E = \frac{(19{,}400 \text{ N})(5.00 \text{ mm})}{2\pi(255 \text{ mm})}$$
$$= 60.5 \text{ N}$$

■ PROBLEMS 8.7

Given $F_R \cdot \text{pitch} = F_E \cdot 2\pi r$, find each missing quantity.

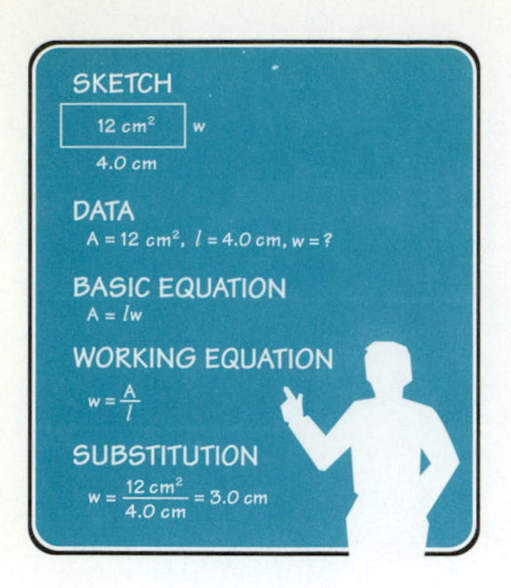

	F_R	F_E	*Pitch*	r
1.	20.7 N	5.30 N	3.70 mm	___ mm
2.	___ lb	17.6 lb	0.130 in.	24.5 in.
3.	234 N	9.80 N	___ mm	53.9 mm
4.	1190 N	___ N	2.97 mm	67.4 mm
5.	$37\bar{0}$ lb	12.0 lb	___ in.	11.2 in.

Given $\text{MA}_{\text{screw}} = \dfrac{2\pi r}{\text{pitch}}$, find each missing quantity.

	MA	r	*Pitch*
6.	7.00	34.0 mm	___ mm
7.	___	3.79 in.	0.812 in.
8.	9.00	___ in.	0.970 in.
9.	___	7.40 mm	1.32 mm
10.	13.0	___ mm	2.10 mm

11. A 3650-lb car is raised using a jackscrew having eight threads to the inch and a handle 15.0 in. long. What effort must be applied?
12. What is the MA in Problem 11?
13. The mechanical advantage of a jackscrew is 97.0. If the handle is 34.5 cm long, what is the pitch?
14. How much weight can be raised by applying an effort of 405 N to the jackscrew in Problem 13?
15. A wood screw with pitch 0.125 in. is advanced into wood using a screwdriver whose handle is 1.50 in. in diameter. What is the mechanical advantage of the screw?
16. What is the resistance of the wood if 15.0 lb of effort is applied on the wood screw in Problem 15?
17. What is the resistance of the wood if 15.0 lb of effort is applied to the wood screw in Problem 15 using a screwdriver whose handle is 0.500 in. in diameter?
18. The handle of a jackscrew is 60.0 cm long. If the mechanical advantage is 78.0, what is the pitch?
19. How much weight can be raised by applying a force of $43\bar{0}$ N to the jackscrew handle in Problem 18?

8.8 THE WEDGE

A **wedge** is an inclined plane in which the plane is moved instead of the resistance. Examples are shown in Fig. 8.18.

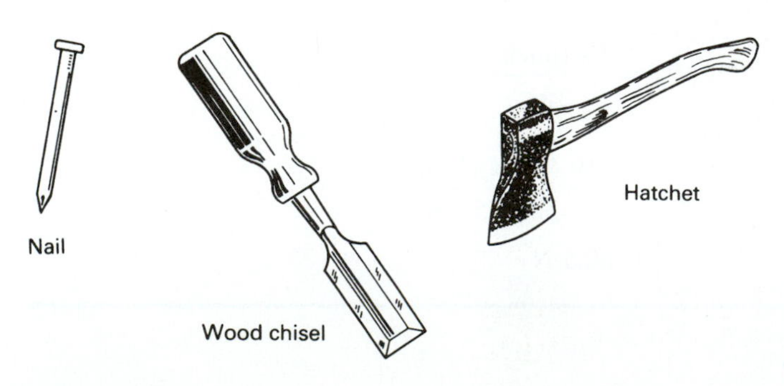

Figure 8.18
Inclined planes where the plane moves instead of the resistance.

Finding the mechanical advantage of a wedge is not practical because of the large amount of friction. A narrow wedge is easier to drive than a thick wedge. Therefore, the mechanical advantage depends on the ratio of its length to its thickness.

8.9 COMPOUND MACHINES

A **compound machine** is a combination of simple machines. Examples are shown in Fig. 8.19. In most compound machines, *the total mechanical advantage is the product of the mechanical advantages of each simple machine.*

$$MA_{\text{compound machine}} = (MA_1)(MA_2)(MA_3)\ldots$$

Figure 8.19
Compound machines multiply mechanical advantage. (Courtesy of Caterpillar Inc.)

EXAMPLE 1

A crate weighing $95\bar{0}0$ N is pulled up the inclined plane using the pulley system shown in Fig. 8.20.

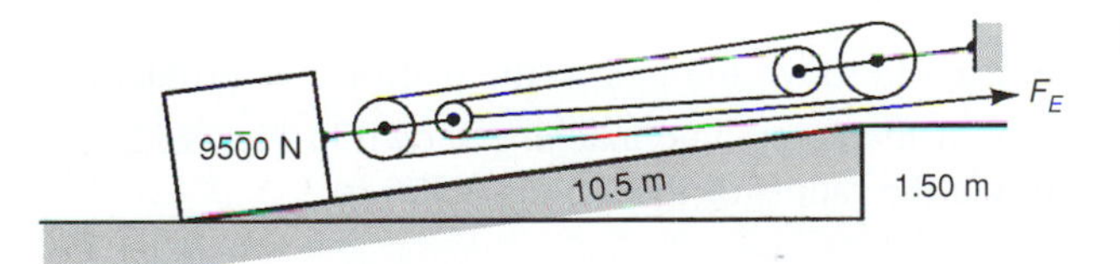

Figure 8.20

(a) Find the mechanical advantage of the total system.

(b) What effort force (F_E) is needed?

(a) First, find the MA of the inclined plane.

$$MA_{\text{inclined plane}} = \frac{\text{length of plane}}{\text{height of plane}} = \frac{10.5\ \cancel{m}}{1.50\ \cancel{m}} = 7.00$$

The MA of the pulley system = 5 (the number of supporting strands).
The MA of the total system (compound machine) is

$$(MA_{\text{inclined plane}})\,(MA_{\text{pulley system}}) = (7.00)(5) = 35.0$$

(b) Data:

$$MA_{\text{compound machine}} = 35.0$$
$$F_R = 95\bar{0}0 \text{ N}$$
$$F_E = ?$$

Basic Equation:

$$MA_{\text{compound machine}} = \frac{F_R}{F_E}$$

Working Equation:

$$F_E = \frac{F_R}{MA_{\text{compound machine}}}$$

Substitution:

$$F_E = \frac{95\bar{0}0 \text{ N}}{35.0}$$
$$= 271 \text{ N}$$

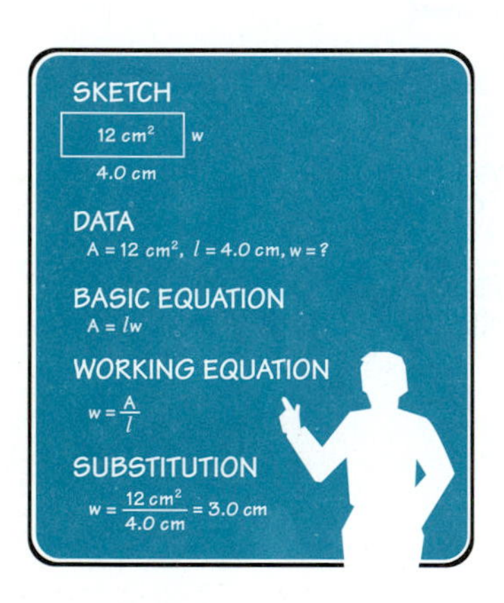

■ PROBLEMS 8.9

1. The box shown in Fig. 8.21 being pulled up an inclined plane using the indicated pulley system (called a block and tackle) weighs 5340 N. If the inclined plane is 6.00 m long and the height of the platform is 2.00 m, find the mechanical advantage of this compound machine.

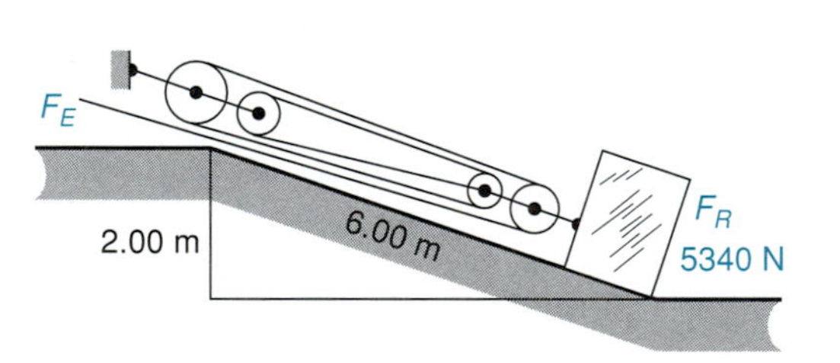

Figure 8.21

2. What effort force must be exerted to move the box to the platform in Problem 1?
3. Find the mechanical advantage of the compound machine shown in Fig. 8.22. The radius of the crank is 1.00 ft and the radius of the axle is 0.500 ft.

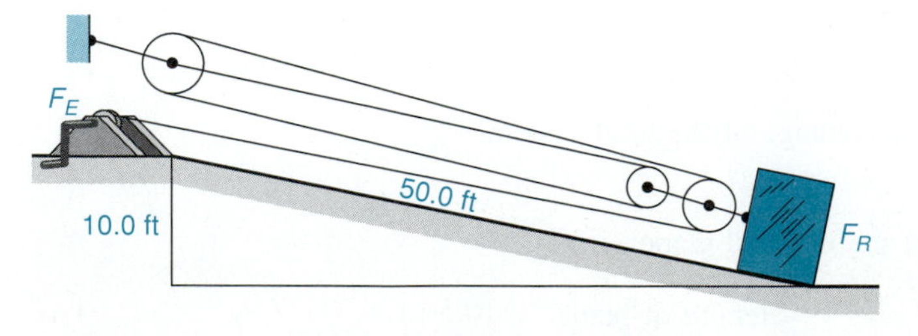

Figure 8.22

4. If an effort of $30\bar{0}$ lb is exerted, what weight can be moved up the inclined plane using the compound machine in Problem 3?
5. What effort is required to move a load of 1.50 tons up the inclined plane using the compound machine in Problem 3? (1 ton = $200\bar{0}$ lb)
6. Find the mechanical advantage of the compound machine in Problem 1 if the inclined plane is 8.00 m long and 2.00 m high.
7. What effort force is needed to move a box of weight $25\bar{0}0$ N to the platform in Problem 6?
8. Find the mechanical advantage of the compound machine in Problem 3 if the radius of the crank is 40.0 cm, the radius of the axle is 12.0 cm, the length of the inclined plane is 12.0 m, and the height of the inclined plane is 50.0 cm.
9. If an effort of $45\bar{0}$ N is exerted in Problem 8, what weight can be moved up the inclined plane?
10. What effort force (in N) is needed to move 2.50 metric tons up the inclined plane in Problem 8? ■

8.10 THE EFFECT OF FRICTION ON SIMPLE MACHINES

The **Law of Simple Machines** has been stated for a particular machine and in the general case in the previous sections. Each has been stated in terms of what is called *Ideal Mechanical Advantage* (IMA), in which we have 100% efficiency. Actually, energy is lost in every machine by overcoming friction through heat loss. This lost energy decreases the efficiency of the machine; that is, more work must be put into a machine than what you get out of the machine. This lost energy is heat energy, which results in machine wear or even burning out certain parts of the machine.

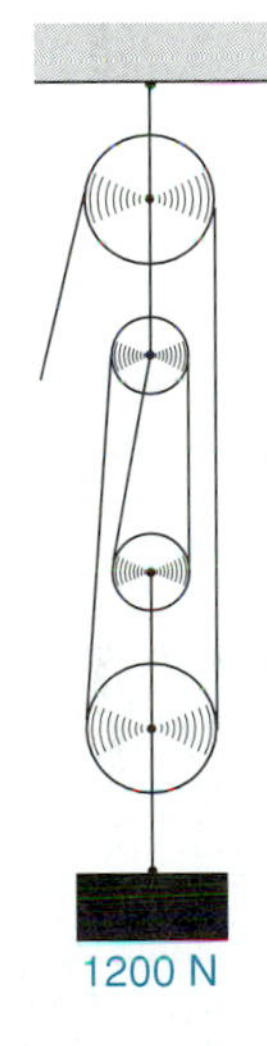

Figure 8.23
Friction has been ignored in our study of pulleys.

Throughout this chapter we have been discussing simple machines, mechanical advantage, resistance force, effort force, resistance distance, and effort distance in the ideal case while ignoring friction. For example, in the pulley system in Fig. 8.23, we find the IMA is 4 to 1. Ideally, it takes 300 N of effort force to lift the resistant force of 1200 N; that is, it ideally takes 1 N of force to raise 4 N of weight. However, it actually takes 400 N of effort to lift the 1200-N weight; the *Actual Mechanical Advantage* (AMA) is then 3 to 1; that is, it actually takes 1 N of force to raise 3 N of weight.

In general the Actual Mechanical Advantage is found by the following formula:

$$\text{AMA} = \frac{F_R}{F_E} = \frac{\text{resistance force}}{\text{effort force}}$$

In Section 4.3, we studied the effects of sliding friction. The actual effects of sliding friction are substantial in inclined plane problems. In Example 1 of Section 8.6, the inclined plane (repeated in Fig. 8.24) has an IMA of 4 to 1. Actu-

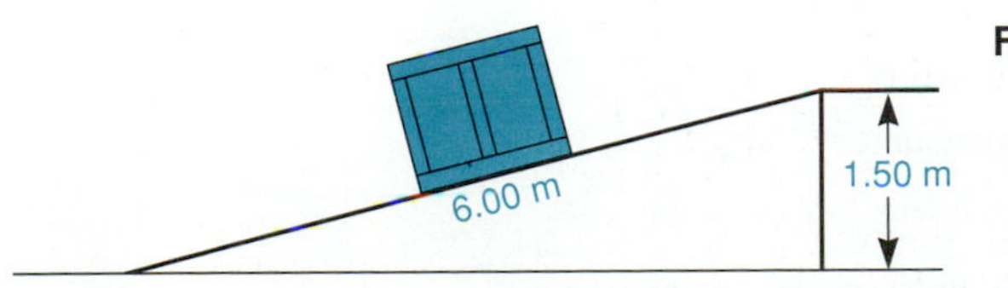

Figure 8.24

ally, it takes 545 N of effort to move the $15\bar{0}0$-N box up the ramp. Therefore, the AMA is

$$\text{AMA} = \frac{F_R}{F_E} = \frac{15\bar{0}0\text{ N}}{545\text{ N}} = 2.75$$

That is, it actually takes 1.00 N of force to push 2.75 N up the ramp.

GLOSSARY

Compound Machine A combination of simple machines. The total mechanical advantage is the product of the mechanical advantage of each simple machine. (p. 211)
Efficiency The ratio of the work output to the work input of a machine. (p. 192)
Effort The force applied to a machine. (p. 190)
Effort Arm The distance from the effort force to the fulcrum of a lever. (p. 193)
Fixed Pulley A pulley that is fastened to a fixed object. (p. 199)
Fulcrum A pivot about which a lever is free to turn. (p. 193)
Inclined Plane A plane surface set at an angle from the horizontal used to raise objects that are too heavy to lift vertically. Gangplanks, chutes, and ramps are examples. (p. 204)
Law of Simple Machines Resistance force × resistance distance = effort force × effort distance. (p. 192)
Lever A rigid bar free to turn on a pivot called a *fulcrum.* (p. 193)
Machine An object or system that is used to transfer energy from one place to another and allows us to do work that could not otherwise be done or could not be done as easily. (p. 190)
Mechanical Advantage The ratio of the resistance force to the effort force. (p. 192)
Movable Pulley A pulley that is fastened to the resistance to be moved. (p. 199)
Pitch The distance a screw advances in one revolution of the screw. Also given by the distance between two successive threads. (p. 207)
Pulley A grooved wheel that turns readily on an axle and is supported in a frame. (p. 199)
Resistance The force overcome by a machine. (p. 190)
Resistance Arm The distance from the resistance force to the fulcrum of a lever. (p. 193)
Screw An inclined plane wrapped around a cylinder. (p. 207)
Wedge An inclined plane in which the plane is moved instead of the resistance. Examples include a hatchet and a wood chisel. (p. 210)
Wheel-and-Axle A large wheel attached to an axle so that both turn together. (p. 197)

FORMULAS

8.1 resistance force × resistance distance = effort force × effort distance

8.2 $\text{MA} = \dfrac{\text{resistance force}}{\text{effort force}}$

$\text{efficiency} = \dfrac{\text{work output}}{\text{work input}} \times 100\% = \dfrac{F_{\text{output}} \times s_{\text{output}}}{F_{\text{input}} \times s_{\text{input}}} \times 100\%$

8.3 $\text{MA}_{\text{lever}} = \dfrac{\text{effort arm}}{\text{resistance arm}} = \dfrac{d_E}{d_R}$

$F_R \cdot d_R = F_E \cdot d_E$

8.4 $\text{MA}_{\text{wheel-and-axle}} = \dfrac{\text{radius of effort}}{\text{radius of resistance}} = \dfrac{r_E}{r_R}$

$F_R \cdot r_R = F_E \cdot r_E$

8.5 $\text{MA}_{\text{pulley}}$ = number of strands holding the resistance

$$MA_{pulley} = \frac{d_E}{d_R}$$

8.6 $$MA_{inclined\ plane} = \frac{\text{length of plane}}{\text{height of plane}}$$

$$F_R \cdot \text{height} = F_E \cdot \text{length}$$

8.7 $$MA_{screw} = \frac{2\pi r}{\text{pitch}}$$

$$F_R \cdot \text{pitch} = F_E \cdot 2\pi r$$

8.9 $$MA_{compound\ machine} = (MA_1)(MA_2)(MA_3) \ldots$$

8.10 $$AMA = \frac{F_R}{F_E} = \frac{\text{resistance force}}{\text{effort force}}$$

REVIEW QUESTIONS

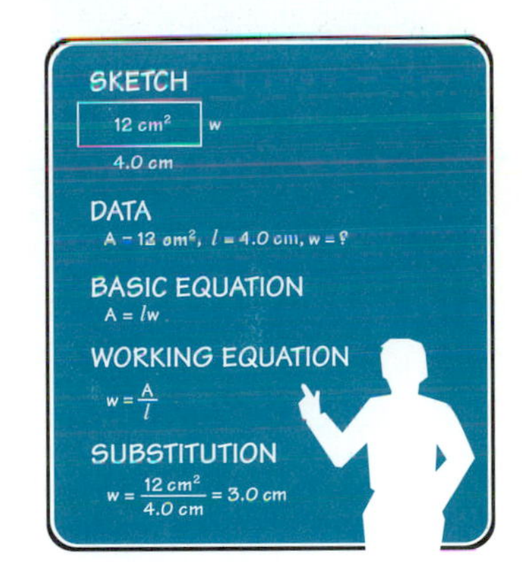

1. Which of the following is not a simple machine?
   (a) Pulley (b) Lever
   (c) Wedge (d) Automobile
2. The force applied to the machine is the
   (a) effort. (b) frictional.
   (c) horizontal. (d) resistance.
3. Efficiency is
   (a) the same as mechanical advantage. (b) a percentage.
   (c) impossible to determine.
4. A second-class lever has
   (a) two fulcrums. (b) two effort arms.
   (c) two resistance arms. (d) a resistance arm shorter than the effort arm.
5. A pulley has eight strands holding the resistance. The mechanical advantage is
   (a) 4. (b) 8.
   (c) 16. (d) 64.
6. The mechanical advantage of a compound machine
   (a) is the sum of the MAs of all the simple machines.
   (b) is the product of the MAs of all the simple machines.
   (c) cannot be found.
   (d) is none of the above.
7. Cite three examples of machines used to multiply speed.
8. What name is given to the force overcome by the machine?
9. State the Law of Simple Machines in your own words.
10. What is the term used for the ratio of the resistance force to the effort force?
11. What is the term used for the ratio of the amount of work obtained from a machine to the amount of work put into the machine?
12. Does a friction-free machine exist?
13. What is the pivot point of a lever called?
14. In your own words, state how to find the MA of a lever.
15. Which type of lever do you think would be most efficient?
16. State the Law of Simple Machines as it is applied to levers.
17. Where is the fulcrum located in a third-class lever?
18. In your own words, explain the Law of Simple Machines as applied to the wheel-and-axle.

19. Does the MA of a wheel-and-axle depend on the force applied?
20. Describe the difference between a fixed pulley and a movable pulley.
21. Does the MA of a pulley depend on the radius of the pulley?
22. How can you find the MA of an inclined plane?
23. In your own words, describe the pitch of a screw.
24. How does the MA of a jackscrew differ from the MA of a screwdriver?
25. How is the total MA of a compound machine found?

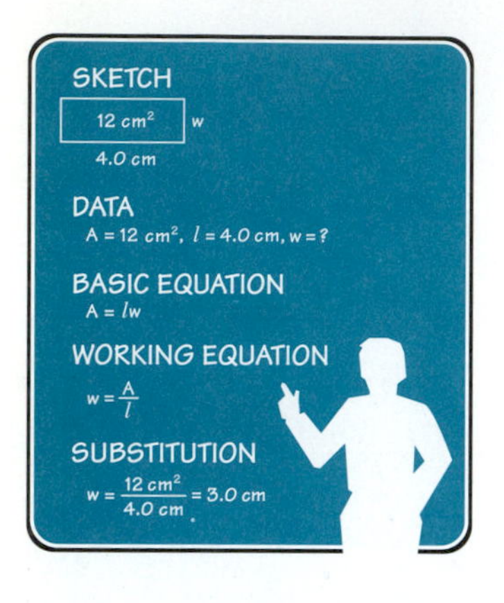

REVIEW PROBLEMS

1. A girl uses a lever to lift a box. The box has a resistance force of $25\bar{0}$ N while she exerts an effort force of 125 N. What is the mechanical advantage of the lever?
2. A bicycle requires 1575 N m of input but only puts out 1150 N m of work. What is the bicycle's efficiency?
3. A lever uses an effort arm of 15.3 m and has a resistance arm of 7.20 m. What is the lever's mechanical advantage?
4. Two people are on a teeter-totter. The person on the left exerts a force of 540 N and is 2.00 m from the fulcrum. If they are to remain balanced, how much force does the other person exert if she is (a) also 2.00 m from the fulcrum? (b) 3.00 m from the fulcrum?
5. A wheel-and-axle has an effort force of 125 N and an effort radius of 17.0 cm. If the resistance force is 325 N, what is the resistance radius?
6. What is the mechanical advantage in Problem 5?
7. What is the mechanical advantage of a pulley system having 12 strands holding the resistance?
8. A pulley system has a mechanical advantage of 5. What is the resistance force if an effort of 135 N is exerted?
9. An inclined plane has a height of 1.50 m and a length of 4.50 m. What effort must be exerted to pull up an 875-N box?
10. What is the mechanical advantage in Problem 9?
11. What height must a 10.0-ft-long inclined plane be to lift a $10\bar{0}0$-lb crate with $23\bar{0}$ lb of effort?
12. A screw has a pitch of 0.0200 cm. An effort force of 29.0 N is used to turn a screwdriver whose handle diameter is 36.0 mm. What is the maximum resistance force?
13. A 945-N resistance force is overcome with a 6.00-N effort using a screwdriver whose handle is 54.0 mm in diameter. What is the pitch of the screw?
14. Find the mechanical advantage of a jackscrew with a 1.50-cm pitch and a handle 36.0 cm long.

Find the mechanical advantage of each.

15.

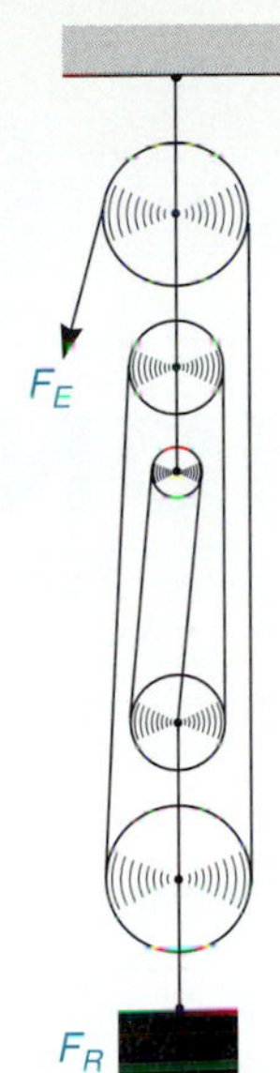

16.

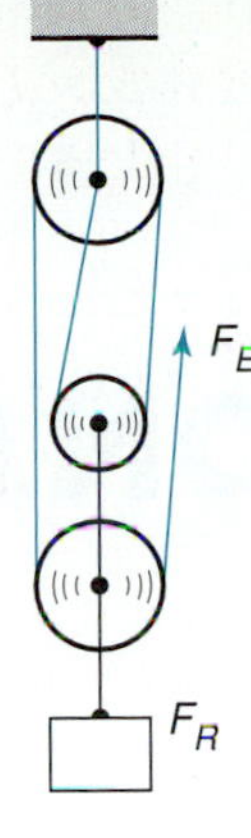

17. (a) Find the mechanical advantage of the compound machine. The radius of the crank is 80.0 cm, and the radius of the axle is 20.0 cm. (b) If an effort force of 75 N is applied to the handle of the crank, what force may be moved up the inclined plane?

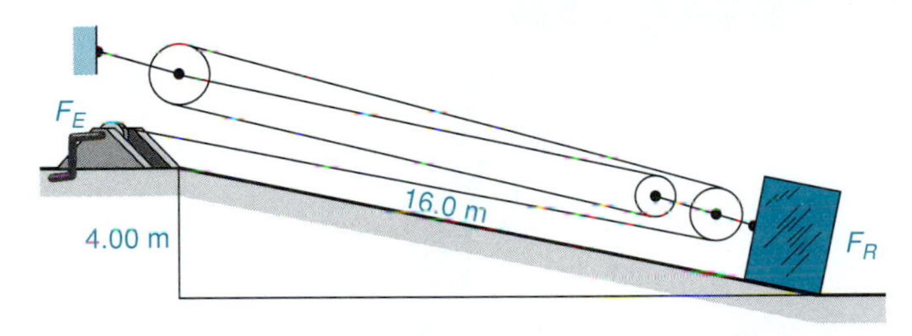

18. If an effort force of 45 N is applied to a simple machine and moves a resistance of 270 N, what is the actual mechanical advantage?

Chapter 9

ROTATIONAL MOTION

To this point we have studied displacement, velocity, acceleration, vectors, and forces in a straight line. Those concepts also apply to motion in a curved path and rotational motion.

OBJECTIVES

The major goals of this chapter are to enable you to:

1. Distinguish between rectilinear, curvilinear, and rotational motion.
2. Apply the torque equation to rotational problems.
3. Calculate the centripetal force of moving objects.
4. Find power in rotational systems.
5. Analyze how gears and gear trains are used to transfer rotational motion.

9.1 MEASUREMENT OF ROTATIONAL MOTION

Until now we have considered only motion in a straight line, called **rectilinear motion.** Technicians are often faced with many problems with motion along a curved path or objects that are rotating about an axis. Although these kinds of motion are similar, we must distinguish between them.

Motion along a curved path is called **curvilinear motion.** A satellite in orbit around the earth is an example of curvilinear motion [Fig. 9.1(a)].

(a) Curvilinear motion of an orbiting satellite.

(b) Rotational motion of earth spinning on its axis.

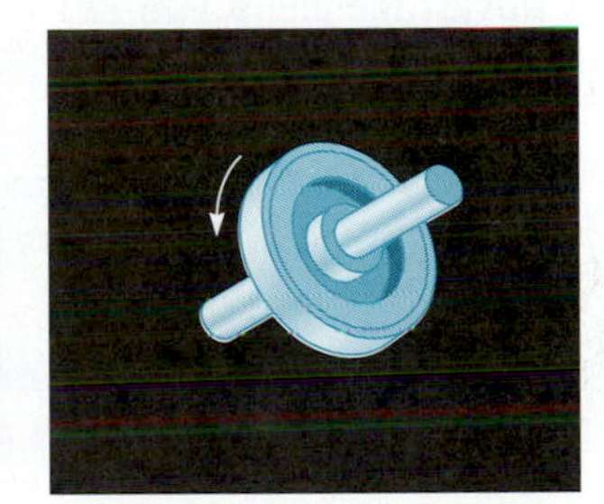

(c) Rotational motion of a wheel spinning on its axle.

Figure 9.1

Rotational motion occurs when the body itself is spinning. Examples of rotational motion are the earth spinning on its axis, a turning wheel, a turning driveshaft, and the turning shaft of an electric motor [Fig. 9.1(b) and (c)].

We can see a wheel turn, but to gather useful information about its motion, we need a system of measurement. There are three basic systems of defining angle measurement. One unit of measurement in rotational motion is the number of rotations—how many times the object goes around. We need to know not only the number of rotations but also the time for each rotation. The unit of rotation (most often used in industry) is the **revolution** (rev). A second system of measurement divides the circle of rotation into 360 degrees (360° = 1 rev).

The **radian** (rad), which is approximately 57.3° or exactly $\left(\frac{360}{2\pi}\right)^{\circ}$, is a third

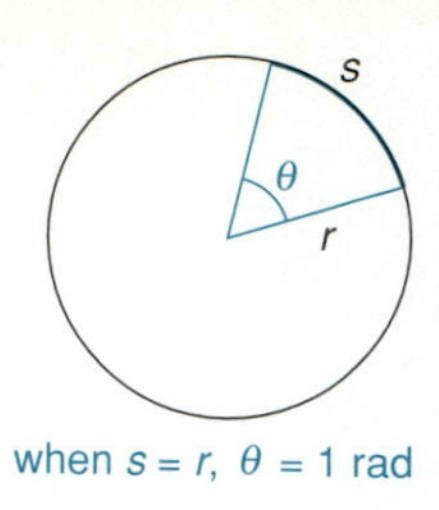

Figure 9.2

angular measure used by scientists. A radian is defined as that angle with its vertex at the center of a circle whose sides cut off an arc on the circle equal to its radius (Fig. 9.2), where $s = r$ and θ (Greek lowercase theta, used as a variable for the angle) = 1 rad.

Stated as a formula.

$$\theta = \frac{s}{r}$$

where θ = angle determined by s and r
s = length of arc of circle
r = radius of the circle

Technically, angle θ measured in radians is defined as the ratio of two lengths: length of arc and radius of a circle. Since the length units in the ratio cancel, the radian is a unitless dimension. As a matter of convenience, though, "rad" is often used to show radian measurement and therefore often appears in formulas. A useful relationship is 2π rad equals one revolution. Therefore,

$$1 \text{ rev} = 360° = 2\pi \text{ rad}$$

You may want to refer to this box when a conversion between systems of measurement is required.

EXAMPLE 1

Convert the angle 10π rad

(a) to rev. (b) to degrees.

Using 1 rev = 360° = 2π rad, form conversion factors, so that the old units are in the denominator and the new units are in the numerator.

$$\text{(a)}\ \theta = (10\pi \cancel{\text{rad}})\left(\frac{1 \text{ rev}}{2\pi \cancel{\text{rad}}}\right) \qquad \text{(b)}\ \theta = (10\pi \cancel{\text{rad}})\left(\frac{360°}{2\pi \cancel{\text{rad}}}\right)$$
$$= 5 \text{ rev} \qquad\qquad = 1800°$$

Angular displacement is the angle through which any point on a rotating body moves. Note that on any rotating body, all points on that body move through the same angle in any given amount of time—even though each may travel different linear distances. Point A on the flywheel shown in Fig. 9.3 travels much father than point B (along a curved line), but during one revolution both travel through the same angle (have equal angular displacements).

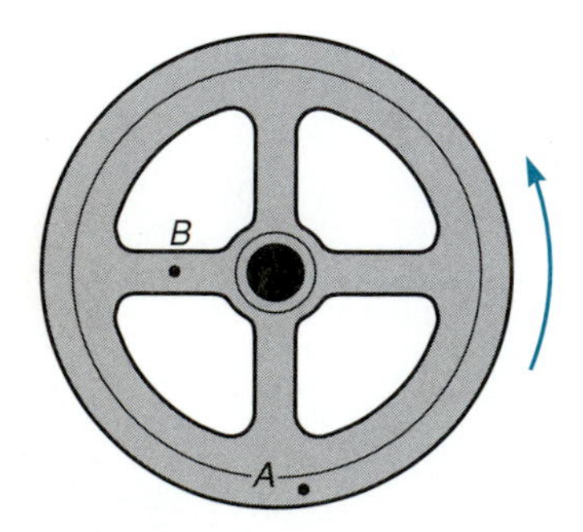

Figure 9.3
The angular displacements of points A and B on the flywheel are always the same.

In the automobile industry, technicians are concerned with the *rate* of rotational motion. Recall that in the linear system, velocity is the rate of motion (displacement/time). Similarly, **angular velocity** in the rotational system is the rate of angular displacement. Angular velocity (designated ω, the Greek lowercase letter omega) is usually measured in rev/min (rpm) for relatively slow rotations (e.g., automobile engines) and rev/s or rad/s for high-speed instruments. We use the term *angular velocity* when referring to a vector and/or knowing the direction of rotation. We use the term *angular speed* in referring to a magnitude when the direction of rotation is either not known or not important.

$$\omega = \text{angular velocity} = \frac{\text{number of revolutions}}{\text{time}} = \frac{\text{angular displacement}}{\text{time}}$$

Written as a formula:

$$\omega = \frac{\theta}{t}$$

where ω = angular velocity or speed (rad/s)
θ = angle (in radians)
t = time (in seconds)

or

ω = angular velocity or speed (rev/min)
θ = angle (in revolutions)
t = time (in minutes)

EXAMPLE 2

A motorcycle wheel turns $36\bar{0}0$ times while being ridden for 6.40 min. What is the angular speed in rev/min?

Sketch: None needed

Data:

$$t = 6.40 \text{ min}$$
$$\text{number of revolutions} = 36\bar{0}0 \text{ rev}$$
$$\omega = ?$$

Basic Equation:

$$\omega = \frac{\text{number of revolutions}}{\text{time}}$$

Working Equation: Same

Substitution:

$$\omega = \frac{36\bar{0}0 \text{ rev}}{6.40 \text{ min}}$$
$$= 563 \text{ rev/min or } 563 \text{ rpm}$$

Formulas for linear speed of a rotating point on a circle and angular speed are related as follows:

We know:

$$(1)\ \theta = \frac{s}{r} \qquad (2)\ v = \frac{s}{t} \qquad (3)\ \omega = \frac{\theta}{t}$$

Therefore, combining and substituting s/r for θ in (3),

$$\omega = \frac{s/r}{t}$$

$$\omega(r) = \frac{(s/r)(r)}{t} \qquad \text{Multiply both sides by } r.$$

$$\omega r = \frac{s}{t}$$

and recalling that $v = s/t$,

$$\omega r = v$$

Thus,

$$v = \omega r$$

where v = linear velocity of a point on the circle
ω = angular speed
r = radius

EXAMPLE 3

A wheel of 1.00 m radius turns at $10\bar{0}0$ rpm.

(a) Express the angular speed in rad/s.
(b) Find the angular displacement in 2.00 s.
(c) Find the linear speed of a point on the rim of the wheel.

Sketch:

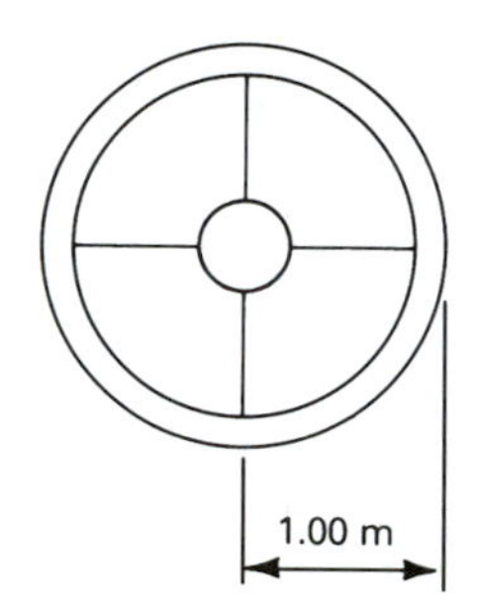

(a) Data:

$$\omega = 10\bar{0}0 \text{ rpm} \qquad \text{(change it to rad/s)}$$

$$\omega = 10\bar{0}0 \frac{\text{rev}}{\text{min}} \times \frac{2\pi \text{ rad}}{1 \text{ rev}} \times \frac{1 \text{ min}}{60 \text{ s}} = 105 \text{ rad/s}$$

(b) Data:

$$\omega = 105 \text{ rad/s}$$
$$t = 2.00 \text{ s}$$
$$\theta = ?$$

Basic Equation:

$$\omega = \frac{\theta}{t}$$

Working Equation:

$$\theta = \omega t$$

Substitution:

$$\theta = (105 \text{ rad/s})(2.00 \text{ s})$$
$$= 21\bar{0} \text{ rad}$$

(c) Data:

$$\omega = 105 \text{ rad/s}$$
$$r = 1.00 \text{ m}$$
$$v = ?$$

Basic Equation:

$$v = \omega r$$

Working Equation: Same

Substitution:

$$v = (105 \text{ rad/s})(1.00 \text{ m})$$
$$= 105 \text{ m/s}$$

(rad/s)(m) = m/s because the rad is a unitless dimension.

A device called a *stroboscope* or strobe light may be used to measure or check the speed of rotation of a shaft or other machinery part. It is used to "slow down" repeating motion to be observed more conveniently. The light flashes rapidly, and the rate of flash can be adjusted to coincide with the rotation of a point or points on the rotating object. Knowing the rate of flashing will also then reveal the rate of rotation. A slight variation in the rate of rotation and flash will cause the observed point to appear to move either forward or backward as the stagecoach wheels in old western movies sometimes appear to do.

■ PROBLEMS 9.1

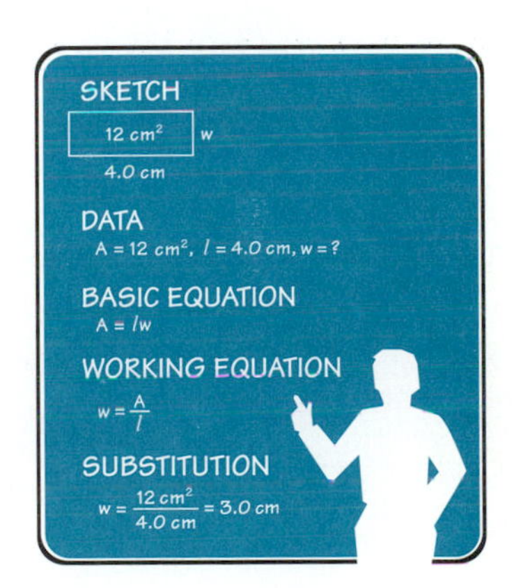

1. Convert $6\frac{1}{2}$ revolutions
   (a) to radians.
   (b) to degrees.
2. Convert 2880°
   (a) to revolutions.
   (b) to radians.
3. Convert 25π rad
   (a) to revolutions.
   (b) to degrees.
4. Convert 12.0 revolutions
   (a) to radians.
   (b) to degrees.

Find the angular speed in Problems 5–10.

5. Number of revolutions = 525
   $t = 3.42$ min
   $\omega =$ _____ rpm
6. Number of revolutions = 7360
   $t = 37.0$ s
   $\omega =$ _____ rev/s
7. Number of revolutions = 4.00
   $t = 3.00$ s
   $\omega =$ _____ rad/s
8. Number of revolutions = 325
   $t = 5.00$ min
   $\omega =$ _____ rpm
9. Number of revolutions = 6370
   $t = 18.0$ s
   $\omega =$ _____ rev/s
10. Number of revolutions = 6.25
    $t = 5.05$ s
    $\omega =$ _____ rad/s
11. Convert 675 rad/s to rpm.
12. Convert 285 rpm to rad/s.
13. Convert 136 rpm to rad/s.
14. Convert 88.4 rad/s to rpm.
15. A motor turns at a rate of 11.0 rev/s. Find its angular speed in rpm.
16. A rotor turns at a rate of 180 rpm. Find its angular speed in rev/s.
17. A rotating wheel completes one revolution in 0.150 s. Find its angular speed
    (a) in rev/s. (b) in rpm. (c) in rad/s.
18. A rotor completes 50.0 revolutions in 3.25 s. Find its angular speed
    (a) in rev/s. (b) in rpm. (c) in rad/s.

19. A flywheel rotates at 1050 rpm.
    (a) How long (in s) does it take to complete one revolution?
    (b) How many revolutions does it complete in 5.00 s?
20. A wheel rotates at 36.0 rad/s.
    (a) How long (in s) does it take to complete one revolution?
    (b) How many revolutions does it complete in 8.00 s?
21. A shaft of radius 8.50 cm rotates 7.00 rad/s. Find its angular displacement (in rad) in 1.20 s.
22. A wheel of radius 0.240 m turns at 4.00 rev/s. Find its angular displacement (in rev) in 13.0 s.
23. A pendulum of length 1.50 m swings through an arc of 5.0°. Find the length of the arc through which the pendulum swings.
24. An airplane circles an airport twice while 5.00 mi from the control tower. Find the length of the arc through which the plane travels.
25. A wheel of radius 27.0 cm has an angular speed of 47.0 rpm. Find the linear speed (in m/s) of a point on its rim.
26. A belt is placed around a pulley that is 30.0 cm in diameter and rotating at 275 rpm. Find the linear speed (in m/s) of the belt. (Assume no belt slippage on the pulley.)
27. A flywheel of radius 25.0 cm is rotating at 655 rpm.
    (a) Express its angular speed in rad/s.
    (b) Find its angular displacement (in rad) in 3.00 min.
    (c) Find the linear distance traveled (in cm) by a point on the rim in one complete revolution.
    (d) Find the linear distance traveled (in m) by a point on the rim in 3.00 min.
    (e) Find the linear speed (in m/s) of a point on the rim.
28. An airplane propeller with blades 2.00 m long is rotating at 1150 rpm.
    (a) Express its angular speed in rad/s.
    (b) Find its angular displacement in 4.00 s.
    (c) Find the linear speed (in m/s) of a point on the end of the blade.
    (d) Find the linear speed (in m/s) of a point 1.00 m from the end of the blade.
29. An automobile is traveling at 60.0 km/h. Its tires have a radius of 33.0 cm.
    (a) Find the angular speed of the tires (in rad/s).
    (b) Find the angular displacement of the tires in 30.0 s.
    (c) Find the linear distance traveled by a point on the tread in 30.0 s.
    (d) Find the linear distance traveled by the automobile in 30.0 s.
30. Find the angular speed (in rad/s) of the following hands on a clock.
    (a) Second hand (b) Minute hand (c) Hour hand
31. A bicycle wheel of diameter 30.0 in. rotates twice each second. Find the linear velocity of a point on the wheel.
32. A flywheel with radius 1.50 ft has a linear velocity of 30.0 ft/s. Find the time to complete 4π rad.
33. The earth rotates on its axis at an angular speed of 1 rev/24 h. Find the linear speed (in km/h)
    (a) of Singapore, which is nearly on the equator.
    (b) of Houston, which is approximately 30.0° north latitude.
    (c) of Minneapolis, which is approximately 45.0° north latitude.
    (d) of Anchorage, which is approximately 60.0° north latitude. ■

9.2 TORQUE

In a linear system, we defined force as a push or a pull. In a rotational system, we have a "twist" which we call torque. **Torque** is the tendency to produce change in rotational motion. It is, in rotational systems, similar to force in the linear system. You may already have studied torque in connection with automobile engines. We first consider the simpler example of pedaling a bicycle [Fig. 9.4(a)]. In pedaling we apply a torque to the sprocket, causing it to rotate. The torque developed depends on two factors:

1. The amount of force applied
2. How far from the axle (shaft) center point the force is applied

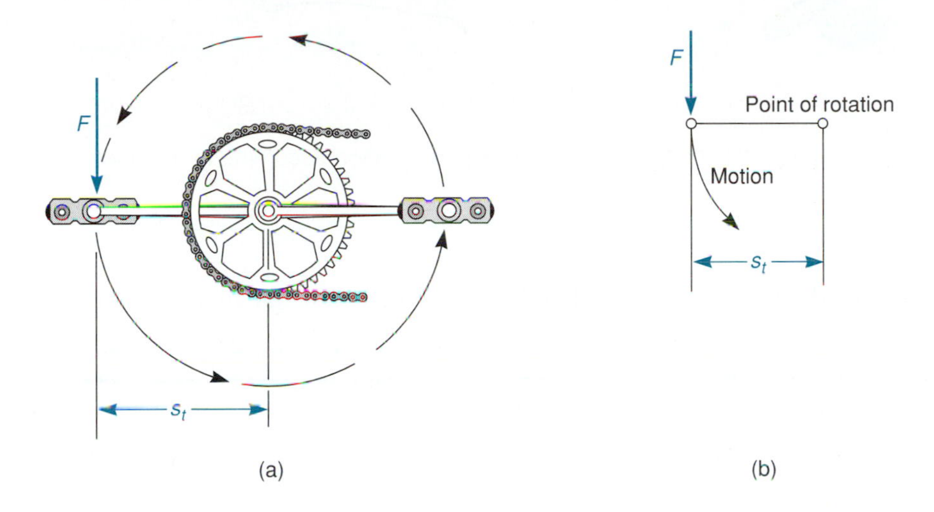

Figure 9.4
Torque produced in pedaling a bicycle.

We can express torque with an equation:

$$T = Fs_t$$

where T = torque (N m or lb ft)
F = applied force (N or lb)
s_t = length of torque arm (m or ft)

Note that s_t, the length of the torque arm, is different from s in the equation defining work *($W = Fs$)*. Recall that s in the work equation is the linear distance over which the force acts.

In all torque problems we are concerned with motion about a point or axis of rotation. The torque arm is the *perpendicular* distance from the point of rotation to the applied force [Fig. 9.4(a)]. In torque problems, s_t is always perpendicular to the force [Fig. 9.4(a)]. Note in Fig. 9.4(a) that s_t is the distance from the axle to the pedal. The units of torque look similar to those of work, but note the difference between s and s_t.

If the force is not exerted tangent to the circle made by the pedal (Fig. 9.5), the length of the torque arm is *not* the length of the pedal arm. The torque arm, s_t, is measured as the perpendicular distance to the force. Since s_t is therefore shorter, the product $F \cdot s_t$ is smaller, and the turning effect, the torque, is less in the pedal position shown in Fig. 9.5. Maximum torque is produced when the pedals are horizontal and the force applied is straight down.

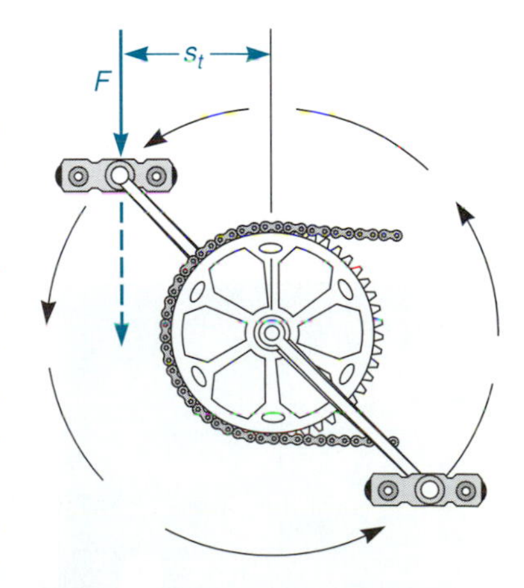

Figure 9.5
Maximum torque is only produced when the pedal reaches a position perpendicular to the applied force.

Torque is a vector quantity that acts along the axis of rotation (not along the force) and points in the direction in which a right-handed screw would advance if turned by the torque as in Fig. 9.6(a). The right-hand rule is often used to determine the direction of the torque as follows: Grasp the axis of rotation with your right hand so that your fingers circle it in the direction that the torque tends to induce rotation. Your thumb will point in the direction of the torque vector [Fig.

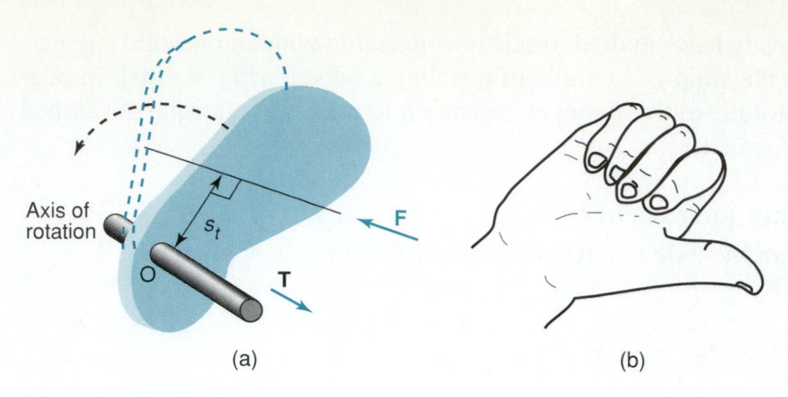

Figure 9.6
Torque is a vector quantity that acts along the axis of rotation according to the right-hand rule.

9.6(b)]. Thus, the torque vector in Fig. 9.6(a) is perpendicular to and points out of the page.

EXAMPLE

A force of 10.0 lb is applied to a bicycle pedal. If the length of the pedal arm is 0.850 ft, what torque is applied to the shaft?

Sketch:

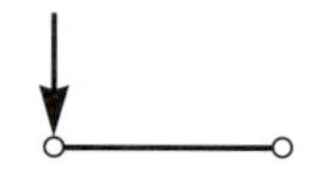

Data:

$$F = 10.0 \text{ lb}$$
$$s_t = 0.850 \text{ ft}$$
$$T = ?$$

Basic Equation:

$$T = Fs_t$$

Working Equation: Same

Substitution:

$$T = (10.0 \text{ lb})(0.850 \text{ ft})$$
$$= 8.50 \text{ lb ft}$$

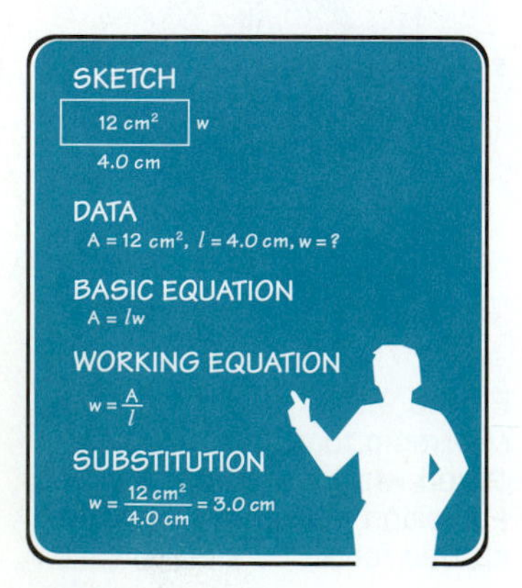

■ PROBLEMS 9.2

Assume that each force is applied perpendicular to the torque arm.

1. Given: $F = 16.0$ lb
   $s_t = 6.00$ ft
   $T = ?$
2. Given: $F = 10\bar{0}$ N
   $s_t = 0.420$ m
   $T = ?$
3. Given: $T = 60.0$ N m
   $F = 30.0$ N
   $s_t = ?$
4. Given: $T = 35.7$ lb ft
   $s_t = 0.0240$ ft
   $F = ?$

5. Given: $T = 65.4$ N m
$s_t = 35.0$ cm
$F = ?$

6. Given: $F = 63\bar{0}$ N
$s_t = 74.0$ cm
$T = ?$

7. If the torque on a shaft of radius 2.37 cm is 38.0 N m (Fig. 9.7), what force is applied to the shaft?

8. If a force of 56.2 lb is applied to a torque wrench 1.50 ft long (Fig. 9.8), what torque is indicated by the wrench?

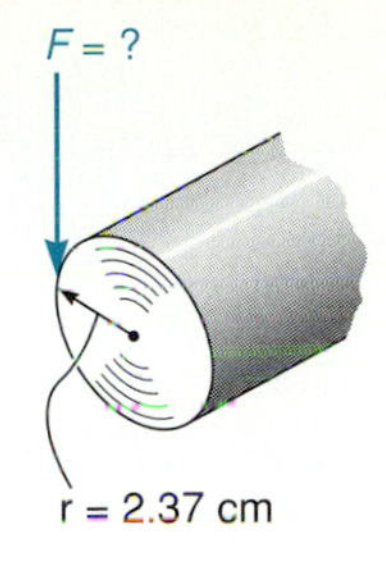

Figure 9.7

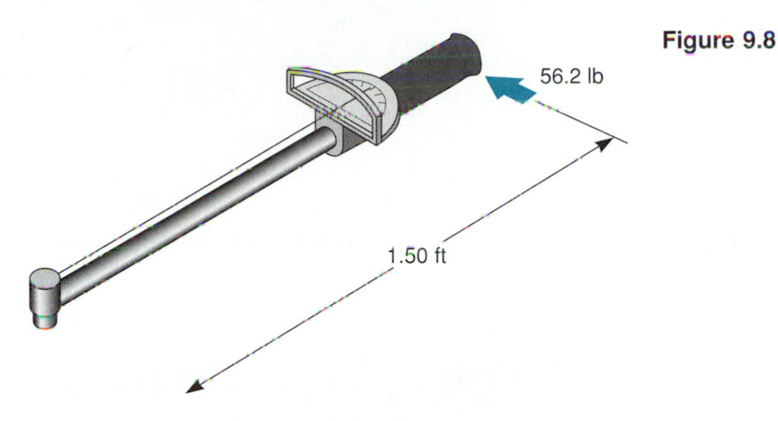

Figure 9.8

9. A motorcycle head bolt is torqued to 25.0 N m. What length shaft do we need on a wrench to exert a maximum force of 70.0 N?

10. A force of 112 N is applied to a shaft of radius 3.50 cm. What is the torque on the shaft?

11. A torque of 175 lb ft is needed to free a large rusted-on nut. The length of the wrench is 1.10 ft. What force must be applied to free it?

12. A torque wrench reads 14.5 N m. If its length is 25.0 cm, what force is being applied to the handle?

13. The torque on a shaft of radius 3.00 cm is 12.0 N m. What force is being applied to the shaft?

14. An automobile bolt is torqued to 27.0 N m. If the length of the wrench is 30.0 cm, what force is applied to the wrench?

15. A torque wrench reads 25 lb ft. (a) If its length is 1.0 ft, what force is being applied to the wrench? (b) What is the force if the length is doubled? Explain the results.

16. If 13 N m of torque is applied to a bolt with an applied force of 28 N, what is the length of the wrench? ■

9.3 MOTION IN A CURVED PATH

Newton's laws of motion apply to motion along a curved path as well as in a straight line. Recall that a moving body tends to continue in a straight line because of inertia. If we are to cause the body to move in a circle, we must constantly apply a force perpendicular to the line of motion of the body. The simplest example is a rock being swung in a circle on the end of a string (Fig. 9.9). By Newton's first law, the rock tends to go in a straight line but the string exerts a constant force on the rock perpendicular to this line of travel. The resulting path of the rock is a circle [Fig. 9.10(a)]. The force of the string on the rock is the *centripetal* (toward the center) *force*. The **centripetal force** acting on a body in circular motion causes it to move in a circular path. This force is exerted in the direction of the center of the circle. If the string should break, however, there would no longer be a centripetal force acting on the rock, and it would fly off tangent to the circle [Fig. 9.10(b)].

Figure 9.9
Rock on a string being swung in a circle.

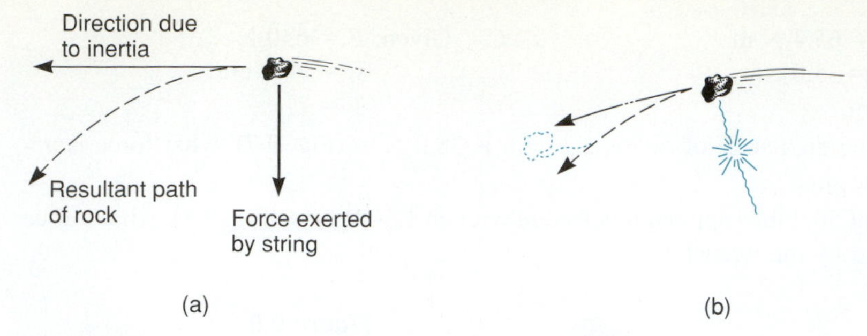

Figure 9.10
Centripetal force on a rock being swung in a circle.

The equation for finding the centripetal force on any body moving along a curved path is

$$F = \frac{mv^2}{r}$$

where F = centripetal force
m = mass of the body
v = velocity of the body
r = radius of curvature of the path of the body

What, then, is the difference between centripetal and centrifugal forces? The term *centrifugal force* is widely misused to mean centripetal force. **Centrifugal force** is the outward force on an object moving along a curved path when viewed from the frame of reference of the object. It is equal in magnitude to the centripetal force but is in the opposite direction.

EXAMPLE

An automobile of mass 1640 kg rounds a curve of radius 25.0 m with a velocity of 15.0 m/s (54.0 km/h). What centripetal force is exerted on the automobile while rounding the curve?

Sketch:

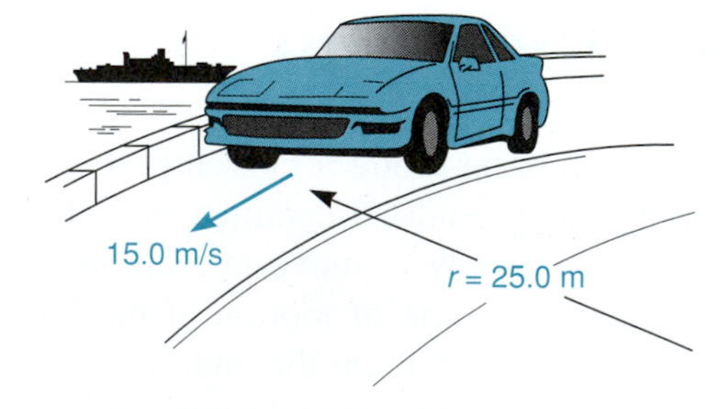

Data:

$$m = 1640 \text{ kg}$$
$$v = 15.0 \text{ m/s}$$
$$r = 25.0 \text{ m}$$
$$F = ?$$

Basic Equation:

$$F = \frac{mv^2}{r}$$

Working Equation: Same

Substitution:

$$F = \frac{(1640 \text{ kg})(15.0 \text{ m/s})^2}{25.0 \text{ m}}$$

$$= 14{,}800 \text{ kg m/s}^2 \qquad (\text{Recall: } 1 \text{ N} = 1 \text{ kg m/s}^2)$$

$$= 14{,}800 \text{ N}$$

■ PROBLEMS 9.3

1. Given: $m = 64.0$ kg
   $v = 34.0$ m/s
   $r = 17.0$ m
   $F =$ _____ N

2. Given: $m = 11.3$ slugs
   $v = 3.00$ ft/s
   $r = 3.24$ ft
   $F =$ _____ lb

3. Given: $F = 25\bar{0}0$ lb
   $v = 47.6$ ft/s
   $r = 72.0$ ft
   $m =$ _____ slugs

4. Given: $F = 587$ N
   $v = 0.780$ m/s
   $m = 67.0$ kg
   $r =$ _____ m

5. Given: $F = 602$ N
   $m = 63.0$ kg
   $r = 3.20$ m
   $v =$ _____ m/s

6. Given: $m = 37.5$ kg
   $v = 17.0$ m/s
   $r = 3.75$ m
   $F =$ _____ N

7. Given: $F = 75.0$ N
   $v = 1.20$ m/s
   $m = 10\bar{0}$ kg
   $r =$ _____ m

8. Given: $F = 80.0$ N
   $m = 43.0$ kg
   $r = 17.5$ m
   $v =$ _____ m/s

9. An automobile of mass 117 slugs follows a curve of radius 79.0 ft with a speed of 49.3 ft/s. What centripetal force is exerted on the automobile while it is rounding the curve?
10. Find the centripetal force exerted on a 7.12-kg mass moving at a speed of 2.98 m/s in a circle of radius 2.72 m.
11. The centripetal force on a car of mass $80\bar{0}$ kg rounding a curve is 6250 N. If its speed is 15.0 m/s, what is the radius of the curve?
12. The centripetal force on a runner is 17.0 lb. If the runner weighs 175 lb and his speed is 14.0 mi/h, find the radius of the curve.
13. An automobile whose mass is 1650 kg is driven around a circular curve of radius $15\bar{0}$ m at 80.0 km/h. Find the centripetal force of the road on the automobile.
14. A cycle of mass 510 kg rounds a curve of radius $4\bar{0}$ m at 95 km/h. What is the centripetal force on the cycle?
15. What is the centripetal force exerted on a rock with mass 3.2 kg moving at 3.5 m/s in a circle of radius 2.1 m?
16. A truck with mass of 215 slugs rounds a curve of radius 53.0 ft with a speed of 62.5 ft/s. (a) What centripetal force is exerted on the truck while rounding the curve? (b) How does the centripetal force change when the velocity is doubled? (c) What is the new force?
17. A 225-kg dirt bike is going through a curve with linear velocity of 35 m/s and an angular speed of 0.25 rad/s. Find the centripetal force exerted on the bike. ■

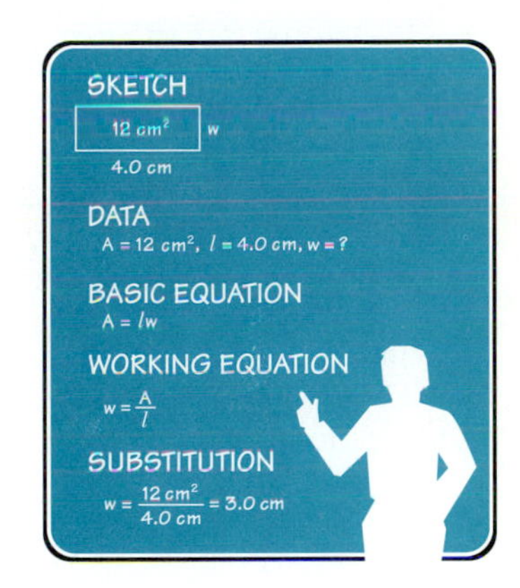

9.4 POWER IN ROTATIONAL SYSTEMS

One of the most important aspects of rotational motion to the technician is the power developed. Recall that torque was discussed in Section 9.2. Power, however, must be considered whenever an engine or motor is used to turn a shaft. Some common examples are the use of winches and drive trains (Fig. 9.11).

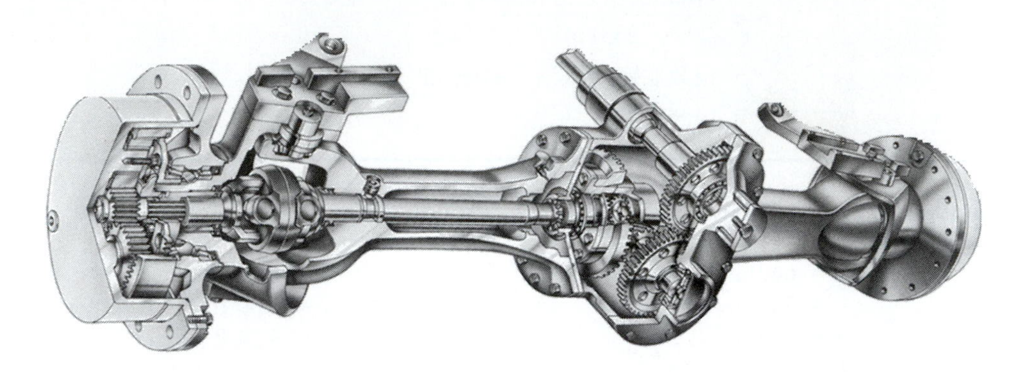

Figure 9.11
This driveshaft connects the engine transmission with the axle to supply power to the wheels and other components of this tractor.(Courtesy of Deere & Company.)

Earlier we learned that

$$\text{power} = \frac{\text{force} \times \text{displacement}}{\text{time}} = \frac{\text{work}}{\text{time}}$$

in the linear system. In the rotational system:

$$\begin{aligned} \text{power} &= \frac{\text{(torque)(angular displacement)}}{\text{time}} \\ &= \text{(torque)(angular velocity)} \\ &= T\omega \end{aligned}$$

Recall that angular displacement is the angle through which the shaft is turned.

In the metric system, angular displacement must be expressed in radians. (*Recall:* 1 rev = 2π radians.)

Substituting symbols and units:

In watts (W):

$$\begin{aligned} \text{power} &= T\omega \\ &= (\text{N m})\left(\frac{1}{\text{s}}\right) = \frac{\text{N m}}{\text{s}} = \frac{\text{J}}{\text{s}} = \text{W} \end{aligned}$$

To find the power in kilowatts (kW), multiply the number of watts by the conversion factor:

$$\frac{1 \text{ kW}}{1000 \text{ W}}$$

Note: In problem solving, the radian unit is a unitless dimension; ω is expressed with the unit /s.

EXAMPLE 1

How many watts of power are developed by a mechanic tightening bolts using 50.0 N m of torque at a rate of 2.50 rad/s? How many kW?

Sketch: None needed

Data:

$$T = 50.0 \text{ N m}$$
$$\omega = 2.50/\text{s}$$
$$P = ?$$

Basic Equation:

$$P = T\omega$$

Working Equation: Same

Substitution:

$$\begin{aligned} P &= (50.0 \text{ N m})(2.50/\text{s}) \\ &= 125 \text{ N m/s} \\ &= 125 \text{ W} \qquad (1 \text{ W} = 1 \text{ N m/s}) \end{aligned}$$

To find the power in kW:

$$125 \cancel{\text{W}} \times \frac{1 \text{ kW}}{1000 \cancel{\text{W}}} = 0.125 \text{ kW}$$

In the English system we measure angular displacement by multiplying the number of revolutions by 2π:

$$\text{angular displacement} = \text{number of revolutions} \times 2\pi$$

For the rotational system

$$\text{power} = \frac{(\text{torque}) \times 2\pi \text{ revolutions}}{\text{time}}$$

When time is in minutes

$$\text{power} = \text{torque} \times 2\pi \times \frac{\text{rev}}{\text{min}} \times \frac{1 \text{ min}}{60 \text{ s}}$$

$$\text{power in } \frac{\text{ft lb}}{\text{s}} = \text{torque in lb ft} \times \frac{\text{number of revolutions}}{\text{min}} \times \underbrace{0.105 \frac{\text{min}}{\text{rev s}}}_{2\pi \times \frac{1 \text{ min}}{60 \text{ s}}}$$

Another common unit of power is the horsepower (hp). The conversion factor between $\frac{\text{ft lb}}{\text{s}}$ and hp is

$$\text{power in hp} = \text{power in } \frac{\text{ft lb}}{\text{s}} \times \frac{\text{hp}}{550 \text{ ft lb/s}}$$

EXAMPLE 2

What power (in ft lb/s) is developed by an electric motor with torque 5.70 lb ft and speed of 425 rpm?

Sketch: None needed

Data:

$$T = 5.70 \text{ lb ft}$$
$$\omega = 425 \text{ rpm}$$
$$P = ?$$

Basic Equation:

$$P = \text{torque} \times \frac{\text{rev}}{\text{min}} \times 0.105 \frac{\text{min}}{\text{rev s}}$$

Working Equation: Same

Substitution:

$$P = (5.70 \text{ lb ft})\left(425 \frac{\cancel{\text{rev}}}{\cancel{\text{min}}}\right)\left(0.105 \frac{\cancel{\text{min}}}{\cancel{\text{rev}} \text{ s}}\right)$$
$$= 254 \text{ ft lb/s}$$

EXAMPLE 3

What horsepower is developed by a racing engine with torque 545 lb ft at $65\bar{0}0$ rpm?

First, find power in ft lb/s and then convert to hp.

Sketch: None needed

Data:

$$T = 545 \text{ lb ft}$$
$$\omega = 65\bar{0}0 \text{ rpm}$$
$$P = ?$$

Basic Equation:

$$P = \text{torque} \times \frac{\text{rev}}{\text{min}} \times 0.105 \frac{\text{min}}{\text{rev s}}$$

Working Equation: Same

Substitution:

$$P = (545 \text{ lb ft})\left(65\bar{0}0 \frac{\cancel{\text{rev}}}{\cancel{\text{min}}}\right)\left(0.105 \frac{\cancel{\text{min}}}{\cancel{\text{rev}} \text{ s}}\right)$$

$$= 372{,}000 \, \frac{\text{ft lb}}{\text{s}} \times \frac{1 \text{ hp}}{550 \, \frac{\text{ft lb}}{\text{s}}}$$

$$= 676 \text{ hp}$$

■ PROBLEMS 9.4

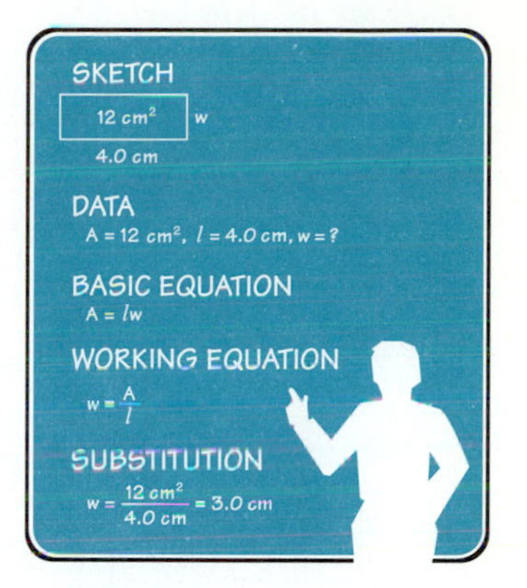

1. Given: $T = 125$ lb ft
   $\omega = 555$ rpm
   $P =$ _____ ft lb/s
2. Given: $T = 39.4$ N m
   $\omega = 6.70$/s
   $P =$ _____ W
3. Given: $T = 372$ lb ft
   $\omega = 264$ rpm
   $P =$ _____ hp
4. Given: $T = 65\bar{0}$ N m
   $\omega = 45.0$/s
   $P =$ _____ kW
5. Given: $P = 8950$ W
   $\omega = 4.80$/s
   $T =$ _____
6. Given: $P = 650$ W
   $T = 540$ N m
   $\omega =$ _____
7. What horsepower is developed by an engine with torque $40\bar{0}$ lb ft at $450\bar{0}$ rpm?
8. What torque must be applied to develop 175 ft lb/s of power in a motor if $\omega =$ 394 rpm?
9. Find the angular velocity of a motor developing 649 W of power with torque 131 N m.
10. A high-speed industrial drill develops 0.500 hp at $16\bar{0}0$ rpm. What torque is applied to the drill bit?
11. An engine has torque of 550 N m at 8.3 rad/s. What power in watts does it develop?
12. Find the angular velocity of a motor developing 33.0 N m/s of power with a torque of 6.0 N m.
13. What power (in hp) is developed by an engine with torque 524 lb ft
    (a) at $30\bar{0}0$ rpm? (b) at $60\bar{0}0$ rpm?
14. Find the angular velocity of a motor developing 650 W of power with a torque of 130 N m.
15. A drill develops 0.500 kW of power at $18\bar{0}0$ rpm. What torque is applied to the drill bit?
16. What power is developed by an engine with torque $75\bar{0}$ N m applied at $45\bar{0}0$ rpm?
17. A tangential force of 150 N is applied to a flywheel of diameter 45 cm to maintain a constant angular velocity of 175 rpm. How much work is done per minute?
18. Find the power developed by an engine with torque 1250 N m applied at $50\bar{0}0$ rpm.
19. Find the angular velocity of a motor developing $10\bar{0}0$ W of power with a torque of $15\bar{0}$ N m.
20. A motor develops 0.75 kW of power at $2\bar{0}00$ revolutions per $1\bar{0}$ min. What torque is applied to the motor shaft?
21. What power is developed when a tangential force of 175 N is applied to a flywheel of diameter 86 cm, causing it to have an angular velocity of 36 revolutions per 6.0 s? ■

9.5 TRANSFERRING ROTATIONAL MOTION

Suppose that two disks are touching each other as in Fig. 9.12. Disk *A* is driven by a motor and turns disk *B* (wheel) by making use of the friction between them.

The relationship between the diameters of the two disks and their number of revolutions is

$$D \cdot N = d \cdot n$$

where D = diameter of the driver disk
d = diameter of the driven disk
N = number of revolutions of the driver disk
n = number of revolutions of the driven disk

Figure 9.12
In the self-propelled lawn mower, disk *A*, driven by the motor, turns disk *B*, the wheel, which results in the mower moving along the ground.

However, using two disks to transfer rotational motion is not very efficient due to slippage that may occur between disks. The most common ways to prevent disk slippage are placing the teeth on the edge of the disk and connecting the disks with a belt. Therefore, instead of using disks, we use gears or belt-driven pulleys to transfer this motion. The teeth on the gears eliminate the slippage; the belt connecting the pulleys helps reduce the slippage and provides for distance between rotating centers (Fig. 9.13).

We can change the equation $D \cdot N = d \cdot n$ to the form $D/d = n/N$ by dividing both sides by dN. The left side indicates the ratio of the diameters of the disks. If the ratio is 2, this means that the larger disk must have a diameter two times the diameter of the smaller disk. The same ratio would apply to gears and pulleys. The ratio of the diameters of the gears must be 2 to 1, and the ratio of the diameters of the pulleys must be 2 to 1. In fact, the ratio of the number of teeth on the gears must be 2 to 1.

The right side of the equation indicates the ratio of the number of revolutions of the two disks. If the ratio is 2, this means that the smaller disk makes two

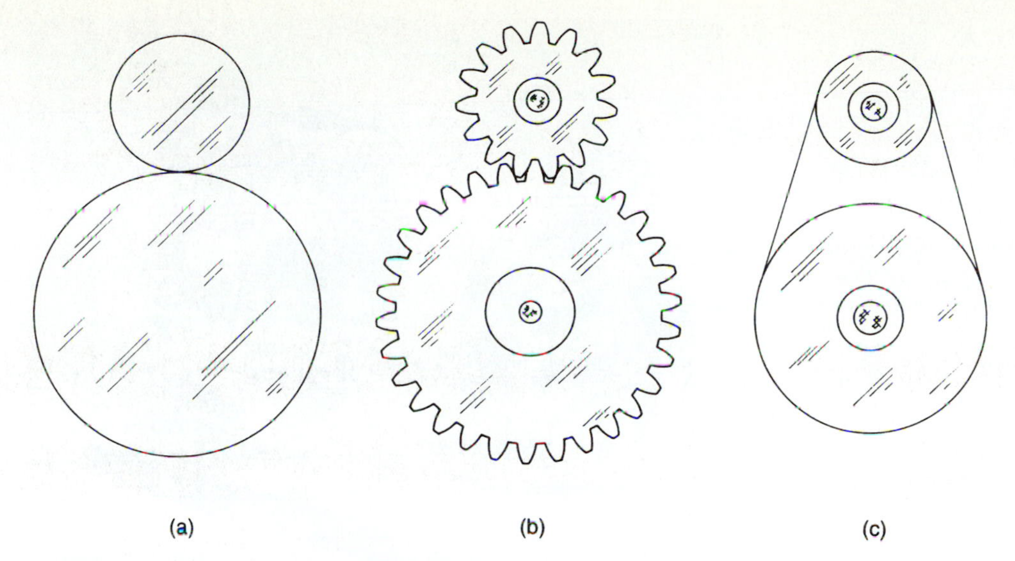

Figure 9.13
Gears and pulleys are used to reduce slippage in transferring rotational motion.

revolutions while the larger disk makes one revolution. The same would be true for gears and for pulleys connected by a belt.

Pulleys and gears are used to increase or reduce the angular velocity of a rotating shaft or wheel. When two gears or pulleys are connected, the speed at which each turns compared to the other is inversely proportional to the diameter of that gear or pulley. The larger the diameter of a pulley or gear, the slower it turns. The smaller the diameter of a pulley or gear, the faster it will turn when connected to a larger one.

9.6 GEARS

Gears are used to transfer rotational motion from one gear to another. The gear that causes the motion is called the *driver gear.* The gear to which the motion is transferred is called the *driven gear.*

There are many different sizes, shapes, and types of gears. A few examples are shown in Fig. 9.14. For any type of gear, we use one basic formula:

$$T \cdot N = t \cdot n$$

where T = number of teeth on the driver gear
N = number of revolutions of the driver gear
t = number of teeth on the driven gear
n = number of revolutions of the driven gear

EXAMPLE 1

A driver gear has 30 teeth. How many revolutions does the driven gear with 20 teeth make while the driver makes 1 revolution?

Data:

$T = 30$ teeth $\quad t = 20$ teeth
$N = 1$ revolution $\quad n = ?$

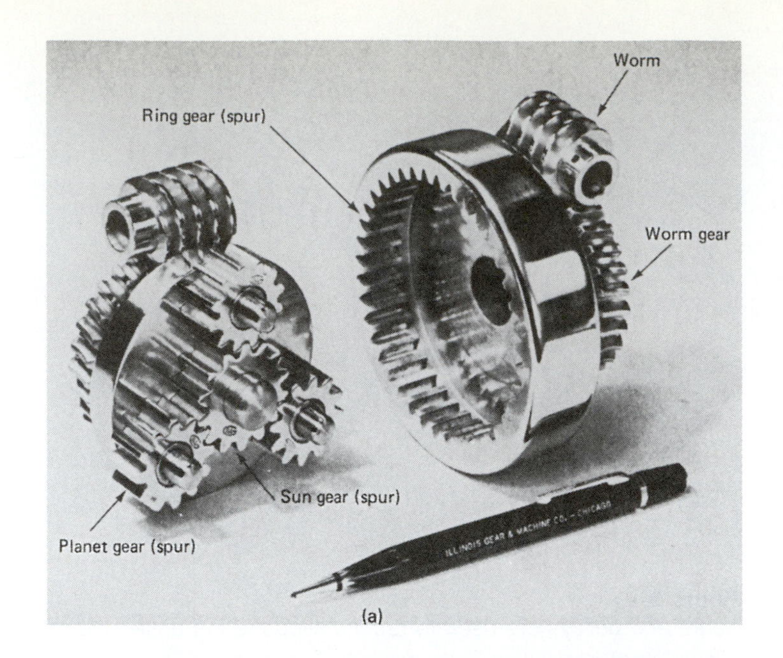

Figure 9.14
Examples of different types of gears. (Courtesy of Illinois Gear, Wallace-Murray Corporation.)

Basic Equation:

$$T \cdot N = t \cdot n$$

Working Equation:

$$n = \frac{T \cdot N}{t}$$

Substitution:

$$n = \frac{(30\ \cancel{\text{teeth}})(1\ \text{rev})}{20\ \cancel{\text{teeth}}}$$
$$= 1.5\ \text{rev}$$

EXAMPLE 2

A driven gear of 70 teeth makes 63.0 revolutions per minute (rpm). The driver gear makes 90.0 rpm. What is the number of teeth required for the driver gear?

Data:

$$N = 90.0\ \text{rpm}$$
$$t = 70\ \text{teeth}$$
$$n = 63.0\ \text{rpm}$$
$$T = ?$$

Basic Equation:

$$T \cdot N = t \cdot n$$

Working Equation:

$$T = \frac{t \cdot n}{N}$$

Substitution:

$$T = \frac{(70\ \text{teeth})(63.0\ \cancel{\text{rpm}})}{90.0\ \cancel{\text{rpm}}}$$
$$= 49\ \text{teeth}$$

Gear Trains

When two gears mesh (Fig. 9.15),* they turn in opposite directions. If gear A turns clockwise, gear B turns counterclockwise. If gear A turns counterclockwise, gear B turns clockwise. If a third gear is inserted between the two (Fig. 9.16), then gears A and B are rotating in the same direction. This third gear is called an *idler*, and such an arrangement of gears is called a **gear train.**

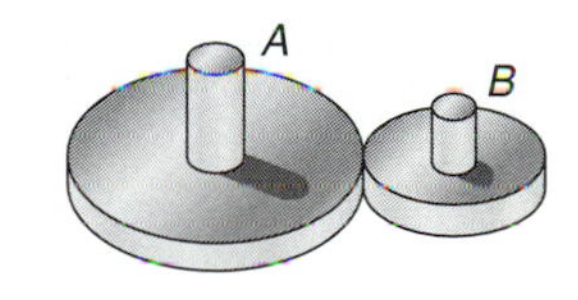

Figure 9.15
Meshed gears shown as cylinders for simplicity.

When the number of shafts in a gear train is odd (such as 1, 3, 5, . . .), the first gear and the last gear rotate in the same direction. When the number of shafts is even, the gears rotate in opposite directions.

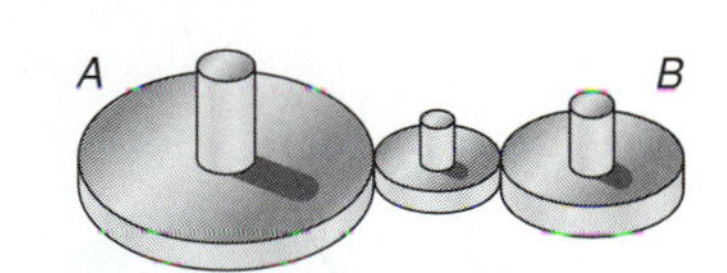

Figure 9.16
Gear train of three gears.

When a complicated gear train is considered, the relationship between revolutions and number of teeth is still present. This relationship is: The number of revolutions of the first driver times the product of the number of teeth of all the driver gears equals the number of revolutions of the final driven gear times the product of the number of teeth on all the driven gears. That is,

$$NT_1T_2T_3T_4 \cdots = nt_1t_2t_3t_4 \cdots$$

*Although gears have teeth, in technical work they are usually shown as cylinders.

where N = number of revolutions of first driver gear
T_1 = teeth on first driver gear
T_2 = teeth on second driver gear
T_3 = teeth on third driver gear
T_4 = teeth on fourth driver gear
n = number of revolutions of last driven gear
t_1 = teeth on first driven gear
t_2 = teeth on second driven gear
t_3 = teeth on third driven gear
t_4 = teeth on four driven gear

EXAMPLE 3

Determine the relative motion of gears A and B in Fig. 9.17.

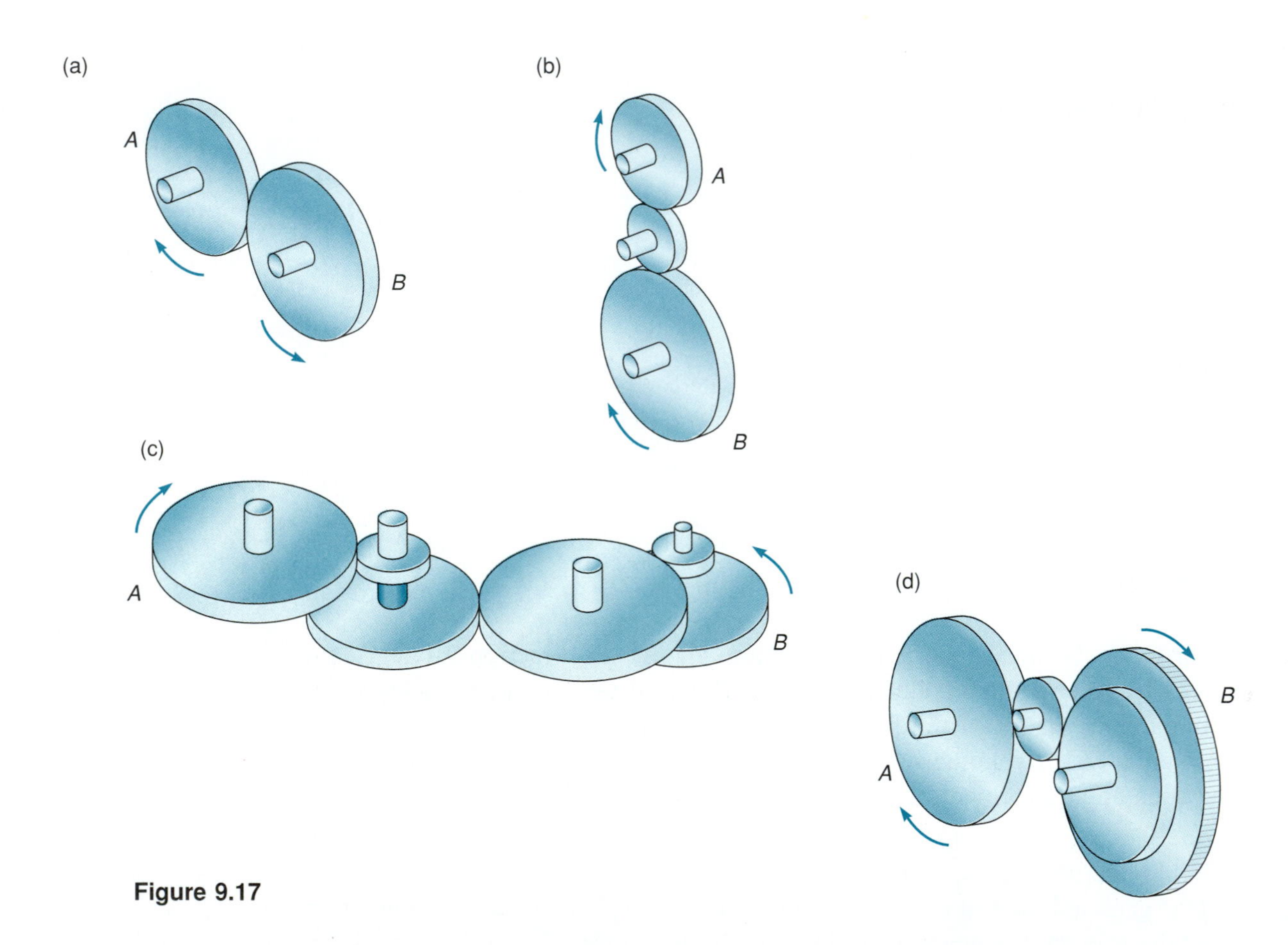

Figure 9.17

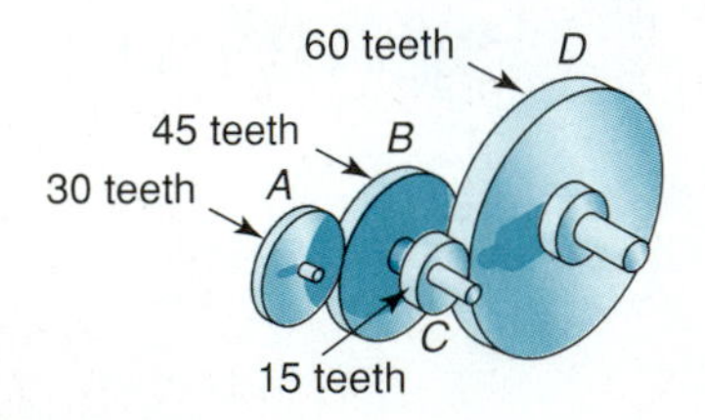

Figure 9.18

EXAMPLE 4

Find the number of revolutions per minute of gear D in Fig. 9.18 if gear A rotates at 20.0 rpm. Gears A and C are drivers and gears B and D are driven.

Data:

$N = 20.0$ rpm $\qquad t_1 = 45$ teeth
$T_1 = 30$ teeth $\qquad t_2 = 60$ teeth
$T_2 = 15$ teeth $\qquad n = ?$

Basic Equation:

$$NT_1T_2 = nt_1t_2$$

Working Equation:

$$n = \frac{NT_1T_2}{t_1t_2}$$

Substitution:

$$n = \frac{(20.0 \text{ rpm})(30 \text{ teeth})(15 \text{ teeth})}{(45 \text{ teeth})(60 \text{ teeth})}$$
$$= 3.33 \text{ rpm}$$

EXAMPLE 5

Find the rpm of gear D in the train shown in Fig. 9.19. Gears A and C are drivers and gears B and D are driven.

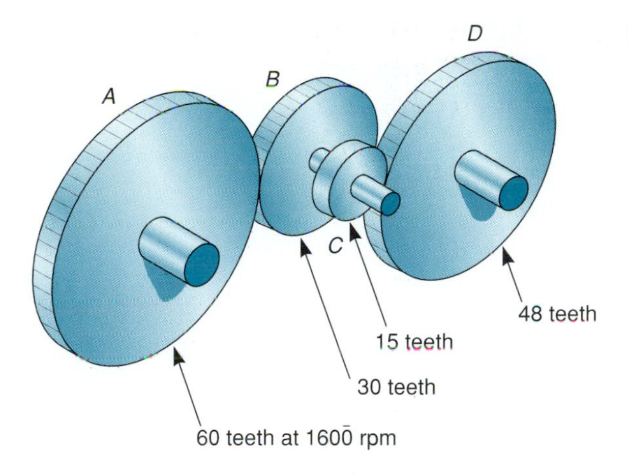

Figure 9.19

Data:

$$N = 16\bar{0}0 \text{ rpm} \qquad t_1 = 30 \text{ teeth}$$
$$T_1 = 60 \text{ teeth} \qquad t_2 = 48 \text{ teeth}$$
$$T_2 = 15 \text{ teeth} \qquad n = ?$$

Basic Equation:

$$NT_1T_2 = nt_1t_2$$

Working Equation:

$$n = \frac{NT_1T_2}{t_1t_2}$$

Substitution:

$$n = \frac{(16\bar{0}0 \text{ rpm})(60 \text{ teeth})(15 \text{ teeth})}{(30 \text{ teeth})(48 \text{ teeth})}$$
$$= 10\bar{0}0 \text{ rpm}$$

EXAMPLE 6

In the gear train shown in Fig. 9.20, find the speed in rpm of gear A.

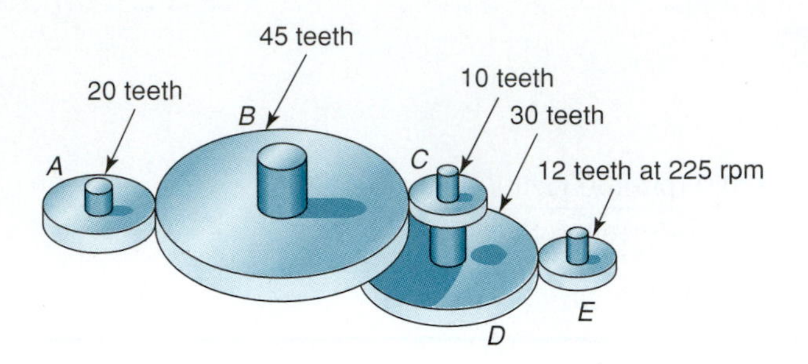

Figure 9.20

Data:

$$
\begin{aligned}
t_1 &= 45 \text{ teeth} & T_1 &= 20 \text{ teeth} \\
t_2 &= 10 \text{ teeth} & T_2 &= 45 \text{ teeth} \\
t_3 &= 12 \text{ teeth} & T_3 &= 30 \text{ teeth} \\
n &= 225 \text{ rpm} & N &= ?
\end{aligned}
$$

Gear B is both a driver and a driven gear.

Basic Equation:

$$NT_1T_2T_3 = nt_1t_2t_3$$

Working Equation:

$$N = \frac{nt_1t_2t_3}{T_1T_2T_3}$$

Substitution:

$$
\begin{aligned}
N &= \frac{(225 \text{ rpm})(45 \text{ teeth})(10 \text{ teeth})(12 \text{ teeth})}{(20 \text{ teeth})(45 \text{ teeth})(30 \text{ teeth})} \\
&= 45.0 \text{ rpm}
\end{aligned}
$$

In a gear train, when a gear is both a driver gear and a driven gear, it may be omitted from the computation.

EXAMPLE 7

The problem in Example 6 could have been worked as follows because gear B is both a driver and a driven.

Basic Equation:

$$NT_1T_3 = nt_2t_3$$

Working Equation:

$$N = \frac{nt_2t_3}{T_1T_3}$$

Substitution:

$$N = \frac{(225 \text{ rpm})(10 \cancel{\text{teeth}})(12 \cancel{\text{teeth}})}{(20 \cancel{\text{teeth}})(30 \cancel{\text{teeth}})}$$

$$= 45.0 \text{ rpm}$$

■ PROBLEMS 9.6

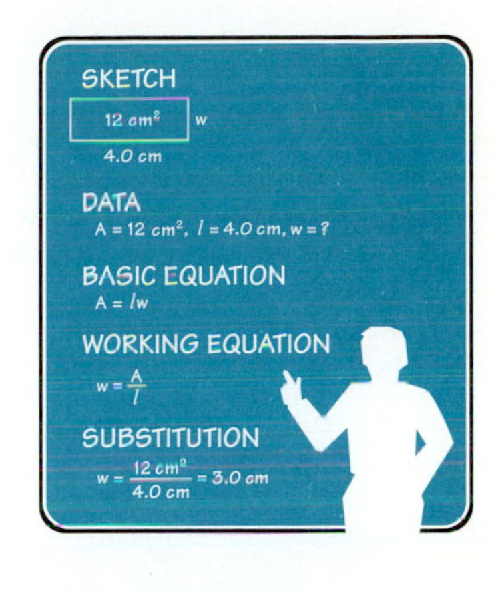

Fill in the blanks.

	Number of Teeth		*rpm*	
	Driver	*Driven*	*Driver*	*Driven*
1.	16	48	156	_____
2.	36	24	_____	225
3.	18	_____	72.0	54.0
4.	_____	64	148	55.5
5.	48	36	_____	276
6.	16	12	144	_____

7. A driver gear has 36 teeth and makes 85.0 rpm. Find the rpm of the driven gear with 72 teeth.
8. A motor turning at 1250 rpm is fitted with a gear having 54 teeth. Find the speed of the driven gear if it has 45 teeth.
9. A gear running at $25\bar{0}$ rpm meshes with another revolving at $10\bar{0}$ rpm. If the smaller gear has 30 teeth, how many teeth does the larger gear have?
10. A driver gear with 40 teeth makes 154 rpm. How many teeth must the driven gear have if it makes $22\bar{0}$ rpm?
11. Two gears have a speed ratio of 4.2 to 1. If the smaller gear has 15 teeth, how many teeth does the larger gear have?
12. What size gear should be mated with a 15-tooth pinion to achieve a speed reduction of 10 to 3?
13. A driver gear has 72 teeth and makes 162 rpm. Find the rpm of the driven gear with 81 teeth.
14. A driver gear with 60 teeth makes 1600 rpm. How many teeth must the driven gear have if it makes 480 rpm?
15. What size gear should be mated with a 20-tooth pinion to achieve a speed reduction of 3 to 1?
16. A motor turning at $15\bar{0}0$ rpm is fitted with a gear having 60 teeth. Find the speed of the driven gear if it has 40 teeth.
17. The larger of two gears in a clock has 36 teeth and turns at a rate of 0.50 rpm. How many teeth does the smaller gear have if it rotates at 1/30 rev/s?
18. How many revolutions does an 88-tooth gear make in 10.0 min when it is mated with a 22-tooth pinion rotating at 44 rpm?

If gear *A* turns in a clockwise motion, determine the motion of gear *B* in each gear train.

19.

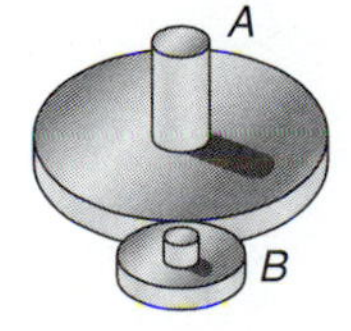

20.

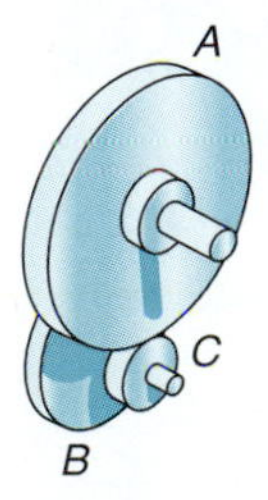

21.

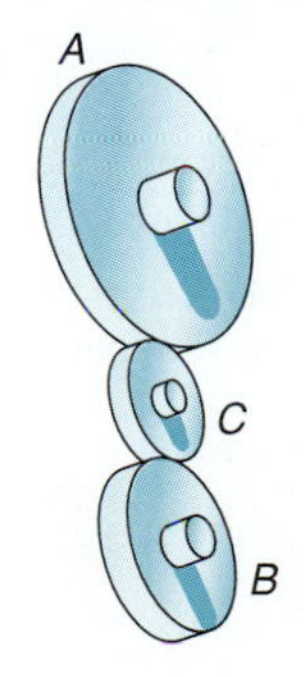

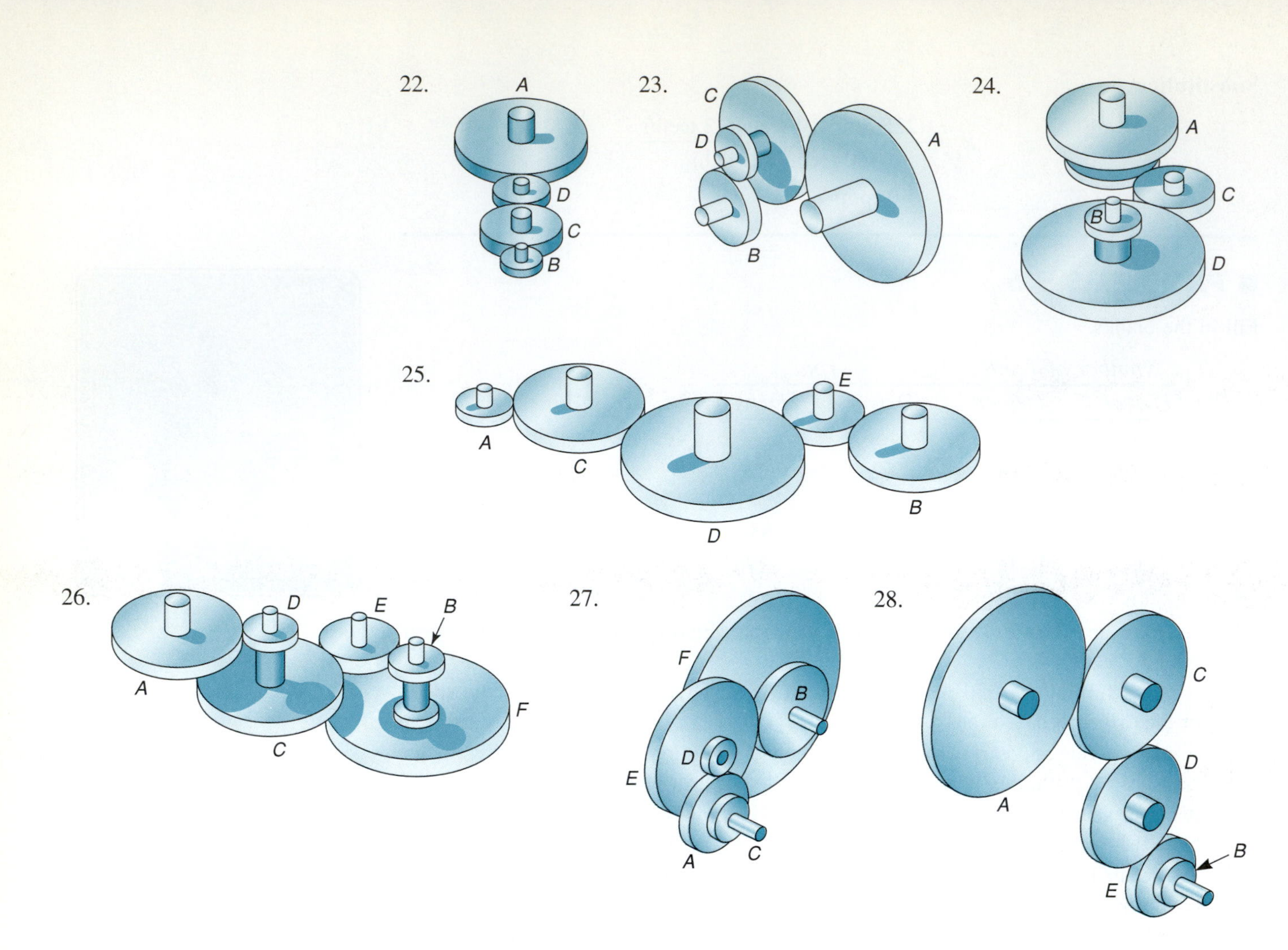

Find the speed in rpm of gear D in each gear train.

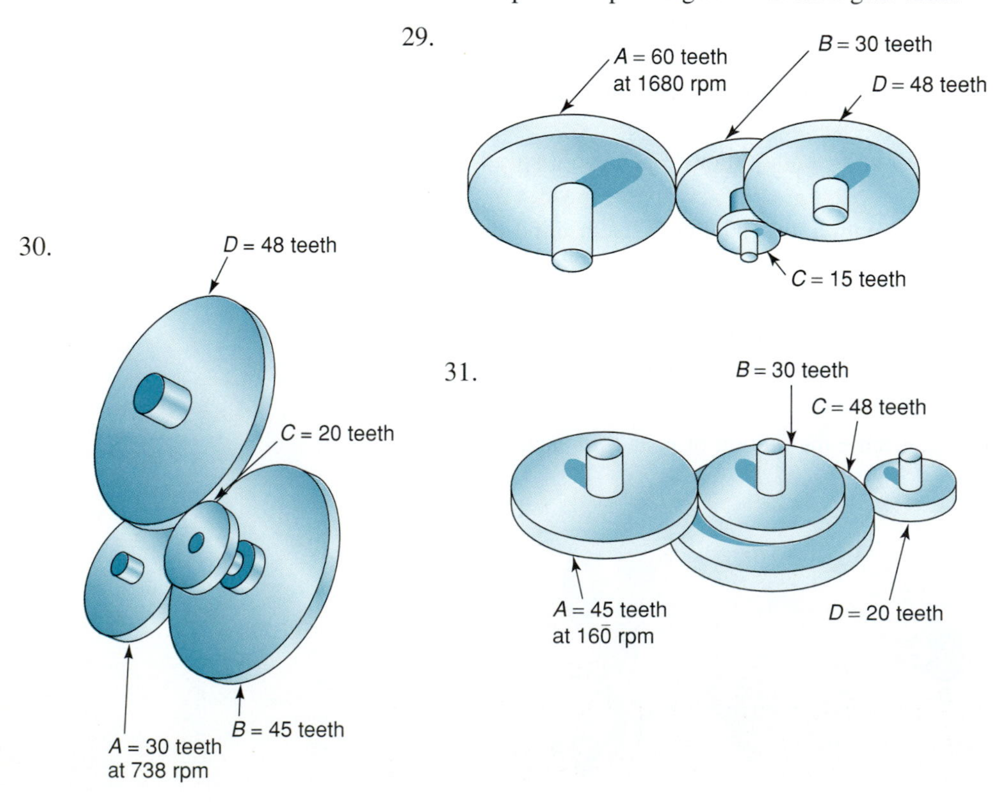

32.

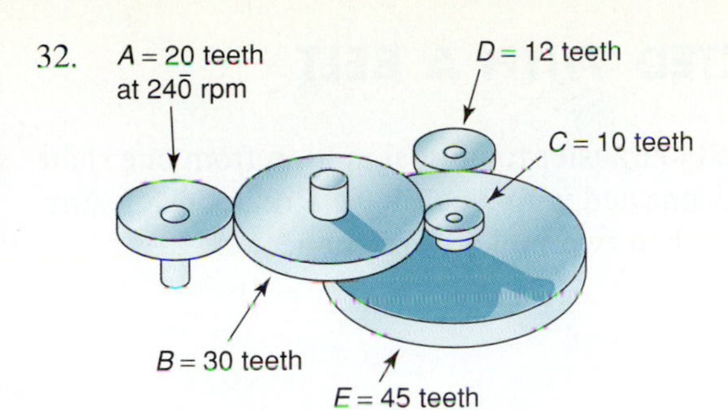

$A = 20$ teeth
at $24\bar{0}$ rpm
$D = 12$ teeth
$C = 10$ teeth
$B = 30$ teeth
$E = 45$ teeth

33.

$C = 45$ teeth
$A = 60$ teeth
at $37\bar{0}$ rpm
$B = 48$ teeth
$D = 15$ teeth

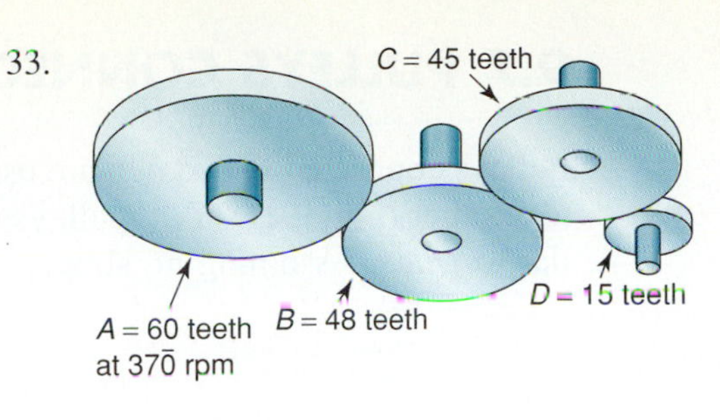

Find the number of teeth for gear D in each gear train.

34.

D: $150\bar{0}$ rpm
$B = 30$ teeth
$C = 15$ teeth
$A = 60$ teeth
at $185\bar{0}$ rpm

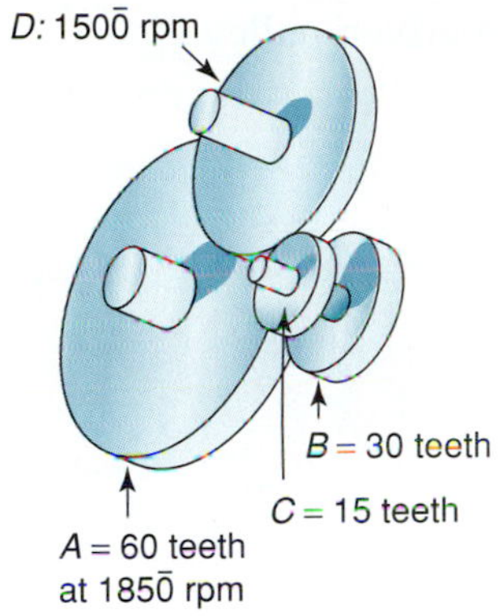

35.

$A = 30$ teeth
at $78\bar{0}$ rpm
$C = 20$ teeth
D: at $26\bar{0}$ rpm
$B = 45$ teeth

36.

$A = 45$ teeth
at $16\bar{0}$ rpm
$B = 30$ teeth
$C = 48$ teeth
D: 576 rpm

37.

D: 1125 rpm
$E = 45$ teeth
$C = 10$ teeth
$B = 30$ teeth
$A = 20$ teeth
at $25\bar{0}$ rpm

38.

D: 555 rpm
$C = 45$ teeth
$A = 60$ teeth
at $37\bar{0}$ rpm
$B = 48$ teeth

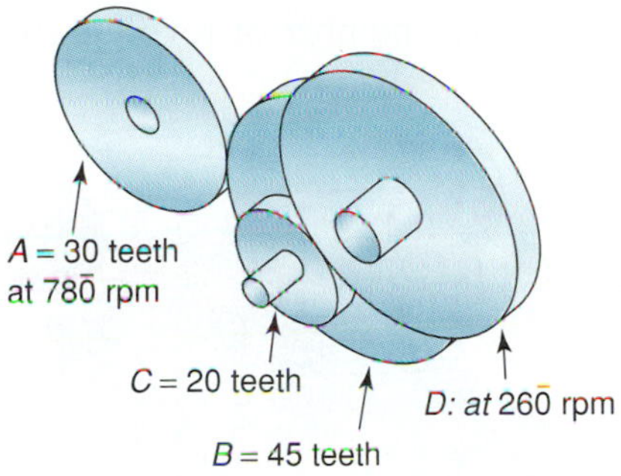

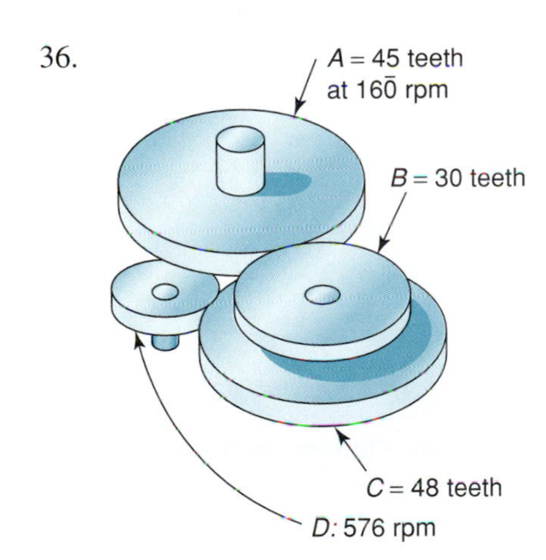

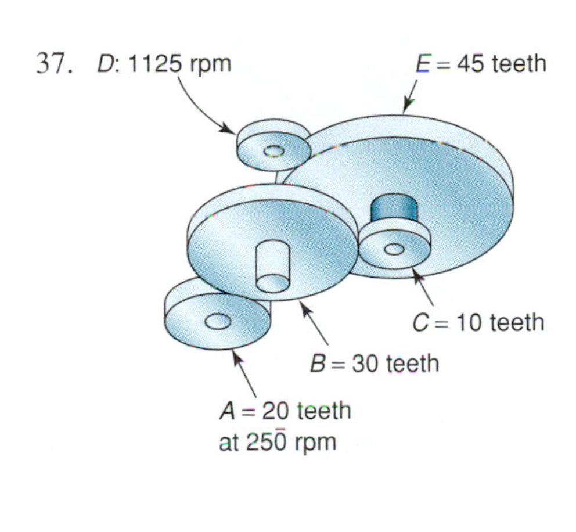

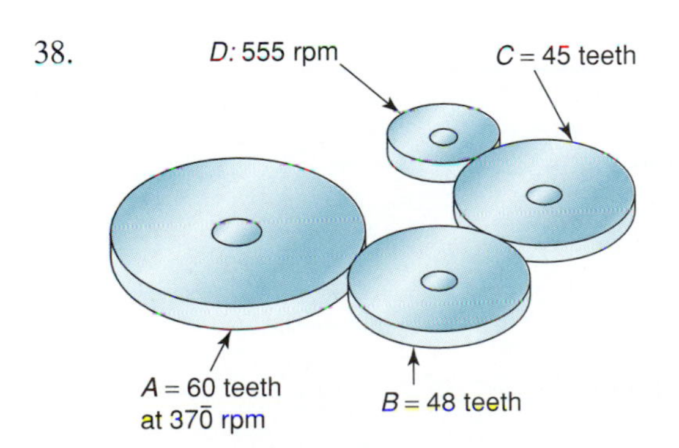

39. Find the direction of rotation of gear B if gear A is turned counterclockwise, in Problems 22 through 28.

40. Find the effect of doubling the number of teeth on gear A in Problem 38. ■

9.7 PULLEYS CONNECTED WITH A BELT

Pulleys connected with a belt are used to transfer rotational motion from one shaft to another (Fig. 9.21). Two pulleys connected with a belt have a relationship similar to gears. Assuming no slippage, when two pulleys are connected

$$D \cdot N = d \cdot n$$

where D = diameter of the driver pulley
N = number of revolutions per minute of the driver pulley
d = diameter of the driven pulley
n = number of revolutions per minute of the driven pulley

Figure 9.21
A single belt drives several components from this engine. (Courtesy of Deere & Company.)

The preceding equation may be generalized in the same manner as for gear trains as follows:

$$ND_1D_2D_3 \cdots = nd_1d_2d_3 \cdots$$

EXAMPLE 1

Find the speed in rpm of pulley A shown in Fig. 9.22.

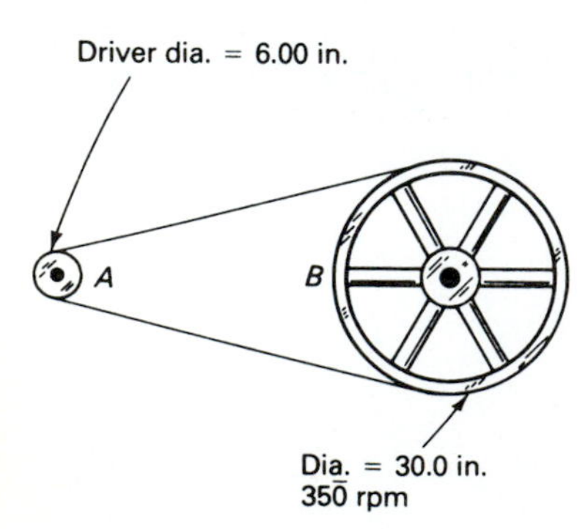

Figure 9.22

Data:

$$D = 6.00 \text{ in.}$$
$$d = 30.0 \text{ in.}$$
$$n = 35\overline{0} \text{ rpm}$$
$$N = ?$$

Basic Equation:

$$D \cdot N = d \cdot n$$

Working Equation:

$$N = \frac{dn}{D}$$

Substitution:

$$N = \frac{(30.0 \text{ in.})(35\overline{0} \text{ rpm})}{6.00 \text{ in.}}$$
$$= 1750 \text{ rpm}$$

When two pulleys are connected with an open-type belt, the pulleys turn in the same direction. When two pulleys are connected with a cross-type belt, the pulleys turn in opposite directions. This is illustrated in Fig. 9.23.

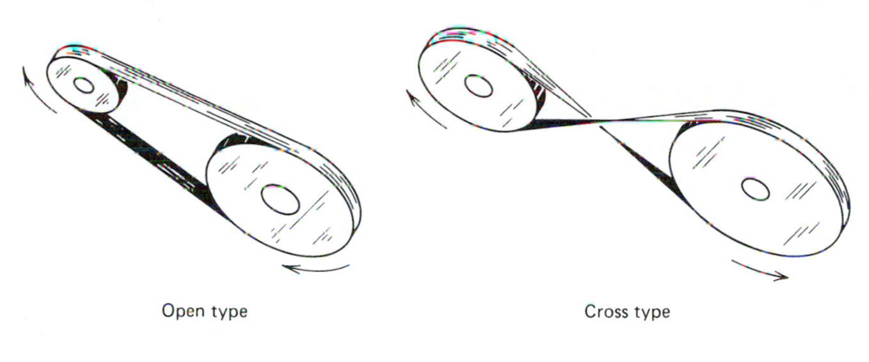

(a) Pulleys rotate in same direction

(b) Pulleys rotate in opposite directions

Figure 9.23
Crossing a belt reverses direction.

■ PROBLEMS 9.7

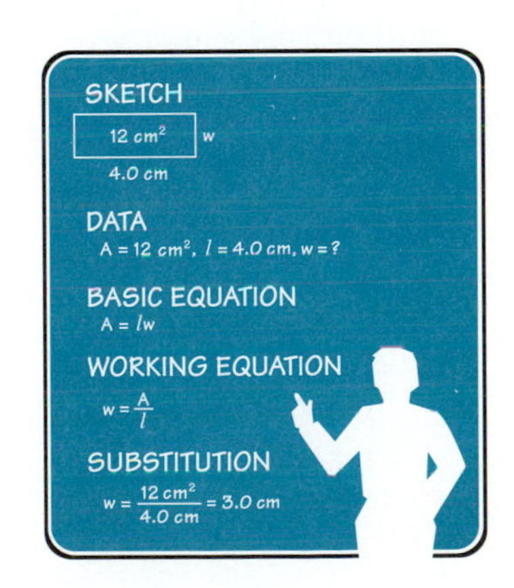

Find each missing quantity using $D \cdot N = d \cdot n$.

	D	N	d	n
1.	18.0	$150\overline{0}$	12.0	____
2.	36.0	____	9.00	972
3.	12.0	$18\overline{0}0$	6.00	____
4.	____	2250	9.00	1125
5.	49.0	1860	____	$62\overline{0}$

6. The diameter of a driving pulley is 6.50 in. and revolves at 1650 rpm. Find the speed of the driven pulley if its diameter is 26.0 in.
7. The diameter of a driving pulley is 25.0 cm and makes $12\overline{0}$ rpm. At what speed will the driven pulley turn if its diameter is 48.0 cm?
8. The diameter of a driving pulley is 36.0 cm and makes $60\overline{0}$ rpm. Find the diameter of the driven pulley if it rotates at $36\overline{0}$ rpm.
9. A driving pulley rotates at $45\overline{0}$ rpm. The diameter of the driven pulley is 15.0 in. and makes 675 rpm. Find the diameter of the driving pulley.
10. The radius of a driving pulley is 10.0 cm and rotates at $12\overline{0}$ rpm. The radius of the driven pulley is 15.0 cm. Find the rpm of the driven pulley.

Determine the direction of pulley *B* in each pulley system.

11.

12.

13.

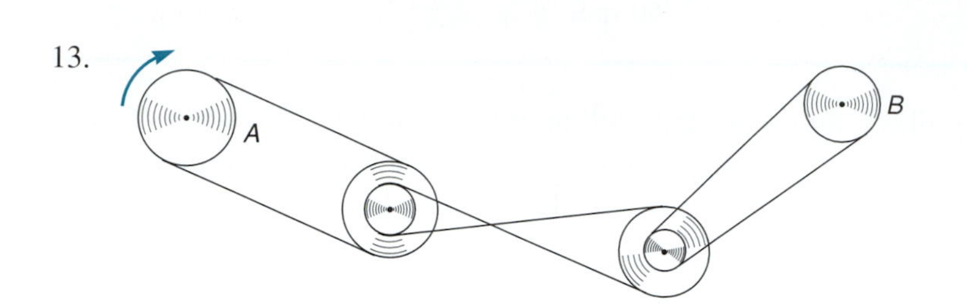

14.

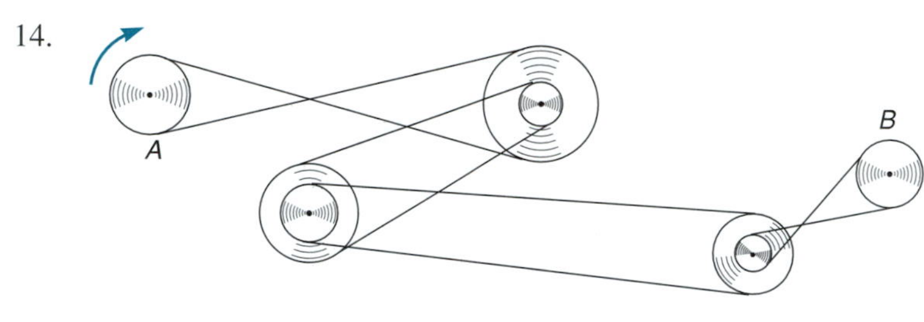

15.

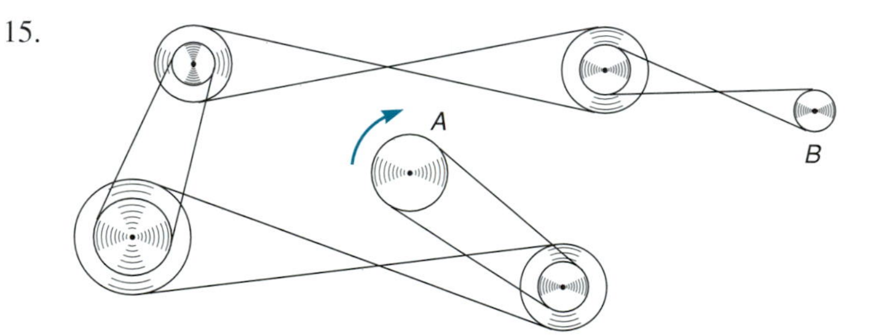

16. What size pulley should be placed on a countershaft turning $15\overline{0}$ rpm to drive a grinder with a 12.0-cm pulley that is to turn at $120\overline{0}$ rpm?
17. If a large pulley and a small pulley are connected, which will turn faster? Why?
18. How do we know the belt connecting two pulleys travels at the same rate while in contact with the different size pulleys? ■

GLOSSARY

Angular Displacement The angle through which any point on a rotating body moves. (p. 220)

Angular Velocity The rate of angular displacement (angular displacement/time). (p. 220)

Centrifugal Force The outward force on an object moving along a curved path when viewed from the frame of reference of the object. Equal in magnitude to the centripetal force but in the opposite direction. (p. 228)

Centripetal Force The force acting on a body in circular motion that causes it to move in a circular path. This force is exerted in the direction of the center of the circle. (p. 227)

Curvilinear Motion Motion along a curved path. (p. 219)

Gear Train A series of gears that transfer rotational motion from one gear to another. (p. 237)

Radian An angular unit of measurement. Defined as that angle with its vertex at the center of a circle whose sides cut off an arc on the circle equal to its radius. Equal to ($360°/2\pi$) or approximately 57.3°. (p. 219)

Rectilinear Motion Motion in a straight line. (p. 219)

Revolution A unit of measurement in rotational motion. One complete rotation of a body. (p. 219)

Rotational Motion Spinning motion of a body. (p. 219)

Torque The tendency to produce change in rotational motion. Equal to the applied force times the length of the torque arm. (p. 224)

FORMULAS

9.1 $\theta = \frac{s}{r}$

$1 \text{ rev} = 360° = 2\pi \text{ rad}$

$\omega = \frac{\theta}{t}$

$v = \omega r$

9.2 $T = Fs_t$

9.3 $F = \frac{mv^2}{r}$

9.4 $\text{Power} = T\omega$

$\text{power in } \frac{\text{ft lb}}{\text{s}} = \text{torque in lb ft} \times \frac{\text{number of revolutions}}{\text{min}} \times 0.105 \frac{\text{min}}{\text{rev s}}$

$\text{power in hp} = \text{power in } \frac{\text{ft lb}}{\text{s}} \times \frac{\text{hp}}{550 \text{ ft lb/s}}$

9.5 $D \cdot N = d \cdot n$

9.6 $T \cdot N = t \cdot n$

9.7 $NT_1T_2T_3T_4 \cdots = nt_1t_2t_3t_4 \cdots$

9.8 $D \cdot N = d \cdot n$

$ND_1D_2D_3 \cdots = nd_1d_1d_3 \cdots$

REVIEW QUESTIONS

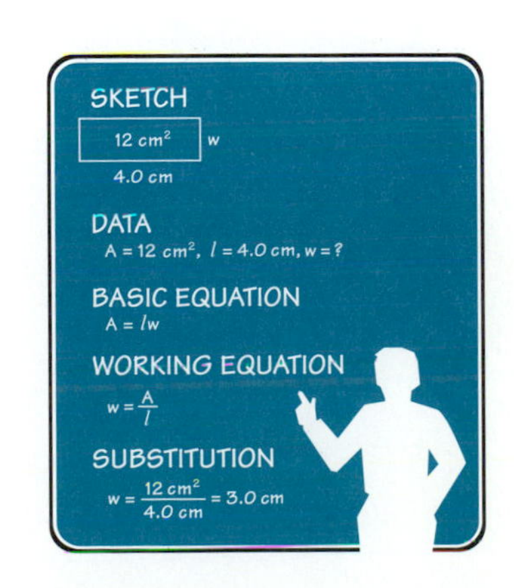

1. Angular velocity is measured in
   (a) revolutions/minute. (b) radians/second.
   (c) revolutions/second. (d) all of the above.
2. Torque is
   (a) applied force in rotational motion.
   (b) the length of the torque arm.
   (c) applied force times the length of the torque arm.
   (d) none of the above.
3. Power in the rotational system
   (a) is found in the same way as in the linear system.
   (b) is found differently than it is in the linear system.
   (c) cannot be determined.
   (d) is always a constant.

4. A gear train has 13 directly connected gears. The first and last gears will
   (a) rotate in opposite directions. (b) rotate in the same direction.
   (c) not rotate.
5. Distinguish between curvilinear motion and rotational motion.
6. Name the two types of measurement of rotation.
7. In your own words, define *radian*.
8. What is angular displacement? In what units is it measured?
9. How is linear velocity of a point on a circle related to angular velocity?
10. Is the length of the pedal necessarily the true length of the torque arm in pedaling a bicycle?
11. Is the tangent to a circle always perpendicular to the radius?
12. Is centripetal force the same as centrifugal force?
13. Will inertia tend to keep a moving body following a curve?
14. Explain the relationship between the number of teeth on two interlocking gears and their relative number of revolutions.
15. How does the presence of an idler gear affect the relationship between a driver gear and a driven gear in a gear train?
16. When the number of directly connected gears in a gear train is four, do the first and last gears in the train rotate in the same or in opposite directions?
17. Why may a gear that is both a driver gear and a driven gear be omitted from a computation?
18. How do pulley combination equations compare to gear train equations?

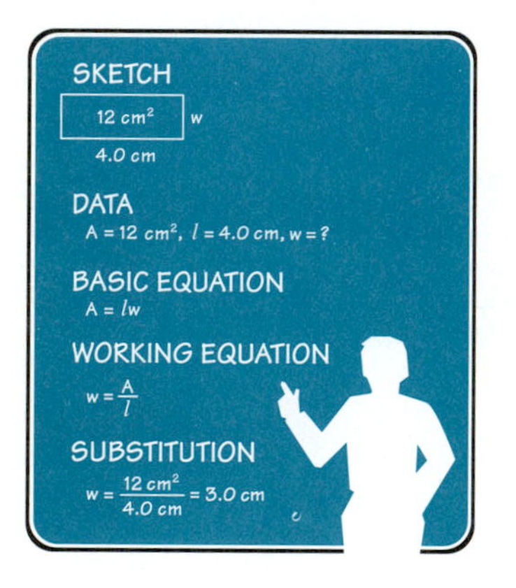

REVIEW PROBLEMS

1. Convert 13 revolutions to (a) radians and (b) degrees.
2. A bicycle wheel turns 25π rad during 45 s. Find the angular velocity of the wheel.
3. A tractor tire turns at 65.0 rpm and has a radius of 13.0 cm. Find the linear speed of the tractor in m/s.
4. A man is changing a flat tire. He is using a tire iron that is 50.0 cm long, and he exerts a force of 53.0 N. How much torque (in N m) does he produce?
5. A torque of 81.0 lb ft is produced by a torque arm of 3.00 ft. What force is being applied?
6. A model plane pulls into a tight curve of a radius of 25.0 m. The $30\overline{0}$ g plane is traveling at 90.0 km/h. What is the plane's centripetal force?
7. A 0.950-kg mass is spun in a circle with a centripetal force of 12.0 N on a string of radius 60.0 cm. At what velocity does it travel?
8. A girl riding her bike creates a torque of 1.20 lb ft with an angular speed of 45.0 rpm. How much power does she produce?
9. A motor generates $30\overline{0}$ W of power. The torque necessary is 50.0 N m. Find the angular velocity.
10. Two rollers are side by side with the large one turning the small. The diameter of the small one is 2.00 cm and turns at 15.0 rpm. The large roller has a diameter of 5.00 cm. How many revolutions does it make in one minute?
11. A clock is driven by a series of gears. The first gear has 30 teeth and revolves 60.0 times a minute. The second gear rotates at 90.0 rpm. How many teeth does it have?
12. Two gears have 13 and 26 teeth, respectively. The first gear turns at 115 rpm. How many times per minute does the second gear rotate?
13. A gear train has 17 directly connected gears. Do the first and last gears rotate in the same direction?

14. A pulley of diameter 14.0 cm is driven by an electric motor to revolve 75.0 rpm. The pulley drives a second one of a diameter of 10.0 cm. How many revolutions does the second pulley make in 1.00 min?

15. A pulley of diameter 5.00 cm is driven at $10\bar{0}$ rpm. Find the diameter of a second pulley if it is driven by the first at $25\bar{0}$ rpm.

16. If gear C turns counterclockwise, in what direction does gear F turn?

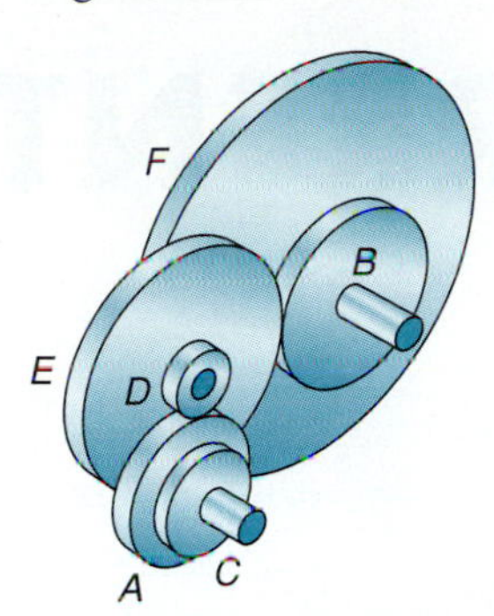

17. Find the speed in rpm of gear D.

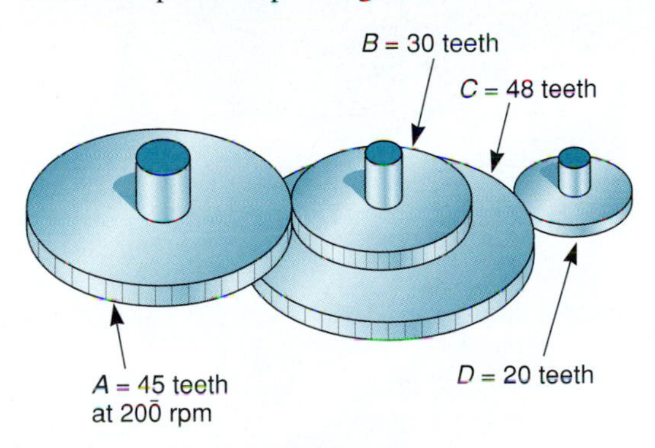

18. Find the number of teeth in gear D.

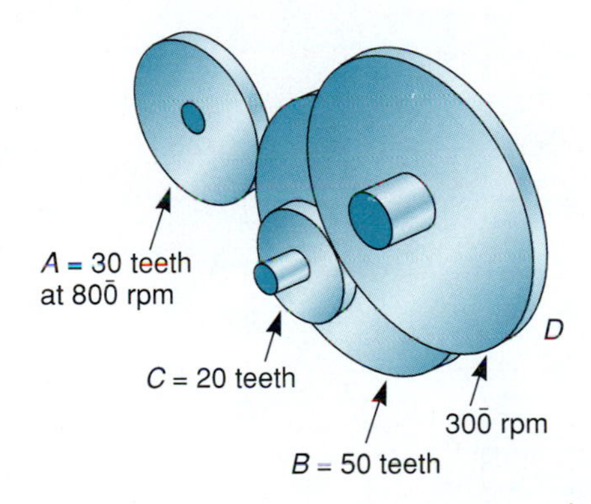

Chapter 10

NONCONCURRENT FORCES

In Chapter 9 we considered rotational motion in terms of rotating shafts and torque. When forces act nonconcurrently (that is, not on the same point), they may also tend to produce rotation. We now consider parallel force problems, equilibrium, and center of gravity.

OBJECTIVES

The major goals of this chapter are to enable you to:

1. Solve parallel force problems.
2. Express the conditions of equilibrium using torque concepts.
3. Understand center of gravity.

10.1 PARALLEL FORCE PROBLEMS

A painter stands 2.00 ft from one end of a 6.00-ft plank that is supported at each end by a scaffold [Fig. 10.1(a)]. How much of the painter's weight must each end of the scaffold support? Problems of this kind are often faced in the construction industry, particularly in the design of bridges and buildings. Using some things we learned about torques and equilibrium, we can now solve problems of this type.

Let's look at the painter problem described above. The force diagram [Fig. 10.1(b)] shows the forces and distances involved. The arrow pointing down represents the weight of the person. The arrows pointing up represent the forces exerted by each end of the scaffold in holding up the plank and painter. (For now, we will neglect the weight of the plank.) We have a condition of equilibrium. The plank and painter are not moving. The sum of the forces exerted by the ends of the scaffold is equal to the weight of the painter (Fig. 10.2). Since these forces are vectors and are parallel, we can show that their sum is zero. Using engineering notation,

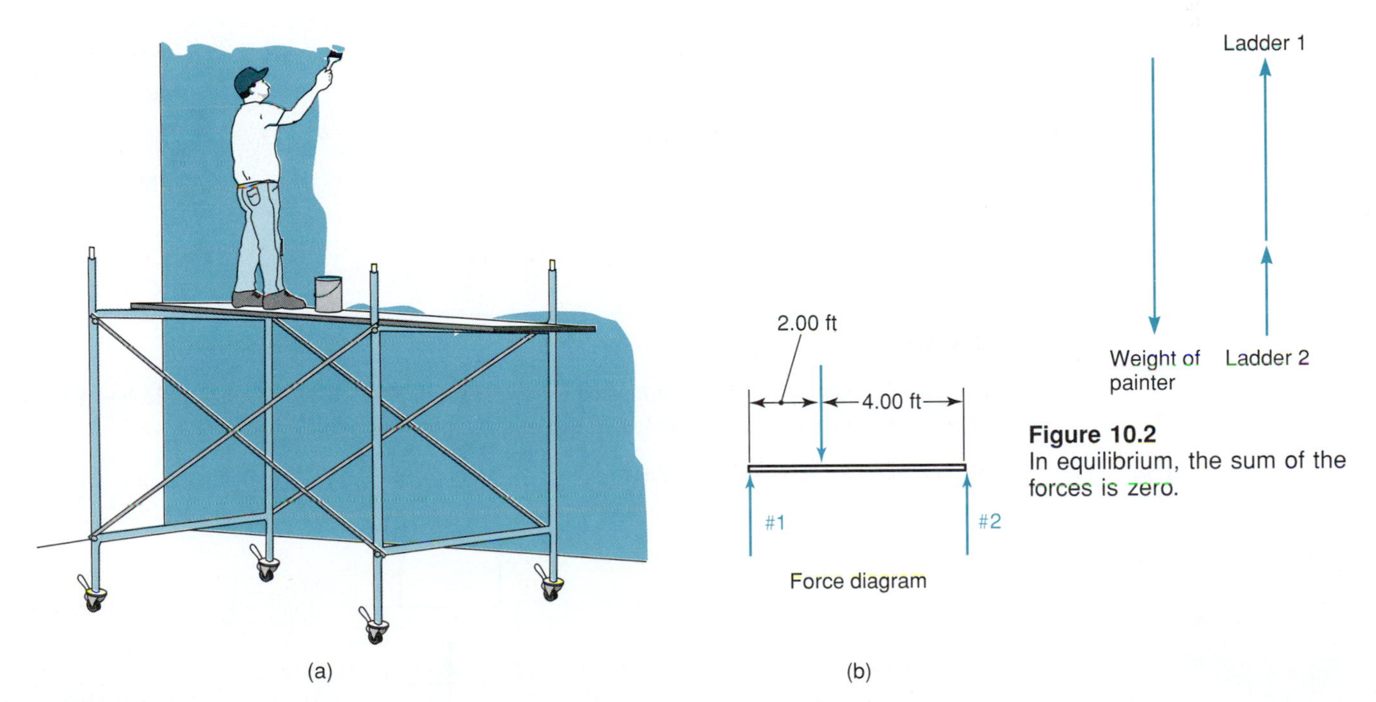

Figure 10.1
Parallel forces shown by painter on a scaffold.

Figure 10.2
In equilibrium, the sum of the forces is zero.

$$\Sigma \mathbf{F} = 0$$

where Σ (Greek capital letter sigma) means summation or "the sum of" and **F** is force, a vector quantity. So $\Sigma \mathbf{F}$ means "the sum of forces," in this case the sum of parallel forces.

> *FIRST CONDITION OF EQUILIBRIUM*
>
> The sum of all parallel forces on an object must be zero.

If the vector sum is not zero (forces up unequal to forces down), we have an unbalanced force tending to cause motion.

Now consider this situation: One end of the scaffold remains firmly in place, supporting the man, and the other is removed. What happens to the painter? The plank, supported only on one end, falls (Fig. 10.3), and the painter has a mess to clean up!

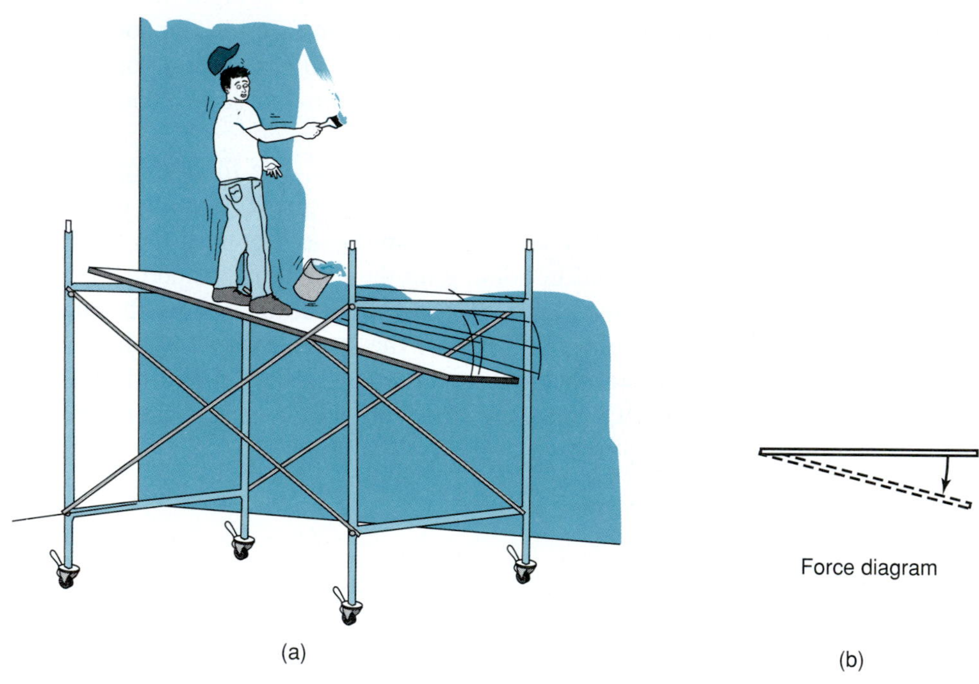

Figure 10.3
The position of the supporting force is important!

EXAMPLE 1

A sign of weight 150$\bar{0}$ lb is supported by two cables (Fig. 10.4). If one cable has a tension of 60$\bar{0}$ lb, what is the tension in the other cable?

Sketch: Draw the free-body diagram.

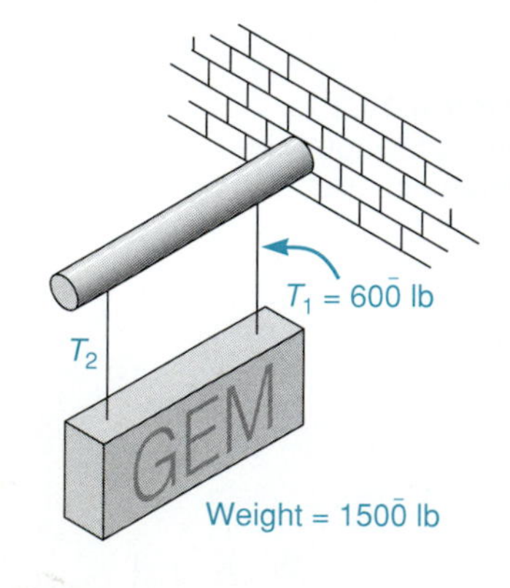

Figure 10.4

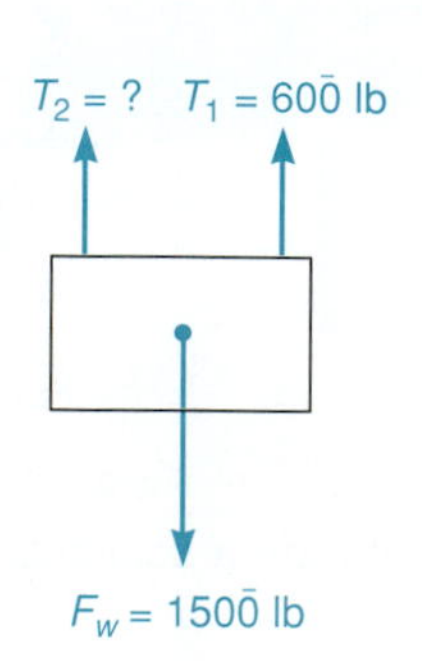

Data:

$$F_w = 150\bar{0} \text{ lb}$$
$$T_1 = 60\bar{0} \text{ lb}$$
$$T_2 = ?$$

Basic Equation:

$$F_+ = F_-$$

Working Equation:

$$T_1 + T_2 = F_w$$
$$T_2 = F_w - T_1$$

Substitution:

$$T_2 = 150\bar{0} \text{ lb} - 60\bar{0} \text{ lb}$$
$$= 90\bar{0} \text{ lb}$$

Not only must the forces balance each other (vector sum = 0), but they must also be positioned so that there is no rotation in the system. To avoid rotation, we can have no unbalanced torques.

Sometimes there will be a natural point of rotation, as in our painter problem. We can, however, choose any point as our center of rotation as we consider the torques present. We will soon see that one of any number of points could be selected. What is necessary, though, is that there be no rotation (no unbalanced torques).

Again, using engineering notation,

$$\Sigma T_{\text{any point}} = 0$$

where $\Sigma T_{\text{any point}}$ is the sum of the torques about any chosen point or,

> *SECOND CONDITION OF EQUILIBRIUM*
>
> The sum of the clockwise torques on an object must equal the sum of the counterclockwise torques about any point.
>
> $$\Sigma T_{\text{clockwise (cw)}} = \Sigma T_{\text{counterclockwise (ccw)}}$$

To illustrate these principles, we will find how much weight each end of the scaffold must support if our painter weighs $15\bar{0}$ lb.

Sketch:

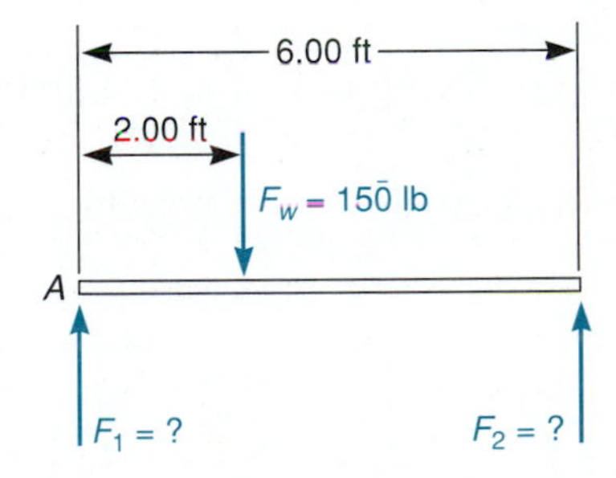

Data:

$$F_w = 15\bar{0} \text{ lb}$$
$$\text{plank} = 6.00 \text{ ft}$$
$$F_w \text{ is 2.00 ft from one end}$$

Basic Equations:

$$(1) \qquad \Sigma F = 0$$

$$\text{sum of forces} = 0$$

$$F_1 + F_2 - F_w = 0$$

(*Note:* F_w is negative because its direction is opposite F_1 and F_2.)

$$\text{or } F_1 + F_2 = F_w$$

$$F_1 + F_2 = 15\bar{0} \text{ lb}$$

$$(2) \qquad \Sigma T_{\text{clockwise}} = \Sigma T_{\text{counterclockwise}}$$

$15\bar{0}$ lb

2.00 ft

A

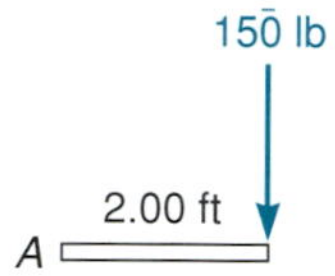

Figure 10.5
Torque arm of painter about point *A*.

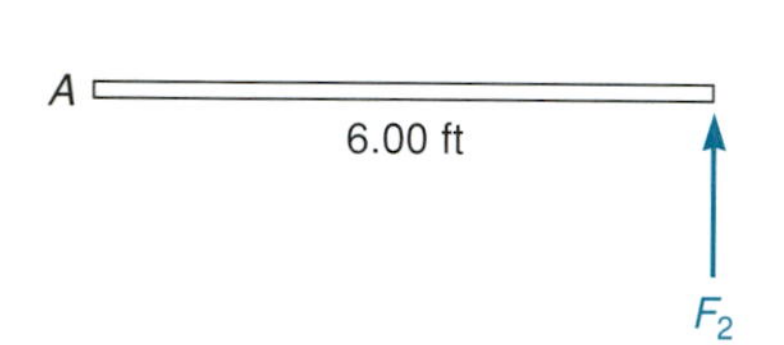

Figure 10.6
Torque arm of F_2 about point *A*.

First, select a point of rotation. Choosing an end is usually helpful in simplifying the calculations. Choose the left end (point *A*) where F_1 acts. What are the clockwise torques about this point?

The force due to the weight of the painter tends to cause clockwise motion. The torque arm is 2.00 ft (Fig. 10.5). Then $T = (15\bar{0} \text{ lb})(2.00 \text{ ft})$. This is the only clockwise torque.

The only counterclockwise torque is F_2 times its torque arm, 6.00 ft (Fig. 10.6). $T = (F_2)(6.00 \text{ ft})$. There is no torque involving F_1 because its torque arm is zero. Setting $\Sigma T_{\text{clockwise}} = \Sigma T_{\text{counterclockwise}}$ we have an equation:

$$(15\bar{0} \text{ lb})(2.00 \text{ ft}) = (F_2)(6.00 \text{ ft})$$

Note that by selecting an end as the point of rotation, we were able to have an equation with just one variable (F_2). Solving for F_2 gives the working equation.

$$F_2 = \frac{(15\bar{0} \text{ lb})(2.00 \not{\text{ft}})}{6.00 \not{\text{ft}}} = 50.0 \text{ lb}$$

Since $\Sigma F = F_1 + F_2 = F_w$, substitute for F_w and F_2 to find F_1.

$$F_1 + 50.0 \text{ lb} = 15\bar{0} \text{ lb}$$

$$F_1 = 15\bar{0} \text{ lb} - 50.0 \text{ lb}$$

$$= 10\bar{0} \text{ lb}$$

To solve parallel force problems:

1. Sketch the problem.
2. Write an equation setting the sums of the opposite forces equal to each other.
3. Choose a point of rotation. Eliminate a variable, if possible (by making its torque arm zero).
4. Write the sum of all clockwise torques.
5. Write the sum of all counterclockwise torques.
6. Set $\Sigma T_{\text{clockwise}} = \Sigma T_{\text{counterclockwise}}$.
7. Solve the equation $\Sigma T_{\text{clockwise}} = \Sigma T_{\text{counterclockwise}}$ for the unknown quantity.
8. Substitute the value found in step 7 back into the equation in step 2 to find the other unknown quantity.

EXAMPLE 2

A bricklayer weighing 175 lb stands on an 8.00-ft scaffold 3.00 ft from one end (Fig. 10.7). He has a pile of bricks, which weighs 40.0 lb, 3.00 ft from the other end. How much weight must each end support?

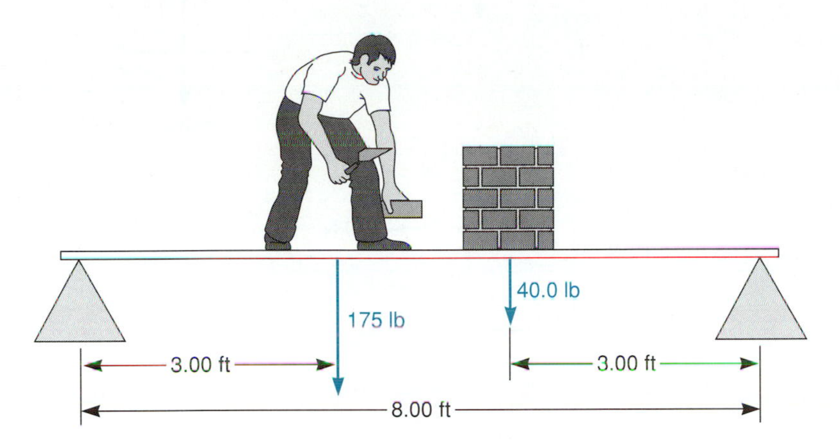

Figure 10.7

1.

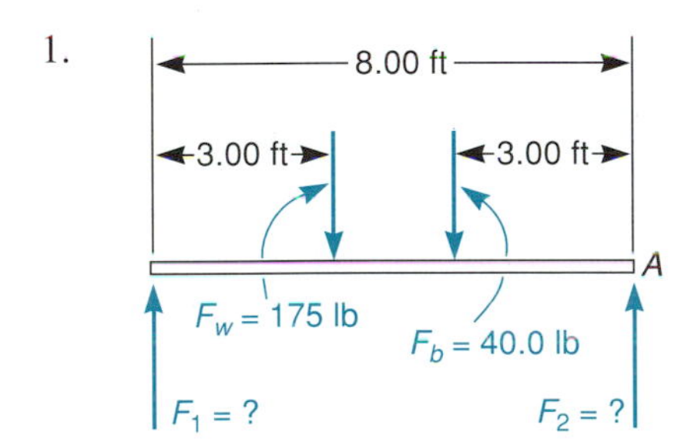

2. $\Sigma F = F_1 + F_2 = 175 \text{ lb} + 40.0 \text{ lb}$
3. Choose a point of rotation. Choose either end to eliminate one of the variables F_1 or F_2. Let us choose the right end and label it A.
4. $\Sigma T_{\text{clockwise}} = (F_1)(8.00 \text{ ft})$
5. $\Sigma T_{\text{counterclockwise}} = (40.0 \text{ lb})(3.00 \text{ ft}) + (175 \text{ lb})(5.00 \text{ ft})$
   Note that there are two counterclockwise torques.
6. $\Sigma T_{\text{clockwise}} = \Sigma T_{\text{counterclockwise}}$
   $F_1(8.00 \text{ ft}) = (40.0 \text{ lb})(3.00 \text{ ft}) + (175 \text{ lb})(5.00 \text{ ft})$
7. $$F_1 = \frac{(40.0 \text{ lb})(3.00 \text{ ft}) + (175 \text{ lb})(5.00 \text{ ft})}{8.00 \text{ ft}}$$
   $$= \frac{12\bar{0} \text{ lb ft} + 875 \text{ lb ft}}{8.00 \text{ ft}} = \frac{995 \text{ lb } \cancel{\text{ft}}}{8.00 \cancel{\text{ft}}} = 124 \text{ lb}$$
8. $$\begin{aligned} F_1 + F_2 &= 175 \text{ lb} + 40.0 \text{ lb} \\ 124 \text{ lb} + F_2 &= 215 \text{ lb} \qquad (F_1 = 124 \text{ lb}) \\ F_2 &= 215 \text{ lb} - 124 \text{ lb} \\ &= 91 \text{ lb} \end{aligned}$$

■ PROBLEMS 10.1

Find the force F that will produce equilibrium for each free-body diagram. Use the same procedure as in Example 1.

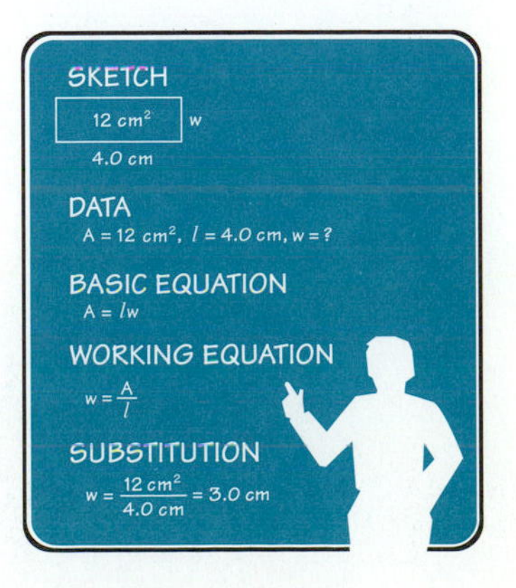

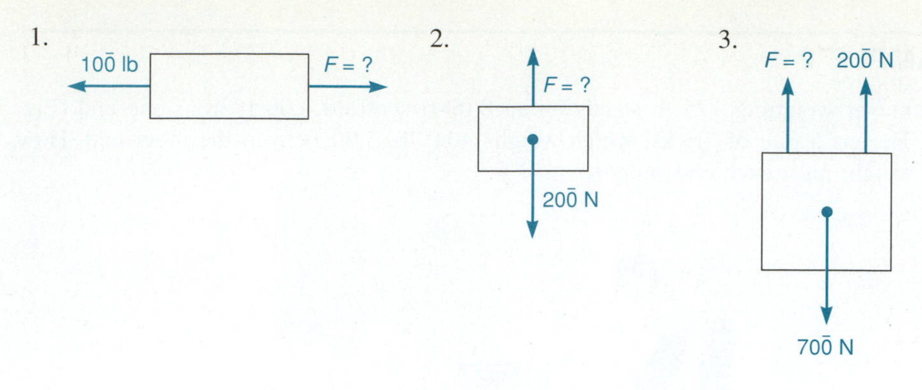

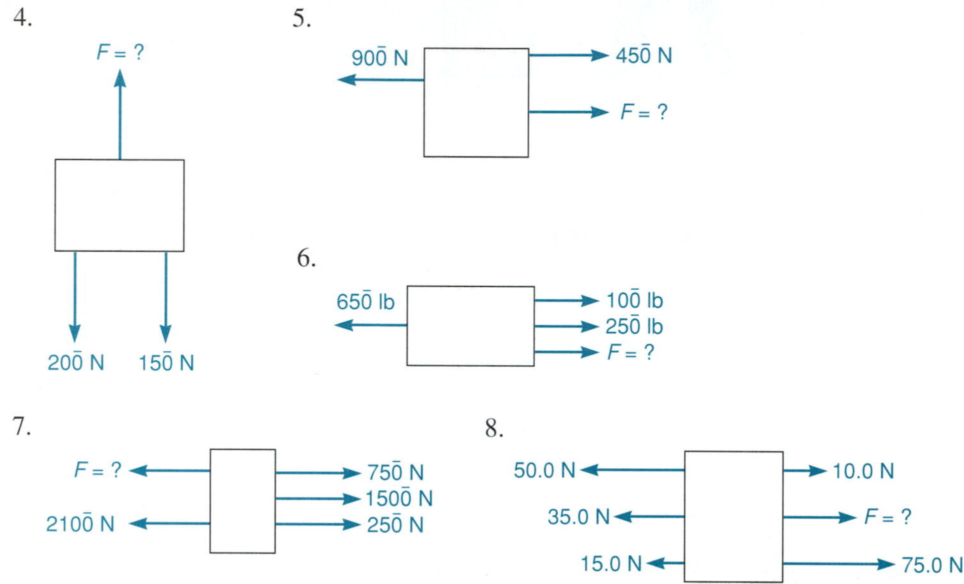

9. A 90.0-kg painter stands 3.00 m from one end of an 8.00-m scaffold. If the scaffold is supported at each end by a stepladder, how much of the weight of the painter must each ladder support?
10. A $50\bar{0}0$-lb truck is 20.0 ft from one end of a 50.0-ft bridge. A $40\bar{0}0$-lb car is 40.0 ft from the same end. How much weight must each end of the bridge support? (Neglect the weight of the bridge.)
11. A $24\bar{0}0$-kg truck is 6.00 m from one end of a 27.0-m-long bridge. A $15\bar{0}0$-kg car is 10.0 m from the same end. How much weight must each end of the bridge support?
12. An auto transmission of mass 165 kg is located 1.00 m from one end of a 2.50-m bench. What weight must each end of the bench support?
13. A bar 8.00 m long supports masses of 20.0 kg on the left end and 40.0 kg on the right end. At what distance from the 40.0-kg mass must the bar be supported for the bar to balance?
14. Two painters each of mass 75.0 kg, stand on a 12.0 m scaffold, 6.00 m apart and 3.00 m from each end. They share a can of paint of mass 1.50 kg in the middle of the scaffold. What mass must be supported by each of the ropes secured to the ends of the scaffold? ■

10.2 CENTER OF GRAVITY

In Section 10.1 we neglected the weight of the plank in the painter example. In practice, the weight of the plank or bridge is extremely important. An engineer must know the weight of the bridge he or she is designing in order to use methods and materials of sufficient strength to support the bridge and the traffic and not collapse.

An important idea in this kind of problem is center of gravity. *The* **center of gravity** *of any object is that point at which all of its weight can be considered to be concentrated.* An object such as a brick or a uniform rod has its center of gravity at its middle or center. The center of gravity of something like an automobile, however, is not at its center or middle because its weight is not evenly distributed throughout. Its center of gravity is located nearer the heavy engine.

The center of gravity of an irregularly shaped uniform thin plate is the point at which it can be supported as in Fig. 10.8(a).

The center of gravity of a uniform thin plate can also be found by suspending it from a point and using a vertical chalkline with a suspended weight as shown in Fig. 10.8(b). The center of gravity is the point of intersection of any two or more such chalklines as in Fig. 10.8(c).

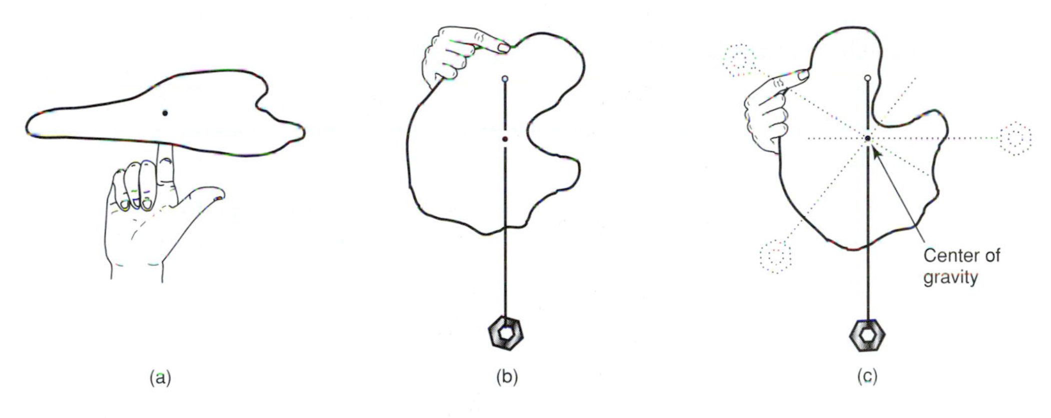

Figure 10.8
The center of gravity of a uniform thin plate can also be found by suspending it from a point and using a vertical chalkline with a suspended weight as shown in part (b). The center of gravity is the point of intersection of any two or more such chalklines as in part (c).

You have probably had the experience of carrying a long board by yourself. If the board was not too heavy, you could carry it yourself by suspending it in about the middle (Fig. 10.9). You didn't have to hold up both ends. You applied the principle of center of gravity and suspended the board at that point.

Figure 10.9
Support of the board at its center of gravity.

We shall represent the weight of an object by a vector through its center of gravity. We use a vector to show the weight (force due to gravity) of the object (Fig. 10.10). It is placed through the center of gravity to show that all the weight of the object may be considered concentrated at that point. If the center of gravity is not at the middle of the object, its location will be given (Fig. 10.10). In solving problems, the weight of the plank or bridge is represented like the other forces by a vector at the center of gravity of the object.

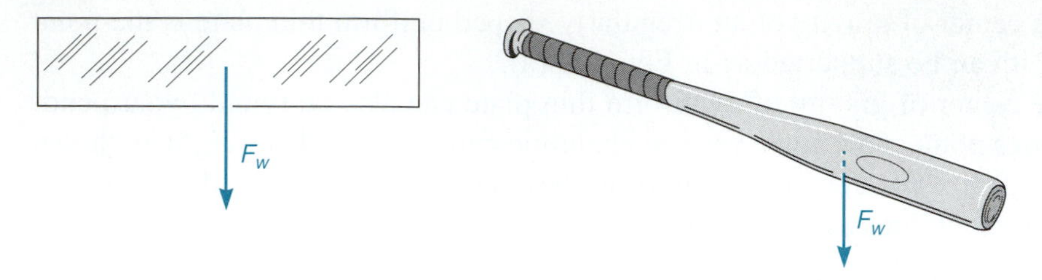

Figure 10.10
Weight can be represented by a vector through the center of gravity.

EXAMPLE 1

A carpenter stands 2.00 ft from one end of a 6.00-ft scaffold that is uniform and weighs 20.0 lb. If the carpenter weighs 165 lb, how much weight must each end support?

1. **Sketch:**

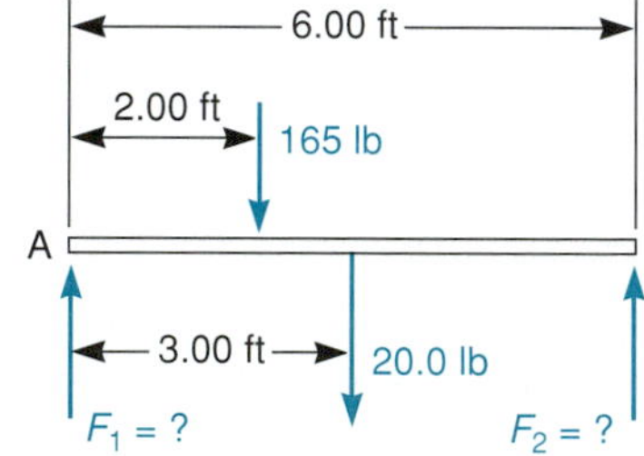

Since the plank is uniform, its center of gravity is at the middle.

2. $\Sigma F = F_1 + F_2 = 165 \text{ lb} + 20.0 \text{ lb}$
3. Choose the left end as the point of rotation and label it A.
4. $\Sigma T_{\text{clockwise}} = (165 \text{ lb})(2.00 \text{ ft}) + (20.0 \text{ lb})(3.00 \text{ ft})$
5. $\Sigma T_{\text{counterclockwise}} = (F_2)(6.00 \text{ ft})$
6. $(165 \text{ lb})(2.00 \text{ ft}) + (20.0 \text{ lb})(3.00 \text{ ft}) = (F_2)(6.00 \text{ ft})$
7. $F_2 = \dfrac{33\bar{0} \text{ lb ft} + 60.0 \text{ lb ft}}{6.00 \text{ ft}} = \dfrac{39\bar{0} \text{ lb } \cancel{\text{ft}}}{6.00 \cancel{\text{ft}}} = 65.0 \text{ lb}$
8. $F_1 + 65.0 \text{ lb} = 165 \text{ lb} + 20.0 \text{ lb}$
$$F_1 = 165 \text{ lb} + 20.0 \text{ lb} - 65.0 \text{ lb}$$
$$= 12\bar{0} \text{ lb}$$

We can also use this method to find the magnitude and position of a parallel force vector that produces equilibrium.

EXAMPLE 2

Find the magnitude, direction, and placement (from point A) of a parallel vector F_6 that will produce equilibrium in the parallel force diagram in Fig. 10.11.

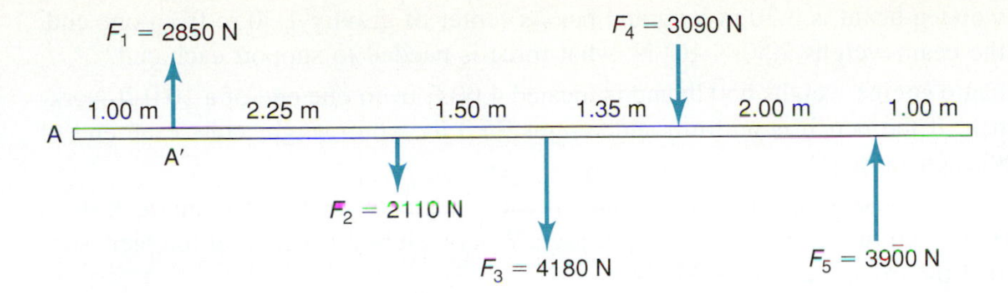

Figure 10.11

1. $\Sigma F = F_1 + F_5 + F_6 = F_2 + F_3 + F_4$
   $2850 \text{ N} + 39\bar{0}0 \text{ N} + F_6 = 2110 \text{ N} + 4180 \text{ N} + 3090 \text{ N}$
   $F_6 = 2630 \text{ N (up)}$
2. Choose A' instead of A as the point of rotation to make the torque arm zero for F_1. Also, let x be the distance of F_6 from point A'.
3. $\Sigma T_{\text{clockwise}} = (2110 \text{ N})(2.25 \text{ m}) + (4180 \text{ N})(3.75 \text{ m}) + (3090 \text{ N})(5.10 \text{ m})$
4. $\Sigma T_{\text{counterclockwise}} = (39\bar{0}0 \text{ N})(7.10 \text{ m}) + (2630 \text{ N})(x)$
5. $\Sigma T_{\text{clockwise}} = \Sigma T_{\text{counterclockwise}}$
6. $(2110 \text{ N})(2.25 \text{ m}) + (4180 \text{ N})(3.75 \text{ m}) + (3090 \text{ N})(5.10 \text{ m}) = (39\bar{0}0 \text{ N})(7.10 \text{ m}) + (2630 \text{ N})(x)$
   $3.23 \text{ m} = x \text{ (from } A')$
   $\text{or } 4.23 \text{ m} = x \text{ (from } A)$

Thus, the equilibrium vector is 2630 N (up) placed at 4.23 m from point A.

■ PROBLEMS 10.2

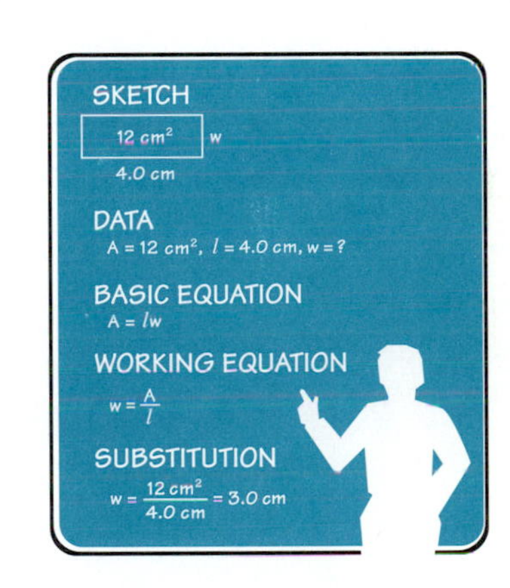

Solve each problem using the methods outlined in this chapter.

1. Solve for F_1:
$$30.0F_1 = (14.0)(18.0) + (25.0)(17.0)$$
2. Solve for F_2:
$$39.0F_2 + (60.0)(55.0) = (20\bar{0})(40.0) + (52.0)(27.0)$$
3. Solve for F_w:
$$(12.0)(15.0) + 45.0F_w = (21.0)(65.0) + (22.0)(32.0)$$
4. Two workers carry a uniform 15.0-ft plank that weighs 22.0 lb (Fig. 10.12). A load of blocks weighing 165 lb is located 7.00 ft from the first worker. What force must each worker exert to hold up the plank and load?

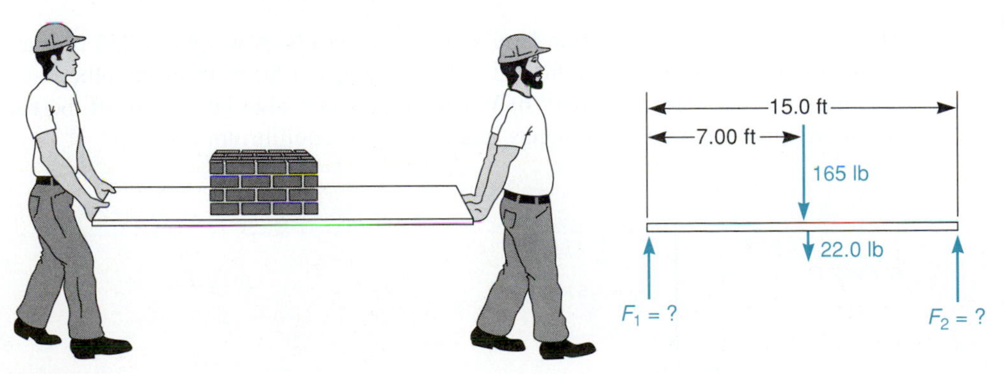

Figure 10.12

5. A wooden beam is 3.30 m long and has its center of gravity 1.30 m from one end. If the beam weighs 2.50×10^4 N, what force is needed to support each end?
6. An auto engine weighs $65\bar{0}$ lb and is located 4.00 ft from one end of a 10.0-ft workbench. If the bench is uniform and weighs 75.0 lb, what weight must each end of the bench support?
7. An old covered bridge across a country stream weighs 89,200 N. A large truck stalls 4.00 m from one end of the 9.00-m bridge. What weight must each of the piers support if the truck weighs $98,\bar{0}00$ N?
8. A window washer's scaffold 12.0 ft long and weighing 75.0 lb is suspended from each end. One washer weighs 155 lb and is 3.00 ft from one end. The other washer is 4.00 ft from the other end. If the force supported by the end near the first washer is $20\bar{0}$ lb, how much does the second washer weigh?
9. A porch swing weighs 29.0 lb. It is 4.40 ft long and has a dog weighing 14.0 lb sleeping on it 1.90 ft from one end. What weight must the support ropes on each end hold up?
10. A wooden plank is 5.00 m long and supports a 75.0-kg block 2.00 m from one end. If the plank is uniform and has a mass of 30.0 kg, how much force is needed to support each end?
11. A bridge has a mass of 1.60×10^4 kg, is 21.0 m long, and has a $35\bar{0}0$-kg truck 7.00 m from one end. What force must each end of the bridge support?
12. A uniform steel beam is 5.00 m long and weighs 3.60×10^5 N. What force is needed to lift one end?
13. A wooden pole is 4.00 m long, weighs 315 N, and has its center of gravity 1.50 m from one end. What force is needed to lift each end?
14. A bridge has a mass of 2.60×10^4 kg, is 32.0 m long, and has a $35\bar{0}0$-kg truck 15.0 m from one end. What force must each end of the bridge support?
15. An auto engine of mass 295 kg is located 1.00 m from one end of a 4.00-m workbench. If the uniform bench has a mass of 45.0 kg, what weight must each end of the bench support?
16. On a 3.00-m scaffold of uniform mass, with supports at each end, there is a pile of bricks 0.500 m from one end. Which support will exert a greater force: the one closer to the bricks or the one farther away?
17. The sign below is 4.00 m long, weighs $155\bar{0}$ N, and is made of uniform material. A weight of 245 N hangs 1.00 m from the end. Find the tension in each support cable.

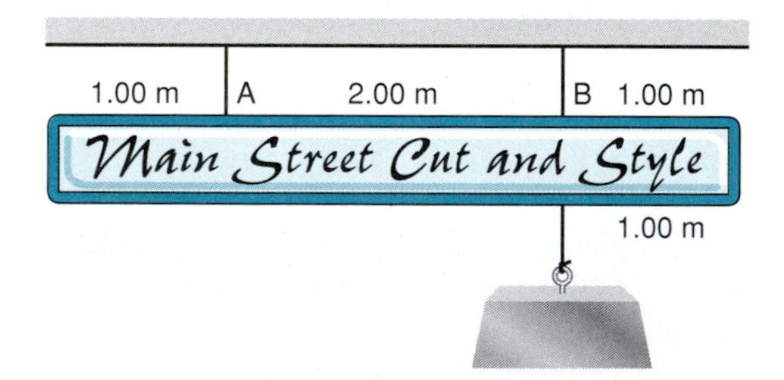

18. The uniform bar below is 5.00 m long and weighs 975 N. A weight of 255 N is attached to one end while a weight of 375 N is attached 1.50 m from the other end. (a) Find the tension in the cable. (b) Where should the cable be tied to lift the bar and its weights so that the bar hangs in a horizontal equilibrium position?

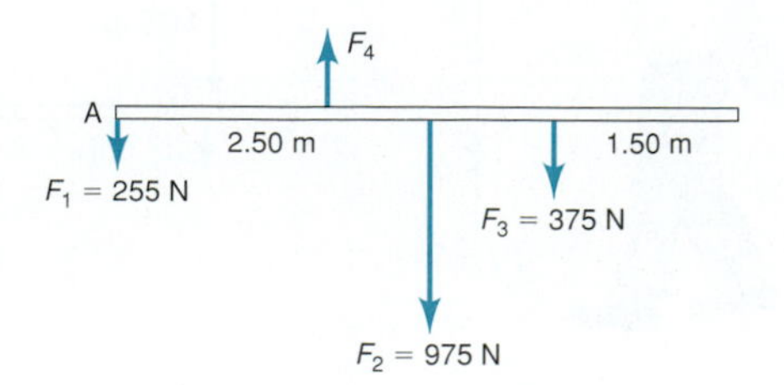

Find the magnitude, direction, and placement (from point A) of a parallel vector F_6 that will produce equilibrium in each force diagram.

19.

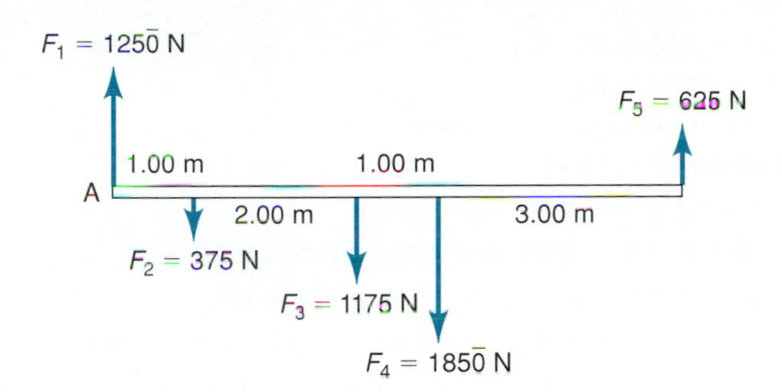

20.

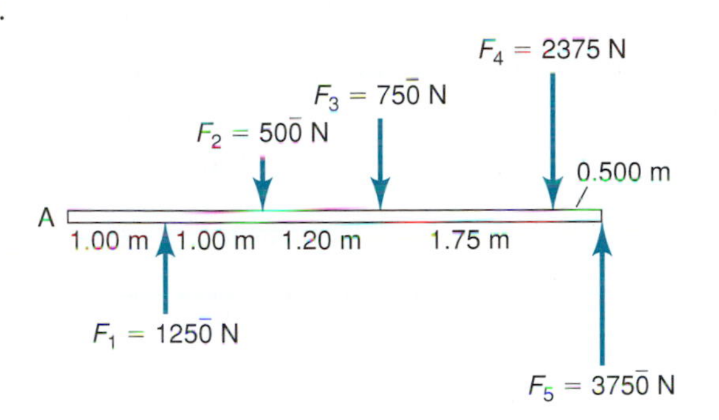

GLOSSARY

First Condition of Equilibrium The sum of all parallel forces on a body in equilibrium must be zero. (p. 252)

Second Condition of Equilibrium The sum of the clockwise torques on a body in equilibrium must be equal to the sum of the counterclockwise torques about any point. (p. 253)

Center of Gravity The point of any object at which all of its weight can be considered to be concentrated. (p. 257)

FORMULAS

10.1 *First condition of equilibrium:* The sum of all parallel forces on an object must be zero.

$$\Sigma F = 0$$

Second condition of equilibrium: The sum of the clockwise torques on an object must equal the sum of the counterclockwise torques.

$$\Sigma T_{\text{clockwise}} = \Sigma T_{\text{counterclockwise}}$$

REVIEW QUESTIONS

1. The first condition of equilibrium states that
   (a) all parallel forces must be zero.
   (b) all perpendicular forces must be zero.
   (c) all frictional forces must be zero.

2. In the second condition of equilibrium,
   (a) clockwise and counterclockwise torques are unequal.
   (b) clockwise and counterclockwise torques are equal.
   (c) there are no torques.
3. The center of gravity of an object
   (a) is always at its geometric center.
   (b) does not have to be at the geometric center.
   (c) exists only in symmetrical objects.
4. Give the expression meaning "the sum of the forces."
5. If the sum of the opposing forces is zero, can there still be rotation?
6. In your own words, explain the second condition of equilibrium.
7. Is it possible to select different points of rotation in an equilibrium problem?
8. What is the primary consideration in the selection of a point of rotation in an equilibrium problem?
9. List five examples from daily life in which you use the concept of center of gravity.
10. Why is weight represented by a vector through the center of gravity of an object?
11. Is the center of gravity of an object always at its geometric center?

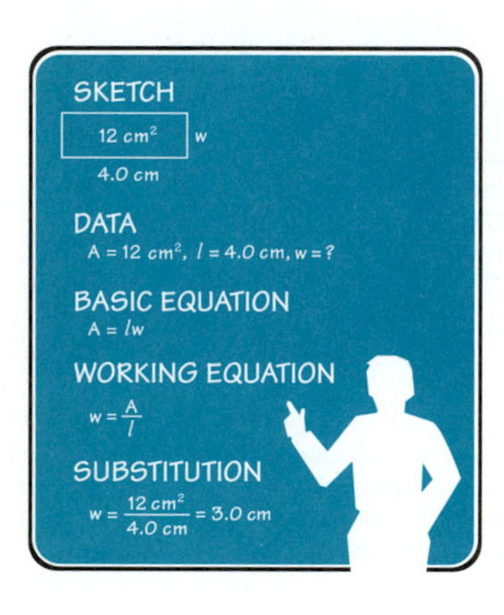

REVIEW PROBLEMS

1. A hanging sign has mass $20\bar{0}$ kg. If the tension in one support cable is 1080 N, what is the tension in the other support cable?
2. A scaffold supports a bricklayer and bricks weighing 450 lb. If the force in one end support is 290 lb, what is the supporting force in the other?
3. Find F in the torque equation $(90.0 \text{ N})(4.00 \text{ m}) = (F)(3.00 \text{ m})$.
4. If $\Sigma F = F_1 + F_2$ when $\Sigma F = 108.0$ N and $F_2 = 76.0$ N, find F_1.
5. If $\Sigma T_{\text{clockwise}} = \Sigma T_{\text{counterclockwise}}$ and $\Sigma T_{\text{clockwise}} = (40.0 \text{ N})(2.50 \text{ m})$, find $\Sigma T_{\text{counterclockwise}}$.
6. Two ladders at the ends of a scaffold support a mass of 90.0 kg each. An 80.0-kg worker is on the scaffold with a pile of bricks. Find the mass of the bricks.
7. Solve for F_w:
   $(12.0 \text{ ft})(6.00 \text{ lb}) + (20.0 \text{ ft})\, F_w = (6.00 \text{ ft})(9.00 \text{ lb}) + (35.0 \text{ ft})(10.0 \text{ lb})$
8. How far from the light end of a 68.0-cm bat would its center of gravity be if it is one-fourth of the length of the bat from the heavy end?
9. A bridge has mass $80\bar{0}0$ kg. If a $32\bar{0}0$-kg truck stops in the middle of the bridge, what mass must each pier support?
10. If the truck in Problem 9 stops 7.00 m from one end of the 26.0-m bridge, what weight must each end support?
11. A uniform 2.20-kg steel bar with length 2.70 m is suspended on each end by a chain. A 40.0-kg child hangs 70.0 cm from one end while a 55.0-kg child hangs 50.0 cm from the other end. An adult pushes up halfway between the two children with a force of 127 N. How much weight does each chain support?
12. Find the vertical force needed to support the 4.00-m-long uniform beam below that weighs 3475 N.

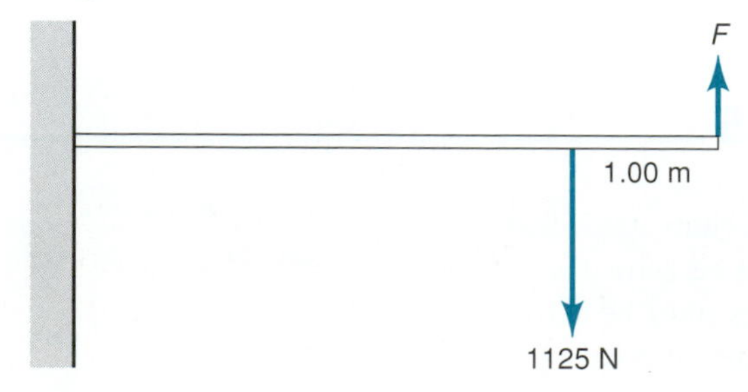

Chapter 11

MATTER

Technology is used to take raw materials and shape, refine, mold, and transform them into products useful to our society. To do this, you need to have some understanding of the nature and properties of matter and basic characteristics of its various forms of solids, liquids, and gases.

OBJECTIVES

The major goals of this chapter are to enable you to:

1. Describe the components of all matter.
2. Learn the properties of solids and the application of Hooke's law in technology.
3. Distinguish the properties of solids, liquids, and gases.
4. Distinguish density and specific gravity.

11.1 PROPERTIES OF MATTER

What are the building blocks of matter? First, matter is anything that occupies space and has mass. Suppose that we take a cube of sugar and divide it into two pieces. Then divide a resulting piece into two pieces. Can we continue this process indefinitely and get smaller and smaller particles of sugar each time? No, at some point the subdivision will result in something different from sugar.

An **element** is a substance that cannot be separated into simpler substances. A **compound** is a substance containing two or more elements.

A **molecule** is the smallest particle of an element that can exist in a free state and still retain the characteristics of that element or compound. Most simple molecules are about 3×10^{-10} m in diameter. An **atom** is the smallest particle of an element that can exist either alone or in combination with other atoms of the same or different elements. The molecules of elements consist of one atom or two or more similar atoms; those molecules of compounds consist of two or more different atoms.

What do we get if we divide the sugar molecule? The resulting particles are carbon, hydrogen, and oxygen atoms. Models of water and sugar molecules are shown in Fig. 11.1. Not all atoms are the same size. The hydrogen atom is the smallest with diameter 6×10^{-11} m and mass 1.67×10^{-27} kg. Uranium is one of the heaviest atoms, 3.96×10^{-25} kg.

What happens if an atom is subdivided? Several particles of the atom have been discovered. Of these, the three most important particles of the atom are the **proton,** the **electron,** and the **neutron.** We limit our discussion here to these three. Table 11.1 provides some basic information about these three particles.

Table 11.1 Properties of Atomic Particles

Particle	*Mass*	*Diameter*	*Charge*
Proton	1.673×10^{-27} kg	8.2×10^{-16} m	+1
Electron	9.109×10^{-31} kg	0*	−1
Neutron	1.675×10^{-27} kg	8.2×10^{-16} m	0

*The zero size of an electron is due to its lack of any internal structure.

Models of the hydrogen atom and the carbon atom are shown in Fig. 11.2. The **nucleus** or center part of an atom is made up of protons and neutrons, while the electrons orbit (circle) the nucleus. Electrons do not really orbit the nuclei like

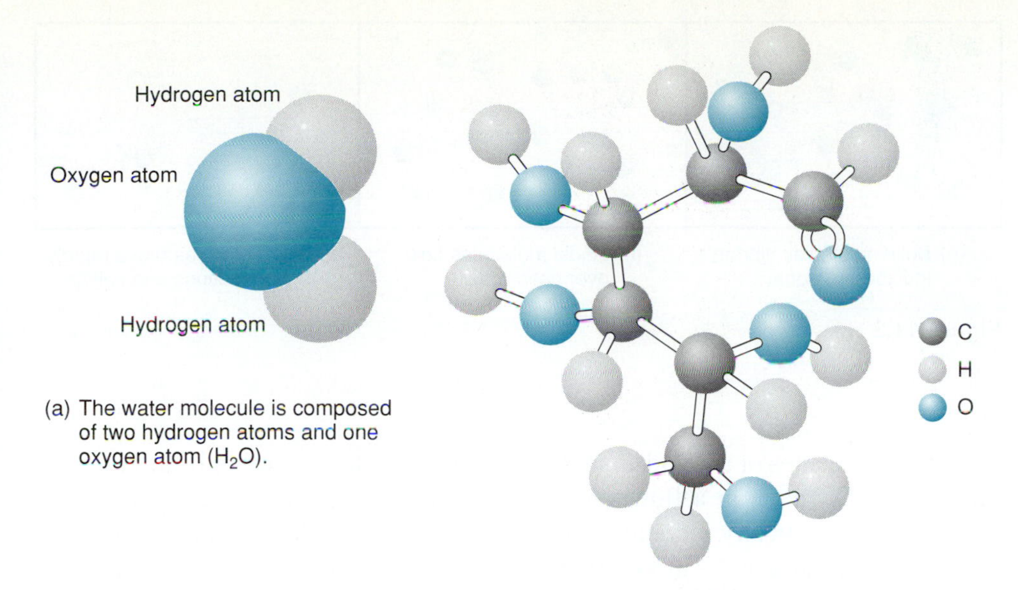

(a) The water molecule is composed of two hydrogen atoms and one oxygen atom (H_2O).

(b) The sugar (glucose) molecule is composed of six carbon atoms, twelve hydrogen atoms, and six oxygen atoms.

Figure 11.1

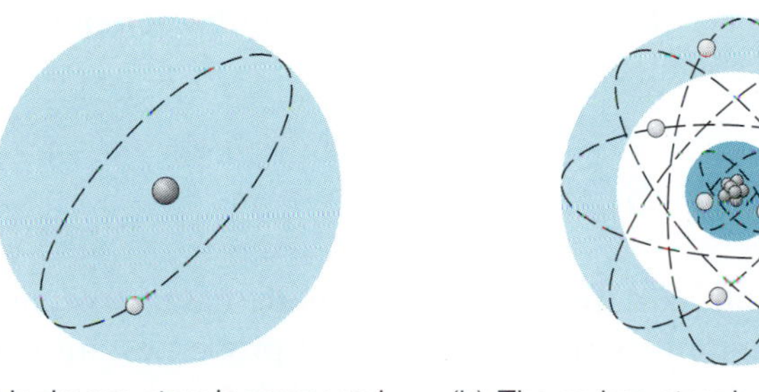

(a) The hydrogen atom is composed of a nucleus which contains one proton. Its one electron moves about or orbits the nucleus.

(b) The carbon atom is composed of a nucleus which contains six protons and six neutrons. Its six electrons move about or orbit the nucleus.

Figure 11.2

planets around the sun; they are more like a cloud around a given nucleus. The atoms of the nuclei are held together by strong nuclear forces. The molecules are held together by electrical forces.

Protons, electrons, and neutrons, formed in various combinations, give us more than 100 known atoms or chemical elements. The atoms, formed in various combinations, give us the very long list of known molecules.

Matter exists in three states: solids, liquids, and gases. A **solid** is a substance that has a definite shape and a definite volume. A **liquid** is a substance that takes the shape of its container and has a definite volume. A **gas** is a substance that takes the shape of its container and has the same volume as its container.

The molecules of a solid are fixed in relation to each other [Fig. 11.3(a)]. They vibrate in a back-and-forth motion. They are so close that a solid can be compressed only slightly. Solids are usually crystalline substances, meaning that their molecules are arranged in a definite pattern. This is why a solid tends to hold its shape and has a definite volume.

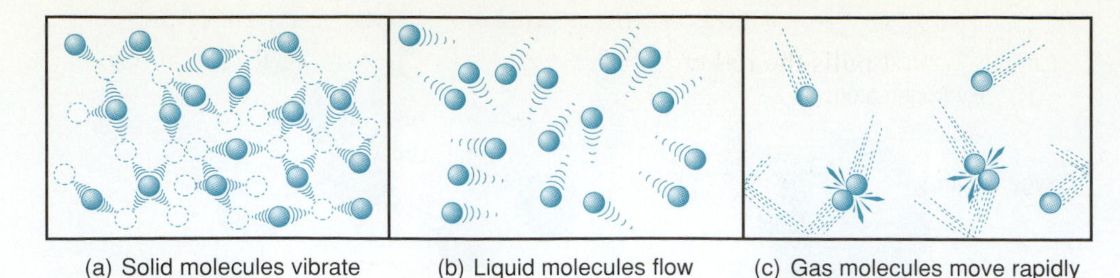

(a) Solid molecules vibrate in fixed positions. (b) Liquid molecules flow over each other. (c) Gas molecules move rapidly in all directions and collide.

Figure 11.3

The molecules of a liquid are not fixed in relation to each other [Fig. 11.3(b)]. They normally move in a flowing type of motion but yet are so close together that they are practically incompressible, thus having a definite volume. Because the molecules move in a smooth flowing motion and not in any fixed manner, a liquid takes the shape of its container.

The molecules of a gas are not fixed in relation to each other and move rapidly in all directions, colliding with each other [Fig. 11.3(c)]. They are much farther apart than molecules in a liquid, and they are extremely far apart when compared to the distance between molecules in solids. The movement of the molecules is limited only by the container. Therefore, a gas takes the shape of its container. Because the molecules are far apart, a gas can easily be compressed, and it has the same volume as its container.

11.2 PROPERTIES OF SOLIDS

Solids have a definite shape and a definite volume. Solids have molecules that are usually arranged in a definite pattern. The following properties are common to most solids.

Cohesion and Adhesion

The molecules of a solid are held together by large internal molecular forces. **Cohesion** is the force of attraction between like molecules. The cohesive forces hold the closely packed molecules of a solid together, which keep its shape and volume from being easily changed.

Cohesion, this force of attraction between like molecules, can also be shown by grinding and polishing the surfaces of two like solids and then sliding their surfaces together. For example, take two pieces of polished plate glass and slide them together. Try to pull them apart. The force of attraction of the like molecules of the two pieces of glass makes it difficult to pull them apart.

Adhesion is the force of attraction between different or unlike molecules. Common examples include glue and wood, adhesive tape and skin, and tar and shoe soles.

Tensile Strength

The **tensile strength** of a solid is a measure of its resistance to being pulled apart. That is, the tensile strength of a solid is a measure of its cohesive forces between adjacent molecules. The tensile strength of a rod or wire is found by putting it in a machine that pulls the rod or wire until it breaks (Fig. 11.4). The tensile strength is the ratio

$$\frac{\text{force required to break the rod or wire}}{\text{cross-sectional area of the rod or wire}}$$

Figure 11.4
This machine determines the tensile strength of a metal rod by finding the force needed to pull the rod until it breaks.

Hardness

The **hardness** of a solid is a measure of the internal resistance of its molecules being forced farther apart or closer together. More commonly, we classify the hardness of a solid in terms of its difficulty in being scratched using a "scratch test." The given material is scratched in a certain way. Its scratch is then compared with a series of standard scratches of materials that form an arbitrary hardness scale from very soft solids to the hardest known substance, diamond.

The **Brinell method** is a common industrial method used to measure the hardness of a metal. A machine is used to press a 10-mm hardened chrome-steel ball with the same force as an equivalent mass of 3000 kg into the metal being tested (Fig. 11.5). The diameter of the resulting impression is used as a measure of the metal's hardness. The Brinell value or number is the ratio

$$\frac{\text{load (in kg)}}{\text{surface area of the impression (in mm}^2\text{)}}$$

This value can also be compared with a scale of the accepted hardnesses of given metals. The larger the Brinell number, the harder the metal.

Steel can be hardened by heating it to a very high temperature, then suddenly cooling it by putting it in water. However, it then becomes brittle. This cooled steel can then be tempered (toughened) by reheating it and allowing it to

Figure 11.5
The Brinell hardness testing machine determines the hardness of a metal.

cool slowly. As the steel cools, it loses hardness and gains toughness. If the steel cools down slowly and completely, we say that it is *annealed*. Annealed steel is soft and tough but not brittle.

Ductility

A metal rod that can be drawn through a die to produce a wire has a property called **ductility.** As the rod is pulled through the die, its diameter is decreased, and its length is increased as it becomes a wire (Fig. 11.6).

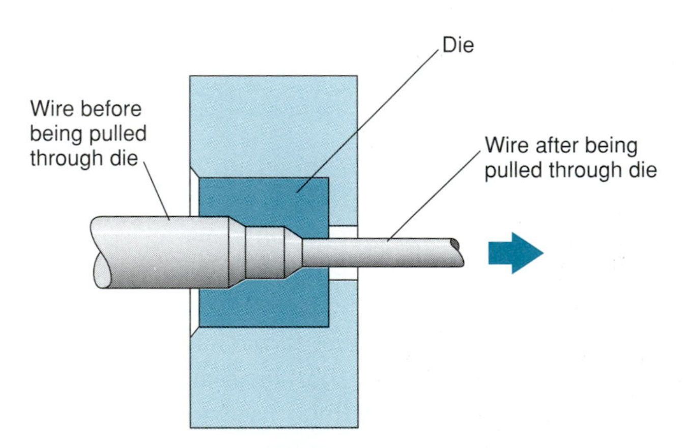

Figure 11.6
Ductility: the ability of a metal to be drawn into a wire.

Malleability

A metal that can be hammered and rolled into sheets has a property called **malleability.** As the metal is hammered or rolled as in Fig. 11.7, its shape or thickness is changed. During this process, the atoms slide over each other and change positions. The cohesive forces are relatively strong; thus, the atoms do not become widely separated during their rearrangement, and the resulting shape remains relatively stable.

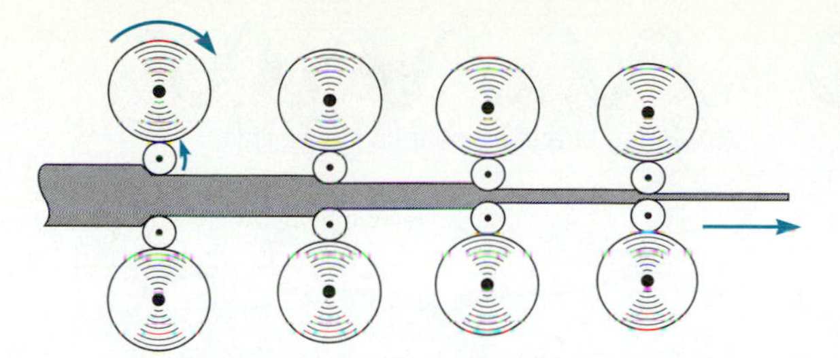

Figure 11.7
Malleability: the ability of a metal to be rolled into a sheet.

Elasticity

An object becomes deformed when outside forces change its shape or size. The object's ability to return to its original size and shape—when the outside forces are removed—is called **elasticity.** When the solid is being deformed, sometimes these molecules attract each other, and sometimes they repel each other. For instance, take a rubber ball and try to pull it apart (Fig. 11.8). You notice that the ball stretches out of shape.

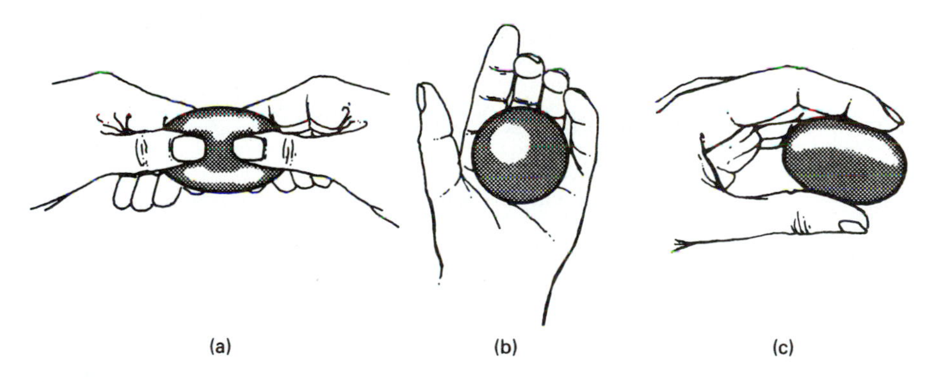

Figure 11.8
Elasticity in a rubber ball.

However, when you release the pulling force, the ball returns to its original shape because the molecules, being farther apart than normal, attract each other. If you squeeze the ball together, the ball will again become out of shape. Now release the pressure and the ball will again return to its original shape because the molecules, being too close together, repel each other. Therefore, we can see that when molecules are slightly pulled out of position, they attract each other. When they are pressed too close together, they repel each other.

Most solids have the property of elasticity; however, some are only slightly elastic. For example, wood and Styrofoam are two solids whose elasticity is small.

Not every elastic object returns to its original shape after being deformed. If too large a deforming force is applied, it will become deformed permanently. Take a spring [Fig. 11.9(a)] and pull it apart [Fig. 11.9(b)]. When you let it go, it should return to its original shape. Next, pull the spring apart as far as you can [Fig. 11.9(c)]. When you let it go this time, it will probably not return to its original shape. When a solid is so deformed, it is said to be deformed past its **elastic limit.** Its molecules have been pulled far enough apart that they slid past one another beyond the point at which the original molecular forces could return the spring to its original shape. If the deforming force is enough greater, the spring breaks apart [Fig. 11.9(d)].

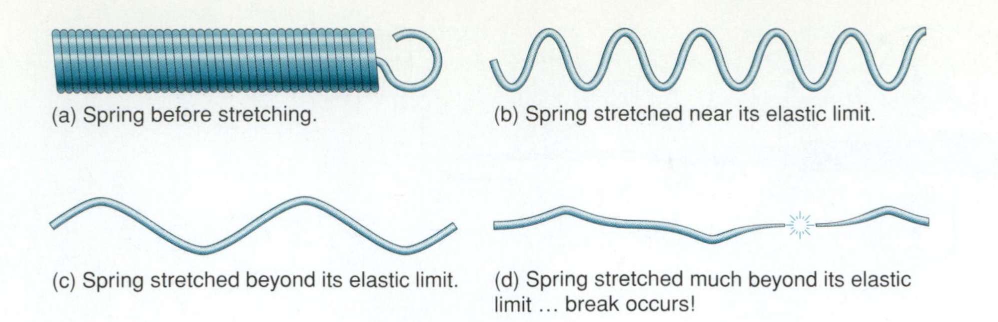

Figure 11.9

Stress is the ratio of the outside applied force, which tends to cause a distortion, to the area over which the force acts. In other words,

$$\text{stress} = \frac{\text{applied force}}{\text{area over which the force acts}}$$

or

$$S = \frac{F}{A}$$

The metric stress unit is usually the pascal (Pa).

$$1 \text{ Pa} = 1 \text{ N/m}^2$$

The English stress unit is lb/in^2. The pascal is discussed more fully in Section 12.1.

There are five basic stresses: tension, compression, shear, bending, and twisting (Fig. 11.10). (Shear, bending, and twisting are not discussed in this book.)

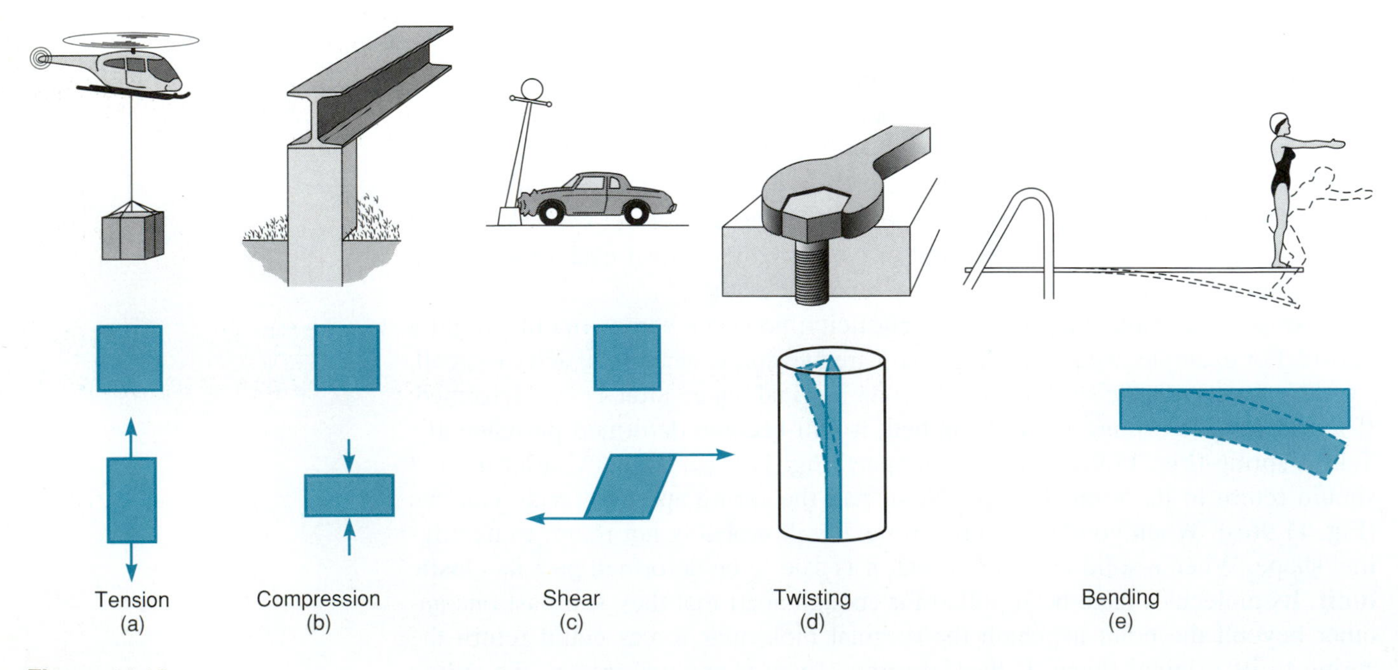

Figure 11.10
Five basic stresses.

EXAMPLE 1

A steel column in a building has a cross-sectional area of 0.250 m^2 and supports a weight of 1.50×10^5 N. Find the stress on the column.

Sketch: None needed

Data:

$$A = 0.250 \text{ m}^2$$
$$F = 1.50 \times 10^5 \text{ N}$$
$$S = ?$$

Basic Equation:

$$S = \frac{F}{A}$$

Working Equation: Same

Substitution:

$$S = \frac{1.50 \times 10^5 \text{ N}}{0.250 \text{ m}^2}$$
$$= 6.00 \times 10^5 \text{ N/m}^2$$
$$= 6.00 \times 10^5 \text{ Pa} \quad \text{or} \quad 60\bar{0} \text{ kPa}$$

Whenever a stress is applied to an object, the object is changed minutely, at least. If you stand on a steel beam, it bends—at least slightly. The resulting deformation is called **strain.** That is, strain is the relative amount of deformation of a body that is under stress. Or, strain is *change in length per unit of length,* change in volume per unit of volume, and so on. Strain is a direct and necessary consequence of stress.

Hooke's Law

One of the most basic principles related to the elasticity of solids is **Hooke's law.** This law is named after Robert Hooke, who first discovered the principle in the seventeenth century while inventing the balance spring for spring-driven clocks.

HOOKE'S LAW

The ratio of the force applied to an object to its change in length (resulting in its being stretched or compressed by the applied force) is constant as long as the elastic limit has not been exceeded.

Or, stated another way: Stress is directly proportional to strain as long as the elastic limit has not been exceeded. In equation form:

$$\frac{F}{\Delta l} = k$$

where F = applied force
Δl = change in length
k = elastic constant

Note: Δ (the Greek letter delta) is often used in mathematics and science to mean "change in."

EXAMPLE 2

A force of 5.00 N is applied to a spring whose elastic constant is 0.250 N/cm. Find its change in length.

Sketch:

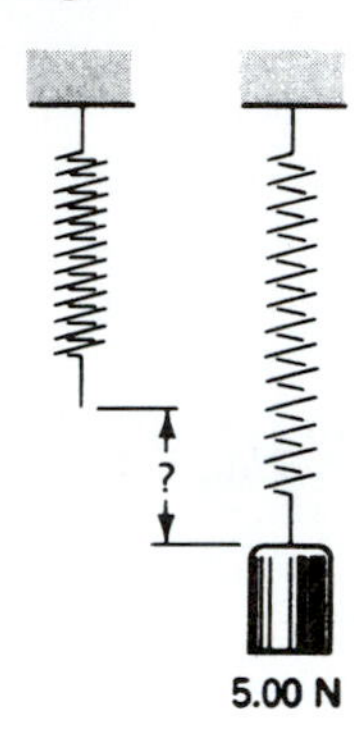

Data:

$$F = 5.00 \text{ N}$$
$$k = 0.250 \text{ N/cm}$$
$$\Delta l = ?$$

Basic Equation:

$$\frac{F}{\Delta l} = k$$

Working Equation:

$$\Delta l = \frac{F}{k}$$

Substitution:

$$\Delta l = \frac{5.00 \text{ N}}{0.250 \text{ N/cm}}$$
$$= 20.0 \text{ cm}$$

$$\frac{\text{N}}{\text{N/cm}} = \text{N} \div \frac{\text{N}}{\text{cm}} = \cancel{\text{N}} \cdot \frac{\text{cm}}{\cancel{\text{N}}} = \text{cm}$$

EXAMPLE 3

A force of 3.00 lb stretches a spring 12.0 in. What force is required to stretch the spring 15.0 in.?

Sketch:

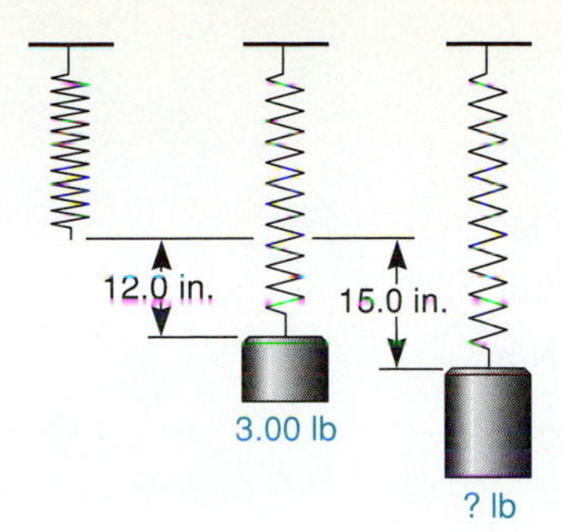

Data:

$$F_1 = 3.00 \text{ lb}$$
$$l_1 = 12.0 \text{ in.}$$
$$l_2 = 15.0 \text{ in.}$$
$$F_2 = ?$$

Basic Equation:

$$\frac{F}{\Delta l} = k$$

Working Equations:

$$\frac{F}{\Delta l} = k \qquad \text{and} \qquad F = k(\Delta l)$$

Substitution: There are two substitutions, one to find k and one to find the second force F_2.

$$\frac{3.00 \text{ lb}}{12.0 \text{ in.}} = k$$
$$0.250 \text{ lb/in.} = k$$

$$F_2 = (0.250 \text{ lb/}\cancel{\text{in.}})(15.0 \ \cancel{\text{in.}})$$
$$= 3.75 \text{ lb}$$

EXAMPLE 4

A support column is compressed 3.46×10^{-4} m under a weight of 6.42×10^5 N. How much is the column compressed under a weight of 5.80×10^6 N?

Sketch: None needed

First find k:

Data:

$$F_2 = 6.42 \times 10^5 \text{ N}$$
$$\Delta l_2 = 3.46 \times 10^{-4} \text{ m}$$
$$k = ?$$

Basic Equation:

$$\frac{F_2}{\Delta l_2} = k$$

Working Equation: Same

Substitution:

$$k = \frac{6.42 \times 10^5 \text{ N}}{3.46 \times 10^{-4} \text{ m}} = 1.86 \times 10^9 \text{ N/m}$$

Then

Data:

$$k = 1.86 \times 10^9 \text{ N/m}$$
$$F_1 = 5.80 \times 10^6 \text{ N}$$
$$\Delta l_1 = ?$$

Basic Equation:

$$\frac{F_1}{\Delta l_1} = k$$

Working Equation:

$$\Delta l_1 = \frac{F_1}{k}$$

Substitution:

$$\Delta l_1 = \frac{5.80 \times 10^6 \not{\text{N}}}{1.86 \times 10^9 \not{\text{N}}\text{/m}} = 3.12 \times 10^{-3} \text{ m or } 3.12 \text{ mm}$$

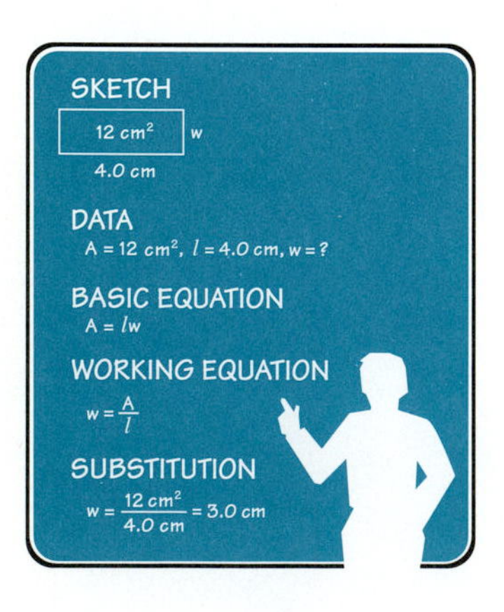

■ PROBLEMS 11.2

1. A spring is stretched 24.0 in. by a force of 54.0 lb. How far will it stretch if a force of 104 lb is applied?
2. What weight will stretch the spring 9.00 in. in Problem 1?
3. A 17.0-N force stretches a wire 0.650 cm. What force will stretch a similar piece of wire 1.87 cm?
4. A force of 21.3 N is applied on a similar piece of wire as in Problem 3. How far will it stretch?
5. A force of 36.0 N stretches a spring 18.0 cm. Find the spring constant (in N/m).
6. A force of 5.00 N is applied to a spring whose spring constant is 0.250 N/cm. Find its change in length (in cm).
7. The vertical steel columns of an office building each support a weight of 1.30×10^5 N at the second floor, with each being compressed 5.90×10^{-3} cm.
   (a) Find the compression in each column if a weight of 5.50×10^5 N is supported.
   (b) If the compression for each steel column is 0.0710 cm, what weight is supported by each column?
8. The vertical steel columns of an office building each support a weight of $30,\bar{0}00$ lb, with each being compressed 0.00234 in. Find the compression of each column if a weight of 125,000 lb is supported.
9. The compression for the steel columns in Problem 8 is 0.0279 in. What weight is supported by each column?

10. A coiled spring is stretched 40.0 cm by a 5.00-N weight. How far is it stretched by a 15.0-N weight?
11. In Problem 10, what weight will stretch the spring 60.0 cm?
12. A 12,$\bar{0}$00-N load is hanging from a steel cable that is 10.0 m long and 16.0 mm in diameter. Find the stress.
13. A rectangular cast-iron column 25.0 cm $\times$ 25.0 cm $\times$ 5.00 m supports a weight of 6.80×10^6 N. Find the stress in pascals on the top portion of the column.
14. In a Hooke's law experiment, the following weights were attached to a spring, resulting in the following elongations:

Weight (N)	*Elongation (cm)*
5$\bar{0}$	2.0
75	3.3
105	4.2
125	5.0
15$\bar{0}$	6.0
175	7.4
225	9.5
275	11.1

  (a) Plot the graph of weight versus elongation.
  (b) Draw the best straight line through the data.
  (c) From the graph, what weight corresponds to an elongation of 7.5 cm?
  (d) From the graph, what elongation corresponds to a weight of 270 N?
  (e) From the graph, determine the spring constant.
15. What was the original length of a spring with spring constant 96.0 N/m that is stretched to 28.0 cm by a 15.0-N weight?
16. Two hanging springs, each 15.0 cm long, with spring constants 0.970 N/cm and 1.45 N/cm, are stretched by a bar weighing 26.0 N that connects them. The bar is 6.00 m long and has a center of mass 2.00 m from the spring with constant 0.970 N/cm. How far does each spring stretch? ■

11.3 PROPERTIES OF LIQUIDS

As noted previously, a liquid is a substance that has a definite volume and takes the shape of its container. The molecules move in a flowing motion, yet are so close together that it is very difficult to compress a liquid. Most liquids share the following common properties.

Cohesion and Adhesion

Cohesion, the force of attraction between like molecules, causes a liquid like molasses to be sticky. Adhesion, the force of attraction between unlike molecules, causes the molasses to also stick to your finger. In the case of water, its adhesive forces are greater than its cohesive forces. Put a plate of glass in a pan of water and pull it out. Some water remains on the glass.

In the case of mercury, the opposite is true. Mercury's cohesive forces are greater than its adhesive forces. If glass is submerged in mercury and then pulled out, virtually no mercury remains on the glass.

A liquid whose adhesive forces are greater than its cohesive forces tends to wet any surface that comes in contact with it. A liquid whose cohesive forces are greater than its adhesive forces tends to leave dry or not wet any surface that comes in contact with it.

Surface Tension

The ability of the surface of water to support a needle is an example of **surface tension.** The water's surface acts like a thin, flexible surface film. The surface tension of water can be reduced by adding soap to the water (Fig. 11.11). Soaps are added to laundry water to decrease the surface tension of water so that the water more easily penetrates the fibers of the clothes being washed.

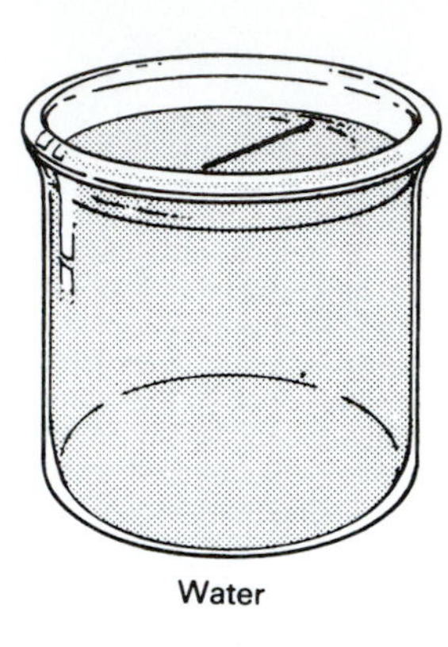

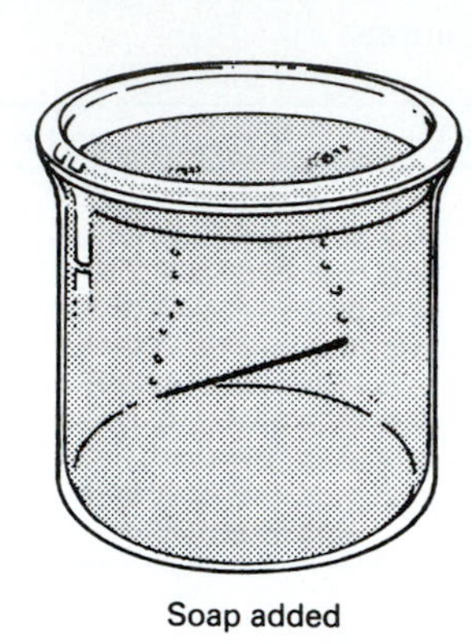

Figure 11.11
The surface tension of water will support a needle. Adding soap reduces surface tension.

Surface tension causes a raindrop to hold together. Surface tension causes a small drop of mercury to keep an almost spherical shape. A liquid drop suspended in space is spherical. A falling raindrop's shape is due to the friction with the air (Fig. 11.12).

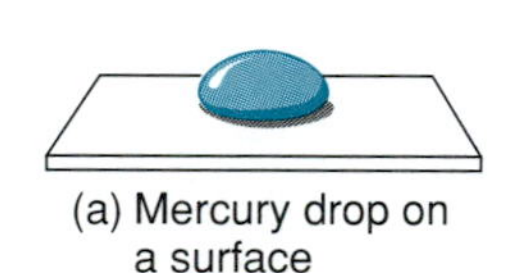

Figure 11.12
(a) Surface tension causes a drop of mercury to be nearly spherical. (b) Surface tension causes a raindrop to hold together. (c) The shape of a falling raindrop is due to the friction with air.

Figure 11.13
Cold oil is more viscous than hot oil.

Viscosity

In liquids there is friction, which is called **viscosity.** The greater the molecular attraction, the more the friction and the greater the viscosity. For example, it takes more force to move a block of wood through oil than through water. This is because oil is more viscous than water.

If a liquid's temperature is increased, its viscosity decreases. For example, the viscosity of oil in a car engine before it is started on a winter morning at $-10.0°C$ is greater than after the engine has been running for an hour (Fig. 11.13).

One common misunderstanding is that higher viscosity means higher density. For example, oil is more viscous, but water is denser as oil floats on water.

Capillary Action

Liquids keep the same level in tubes whose ends are submerged in a liquid if the tubes have a large enough diameter (Fig. 11.14). In tubes of different small diameters, water does not stand at the same level. The smaller the diameter, the

Figure 11.14
A liquid keeps the same level in tubes that are connected. Neither the shape nor the size of the containers makes a difference.

higher the water rises. If mercury is used, it does not stand at the same level either, but instead of rising up the tube, the mercury level falls, or its level is depressed. The smaller the diameter, the lower its level is depressed. This behavior of liquids in small tubes is called **capillary action** (Fig. 11.15).

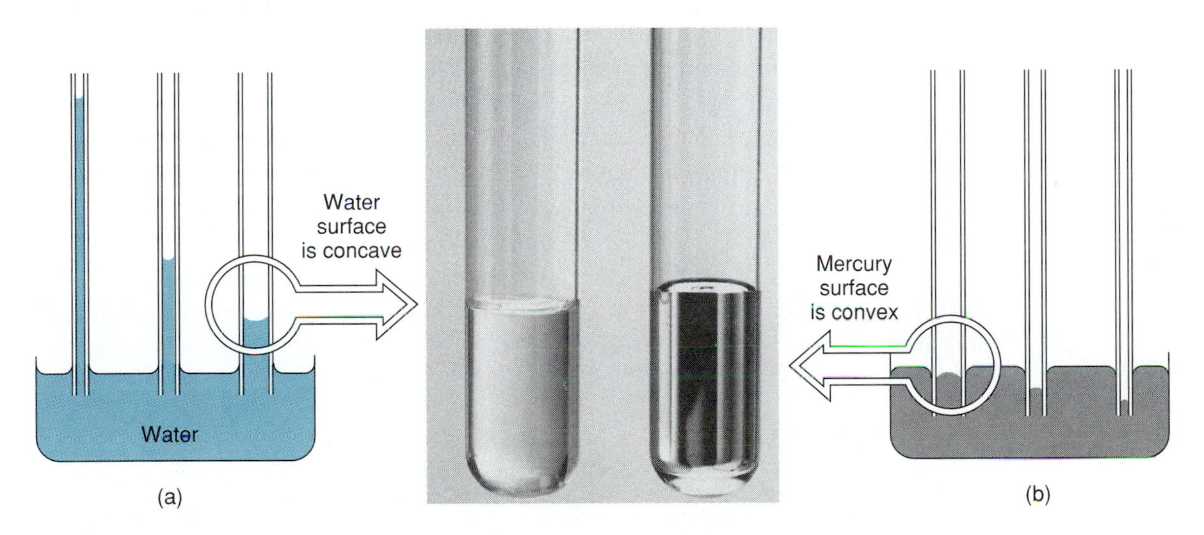

Figure 11.15
Capillary action. In small tubes, water stands at higher levels for small diameters. The water surface is concave. In small tubes, mercury stands at lower levels for small diameters. The mercury surface is convex. (The tube diameters and differences in tube levels are exaggerated in a and b.)

Capillary action is due to both adhesion of the liquid molecules with the tube and the surface tension of the liquid. In the case of water, the adhesive forces are greater than the cohesive molecular forces. Thus, water creeps up the sides of the tube and produces a concave water surface. The surface tension of the water tends to flatten the concave surface. Together, these two forces raise the water up the tubes until it is counterbalanced by the weight of the water column itself.

In the case of mercury, the cohesive molecular forces are greater than the adhesive forces and produce a convex mercury surface. The surface tension of the mercury tends to further hold down the mercury level.

This crescent-shaped surface of a liquid column, whether concave or convex, is called a **meniscus.** To measure the height of a liquid in a tube, measure to the lowest point of a concave meniscus or to the highest point of a convex meniscus.

Experimentally, scientists have found that:

1. Liquids rise in capillary tubes they tend to wet and are depressed in tubes they tend not to wet.
2. Elevation or depression in the tube is inversely proportional to the diameter of the tube.
3. The elevation or depression decreases as the temperature increases.

Capillary action causes the rise of oil (or kerosene) in the wick of an oil lamp. Towels also absorb water because of capillary action.

11.4 PROPERTIES OF GASES

Because of the rapid random movement of its molecules, a gas spreads to completely occupy the volume of its container. This property is called **expansion.**

Diffusion of a gas is the process by which molecules of the gas mix with the molecules of a solid, liquid, or another gas. If you remove the cap from a can of gasoline, you soon smell the fumes. The air molecules and the gasoline molecules mix throughout the room because of diffusion.

A balloon inflates due to the pressure of the air molecules on the inside surface of the balloon. This pressure is caused by the bombardment on the walls by the moving molecules. The pressure may be increased by increasing the number of molecules by blowing more air into the balloon. Pressure may also be increased by heating the air molecules already in the balloon. Heat increases the velocity of the molecules.

The behavior of liquids and gases is very similar in many cases. The term **fluid** is used when discussing principles and behaviors common to both liquids and gases.

11.5 DENSITY

Density is a property of all three states of matter. **Mass density,** D_m, is defined as mass per unit volume. **Weight density,** D_w, is defined as weight per unit volume, or,

$$D_m = \frac{m}{V} \qquad D_w = \frac{w}{V}$$

where D_m = mass density $\qquad D_w$ = weight density
m = mass $\qquad w$ = weight
V = volume $\qquad V$ = volume

Although either mass density or weight density can be expressed in both the metric system and the English system, mass density is usually given in the metric units kg/m^3, and weight density is usually given in the English units lb/ft^3 (Table 11.2).

Table 11.2 Densities for Various Substances

Substance	*Mass Density* (kg/m^3)	*Weight Density* (lb/ft^3)
Solids		
Copper	8,890	555
Iron	7,800	490
Lead	11,300	708
Aluminum	2,700	169
Ice	917	57
Wood, white pine	420	26
Concrete	2,300	145
Cork	240	15
Liquids		
Water	$1,0\bar{0}0$	62.4
Seawater	1,025	64.0
Oil	870	54.2
Mercury	13,600	846
Alcohol	790	49.4
Gasoline	680	42.0
Gases*	At 0°C and 1 atm pressure	At 32°F and 1 atm pressure
Air	1.29	0.081
Carbon dioxide	1.96	0.123
Carbon monoxide	1.25	0.078
Helium	0.178	0.011
Hydrogen	0.0899	0.0056
Oxygen	1.43	0.089
Nitrogen	1.25	0.078
Ammonia	0.760	0.047
Propane	2.02	0.126

*The density of a gas is found by pumping the gas into a container, by measuring its volume and mass or weight, and then by using the appropriate density formula.

The mass density of water is $10\bar{0}0$ kg/m^3; that is, 1 cubic metre of water has a mass of $10\bar{0}0$ kg. The weight density of water is 62.4 lb/ft^3; that is, 1 cubic foot of water weighs 62.4 lb. (A suggested project is to take a container 1 cubic foot in volume, pour it full of water, and find the weight of the water. If you fill the container with a gallon container, you will also find that 1 ft^3 is approximately 7.5 gal.)

In nearly all forms of matter, the density usually decreases as the temperature increases and increases as the temperature decreases. Water does not follow the usual pattern of increasing density at lower temperatures; ice is actually less dense than liquid water. This phenomenon is discussed more fully in Section 13.7.

Note: Conversion factors must often be used to obtain the desired units.

EXAMPLE 1

Find the weight density of a block of wood 3.00 in. × 4.00 in. × 5.00 in. with weight 0.700 lb.

Sketch:

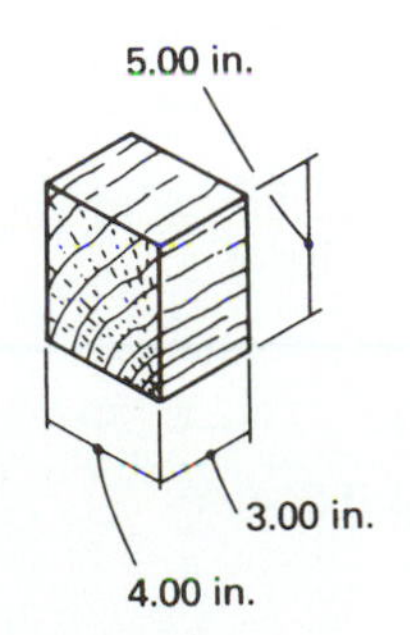

Data:

$$l = 4.00 \text{ in.}$$
$$w = 3.00 \text{ in.}$$
$$h = 5.00 \text{ in.}$$
$$w = 0.700 \text{ lb}$$
$$D_w = ?$$

Basic Equations:

$$V = lwh \qquad \text{and} \qquad D_w = \frac{w}{V}$$

Working Equations: Same

Substitutions:

$$V = (4.00 \text{ in.})(3.00 \text{ in.})(5.00 \text{ in.}) = 60.0 \text{ in}^3$$

$$D_w = \frac{0.700 \text{ lb}}{60.0 \text{ in}^3} = 0.0117 \frac{\text{lb}}{\text{in}^3} \times \frac{1728 \text{ in}^3}{1 \text{ ft}^3} = 20.2 \text{ lb/ft}^3$$

EXAMPLE 2

Find the mass density of a ball bearing with mass 22.0 g and radius 0.875 cm.

Data:

$$r = 0.875 \text{ cm}$$
$$m = 22.0 \text{ g}$$
$$D_m = ?$$

Basic Equations:

$$V = \tfrac{4}{3}\pi r^3 \qquad \text{and} \qquad D_m = \frac{m}{V}$$

Working Equations: Same

Substitutions:

$$V = \tfrac{4}{3}\pi(0.875 \text{ cm})^3 = 2.81 \text{ cm}^3$$

$$D_m = \frac{22.0 \text{ g}}{2.81 \text{ cm}^3} = 7.83 \text{ g/cm}^3 = 7.83 \frac{\text{g}}{\text{cm}^3} \times \frac{10^6 \text{ cm}^3}{1 \text{ m}^3} \times \frac{1 \text{ kg}}{10^3 \text{ g}} = 7830 \text{ kg/m}^3$$

EXAMPLE 3

Find the weight density of a gallon of water weighing 8.34 lb.

Data:

$$w = 8.34 \text{ lb}$$
$$V = 1 \text{ gal} = 231 \text{ in}^3$$
$$D_w = ?$$

Basic Equation:

$$D_w = \frac{w}{V}$$

Working Equation: Same

Substitution:

$$D_w = \frac{8.34 \text{ lb}}{231 \text{ in}^3}$$
$$= 0.0361 \frac{\text{lb}}{\cancel{\text{in}^3}} \times \frac{1728 \cancel{\text{in}^3}}{1 \text{ ft}^3}$$
$$= 62.4 \text{ lb/ft}^3$$

EXAMPLE 4

Find the weight density of a can of oil (1 quart) weighing 1.90 lb.

Data:

$$V = 1 \text{ qt} = \tfrac{1}{4} \text{ gal} = \tfrac{1}{4}(231 \text{ in}^3) = 57.8 \text{ in}^3$$
$$w = 1.90 \text{ lb}$$
$$D_w = ?$$

Basic Equation:

$$D_w = \frac{w}{V}$$

Working Equation: Same

Substitution:

$$D_w = \frac{1.90 \text{ lb}}{57.8 \text{ in}^3}$$
$$= 0.0329 \frac{\text{lb}}{\cancel{\text{in}^3}} \times \frac{1728 \cancel{\text{in}^3}}{1 \text{ ft}^3}$$
$$= 56.9 \text{ lb/ft}^3$$

EXAMPLE 5

A quantity of gasoline weighs 5.50 lb with weight density of 42.0 lb/ft^3. What is its volume?

Data:

$$D_w = 42.0 \text{ lb/ft}^3$$
$$w = 5.50 \text{ lb}$$
$$V = ?$$

Basic Equation:

$$D_w = \frac{w}{V}$$

Working Equation:

$$V = \frac{w}{D_w}$$

Substitution:

$$V = \frac{5.50\ \cancel{\text{lb}}}{42.0\ \cancel{\text{lb}}/\text{ft}^3}$$
$$= 0.131\ \text{ft}^3$$

The density of an irregular solid (rock) cannot be found directly because of the difficulty of finding its volume. However, we could find the amount of water the solid displaces, which is the same as the volume of the irregular solid (Fig. 11.16). The volume of water in the small beaker equals the volume of the rock.

Figure 11.16
The volume of this rock can be found by measuring the volume of the liquid it displaces.

EXAMPLE 6

A rock of mass 10.8 kg displaces $32\bar{0}0\ \text{cm}^3$ of water. What is the mass density of the rock?

Data:

$$m = 10.8\ \text{kg}$$
$$V = 32\bar{0}0\ \text{cm}^3$$
$$D_m = ?$$

Basic Equation:

$$D_m = \frac{m}{V}$$

Working Equatoin: Same

Substitution:

$$D_m = \frac{10.8\ \text{kg}}{32\bar{0}0\ \cancel{\text{cm}^3}} \times \frac{10^6\ \cancel{\text{cm}^3}}{1\ \text{m}^3}$$
$$= 3380\ \text{kg/m}^3$$

EXAMPLE 7

A rock displaces 3.00 gal of water and has a weight density of 156 lb/ft^3. What is its weight?

Data:

$$D_w = 156 \text{ lb/ft}^3$$
$$V = 3.00 \text{ gal}$$
$$w = ?$$

Basic Equation:

$$D_w = \frac{w}{V}$$

Working Equation:

$$w = D_w V$$

Substitution:

$$w = 156 \frac{\text{lb}}{\cancel{\text{ft}^3}} \times 3.00 \cancel{\text{gal}} \times \frac{231 \cancel{\text{in}^3}}{1 \cancel{\text{gal}}} \times \frac{1 \cancel{\text{ft}^3}}{1728 \cancel{\text{in}^3}}$$
$$= 62.6 \text{ lb}$$

To compare the densities of two materials, we compare each with the density of water. The ratio of the density of any material to the density of water is called its **specific gravity.** That is,

$$\text{specific gravity (sp gr)} = \frac{D_{\text{material}}}{D_{\text{water}}}$$

Note that specific gravity is a unitless quantity.

EXAMPLE 8

The density of iron is 7830 kg/m^3. Find its specific gravity.

Data:

$$D_{\text{material}} = 7830 \text{ kg/m}^3$$
$$D_{\text{water}} = 10\bar{0}0 \text{ kg/m}^3$$
$$\text{sp gr} = ?$$

Basic Equation:

$$\text{sp gr} = \frac{D_{\text{material}}}{D_{\text{water}}}$$

Working Equation: Same

Substitution:

$$\text{sp gr} = \frac{7830 \text{ kg/m}^3}{10\bar{0}0 \text{ kg/m}^3}$$
$$= 7.83$$

This means that iron is 7.83 times as dense as water, and thus it sinks in water.

EXAMPLE 9

The density of oil is 54.2 lb/ft^3. Find its specific gravity.

$$D_{\text{material}} = 54.2 \text{ lb/ft}^3$$
$$D_{\text{water}} = 62.4 \text{ lb/ft}^3$$
$$\text{sp gr} = ?$$

Basic Equation:

$$\text{sp gr} = \frac{D_{\text{material}}}{D_{\text{water}}}$$

Working Equation: Same

Substitution:

$$\text{sp gr} = \frac{54.2 \text{ lb/ft}^3}{62.4 \text{ lb/ft}^3}$$
$$= 0.869$$

This means oil is 0.869 times as dense as water and that it floats on water.

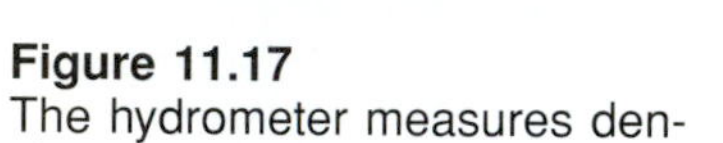

Figure 11.17
The hydrometer measures density of a liquid.

In general, the specific gravity of

water = 1,

a material denser than water > 1, and

a material less dense than water < 1.

When we check the antifreeze in a radiator in winter, we are really finding the specific gravity of the liquid. Specific gravity is a comparison of the density of a substance to that of water. Because the density of antifreeze is different from the density of water, we find the concentration of antifreeze (and thus the amount of protection from freezing) by measuring the specific gravity of the solution in the radiator.

A *hydrometer* is a sealed glass tube weighted at one end so that it floats vertically in a liquid (Fig. 11.17). Its sinks in the liquid until it displaces an amount of liquid equal to its own weight. The densities of the displaced liquids are inversely proportional to the depths to which the tube sinks. That is, the greater the density of the liquid, the less the tube sinks; the less the density of the liquid, the more the tube sinks. A hydrometer usually has a scale inside the tube and is calibrated so that it floats in water at the 1.000 mark. Anything with a specific gravity greater than 1 sinks in water. A substance with a specific gravity less than 1 floats in water; its specific gravity indicates the fractional volume that is under water.

Hydrometers are commonly used to measure the specific gravities of battery acid and antifreeze in radiators (Fig. 11.18). In a lead storage battery, the electrolyte is a solution of sulfuric acid and water, and the specific gravity of the solution varies with the amount of charge of the battery. Table 11.3 gives common specific gravities of conditions of a lead storage battery. Table 11.4 gives various specific gravities and the corresponding temperatures below which the antifreeze and water solution will freeze.

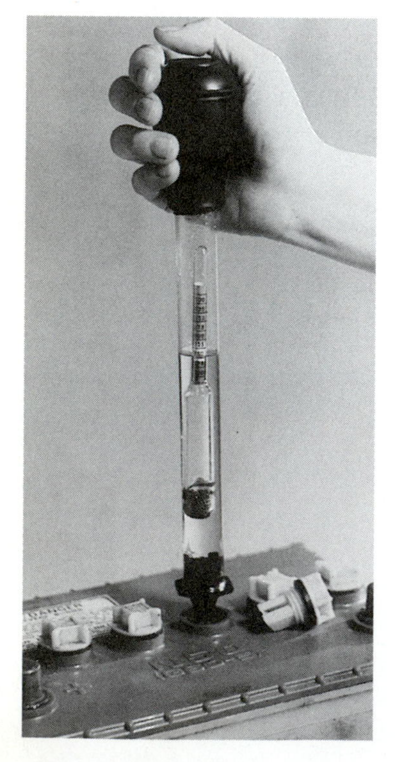

Figure 11.18
A common hydrometer.

Table 11.3 Specific Gravities for a Lead Storage Battery

Condition	*Specific Gravity*
New (fully charged)	1.30
Old (discharged)	1.15

Table 11.4 Specific Gravities for Antifreeze and Water Solution

Temperature (°C)	*Specific Gravity*
−1.24	1.00
−2.99	1.01
−6.89	1.02
−19.82	1.05
−44.83	1.07
−51.23	1.08

One other factor must be considered in the use of the hydrometer—that of temperature. Significant differences in readings will occur over a range of temperatures. Specific gravities of some common liquids at room temperature are given in Table 11.5.

Table 11.5 Specific Gravities of Common Liquids at Room Temperature (20°C or 68°F)

Liquid	*Specific Gravity*
Benzene	0.9
Ethyl alcohol	0.79
Gasoline	0.68
Kerosene	0.82
Mercury	13.6
Seawater	1.025
Sulfuric acid	1.84
Turpentine	0.87
Water	1.000

PROBLEMS 11.5

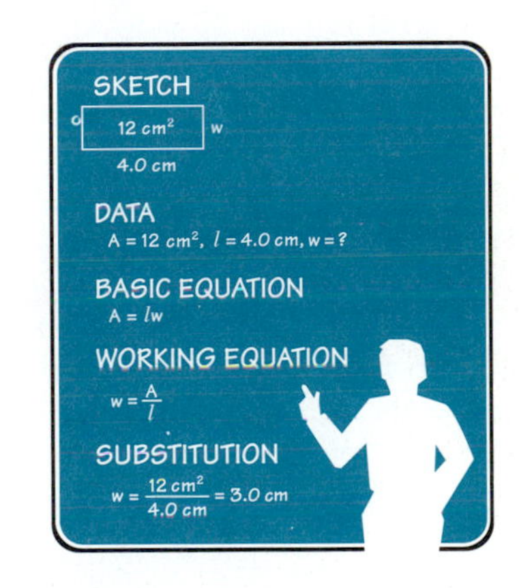

Express mass density in kg/m^3 and weight density in lb/ft^3.

1. Find the mass density of a chunk of rock of mass 215 g that displaces a volume of 75.0 cm^3 of water.
2. A block of wood is 55.9 in. × 71.1 in. × 25.4 in. and weighs 1810 lb. Find its weight density.
3. If a block of wood of the size in Problem 2 has a weight density of 30.0 lb/ft^3, what does it weigh?
4. Find the volume (in cm^3) of 1350 g of mercury.
5. Find the volume (in cm^3) of 1350 g of cork.
6. Find the volume (in m^3) of 1350 g of nitrogen at 0°C and 1 atmosphere pressure.
7. A block of gold 9.00 in. × 8.00 in. × 6.00 in. weighs 302 lb. Find its weight density.
8. A cylindrical piece of copper is 9.00 in. tall and 1.40 in. in radius. How much does it weigh?
9. A piece of aluminum with mass 6.24 kg displaces water that fills a container 12.0 cm × 12.0 cm × 16.0 cm. Find its mass density.
10. If 1.00 pint of turpentine weighs 0.907 lb, what is its weight density?
11. Find the mass density of gasoline if 106 g occupies 155 cm^3.
12. How much does 1.00 gal of gasoline weigh?
13. Determine the volume in cubic metres of 3045 kg of oil.

14. How many ft^3 will 573 lb of water occupy?
15. If 20.4 in^3 of linseed oil weighs 0.694 lb, what is its weight density?
16. If 108 in^3 of ammonia gas weighs 0.00301 lb, what is its weight density?
17. Find the volume of 3.00 kg of propane at 0°C and 1 atm pressure.
18. Granite has a mass density of 2650 kg/m^3. Find its weight density in lb/ft^3.
19. Find the mass density of a metal block 18.0 cm × 24.0 cm × 8.00 cm with mass 9.76 kg.
20. Find the mass (in kg) of 1.00 m^3 of
    (a) water. (b) gasoline.
    (c) copper. (d) mercury.
    (e) air at 0°C and 1 atm pressure.
21. What size tank (in litres) is needed for $10\bar{0}0$ kg of
    (a) water? (b) gasoline? (c) mercury?
22. Copper has a mass density of 8890 kg/m^3. Find its mass density in g/cm^3.

From Table 11.2, find the specific gravity of each material.

23. Ice
24. Concrete
25. Iron
26. Air
27. Gasoline
28. Cork
29. The specific gravity of material X is 0.82. Does it sink or float on water?
30. The specific gravity of material Y is 1.7. Does it sink or float on water?
31. The specific gravity of material Z is 0.52. Does it sink or float on gasoline?
32. The specific gravity of material W is 11.5. Does it sink or float in a tank of mercury?
33. A proton has a mass of 1.67×10^{-27} kg and a diameter of 8.2×10^{-16} m. Find its specific gravity.
34. Find the mass density of a 315-g object that displaces 0.275 m^3 of water. ■

GLOSSARY

Adhesion The force of attraction between different or unlike molecules. (p. 266)

Atom The smallest particle of an element that can exist in a stable or independent state. (p. 264)

Brinell Method A common industrial method used to measure the hardness of a metal. A hardened chrome-steel ball is forced into the metal with an equivalent mass of 3000 kg. The diameter of the resulting impression is used as a measure of the metal's hardness. (p. 267)

Capillary Action The behavior of liquids in small tubes that causes the liquid level in the tube to be higher or lower than in larger tubes. This behavior is due both to adhesion of the liquid molecules with the tube and to the surface tension of the liquid. (p. 277)

Cohesion The force of attraction between like molecules. Holds the closely packed molecules of a solid together. (p. 266)

Compound A substance containing two or more elements. (p. 264)

Diffusion The process by which molecules of a gas mix with the molecules of a solid, liquid, or another gas. (p. 278)

Ductility A property of a metal that can be drawn through a die to produce a wire. (p. 268)

Elastic Limit The point beyond which a deformed object cannot return to its original shape. (p. 269)

Elasticity A measure of a deformed object's ability to return to its original size and shape once the deforming force is removed. (p. 269)

Electron One of the particles that make up atoms. Has a negative charge. (p. 264)

Element A substance that cannot be separated into simpler substances. (p. 264)

Expansion Property of a gas in which the rapid random movement of its molecules causes the gas to occupy the volume of its container. (p. 278)

Fluid A substance that takes the shape of its container. Either a liquid or a gas. (p. 278)

Gas A substance that takes the shape of its container and has the same volume as its container. (p. 265)

Hardness A measure of the internal resistance of the molecules of a solid to being forced farther apart or closer together. (p. 267)

Hooke's Law A principle of elasticity in solids that indicates that the ratio of the change in length of an object that is stretched or compressed to the force causing this change is constant as long as the elastic limit has not been exceeded. (p. 271)

Liquid A substance that takes the shape of its container and has a definite volume. (p. 265)

Malleability A property of a metal that can be hammered and rolled into sheets. (p. 268)

Mass Density The mass per unit volume of a substance. (p. 278)

Meniscus The crescent-shaped surface of a liquid column in a tube. (p. 278)

Molecule The smallest particle of a substance that exists in a stable and independent state. (p. 264)

Neutron One of the particles that makes up atoms. Does not carry an electric charge. (p. 264)

Nucleus The center part of an atom that is made up of protons and neutrons. (p. 264)

Proton One of the particles that makes up atoms. Has a positive charge. (p. 264)

Solid A substance that has a definite shape and a definite volume. (p. 265)

Specific Gravity The ratio of the density of any material to the density of water. (p. 283)

Strain The deformation of an object due to an applied force. (p. 271)

Stress The ratio of an outside applied distorting force to the area over which the force acts. (p. 270)

Surface Tension The ability of the surface of a liquid to act like a thin, flexible film. (p. 276)

Tensile Strength A measure of a solid's resistance to being pulled apart. (p. 266)

Viscosity The friction between the molecules of a liquid that resists motion of the molecules across each other. (p. 276)

Weight Density The weight per unit volume of a substance. (p. 278)

FORMULAS

11.2 $S = \dfrac{F}{A}$

$$\frac{F}{\Delta l} = \text{k}$$

11.5 $D_m = \dfrac{m}{V}$

$$D_w = \frac{w}{V}$$

$$\text{specific gravity} = \frac{D_{\text{material}}}{D_{\text{water}}}$$

REVIEW QUESTIONS

1. The most important particles that make up atoms include which of the following?
   (a) Neutron (b) Molecule (c) Electron
   (d) Hydrogen (e) Proton
2. Matter exists in which of the following states?
   (a) Gas (b) Neutrons (c) Electrons
   (d) Solid (e) Liquid

3. The common industrial method used to measure the hardness of a metal is
   (a) the Bernouilli method. (b) Hooke's method.
   (c) the capillary method. (d) the Brinell method.
   (e) none of the above.
4. Density is a property of
   (a) gases. (b) liquids. (c) solids.
   (d) all of the above. (e) none of the above.
5. The process by which molecules of a gas mix with the molecules of a solid, liquid, or gas is called
   (a) expansion. (b) contraction. (c) capillary action.
   (d) diffusion. (e) none of the above.
6. Capillary action refers to
   (a) the mixing of molecules of different types.
   (b) the behavior of liquids in small tubes.
   (c) the attractive force between molecules.
   (d) stretching beyond the elastic limit.
7. The relationship of the change in length of a stretched or compressed object to the force causing the change is given by
   (a) Pascal's law. (b) Brinell's law. (c) the elastic limit.
   (d) Hooke's law. (e) none of the above.
8. The ability of the surface of water to support a needle is an example of
   (a) mass density. (b) Hooke's law. (c) diffusion.
   (d) stress. (e) surface tension.
9. In your own words, describe the difference between mass density and weight density.
10. Would the mass density of an object be the same if the object were on the moon rather than on earth? Would the weight density be the same?
11. In your own words, describe capillary action.
12. What is the difference between adhesion and cohesion?
13. Give one example of the effects of surface tension that are not described in this book.
14. A proton is approximately _____ times heavier in mass than an electron.
15. The applied force divided by the area over which the force acts is called _____.
16. In your own words, state Hooke's law.
17. The commonly used unit of stress in the metric system is the _____.
18. Describe how to find the specific gravity of an object.
19. What is the ratio of mass to volume called?
20. What is friction in liquids called?
21. A spring that has been permanently deformed is said to have been deformed past its _____ _____.
22. List the three states of matter.
23. Distinguish between a molecule and an atom.
24. Distinguish between a neutron and a proton.
25. List the five basic stresses.
26. Explain how a hydrometer measures the charge in a lead storage battery. Does the temperature affect the measurement?

REVIEW PROBLEMS

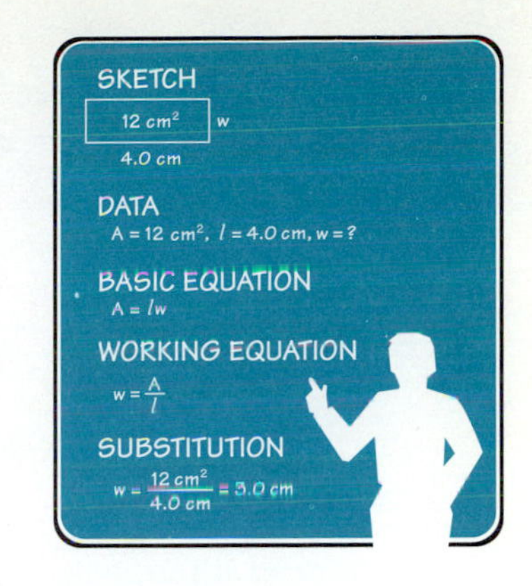

1. A force of 32.5 N stretches a wire 0.470 cm. What force will stretch a similar piece of wire 2.39 cm?
2. A force of 7.33 N is applied to a spring whose spring constant is 0.298 N/cm. Find its change in length.
3. The vertical steel columns of an office building each support a weight of 42,100 lb, with each being compressed 0.0258 in. What is the spring constant of the steel? Find the compression if the weight were 51,700 lb.
4. A rectangular cast-iron column 16.0 cm × 16.0 cm × 4.50 m supports a weight of 7.95×10^6 N. Find the stress on the top of the column.
5. Find the weight density of a block of metal 7.00 in. × 6.50 in. × 8.00 in. that weighs 425 lb.
6. A cylindrical piece of aluminum is 4.25 cm tall and 1.95 cm in radius. How much does it weigh?
7. A piece of metal has a mass of 8.36 kg. If it displaces water that fills a container 9.34 cm × 10.0 cm × 10.0 cm, what is the mass density of the metal?
8. A block of wood is 27.7 in. × 36.3 in. × 12.4 in. and weighs 602 lb. Find its weight density.
9. Find the volume (in cm^3) of 759 g of mercury.
10. Find the volume (in m^3) of 1970 g of hydrogen at 0° C and 1 atm.
11. Find the mass of 1510 m^3 of oxygen at 0°C and 1 atm.
12. Find the weight of 951 ft^3 of water.
13. Find the weight density of a block of material 4.27 in. × 3.87 in. × 5.44 in. and weight 0.982 lb.
14. Find the weight density of 2.00 quarts of liquid weighing 3.67 lb.
15. A quantity of liquid weighs 4.65 lb with a weight density of 39.8 lb/ft^3. What is its volume?
16. The density of a metal is 694 kg/m^3. Find its specific gravity.
17. A solid displaces 4.30 gal of water and has a weight density of 135 lb/ft^3. What is its weight?
18. Find the mass of a rectangular gold bar 4.00 cm × 6.00 cm × 20.00 cm. The mass density of gold is 19,300 kg/m^3.
19. Find the mass density of a chunk of rock using only a scale knowing the following information: mass of rock is 225 g; mass of water the rock displaces is 75.9 g.
20. The specific gravity of an unknown substance is 0.80. Will it float or sink in gasoline?

Chapter 12

FLUIDS

Substances that flow are called fluids. Because liquids and gases behave in much the same manner, they will be presented together. From ships to airplanes to automobile brake systems to balloons, the utilization of fluids and their properties is important to technology and everyday life. We will now study hydraulics, pressure, buoyancy, and flow.

OBJECTIVES

The major goals of this chapter are to enable you to:

1. Describe the behavior of fluids.
2. Determine pressure using the hydraulic principle.
3. Distinguish between gauge pressure and absolute pressure.
4. Calculate buoyancy using Archimedes' principle.
5. Understand fluid flow and Bernoulli's principle.

12.1 PRESSURE

Since liquids and gases behave in much the same manner, they are often studied together as fluids. The gas and water piped to your home are fluids having several common characteristics.

If you press your hand against the table, you apply a force to the table. You are also applying pressure to the table. Is there a difference? The difference is an important one. **Pressure** is the force applied to a unit area. It is the concentration of the force.

$$P = \frac{F}{A}$$

where P = pressure, usually in N/m^2 (Pa) or lb/in^2 (psi)
F = force applied, N or lb, perpendicular to the surface to which it is applied
A = area, m^2 or in^2

Since the SI metric unit for force is the newton (N) and for area is the square metre (m^2), the corresponding pressure unit is N/m^2. This unit is given the special name *pascal* (Pa), named after Blaise Pascal, a French physicist (1623–1662), who made important discoveries in the study of pressure.

$$1 \text{ N/m}^2 = 1 \text{ Pa}$$

Pressure of Solids

Imagine a brick weighing 12.0 N first lying on its side on a table and then standing on one end (Fig. 12.1). The weight of the brick is the same no matter what its position, so the total force (the weight of the brick) on the table is the same in both cases. However, the position of the brick does make a difference in the pressure exerted on the table. In which case is the pressure greater? When standing on end, the brick exerts a greater pressure on the table. The reason is that the area of contact on the end is *smaller* than on the side. Using $P = F/A$, find the pressure in each case.

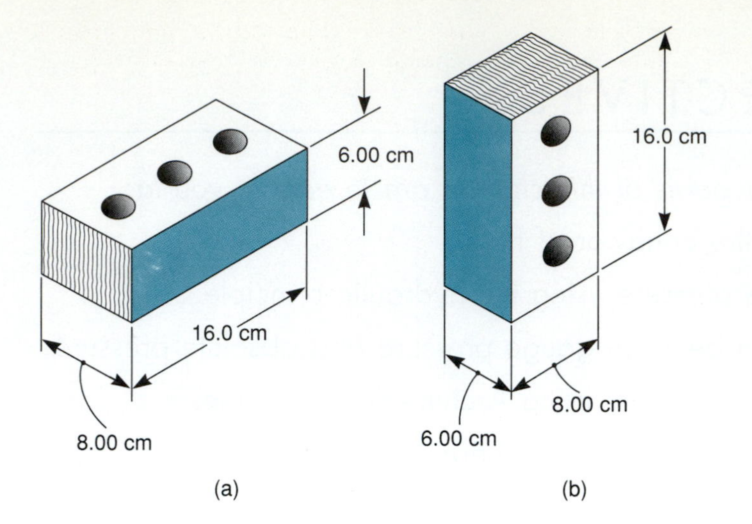

Figure 12.1
The weight of the brick is constant, but the pressure in part (b) is greater.

Case 1	*Case 2*
$F = 12.0 \text{ N}$	$F = 12.0 \text{ N}$
$A = 8.00 \text{ cm} \times 16.0 \text{ cm}$	$A = 6.00 \text{ cm} \times 8.00 \text{ cm}$
$A = 128 \text{ cm}^2$	$A = 48.0 \text{ cm}^2$
$P = \frac{F}{A} = \frac{12.0 \text{ N}}{128 \cancel{\text{cm}^2}} \times \frac{10^4 \cancel{\text{cm}^2}}{1 \text{ m}^2}$	$P = \frac{F}{A} = \frac{12.0 \text{ N}}{48.0 \cancel{\text{cm}^2}} \times \frac{10^4 \cancel{\text{cm}^2}}{1 \text{ m}^2}$
$P = 938 \text{ N/m}^2 = 938 \text{ Pa}$	$P = 25\bar{0}0 \text{ N/m}^2 = 25\bar{0}0 \text{ Pa}$

This shows that when the same force is applied to a smaller area, the pressure is greater. From the discussion so far, would you rather a woman step on your foot with a pointed-heel shoe or with a flat-heel shoe? (See Fig. 12.2.) This was a serious problem for the aircraft industry. They must design and construct

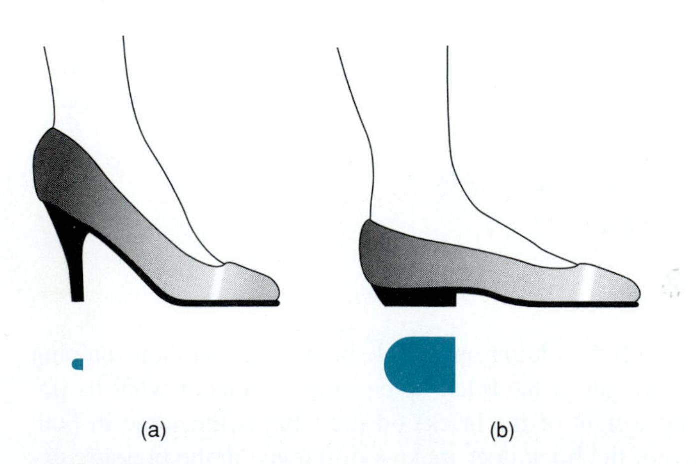

Figure 12.2
The pressure of the heel in part (a) is greater because the weight of the woman rests on a smaller area.

airplane floors light in weight but strong enough to withstand the pressure of pointed-heel shoes. For example, if a 160-lb woman rests her weight on a 4.0-in^2 heel, the pressure is

$$P = \frac{F}{A} = \frac{160 \text{ lb}}{4.0 \text{ in}^2} = 4\bar{0} \text{ lb/in}^2$$

But if she rests her weight on a pointed heel of $\frac{1}{4}$ in^2, which is a common area of a pointed heel, the pressure is

$$P = \frac{F}{A} = \frac{160 \text{ lb}}{\frac{1}{4} \text{ in}^2} = 640 \text{ lb/in}^2$$

A similar comparison may be shown using metric units. If a 65-kg woman rests her weight on a 25-cm^2 heel, the pressure is

$$P = \frac{F}{A}$$

$$P = \frac{(65 \text{ kg})(9.80 \text{ m/s}^2)}{25 \text{ cm}^2 \times \dfrac{1 \text{ m}^2}{10^4 \text{ cm}^2}} = 2.5 \times 10^5 \text{ N/m}^2 \qquad (1 \text{ N} = 1 \text{ kg m/s}^2)$$

$$= 2.5 \times 10^5 \text{ Pa} = 250 \text{ kPa}$$

If she rests her weight on a pointed heel of 1.0 cm^2, the pressure is

$$P = \frac{F}{A}$$

$$P = \frac{(65 \text{ kg})(9.80 \text{ m/s}^2)}{1.0 \text{ cm}^2 \times \dfrac{1 \text{ m}^2}{10^4 \text{ cm}^2}} = 6.4 \times 10^6 \text{ N/m}^2$$

$$= 6.4 \times 10^6 \text{ Pa} = 6400 \text{ kPa}$$

Since the pascal is a relatively small unit, the kilopascal (kPa) is a commonly used unit of pressure.

Pressure of Liquids

Liquids present a slightly different situation. As you probably know, the pressure increases as you go deeper in water. Liquids are different in this respect from solids in that, whereas solids exert only a downward force due to gravity, the force exerted by liquids is in all directions. The pressure a liquid exerts on a submerged object is called **hydrostatic pressure** (Fig. 12.3).

The pressure in a liquid depends *only* on the depth and weight density of the liquid and *not* on the surface area. Because the pressure exerted by water increases with depth, dams are built much thicker at the base than at the top (Fig. 12.4).

To find the pressure at a given depth in a liquid, use the formula

$$P = hD_w$$

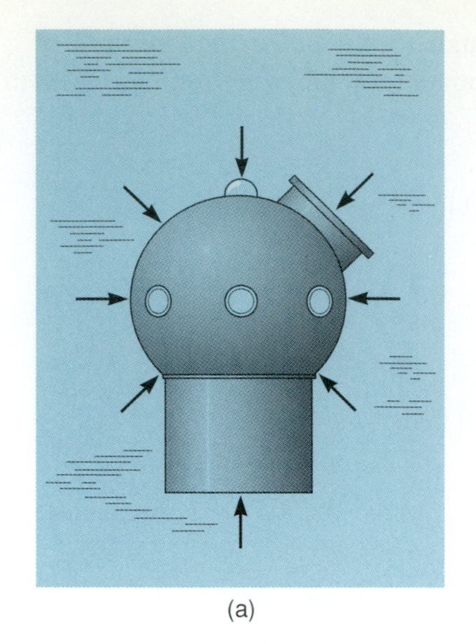

(a)

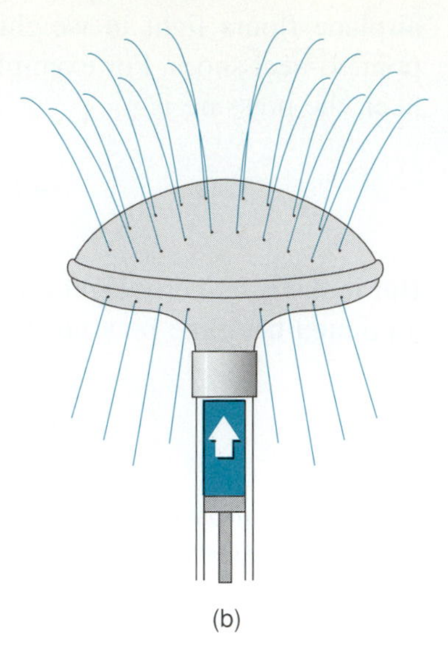

(b)

Figure 12.3
(a) The pressure on a submerged object is exerted by the liquid in all directions. The pressure on the ocean floor is so great that very strong containers must be built to withstand the enormous pressures. (b) When a force is exerted on a liquid in a container, the force is exerted uniformly in all directions. This device has holes, which show that the liquid squirts out uniformly in all directions.

Figure 12.4
Dams must be built much thicker at the base because the pressure exerted by the water increases with depth.

where P = pressure
h = height (or depth)
D_w = weight density of the liquid

EXAMPLE 1

Find the pressure at the bottom of a water-filled drum 4.00 ft high.

Sketch:

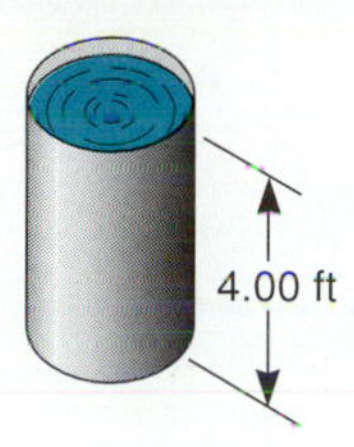

Data:

$$h = 4.00 \text{ ft}$$
$$D_w = 62.4 \text{ lb/ft}^3$$
$$P = ?$$

Basic Equation:

$$P = hD_w$$

Working Equation: Same

Substitution:

$$P = (4.00 \text{ ft})\left(62.4 \frac{\text{lb}}{\text{ft}^3}\right)$$
$$= 25\bar{0} \frac{\text{lb}}{\cancel{\text{ft}^2}} \times \frac{1 \cancel{\text{ft}^2}}{144 \text{ in}^2}$$
$$= 1.74 \text{ lb/in}^2$$

Note: The pressure depends only on the height, not the width or area of the container.

EXAMPLE 2

Find the depth in a lake at which the pressure is 105 lb/in^2.

Sketch: None needed

Data:

$$P = 105 \text{ lb/in}^2$$
$$D_w = 62.4 \text{ lb/ft}^3$$
$$h = ?$$

Basic Equation:

$$P = hD_w$$

Working Equation:

$$h = \frac{P}{D_w}$$

Substitution:

$$h = \frac{105 \text{ lb/in}^2}{62.4 \text{ lb/ft}^3}$$

$$= 1.68 \frac{\text{ft}^3}{\text{in}^2}$$

$$\frac{\text{lb/in}^2}{\text{lb/ft}^3} = \frac{\text{lb}}{\text{in}^2} \div \frac{\text{lb}}{\text{ft}^3} = \frac{\cancel{\text{lb}}}{\text{in}^2} \times \frac{\text{ft}^3}{\cancel{\text{lb}}} = \frac{\text{ft}^3}{\text{in}^2}$$

$$= 1.68 \frac{\overset{\text{ft}}{\cancel{\text{ft}^3}}}{\cancel{\text{in}^2}} \times \frac{144\ \cancel{\text{in}^2}}{1\ \cancel{\text{ft}^2}}$$

$$= 242 \text{ ft}$$

EXAMPLE 3

Find the height of a water column where the pressure at the bottom of the column is $40\bar{0}$ kPa and the weight density of water is $98\bar{0}0$ N/m^3.

Sketch: None needed

Data:

$$P = 40\bar{0} \text{ kPa}$$
$$D_w = 98\bar{0}0 \text{ N/m}^3$$
$$h = ?$$

Basic Equation:

$$P = hD_w$$

Working Equation:

$$h = \frac{P}{D_w}$$

Substitution:

$$h = \frac{40\bar{0} \text{ kPa}}{98\bar{0}0 \text{ N/m}^3}$$

$$= \frac{40\bar{0} \text{ kPa}}{98\bar{0}0 \text{ N/m}^3} \times \frac{10^3 \text{ N/m}^2}{1 \text{ kPa}} \quad (\textit{Recall: } 1 \text{ kPa} = 10^3 \text{ N/m}^2)$$

$$= 40.8 \text{ m}$$

$$\frac{\cancel{\text{kPa}}}{\text{N/m}^3} \times \frac{\text{N/m}^2}{\cancel{\text{kPa}}} = \text{N/m}^2 \div \text{N/m}^3 = \frac{\cancel{\text{N}}}{\text{m}^2} \times \frac{\text{m}^3}{\cancel{\text{N}}} = \text{m}$$

Total Force Exerted by Liquids

The *total force* exerted by a liquid on a horizontal surface (such as the bottom of a barrel) depends on the area of the surface, the depth of the liquid, and the weight density of the liquid. By formula:

$$F_t = AhD_w$$

where F_t = total force
A = area of bottom or horizontal surface
h = height or depth of the liquid
D_w = weight density

EXAMPLE 4

Find the total force on the bottom of a rectangular tank 10.0 ft by 5.00 ft by 4.00 ft deep filled with water.

Sketch:

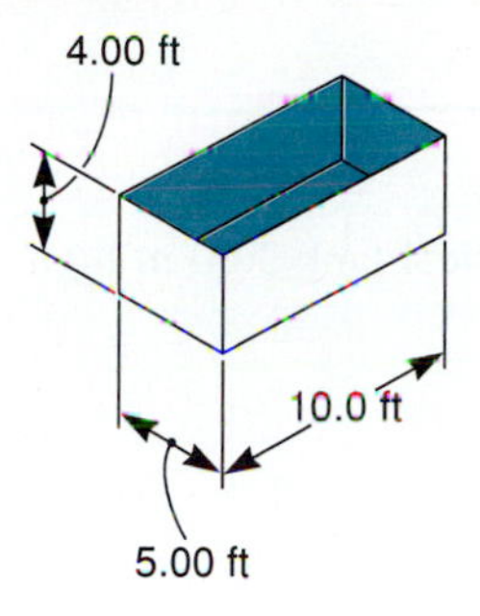

Data:

$$A = lw = (10.0 \text{ ft})(5.00 \text{ ft}) = 50.0 \text{ ft}^2$$
$$h = 4.00 \text{ ft}$$
$$D_w = 62.4 \text{ lb/ft}^3$$
$$F_t = ?$$

Basic Equation:

$$F_t = AhD_w$$

Working Equation: Same

Substitution:

$$F_t = (50.0 \cancel{\text{ft}^2})(4.00 \cancel{\text{ft}})\left(62.4 \frac{\text{lb}}{\cancel{\text{ft}^3}}\right)$$
$$= 12{,}500 \text{ lb}$$

The total force on a vertical surface F_s (such as the *side* of a tank) is found by using half the vertical height (average height):

$$F_s = \tfrac{1}{2}AhD_w$$

where A is the area of the side or vertical surface.

EXAMPLE 5

Find the total force on the end of the rectangular tank in Example 4.

Data:

$$A = lw = (5.00 \text{ ft})(4.00 \text{ ft}) = 20.0 \text{ ft}^2$$
$$h = 4.00 \text{ ft}$$
$$D_w = 62.4 \text{ lb/ft}^3$$
$$F_s = ?$$

Basic Equation:

$$F_s = \tfrac{1}{2}AhD_w$$

Working Equation: Same

Substitution:

$$F_s = \frac{1}{2}(20.0 \text{ ft}^2)(4.00 \text{ ft})(62.4 \text{ lb/ft}^3)$$
$$= 25\bar{0}0 \text{ lb}$$

EXAMPLE 6

Find the total force on the side of a water-filled cylindrical tank 3.00 m high with radius 5.00 m. The weight density of water is $98\bar{0}0$ N/m^3.

Sketch:

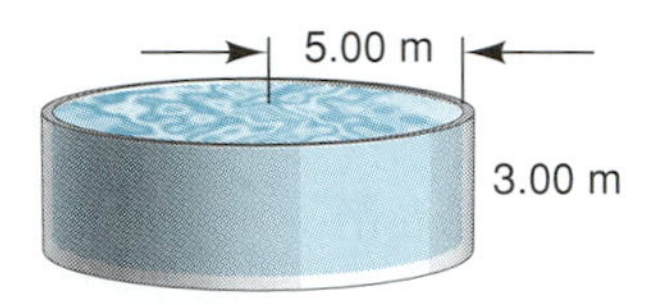

Data:

$A = 2\pi rh = 2\pi(5.00 \text{ m})(3.00 \text{ m})$ — ***Note:*** the area of the vertical surface is the lateral surface of the cylinder.

$h = 3.00$ m
$D_w = 98\bar{0}0$ N/m^3
$F_s = ?$

Basic Equation:

$$F_s = \frac{1}{2}AhD_w$$

Working Equation: Same

Substitution:

$$F_s = \frac{1}{2}[2\pi(5.00 \text{ m})(3.00 \text{ m})](3.00 \text{ m})(98\bar{0}0 \text{ N/m}^3)$$
$$= 1.39 \times 10^6 \text{ N}$$

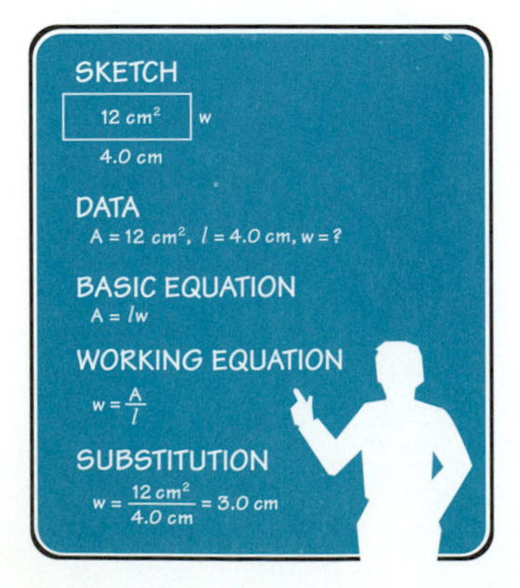

■ PROBLEMS 12.1

1. A packing crate 2.50 m × 0.80 m × 0.45 m weighs 1.41×10^5 N. Find the pressure (in kPa) exerted by the crate on the floor in each of its three possible positions.
2. A packing crate 2.50 m × 20.0 cm × 30.0 cm has a mass of 975 kg. Find the pressure (in kPa) exerted by the crate on the floor in each of its three possible positions.
3. Find the pressure (in lb/in^2) at the bottom of a tower with water 50.0 ft deep.
4. Find the height of a column of water where the pressure at the bottom of the column is 20.0 lb/in^2.
5. Find the density of a liquid that exerts a pressure of 0.400 lb/in^2 at a depth of 42.0 in.
6. Find the total force on the bottom of a water-filled circular cattle tank 0.750 m high with a radius of 1.30 m where the weight density of water is $98\bar{0}0$ N/m^3.
7. Find the total force on the side of the tank in Problem 6.

8. What must the water pressure be to supply water to the third floor of a building (35.0 ft up) with a pressure of 40.0 lb/in^2 at that level?
9. A small tank 5.00 in. by 9.00 in. is filled with mercury. If the total force on the bottom of the tank is 165 lb, how deep is the mercury? (Weight density of mercury = 0.490 lb/in^3.)
10. Find the total force on the largest side of the tank in Problem 9.
11. Find the water pressure (in kPa) at the 25.0-m level of a water tower containing water 50.0 m deep.
12. Find the height of a column of water where the pressure at the bottom is 115 kPa.
13. What is the mass density of a liquid that exerts a pressure of 178 kPa at a depth of 24.0 m?
14. Find the total force on the bottom of a cylindrical gasoline storage tank 15.0 m high with a radius of 23.0 m.
15. Find the total force on the side of the tank in Problem 14.
16. What must the water pressure be to supply the second floor (18.0 ft up) with a pressure of 50.0 lb/in^2 at that level?
17. Find the water pressure at ground level to supply water to the third floor of a building 8.00 m high with a pressure of 325 kPa at the third-floor level.
18. What pressure must a pump supply to pump water up to the thirtieth floor of a skyscraper with a pressure of 25 lb/in^2? Assume that the pump is located on the first floor and that there are 16.0 ft between floors.
19. A submarine is submerged to a depth of 3550 m in the Pacific Ocean. What air pressure (in kPa) is needed to blow water out of the ballast tanks?

A filled water tower sits on the top of the highest hill in a town (use Fig. 12.5 for Problems 20–24). The cylindrical tower has a radius of 12.0 m and a height of 50.0 m.

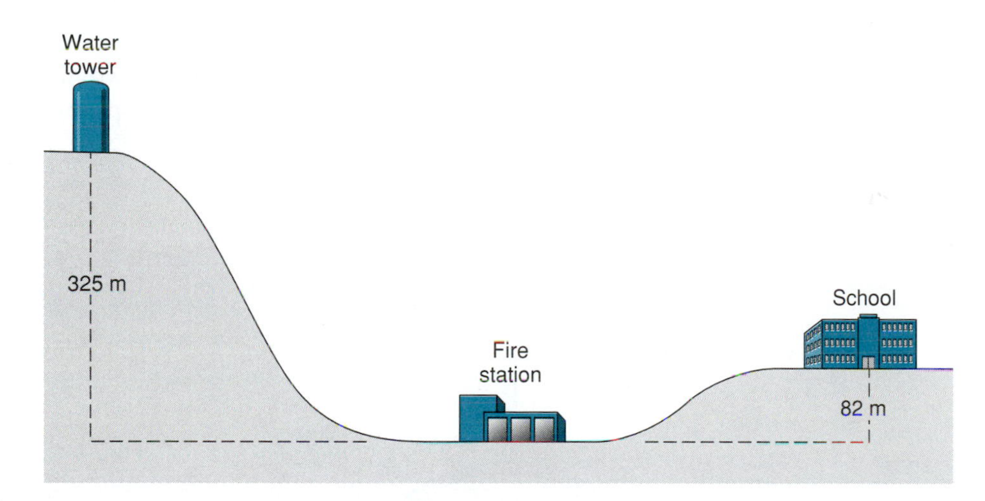

Figure 12.5

20. Find the total force on the bottom of the water tower.
21. Find the total force on the sides of the water tower.
22. Find the pressure (in kPa) on the bottom of the water tower.
23. What is the water pressure (in kPa) at the fire station?
24. What is the water pressure (in kPa) at the school?
25. Find the height of a 20.0-cm-diameter pipe filled with water closed at one end with a force of 115 N applied to the other.
26. A cylindrical grain bin 24.0 ft in diameter is filled with corn whose weight density is 451 lb/ft^3. How tall can the bin be for the floor to support 94.0 lb/in^2 of pressure?

12.2 HYDRAULIC PRINCIPLE

Put a stopper in one end of a metal pipe. Fill it with a fluid such as water. Then put a second stopper into the open end. Put the pipe in a horizontal position as in Fig. 12.6 and push on stopper *A*. What happens? Stopper *B* is pushed out. This illustrates a basic principle of hydraulics: The liquid in the pipe transmits the pressure from one stopper or piston to another without measurable loss.

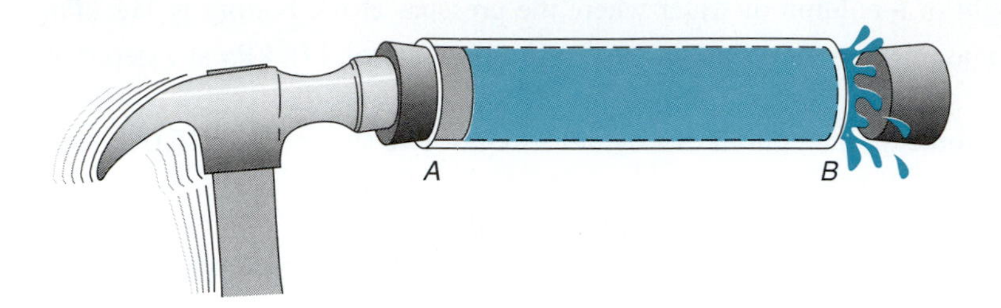

Figure 12.6
Pressure on a confined liquid is transmitted in all directions without measurable loss.

> *HYDRAULIC PRINCIPLE (PASCAL'S PRINCIPLE)*
>
> Pressure applied to a confined liquid is transmitted without measurable loss throughout the entire liquid to all inner surfaces of the container.

The hydraulic brake system of an automobile (Fig. 12.7) is an application of **Pascal's principle.** When the driver pushes the brake pedal, the pressure on a piston on the master cylinder is transmitted through the brake fluid to the two pistons in the brake cylinder. This transmitted pressure then forces the two pistons

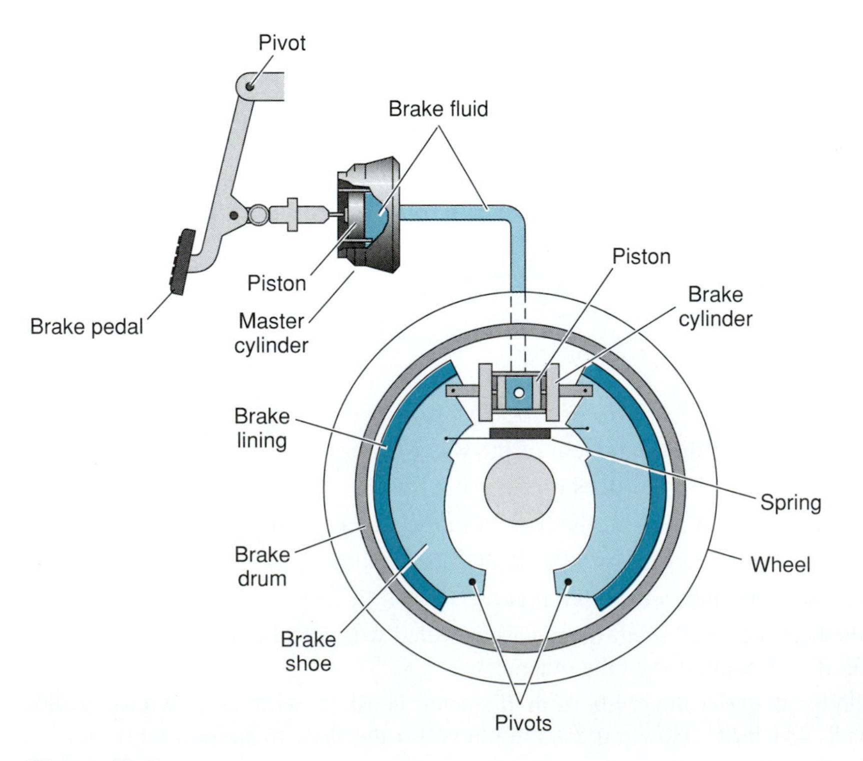

Figure 12.7
Hydraulic brake system of an automobile.

to move outward, forcing the brake shoes against the brake drum and stopping the automobile. Releasing the brake pedal releases the pressure on the pistons in the brake cylinder. The spring pulls the brake shoes away from the brake drum, which allows the wheel to turn freely again.

The hydraulic jack and press are applications of how hydraulics can be used as a simple machine to multiply force. If we apply a force to the small piston of the hydraulic lift in Fig. 12.8, the pressure is transmitted without measurable loss in all directions. The reason for this is the virtual noncompressibility of liquids. The *pressure* on the large piston is the same as the pressure on the small piston; however, the *total force* on the large piston is greater because of its larger surface area.

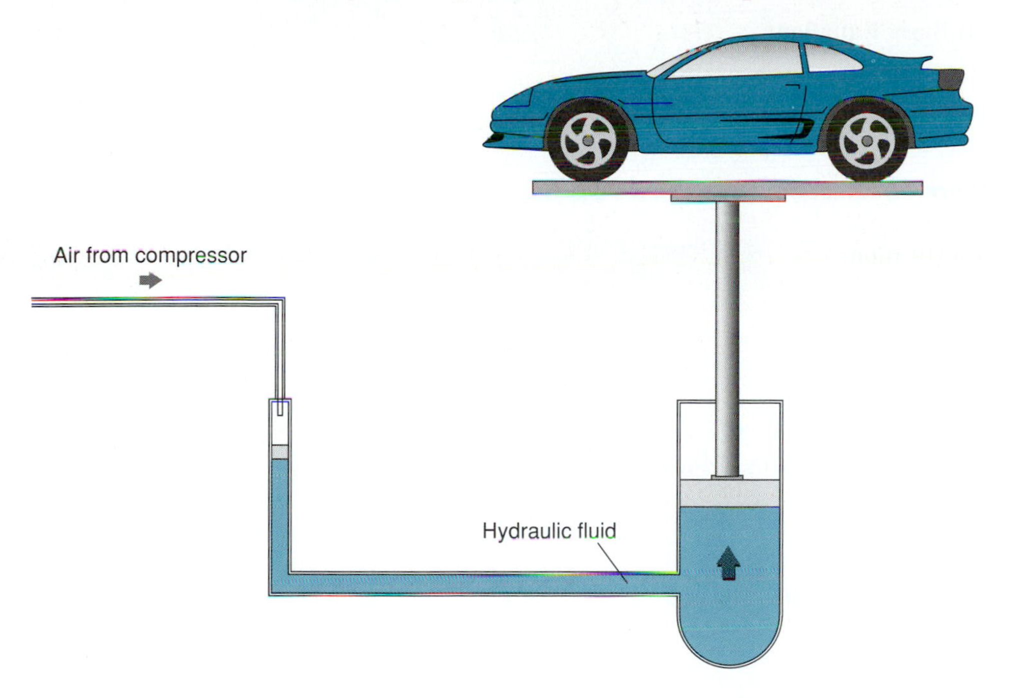

Figure 12.8
Hydraulic lift.

EXAMPLE 1

From the diagram of the hydraulic jack in Fig. 12.9, find

(a) the pressure on the small piston.
(b) the pressure on the large piston.
(c) the total force on the large piston.
(d) the mechanical advantage of the jack.

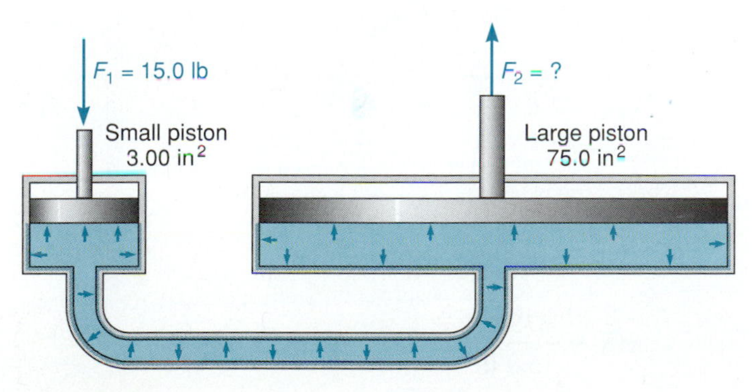

Figure 12.9
Hydraulic jack.

Data:

$$F_1 = 15.0 \text{ lb}$$
$$A_{\text{small piston}} = 3.00 \text{ in}^2$$
$$A_{\text{large piston}} = 75.0 \text{ in}^2$$
$$P_1 = ?$$
$$P_2 = ?$$
$$F_2 = ?$$
$$\text{MA} = ?$$

(a) Basic Equation:

$$P_1 = \frac{F_1}{A}$$

Working Equation: Same

Substitution:

$$P_1 = \frac{15.0 \text{ lb}}{3.00 \text{ in}^2}$$
$$= 5.00 \text{ lb/in}^2$$

(b) Applying Pascal's Principle:

$$P_2 = P_1 = 5.00 \text{ lb/in}^2$$

(c) Basic Equation:

$$P_2 = \frac{F_2}{A}$$

Working Equation:

$$F_2 = P_2 A$$

Substitution:

$$F_2 = \left(5.00 \frac{\text{lb}}{\cancel{\text{in}^2}}\right)(75.0 \cancel{\text{in}^2})$$
$$= 375 \text{ lb}$$

(d) Basic Equation:

$$\text{MA} = \frac{F_R}{F_E}$$

Working Equation: Same

Substitution:

$$\text{MA} = \frac{375 \cancel{\text{lb}}}{15.0 \cancel{\text{lb}}}$$
$$= 25.0$$

EXAMPLE 2

The small piston of a hydraulic press has an area (A_1) of 10.0 cm^2. If the applied force (F_1) is 50.0 N, what must the area of the large piston (A_2) be to exert a pressing force of 4800 N (F_2)?

Sketch: None needed

Data:

$$A_1 = 10.0 \text{ cm}^2$$
$$F_1 = 50.0 \text{ N}$$
$$F_2 = 4800 \text{ N}$$
$$A_2 = ?$$

Basic Equations:

$$P_1 = \frac{F_1}{A_1}, \qquad P_2 = \frac{F_2}{A_2}, \qquad \text{and since } P_1 = P_2, \qquad \frac{F_1}{A_1} = \frac{F_2}{A_2}$$

Working Equation:

$$A_2 = \frac{A_1 F_2}{F_1}$$

Substitution:

$$A_2 = \frac{(10.0 \text{ cm}^2)(4800 \cancel{\text{N}})}{50.0 \cancel{\text{N}}}$$
$$= 960 \text{ cm}^2$$

■ PROBLEMS 12.2

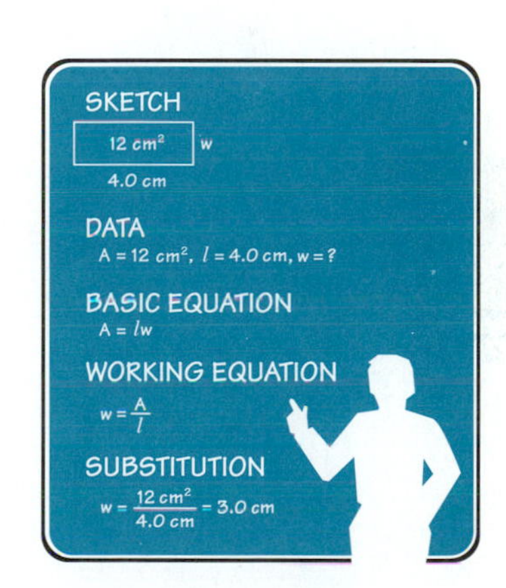

1. The area of the small piston in a hydraulic jack is 0.750 in^2. The area of the large piston is 3.00 in^2. If a force of 15.0 lb is applied to the small piston, what weight can be lifted by the large one?
2. The mechanical advantage of a hydraulic press is 25. What applied force is necessary to produce a pressing force of 2400 N?
3. Find the mechanical advantage of a hydraulic press that produces a pressing force of 8250 N when the applied force is 375 N.
4. The mechanical advantage of a hydraulic press is 18. What applied force is necessary to produce a pressing force of 990 lb?
5. Find the mechanical advantage of a hydraulic press that produces a pressing force of 1320 N when the applied force is 55.0 N.
6. The small piston of a hydraulic press has an area of 8.00 cm^2. If the applied force is 25.0 N, what must the area of the large piston be to exert a pressing force of $36\bar{0}0$ N?
7. The MA of a hydraulic jack is 250. What force must be applied to lift an automobile weighing 12,000 N?
8. The small piston of a hydraulic press has an area of 4.00 in^2. If the applied force is 10.0 lb, what must the area of the large piston be to exert a pressing force of 865 lb?
9. The MA of a hydraulic jack is 420. Find the weight of the heaviest automobile that can be lifted by an applied force of 55 N.

10. The mechanical advantage of a hydraulic jack is 450. Find the weight of the heaviest automobile that can be lifted by an applied force of 60.0 N.
11. The pistons of a hydraulic press have radii of 2.00 cm and 12.0 cm.
    (a) What force must be applied to the smaller piston to exert a force of 5250 N on the larger?
    (b) What is the pressure (in N/cm^2) on each piston?
    (c) What is the mechanical advantage of the press?
12. The small circular piston of a hydraulic press has an area of $8.00\ cm^2$. If the applied force is 25.0 N, what must the area of the large piston be to exert a pressing force of 3650 N?
13. The large piston on a hydraulic lift has radius 40.0 cm. The small piston has radius 5.00 cm to which a force of 75.0 N is applied. Find
    (a) the force exerted by the large piston.
    (b) the pressure on the large piston.
    (c) the pressure on the small piston.
    (d) the mechanical advantage of the lift.
    (e) What happens when the area of the small piston is half as large?
    (f) What happens when the radius of the small piston is half as large? ■

12.3 AIR PRESSURE

Since air has weight, as any fluid, it must exert pressure. The atmosphere exerts pressure on objects on the surface of the earth. This atmospheric pressure can be illustrated by using a bell jar with a hole in the top over which a thin rubber membrane is stretched as in Fig. 12.10(a). As air is pumped out of the bell jar, the inside air pressure is reduced by removing a number of air molecules. Thus, there are fewer molecular bombardments on the inside surface of the rubber membrane than there are on its outside surface. The outside air pressure, now greater than the inside air pressure, pushes the rubber membrane down into the bell jar. When

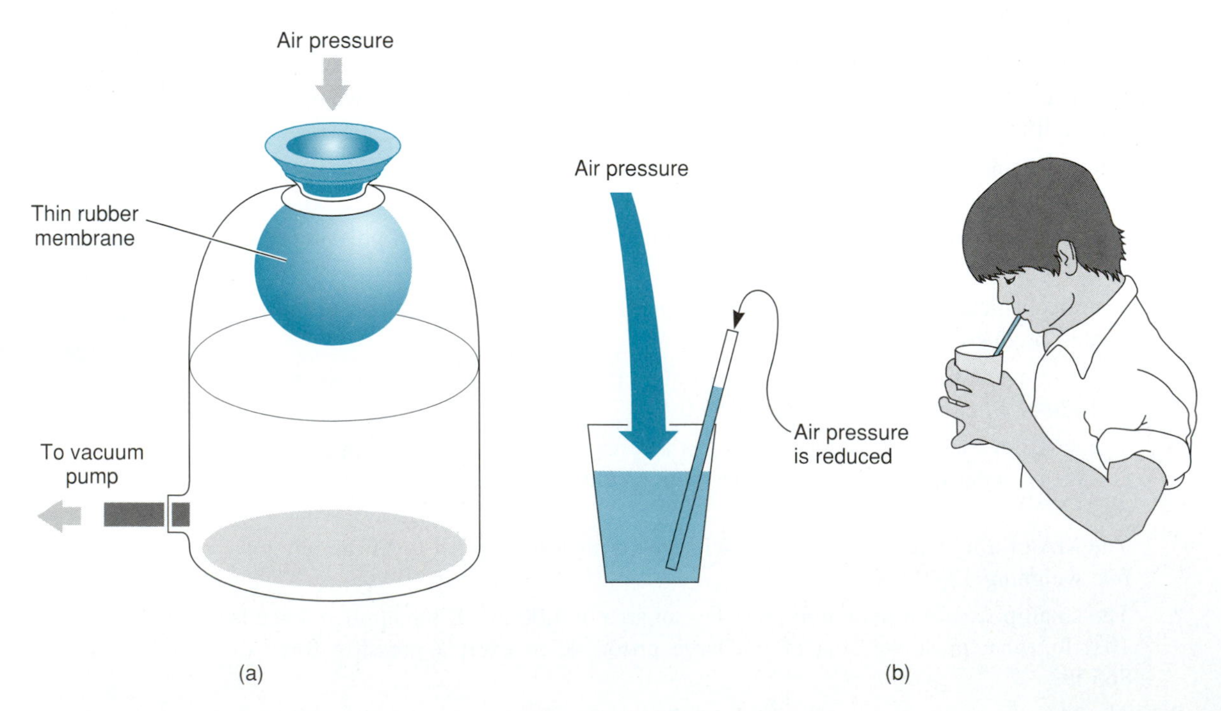

Figure 12.10
Effects of air pressure.

a straw is used to drink, the air pressure inside the straw is reduced [Fig. 12.10(b)]. As a result, the outside air pressure is higher than the pressure in the straw, which forces the fluid up the straw.

In Section 12.1 we saw that the pressure on a submerged body increases as the body goes deeper into the liquid. Some creatures live near the bottom of the ocean, where the pressure of the water above is so great that it would collapse any human body and most submarines, but through the process of evolution, such creatures have adapted to this tremendous pressure. Similarly, we on earth live at the bottom of a fluid, air, that is several miles deep. The pressure from this fluid is normally 14.7 lb/in^2 or 101.32 kPa at sea level. We do not feel this pressure because it normally is almost the same from all directions and also because our bodies have become accustomed to it. **Atmospheric pressure** is the pressure felt resulting from the weight of the air in the atmosphere. When the air pressure becomes unequal, its force becomes quite evident in the form of wind. This wind may be a cool summer breeze or the tremendous concentrated force of a tornado.

What is the *pressure of our atmosphere* equivalent to? Experiments have shown that the atmosphere supports a column of water 33.9 ft high in a tube in which the air has been removed. The atmosphere supports 29.9 in. of mercury in a similar tube (Fig. 12.11). This is not surprising; mercury is 13.6 times as dense as water and

$$\frac{1}{13.6} \times 33.9 \text{ ft} \times \frac{12.0 \text{ in.}}{1.00 \text{ ft}} = 29.9 \text{ in.}$$

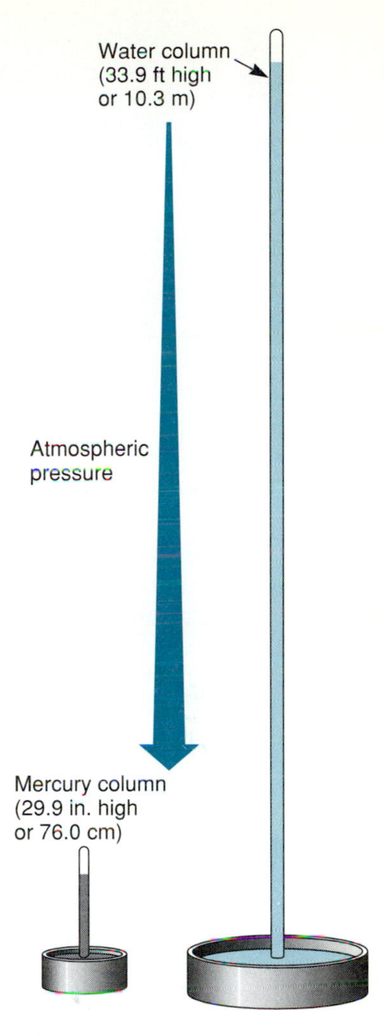

Figure 12.11
The barometer measures air pressure.

The height of the mercury column in a barometer is independent of the width (or diameter or cross-sectional area) of the barometer tube. This "inches of mercury" measurement has been standard for many years on TV weather programs but is increasingly being replaced by the metric standard measurement in kilopascals (kPa).

The pressure of the atmosphere can be expressed in terms of the pressure of an equivalent column of mercury. Air pressure at sea level is normally 29.9 in. or 76.0 cm or 76$\bar{0}$ mm of mercury. How do we arrive at the 14.7 lb/in^2 measurement? In terms of mercury, its height is 29.9 in. or 2.49 ft. Its density is 13.6 × 62.4 lb/ft^3 or 849 lb/ft^3. Therefore,

$$P = hD_w$$

$$P = 2.49 \text{ ft} \times 849 \frac{\text{lb}}{\text{ft}^3} \times \frac{1.00 \text{ ft}^2}{144 \text{ in}^2}$$

$$= 14.7 \text{ lb/in}^2 \text{ at standard temperature}$$

The pressure of 2 atm would be 29.4 lb/in^2 or 202.64 kPa. If the pressure is $\frac{1}{2}$ atm at one point in the sky, it would be 7.35 lb/in^2 or 50.66 kPa.

Different types of gauges read either atmospheric or gauge pressure. When we purchase bottled gas, the amount of gas and its density vary with the pressure. If the pressure is low, the amount of gas in the bottle is low. If the pressure is high, the bottle is "nearly full." The *gauge* that is usually used for checking the pressure in bottles and tires shows a reading of zero at normal atmospheric pressure. The pressure of the atmosphere is not included in this reading. Thus, **gauge pressure** is the amount of air pressure excluding the normal atmospheric pressure. The actual pressure, called **absolute pressure,** is the gauge pressure reading plus the normal atmospheric pressure, 101.32 kPa or 14.7 lb/in^2. That is,

$$\text{absolute pressure} = \text{gauge pressure} + \text{atmospheric pressure}$$

or

$$P_{abs} = P_{ga} + P_{atm}$$

where $P_{atm} = 101.32$ kPa or 14.7 lb/in^2 at standard temperature.

EXAMPLE

What is the absolute pressure in a tire inflated to 32.0 lb/in^2

(a) in lb/in^2?

(b) in kPa?

(a) Sketch: None needed

Data:

$$P_{ga} = 32.0 \text{ lb/in}^2$$
$$P_{atm} = 14.7 \text{ lb/in}^2$$
$$P_{abs} = ?$$

Basic Equation:

$$P_{abs} = P_{ga} + P_{atm}$$

Working Equation: Same

Substitution:

$$P_{abs} = 32.0 \text{ lb/in}^2 + 14.7 \text{ lb/in}^2$$
$$= 46.7 \text{ lb/in}^2$$

(b) We use the conversion factor:

$$101.32 \text{ kPa} = 14.7 \text{ lb/in}^2$$

Therefore,

$$P_{abs} = 46.7 \cancel{\text{lb/in}^2} \times \frac{101.32 \text{ kPa}}{14.7 \cancel{\text{lb/in}^2}}$$
$$= 322 \text{ kPa}$$

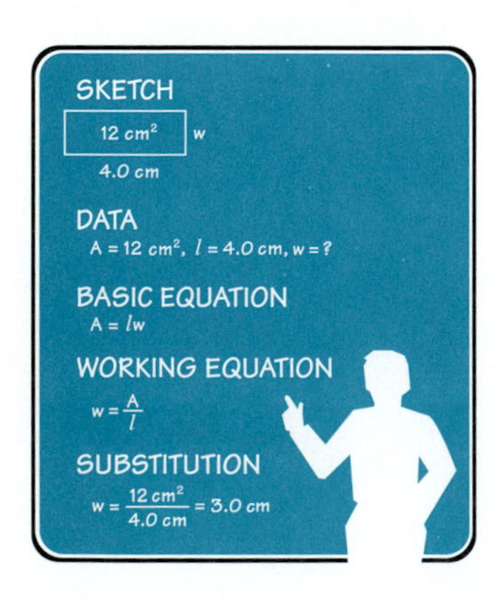

■ PROBLEMS 12.3

1. Change 815 kPa to lb/in^2.
2. Change 64.3 lb/in^2 to kPa.
3. Change 42.5 lb/in^2 to kPa.
4. Change 215 kPa to lb/in^2.
5. Find the pressure of
   (a) 3 atm (in kPa).
   (b) 2 atm (in kPa).
   (c) 6 atm (in lb/in^2).
   (d) 5 atm (in kPa).
   (e) $\frac{1}{3}$ atm (in kPa).
   (f) $\frac{1}{4}$ atm (in kPa).
6. A barometer in the Rocky Mountains reads 516 mm of mercury. Find this pressure (a) in kPa and (b) in lb/in^2.
7. Find the absolute pressure in a bicycle tire with a gauge pressure of 485 kPa.
8. Find the absolute pressure of a motorcycle tire with a gauge pressure of 255 kPa.
9. Find the gauge pressure of a tire with an absolute pressure of 45.0 lb/in^2.
10. Find the gauge pressure of a tire with an absolute pressure of 425 kPa.
11. Find the absolute pressure of a tire gauge that reads 205 kPa.
12. Find the absolute pressure of a tank whose gauge pressure reads 362 lb/in^2.
13. Find the gauge pressure of a tank whose absolute pressure is 1275 kPa.
14. Find the gauge pressure of a tank whose absolute pressure is 218 lb/in^2.

15. Find the absolute pressure of a cycle tire with gauge pressure 3.00×10^5 Pa.
16. Find the absolute pressure in a hydraulic jack with a small piston of area 23.0 cm^2 when a force of 125 N is applied. The area of the large piston is 46.0 cm^2. ■

12.4 BUOYANCY

A floating boat displaces an amount of water equal to its own weight. Three people in a boat displace more water than when only one person is in the boat. The boat rides lower due to the increased weight (Fig. 12.12).

Figure 12.12
A floating boat displaces an amount of water equal to its weight.

Archimedes, a Greek, was one of the first to study fluids and formulated what is now called **Archimedes' principle.**

ARCHIMEDES' PRINCIPLE

Any object placed in a fluid apparently loses weight equal to the weight of the displaced fluid.

In Fig. 12.13, the object floats on water. What happens to the weight of a brick or some other object that sinks in water? First, weigh the brick in air. Then, lower the brick under the water and weigh it again. It weighs less. The difference between the two weights is the buoyant (upward) force of the water. That is, the **buoyant force** is the upward force exerted on a submerged or partially submerged object. This is illustrated in the following example.

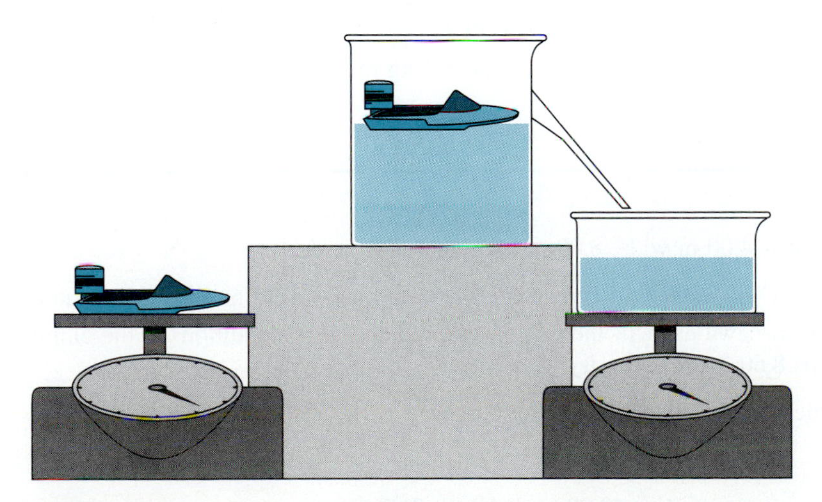

Figure 12.13
Weight of object = weight of displaced water.

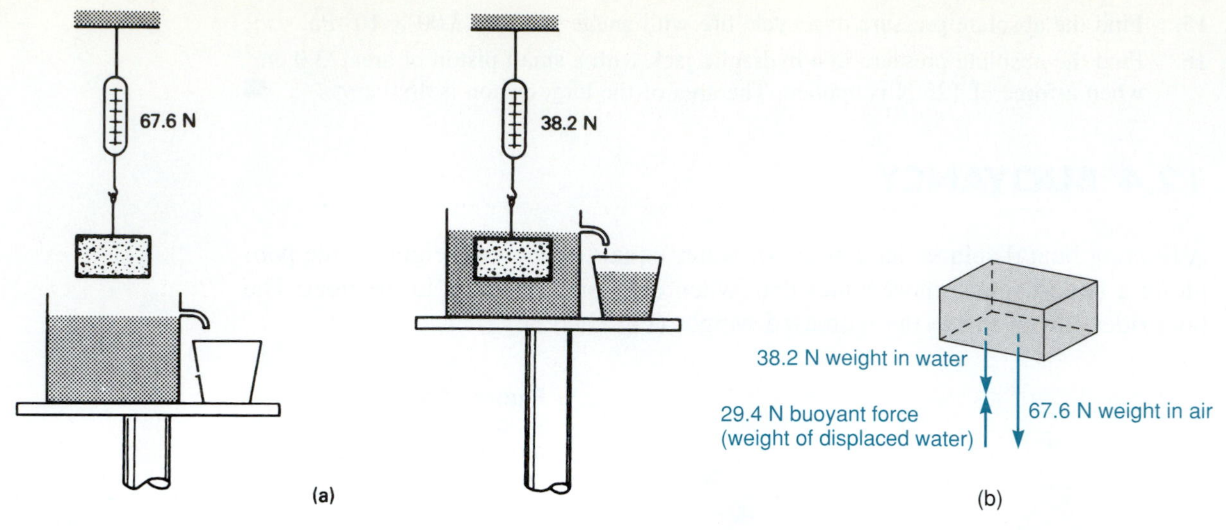

Figure 12.14
Weight in water = weight in air − buoyant force (weight of displaced water).

EXAMPLE 1

A solid concrete block 15.0 cm × 20.0 cm × 10.0 cm weighs 67.6 N in air (Fig. 12.14). When it is lowered into the water, it weighs 38.2 N. The buoyant force is 67.6 N − 38.2 N = 29.4 N. The volume of the displaced water is

$$V = lwh$$
$$V = 15.0 \text{ cm} \times 20.0 \text{ cm} \times 10.0 \text{ cm}$$
$$= 30\bar{0}0 \text{ cm}^3 = 3.00 \times 10^{-3} \text{ m}^3$$

The mass of the displaced water is

$$m = D_m V$$
$$m = (100\bar{0} \text{ kg/}\cancel{\text{m}^3})(3.00 \times 10^{-3} \cancel{\text{m}^3})$$
$$= 3.00 \text{ kg}$$

The weight of the displaced water is then

$$F_w = mg$$
$$F_w = (3.00 \text{ kg})(9.80 \text{ m/s}^2)$$
$$= 29.4 \text{ N} \qquad (1 \text{ N} = 1 \text{ kg m/s}^2)$$

which equals the buoyant force.

EXAMPLE 2

A rectangular boat is 4.00 m wide, 8.00 m long, and 3.00 m deep.

(a) How many m^3 of water will it displace if the top stays 1.00 m above the water?

(b) What load (in newtons) will the boat contain under these conditions if the empty boat weighs 8.60×10^4 N in dry dock?

(a) The volume of water displaced by the boat is

$$V = lwh$$
$$V = (8.00 \text{ m})(4.00 \text{ m})(2.00 \text{ m})$$
$$= 64.0 \text{ m}^3$$

(b) The load of the boat is the buoyant force of the displaced water (D_wV) minus the weight of the boat in dry dock.

$$(98\bar{0}0 \text{ N/m}^3)(64.0 \text{ m}^3) - (8.60 \times 10^4 \text{ N}) = 5.41 \times 10^5 \text{ N}$$

Note: The weight density of water is $98\bar{0}0$ N/m^3.

Archimedes' principle applies to gases as well as liquids. Lighter-than-air craft (such as the Goodyear blimps) operate on this principle. Since they are filled with a gas lighter than air (helium), the buoyant force on them causes them to be supported by the air. Being "submerged" in the air, a blimp is buoyed up by the weight of the air it displaces, which equals the buoyant force of the air on the balloon.

PROBLEMS 12.4

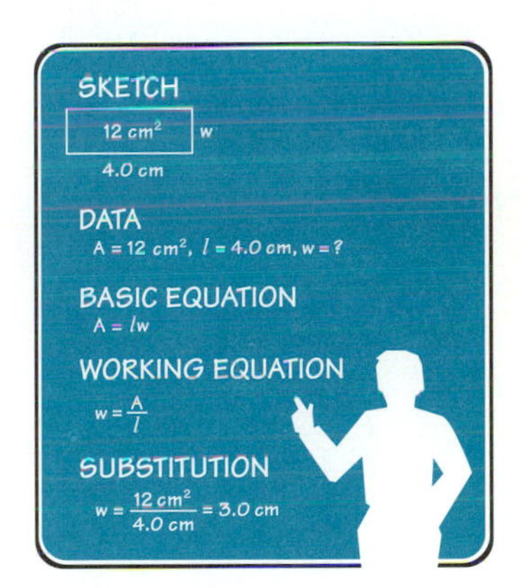

1. A metal alloy weighs 81.0 lb in air and 68.0 lb when under water. Find the buoyant force of the water.
2. A piece of metal weighs 67.0 N in air and 62.0 N in water. Find the buoyant force of the water.
3. A rock weighs 25.7 N in air and 21.8 N in water. What is the buoyant force of the water?
4. A metal bar weighs 455 N in air and 437 N in water. What is the buoyant force of the water?
5. A rock displaces 1.21 ft^3 of water. What is the buoyant force of the water?
6. A metal displaces 16.8 m^3 of water. Find the buoyant force of the water.
7. A metal casting displaces 327 cm^3 of water. Find the buoyant force of the water.
8. A piece of metal displaces 657 cm^3 of water. Find the buoyant force of the water.
9. A metal casting displaces 2.12 ft^3 of alcohol. Find the buoyant force of the alcohol.
10. A metal cylinder displaces 515 cm^3 of gasoline. Find the buoyant force of the gasoline.
11. A 75.0-kg rock lies at the bottom of a pond. Its volume is 3.10×10^4 cm^3. How much force is needed to lift the rock?
12. A 125-lb rock lies at the bottom of a pond. Its volume is 0.800 ft^3. How much force is needed to lift the rock?
13. A flat-bottom river barge is 30.0 ft wide, 85.0 ft long, and 15.0 ft deep.
    (a) How many ft^3 of water will it displace while the top stays 3.00 ft above the water?
    (b) What load in tons will the barge contain under these conditions if the empty barge weighs $16\bar{0}$ tons in dry dock?
14. A flat-bottom river barge is 12.0 m wide, 30.0 m long, and 6.00 m deep.
    (a) How many m^3 of water will it displace while the top stays 1.00 m above the water?
    (b) What load (in newtons) will the barge contain under these conditions if the empty barge weighs 3.55×10^6 N in dry dock?
15. What is the volume (in m^3) of the water displaced by a submerged air tank that is acted on by a buoyant force of 7.50×10^4 N?

12.5 FLUID FLOW

Think for a minute about the motion of water flowing down a fast-moving mountain stream that contains boulders and rapids and about the motion of the air during a thunderstorm or during a tornado. These types of motion are complex, indeed. Our discussion will focus on the simpler examples of fluid flow.

Streamline flow, also known as *laminar flow,* is the smooth flow of a fluid through a tube (Fig. 12.15). By smooth flow we mean that all particles of the fluid follow the same uniform path. **Turbulent flow,** also known as *nonlaminar flow,* is the erratic, unpredictable flow of a fluid resulting from excessive speed of the flow or sudden changes in direction or size of the tube.

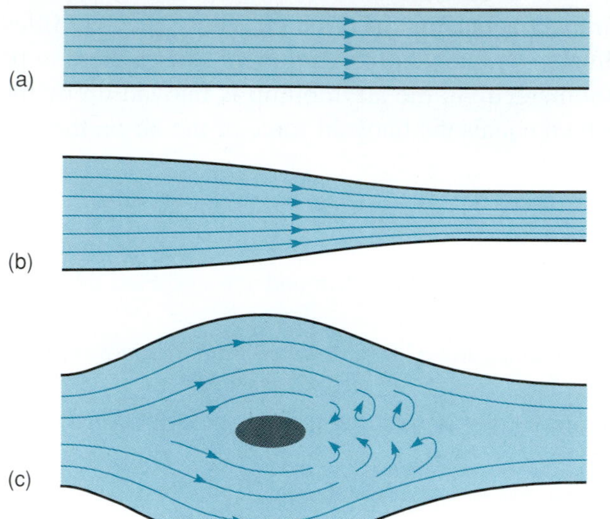

Figure 12.15 Streamline flow of a fluid through a smooth tube or pipe is shown in (a) and (b). Water flowing in a creek or a mountain stream over and around rocks, resulting in sudden changes in direction and speed, is a common example of turbulent flow as in (c).

The **flow rate** of a fluid is the volume of fluid flowing past a given point in a pipe per unit time. Assume that we have a streamline flow through a straight section of pipe at speed v. During a time interval of t, each particle of fluid travels a distance vt. If A is the cross-sectional area of the pipe, the volume of fluid passing a given point during the time interval t is vtA. Thus, the flow rate, Q, is given by

$$Q = \frac{vtA}{t}$$

or

$$Q = vA$$

where Q = flow rate
v = velocity of the fluid
A = cross-sectional area of the tube or pipe

EXAMPLE

Water flows through a fire hose of diameter 6.40 cm at a velocity of 5.90 m/s. Find the flow rate of the fire hose in L/min.

Data:

$$v = 5.90 \text{ m/s}$$

$$r = 3.20 \text{ cm} = 0.0320 \text{ m}$$

$$A = \pi r^2 = \pi(0.0320 \text{ m})^2$$

$$Q = ?$$

Basic Equation:

$$Q = vA$$

Working Equation: Same

Substitution:

$$Q = \left(5.90 \frac{\cancel{m}}{\cancel{s}}\right) \pi(0.0320 \cancel{m})^2 \times \frac{10^3 \text{ L}}{1 \cancel{m^3}} \times \frac{60 \cancel{s}}{1 \text{ min}}$$

$$= 1140 \text{ L/min}$$

For an incompressible fluid, the flow rate is constant throughout the pipe. If the cross-sectional area of the pipe changes and streamline flow is maintained, the flow rate is the same all along the pipe. That is, as the cross-sectional area increases, the velocity decreases, and vice versa (Fig. 12.16).

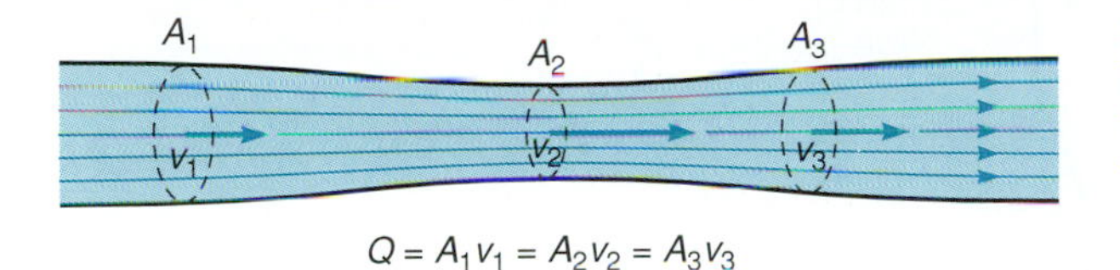

Figure 12.16
For an incompressible fluid, the flow rate is constant throughout.

What happens to the pressure as the cross-sectional area of the pipe changes? This concept can be illustrated by use of a Venturi meter (Fig. 12.17). Here the vertical tubes act like pressure gauges; the higher the column, the higher the pressure. As you can see, the higher the speed, the lower the pressure, and vice versa. This change in pressure of a fluid in streamline flow was first explained by Daniel Bernoulli (1700–1782).

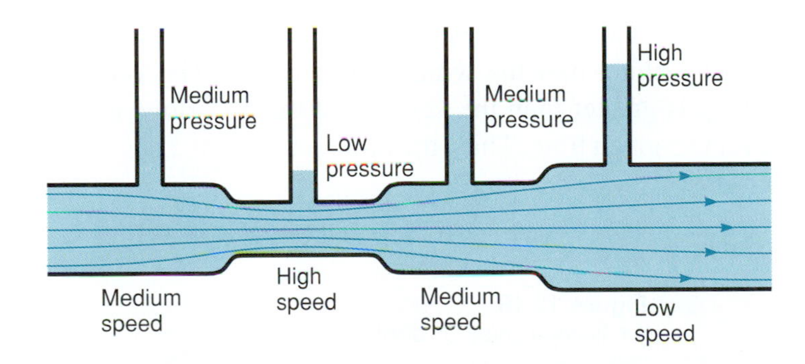

Figure 12.17
A Venturi meter shows that the higher the speed of a fluid through a tube, the lower the pressure; the lower the speed of a fluid, the higher the pressure.

Do you recall walking into the wind in winter on a city street lined with several tall buildings when the wind seemed stronger than usual? The wind was actually stronger because it was acting as a fluid as per the Venturi meter principle in Fig. 12.17. That is, the speed of the wind increased as it was forced to flow between the tall buildings along the street.

BERNOULLI'S PRINCIPLE

For the horizontal flow of a fluid through a tube, the sum of the pressure and energy of motion (kinetic energy) per unit volume of the fluid is constant.

One application of **Bernoulli's principle** involves an automobile carburetor (Fig. 12.18). The volume of airflow is determined by the position of the butterfly valve. As the air flows through the throat, the air gains speed and loses pressure. The pressure in the fuel bowl equals the pressure above the throat. Due to the difference in pressure between the fuel bowl and the throat, gasoline is drawn into and is mixed with the airstream. The reduced pressure in the throat also helps the gasoline to vaporize.

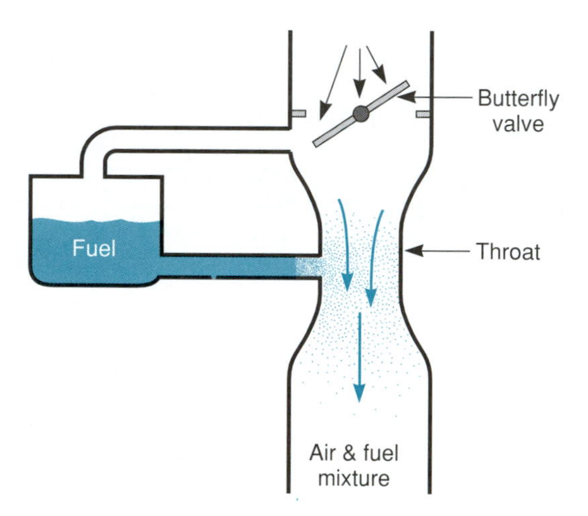

Figure 12.18 Automobile carburetor.

Another application of Bernoulli's principle involves airplane travel. Figure 12.19 shows the flow of air rushing past the wing of an airplane. The velocity, v_1, of the air above the wing is greater than the velocity of the air below, v_2, because it has farther to travel in a given time. Thus, the pressure at point 2 is greater, which causes lift on the wing.

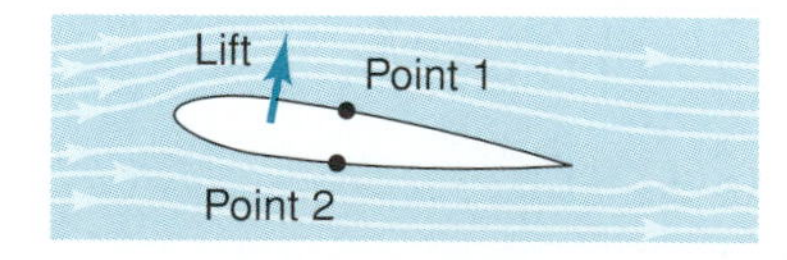

Figure 12.19 Air flowing past an airplane wing creates lift.

An airplane requires a longer distance for takeoff in summer than in winter. Why? Hot air is less dense and has fewer air molecules to lift the plane as it moves down the runway.

Two other examples that illustrate Bernoulli's principle are a curving baseball and a paint spray gun. The spin of the ball causes the air on either side of the ball to be moving past the ball at different speeds. The air pressure on the faster side is reduced, so the ball curves that way (Fig. 12.20).

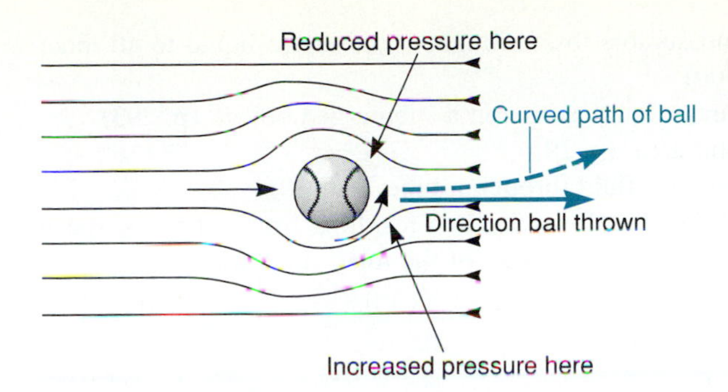

Figure 12.20
Bernoulli's principle explains why a baseball curves.

When the air in a spray gun is accelerated through a narrowing in the line, the pressure is reduced, and paint is drawn into the airstream and is then forced from the gun.

PROBLEMS 12.5

1. Water flows through a hose of diameter 3.90 cm at a velocity of 5.00 m/s. Find the flow rate of the hose in L/min.
2. Water flows through a 15.0-cm fire hose at a rate of 5.00 m/s.
   (a) Find the rate of flow through the hose in L/min.
   (b) How many litres pass through the hose in 30.0 min?
3. Water flows from a pipe at 650 L/min.
   (a) What is the diameter (in cm) of the pipe if the velocity of the water is 1.5 m/s?
   (b) Find the velocity (in m/s) of the water if the diameter of the pipe is 20.0 cm.
4. Water flows through a pipe of diameter 8.00 cm at 45.0 m/min. Find the flow rate (a) in m^3/min and (b) in L/s.
5. A pump is rated to deliver 50.0 gal/min. Find the velocity of water in (a) a 6.00-in.-diameter pipe and (b) a 3.00-in.-diameter pipe.
6. What size pipe needs to be attached to a pump rated at 36.0 gal/min if the desired velocity is 10.0 ft/min? Give the inner diameter in inches.
7. What is the diameter of a pipe in which water travels 32.0 m in 16.0 s and has a flow rate of 2620 L/min?

SKETCH
12 cm^2 w
4.0 cm
DATA
$A = 12\ cm^2$, $l = 4.0\ cm$, $w = ?$
BASIC EQUATION
$A = lw$
WORKING EQUATION
$w = \frac{A}{l}$
SUBSTITUTION
$w = \frac{12\ cm^2}{4.0\ cm} = 3.0\ cm$

GLOSSARY

Absolute Pressure The actual air pressure that is given by the gauge reading plus the normal atmospheric pressure. (p. 305)

Archimedes' Principle States that any object placed in a fluid apparently loses weight equal to the weight of the displaced fluid. (p. 307)

Atmospheric Pressure is the pressure felt resulting from the weight of the air in the atmosphere. (p. 305)

Bernoulli's Principle States that for the horizontal flow of a fluid through a tube, the sum of the pressure and energy of motion (kinetic energy) per unit volume of the fluid is constant. (p. 312)

Buoyant Force The upward force exerted on a submerged or partially submerged object. (p. 307)

Flow Rate The volume of fluid flowing past a given point in a pipe per unit time. (p. 310)

Gauge Pressure The amount of air pressure excluding the normal atmospheric pressure. (p. 305)

Hydraulic Principle (Pascal's Principle) States that the pressure applied to a confined liquid is transmitted without measurable loss throughout the entire liquid to all inner surfaces of the container. (p. 300)

Hydrostatic Pressure The pressure a liquid exerts on a submerged object. (p. 293)

Pressure The force applied per unit area. (p. 291)

Streamline Flow The smooth flow of a fluid through a tube. (p. 310)

Turbulent Flow The erratic, unpredictable flow of a fluid resulting from excessive speed of the flow or sudden changes in direction or size of the tube. (p. 310)

FORMULAS

12.1 $P = \frac{F}{A}$

$P = hD_w$

$F_t = AhD_w$

$F_s = \frac{1}{2}AhD_w$

12.3 $P_{abs} = P_{ga} + P_{atm}$

12.5 $Q = vA$

REVIEW QUESTIONS

1. The force applied to a unit area is called
   (a) strain. (b) total force.
   (c) pressure. (d) none of the above.
2. The statement that the pressure applied to a confined liquid is transmitted without measurable loss throughout the entire liquid to all inner surfaces of the container is called
   (a) Hooke's law. (b) Pascal's principle.
   (c) Archimedes' principle. (d) none of the above.
3. For an incompressible fluid, the flow rate is
   (a) equal for all surfaces. (b) constant throughout the pipe.
   (c) greater for the larger parts of the pipe. (d) none of the above.
4. Bernoulli's principle states that for horizontal flow of a fluid through a tube, the sum of the pressure and energy of motion per unit volume is
   (a) increasing with time. (b) decreasing with time.
   (c) constant. (d) none of the above.
5. Bernoulli's principle explains
   (a) curving baseballs. (b) the hydraulic principle.
   (c) absolute pressure. (d) buoyant forces.
   (e) none of the above.
6. What is the metric unit for pressure?
7. In your own words, define *pressure*.
8. In your own words, state how to find the force exerted on the vertical side of a rectangular water tank.
9. In your own words, state the hydraulic principle.
10. Describe why a ship floats.
11. Describe how a rotating baseball follows a curved path.
12. How does an airplane wing provide lift?
13. What is the difference between streamline and turbulent flow?

14. Give an example of how Archimedes' principle applies to gases.
15. Describe the difference between absolute and gauge pressure. Which do you use when you measure the pressure on your automobile tires?
16. Is the pressure on a small piston different from the pressure on a large piston in the same hydraulic system? Are the forces on the two pistons the same?
17. On what does the total force exerted by a liquid on a horizontal surface depend?
18. Why must the thickness of a dam be greater at the bottom than at the top?
19. Is the hydraulic piston in the master brake cylinder in an automobile larger or smaller than the piston in the brake cylinder at the wheels? Why?
20. Would a drinking straw work in space where there is no gravity? Explain.

REVIEW PROBLEMS

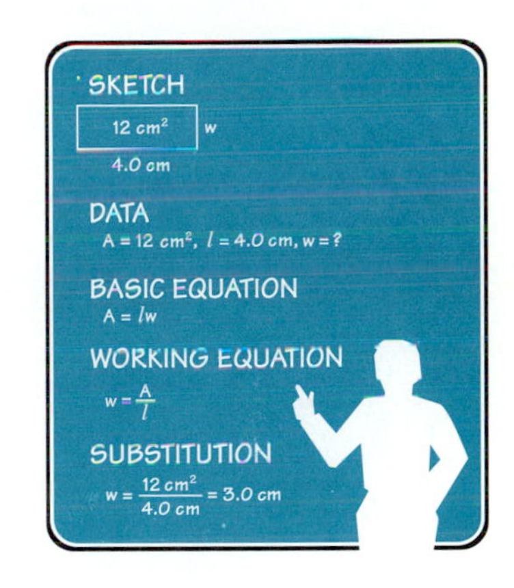

1. Find the pressure (in kPa) at the bottom of a water-filled drum 3.24 m high.
2. Find the depth in a lake at which the pressure is 197 lb/in^2.
3. Find the height of a water column when the pressure at the bottom of the column is 297 kPa.
4. What is the total force exerted on the bottom of a rectangular tank 8.67 ft by 4.83 ft by 3.56 ft deep?
5. Find the water pressure (in kPa) at a point 35.0 m from the bottom of a 55.0-m-tall full water tower.
6. Find the total force on the bottom of a cylindrical water tower 55.0 m high and 7.53 m in radius.
7. Find the total force on the side of a cylindrical water tower 55.0 m high and 7.53 m in radius.
8. Find the total force on the side of a rectangular water trough 1.25 m high by 1.55 m by 2.95 m.
9. What must the water pressure (in kPa) be on the ground to supply a water pressure of 252 N/cm^2 on the third floor, which is 9.00 m above the ground?
10. What water pressure must a pump that is located on the first floor supply to have water on the twenty-fifth floor of a building with a pressure of 26 lb/in^2? Assume that the distance between floors is 16.0 ft.
11. A submarine is submerged to a depth of 3150 ft in the Atlantic Ocean. What air pressure (in kPa) is needed to blow water out of the ballast tanks?
12. The area of the large piston in a hydraulic jack is 4.75 in^2. The area of the small piston is 0.564 in^2. What force must be applied to the small piston if a weight of $65\bar{0}$ lb is to be lifted?
13. What is the mechanical advantage of the hydraulic jack in Problem 12?
14. The MA of a hydraulic jack is 324. What force must be applied to lift an automobile weighing 11,500 N?
15. The pistons of a hydraulic press have radii of 0.543 cm and 3.53 cm. What force must be applied to the smaller piston to exert a force of 4350 N on the larger?
16. What is the pressure (in N/cm^2) on each piston in Problem 15?
17. What is the mechanical advantage of the press in Problem 15?
18. Find the absolute pressure in a bicycle tire with a gauge pressure of 202 kPa.
19. Find the gauge pressure of a tire with an absolute pressure of 655 kPa.
20. Find the gauge pressure of a tank whose absolute pressure is 314 lb/in^2.
21. A rock weighs 55.4 N in air and 52.1 N in water. What is the buoyant force on the rock?
22. A metal displaces 643 cm^3 of water. Find the buoyant force of the water.

23. A rock displaces 314 cm^3 of alcohol. Find the buoyant force on the rock.

24. A flat-bottom barge is 22.3 ft wide, 87.5 ft long, and 16.5 ft deep. How many ft^3 of water will it displace while the top stays 3.20 ft above the water? What load in tons will the barge contain if the barge weighs 157 tons in dry dock?

25. Water flows through a hose of diameter 3.00 cm at a velocity of 4.43 m/s. Find the flow rate of the hose in L/min.

26. Water flows through a 13.0-cm-diameter fire hose at a rate of 4.53 m/s. What is the rate of flow through the hose in L/min? How many litres pass through the hose in 25.0 min?

27. What is the weight density of a liquid that exerts a total force of 433 N on the sides of a 3.00-m-tall cylindrical tank? The radius of the tank is 0.913 m, and it is filled to 1.75 m. What liquid might this be?

Chapter 13

Temperature and Heat

Almost all forms of technology have concerns about temperature and heat. The concern may be direct, as in refrigeration, or indirect, as in the thermal expansion of highways. Being able to measure the effect of heat can mean the difference between success and failure of many things, from steam heat to space travel.

OBJECTIVES

The major goals of this chapter are to enable you to:

1. Distinguish between temperature and heat.
2. Express temperature using different scales.
3. Understand heat and heat transfer.
4. Determine final temperature using the method of mixtures.
5. Relate heat to the expansion of solids and liquids.
6. Calculate the heat required for change of state of solids, liquids, and gases.

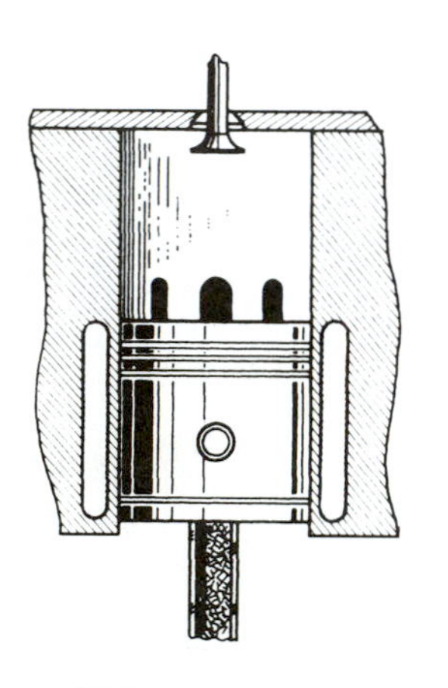

Figure 13.1
Force on piston produced by hot expanding gas.

13.1 TEMPERATURE

An understanding of temperature and heat is very important to all technicians. An automotive technician is concerned with the heat energy released by a fuel mixture in a combustion chamber (Fig. 13.1). The excess heat produced by an engine must be transferred to the atmosphere. The highway technician is concerned with the expansion and contraction of bridges and roads when the temperature changes.

Basically, **temperature** is a measure of the hotness or coldness of an object. Temperature could be measured in a simple way by using your hand to sense the hotness or coldness of an object. However, the range of temperatures that your hand can withstand is too small, and your hand is not precise enough to measure temperature adequately. Therefore, other methods are used for measuring temperature.

We use the fact that certain properties of matter vary with their temperature. For example, when objects are heated, they give off light of different colors. When an object is heated, in the absence of chemical reactions, it first gives off red light. As it is heated more, it appears white.

Chemical reactions sometimes cause different colors to be given off. When carbon steel is heated and exposed to air, several colors are given off before the rod appears red (see Fig. 13.2). This is due to a chemical reaction involving the carbon. If we could measure the color of the light given off, we could then determine the temperature. Although this works only for high temperatures, it is used in the production of metal alloys. The temperature of hot molten metals is determined this way.

Another property of matter that we use to find temperature is that the volume of a liquid or a solid changes as its temperature changes. The liquid in glass thermometers is an example. This type of thermometer (Fig. 13.3) consists of a hollow glass bulb and a hollow glass tube joined together. A small amount of liq-

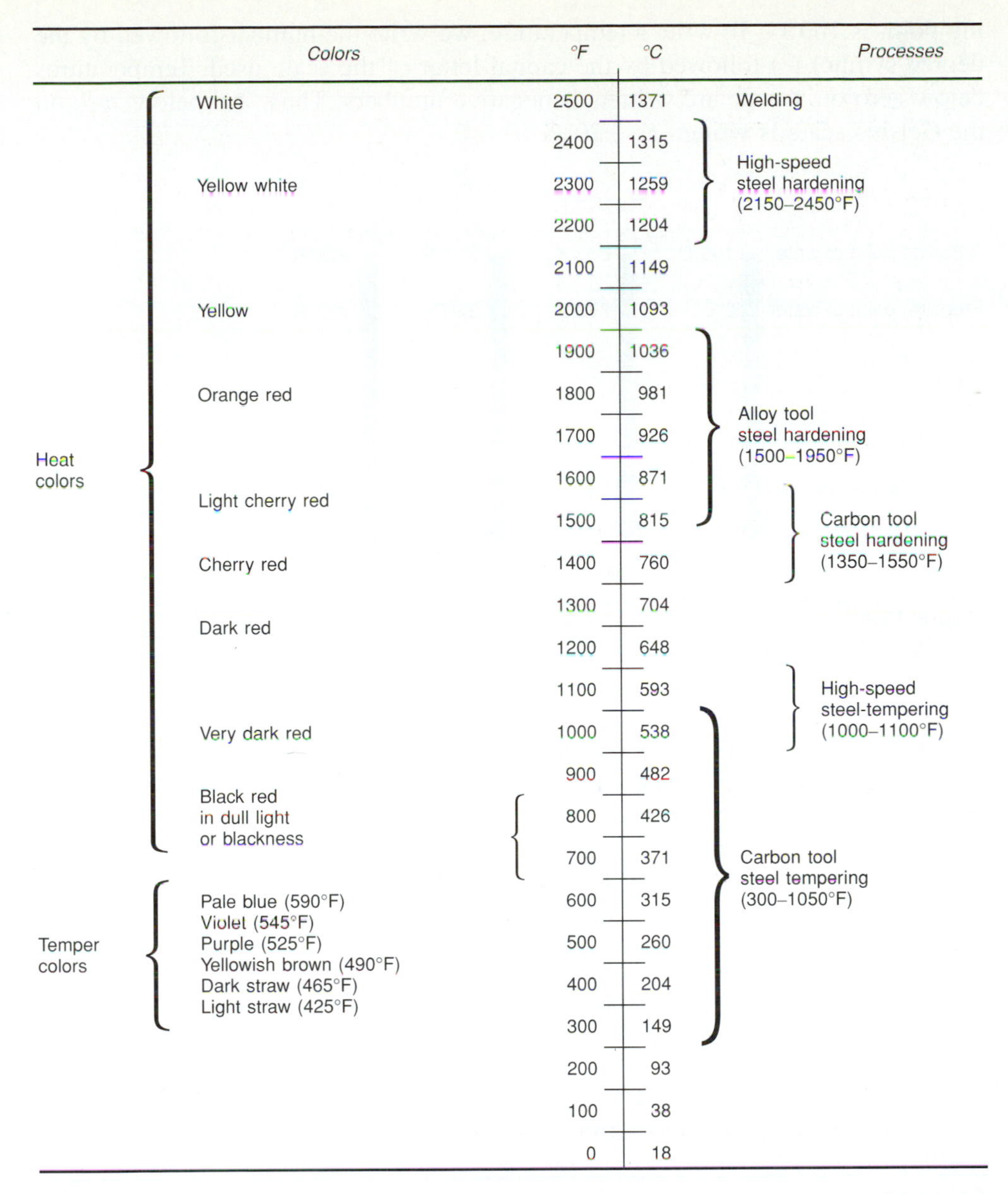

Figure 13.2
Metallurgy and heat treatment: temperatures, steel colors, and related processes. (Courtesy of Allegheny Ludlum Steel Corp. Reprinted by permission.)

uid mercury or alcohol is placed in the bulb. The air is removed from the tube. When the liquid is heated, it expands and rises up the glass tube. The height to which the liquid rises indicates the temperature.

The thermometer is standardized by marking two points on the glass which indicate the liquid level at two known temperatures. The temperatures used are the melting point of ice (called the **ice point**) and the **boiling point** of water at sea level. The distance between these marks is then divided up into equal segments called **degrees.**

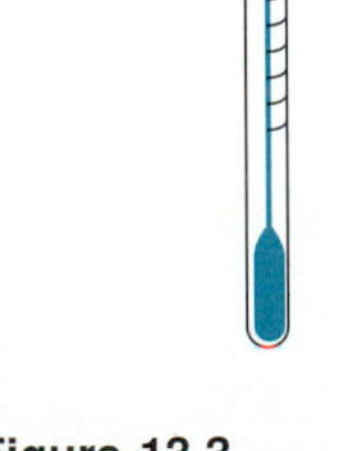

Figure 13.3
Common thermometer.

We will study the four temperature scales shown in Fig. 13.4. The common metric temperature scale is the **Celsius scale.** The ice point is 0°C and the boiling point is 100°C. To write a temperature, we write the number followed by the degree symbol (°) followed by the capital letter of the scale used. Temperatures below zero on a scale are written as negative numbers. Thus, 20° below zero on the Celsius scale is written as −20°C.

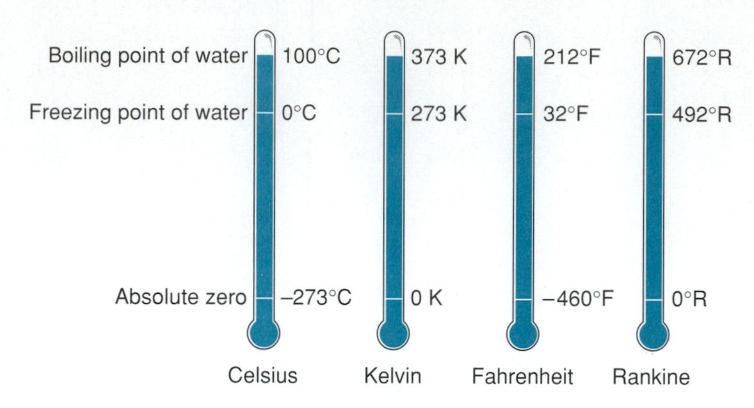

Figure 13.4
Four basic temperature scales.

On the **Fahrenheit scale,** the ice point is 32°F and the boiling point is 212°F. The relationship between Fahrenheit temperatures (T_F) and Celsius temperatures (T_C) is given by

$$T_C = \frac{5}{9}(T_F - 32°)$$

$$T_F = \frac{9}{5}T_C + 32°$$

where T_C = Celsius temperature
T_F = Fahrenheit temperature

EXAMPLE 1

The average human body temperature is 98.6°F. What is it in Celsius?

Data:

$$T_F = 98.6°\text{F}$$
$$T_C = ?$$

Basic Equation:

$$T_C = \frac{5}{9}(T_F - 32°)$$

Working Equation: Same

Substitution:

$$T_C = \frac{5}{9}(98.6° - 32°)$$

$$= \frac{5}{9}(66.6°)$$
$$= 37.0°C$$

In some technical work it is necessary to use the *absolute temperature scales,* which are the Kelvin scale and the Rankine scale. These are called absolute scales because 0 on either scale refers to the lowest limit of temperature, called *absolute zero.*

The **Kelvin scale** is closely related to the Celsius scale. The relationship is

$$T_K = T_C + 273^*$$

The **Rankine scale** is closely related to the Fahrenheit scale. The relationship is

$$T_R = T_F + 46\bar{0}°$$

EXAMPLE 2

Change 18°C to Kelvin.

Data:

$$T_C = 18°C$$
$$T_K = ?$$

Basic Equation:

$$T_K = T_C + 273$$

Working Equation: Same

Substitution:

$$T_K = 18 + 273$$
$$= 291 \text{ K}$$

EXAMPLE 3

Change 535°R to degrees Fahrenheit.

Data:

$$T_R = 535°R$$
$$T_F = ?$$

Basic Equation:

$$T_R = T_F + 46\bar{0}°$$

Working Equation:

$$T_F = T_R - 46\bar{0}°$$

*The degree symbol (°) is not used when writing a temperature on the Kelvin scale.

Substitution:

$$T_F = 535° - 46\bar{0}°$$
$$= 75°F$$

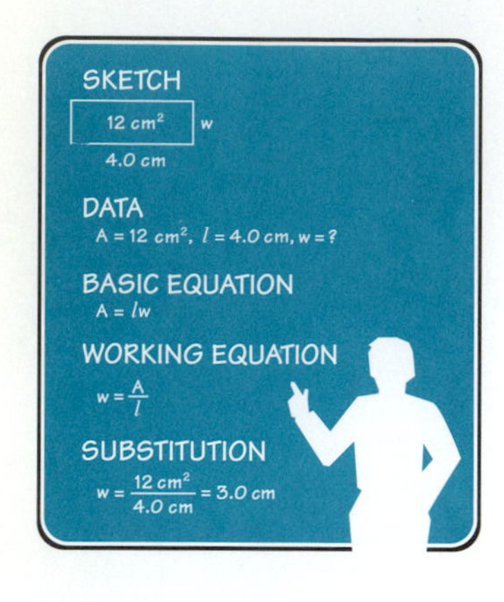

■ PROBLEMS 13.1

Find each temperature as indicated.

1. $T_F = 77°F$, $T_C =$ _____
2. $T_F = 113°F$, $T_C =$ _____
3. $T_F = 257°F$, $T_C =$ _____
4. $T_C = 15°C$, $T_F =$ _____
5. $T_C = 145°C$, $T_F =$ _____
6. $T_C = 35°C$, $T_F =$ _____
7. $T_F = 1\bar{0}°F$, $T_C =$ _____
8. $T_F = 2\bar{0}°F$, $T_C =$ _____
9. $T_C = 95°C$, $T_F =$ _____
10. $T_F = -5\bar{0}°F$, $T_C =$ _____
11. $T_C = 25°C$, $T_K =$ _____
12. $T_F = -45°F$, $T_R =$ _____
13. $T_K = 406$ K, $T_C =$ _____
14. $T_C = 75°C$, $T_K =$ _____
15. $T_C = -5\bar{0}°C$, $T_K =$ _____
16. $T_K = 175$ K, $T_C =$ _____
17. $T_K = 600\bar{0}$ K, $T_C =$ _____
18. The melting point of pure iron is 1505°C. What Fahrenheit temperature is this?
19. The melting point of mercury is −38.0°F. What Celsius temperature is this?
20. A welding white heat is approximately $14\bar{0}0$°C. Find this temperature expressed in degrees Fahrenheit.
21. The temperature in a crowded room is 85°F. What is the Celsius reading?
22. The temperature of an iced tea drink is 5°C. What is the Fahrenheit reading?
23. The boiling point of liquid nitrogen is −196°C. What is the Fahrenheit reading?
24. The melting point of ethyl alcohol is −179°F. What is the Celsius reading?

During the forging and heat-treating of steel, the color of heated steel is used to determine its temperature. Complete the following table, which shows the color of heat-treated steel and the corresponding approximate temperatures in Celsius and Fahrenheit. (Round to three significant digits.)

	Color	*°C*	*°F*
25.	White	—	2200
26.	Yellow	1100	—
27.	Orange	—	1725
28.	Cherry red	718	—
29.	Dark red	635	—
30.	Faint red	—	$90\bar{0}$
31.	Pale blue	310	—

■

13.2 HEAT

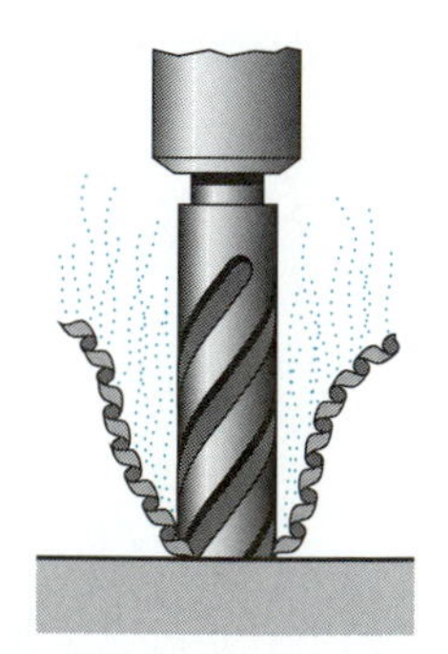

Figure 13.5
Friction causes a rise in temperature of the drill and plate.

When a machinist drills a hole in a metal block (Fig. 13.5), it becomes very hot. As the drill does mechanical work on the metal, the temperature of the metal increases. How can we explain this? Note the difference between the metal at low temperatures and at high temperatures. At high temperatures the atoms in the metal vibrate more rapidly than at low temperatures. Their velocity is higher at high temperatures, and thus their kinetic energy ($KE = \frac{1}{2}mv^2$) is greater. To raise the temperature of a material, we must speed up the atoms; that is, we must add energy to them. **Heat** is the name given to this energy that is added to or taken from a material.

Drilling a hole in a metal block causes a temperature increase. As the drill turns, it collides with atoms of the metal, causing them to speed up. This mechanical work done on the metal has caused an increase in the energy (speed) of the atoms. For this reason, any friction between two surfaces results in a temperature rise of the materials.

Since heat is a form of energy, we could measure it in ft lb or joules, which are energy units. However, before it was known that heat is a form of energy, special units for heat were developed, which are still in use. These units are the calorie and the kilocalorie in the metric system and the Btu (British thermal unit) in the English system. The **kilocalorie** (kcal) is the amount of heat necessary to raise the temperature of 1 kilogram of water 1°C. ***Note:*** One food calorie is the same as 1 kilocalorie. The **Btu** is the amount of heat (energy) necessary to raise the temperature of 1 lb of water 1°F. The **calorie** (cal) is the amount of heat (energy) necessary to raise the temperature of 1 gram of water 1°C at its maximum density.

To lower the temperature of a substance, we need to remove some of the energy of motion of the molecules (heat). When we have removed all the heat possible (when the molecules are moving as slowly as possible), we have reached the lowest possible temperature, called **absolute zero.** Lower temperatures cannot be reached because all the heat has been removed. However, there is no upper limit on temperature because we can always add more heat (energy) to a substance to increase its temperature.

As mentioned before, heat and work are somehow related. James Prescott Joule (1818–1889), an English scientist, determined by experiments the relationship between heat and work. He found that

1. 1 cal of heat is produced by 4.19 J of work.
2. 1 kcal of heat is produced by 4190 J of work.
3. 1 Btu of heat is produced by 778 ft lb of work.

These relationships are known as the **mechanical equivalent of heat.**

The following are some examples by which heat is converted into useful work:

1. *In our bodies.* When food is oxidized, heat energy is produced, which can be converted into muscular energy, which in turn can be turned into work. Experiments have shown that only about 25% of the heat energy from our food is converted into muscular energy. That is, our bodies are about 25% efficient.
2. *By burning gases.* When a gas is burned, the gas expands and builds up a tremendous pressure that may convert heat to work by exerting a force to move a piston in an engine or turn the blades of a turbine. Since the burning of the fuel occurs within the cylinder or turbine, such engines are called *internal combustion engines.*
3. *By steam.* Heat from burning oil, coal, or wood may be used to generate steam. When water changes to steam under normal atmospheric pressure, it expands about 1700 times. When confined to a boiler, the pressure exerts a force against the piston in a steam engine or against the blades of a steam turbine. Since the fuel burns outside the engine, most steam engines or steam turbines are *external combustion engines.*

Technically, what is the difference between temperature and heat? *Temperature* is a measure of the average velocity of the molecules in a substance. *Heat* is the total thermal energy (kinetic and potential) that is added to or taken away from the molecules in a substance. There are two basic ways of changing the temperature of an object:

1. By doing work *on* the substance, such as the work done by the drill on the metal block in Fig. 13.5
2. By supplying energy *to* the object, such as mechanical, chemical, or electrical

EXAMPLE 1

Find the amount of work (in J) that is equivalent to 4850 cal of heat.

$$4850\ \cancel{\text{cal}} \times \frac{4.19\ \text{J}}{1\ \cancel{\text{cal}}} = 20{,}300\ \text{J} \quad \text{or} \quad 20.3\ \text{kJ}$$

EXAMPLE 2

How much work must a person do to offset eating a 775-calorie breakfast?

First, note that one food calorie equals one kilocalorie.

$$775\ \cancel{\text{kcal}} \times \frac{4190\ \text{J}}{1\ \cancel{\text{kcal}}} = 3.25 \times 10^6\ \text{J} \qquad \text{or} \qquad 3.25\ \text{MJ}$$

EXAMPLE 3

A given coal gives off 7150 kcal/kg of heat when burned. How many joules of work result from burning one metric ton, assuming that 35.0% of the heat is lost?

First, note that one metric ton equals 1000 kg.

$$7150\ \frac{\cancel{\text{kcal}}}{\cancel{\text{kg}}} \times \frac{4190\ \text{J}}{\cancel{\text{kcal}}} \times 1000\ \cancel{\text{kg}} \times 0.350 = 1.05 \times 10^{10}\ \text{J}$$

■ PROBLEMS 13.2

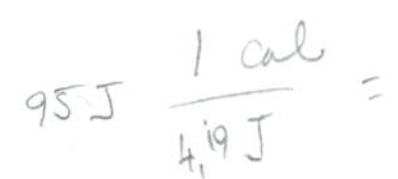

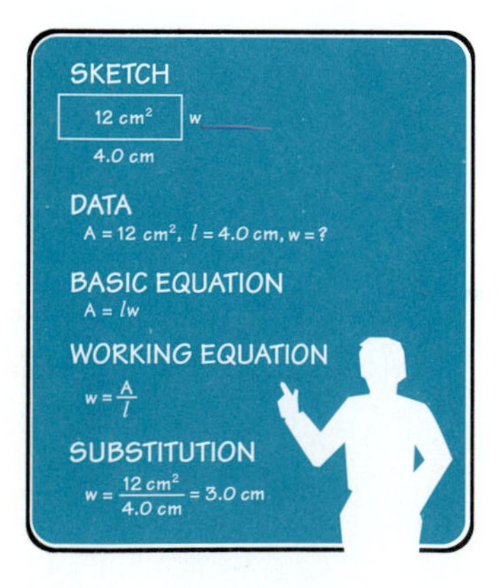

1. Find the amount of heat in cal generated by 95 J of work.
2. Find the amount of heat in kcal generated by 7510 J of work.
3. Find the amount of work that is equivalent to 1550 Btu.
4. Find the amount of work that is equivalent to 3850 kcal.
5. Find the mechanical work equivalent (in J) of 765 kcal of heat.
6. Find the mechanical work equivalent (in J) of 8550 cal of heat.
7. Find the heat equivalent (in Btu) of 3.46×10^6 ft lb of work.
8. Find the heat equivalent (in kcal) of 7.63×10^5 J of work.
9. How much work must a person do to offset eating a piece of cake containing 625 cal?
10. How much work must a person do to offset eating a $2\bar{0}0$-g bag of potato chips if 28 g of chips contain 150 calories?
11. A given gasoline yields 1.15×10^4 cal/g when burned. How many joules of work are obtained by burning 875 g of gasoline?
12. A coal sample yields 1.25×10^4 Btu/lb. How many foot-pounds of work result from burning 1.00 ton of this coal?
13. Natural gas burned in a gas turbine has a heating value of 1.10×10^5 cal/g. If the turbine is 24.0% efficient and 2.50 g of gas is burned each second, find (a) how many joules of work are obtained and (b) the power output in kilowatts.

14. Find the amount of heat energy that must be produced by the body to be converted into muscular energy and then into 1000 ft lb of work. Assume that the body is only 25% efficient. ■

13.3 HEAT TRANSFER

The movement of heat from a hot engine to the air is necessary to keep the engine from overheating. The heat produced by a furnace must be transferred to the various rooms in a house. The movement of heat is a major technical problem.

The transfer of heat from one object to another is always from the warmer object to the colder one or from the warmer part of an object to a colder part (Fig. 13.6). There are three methods of heat transfer: *conduction*, *convection*, and *radiation*. The usual method of heat transfer in solids is **conduction.** When one end of a metal rod is heated, the molecules in that end move faster than before. These molecules collide with other molecules and cause them to move fast also. In this way the heat is transferred from one end of the metal to the other (Fig. 13.7). Another example of conduction is the transfer of the excess heat produced in the combustion chamber of an engine through the engine block into the water coolant (Fig. 13.8).

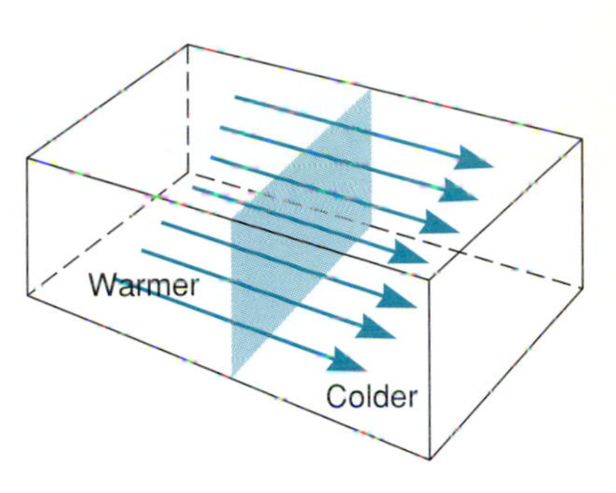

Figure 13.6
Transfer of heat from warmer to colder.

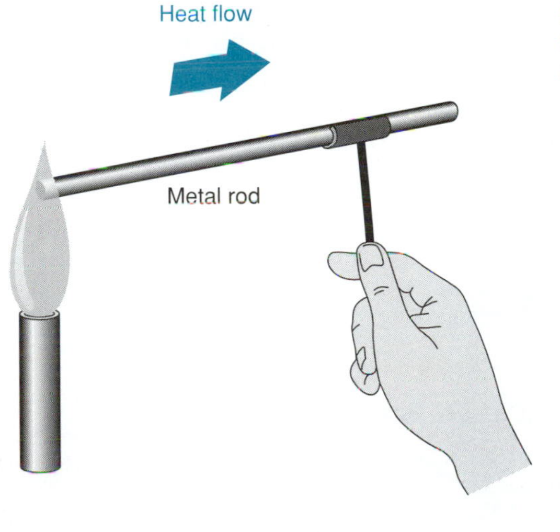

Figure 13.7
Heat flows by conduction in the metal rod.

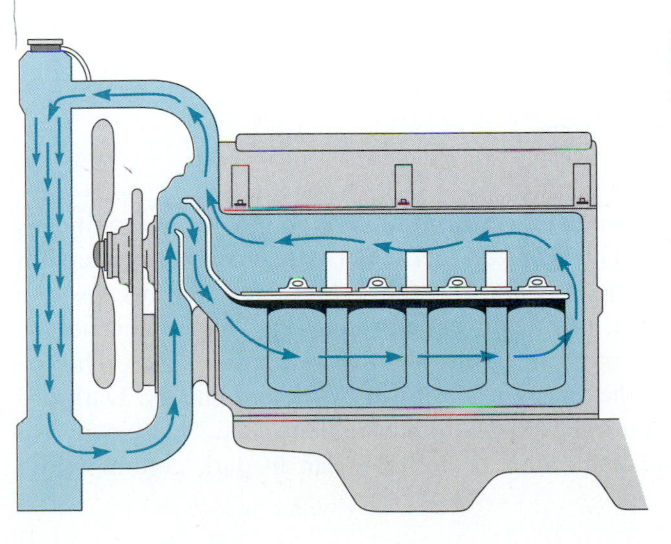

Figure 13.8
Heat conduction in an auto engine cooling system.

The conduction of heat through some materials is better than through others. A poor conductor of heat is called an *insulator.* A list of good conductors and poor conductors is given in Table 13.1.

Table 13.1 Good and Poor Heat Conductors

Good Heat Conductors	*Poor Heat Conductors*
Copper	Asbestos
Aluminum	Glass
Steel	Wood
	Air

Another method of heat transfer is called **convection.** This is the movement of warm gases or liquids from one place to another. The wind carries heat along with it. The water coolant in an engine carries hot water from the engine block to the radiator by a convection process. The wind is a natural convection process. The engine coolant is a forced convection process because it depends on a pump.

Convection currents are caused by the expansion of liquids or gases as they are heated or cooled. This expansion makes the hot gas or liquid less dense than the surrounding fluid. The lighter fluid is then forced upward by the heavier surrounding fluid, which then flows in to replace it (Fig. 13.9). This type of behavior occurs in a fireplace as hot air goes up the chimney and is replaced by cool air from the adjacent room. The cool air draft, as this is called, is eventually supplied from outside air. This is why a fireplace is not very effective in heating a house. An airtight woodburning stove, however, draws little air from the inside of a house and is therefore much more efficient at heating the house.

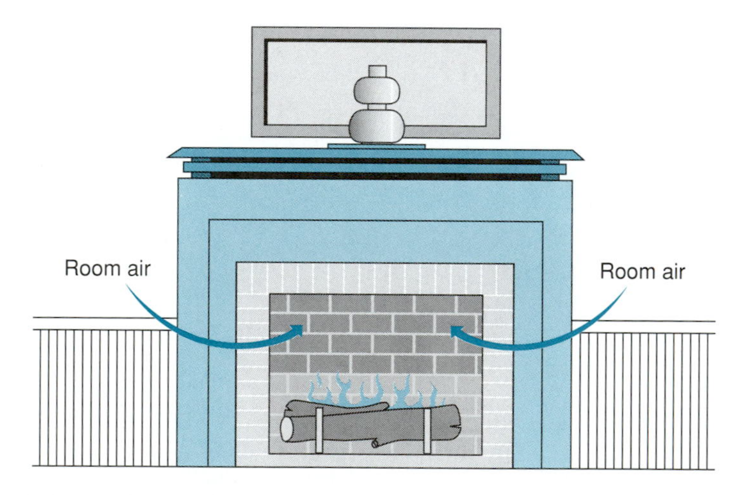

Figure 13.9
Fireplace draws room air up into chimney.

The third method of heat transfer is **radiation.** Put your hand several inches from a hot iron (Fig. 13.10). The heat you feel is not transferred by conduction because air is a poor conductor. It is not transferred by convection because the hot air rises. This kind of heat transfer is called radiation. This radiant heat is similar to light and passes through air, glass, and the vacuum of space. The energy that comes to us from the sun is in the form of radiant energy. Dark objects absorb radiant heat, and light objects reflect radiant heat. This is why we feel cooler on a hot day in light-colored clothing than in dark clothing (Fig. 13.11).

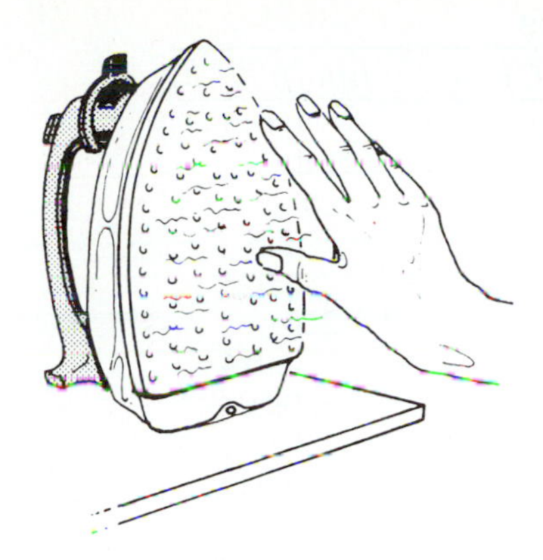

Figure 13.10
Heat radiation.

Figure 13.11
Dark objects absorb more radiant heat than light objects.

Calculations of heat flow are of great importance because of concern about energy conservation. All three of the heat-transfer mechanisms discussed here must be accounted for in any estimation of heat loss from a building. In addition, infiltration losses that arise from leakage through cracks and openings near doors, windows, and other such areas must also be considered. Infiltration losses are, of course, a form of convective transfer but are often treated separately. We will discuss methods for calculating heat loss by conduction.

The equations describing the flow of heat through an object are very similar to those for the flow of electricity, which are developed in later chapters. The driving potential for heat flow is the temperature difference between the hot and cold sides of the object. Heat flow is similar to the flow of electrical charge.

The ability of a material to transfer heat by conduction depends on its **thermal conductivity.** Metals are good conductors of heat. Glass and air are poor conductors. The rate at which heat is transferred through an object depends on the following factors:

1. The thermal conductivity K
2. The cross-sectional area A through which the heat flows
3. The thickness of the material L
4. The temperature difference between the two sides of the material

The total amount of heat transferred is given by the equation

$$Q = \frac{KAt(T_2 - T_1)}{L}$$

where Q = heat transferred in Btu or J
K = thermal conductivity (from Table 13.2)
A = cross-sectional area
t = total time
T_2 = temperature of the hot side
T_1 = temperature of the cool side
L = thickness of the material

Table 13.2 gives the thermal conductivities of some common materials.

Table 13.2 Thermal Conductivities

Substance	*J/(s m °C)*	*Btu/(ft °F h)*
Air	0.025	0.015
Aluminum	230	140
Brass	120	68
Brick/concrete	0.84	0.48
Cellulose fiber (loose fill)	0.039	0.023
Copper	380	220
Corkboard	0.042	0.024
Glass	0.75	0.50
Gypsum board (sheetrock)	0.16	0.092
Mineral wool	0.045	0.026
Plaster	0.14	0.083
Polystyrene foam	0.035	0.020
Polyurethane (expanded)	0.024	0.014
Steel	45	26

EXAMPLE 1

Find the heat flow in an 8.0-h period through a 36 in. × 36 in. pane of glass (0.125 in. thick) if the temperature of the inner surface of the glass is 65°F and the temperature of the outer surface is 15°F.

Data:

$$K = 0.50 \text{ Btu/(ft °F h)}$$
$$A = 36 \text{ in.} \times 36 \text{ in.} = 3.0 \text{ ft} \times 3.0 \text{ ft} = 9.0 \text{ ft}^2$$
$$t = 8.0 \text{ h}$$
$$T_2 = 65°\text{F}$$
$$T_1 = 15°\text{F}$$
$$L = 0.125 \text{ in.} \times \left(\frac{1 \text{ ft}}{12 \text{ in.}}\right) = 0.0104 \text{ ft}$$
$$Q = ?$$

Basic Equation:

$$Q = \frac{KAt(T_2 - T_1)}{L}$$

Working Equation: Same

Substitution:

$$Q = \frac{[0.50 \text{ Btu/(ft °F h)}](9.0 \text{ ft}^2)(8.0 \text{ h})(65°\text{F} - 15°\text{F})}{0.0104 \text{ ft}}$$
$$= 1.7 \times 10^5 \text{ Btu}$$

The insulation value of construction material is often expressed in terms of the *R value,* which indicates the ability of the material to resist the flow of heat using English units. The *R* value is inversely proportional to the thermal conductivity and directly proportional to the thickness. Low thermal conductivity is characteristic of good insulators. This is described by the equation

$$R = \frac{L}{K}$$

where R = R value (in ft^2 °F/Btu/h)
K = thermal conductivity
L = thickness of the material (in ft)

EXAMPLE 2

Calculate the R value of 6.0 in. of mineral wool insulation.

Data:

$$L = 6.0 \text{ in.} = 0.50 \text{ ft}$$
$$K = 0.026 \text{ Btu/(ft °F h)}$$
$$R = ?$$

Basic Equation:

$$R = \frac{L}{K}$$

Working Equation: Same

Substitution:

$$R = \frac{0.50 \text{ ft}}{0.026 \text{ Btu/(ft °F h)}}$$
$$= 19 \text{ ft}^2 \text{ °F/Btu/h}$$

$$\frac{\text{ft}}{\dfrac{\text{Btu}}{\text{ft °F h}}} = \text{ft} \div \frac{\text{Btu}}{\text{ft °F h}} = \text{ft} \cdot \frac{\text{ft °F h}}{\text{Btu}} = \frac{\text{ft}^2 \text{ °F h}}{\text{Btu}} = \frac{\text{ft}^2 \text{ °F}}{\text{Btu/h}}$$

This result could also have been written R-19. There is no equivalent in the metric system.

■ PROBLEMS 13.3

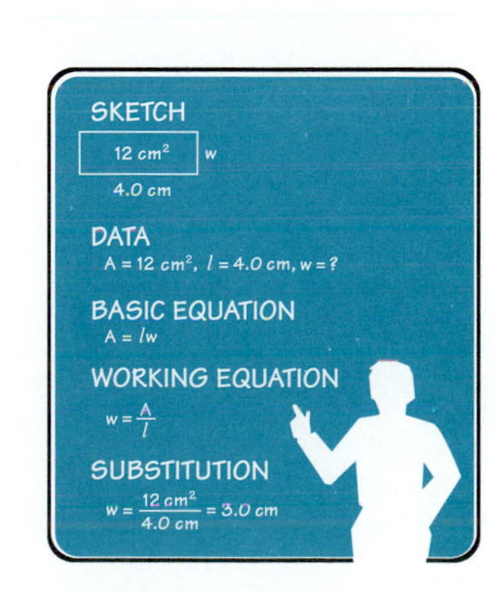

1. Find the R value of a pane of 0.125-in.-thick glass.
2. Find the R value of a brick wall 4.0 in. thick.
3. Find the R value of 0.50-in.-thick sheetrock.
4. Find the thermal conductivity of a piece of building material 0.25 in. thick that has an R value of 1.6 ft^2 °F/Btu/h.
5. Find the R value of 0.50-in.-thick-corkboard.
6. Find the amount of heat conducted through the walls of a building in 24 h if the R value of the walls is 11 ft^2 °F/Btu/h. Assume the dimensions of each of the building's four walls are $2\bar{0}$ ft × $10\bar{0}$ ft. Also assume that the average outer wall temperature is $2\bar{0}$°F and that the average inner wall temperature is 55°F.
7. Find the heat flow during 30.0 days through a glass window of thickness 0.20 in. with an area of 15 ft^2 if the average outer surface temperature is 25°F and the average inner glass surface temperature is $5\bar{0}$°F.
8. Find the heat flow in 30.0 days through a 0.25-cm-thick steel plate with a cross section of 45 cm × 75 cm. Assume a temperature differential of 95°C.

9. Find the heat flow through a steel rod in 75 s if the length is 85 cm and the diameter is 0.50 cm. Assume that the temperature of the hot end of the rod is $11\bar{0}$°C and the temperature of the cool end is −25°C.
10. Find the heat flow in 15 min through a 0.10-cm-thick copper plate with a cross-sectional area of 150 cm^2 if the temperature of the hot side is $99\bar{0}$°C and the temperature of the cool side is 5°C.
11. Find the heat flow in 24 h through a refrigerator door 30.0 in. × 58.0 in. insulated with cellulose fiber 2.0 in. thick. The temperature inside the refrigerator is 38°F. Room temperature is 72°F.
12. Find the heat flow in 30.0 days through a freezer door 30.0 in. × 58.0 in. insulated with cellulose fiber 2.0 in. thick. The temperature inside the freezer is $-1\bar{0}$°F. Room temperature is 72°F.
13. Find the heat flow in 24 h through a refrigerator door 76.0 cm × 155.0 cm insulated with cellulose fiber 5.0 cm thick. The temperature inside the refrigerator is 3°C. Room temperature is 21°C.
14. Find the heat flow in 30.0 days through a freezer door 76.0 cm × 155.0 cm insulated with cellulose fiber 5.0 cm thick. The temperature inside the freezer is −18°C. Room temperature is 21°C.
15. Find the heat flow through the sides of an 18-cm-tall glass of ice water in 45 s. The glass is 6.00 mm thick, the temperature inside is 28.0°C. The temperature outside is 43.3°C and the radius is 7.0 cm. ■

13.4 SPECIFIC HEAT

If we placed a piece of steel and a pan of water in the direct summer sunlight, we would find that the water becomes only slightly warmer whereas the steel gets quite hot. Why should one get so much hotter than the other? If equal masses of steel and water were placed over the same flame for 1 minute, the temperature of the steel would increase almost 10 times more than that of the water. The water has a greater capacity to absorb heat.

The specific heat of a substance is a measure of its capacity to absorb or give off heat per degree change in temperature. This property of water to absorb or give off large amounts of heat makes it an effective substance to transfer heat in industrial processes.

The **specific heat** of a substance is the amount of heat necessary to change the temperature of 1 lb 1°F (1 kg 1°C in the metric system). By formula.

$$c = \frac{Q}{m\Delta T} \quad \text{(metric)} \qquad c = \frac{Q}{w\Delta T} \quad \text{(English)}$$

These equations can be rearranged to give the amount of heat added or taken away from a material to produce a certain temperature change.

$$Q = cm\,\Delta T \quad \text{(metric)} \qquad Q = cw\,\Delta T \quad \text{(English)}$$

where c = specific heat
Q = heat
m = mass
w = weight
ΔT = change in temperature

A table of specific heat is found in Table 15 of Appendix C.

EXAMPLE 1

How many kilocalories of heat must be added to 10.0 kg of steel to raise its temperature $15\bar{0}$°C?

Data:

$$m = 10.0 \text{ kg}$$
$$\Delta T = 15\bar{0}°\text{C}$$
$$c = 0.115 \text{ kcal/kg °C} \qquad \text{(from Table 15 of Appendix C)}$$
$$Q = ?$$

Basic Equation:

$$Q = cm\,\Delta T$$

Working Equation: Same

Substitution:

$$Q = 0.115 \frac{\text{kcal}}{\cancel{\text{kg}}\ °\cancel{\text{C}}} \times 10.0\ \cancel{\text{kg}} \times 15\bar{0}°\cancel{\text{C}}$$
$$= 173 \text{ kcal}$$

EXAMPLE 2

How many joules of heat must be absorbed to cool 5.00 kg of water from 75.0°C to 10.0°C?

Data:

$$m = 5.00 \text{ kg}$$
$$\Delta T = 75.0°\text{C} - 10.0°\text{C} = 65.0°\text{C}$$
$$c = 4190 \text{ J/kg °C} \qquad \text{(from Table 15 of Appendix C)}$$
$$Q = ?$$

Basic Equation:

$$Q = cm\,\Delta T$$

Working Equation: Same

Substitution:

$$Q = 4190 \frac{\text{J}}{\cancel{\text{kg}}\ °\cancel{\text{C}}} \times 5.00\ \cancel{\text{kg}} \times 65.0°\cancel{\text{C}}$$
$$= 1.36 \times 10^6 \text{ J} \qquad \text{or} \qquad 1.36 \text{ MJ}$$

■ PROBLEMS 13.4

Find Q for each material.

1. Steel, $w = 3.00$ lb, $\Delta T = 50\bar{0}$°F, $Q =$ _____ Btu
2. Copper, $m = 155$ kg, $\Delta T = 170$°C, $Q =$ _____ kcal
3. Water, $w = 19.0$ lb, $\Delta T = 20\bar{0}$°F, $Q =$ _____ Btu
4. Water, $m = 25\bar{0}$ g, $\Delta T = 17.0$°C, $Q =$ _____ cal
5. Ice, $m = 5.00$ kg, $\Delta T = 2\bar{0}$°C, $Q =$ _____ J

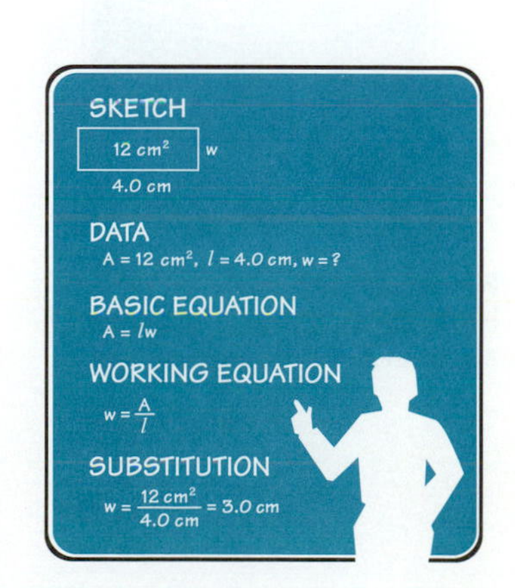

6. Steam, $w = 5.00$ lb, $\Delta T = 4\bar{0}$°F, $Q =$ _____ Btu
7. Aluminum, $m = 79.0$ g, $\Delta T = 16$°C, $Q =$ _____ cal
8. Brass, $m = 750$ kg, $\Delta T = 125$°C, $Q =$ _____ J
9. Steel, $m = 1250$ g, $\Delta T = 50.0$°C, $Q =$ _____ J
10. Aluminum, $m = 85\bar{0}$ g, $\Delta T = 115$°C, $Q =$ _____ kcal
11. Water, $m = 80\bar{0}$ g, $\Delta T = 80.0$°C, $Q =$ _____ kcal
12. Lead, $m = 475$ kg, $\Delta T = 245$°C, $Q =$ _____ J
13. How many Btu of heat must be added to 1200 lb of copper to raise its temperature from $10\bar{0}$°F to $45\bar{0}$°F?
14. How many Btu of heat are given off by $50\bar{0}$ lb of aluminum when it cools from $65\bar{0}$°F to 75°F?
15. How many kcal of heat must be added to 1250 kg of copper to raise its temperature from 25°C to 275°C?
16. How many joules of heat are absorbed by an electric freezer in lowering the temperature of 1850 g of water from 80.0°C to 10.0°C?
17. How many joules of heat are required to raise the temperature of $75\bar{0}$ kg of water from 15.0°C to 75.0°C?
18. How many kilocalories of heat must be added to $75\bar{0}$ kg of steel to raise its temperature from 75°C to $30\bar{0}$°C?
19. How many joules of heat are given off when 125 kg of steel cools from 1425°C to 82°C?
20. A 525-kg steam boiler is made of steel and contains 315 kg of water at 40.0°C. Assuming that 75% of the heat is delivered to the boiler and water, how many kilocalories are required to raise the temperature of both the boiler and water to 100.0°C?
21. Find the initial temperature of a 49.0-N cube of zinc, 16.0 cm to a side, that gives off 6.30×10^6 J of heat while cooling to 80.0°C.
22. If a coolant lowers the temperature in a steel engine weighing 16,250 N by 13°C that is running at a temperature of $11\bar{0}$°C, what is the heat reduction (in joules) in the steel engine?

13.5 METHOD OF MIXTURES

Figure 13.12
Heat flows from the warmer substance to the cooler.

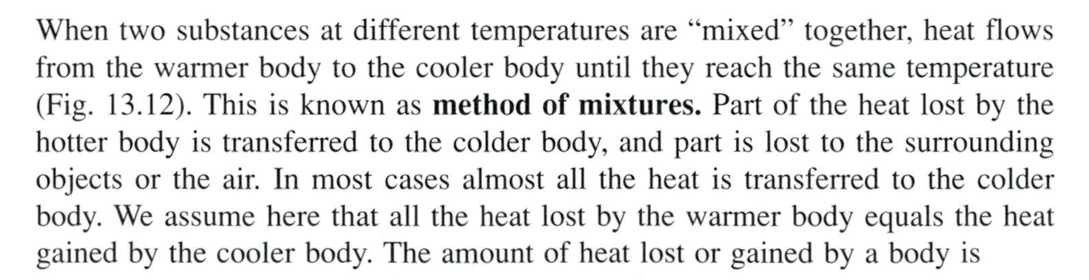

When two substances at different temperatures are "mixed" together, heat flows from the warmer body to the cooler body until they reach the same temperature (Fig. 13.12). This is known as **method of mixtures.** Part of the heat lost by the hotter body is transferred to the colder body, and part is lost to the surrounding objects or the air. In most cases almost all the heat is transferred to the colder body. We assume here that all the heat lost by the warmer body equals the heat gained by the cooler body. The amount of heat lost or gained by a body is

$$Q = cm\,\Delta T \qquad \text{or} \qquad Q = cw\,\Delta T$$

By formula,

$$Q_{\text{lost}} = Q_{\text{gained}}$$
$$c_l w_l(T_l - T_f) = c_g w_g(T_f - T_g)$$

where the subscript l refers to the warmer body, which *loses* heat, the subscript g refers to the cooler body, which *gains* heat, and T_f is the final temperature of the mixture.

EXAMPLE 1

A 10.0-lb piece of hot copper is dropped into 30.0 lb of water at $5\overline{0}$°F. If the final temperature of the mixture is 65°F, what was the initial temperature of the copper?

Data:

$$
\begin{aligned}
w_l &= 10.0 \text{ lb} & w_g &= 30.0 \text{ lb} \\
c_l &= 0.093 \text{ Btu/lb °F} & c_g &= 1 \text{ Btu/lb °F} \\
T_l &= ? & T_g &= 5\overline{0}\text{°F} \\
T_f &= 65\text{°F}
\end{aligned}
$$

Basic Equation:

$$c_l w_l (T_l - T_f) = c_g w_g (T_f - T_g)$$

Working Equation:

$$T_l = \frac{c_g w_g}{c_l w_l} (T_f - T_g) + T_f$$

Substitution:

$$
\begin{aligned}
T_l &= \frac{(1 \text{ Btu/lb °F})(30.0 \text{ lb})}{(0.093 \text{ Btu/lb °F})(10.0 \text{ lb})} (65\text{°F} - 5\overline{0}\text{°F}) + 65\text{°F} \\
&= 550\text{°F}
\end{aligned}
$$

Some find it easier to find T_l using a second method. Substitute the data directly into the basic equation. Then solve for T_l as follows:

$$
\begin{aligned}
\left(0.093 \frac{\text{Btu}}{\text{lb °F}}\right)(10.0 \text{ lb})(T_l - 65\text{°F}) &= \left(1 \frac{\text{Btu}}{\text{lb °F}}\right)(30.0 \text{ lb})(65\text{°F} - 5\overline{0}\text{°F}) \\
0.93 T_l \text{ Btu/°F} - 6\overline{0} \text{ Btu} &= 450 \text{ Btu} \\
0.93 T_l \text{ Btu/°F} &= 510 \text{ Btu} \\
T_l &= \frac{510 \cancel{\text{Btu}}}{0.93 \cancel{\text{Btu}}\text{/°F}} \\
T_l &= 550\text{°F}
\end{aligned}
$$

EXAMPLE 2

If $20\overline{0}$ g of steel at $22\overline{0}$°C is added to $50\overline{0}$ g of water at 10.0°C, find the final temperature of this mixture.

Data:

$$
\begin{aligned}
c_l &= 0.115 \text{ cal/g°C} & c_g &= 1.00 \text{ cal/g°C} \\
m_l &= 20\overline{0} \text{ g} & m_g &= 50\overline{0} \text{ g} \\
T_l &= 22\overline{0}\text{°C} & T_g &= 10.0\text{°C} \\
& & T_f &= ?
\end{aligned}
$$

Basic Equation:

$$c_l m_l (T_l - T_f) = c_g m_g (T_f - T_g)$$

Working Equation:

$$T_f = \frac{c_l m_l T_l + c_g m_g T_g}{c_l m_l + c_g m_g}$$

Substitution:

$$T_f = \frac{(0.115 \text{ cal/g °C})(20\bar{0} \text{ g})(22\bar{0}\text{°C}) + (1.00 \text{ cal/g °C})(50\bar{0} \text{ g})(10.0\text{°C})}{(0.115 \text{ cal/g °C})(20\bar{0} \text{ g}) + (1.00 \text{ cal/g °C})(50\bar{0} \text{ g})}$$

$$= 19.2\text{°C}$$

To find T_f by the second method, substitute the data directly into the basic equation. Then, solve for T_f as follows:

$$\left(0.115 \frac{\text{cal}}{\text{g °C}}\right)(20\bar{0} \text{ g})(22\bar{0}\text{°C} - T_f) = \left(1.00 \frac{\text{cal}}{\text{g °C}}\right)(50\bar{0} \text{ g})(T_f - 10.0\text{°C})$$

$$5060 \text{ cal} - 23.0 \frac{\text{cal}}{\text{°C}} T_f = 50\bar{0} \frac{\text{cal}}{\text{°C}} T_f - 50\bar{0}0 \text{ cal}$$

$$10{,}060 \text{ cal} = 523 \frac{\text{cal}}{\text{°C}} T_f$$

$$\frac{10{,}060 \text{ cal}}{523 \text{ cal/°C}} = T_f$$

$$19.2\text{°C} = T_f$$

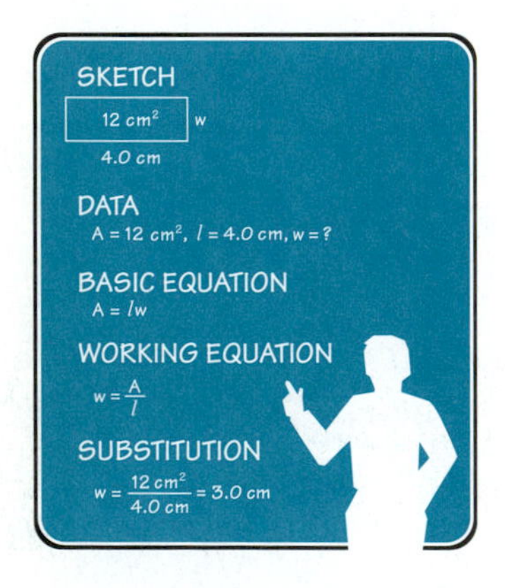

PROBLEMS 13.5

Refer to Table 15, "Heat Constants," of Appendix C.

1. A 2.50-lb piece of steel is dropped into 11.0 lb of water at 75.0°F. The final temperature is 84.0°F. What was the initial temperature of the steel?
2. Mary mixes 5.00 lb of water at $20\bar{0}$°F with 7.00 lb of water at 65.0°F. Find the final temperature of the mixture.
3. A $25\bar{0}$-g piece of tin at 99°C is dropped in $10\bar{0}$ g of water at $1\bar{0}$°C. If the final temperature of the mixture is $2\bar{0}$°C, what is the specific heat of the tin?
4. How many grams of water at $2\bar{0}$°C are necessary to change $80\bar{0}$ g of water at $9\bar{0}$°C to $5\bar{0}$°C?
5. A 159-lb piece of aluminum at $50\bar{0}$°F is dropped into $40\bar{0}$ lb of water at $6\bar{0}$°F. What is the final temperature?
6. A 42.0-lb piece of steel at $67\bar{0}$°F is dropped into $10\bar{0}$ lb of water at 75.0°F. What is the final temperature of the mixture?
7. If 1250 g of copper at 20.0°C is mixed with $50\bar{0}$ g of water at 95.0°C, find the final temperature of the mixture.
8. If $50\bar{0}$ g of brass at $20\bar{0}$°C and $30\bar{0}$ g of steel at $15\bar{0}$°C are added to $90\bar{0}$ g of water in a $15\bar{0}$-g aluminum pan and both are at 20.0°C, what is the final temperature of this mixture, assuming no loss of heat to the surroundings?
9. The following data were collected in the laboratory to determine the specific heat of an unknown metal:

Mass of copper calorimeter	153 g
Specific heat of calorimeter	0.092 cal/g °C
Mass of water	275 g
Specific heat of water	1.00 cal/g °C
Mass of metal	236 g

Initial temperature of water and calorimeter	16.2°C
Initial temperature of metal	99.6°C
Final temperature of calorimeter, water, and metal	22.7°C

Find the specific heat of the unknown metal. ***Note:*** A calorimeter is usually a metal cup inside another metal cup that is insulated by the air between them (Fig. 13.13).

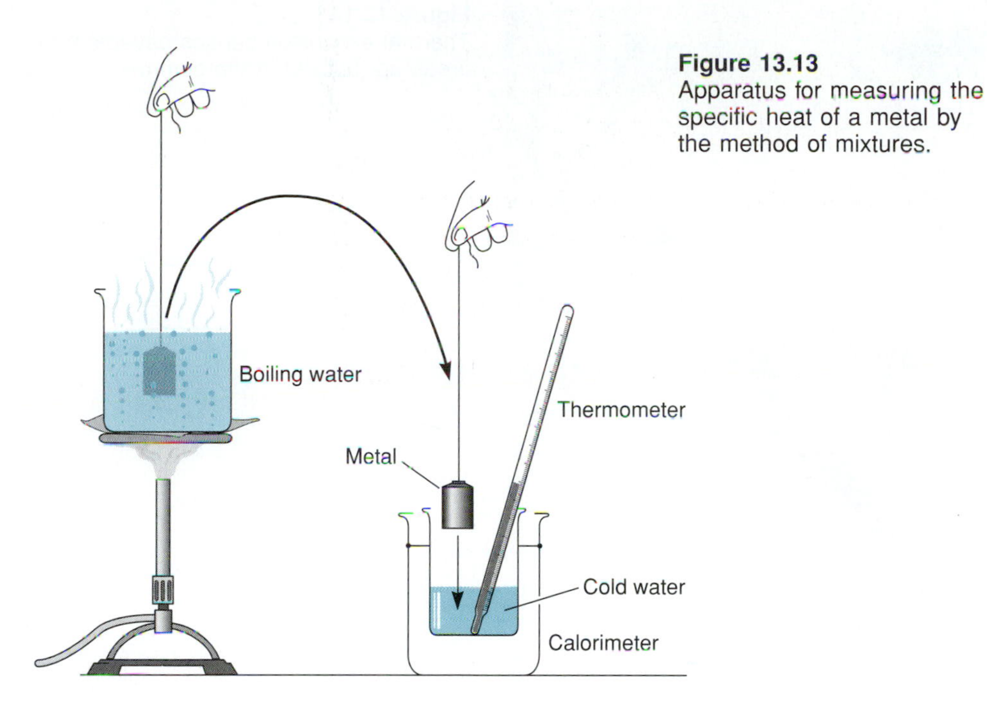

Figure 13.13
Apparatus for measuring the specific heat of a metal by the method of mixtures.

10. The following data were collected in the laboratory to determine the specific heat of an unknown metal:

Mass of aluminum calorimeter	132 g
Specific heat of calorimeter	$92\bar{0}$ J/kg °C
Mass of water	285 g
Specific heat of water	4190 J/kg °C
Mass of metal	215 g
Initial temperature of water and calorimeter	12.6°C
Initial temperature of metal	99.1°C
Final temperature of calorimeter, water, and metal	18.6°C

Find the specific heat of the unknown metal.

11. Determine the original temperature of a $56\bar{0}$-g piece of lead placed in a 165-g brass calorimeter that contains 325 g of water. The initial temperature of the water and calorimeter was 18.0°C. The final temperature of the lead, calorimeter, and water is 31.0°C. ■

13.6 EXPANSION OF SOLIDS

Most solids expand when heated and contract when cooled. They expand or contract in all three dimensions—length, width, and thickness. When a solid is heated, the expansion is due to the increased length of the vibrations of the atoms and molecules. This results in the solid expanding in all directions. This increase in volume results in a decrease in weight density, which was discussed in Chapter 11. Engineers, technicians, and designers must know the effects of thermal expansion. You have no doubt heard of highway pavements buckling on a hot summer day (Fig. 13.14). Bridges are built with special joints that allow for expansion and contraction of the bridge deck (Fig. 13.15). The clicking noise of a

Figure 13.14
Thermal expansion causes pavement to break up (buckle) in the summer.

Figure 13.15
Thermal expansion joint in a roadway.

train's wheels passing over the rails can be heard more in winter than in summer. The space between rails is larger in winter than in summer (Fig. 13.16). If the rails were placed snugly end to end in the winter, they would buckle in the summer. Similarly, TV towers, pipelines, and buildings must be designed and built to allow for this expansion and contraction.

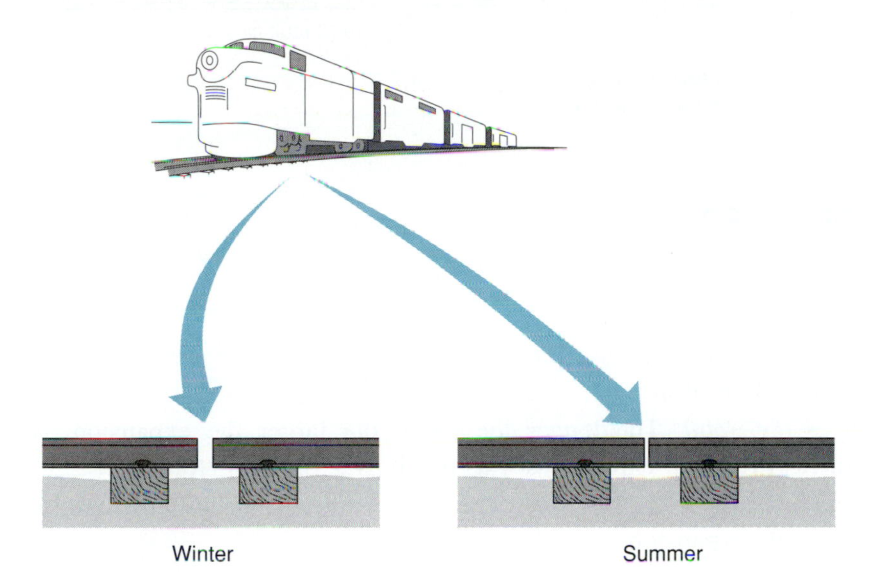

Figure 13.16
Thermal expansion in rails.

There are some advantages to solids expanding. A bimetallic strip is made by fusing two different metals together side by side as illustrated in Fig. 13.17. When heated, the brass expands more than the steel, which makes the strip curve. If the bimetallic strip is cooled below room temperature, the brass will contract more than the steel, forcing the strip to curve in the opposite direction. The thermostat operates on this principle. As shown in Fig. 13.18, the basic parts of a thermostat are a bimetallic strip on the right and a regular metal strip on the left. The bimetallic strip of brass and steel bends with the temperature. The regular metal strip is moved by hand to set the temperature desired. This particular bimetallic strip is made and placed so that it bends to the left when cooled. As a result, when it comes in contact with the strip on the left, it completes a circuit, which turns on the furnace. When the room warms to the desired temperature, the bimetallic strip moves back to the right, which opens the contacts and shuts off the heat. Bimetallic strips are in spiral form in some thermostats (Fig. 13.19).

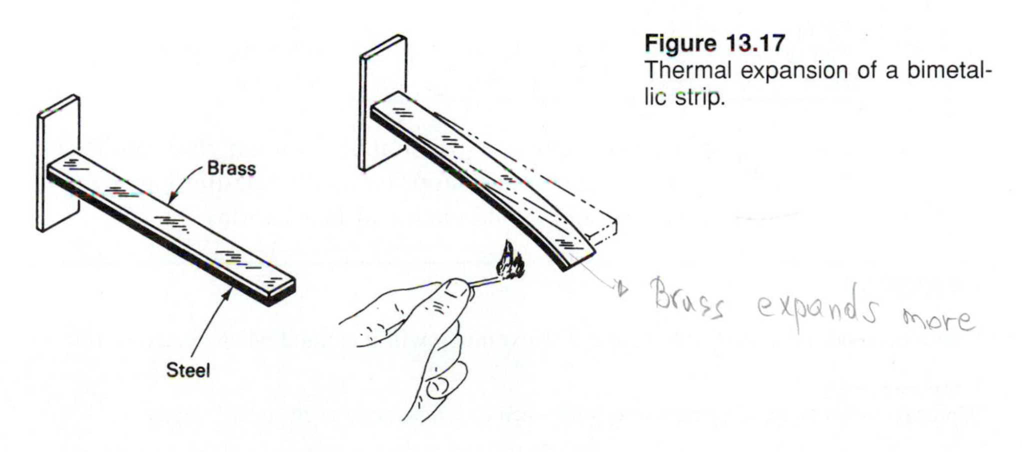

Figure 13.17
Thermal expansion of a bimetallic strip.

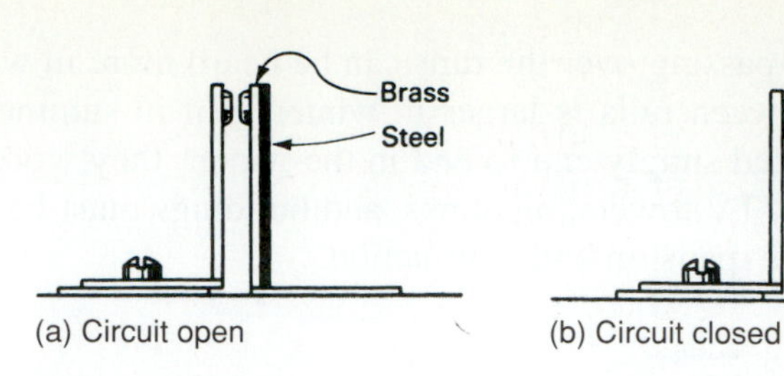

Figure 13.18
Simple thermostat

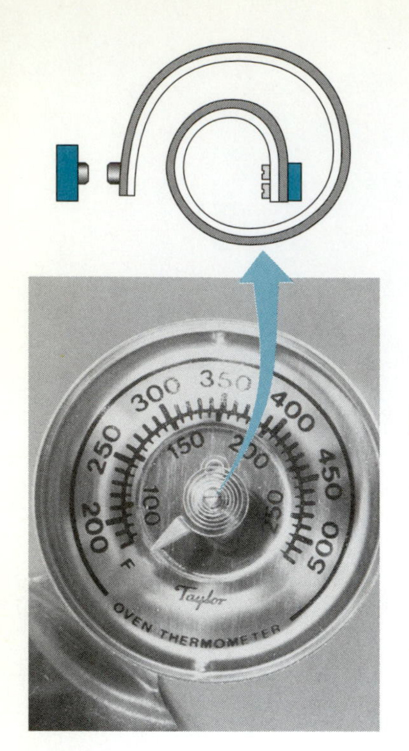

Figure 13.19
Thermostat using spiral bimetallic nut.

Linear Expansion

The amount that a solid expands depends on the following:

1. *Material.* Different materials expand at different rates. Steel expands at a rate less than that of brass.
2. *Length of the solid.* The longer the solid, the larger the expansion. A 20.0-cm steel rod will expand twice as much as a 10.0-cm steel rod.
3. *Amount of change in temperature.* The greater the change in temperature, the greater the expansion.

This can be written by formula:

$$\Delta l = \alpha l \, \Delta T$$

where Δl = change in length
α = a constant called the **coefficient of linear expansion***
l = original length
ΔT = change in temperature

Table 13.3 lists the coefficients of linear expansion for some common solids.

Table 13.3 Coefficients of Linear Expansion

Material	α *(metric)*	α *(English)*
Aluminum	2.3×10^{-5}/C°	1.3×10^{-5}/F°
Brass	1.9×10^{-5}/C°	1.0×10^{-5}/F°
Concrete	1.1×10^{-5}/C°	6.0×10^{-6}/F°
Copper	1.7×10^{-5}/C°	9.5×10^{-6}/F°
Glass	9.0×10^{-6}/C°	5.1×10^{-6}/F°
Pyrex	3.0×10^{-6}/C°	1.7×10^{-6}/F°
Steel	1.3×10^{-5}/C°	6.5×10^{-6}/F°
Zinc	2.6×10^{-5}/C°	1.5×10^{-5}/F°

Comparing the coefficients of linear expansion of common glass and Pyrex, we can see that Pyrex expands and contracts approximately one-third as much as glass. This is why it is used in cooking and chemical laboratories.

EXAMPLE 1

A steel railroad rail is 40.0 ft long at 0°F. How much will it expand when heated to $10\bar{0}$°F?

*Defined as change in unit length of a solid when its temperature is changed 1 degree.

Data:

$$l = 40.0 \text{ ft}$$
$$\Delta T = 10\bar{0}°\text{F} - 0°\text{F} = 10\bar{0}°\text{F}$$
$$\alpha = 6.5 \times 10^{-6}/\text{F}°$$
$$\Delta l = ?$$

Basic Equation:

$$\Delta l = \alpha l \, \Delta T$$

Working Equation: Same

Substitution:

$$\Delta l = (6.5 \times 10^{-6}/\text{F}°)(40.0 \text{ ft})(10\bar{0}°\text{F})$$
$$= 0.026 \text{ ft} \quad \text{or} \quad 0.31 \text{ in.}$$

Pipes that undergo large temperature changes are usually installed to allow for expansion and contraction (Fig. 13.20).

Figure 13.20
Thermal expansion joints in pipes.

EXAMPLE 2

What allowance for expansion must be made for a steel pipe $12\bar{0}$ m long that handles coolants and must undergo temperature changes of $20\bar{0}$°C?

Data:

$$\alpha = 1.3 \times 10^{-5}/\text{C}°$$
$$l = 12\bar{0} \text{ m}$$
$$\Delta T = 20\bar{0}°\text{C}$$
$$\Delta l = ?$$

Basic Equation:

$$\Delta l = \alpha l \, \Delta T$$

Working Equation: Same

Substitution:

$$\begin{aligned}\Delta l &= (1.3 \times 10^{-5}/\cancel{C°})(12\bar{0} \text{ m})(20\bar{0}\cancel{°C}) \\ &= 0.31 \text{ m} \quad \text{or} \quad 31 \text{ cm}\end{aligned}$$

Area and Volume Expansion of Solids

Solids expand in width and thickness as well as in length when heated. The area of a hole cut out of a metal sheet will expand in the same way as the surrounding material. To allow for this expansion the following formulas are used:

Area expansion: $\Delta A = 2\alpha A \, \Delta T$

Volume expansion: $\Delta V = 3\alpha V \, \Delta T$

where A = original area
V = original volume

EXAMPLE 3

The top of a circular copper disk has an area of 64.2 in^2 at 2$\bar{0}$°F. What is the change in area when the temperature is increased to 15$\bar{0}$°F?

Data:

$$\begin{aligned}\alpha &= 9.5 \times 10^{-6}/\text{F°} \\ A &= 64.2 \text{ in}^2 \\ \Delta T &= 15\bar{0}\text{°F} - 2\bar{0}\text{°F} = 13\bar{0}\text{°F} \\ \Delta A &= ?\end{aligned}$$

Basic Equation:

$$\Delta A = 2\alpha A \, \Delta T$$

Working Equation: Same

Substitution:

$$\begin{aligned}\Delta A &= 2(9.5 \times 10^{-6}/\cancel{F°})(64.2 \text{ in}^2)(13\bar{0}\cancel{°F}) \\ &= 0.16 \text{ in}^2\end{aligned}$$

EXAMPLE 4

A section of concrete measures 6.00 m × 12.0 m × 30.0 m at 38°C. What allowance for change in volume is necessary for a temperature of −15°C?

Data:

$$\begin{aligned}V &= (6.00 \text{ m})(12.0 \text{ m})(30.0 \text{ m}) = 2160 \text{ m}^3 \\ \alpha &= 1.1 \times 10^{-5}/\text{C°} \\ \Delta T &= 38\text{°C} - (-15\text{°C}) = 53\text{°C}\end{aligned}$$

Basic Equation:

$$\Delta V = 3\alpha V\,\Delta T$$

Working Equation: Same

Substitution:

$$\begin{aligned}\Delta V &= 3(1.1 \times 10^{-5}/\text{C}°)(2160\ \text{m}^3)(53\ \text{C}°)\\ &= 3.8\ \text{m}^3\end{aligned}$$

■ PROBLEMS 13.6

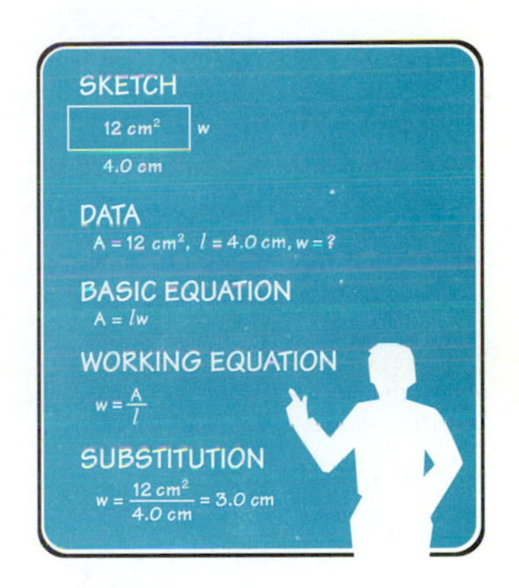

1. Find the increase in length of copper tubing 200.0 ft long at 40.0°F when it is heated to 200.0°F.
2. Find the increase in length of a zinc rod 50.0 m long at 15.0°C when it is heated to 130.0°C.
3. Compute the increase in length of 300.00 m of copper wire when its temperature changes from 14°C to 34°C.
4. A steel pipe 8.25 m long is installed at 45°C. Find the decrease in length when coolants at $-6\bar{0}$°C pass through the pipe.
5. A steel tape measures 200.00 m at 15°C. What is its length at 55°C?
6. A brass rod 1.020 m long expands by 3.0 mm when it is heated. Find the temperature change.
7. The road bed on a bridge 500.0 ft long is made of concrete. What allowance is needed for temperatures of $-4\bar{0}$°F in winter and $14\bar{0}$°F in summer?
8. An aluminum plug has a diameter of 10.003 cm at 40.0°C. At what temperature will it fit precisely into a hole of constant diameter 10.000 cm?
9. The diameter of a steel drill at 45°F is 0.750 in. Find its diameter at 375°F.
10. A brass ball with diameter 12.000 cm is 0.011 cm too large to pass through a hole in a copper plate when the ball and plate are at a temperature of 20.0°C. What is the temperature of the ball when it will just pass through the plate, assuming that the temperature of the plate does not change? What is the temperature of the plate when the ball will just pass through, assuming that the temperature of the ball does not change?
11. A brass cylinder has a cross-sectional area of 482 cm^2 at −5°C. Find its change in area when heated to 95°C.
12. The volume of the cylinder in Problem 11 is 4820 cm^3 at 240.0°C. Find its change in volume when cooled to −75.0°C.
13. An aluminum pipe has a cross-sectional area of 88.40 cm^2 at 15°C. What is its cross-sectional area when the pipe is heated to 155°C?
14. A steel pipe has a cross-sectional area of 127.20 in^2 at 25°F. What is its cross-sectional area when the pipe is heated to 175°F?
15. A glass plug has a volume of 60.00 cm^3 at 12°C. What is its volume at 76°C?
16. The diameter of a hole drilled through brass at 21°C measures 6.500 cm. Find the diameter and area of the hole when the brass is heated to 175°C.
17. Steel rails 15.000 m long are laid at 10.0°C. How much space should be left between them if they are to just touch at 35.0°C?
18. Steel beams 120.000 ft long are placed in a highway overpass to allow for expansion and contraction. The temperature range allowance is $-3\bar{0}$°F to $13\bar{0}$°F.
    (a) Find the space allowance (in inches) between the beams at $-3\bar{0}$°F if the beams touch at $13\bar{0}$°F.
    (b) Find the space allowance between the beams if placed at 75°F.

19. The spaces between 13.00-m steel rails are 0.711 cm at −15°C. If the rails touch at 35.5°C, what is the coefficient of linear expansion?
20. A section of concrete dam is a rectangular solid 20.0 ft by 50.0 ft by 80.0 ft at 115°F. What allowance for change in volume is necessary for a temperature of −15°F?
21. A glass ball has a radius of 12.000 cm at 6.0°C. Find its change in volume when the temperature is increased to 81.0°C.
22. Find the final height of a concrete column that is 1.250 m × 1.250 m × 4.250 m at 0.0°C when the column is heated to 45.0°C.
23. What is the final volume of a glass right circular cylinder with original height 1.200 m and radius 30.00 cm that is heated from 13.0°C to 56.0°C? ■

13.7 EXPANSION OF LIQUIDS

Liquids also generally expand when heated and contract when cooled. The thermometer is made using this principle. When a thermometer is placed under your tongue, the heat from your mouth causes the mercury in the bottom of the thermometer to expand. Mercury is then forced to rise up the thin calibrated tube (Fig. 13.21). The formula for *volume expansion of liquids* is

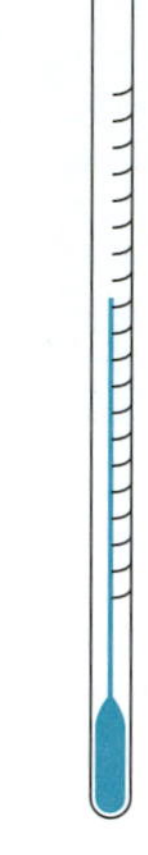

Figure 13.21
Liquid expansion in a thermometer.

$$\Delta V = \beta V \, \Delta T$$

where β = coefficient of volume expansion for liquids
V = original volume

Table 13.4 lists the coefficients of volume expansion for some common liquids.

Table 13.4 Coefficients of Volume Expansion

Liquid	β *(metric)*	β *(English)*
Acetone	1.49×10^{-3}/C°	8.28×10^{-4}/F°
Alcohol, ethyl	1.12×10^{-3}/C°	6.62×10^{-4}/F°
Carbon tetrachloride	1.24×10^{-3}/C°	6.89×10^{-4}/F°
Mercury	1.8×10^{-4}/C°	1.0×10^{-4}/F°
Petroleum	9.6×10^{-4}/C°	5.33×10^{-4}/F°
Turpentine	9.7×10^{-4}/C°	5.39×10^{-4}/F°
Water	2.1×10^{-4}/C°	1.17×10^{-4}/F°

EXAMPLE 1

If petroleum at 0°C occupies $25\bar{0}$ L, what is its volume at $5\bar{0}$°C?

Data:

$$\beta = 9.6 \times 10^{-4}/\text{C}^\circ$$
$$V = 25\bar{0} \text{ L}$$
$$\Delta T = 5\bar{0}^\circ\text{C}$$
$$\Delta V = ?$$

Basic Equation:

$$\Delta V = \beta V \, \Delta T$$

Working Equation: Same

Substitution:

$$\Delta V = (9.6 \times 10^{-4}/\cancel{C}°)(25\bar{0}\text{ L})(5\bar{0}°\cancel{C})$$
$$= 12\text{ L}$$
$$\text{volume at }5\bar{0}°\text{C} = V + \Delta V$$
$$= 25\bar{0}\text{ L} + 12\text{ L} = 262\text{ L}$$

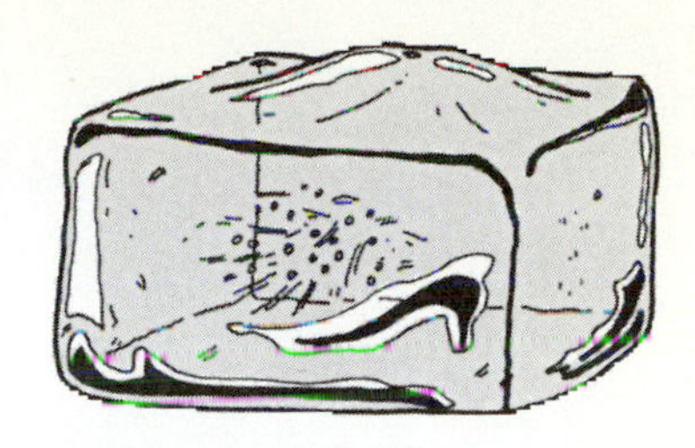

Figure 13.22
Expansion of water in change of state from liquid to solid.

EXAMPLE 2

Find the increase in volume of 18.2 in^3 of water when the water is heated from 4$\bar{0}$°F to 18$\bar{0}$°F.

Data:

$$\beta = 1.17 \times 10^{-4}/\text{F}°$$
$$V = 18.2\text{ in}^3$$
$$\Delta T = 18\bar{0}°\text{F} - 4\bar{0}°\text{F} = 14\bar{0}°\text{F}$$
$$\Delta V = ?$$

Basic Equation:

$$\Delta V = \beta V\,\Delta T$$

Working Equation: Same

Substitution:

$$\Delta V = (1.17 \times 10^{-4}/\cancel{F}°)(18.2\text{ in}^3)(14\bar{0}°\cancel{F})$$
$$= 0.298\text{ in}^3$$

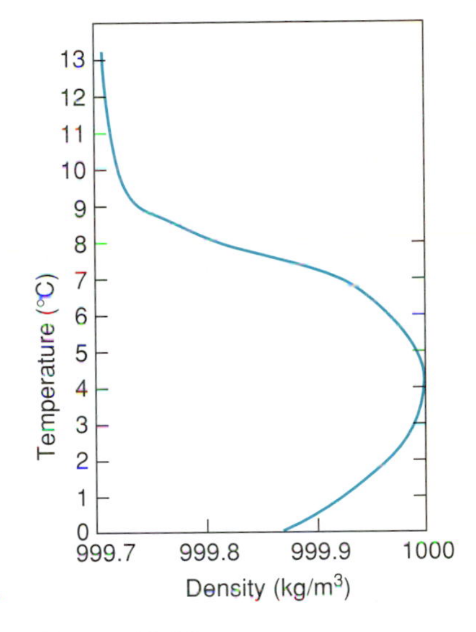

Figure 13.23
Change in density of water with change in temperature.

Expansion of Water

Water is unusual in its expansion characteristics. Recall the mound in the middle of each ice cube in ice cube trays (Fig. 13.22). This evidence shows the expansion of water during its change of state from liquid to solid form.

Nearly all liquids are densest at their lowest temperature before a change of state to become solids. As the temperature drops, the molecular motion slows, and the substance becomes denser. Water does not follow this general rule. Because of its unusual structural characteristics, water is densest at 4°C or 39.2°F instead of 0°C or 32°F. A graph of its change in density with an increase in temperature is shown in Fig. 13.23. As ice melts and the water temperature is slightly increased, there are still groups of molecules that have the open crystallographic structure of ice, which is less dense than water. As the water is heated to 4°C, these groupings disappear and the water becomes denser. Above 4°C, water then expands normally as the temperature is raised.

When ice melts at 0°C or 32°F, the water formed *contracts* as the temperature is raised to 4°C or 39.2°F. Then it begins to *expand,* as do most other liquids.

■ PROBLEMS 13.7

1. A quantity of carbon tetrachloride occupies 625 L at 12°C. Find its volume at 48°C.
2. Some mercury occupies 157 in^3 at −3$\bar{0}$°F. What is the change in volume when heated to 9$\bar{0}$°F?

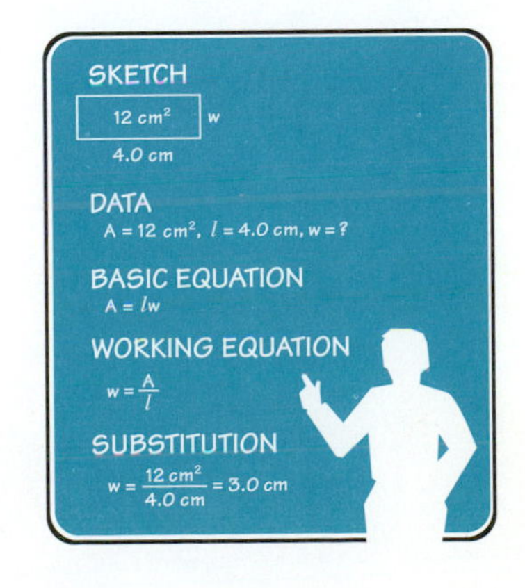

3. Some petroleum occupies 11.7 m^3 at -17°C. Find its volume at 28°C.
4. Find the increase in volume of 35 L of acetone when heated from 28°C to 38°C.
5. Some water at $18\overline{0}$°F occupies $378\overline{0}$ ft^3. What is its volume at 122°F?
6. A $120\overline{0}$-L tank of petroleum is completely filled at 9°C. How much spills over if the temperature rises to 45°C?
7. Find the increase in volume of 215 cm^3 of mercury when its temperature increases from $1\overline{0}$°C to 25°C.
8. Find the decrease in volume of alcohol in a railroad tank car that contains $200\overline{0}$ ft^3 if the temperature drops from 75°F to 54°F.
9. A gasoline service station owner receives a truckload of $33{,}00\overline{0}$ L of gasoline at 32°C. It cools to 15°C in the underground tank. At 40 cents/L, how much money is lost as a result of the contraction of the gasoline?
10. A Pyrex container is completely filled with 275 cm^3 of mercury at 10.0°C. How much mercury spills over when heated to 75.0°C?
11. What was the temperature of $18\overline{0}$ mL of acetone before it was heated to 98°C and increased to a volume of $20\overline{0}$ mL? ■

13.8 CHANGE OF STATE

Many industries are concerned with a change of state in the materials they use. In foundries the principal activity is to change that state of solid metals to liquid, pour the liquid metal into molds, and allow it to become solid again (Fig. 13.24). **Change of state** is a change in a substance from one form of matter to another form of matter.

Figure 13.24
Molten pig iron from the blast furnace is poured into an open hearth furnace, refined, and purified into steel at temperatures of about 2900°F.

Fusion

The change of state from solid to liquid is called **melting** or **fusion.** The change from the liquid to the solid state is called **freezing** or **solidification.** Most solids have a crystalline structure and a definite melting point at any given pressure. Melting and solidification of these substances occur at the same temperature. For example, water at 0°C (32°F) changes to ice and ice changes to water at the same temperature. There is no temperature change during change of state. Ice at 0°C changes to water at 0°C. Only a few substances, such as butter and glass, have no particular melting temperature but change state gradually.

Although there is no temperature change during a change of state, *there is a transfer of heat.* A melting solid *absorbs* heat, and a solidifying liquid *gives off* heat. When 1 g of ice at 0°C melts, it absorbs $8\overline{0}$ cal of heat. Similarly, when 1 g of water freezes at 0°C, ice at 0°C is produced, and $8\overline{0}$ cal of heat is released.

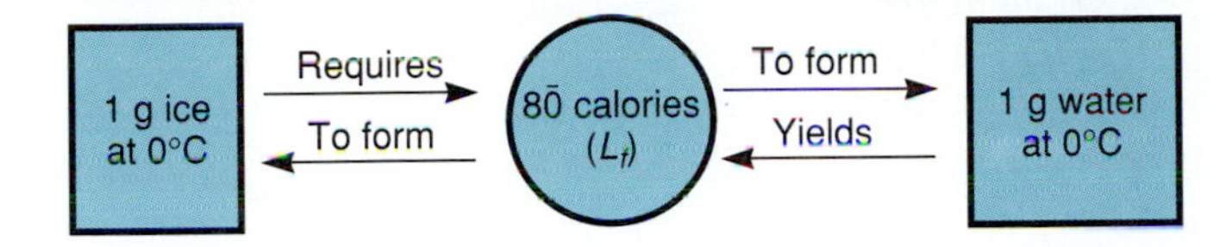

When 1 kg of ice at 0°C melts, it absorbs $8\overline{0}$ kcal of heat. Similarly, when 1 kg of water freezes at 0°C, ice at 0°C is produced and $8\overline{0}$ kcal of heat is released.

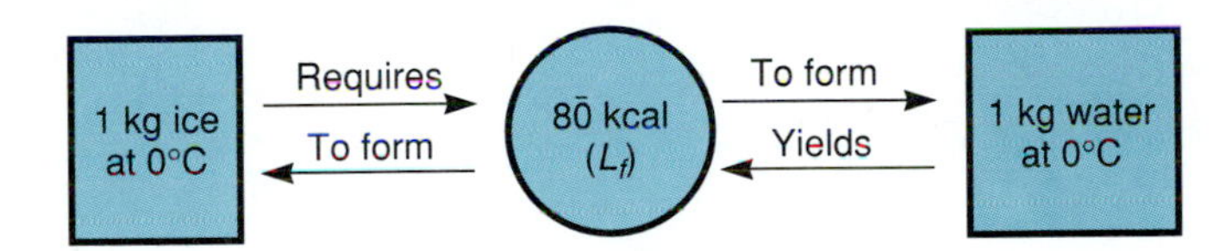

Or, when 1 kg of ice at 0°C melts, it absorbs 335 kilojoules (kJ) of heat. Then, when 1 kg of water freezes at 0°C, ice at 0°C is produced and 335 kJ of heat is released.

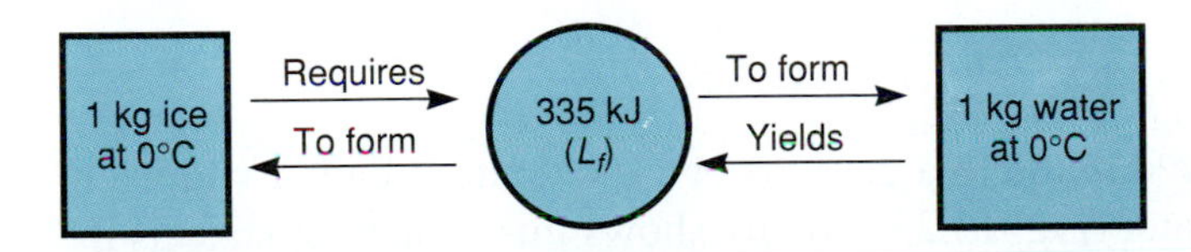

When 1 lb of ice at 32°F melts, it absorbs 144 Btu of heat. Similarly, when 1 lb of water freezes at 32°F, ice at 32°F is produced and 144 Btu of heat is released.

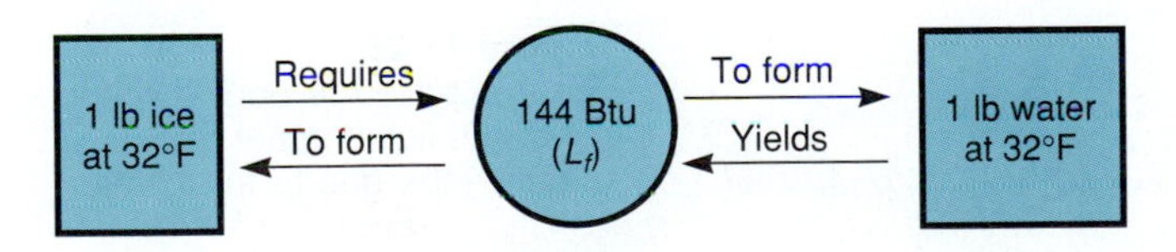

The amount of heat required to melt 1 g or 1 kg or 1 lb of a liquid is called its **heat of fusion,** designated L_f.

$$L_f = \frac{Q}{m} \quad \text{(metric)} \qquad L_f = \frac{Q}{w} \quad \text{(English)}$$

where L_f = heat of fusion (see Table 15 in Appendix C)
Q = quantity of heat

m = mass of substance (metric system)
w = weight of substance (English system)

EXAMPLE 1

If 1340 kJ of heat is required to melt 4.00 kg of ice at 0°C into water at 0°C, what is the heat of fusion of water?

Data:

$$Q = 1340 \text{ kJ}$$
$$m = 4.00 \text{ kg}$$
$$L_f = ?$$

Basic Equation:

$$L_f = \frac{Q}{m}$$

Working Equation: Same

Substitution:

$$L_f = \frac{1340 \text{ kJ}}{4.00 \text{ kg}}$$
$$= 335 \text{ kJ/kg}$$

In general, heat of fusion (water) = $8\overline{0}$ cal/g, or $8\overline{0}$ kcal/kg, or 335 kJ/kg, or 144 Btu/lb.

Vaporization

The change of state from a liquid to a gaseous or vapor state is called **vaporization.** A pan of boiling water (Fig. 13.25) vividly shows this change of state as the steam evaporates and leaves the liquid. Note that vaporization requires that heat be supplied; in this case heat is required to boil the water. The reverse process (change from a gas to a liquid) is called **condensation.** As steam condenses in radiators (Fig. 13.26), large amounts of heat are released.

Figure 13.25
Heat supplied to boiling water changes liquid water into steam—the gas form of water.

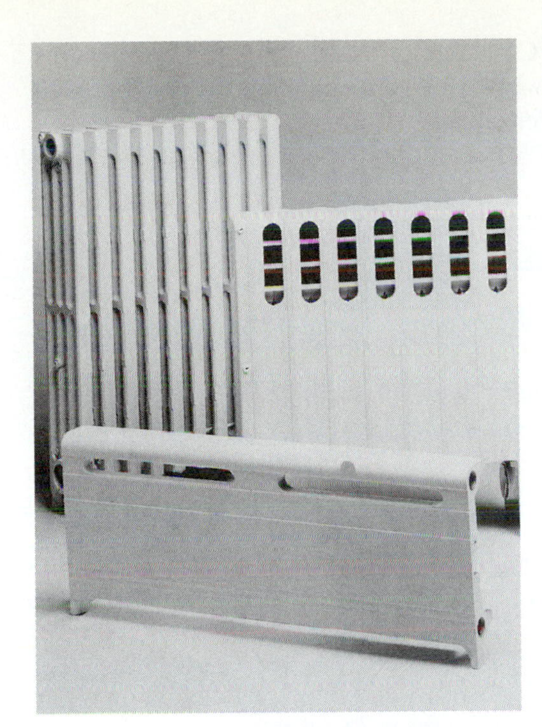

Figure 13.26
A large amount of heat is released by condensation of steam in a radiator.

At the point of condensation, the vapor becomes *saturated;* that is, the vapor cannot hold any more moisture. For example, water vapor is always present in some amount in the earth's atmosphere. The weather term, **relative humidity,** is the ratio of the actual amount of vapor in the atmosphere to the amount of vapor required to reach 100% of saturation at the existing temperature. As the temperature of the air decreases without change in pressure or vapor content, the relative humidity increases until it reaches 100% at saturation. The temperature at which saturation is reached is called the *dew point.* Once saturation is reached and the temperature continues to decrease, condensation occurs in the form of dew, fog, mist, clouds, rain, or other forms of precipitation.

While a liquid is boiling, the temperature of the liquid does not change. However, there is a transfer of heat. A liquid being vaporized (boiled) *absorbs* heat. As a vapor condenses, heat is given off.

The amount of heat required to vaporize 1 g or 1 kg or 1 lb of a liquid is called its **heat of vaporization,** designated L_v. So, when 1 g of water at $10\bar{0}$°C changes to steam at $10\bar{0}$°C, it absorbs $54\bar{0}$ cal; when 1 g of steam at $10\bar{0}$°C condenses to water at $10\bar{0}$°C, $54\bar{0}$ cal of heat is given off.

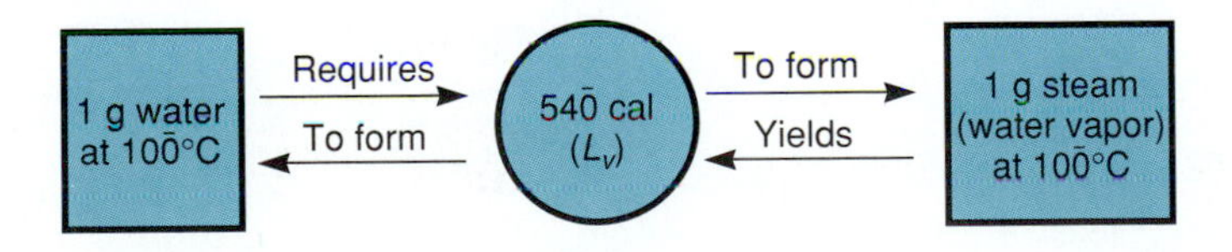

When 1 kg of water at $10\bar{0}$°C changes to steam at $10\bar{0}$°C, it absorbs $54\bar{0}$ kcal of heat. Similarly, when 1 kg of steam at $10\bar{0}$°C condenses to water at $10\bar{0}$°C, $54\bar{0}$ kcal of heat is given off.

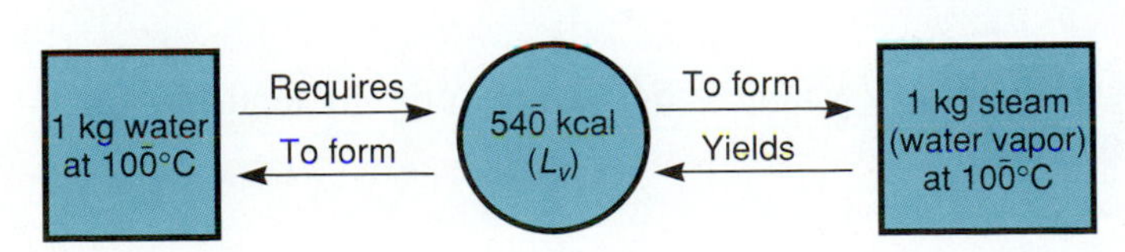

Or, when 1 kg of water at $10\bar{0}°C$ changes to steam at $10\bar{0}°C$, it absorbs 2.26 MJ (2.26×10^6 J) of heat. Then, when 1 kg of steam at $10\bar{0}°C$ condenses to water at $10\bar{0}°C$, 2.26 MJ of heat is given off.

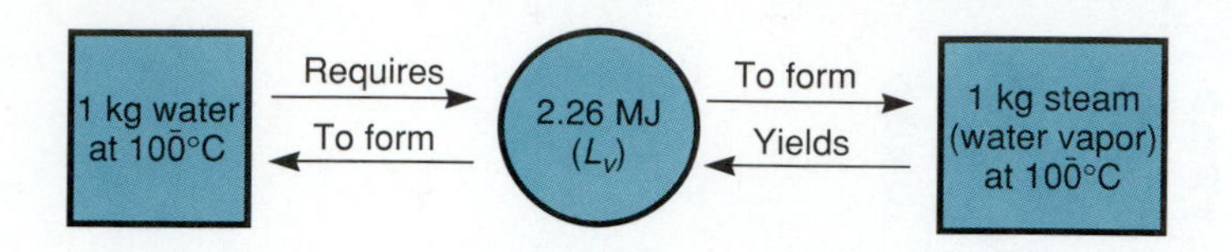

When 1 lb of water at 212°F changes to steam at 212°F, $97\bar{0}$ Btu of heat is absorbed; when 1 lb of steam at 212°F condenses to water at 212°F, $97\bar{0}$ Btu of heat is given off.

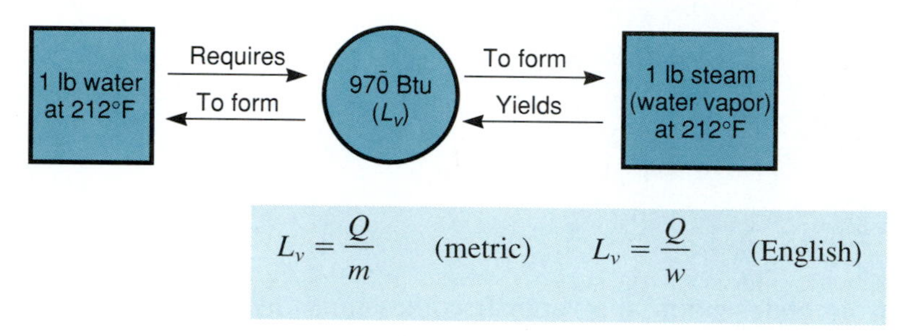

$$L_v = \frac{Q}{m} \quad \text{(metric)} \qquad L_v = \frac{Q}{w} \quad \text{(English)}$$

where L_v = heat of vaporization (see Table 15 in Appendix C)
Q = quantity of heat
m = mass of substance (metric system)
w = weight of substance (English system)

EXAMPLE 2

If 135,000 cal of heat is required to vaporize $25\bar{0}$ g of water at $10\bar{0}°C$, what is the heat of vaporization of water?

Data:

$$Q = 135{,}000 \text{ cal}$$
$$m = 25\bar{0} \text{ g}$$
$$L_v = ?$$

Basic Equation:

$$L_v = \frac{Q}{m}$$

Working Equation: Same

Substitution:

$$L_v = \frac{135{,}000 \text{ cal}}{25\bar{0} \text{ g}} = 54\bar{0} \text{ cal/g}$$

In general, heat of vaporization (water) = $54\bar{0}$ cal/g, or $54\bar{0}$ kcal/kg, or 2.26 MJ/kg, or $97\bar{0}$ Btu/lb.

EXAMPLE 3

If 15.8 MJ of heat is required to vaporize 18.5 kg of ethyl alcohol at 78.5°C (its boiling point), what is the heat of vaporization of ethyl alcohol?

Data:

$$Q = 15.8 \text{ MJ}$$
$$m = 18.5 \text{ kg}$$
$$L_v = ?$$

Basic Equation:

$$L_v = \frac{Q}{m}$$

Working Equation: Same

Substitution:

$$L_v = \frac{15.8 \text{ MJ}}{18.5 \text{ kg}}$$
$$= 0.854 \text{ MJ/kg} \quad \text{or} \quad 854 \text{ kJ/kg} \quad \text{or} \quad 8.54 \times 10^5 \text{ J/kg}$$

Figures 13.27 through 13.29 show the heat gained by one unit of ice at a temperature below its melting point as it warms to its melting point, changes to water, warms to its boiling point, changes to steam, and then is heated above its boiling point in Btu, calories, and joules. Note that during changes of state there are no temperature changes. Recall the basic shape of these graphs because we will use it to find the amount of heat gained or lost when a quantity of material goes through one or both changes of state. Refer to Figure 13.30 to do such problems. See Table 15 of Appendix C for heat constants of some common substances.

Figure 13.27

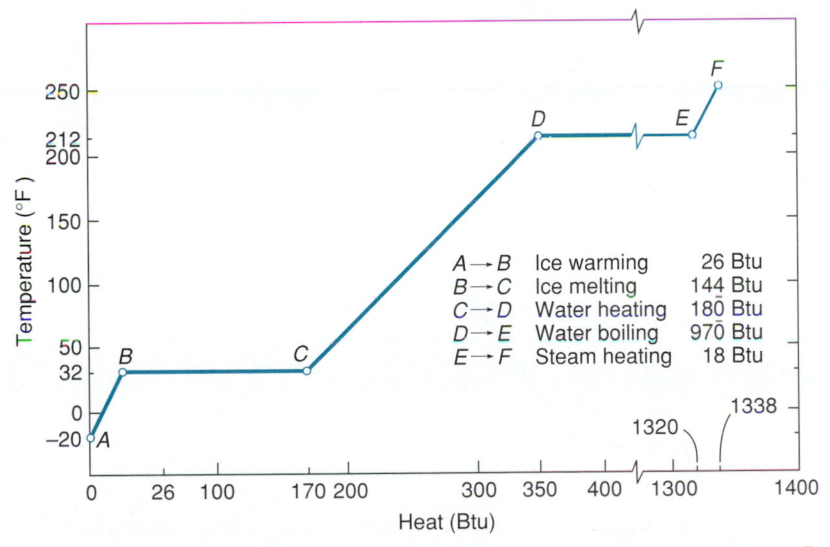

Heat gained by one pound of ice at −20°F as it is converted to steam at 250°F.

EXAMPLE 4

How many Btu of heat are released when 4.00 lb of steam at 222°F is cooled to water at 82°F?

Figure 13.28

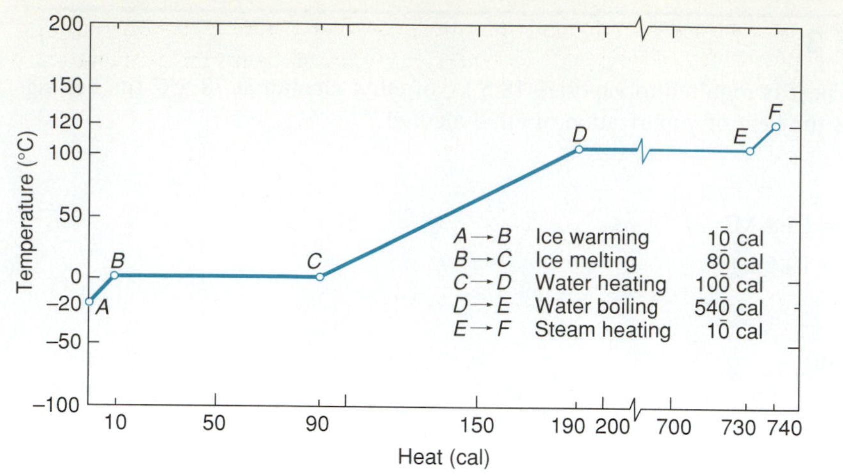

Heat gained by one gram of ice at −20°C as it is converted to steam at 120°C.

Figure 13.29

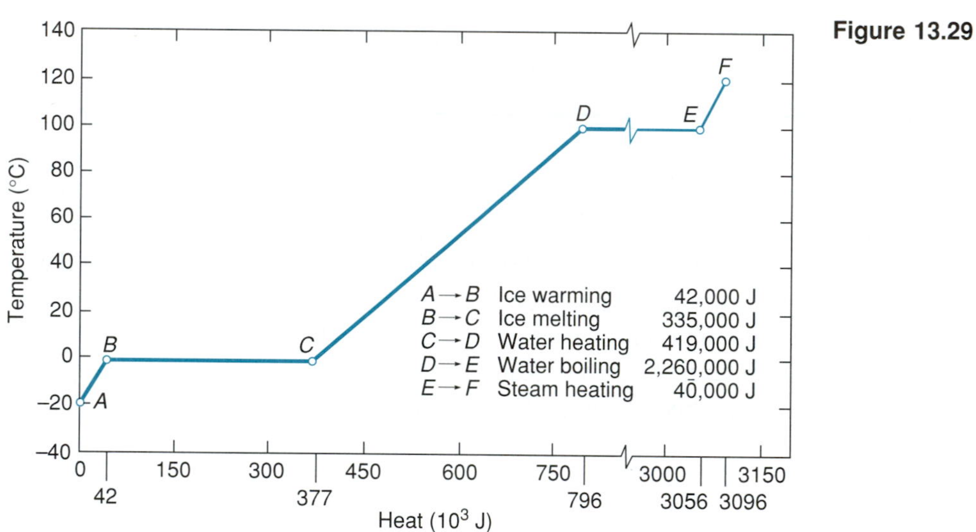

Heat gained by one kilogram of ice at −20°C as it is converted to steam at 120°C.

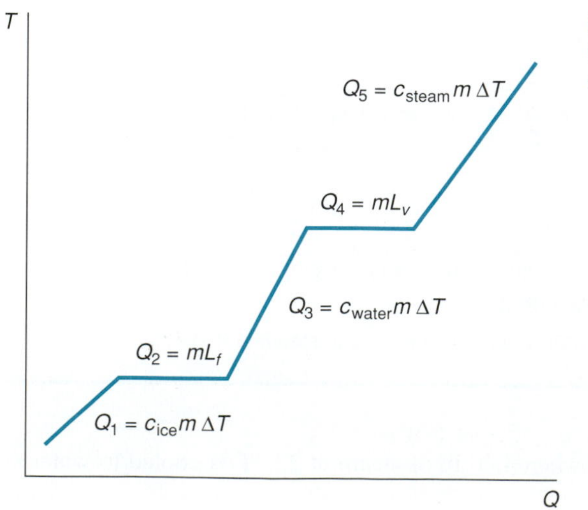

Figure 13.30
Graph of heat transfer during change of state.

To find the amount of heat released when steam at a temperature above its vaporization point is cooled to water below its boiling point, we need to consider three amounts (see Fig. 13.31).

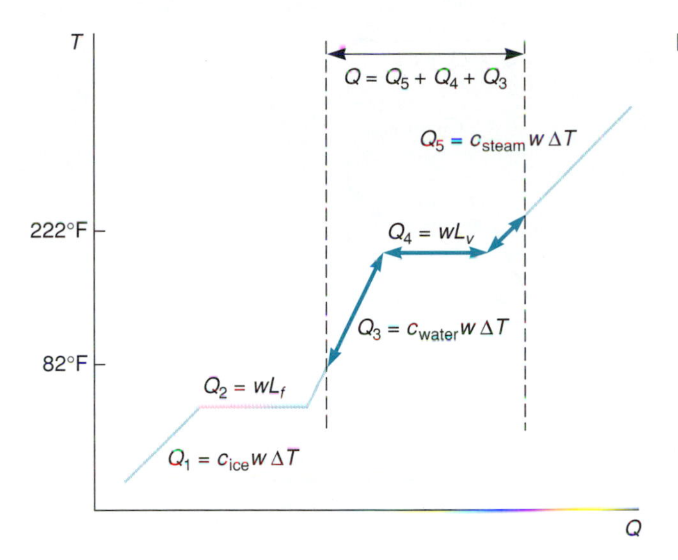

Figure 13.31

$Q_5 = c_{\text{steam}}w\ \Delta T$	(amount of heat released as the steam changes temperature from 222°F to 212°F)
$Q_4 = wL_v$	(amount of heat released as the steam changes to water)
$Q_3 = c_{\text{water}}w\ \Delta T$	(amount of heat released as the water changes temperature from 212°F to 82°F)

So the total amount of heat released is

$$Q = Q_5 + Q_4 + Q_3$$

Data:

$$w = 4.00 \text{ lb}$$
$$T_i \text{ of steam} = 222°\text{F}$$
$$T_f \text{ of water} = 82°\text{F}$$
$$Q = ?$$

Basic Equation:

$$Q = Q_5 + Q_4 + Q_3$$

Working Equation:

$$Q = c_{\text{steam}}w\ \Delta T + wL_v + c_{\text{water}}w\ \Delta T$$

Substitution:

$$Q = \left(0.48\ \frac{\text{Btu}}{\cancel{\text{lb}}\ \cancel{°\text{F}}}\right)(4.00\ \cancel{\text{lb}})(1\bar{0}\cancel{°\text{F}}) + (4.00\ \cancel{\text{lb}})\left(97\bar{0}\ \frac{\text{Btu}}{\cancel{\text{lb}}}\right)$$
$$+ \left(1\ \frac{\text{Btu}}{\cancel{\text{lb}}\ \cancel{°\text{F}}}\right)(4.00\ \cancel{\text{lb}})(13\bar{0}\cancel{°\text{F}})$$
$$= 4420 \text{ Btu}$$

EXAMPLE 5

How many joules of heat are needed to change 3.50 kg of ice at $-15.0°C$ to steam at $120.0°C$?

Sketch:

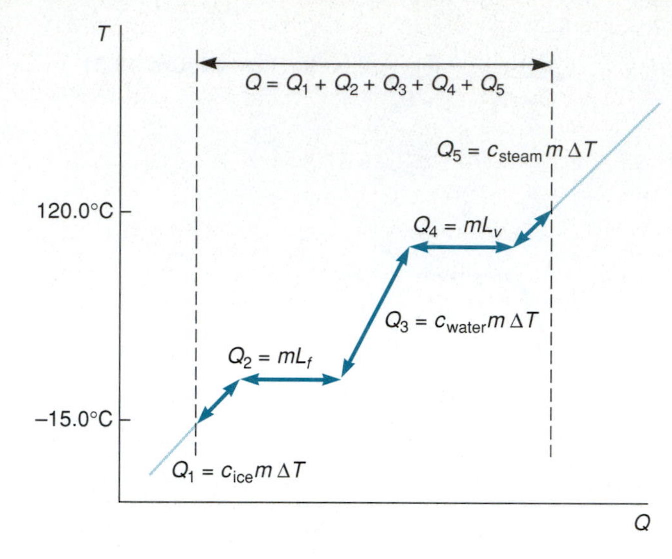

Data:

$$m = 3.50 \text{ kg}$$
$$T_i \text{ of ice} = -15.0°\text{C}$$
$$T_f \text{ of steam} = 120.0°\text{C}$$
$$Q = ?$$

Basic Equation:

$$Q = Q_1 + Q_2 + Q_3 + Q_4 + Q_5$$

Working Equation:

$$Q = c_{ice} m\, \Delta T + mL_f + c_{water} m\, \Delta T + mL_v + c_{steam} m\, \Delta T$$

Substitution:

$$Q = \left(2100 \frac{\text{J}}{\cancel{\text{kg}}\cancel{°\text{C}}}\right)(3.50\ \cancel{\text{kg}})(15.0\cancel{°\text{C}}) + (3.50\ \cancel{\text{kg}})\left(335 \frac{\text{kJ}}{\cancel{\text{kg}}}\right)$$
$$+ \left(4190 \frac{\text{J}}{\cancel{\text{kg}}\cancel{°\text{C}}}\right)(3.50\ \cancel{\text{kg}})(100.0\cancel{°\text{C}}) + (3.50\ \cancel{\text{kg}})\left(2.26 \frac{\text{MJ}}{\cancel{\text{kg}}}\right)$$
$$+ \left(2\bar{0}00 \frac{\text{J}}{\cancel{\text{kg}}\cancel{°\text{C}}}\right)(3.50\ \cancel{\text{kg}})(20.0\cancel{°\text{C}})$$
$$= 1.080 \times 10^7 \text{ J} \quad \text{or} \quad 10.80 \text{ MJ}$$

Evaporation as a Cooling Process

Evaporation, which is the change of a liquid to a gas, helps keep your body cool. When you become too warm, your sweat glands produce water, which evaporates from your skin. As the water evaporates, your body loses heat at the rate of $\frac{Q}{m} = L_v$. In a cool summer breeze, the perspiration evaporates more rapidly, which cools you faster. On a hot, humid day, you tend to remain hot because the perspiration does not evaporate as quickly.

Years ago when a person, especially a child, had a very high body temperature, the common medical practice was to rub the body with rubbing alcohol because it quickly evaporates from the skin. As it evaporates, it removes heat from the body. This practice is not recommended now because we know the alcohol is also absorbed in the body.

During evaporation the molecules of a liquid continually leave the surface of the liquid. Some molecules have enough energy to leave, freeing themselves from the liquid's surface. The rest, not having enough energy to leave, fall back and remain as part of the liquid. The rate of evaporation of a liquid depends on the following:

1. *Amount of surface area.* The larger the surface area, the more molecules that have a chance to escape from the surface.
2. *Temperature.* The higher the temperature, the higher the molecular energy of the molecules, which allows more molecules to escape.
3. *Surface currents.* Air currents blowing over the liquid's surface remove many of the molecules that have evaporated before they fall back into the liquid, which is why a cool summer breeze "feels so good."
4. *Volatility.* The **volatility** of a liquid is a measure of its ability to vaporize. Examples of highly volatile liquids are rubbing alcohol and gasoline. The more volatile the liquid, the greater its rate of evaporation.
5. *Pressure on or above the liquid.* The lower the pressure, the greater the rate of evaporation. Under a partial vacuum, there are fewer molecules around with which the liquid molecules may collide, allowing for a higher rate of escape and a higher rate of evaporation.
6. *Humidity.* If the liquid is exposed to the atmosphere, lower humidity values will provide for greater evaporation.

Heat Pump

A **heat pump** transfers heat from a lower-temperature source to a higher-temperature source or vice versa. It is often used for heating in the winter and cooling in the summer.

A heat pump contains a vapor, usually called a *refrigerant,* that is easily condensed to a liquid when under pressure. The liquid refrigerant gives up the heat gained during compression to the higher-temperature source. The liquid is then released to a low-pressure part of the heat pump, where it quickly evaporates and takes its heat of vaporization from the lower-temperature heat source. The vapor is then compressed again, and the cycle repeats. Work is done on the vapor for the heat pump to transfer the heat from a lower-temperature source to a higher-temperature one.

In winter, a house may be heated by a heat pump that extracts heat from the outside air and transfers it into the house. In summer, the heat pump is reversed and extracts heat from the inside air and transfers it to the outdoors (Fig. 13.32).

Refrigerators and freezers are forms of heat pumps. Heat is transferred from the appliance to the room.

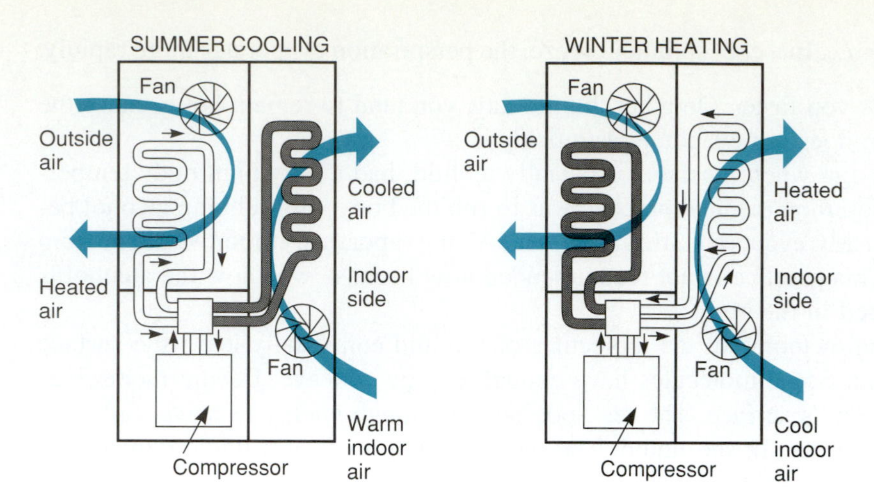

Figure 13.32
The heat pump is used to heat a home in winter and cool it in summer.

Effects of Pressure and Impurities on Change of State

Automobile cooling systems present important problems concerning change of state. Most substances contract on solidifying. However, water and a few other substances expand. The tremendous force exerted by this expansion is shown by the number of cracked automobile blocks and burst radiators suffered by careless motorists every winter.

Impurities in water tend to *lower* the freezing point. Alcohol has a lower freezing point than water and is used in some types of antifreeze. By mixing antifreeze with water in the cooling system, the freezing point of the water may be lowered to avoid freezing in winter. Automobile engines may also be ruined in winter by overheating if the water in the radiator is frozen, preventing the engine from being cooled by circulation in the system.

An increase in the pressure on a liquid *raises* the boiling point. Automobile manufacturers utilize this fact by pressurizing their cooling systems and thereby raising the boiling point of the coolant used.

A decrease in the pressure on a liquid *lowers* the boiling point. Frozen concentrated orange juice is produced by subjecting the pure juice to very low pressures at which the water in the juice is evaporated. Then the consumer must restore the lost water before serving the juice.

■ PROBLEMS 13.8

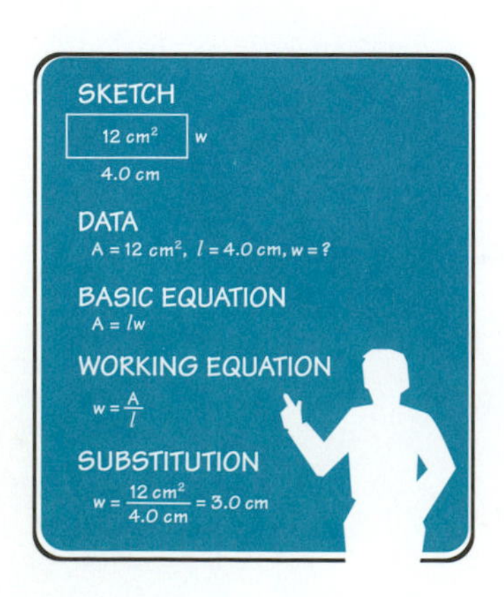

1. How many calories of heat are required to melt 14.0 g of ice at 0°C?
2. How many pounds of ice at 32°F can be melted by the addition of 635 Btu of heat?
3. How many Btu of heat are required to vaporize 11.0 lb of water at 212°F?
4. How many grams of steam in a boiler at $10\bar{0}$°C can be condensed to water at $10\bar{0}$°C by the removal of 1520 cal of heat?
5. How many calories of heat are required to melt 320 g of ice at 0°C?
6. How many calories of heat are given off when 3250 g of steam is condensed to water at $10\bar{0}$°C?
7. How many joules of heat are required to melt 20.0 kg of ice at 0°C?
8. How many kilocalories of heat are required to melt 20.0 kg of ice at 0°C?
9. How many joules of heat need to be removed to condense 1.50 kg of steam at $10\bar{0}$°C?

10. How many litres of water at $10\bar{0}$°C are vaporized by the addition of 5.00 MJ of heat?
11. How many Btu of heat are required to melt 33.0 lb of ice at 32°F and to raise the temperature of the melted ice to 72°F?
12. How many Btu of heat are liberated when 20.0 lb of water at $8\bar{0}$°F is cooled to 32°F and then frozen in an ice plant?
13. How many Btu of heat are required to change 9.00 lb of ice at $1\bar{0}$°F to steam at 232°F?
14. How many calories of heat are liberated when $20\bar{0}$ g of steam at $12\bar{0}$°C is changed to ice at −12°C?
15. How many kilocalories of heat are required to melt 50.0 kg of ice at 0°C and to raise the temperature of the melted ice to $2\bar{0}$°C?
16. How many joules of heat are required to melt 15.0 kg of ice at 0°C and to raise the temperature of the melted ice to 75°C?
17. How many joules of heat need to be removed from 1.25 kg of steam at 115°C to condense it to water and cool the water to $5\bar{0}$°C?
18. How many kcal of heat are needed to vaporize 5.00 kg of water at $10\bar{0}$°C and raise the temperature of the steam to 145°C?
19. How many calories of heat are needed to change 625 g of ice at −24.0°C to steam at 132.0°C?
20. How many Btu of heat must be withdrawn from 5.65 lb of steam at 236.0°F to change it to ice at 12.0°F?
21. How many kilocalories of heat are needed to change 143 N of ethyl alcohol at 65.0°C to vapor?
22. How many joules of heat does 620 g of mercury require to go from a solid at −38.9°C to vapor? ■

GLOSSARY

Absolute Zero The lowest possible temperature. (p. 323)
Boiling Point The temperature at which water boils at sea level. (p. 319)
Btu The amount of heat (energy) necessary to raise the temperature of 1 lb of water 1°F. (p. 323)
Calorie The amount of heat necessary to raise the temperature of 1 gram of water 1°C at its maximum density. (p. 323)
Celsius Scale The metric temperature scale on which ice melts at 0° and water boils at 100°. (p. 319)
Change of State A change in a substance from one form of matter (solid, liquid, or gas) to another form. (p. 344)
Coefficient of Linear Expansion A constant that indicates the amount by which a material expands or contracts when heated. (p. 338)
Condensation The change of state from gas or vapor to the liquid state. (p. 346)
Conduction A form of heat transfer from a warmer part of a substance to a cooler part as a result of molecular collisions, which cause the slower-moving molecules to move faster. (p. 325)
Convection A form of heat transfer in which the movement of warm gases or liquids from one place to another carries heat along with it. Caused by the expansion of liquids or gases as they are heated or cooled. This expansion makes the warmer gas or liquid less dense than the surrounding fluid. The lighter fluid is then forced upward by the heavier surrounding fluid. (p. 326)
Degree A unit of measurement on a temperature scale. (p. 319)
Evaporation The process by which molecules of a liquid continually leave the surface of a liquid. The most energetic molecules escape from the liquid. (p. 352)

Fahrenheit Scale The temperature scale on which ice melts at 32° and water boils at 212°. (p. 320)

Freezing The change of state from liquid to solid. Also called *solidification*. (p. 344)

Fusion The change of state from solid to liquid. Also called *melting*. (p. 344)

Heat The name given to the energy added to or taken away from an object when the internal kinetic or potential energy is increased or decreased. A temperature increase or change in phase (solid to a liquid, etc.) is often associated with the addition of heat to an object. (p. 322)

Heat of Fusion The heat required to melt 1 g or 1 kg or 1 lb of a liquid. (p. 345)

Heat of Vaporization The amount of heat required to vaporize 1 g or 1 kg or 1 lb of a liquid. (p. 347)

Heat Pump A pump containing a vapor (refrigerant) that is easily condensed to a liquid when under pressure. Produces heat during compression and cools during vaporization. (p. 353)

Ice Point The temperature at which ice melts. (p. 319)

Kelvin Scale The absolute temperature scale on which absolute zero is 0 and the units are the same as on the Celsius scale. (p. 321)

Kilocalorie The amount of heat necessary to raise the temperature of 1 kilogram of water 1°C. (p. 323)

Mechanical Equivalent of Heat The relationships that show that heat and mechanical work are related. Conversion factors can be used to relate amounts of heat and work. (p. 323)

Melting The change of state from solid to liquid. Also called *fusion*. (p. 344)

Method of Mixtures When two substances at different temperatures are "mixed together," heat flows from the warmer body to the cooler body until they reach the same temperature. Heat lost by the warmer body is transferred to the cooler body and to surrounding objects. If the two substances are well insulated from surrounding objects, the heat lost by the warmer body is equal to the heat gained by the cooler body. (p. 332)

Radiation A method of heat transfer in which heat is transferred in the form of radiant energy. Radiant energy is electromagnetic radiation such as infrared or visible light. (p. 326)

Rankine Scale The absolute temperature scale on which absolute zero is 0° and the degree units are the same as on the Fahrenheit scale. (p. 321)

Relative Humidity Ratio of the actual amount of vapor in the atmosphere to the amount of vapor required to reach 100% of saturation at the existing temperature. (p. 347)

Solidification The change of state from liquid to solid. Also called *freezing*. (p. 344)

Specific Heat The amount of heat necessary to change the temperature of 1 lb by 1°F in the English system or 1 kg by 1°C in the metric system. (p. 330)

Temperature A measure of the average velocity of the molecules in a substance. (p. 318)

Thermal Conductivity The ability of a material to transfer heat by conduction. (p. 327)

Vaporization The change of state from liquid to the gaseous or vapor state. (p. 346)

Volatility A measure of a liquid's ability to vaporize. The more volatile the liquid, the greater its rate of evaporation. Gasoline and rubbing alcohol are highly volatile liquids. (p. 353)

FORMULAS

13.1 $T_F = \frac{9}{5} T_C + 32°$

$T_C = \frac{5}{9}(T_F - 32°)$

$T_R = T_F + 46\bar{0}°$

$T_K = T_C + 273$

13.3 $Q = \dfrac{KAt(T_2 - T_1)}{L}$

$R = \dfrac{L}{K}$

13.4 $Q = cw\ \Delta T$

$Q = cm\ \Delta T$

13.5 $Q_{\text{lost}} = Q_{\text{gained}}$

$c_l w_l(T_l - T_f) = c_g w_g(T_f - T_g)$

13.6 $\Delta l = \alpha l\ \Delta T$

$\Delta A = 2\alpha A\ \Delta T$

$\Delta V = 3\alpha V\ \Delta T$

13.7 $\Delta V = \beta V\ \Delta T$

13.8 $L_f = \dfrac{Q}{w}$ $\quad L_f = \dfrac{Q}{m}$

$L_v = \dfrac{Q}{w}$ $\quad L_v = \dfrac{Q}{m}$

REVIEW QUESTIONS

1. Which of the following are methods of heat transfer?
   (a) Convection (b) Conduction
   (c) Temperature (d) Radiation
   (e) Potential energy
2. Which of the following are good conductors of heat?
   (a) Air (b) Copper (c) Steel
   (d) Aluminum (e) Brick (f) Mineral wool
3. The amount that a solid expands when heated depends on
   (a) the type of material. (b) the length of the solid.
   (c) the density of the solid. (d) the amount of temperature change.
   (e) all of the above.
4. The rate of evaporation from the surface of a liquid depends on
   (a) the temperature. (b) the volatility of the liquid.
   (c) the mass density of the liquid. (d) the air pressure above the liquid.
5. The amount of heat required to melt 1 kg of a solid is called its
   (a) heat of vaporization. (b) mass density.
   (c) weight density. (d) heat of fusion.
   (e) volume expansion.
6. The operation of a simple thermostat depends on
   (a) the mechanical equivalent of heat. (b) specific heat.
   (c) thermal expansion. (d) the R value.
7. In your own words, describe the method of mixtures.
8. What is the mechanical equivalent of heat in the English system?
9. Which other temperature scale is closely related to the Fahrenheit scale?
10. Which other temperature scale is closely related to the Celsius scale?
11. Distinguish between the Celsius and Fahrenheit temperature scales.
12. Distinguish between heat and temperature.
13. Give three examples of the conversion of heat into useful work.

14. Give three examples of the conversion of work into heat.
15. Should you wear light- or dark-colored clothing on a hot sunny summer day? Explain.
16. Does the area of a hole cut out of a metal block increase or decrease as the metal is heated? Explain.
17. At what temperature does water have its highest density? How does water differ from other liquids in this regard?
18. What is the impact of the characteristic of water discussed in Question 17?
19. Which would cool a hot object better: 10 kg of water at 0° C or 10 kg of ice at 0° C? Explain.
20. Steam can cause much more severe burns than hot water. Explain.
21. Why are ice cubes often observed to have a slight mound on the top of the cube?
22. In your own words, describe each method of heat transfer.
23. Describe why automotive cooling systems are designed to operate at elevated pressures.
24. Explain how a heat pump works to heat in the winter and cool in the summer.

REVIEW PROBLEMS

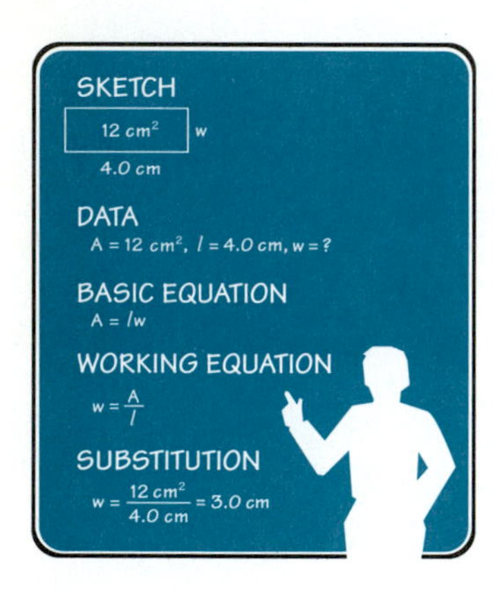

1. Change 344 K to degrees Celsius.
2. Change 24°C to Kelvin.
3. Change 5110°C to degrees Fahrenheit.
4. Change 635°F to degrees Celsius.
5. Find the amount of heat in cal generated by 43.0 J of work.
6. Find the amount of heat in kcal generated by 6530 J of work.
7. Find the amount of work equivalent to 435 Btu.
8. Find the heat flow during 4.10 h through a glass window of thickness 0.15 in. with an area of 33 ft^2 if the average outer surface temperature is 22°F and the average inner glass surface is 48°F.
9. Find the heat flow in 25.0 days through a freezer door 80.0 cm × 144 cm insulated with cellulose fiber 4.0 cm thick. The temperature inside the freezer is −14°C. Room temperature is 22°C.
10. How many Btu of heat must be added to 835 lb of steel to raise its temperature from 20.0°F to 455°F?
11. How many kcal of heat must be added to 148 kg of aluminum to raise its temperature from 21.5°C to 485°C?
12. A 161-kg steam boiler is made of steel and contains 1127 N of water at 8.9°C. How much heat is required to raise the temperature of both the boiler and water to $10\bar{0}$°C?
13. A 3.80-lb piece of copper is dropped into 8.35 lb of water at 48.0°F. The final temperature is 98.2°F. What was the initial temperature of the copper?
14. A 355-g piece of metal at 48.0°C is dropped into 111 g of water at 15.0°C. If the final temperature of the mixture is 32.5°C, what is the specific heat of the metal?
15. A brass rod 45.2 cm long expands 0.734 mm when heated. Find the temperature change.
16. The length of a steel rod at 5°C is 12.500 m. What is its length when heated to 154°C?
17. The diameter of a hole drilled through aluminum at 22°C is 7.50 mm. Find the diameter and the area of the hole at a temperature of 89°C.
18. A steel ball has a radius of 1.54 cm at 35°C. Find its change in volume when the temperature is increased to 84.5°C.

19. Find the increase in volume of 44.8 L of acetone when it is heated from 37.0°C to 75.5°C.
20. What is the decrease in volume of alcohol in a railroad tank car that contains 3450 ft^3 if the temperature drops from 87.0°F to 33.0°F?
21. How many kcal of heat are required to vaporize 21.5 kg of water at $10\bar{0}$°C?
22. How many Btu of heat are required to melt 8.35 lb of ice at 32°F?
23. How many kilocalories of heat must be withdrawn from 4.56 kg of steam at 125°C to change it to ice at −44.5°C?
24. How many joules of heat are required to change 336 g of ethyl alcohol from a solid at −117°C to a vapor at 78.5°C?

Chapter 14

GAS LAWS

Gas laws concern the behavior of "ideal" gases at standard conditions of temperature and pressure. The focus of this chapter is on the behavior of gases as temperature and pressure are varied.

OBJECTIVES

The major goals of this chapter are to enable you to:

1. Use Charles' law to determine thermal expansion.
2. Apply Boyle's law to calculate volume changes of gases.
3. Relate gas density to pressure and temperature.

14.1 CHARLES' LAW

Before making a long summer trip, you notice that the tires are low. You stop at a gas station around the corner and have the attendant add air to 28 lb/in^2 gauge pressure. Later in the afternoon you stop for gas. Since your tires were low that morning, you decide to check them again. Now you notice that they look a bit larger, and the gauge pressure is 40 lb/in^2. What happened? When a gas is heated, the increased kinetic energy causes the volume to increase, the pressure to increase, or both to increase.

CHARLES' LAW

If the pressure on a gas is constant, the volume is directly proportional* to its absolute (Kelvin or Rankine) temperature.

By formula,

$$\frac{V}{T} = \frac{V'}{T'} \quad \text{or} \quad VT' = V'T$$

where V = original volume
T = original temperature
V' = final volume
T' = final temperature

EXAMPLE 1

A gas occupies $45\bar{0}$ cm^3 at $3\bar{0}$°C. At what temperature will the gas occupy $48\bar{0}$ cm^3?

Data:

$V = 45\bar{0} \text{ cm}^3$
$T = 3\bar{0}° + 273 = 303 \text{ K}$ (***Note:*** We must use Kelvin temperature.)
$V' = 48\bar{0} \text{ cm}^3$
$T' = ?$

*Directly proportional means that as temperature increases, volume increases, and as temperature decreases, volume decreases.

Basic Equation:

$$\frac{V}{T} = \frac{V'}{T'}$$

Working Equation:

$$T' = \frac{TV'}{V}$$

Substitution:

$$T' = \frac{(303\text{ K})(48\bar{0}\text{ cm}^3)}{45\bar{0}\text{ cm}^3}$$
$$= 323\text{ K} \quad \text{or} \quad 5\bar{0}°\text{C}$$

EXAMPLE 2

At $4\bar{0}$°F, some helium occupies 15.0 ft^3. What will be the volume of this gas at $9\bar{0}$°F?

Data:

$V = 15.0\text{ ft}^3$

$T = 4\bar{0}° + 46\bar{0}° = 50\bar{0}°\text{R}$ (***Note:*** We must use Rankine temperature.)

$T' = 9\bar{0}° + 46\bar{0}° = 55\bar{0}°\text{R}$

$V' = ?$

Basic Equation:

$$\frac{V}{T} = \frac{V'}{T'}$$

Working Equation:

$$V' = \frac{VT'}{T}$$

Substitution:

$$V' = \frac{(15.0\text{ ft}^3)(55\bar{0}°\text{R})}{50\bar{0}°\text{R}}$$
$$= 16.5\text{ ft}^3$$

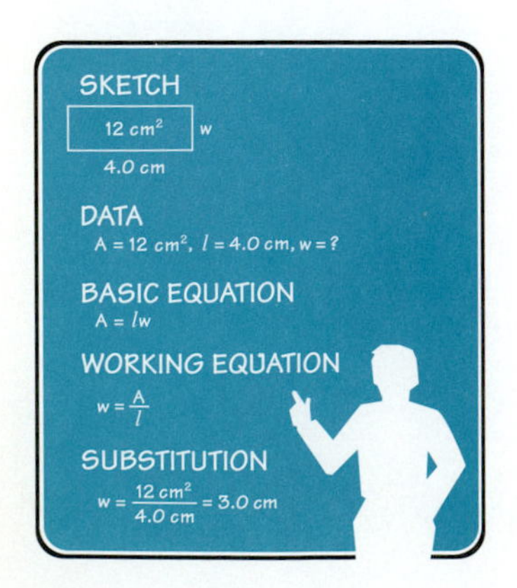

■ PROBLEMS 14.1

1. Change 15°C to K.
2. Change −14°C to K.
3. Change 317 K to °C
4. Change 235 K to °C.
5. Change 72°F to °R.
6. Change −55°F to °R.
7. Change $55\bar{0}$°R to °F.
8. Change 375°R to °F.

Use $\dfrac{V}{T} = \dfrac{V'}{T'}$ to find each quantity:

9. $T = 315$ K, $V' = 225$ cm^3, $T' = 275$ K, find V.
10. $T = 615$°R, $V = 60.3$ in^3, $T' = 455$°R, find V'.
11. $V = 20\bar{0}$ ft^3, $T' = 95$°F, $V' = 25\bar{0}$ ft^3, find T.
12. $V = 19.7$ L, $T = 51$°C, $V' = 25.2$ L, find T'.
13. Some gas occupies a volume of 325 m^3 at 41°C. What is its volume at 94°C?

14. Some oxygen occupies 275 in^3 at 35°F. Find its volume at 95°F.
15. Some methane occupies 1575 L at 45°C. Find its volume at 15°C.
16. Some helium occupies $120\bar{0}$ ft^3 at $7\bar{0}$°F. At what temperature will its volume be $60\bar{0}$ ft^3?
17. Some nitrogen occupies 14,300 cm^3 at 25.6°C. What is the temperature when its volume is 10,250 cm^3?
18. Some propane occupies 1270 cm^3 at 18.0°C. What is the temperature when its volume is 1530 cm^3?
19. Some carbon dioxide occupies 34.5 L at 49.0°C. Find its volume at 12.0°C.
20. Some oxygen occupies 28.7 ft^3 at 11.0°F. Find its temperature when its volume is 18.5 ft^3.
21. A balloon contains 26.0 L of hydrogen at 40.0°F. What is the Kelvin temperature change needed to make the balloon expand to 36.0 L?
22. Using Charles' law, what is the effect
    (a) on the temperature of a gas when the volume is doubled?
    (b) on the temperature of a gas when the volume is tripled?
    (c) on the volume when temperature is doubled?
    (d) Explain the relationship between the volume and temperature. ■

14.2 BOYLE'S LAW

BOYLE'S LAW

If the temperature of a gas is constant, the volume is inversely proportional* to the absolute pressure.

By formula,

$$\frac{V}{V'} = \frac{P'}{P} \quad \text{or} \quad VP = V'P'$$

where V = original volume
V' = final volume
P = original pressure
P' = final pressure

Note: The pressure must be expressed in terms of *absolute pressure*.

EXAMPLE 1

Some oxygen occupies $50\bar{0}$ in^3 at an absolute pressure of 40.0 lb/in^2 (psi). What is its volume at an absolute pressure of $10\bar{0}$ psi?

Data:

$$V = 50\bar{0} \text{ in}^3$$
$$P = 40.0 \text{ psi}$$
$$P' = 10\bar{0} \text{ psi}$$
$$V' = ?$$

*Inversely proportional means that as volume increases, pressure decreases, and as volume decreases, pressure increases.

Basic Equation:

$$\frac{V}{V'} = \frac{P'}{P}$$

Working Equation:

$$V' = \frac{VP}{P'}$$

Substitution:

$$V' = \frac{(50\bar{0} \text{ in}^3)(40.0 \text{ psi})}{10\bar{0} \text{ psi}}$$
$$= 20\bar{0} \text{ in}^3$$

EXAMPLE 2

Some nitrogen occupies 20.0 m^3 at a gauge pressure of 274 kPa. Find the absolute pressure when its volume is 30.0 m^3.

Data:

$$V = 20.0 \text{ m}^3$$
$$P_{abs} = 274 \text{ kPa} + 101 \text{ kPa} = 375 \text{ kPa}$$
$$V' = 30.0 \text{ m}^3$$
$$P' = ?$$

Basic Equation:

$$\frac{V}{V'} = \frac{P'}{P}$$

Working Equation:

$$P' = \frac{VP}{V'}$$

Substitution:

$$P' = \frac{(20.0 \text{ m}^3)(375 \text{ kPa})}{30.0 \text{ m}^3}$$
$$= 25\bar{0} \text{ kPa}$$

Density and Pressure

If the pressure of a given amount (constant volume) of gas is increased, its density increases as the gas molecules are forced closer together. (Recall that density is discussed in Section 11.5.) Also, if the pressure is decreased, the density decreases. That is, the *density of a gas is directly proportional to its pressure* as long as there is no change in state. In equation form,

$$\frac{D}{D'} = \frac{P}{P'} \quad \text{or} \quad DP' = D'P$$

where D = original density
D' = final density
P = original pressure (absolute)
P' = final pressure (absolute)

EXAMPLE 3

Some carbon dioxide has a density of 1.60 kg/m^3 at an absolute pressure of 95.0 kPa. What is the density when the pressure is decreased to 80.0 kPa?

Data:

$$D = 1.60 \text{ kg/m}^3$$
$$P = 95.0 \text{ kPa}$$
$$P' = 80.0 \text{ kPa}$$
$$D' = ?$$

Basic Equation:

$$\frac{D}{D'} = \frac{P}{P'}$$

Working Equation:

$$D' = \frac{DP'}{P}$$

Substitution:

$$D' = \frac{(1.60 \text{ kg/m}^3)(80.0 \cancel{\text{kPa}})}{95.0 \cancel{\text{kPa}}}$$
$$= 1.35 \text{ kg/m}^3$$

EXAMPLE 4

A gas has a density of 2.00 kg/m^3 at a gauge pressure of $16\overline{0}$ kPa. What is the density at a gauge pressure of $30\overline{0}$ kPa?

Data:

$$D = 2.00 \text{ kg/m}^3$$
$$P = 16\overline{0} \text{ kPa} + 101 \text{ kPa} = 261 \text{ kPa}$$
$$P' = 30\overline{0} \text{ kPa} + 101 \text{ kPa} = 401 \text{ kPa}$$
$$D' = ?$$

Basic Equation:

$$\frac{D}{D'} = \frac{P}{P'}$$

Working Equation:

$$D' = \frac{DP'}{P}$$

Substitution:

$$D' = \frac{(2.00 \text{ kg/m}^3)(401 \cancel{\text{kPa}})}{261 \cancel{\text{kPa}}}$$
$$= 3.07 \text{ kg/m}^3$$

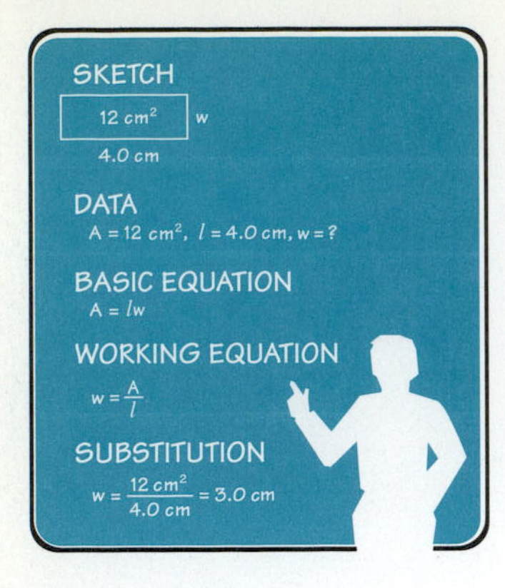

■ PROBLEMS 14.2

Use $\frac{V}{V'} = \frac{P'}{P}$ or $\frac{D}{D'} = \frac{P}{P'}$ to find each quantity. (All pressures are absolute unless otherwise stated.)

1. $V' = 315\text{ cm}^3$, $P = 101$ kPa, $P' = 85.0$ kPa; find V.
2. $V = 45\bar{0}$ L, $V' = 70\bar{0}$ L, $P = 75\bar{0}$ kPa; find P'.
3. $V = 76.0\text{ m}^3$, $V' = 139\text{ m}^3$, $P' = 41.0$ kPa; find P.
4. $V = 439\text{ in}^3$, $P' = 38.7$ psi, $P = 47.1$ psi; find V'.
5. $D = 1.80\text{ kg/m}^3$, $P = 108$ kPa, $P' = 125$ kPa; find D'.
6. $D = 1.65\text{ kg/m}^3$, $P = 87.0$ kPa, $D' = 1.85\text{ kg/m}^3$; find P'.
7. $P = 51.0$ psi, $P' = 65.3$ psi, $D' = 0.231\text{ lb/ft}^3$; find D.
8. Some air at 22.5 psi occupies $14\bar{0}0\text{ in}^3$. What is its volume at 18.0 psi?
9. Some nitrogen at a pressure of 110.0 kPa occupies 185 m^3. Find its pressure if its volume is changed to 225 m^3.
10. Some methane at 185.0 kPa occupies 65.0 L. What is its volume at a pressure of 95.0 kPa?
11. Some carbon dioxide has a density of 3.75 kg/m^3 at 815 kPa. What is its density if the pressure is decreased to 725 kPa?
12. Some oxygen has a density of 1.75 kg/m^3 at normal atmospheric pressure. What is its pressure (in kPa) when the density is changed to 1.45 kg/m^3?
13. Some methane at $50\bar{0}$ kPa gauge pressure occupies $75\bar{0}\text{ m}^3$. What is its gauge pressure if its volume is $50\bar{0}\text{ m}^3$?
14. Some helium at 15.0 psi gauge pressure occupies 20.0 ft^3. Find its volume at 20.0 psi gauge pressure.
15. Some nitrogen at 80.0 psi gauge pressure occupies 13.0 ft^3. Find its volume at 50.0 psi gauge pressure.
16. Some carbon dioxide has a density of 6.35 kg/m^3 at 685 kPa gauge pressure. What is the density when the gauge pressure is 455 kPa?
17. Some propane has a density of 48.5 oz/ft^3 at 265 psi gauge pressure. What is the gauge pressure when the density is 30.6 oz/ft^3?
18. Some air occupies 4.5 m^3 at a gauge pressure of 46 kPa. What is the volume at a gauge pressure of 13 kPa?
19. Some oxygen at 87.6 psi (absolute) occupies 75.0 in^3. Find its volume if its absolute pressure is (a) doubled; (b) tripled; (c) halved.
20. A gas at $30\bar{0}$ kPa (absolute) occupies 40.0 m^3. Find its absolute pressure if its volume is (a) doubled; (b) tripled; (c) halved.
21. A volume of 58.0 L of hydrogen is heated from 33°C to 68°C. If its original density is 4.85 kg/m^3 and its original absolute pressure is $12\bar{0}$ kPa, what is the resulting density? ■

14.3 CHARLES' AND BOYLE'S LAWS COMBINED

Most of the time it is very difficult to keep the pressure constant or the temperature constant. In this case we combine Charles' law and Boyle's law as follows:

$$\frac{VP}{T} = \frac{V'P'}{T'} \quad \text{or} \quad VPT' = V'P'T$$

Note: Both pressure and temperature must be absolute.

EXAMPLE

We have $50\bar{0}$ in^3 of acetylene at $4\bar{0}$°F at $200\bar{0}$ psi (absolute). What is the pressure if its volume is changed to $80\bar{0}$ in^3 at $10\bar{0}$°F?

Data:

$$V = 50\bar{0} \text{ in}^3$$
$$P = 200\bar{0} \text{ psi}$$
$$T = 4\bar{0}° + 46\bar{0}° = 50\bar{0}°\text{R}$$
$$V' = 80\bar{0} \text{ in}^3$$
$$T' = 10\bar{0}° + 46\bar{0}° = 56\bar{0}°\text{R}$$
$$P' = ?$$

Basic Equation:

$$\frac{VP}{T} = \frac{V'P'}{T'}$$

Working Equation:

$$P' = \frac{VPT'}{TV'}$$

Substitution:

$$P' = \frac{(50\bar{0} \text{ in}^3)(200\bar{0} \text{ psi})(56\bar{0}°\text{R})}{(50\bar{0}°\text{R})(80\bar{0} \text{ in}^3)}$$
$$= 14\bar{0}0 \text{ psi}$$

The gas laws are reasonably accurate except at very low temperatures and under extreme pressures.

A commonly used reference in gas laws is called **standard temperature and pressure** (STP). Standard temperature is the freezing point of water, 0°C or 32°F. Standard pressure is equivalent to atmospheric pressure, 101.32 kPa or 14.7 lb/in^2.

Vapor Pressure and Humidity

When a liquid evaporates, molecules of water pass from the surface of the water into the air above the liquid. This increase in the number of molecules in the air causes an increase in the pressure above the liquid. This increase in pressure is called *vapor pressure*.

Similarly, some molecules of water vapor in the air will return to the liquid state when moist air is cooled. This is called *condensation*. Condensation occurs at a point in temperature called the *dew point*. Dew forms when moist air is cooled by the earth's surface. It can be closely reproduced in the laboratory by taking a container partly filled with water and adding pieces of ice. When the water and container reach the dew point in temperature, water will condense from the air on the outside of the glass.

The maximum amount of water vapor that air will hold at a given temperature is *absolute humidity*. Any increase becomes rainfall. *Relative humidity*, in contrast, is the comparison of, or ratio of the amount of water vapor a sample of air holds as compared to the maximum amount it can hold if it is saturated.

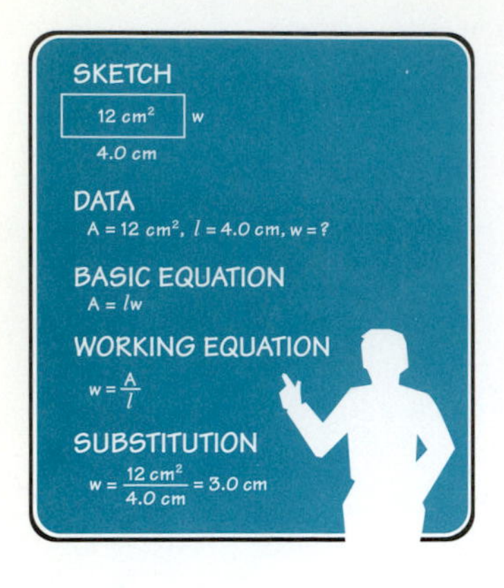

■ PROBLEMS 14.3

Use $\dfrac{VP}{T} = \dfrac{V'P'}{T'}$ to find each quantity. (All pressures are absolute unless otherwise stated.)

1. $P = 825$ psi, $T = 575°$R, $V' = 1550$ in^3, $P' = 615$ psi, $T' = 525°$R; find V.
2. $V = 50\bar{0}$ in^3, $T = 50\bar{0}°$R, $V' = 80\bar{0}$ in^3, $P' = 80\bar{0}$ psi, $T' = 45\bar{0}°$R; find P.
3. $V = 90\bar{0}$ m^3, $P = 105$ kPa, $T = 30\bar{0}$ K, $P' = 165$ kPa, $T' = 265$ K; find V'.
4. $V = 18.0$ m^3, $P = 112$ kPa, $V' = 15.0$ m^3, $P' = 135$ kPa, $T' = 235$ K; find T.
5. $V = 532$ m^3, $P = 135$ kPa, $T = 87°$C, $V' = 379$ m^3, $P' = 123$ kPa; find T'.
6. We have $60\bar{0}$ in^3 of oxygen at $150\bar{0}$ psi at 65°F. What is the volume at $120\bar{0}$ psi at $9\bar{0}$°F?
7. We have $80\bar{0}$ m^3 of natural gas at 235 kPa at $3\bar{0}$°C. What is the temperature if the volume is changed to $120\bar{0}$ m^3 at 215 kPa?
8. We have $140\bar{0}$ L of nitrogen at 135 kPa at 54°C. What is the temperature if the volume changes to $80\bar{0}$ L at 275 kPa?
9. An acetylene welding tank has a pressure of $20\bar{0}0$ psi at $4\bar{0}$°F. If the temperature rises to $9\bar{0}$°F, what is the new pressure?
10. What is the new pressure in Problem 9 if the temperature falls to $-3\bar{0}$°F?
11. An ideal gas occupies a volume of 5.00 L at STP. What is its gauge pressure (in kPa) if the volume is halved and its temperature increases to $40\bar{0}$°C?
12. An ideal gas occupies a volume of 5.00 L at STP.
    (a) What is its temperature if its volume is halved and its absolute pressure is doubled?
    (b) What is its temperature if its volume is doubled and its absolute pressure is tripled?
13. Some propane occupies 2.00 m^3 at 18.0°C at an absolute pressure of 3.50×10^5 N/m^2. Find the absolute pressure (in kPa) at the same temperature when the volume is halved.
14. Find the new temperature in Problem 13 when the absolute pressure is doubled and the volume is doubled.
15. Find the new volume in Problem 13 when the absolute pressure is halved and the temperature is decreased to -12.0°C.
16. Find the new volume in Problem 13 if the absolute pressure is 1.30×10^6 N/m^2 and the temperature is 31.0°C.
17. A balloon with volume $32\bar{0}0$ mL of Xenon gas is at a gauge pressure of 122 kPa and a temperature of 27°C. What is the volume when the balloon is heated to 65°C and the gauge pressure is decreased to 112 kPa? ■

GLOSSARY

Boyle's Law States that if the temperature of a gas is constant, the volume is inversely proportional to the absolute pressure, $V/V' = P'/P$. (p. 363)

Charles' Law States that if the pressure on a gas is constant, the volume is directly proportional to its Kelvin or Rankine temperature, $V/T = V'/T'$. (p. 361)

Standard Temperature and Pressure (STP) A commonly used reference in gas laws. Standard temperature is the freezing point of water. Standard pressure is equivalent to atmospheric pressure. (p. 367)

FORMULAS

14.1 Charles' Law: $\frac{V}{T} = \frac{V'}{T'}$

14.2 Boyle's law: $\frac{V}{V'} = \frac{P'}{P}$

$$\frac{D}{D'} = \frac{P}{P'}$$

14.3 $\frac{VP}{T} = \frac{V'P'}{T'}$

REVIEW QUESTIONS

1. The gas law that relates volume and temperature is called
   (a) Boyle's law. (b) Hooke's law.
   (c) Charles' law. (d) none of the above.
2. The gas law that relates volume and pressure is called
   (a) Boyle's law. (b) Hooke's law.
   (c) Charles' law. (d) none of the above.
3. If the temperature of a gas is constant and the volume is decreased, the pressure will
   (a) stay the same. (b) decrease.
   (c) increase. (d) increase or decrease, depending on the gas.
4. If the temperature of a gas is constant and the pressure is decreased, the volume will
   (a) stay the same. (b) decrease.
   (c) increase. (d) increase or decrease, depending on the gas.
5. If the pressure on a gas is constant and the temperature is decreased, the volume will
   (a) stay the same. (b) decrease.
   (c) increase. (d) increase or decrease, depending on the gas.
6. If the pressure on a gas is constant and the volume is decreased, the temperature will
   (a) stay the same. (b) decrease.
   (c) increase. (d) increase or decrease, depending on the amount of gas.
7. Describe the conditions of standard temperature and pressure.
8. Describe what happens to the volume of a gas if its temperature and pressure increase.
9. Describe what happens to the temperature of a gas if its volume and pressure increase.
10. What causes the tendency of the volume and pressure of a gas to increase when it is heated?
11. What causes the tendency of the temperature of a gas to increase when it is compressed?
12. What causes the tendency of the pressure of a gas to decrease when the volume is increased?

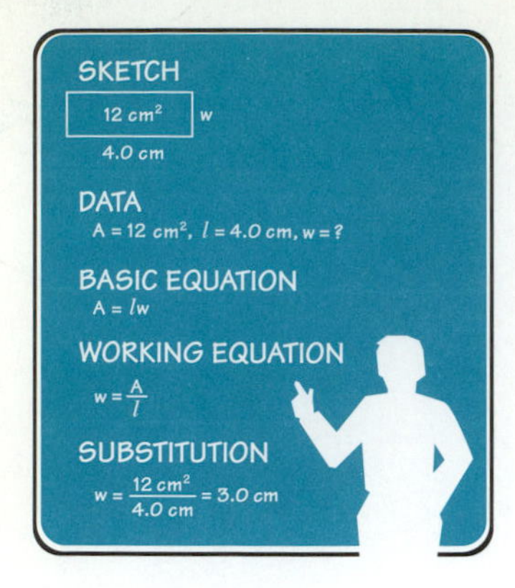

REVIEW PROBLEMS

(All pressures are absolute unless otherwise stated.)

1. A gas occupies 13.5 ft^3 at 35.8°F. What will the volume of this gas be at 88.6°F if the pressure is constant?
2. A gas occupies 3.45 m^3 at 18.5°C. What will the volume of this gas be at 98.5°C if the pressure is constant?
3. Some hydrogen occupies 115 ft^3 at 54.5°F. What is the temperature when the volume is 132 ft^3 if the pressure is constant?
4. Some carbon dioxide occupies 45.3 L at 38.5°C. What is the temperature when the volume is 44.2 L if the pressure is constant?
5. Some propane occupies 145 cm^3 at 12.4°C. What is the temperature when the volume is 156 cm^3 if the pressure is constant?
6. Some air at 276 kPa occupies 32.4 m^3. What is its absolute pressure if its volume is doubled at constant temperature?
7. Some helium at 17.5 psi gauge pressure occupies 35.0 ft^3. What is the volume at 32.4 psi if the temperature is constant?
8. Some carbon dioxide has a density of 6.35 kg/m^3 at 685 kPa gauge pressure. What is the density when the gauge pressure is 355 kPa if the temperature is constant?
9. We have 435 in^3 of nitrogen at 1340 psi gauge pressure at 75°F. What is the volume at 1150 psi gauge pressure at 45°F?
10. We have 755 m^3 of carbon dioxide at 344 kPa at 25°C. Find the temperature if the volume is changed to 1330 m^3 at 197 kPa.
11. A welding tank has a gas pressure of 1950 psi at 38°F. What is the new pressure if the temperature rises to 98°F?
12. What is the temperature if the gas pressure in Problem 11 falls to 1870 psi?
13. An ideal gas occupies a volume of 4.50 L at STP. What is its gauge pressure (in kPa) if the volume is halved and the absolute temperature is doubled?
14. An ideal gas occupies a volume of 5.35 L at STP. What is its gauge pressure (in kPa) if the volume is doubled and the temperature is increased by 45.5°C?
15. A volume of 1120 L of helium at $40\bar{0}0$ Pa is heated from 45°C to 77°C, increasing the pressure to $60\bar{0}0$ Pa. What is the resulting volume of the gas?
16. In a 47-cm-tall cylinder (radius 7.0 cm), hydrogen of density 2.50 kg/m^3 is at a gauge pressure of 327 kPa. What is the density when the absolute pressure is changed to 525 kPa?

Chapter 15

WAVE MOTION AND SOUND

The importance of radio and television signals and other forms of electromagnetic waves cannot be overstated. Communication using these phenomena is the backbone of modern civilization. We will first study mechanical waves to see how waves in general behave and then study electromagnetic radiation and sound.

OBJECTIVES

The major goals of this chapter are to enable you to:

1. Determine characteristics of mechanical waves.
2. Describe electromagnetic waves and the electromagnetic spectrum.
3. Analyze sound waves and explain the Doppler effect.

15.1 CHARACTERISTICS OF WAVES

Energy may be transferred by the motion of particles. Electricity, for example, is conducted along a wire by the motion of electrons. Heat is conducted by the motion of atoms and molecules. Tides and winds are examples of transfer of energy by the motion of fluids.

Another means by which energy transfer may occur is *wave motion.* The sun's energy is transported to the earth by light waves. Radio waves are an illustration of energy transfer for communications. Light waves produced by lasers are being used for voice and data transmission in optical fibers. Sound waves are yet another method by which energy may be transferred.

A **wave** is a disturbance that moves through a medium or through space. This disturbance may be a displacement of atoms away from their equilibrium positions in an elastic medium, a pulse in a spring, a change in pressure of a gas, or a variation in light intensity. There is a transfer of energy in the direction of propagation of the disturbance for each type of wave.

The elastic medium through which a wave travels or propagates is in many respects similar to a chain of particles connected by a series of springs like those shown in Fig. 15.1. If particle A is pulled to the left away from its equilibrium position, the neighboring spring exerts a force that tends to return A to its equilibrium position. The same spring exerts an equal but opposite force on particle B, which also tends to displace B to the left. As particle B moves to the left, the next particle experiences a force to the left, and so on until each particle experiences a displacement. If particle A is returned to its equilibrium position, the other particles will return to their equilibrium positions at a later time.

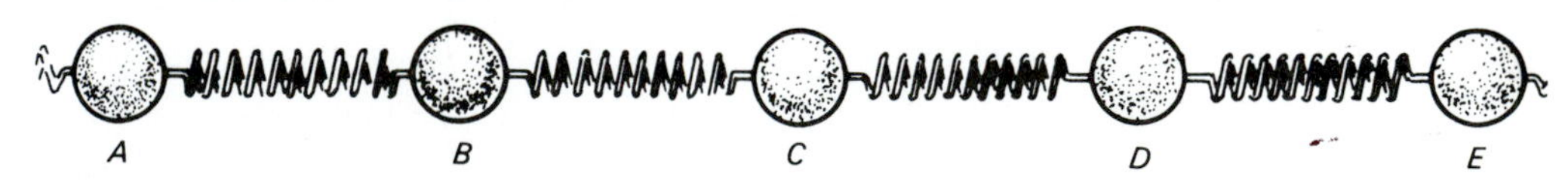

Figure 15.1
Springs in series are a form of elastic medium.

If particle A is forced to oscillate about its equilibrium position, all the other particles will also oscillate about their equilibrium positions. The kinetic energy given to the first particle is transmitted to each successive particle in the system. Although energy is transferred through the connecting springs, there is no transfer of particles from position A to E. This energy transfer without matter transfer is typical of all types of wave motion.

Another type of wave motion is shown in Fig. 15.2. In this case the elastic medium is a long spring. If the left end of the spring is rapidly lifted up and then returned to its starting position, a crest is formed which travels to the right [Fig. 15.2(a)]. If the left end is displaced downward and rapidly returned to its original position, a trough is formed which travels to the right [Fig. 15.2(b)].

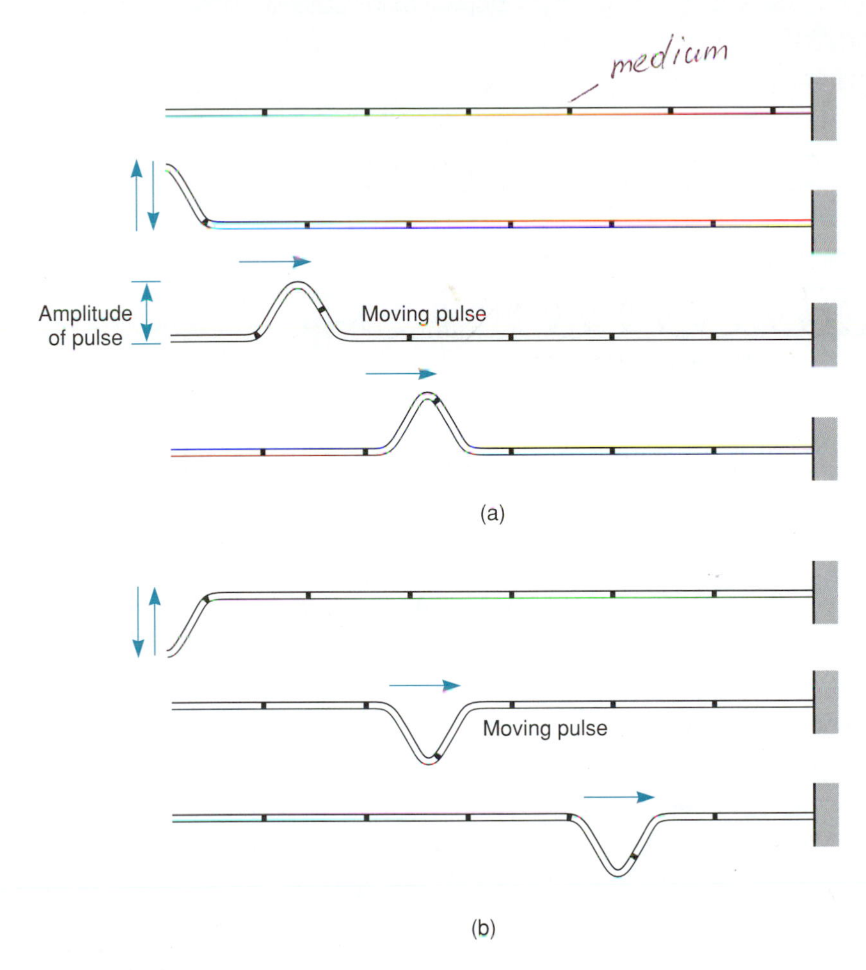

Figure 15.2
Pulses in a long spring.

These nonrepeated disturbances are called **pulses.** If they are repeated periodically, then a series of crests and troughs will travel through the medium, creating a traveling wave. The displacement of the particles is perpendicular to the direction of wave motion. This is referred to as a **transverse wave** (Fig. 15.3). Water waves are another example of transverse waves.

If a spring is compressed at the left end as shown in Fig. 15.4 and then released, the compression will travel to the right. Similarly, if the spring is stretched, a rarefaction is formed that will propagate or travel to the right. In this case the particle motion is along the direction of the wave travel and is referred to as **longitudinal wave** motion. Sound is another example of a longitudinal wave.

The **wavelength** λ is the minimum distance between particles that have the same displacement and are moving in the same direction (Fig. 15.5).

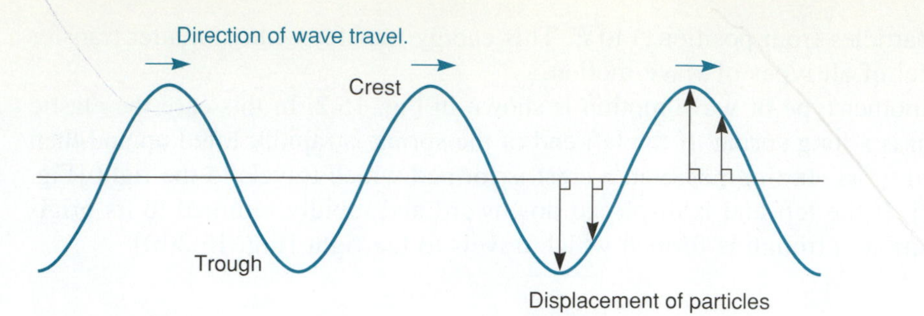

Figure 15.3
Transverse waves.

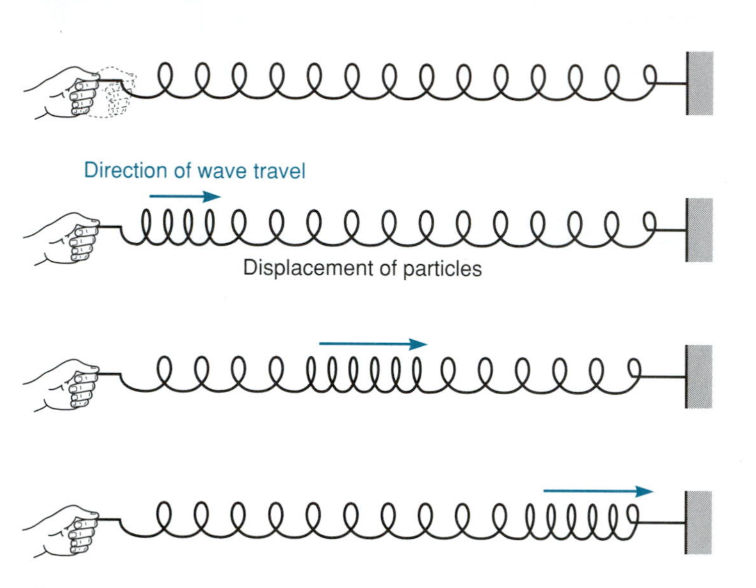

Figure 15.4
Longitudinal wave in a spring.

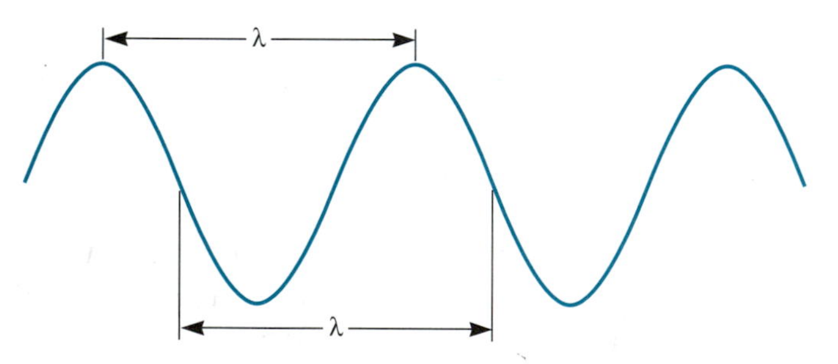

Figure 15.5
The wavelength of a wave is the distance between corresponding points on a uniformly repeated wave.

The **period** is the time required for a single wave to pass a given point. The **frequency** is the number of complete waves passing a given point per unit time. It is common to use the unit hertz (Hz) for frequency, where one oscillation per second is equal to 1 hertz (1 Hz = 1/s). The period and frequency are related by

$$f = \frac{1}{T}$$

where f = frequency
T = period

The **propagation velocity** v of a wave is given by the distance traveled by the wave in one period divided by the period, or

$$v = \frac{\lambda}{T} = \lambda f$$

where v = velocity
λ = wavelength
T = period
f = frequency

EXAMPLE

Find the velocity of a wave with wavelength 2.5 m and frequency 44 Hz.

Data:

$$\lambda = 2.5 \text{ m}$$
$$f = 44 \text{ Hz} = 44/\text{s}$$
$$v = ?$$

Basic Equation:

$$v = \lambda f$$

Working Equation: Same

Substitution:

$$v = (2.5 \text{ m})(44/\text{s})$$
$$= 110 \text{ m/s}$$

Superposition of Waves

Two waves of similar type can pass through the same medium. This **superposition of waves** forms a new wave created by adding their amplitudes together. The shape of the resultant wave is given by the sum of their amplitudes (Fig. 15.6). Where the waves add together to form a larger disturbance, as at point A, **constructive interference** occurs. If the waves oppose each other, producing a smaller disturbance as at point B, **destructive interference** occurs.

At any given point:

Location (1) amplitude of wave (d)	↑↑ ↑	amplitude of wave (a) amplitude of wave (b)	constructive interference
Location (2) amplitude of wave (d)	↓ ↑	amplitude of wave (a) amplitude of wave (b)	destructive interference

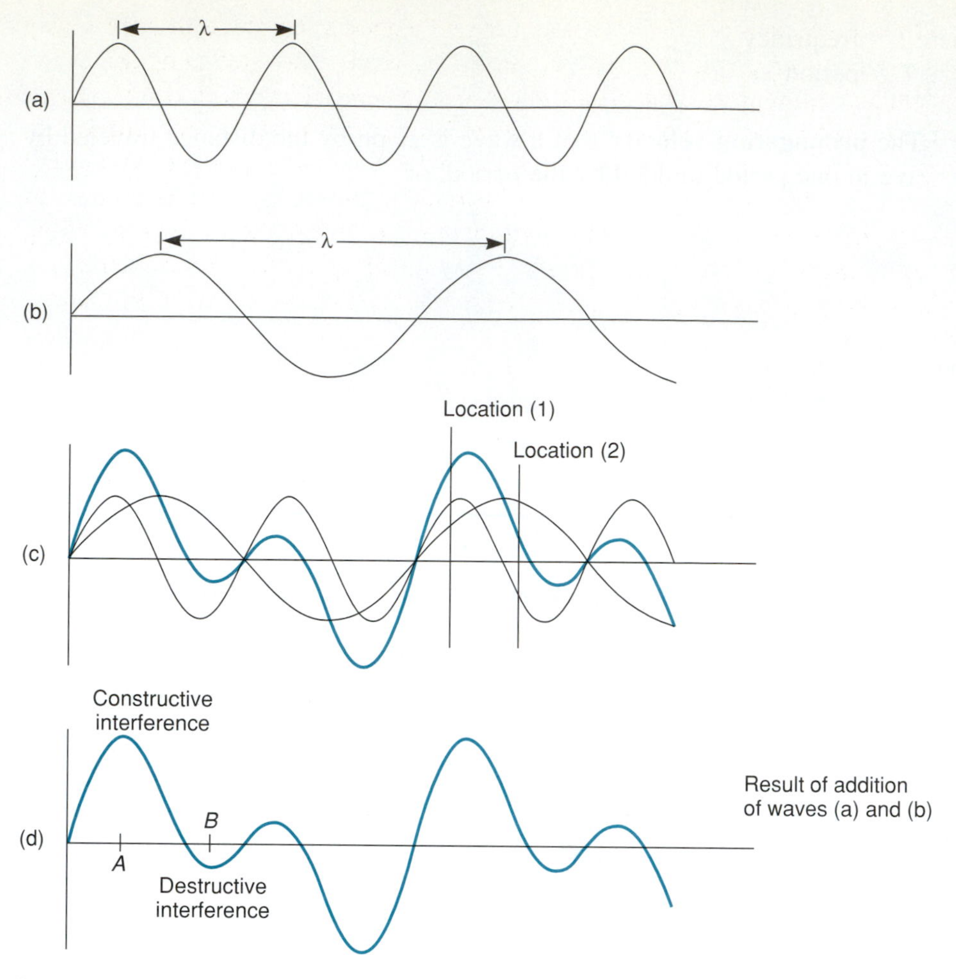

Figure 15.6
Superposition of two waves to form a new wave by adding their amplitudes.

Standing Waves

When a transverse pulse reaches the end of a spring that is fastened to a rigid support (Fig. 15.7), the pulse is reflected back along the spring with the displacement of the reflected wave opposite in direction to that of the incident wave. A traveling wave is also reflected at the rigid end of a spring, producing two waves moving in opposite directions. Reflections from a free end are not inverted.

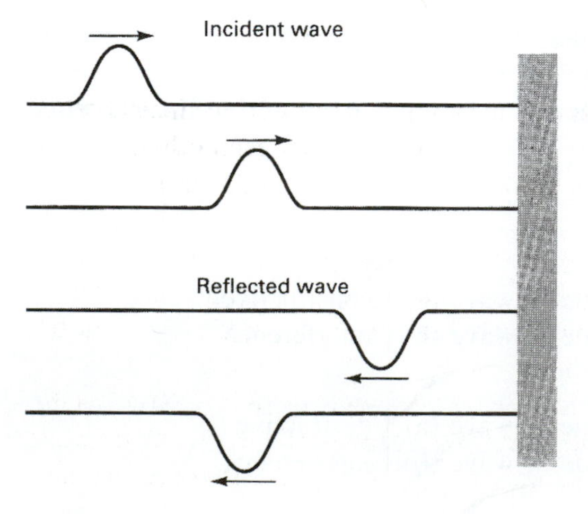

Figure 15.7
The reflected wave in the spring has the same amplitude as the incident wave.

In one special case it is possible for two waves to combine in such a way that there is no propagation of energy along the wave. The wave amplitudes are constant and motionless. This is called a **standing wave** (Fig. 15.8). It can be considered as two waves of equal amplitude and wavelength that are moving in opposite directions. Figure 15.8 shows an example of a standing wave on a string that could produce sound on a musical instrument. Note that there is no motion of the string at the end points. Although there is no propagation of energy along the string, there may be energy transfer from the string to the air surrounding it, producing sound waves at the same frequency of oscillation as that of the vibrating string.

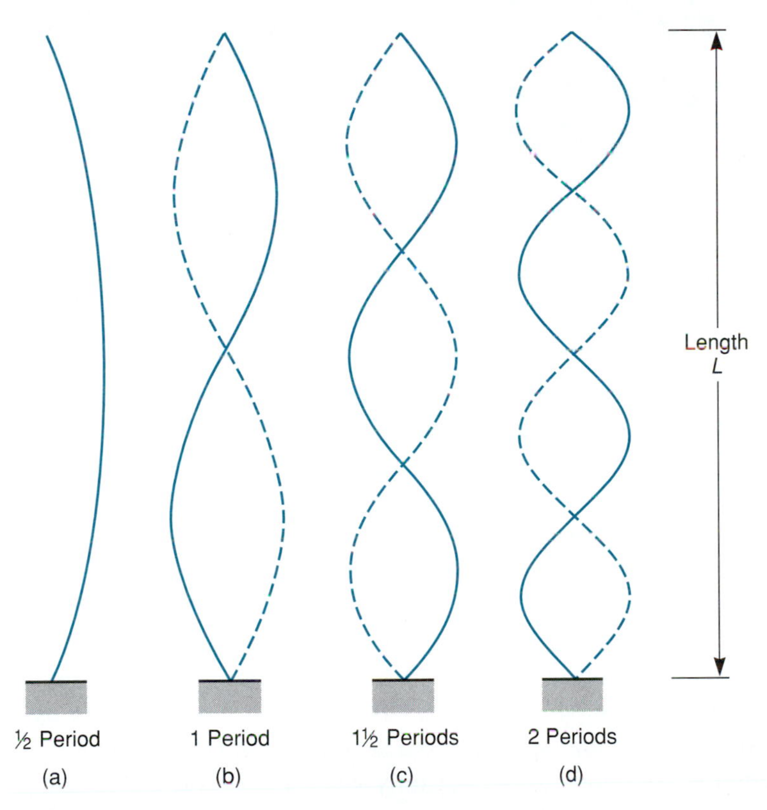

Figure 15.8
Standing waves in a string generated by increasing the frequency of vibration.

Interference and Diffraction

If two rocks are simultaneously dropped into a pool, each will produce a set of waves or ripples (Fig. 15.9). Wherever two wave crests cross each other, the water height is higher than for either crest alone. Where two troughs cross, the water level is lower than for one alone. If a trough crosses a crest, the water level is nearly undisturbed. This is an example of wave **interference.** Constructive interference occurs when two crests or troughs meet, giving a larger disturbance than for either wave alone. Destructive interference occurs when a wave and a trough meet and cancel each other out.

Interference of waves can occur for any type of wave, including sound, light, or radio waves. Some directional radio broadcast antennas rely upon constructive interference between waves from different parts of the antenna to direct the sig-

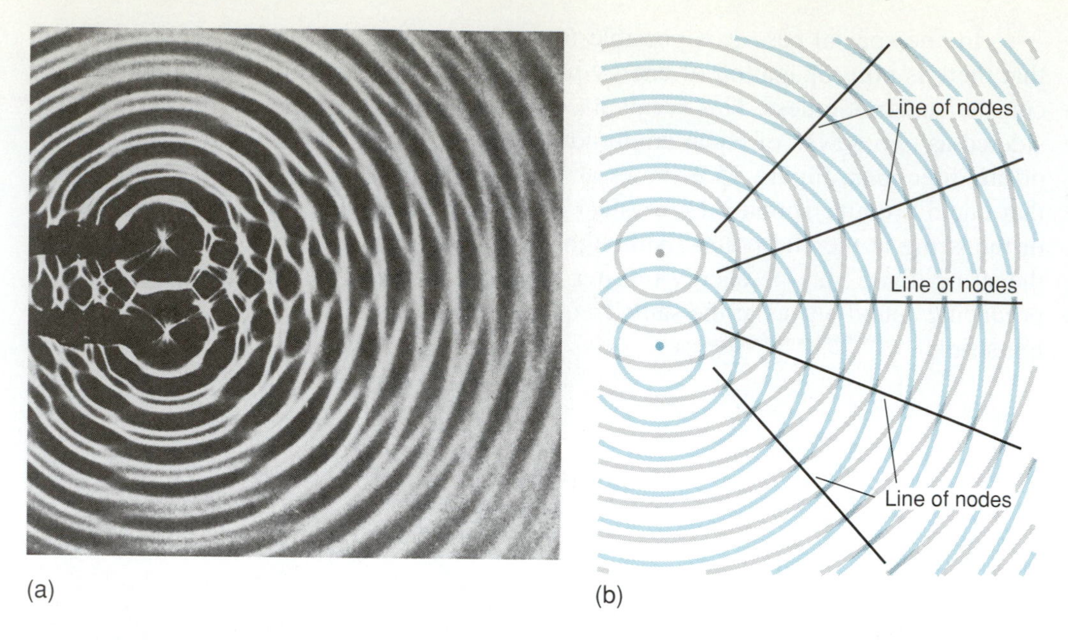

Figure 15.9
Interference of waves from two sources. The waves combine to form larger waves and cancel each other out as the wave fronts meet.

nal in the desired direction. Destructive interference prevents the signal from propagating in undesired directions. "Dead spots" in an auditorium are caused by destructive interference of waves direct from the sound source, with sound waves reflected from the walls, ceiling, or floor of the room. Proper choice of room shape and proper placement of materials that absorb sound well can lead to a room with good acoustical characteristics.

Another interesting property of waves is that they need not travel in straight lines. Waves can bend around obstacles in their path. This is called **diffraction.** Water waves bend around the supports of a pier (Fig. 15.10). Sound waves pass from one room through a door and spread into a second room (Fig. 15.11).

Wave diffraction is commonly observed only when the obstacle or opening is nearly the same size as the wavelength. Water waves and sound waves often have wavelengths in an easily observed range. Light, however, has a wavelength approximately 5×10^{-7} m. For this reason diffraction of light is not as easily rec-

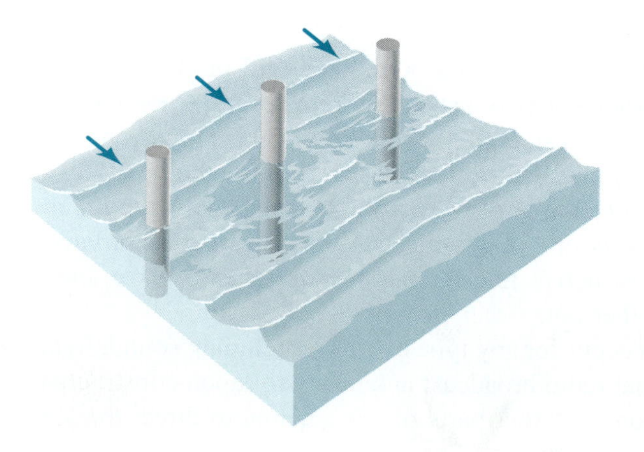

Figure 15.10
Water waves bend around obstacles and pass into the region behind the pier supports. This property is called diffraction.

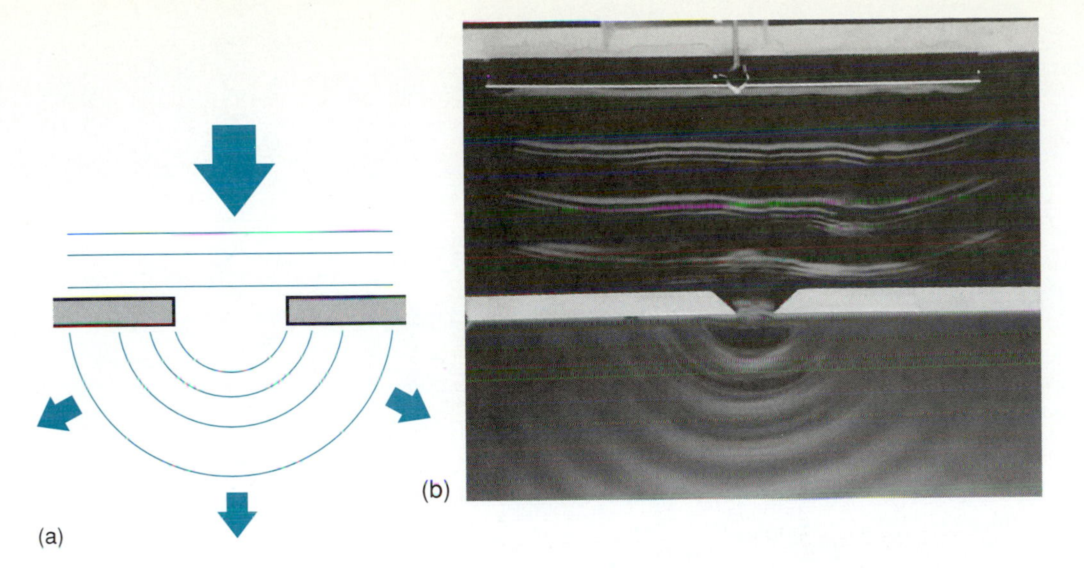

Figure 15.11
Diffraction of (a) sound waves entering a room through a door or (b) water waves through a small opening.

ognized as some other waves. The light waves from the sun when they encounter obstacles (air molecules in the atmosphere) are diffracted around them. The diffracted wavefronts are more or less spherical, and spread out or *scattered.*

Blue sky and a red setting sun are accounted for by scattering. When we look up at the sky, we see scattered light. Those colors with the shortest wavelengths (blue) are scattered the most; the longer wavelengths (red and yellow) are transmitted with very little scattering.

15.2 ELECTROMAGNETIC WAVES

An **electromagnetic wave** is radiated by any periodic motion of charge, is composed of electric and magnetic fields, and moves with velocity $v = c =$ **speed of light** $= 3.00 \times 10^8$ m/s. So, for electromagnetic waves

$$c = \lambda f$$

where c = speed of light (3.00×10^8 m/s)
λ = wavelength
f = frequency

Note that λ and f are inversely proportional. That is, when the frequency increases, the wavelength decreases; and when the frequency decreases, the wavelength increases.

EXAMPLE

The FM band of a radio is centered around a frequency of $10\overline{0}$ megahertz (MHz). Find the length of an FM antenna if each arm must be a quarter wavelength.

First, find the wavelength, λ.

Data:

$$f = 10\overline{0} \text{ MHz} = 10\overline{0} \times 10^6 \text{ Hz} = 1.00 \times 10^8/\text{s} \qquad (1 \text{ Hz} = 1/\text{s})$$
$$c = 3.00 \times 10^8 \text{ m/s}$$
$$\lambda = ?$$

Basic Equation:

$$c = \lambda f$$

Working Equation:

$$\lambda = \frac{c}{f}$$

Substitution:

$$\lambda = \frac{3.00 \times 10^8 \text{ m/s}}{1.00 \times 10^8\text{/s}}$$
$$= 3.00 \text{ m}$$

Therefore,

$$\frac{\lambda}{4} = \frac{3.00 \text{ m}}{4} = 0.750 \text{ m}$$

The classification of electromagnetic waves according to frequency is called the **electromagnetic spectrum** (Fig. 15.12). The radio broadcast band is in the region of 1 MHz. The VHF television band starts in the region of 50 MHz; the UHF band is even higher. The highest-frequency waves generated by electronic oscillators are microwaves.

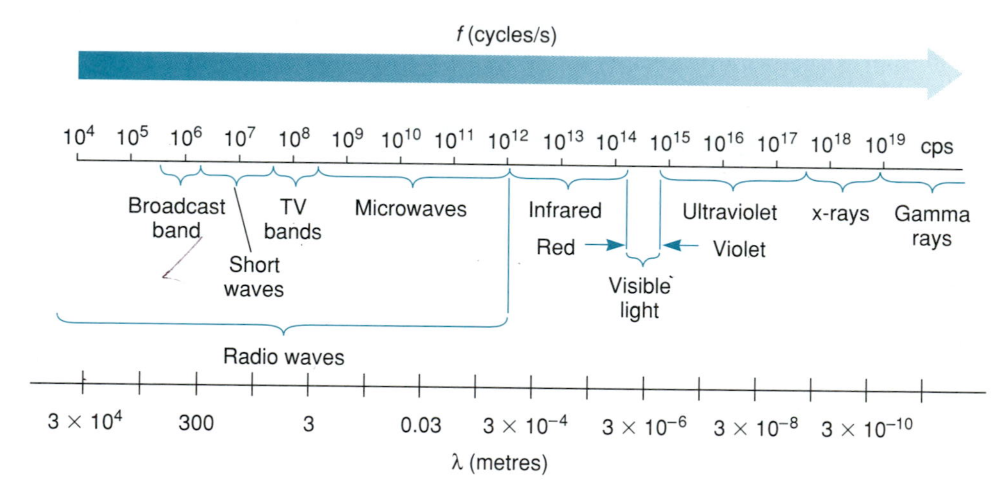

Figure 15.12
The electromagnetic spectrum.

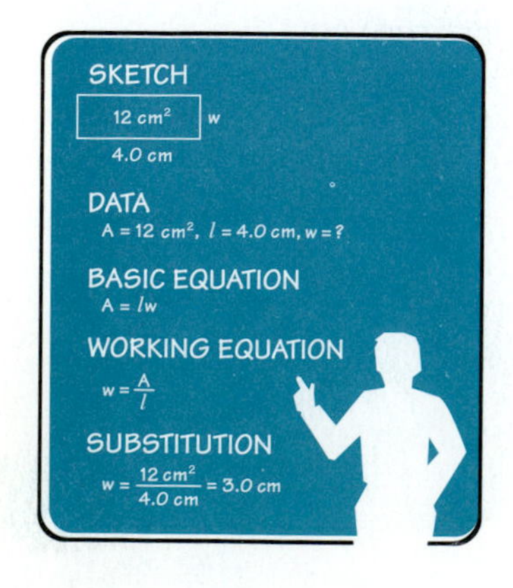

Molecular and atomic oscillations create waves of even higher frequencies, including infrared, visible light, ultraviolet, and X-ray waves. Visible light is electromagnetic radiation in the range 4.3×10^{14} to 7×10^{14} Hz.

■ PROBLEMS 15.2

1. Find the period of a wave whose frequency is $50\bar{0}$ Hz.
2. Find the frequency of a wave whose period is 0.550 s.
3. Find the velocity of a wave with wavelength 2.00 m and frequency $40\bar{0}$ Hz.
4. What is the frequency of a light wave with wavelength 5.00×10^{-7} m and velocity 3.00×10^8 m/s?

5. Find the period of the wave in Problem 4.
6. What is the speed of a wave with frequency 3.50 Hz and wavelength 0.550 m?
7. Find the wavelength of water waves with frequency 0.650 Hz and velocity 1.50 m/s.
8. What is the wavelength of longitudinal waves in a coil spring with frequency 7.50 Hz and velocity 6.10 m/s?
9. A wave generator produces 20 pulses in 3.50 s.
   (a) What is its period?
   (b) What is its frequency?
10. Find the frequency of a wave produced by a generator that emits 30 pulses in 2.50 s.
11. What is the wavelength of an electromagnetic wave with frequency 50.0 MHz?
12. Find the frequency of an electromagnetic wave with wavelength 1.50 m.
13. Find the wavelength of a wave traveling at 2.68×10^6 m/s with a period of 0.0125 s.
14. Find the wavelength of a wave traveling twice the speed of sound (speed of sound = 331 m/s) that is produced by an oscillator emitting 63 pulses every 8.3×10^{-6} min. ■

15.3 SOUND WAVES

We are exposed to sounds of many kinds daily and communicate with other people using the medium of sound. Whether the sound is pleasant music, the voice of another person, or a loud siren, there are three requirements for the detection of sound in a physiological sense. There must be a source of sound, a medium (such as air) for transmitting it, and an ear to receive it. In a physical sense, sound is a vibratory disturbance in an elastic medium which may produce the sensation of sound. The frequency range over which the human ear responds is approximately 20 to 20,000 Hz. Ultrasonic waves have a frequency higher than 20,000 Hz.

Sound is produced by the vibration of a source. A ringing bell, the vibrating head of a drum, and a tuning fork (Fig. 15.13) are common examples of vibrating sources of sound. Other vibrating sources are not as easily recognized. Vibrating vocal cords produce speech. The notes of a clarinet originate with a vibrating reed. An auto horn uses an electrically driven vibrating diaphragm.

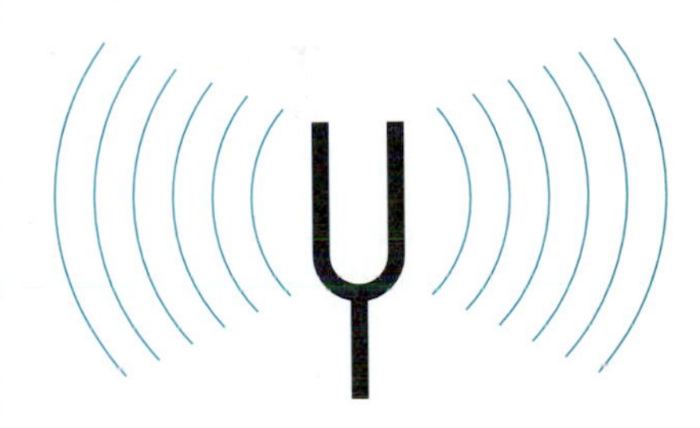

Figure 15.13
Sound waves produced by a vibrating tuning fork.

The most common medium for the propagation of sound is air. Sound will also propagate in solids or liquids. A vacuum will not carry sound. A mechanic may listen to the sounds of a running engine by placing one end of a metal rod against the engine and the other end against his ear. Sounds transmitted through water are utilized by passive sonar receivers aboard ships or submarines to identify other ships nearby.

Sound waves transmitted through the earth may be detected by an instrument called a *seismograph,* which can detect small motions of the earth's crust. An earthquake produces both longitudinal and transverse waves, which propagate with different velocities through the earth's crust. The distance to the source may be determined by measuring the time interval between the arrival of the two types of waves. Comparison of such data from seismographs at several points on the surface of the earth allows the location of the epicenter of the earthquake. Waves may be intentionally set off by exploding buried dynamite. Reflections of these sound waves off different rock formations are recorded by seismographs. These recordings allow geologists to determine the underlying structure of the earth and predict the location of possible oil- or gas-producing regions.

Musical instruments utilize vibrating strings and vibrating columns of air to produce regular sounds. Wind instruments use an enclosed or partially enclosed column of air where waves are produced in the confined air by a pressure change (usually, blowing). Standing waves can then be produced in a column of proper length.

You may have watched distant lightning and noticed the considerable time before the sound of thunder reaches you. This is an example of the relatively slow speed of sound compared to the **speed of light** (3.00×10^8 m/s). The **speed of sound** in dry air at 1 atm pressure and 0°C is 331 m/s. Changes in humidity and temperature cause a variation in the speed of sound. The speed of sound increases with temperature at the rate of 0.61 m/s/°C. The speed of sound in dry air at 1 atm pressure is then given by

$$v = 331 \text{ m/s} + \left(0.61 \frac{\text{m/s}}{°\text{C}}\right)T$$

$$v = 1087 \text{ ft/s} + \left(1.1 \frac{\text{ft/s}}{°\text{F}}\right)(T - 32°\text{F})$$

where v = speed of sound in air
T = air temperature

EXAMPLE 1

Find the speed of sound in dry air at 1 atm pressure if the temperature is 23°C.

Data:

$$T = 23°\text{C}$$
$$v = ?$$

Basic Equation:

$$v = 331 \text{ m/s} + \left(0.61 \frac{\text{m/s}}{°\text{C}}\right)T$$

Working Equation: Same

Substitution:

$$v = 331 \text{ m/s} + \left(0.61 \frac{\text{m/s}}{°\text{C}}\right)(23°\text{C})$$
$$= 345 \text{ m/s}$$

EXAMPLE 2

What is the time required for the sound from an explosion to reach an observer $190\overline{0}$ m away for the conditions of Example 1?

Data:

$$v = 345 \text{ m/s}$$
$$s = 190\overline{0} \text{ m}$$
$$t = ?$$

Basic Equation:

$$s = vt \qquad \text{(from Section 3.4)}$$

Working Equation:

$$t = \frac{s}{v}$$

Substitution:

$$t = \frac{190\bar{0}\text{ m}}{345\text{ m/s}}$$

$$= 5.51\text{ s}$$

$$\frac{\text{m}}{\text{m/s}} = \text{m} \div \frac{\text{m}}{\text{s}} = \cancel{\text{m}} \cdot \frac{\text{s}}{\cancel{\text{m}}} = \text{s}$$

Sound propagates faster in a dense medium such as water than it does in a less dense medium such as air. A list of the speed of sound in various media is given in Table 15.1.

Table 15.1 Speed of Sound in Various Media

	Speed	
Medium	*m/s*	*ft/s*
Aluminum	6,420	21,100
Brass	4,7$\bar{0}$0	15,400
Steel	5,960	19,500
Granite	6,0$\bar{0}$0	19,700
Alcohol	1,210	3,970
Water (25°C)	1,5$\bar{0}$0	4,920
Air, dry (0°C)	331	1,090
Vacuum	0	0

All sounds have characteristics which we associate only with that sound. A siren is loud. A whisper is soft. Music can be loud or soft. The physical properties that differ for these sounds are intensity and frequency. The physiological characteristics of these sounds are loudness and pitch. Sound quality is related to the number of frequencies present. Pitch and frequency are closely related terms.

Intensity is the energy transferred by sound per unit time through unit area, thus $\frac{\text{power}}{\text{area}}$. **Loudness** describes how strong or faint the sensation of sound seems to an observer. The ear does not respond equally to all frequencies. Sound must reach a certain intensity before it can be heard. The human ear normally detects sounds ranging in intensity from 10^{-12} W/m^2 (the threshold of hearing) to 10^0 W/m^2 or 1 (the threshold of pain). Levels of intensity are also measured on a logarithmic scale in decibels (dB); the unit "bel" is named after Alexander Graham Bell, the inventor of the telephone. Table 15.2 shows a range of familiar sounds.

Sound waves from two sources arriving at an observer at the same time will constructively or destructively interfere with each other. Identical waves from two separate sources such as stereo loudspeakers will enhance or reduce each other at different points in space. Sound waves of slightly different frequencies will also enhance or reduce each other at different points in time. The alternating periods of increasing and decreasing volume are called *beats*. The number of times per second the sound reaches a maximum is known as the *beat frequency* and is equal to the difference in frequencies of the two sound waves. This phenomenon is used for tuning musical instruments against a set of standard-frequency tuning forks. When the beat frequency is small, the instrument is well matched with the tuning fork.

Table 15.2 Range of Sound Levels and Intensities

Situation	*Sound Level (dB)*	*Intensity (W/m^2)*	*Sound*
Threshold of hearing	0	10^{-12}	Scarcely audible
Minimal sounds	10	10^{-11}	
Ticking watch	20	10^{-10}	
Whisper, library	30	10^{-9}	Faint
Leaves rustling, refrigerator	40	10^{-8}	
Average home, neighborhood street	50	10^{-7}	Moderate
Normal conversation, microwave	60	10^{-6}	
Car, alarm clock, city traffic	70	10^{-5}	Loud
Garbage disposal, vacuum cleaner, outboard motor, noisy restaurant	80	10^{-4}	Very loud
Factory, electric shaver, screaming child	85		
Lawn mower, passing motorcycle, convertible ride on a freeway	90	10^{-3}	
Blow dryer, diesel truck, chain saw, helicopter, subway train	100	10^{-2}	
Car horn, snowblower	110	10^{-1}	Deafening
Rock concert, prop plane	120	10^{0}	Painful
100 ft from jet engine, air raid siren	130	10^{1}	
Shotgun blast	140	10^{2}	

Sources: Better Hearing Institute, *Self-Help for Hard of Hearing People*. Each increase of 10 dB is a tenfold increase in sound intensity; that is, 90 dB is ten times noisier than 80 dB. The U.S. Government advises wearing earplugs or other hearing protection whenever anyone is exposed to 85 dB for a period of more than a few hours. Some hearing experts have found that hearing damage is done at sound levels as low as 70 dB. Federal regulations require a hearing conservation program in the workplace where employees are exposed to 85 dB or more during an eight-hour work period.

15.4 THE DOPPLER EFFECT

As an automobile or a train passes by at a high speed sounding its horn or whistle, the frequency or pitch of the sound drops noticeably as it passes the observer. This variation in pitch is called the **Doppler effect.** Radar units employed by police to monitor highway speeds employ this principle. A high-frequency radio signal is transmitted by the radar unit toward an oncoming car. Because of the motion of the auto, the reflected signal has a different frequency (Fig. 15.14). This signal is received by the radar unit, which electronically measures the frequency shift to determine the speed of the auto.

Motion of a source of sound toward an observer increases the rate at which he or she receives the vibrations. The velocity of each vibration is the speed of sound whether the source is moving or not. Each vibration from an approaching source has a shorter distance to travel. The wavelength is shortened when the source is moving toward the observer and is lengthened when the source is moving away from the observer. The vibrations are therefore received at a higher frequency than they are sent. Similarly, sound waves from a receding source are received at a lower frequency than that at which they are sent.

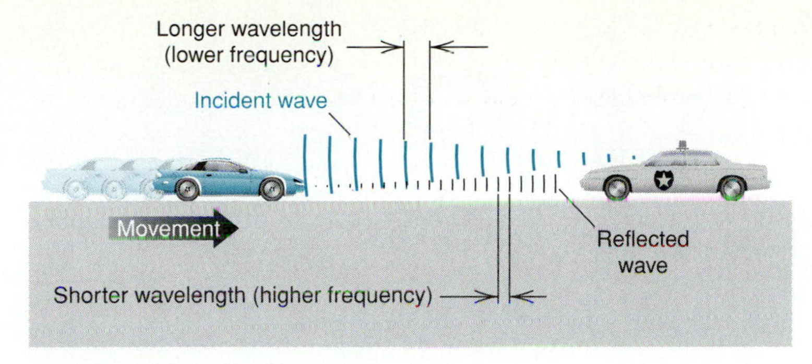

Figure 15.14
Use of reflected radar waves to determine the speed of an automobile. The black reflected wave has a shorter wavelength (higher frequency) than the blue incident wave because of the car's movement toward the radar gun.

The apparent Doppler-shifted frequency for sound is given by the equation

$$f' = f\left(\frac{v}{v \pm v_s}\right)$$

where f' = Doppler-shifted frequency
f = actual source frequency
v = speed of sound
v_s = speed of the source

The + sign in the denominator is used when the source is moving away from the observer. The − sign is used when the source is moving toward the observer.

EXAMPLE

An automobile sounds its horn while passing an observer at 25 m/s. The actual horn frequency is $40\overline{0}$ Hz.

(a) What is the frequency heard by the observer while the car is approaching?

(b) What is the frequency heard when the car is leaving?
Assume that the speed of sound is 345 m/s.

(a) Data:

$f = 40\overline{0}$ Hz
$v = 345$ m/s
$v_s = 25$ m/s toward observer
$f' = ?$

Basic Equation:

$$f' = f\left(\frac{v}{v - v_s}\right)$$

Working Equation: Same

Substitution:

$$f' = (40\bar{0}\ \text{Hz})\left(\frac{345\ \text{m/s}}{345\ \text{m/s} - 25\ \text{m/s}}\right)$$
$$= 431\ \text{Hz}$$

(b) We simply change the sign from − to + in the basic equation of part (a). All other data remain the same. We then find

$$f' = (40\bar{0}\ \text{Hz})\left(\frac{345\ \text{m/s}}{345\ \text{m/s} + 25\ \text{m/s}}\right)$$
$$= 373\ \text{Hz}$$

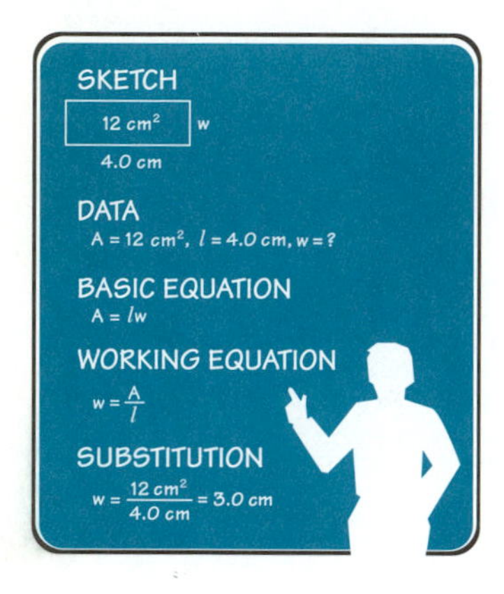

■ PROBLEMS 15.4

1. Find the speed of sound in m/s at $1\bar{0}$°C at 1 atm pressure in dry air.
2. Find the speed of sound in m/s at 35°C at 1 atm pressure in dry air.
3. Find the speed of sound in m/s at −23°C at 1 atm pressure in dry air.
4. How long will it take a sound wave to propagate 21.0 m for the conditions of Problem 1?
5. How long will it take a sound wave to propagate through $750\bar{0}$ m of water at 25°C?
6. A sound wave is transmitted through water from one submarine, is reflected off another submarine 15 km away, and returns to the sonar receiver on the first sub. What is the round-trip transit time for the sound wave? Assume that the water temperature is 25°C.
7. A sonar receiver detects a reflected sound wave from another ship 3.52 s after the wave was transmitted. How far away is the other ship? Assume that the water temperature is 25°C.
8. A woman is swimming when she hears the underwater sound wave from an exploding ship across the harbor. She immediately lifts her head out of the water. The sound wave from the explosion propagating through the air reachers her 4.00 s later. How far away is the ship? Assume that the water temperature is 25°C and the air temperature is 23°C.
9. A train traveling at a speed of $4\bar{0}$ m/s approaches an observer at a station and sounds a $55\bar{0}$-Hz whistle. What frequency will be heard by the observer? Assume that the sound velocity in air is 345 m/s.
10. What frequency is heard by an observer who hears the $45\bar{0}$-Hz siren on a police car traveling at 35 m/s away from her? Assume that the velocity of sound in air is 345 m/s.
11. A car is traveling toward you at 40.0 mi/h. The car horn produces a sound at a frequency of $48\bar{0}$ Hz. What frequency do you hear? Assume that the sound velocity in air is 1090 ft/s.
12. A car is traveling away from you at 40.0 mi/h. The car horn produces a sound at a frequency of $48\bar{0}$ Hz. What frequency do you hear? Assume that the sound velocity in air is 1090 ft/s.
13. A jet airplane taxiing on the runway at 13.0 km/h is moving away from you, the traffic controller. The engine produces a frequency of $66\bar{0}$ Hz in −6.0°C air. What frequency do you hear?
14. While snorkeling you hear a dolphin's sound as it approaches at 5.00 m/s. If the perceived frequency is $85\bar{0}$ Hz, what is the actual frequency being emitted? ■

15.5 RESONANCE

When a tuning fork is struck with a rubber hammer, it vibrates at its **natural frequency.** This frequency depends on the length, thickness, and the material from which the tuning fork is made. Strings on a guitar also vibrate at a natural frequency. The sounding boards of a guitar are forced to vibrate at the same frequency as the strings because of energy transfer from the strings to the sounding board (Fig. 15.15). This is an example of *forced vibration.* The natural frequency of the board is typically different from that of the strings or tuning fork. Because the area of the sounding board is large, energy transfer into sound waves is very efficient. Therefore, the vibrating string or tuning fork loses its energy or dies out more rapidly if in contact with a sounding board.

Figure 15.15
Forced vibration of a guitar sounding board.

Consider two objects such as tuning forks with the same natural frequency that are set close together (Fig. 15.16). One is set into vibration and then stopped after a few seconds. It is found that the other tuning fork is weakly vibrating. The sound waves of the first fork cause the second to vibrate. This is called *sympathetic vibration* or *resonance.* **Resonance** occurs when the natural vibration rates of two objects are the same. Energy transfer into vibrations of the second fork is found to be much more efficient when both forks have the same frequency than when they have different frequencies. Large vibrations can be set up if the driving force is at the natural frequency of a system. Auto body rattles sometimes occur at certain speeds and disappear for small speed changes. Radio receivers operate on the principle of resonance. The natural frequency of vibration of electrical currents in a circuit may be tuned to that of an incoming radio signal, which is then amplified and converted into sound.

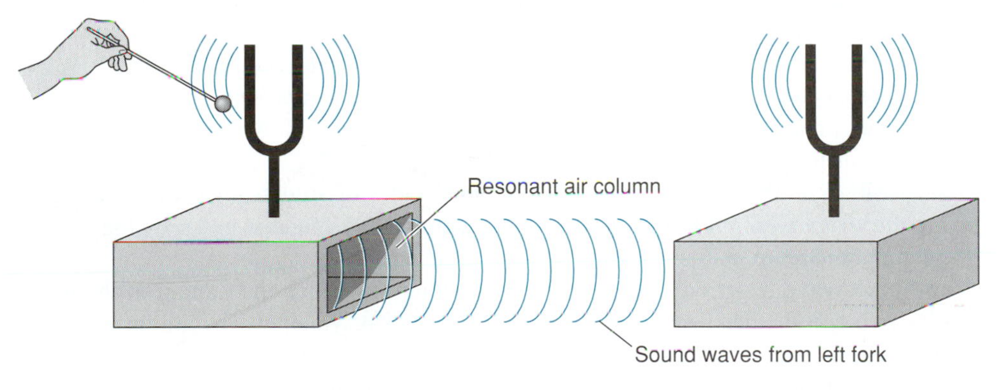

Figure 15.16
Identical tuning forks. Resonant air column. Sound waves of the left fork cause the right fork to vibrate.

GLOSSARY

Constructive Interference The superposition of waves to form a larger disturbance (wave) in a medium. Occurs when two crests or troughs of superimposed waves meet. (p. 375)

Destructive Interference The superposition of waves to form a smaller disturbance (wave) in a medium. (p. 375)

Diffraction The property of a wave that describes its ability to bend around obstacles in its path. (p. 378)

Doppler Effect The variation in frequency or pitch of a wave caused by the motion of the source. (p. 384)

Electromagnetic Spectrum The entire range of electromagnetic waves classified according to frequency. Radio, light, and X-rays are examples of different-frequency electromagnetic waves. (p. 380)

Electromagnetic Wave A type of wave that is created by the periodic motion of charge and can travel through vacuum and many materials carrying energy in the form of electric and magnetic fields. Moves at the velocity of light (3.00×10^8 m/s in a vacuum). (p. 379)

Frequency The number of complete waves passing by a given point per unit time. It is common to use the unit hertz (Hz) for frequency. (p. 374)

Intensity The energy transferred by sound per unit time through unit area. (p. 383)

Interference The effect of two intersecting waves resulting in a loss of amplitude in certain areas and an increase in amplitude in others. (p. 377)

Longitudinal Wave A periodic disturbance in a medium in which the motion of the particles is along the direction of the wave travel. Sound is an example. (p. 373)

Loudness The strength of the sensation of sound to an observer. (p. 383)

Natural Frequency The frequency at which an object vibrates when struck by another object, such as a rubber hammer. (p. 387)

Period (T) The time required for a single wave to pass a given point. (p. 374)

Propagation Velocity The velocity of energy transfer of a wave. Given by the distance traveled by the wave in one period divided by the period. (p. 375)

Pulses Nonrepeated disturbances that carry energy through a medium or through spaces. (p. 373)

Resonance A sympathetic vibration of an object caused by the transfer of energy from another object vibrating at the natural frequency of vibration of the first object. (p. 387)

Sound A wave traveling through a medium such as air or water that is produced by the vibration of an object. (p. 381)

Speed of Light The speed with which light and other forms of electromagnetic radiation travel: 3.00×10^8 m/s in vacuum. (pp. 379, 382)

Speed of Sound The speed with which sound waves travel in a medium: 331 m/s in dry air at 1 atm pressure and 0°C. (p. 382)

Standing Waves A special case of superposition of two waves such that there is no energy propagation along the wave. Can be considered to be two waves of equal amplitude and wavelength that are moving in opposite directions. A vibrating string on a musical instrument is an example. (p. 377)

Superposition of Waves The addition of the amplitudes of similar types of waves passing through a medium to form a new wave pattern. The shape is given by the sum of the amplitudes. (p. 375)

Transverse Wave A periodic disturbance in a medium in which the motion of the particles is perpendicular to the direction of wave motion. Water waves are an example. (p. 373)

Wave A disturbance that moves through a medium or through space in which a transfer of energy usually occurs. (p. 372)

Wavelength (λ) The minimum distance between particles in a wave that have the same displacement and are moving in the same direction. (p. 373)

FORMULAS

15.1 $f = \frac{1}{T}$

$$v = \lambda f$$

15.2 $c = \lambda f$

15.3 $v = 331 \text{ m/s} + \left(0.61 \frac{\text{m/s}}{°\text{C}}\right)T$

$$v = 1087 \text{ ft/s} + \left(1.1 \frac{\text{ft/s}}{°\text{F}}\right)(T - 32°\text{F})$$

15.4 $f' = f\left(\frac{v}{v \pm v_s}\right)$

REVIEW QUESTIONS

1. Which of the following are methods of energy transfer?
   (a) Conduction (b) Radiation
   (c) Wave motion (d) None of the above
2. The minimum distance between particles in a wave that have the same displacement and are moving in the same direction is called
   (a) the period. (b) the frequency.
   (c) the wavelength. (d) none of the above.
3. Which of the following refers to the time required for a single wave to pass a given point?
   (a) The period (b) The frequency
   (c) The wavelength (d) None of the above
4. Which of the following refers to the number of complete waves passing a given point per unit time?
   (a) The period (b) The frequency
   (c) The wavelength (d) None of the above
5. An example of a transverse wave is
   (a) sound waves. (b) water waves.
   (c) interference. (d) none of the above.
6. Which of the following are examples of a longitudinal wave?
   (a) Sound waves (b) Water waves
   (c) Interference (d) None of the above
7. Which of the following are electromagnetic waves?
   (a) Sound (b) Water waves
   (c) Radar waves (d) X-rays
   (e) All of the above
8. Explain the difference between interference and diffraction.
9. Explain the difference between constructive and destructive interference.
10. If waves did not exhibit the property of diffraction, under what conditions would your stereo system sound different?
11. Give an example of diffraction of water waves.
12. What happens to the frequency of a vibrating string on a guitar if the length of the string is decreased?
13. Explain the difference between a wave and a pulse.
14. Give an example of a pulse.
15. Why does the setting sun appear reddish in color?
16. Why does the sky appear blue?
17. What happens to the speed of sound when the temperature increases? Explain why this might happen.
18. Explain how a seismograph works.
19. How does the speed of sound differ in water and air? Explain the reason for this difference.
20. Explain the importance of the Doppler effect in traffic control.
21. In your own words, explain the Doppler effect.
22. Distinguish between sympathetic and forced vibration.

23. In your own words, explain resonance.
24. State a reason that might explain why many stars appear to have their light shifted to the red (longer wavelength) part of the electromagnetic spectrum when viewed from earth.

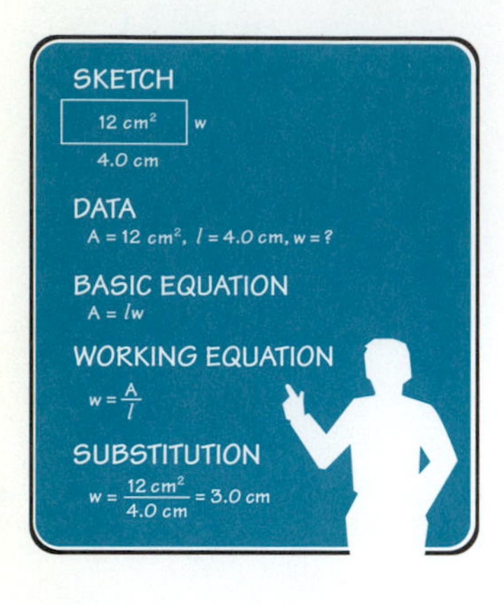

REVIEW PROBLEMS

1. Find the period of a wave with frequency 355 kHz.
2. Find the frequency of a wave with period 0.320 s.
3. What is the frequency of a light wave with wavelength 4.50×10^{-7} m and velocity 3.00×10^{8} m/s?
4. Find the period of the wave in Problem 3.
5. Find the speed of a wave with frequency 8.97 Hz and wavelength 0.654 m.
6. What is the wavelength of longitudinal waves in a coil spring with frequency 4.65 Hz and velocity 5.78 m/s?
7. Find the frequency of a wave produced by a generator that emits 85 pulses in 1.3 s.
8. What is the wavelength of an electromagnetic wave with frequency 65.5 MHz?
9. Find the speed of sound in m/s at 85°C at 1 atm pressure in dry air.
10. Find the speed of sound in m/s at −35°C at 1 atm pressure in dry air.
11. How long will it take a sound wave to propagate through 1450 m of water at 25°C?
12. A sound wave is transmitted through water from one ship, is reflected off another ship 22 km away, and returns to the sonar receiver on the first ship. What is the round-trip transit time for the sound wave if the water temperature is 23°C?
13. A train traveling at a speed of 95 mi/h approaches an observer at a station and sounds a 525-Hz whistle. What frequency will be heard by the observer? Assume that the sound velocity in air is 1090 ft/s.
14. A car is traveling toward you at 95 km/h. The car horn produces a sound at frequency 4950 Hz. What frequency do you hear? Assume that the sound velocity in air is 345 m/s.
15. What frequency do you hear if the car in Problem 14 is traveling away from you?
16. What is the frequency of the sound waves being emitted from a train whistle while approaching at 45 m/s, in air that is 11°C? The perceived frequency is 425 Hz.
17. The tail light on a car produces light with wavelength 5.00×10^{-7} m. What frequency do you observe when the car is departing at 24 m/s at 0°C?

Chapter 16

STATIC ELECTRICITY

From the beginning of time, lightning has been a source of awe and terror.

Electricity remains one of the most useful but mysterious servants of human beings. We will try to reduce its mystery by looking at the nature of electrical charge, electric fields, and static electricity.

OBJECTIVES

The major goals of this chapter are to enable you to:

1. Describe the nature of electrical charges.
2. Distinguish conduction and induction.
3. Use Coulomb's law to find the force between charges.

16.1 ELECTRICAL CHARGES

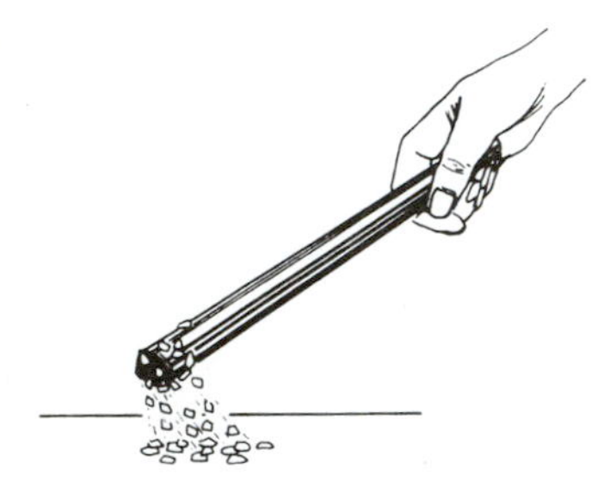

Figure 16.1
Amber rod attracting bits of paper after being rubbed.

Electrification was first studied 2500 years ago in ancient Greece when an amber rod was rubbed with a wool cloth and the rod attracted small objects (Fig. 16.1). When two objects are rubbed together, they become electrified.

When you slide rubber-soled shoes on a wool rug on a dry day, you become electrified. We say you have acquired a static charge. This static charge is usually lost when you touch an object at a different potential as in Fig. 16.2. Part of the static charge may be lost when you touch an object, such as another person, to which charge is transferred. Trucks that carry flammable liquids prevent a buildup of static charge by dragging a chain on the pavement (Fig. 16.3). Otherwise, a spark caused by a discharge could cause an explosion.

Figure 16.2
Stored charge.

Figure 16.3
Dissipation of static electricity.

To understand electricity, we need to know more about the structure of matter. We have seen that all matter is made up of atoms. These atoms are made of **electrons, protons,** and **neutrons.** Each proton has one unit of positive charge, and each electron has one unit of negative charge. The neutron has no charge. The protons and neutrons are tightly packed into what is called the *nucleus*. Electrons may be thought of as small charged clouds that occupy certain "orbits" around the nucleus of atoms (Fig. 16.4). An atom normally has the same number of elec-

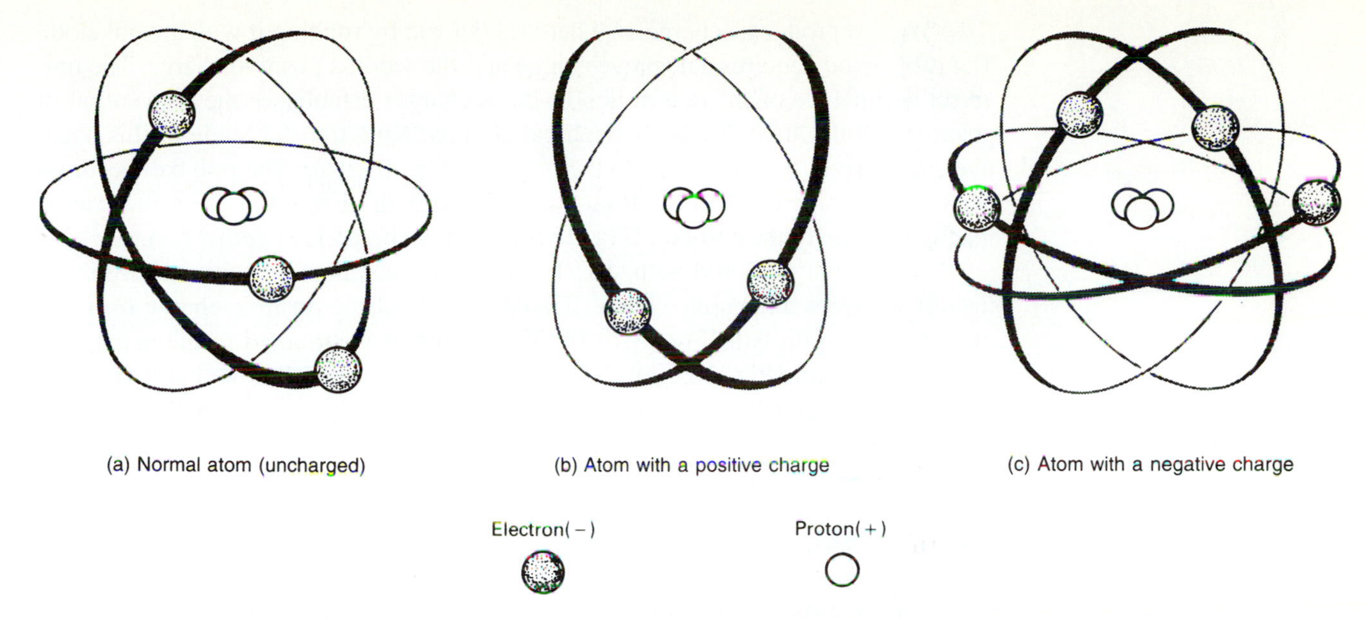

(a) Normal atom (uncharged)
(b) Atom with a positive charge
(c) Atom with a negative charge

Figure 16.4

trons as protons and thus is uncharged. If an electron is removed, the atom is left with a *positive charge* (+)—that is, an excess of protons. If an extra electron is added, the atom has a *negative charge* (−)—that is, an excess of electrons. The study of electric charges and the forces between them is referred to as *electrostatics*.

When two materials are rubbed together, the atoms on the two surfaces move across each other and brush off electrons. The electrons are transferred from one surface to the other. One surface is then left with a *positive charge*, and the other is *negative*. This is the process we have called *electrification*.

The two types of electrical charges can be observed indirectly by using an electroscope. A very simple electroscope is a ball of wood pith on a silk thread [Fig. 16.5(a)].

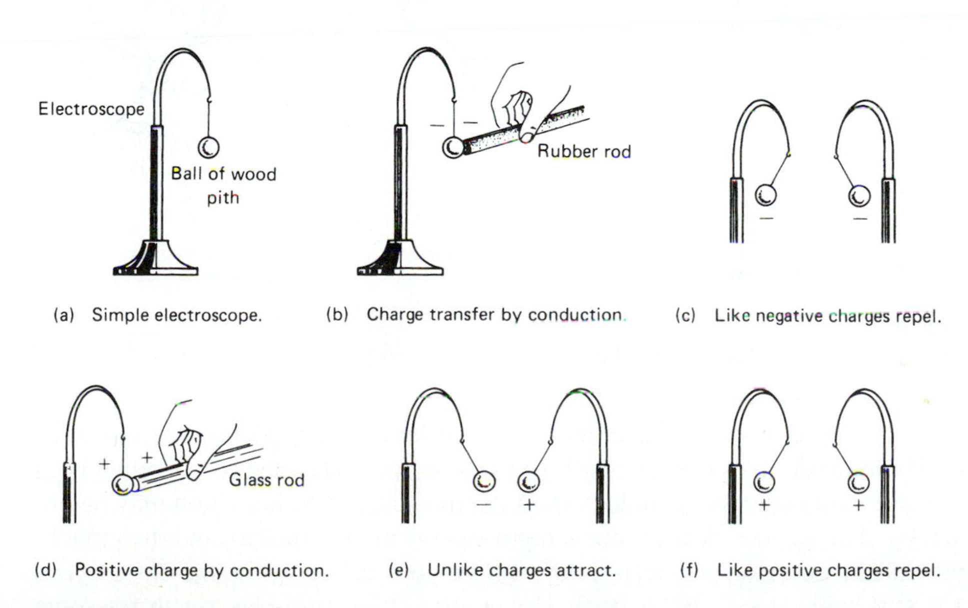

(a) Simple electroscope.
(b) Charge transfer by conduction.
(c) Like negative charges repel.
(d) Positive charge by conduction.
(e) Unlike charges attract.
(f) Like positive charges repel.

Figure 16.5
Charging a simple electroscope.

We can produce a charge on a hard rubber rod by rubbing it with a wool cloth. The rubber rod acquires a negative charge and the wool, a positive charge. The universal acceptance of the description of these charges establishes the convention of positive and negative charge in electrical circuits. Now transfer some of this negative charge from the rubber rod to the pith ball [Fig. 16.5(b)]. The pith ball becomes negatively charged by **conduction.** Another pith ball charged in the same way is repelled by the other pith ball [Fig. 16.5(c)]. This charge is *negative* (−).

Now rub a glass rod with silk. The glass rod acquires a positive charge, and the silk acquires a negative charge. Transfer some of the positive charge from the glass rod to a pith ball [Fig. 16.5(d)]. This pith ball is attracted to the negatively charged pith ball [Fig. 16.5(e)]. The charge produced by glass and silk is called a *positive charge* (+). Two pith balls that are positively charged will repel each other [Fig. 16.5(f)].

16.2 INDUCTION

The leaf electroscope is more sensitive than the pith ball type and can also be used to show electrification or charging by **induction.** The leaf electroscope [Fig. 16.6(a)] consists of thin strips of usually gold foil leaf hanging from a metal rod with a metal ball on the other end. The delicate leaves are enclosed to protect them and the rod is insulated from the enclosure. When charge is placed on the leaves, they diverge because of the force of repulsion of their similar charge.

Electroscopes may be charged by *conduction* by touching a charged object to the metal ball. Electroscopes are charged by *induction* by bringing a charged object near to, but not touching, the metal ball. Charges of the same sign as the charged object are repelled to the leaves where they repel each other and cause the leaves to separate [Fig. 16.6(b) and (c)].

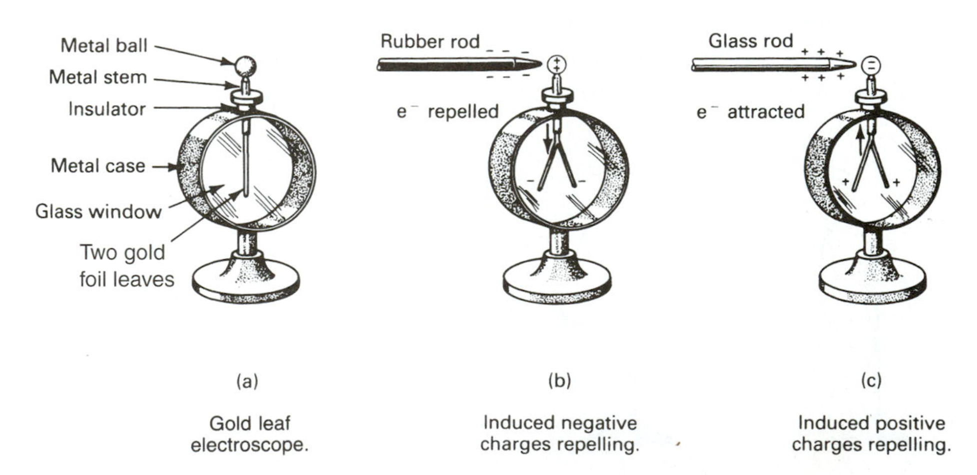

Figure 16.6
Charging an electroscope by induction.

When the charged object is removed, the leaves close as the free electrons redistribute themselves over the ball, rod, and leaves. The electroscope has been only temporarily charged by induction. A residual charge by induction may be obtained by charging the electroscope temporarily as in Fig. 16.7(a), and then touching the electroscope with a neutral object (like your finger) while the charged object is still held close [Fig. 16.7(b)]. The neutral object provides a path for some of the induced charge on the electroscope to escape. When everything is removed, a residual charge by induction remains on the electroscope [Fig. 16.7(d)].

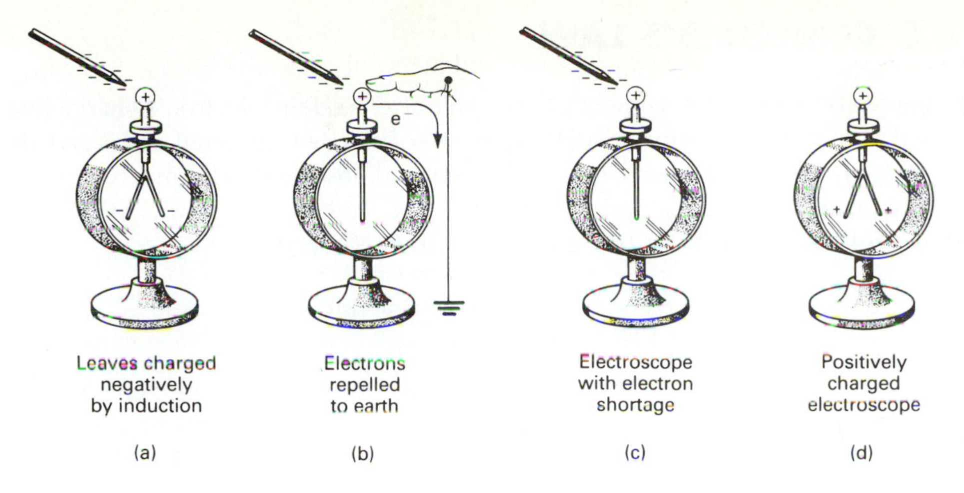

Figure 16.7
Residual charge by induction.

In summary,

> Like charges repel each other, and unlike charges attract each other.

The *van de Graaff generator* is a laboratory machine that is used to produce static electricity and transfer it to a metal sphere by conduction. The van de Graaff consists of an electron source, a rubber belt driven by a motor, and a metal sphere supported by an insulating stand (Fig. 16.8). The electron source charges the rubber belt as it passes by and carries the electrons up to the sphere and deposits them there. This builds up a high potential difference (several hundred thousand volts) between the sphere and the ground. Note that there is no charge on the inside surface of the sphere.

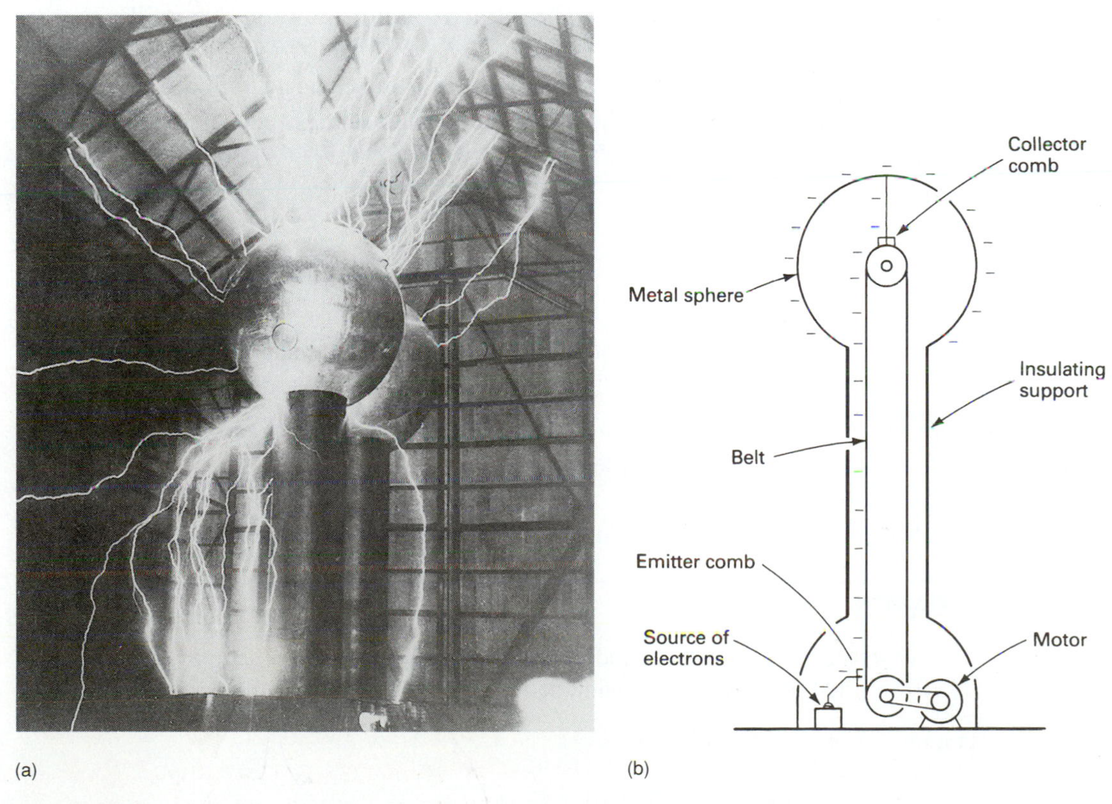

Figure 16.8
van de Graaff generator.

16.3 COULOMB'S LAW

The charge on a proton is denoted by the symbol e^+ and the electron's charge by e^-. In the study of electricity, a much larger unit of charge is required and is called the *coulomb*, C, named after Charles Coulomb (1736–1806). The measurement of the unit charges gives $e^+ = +1.60 \times 10^{-19}$ C and $e^- = -1.60 \times 10^{-19}$ C. Thus, a collection of 6.25×10^{18} electrons, which is found by

$$\frac{1}{-1.60 \times 10^{-19}} = -6.25 \times 10^{18}$$

has a total charge of -1.00 C. We use q to denote amount of electric charge.

In 1789, Charles Coulomb made a scientific study of the forces of attraction and repulsion between charged objects using a very sensitive torsion balance. From his experiments he determined the existence of an *inverse square law* for charged particles that can be used to calculate the forces of attraction or repulsion between charged objects. The inverse square law behavior was also found to apply to the much weaker gravitational force studied in Chapter 4. While the gravitational force is always attractive, the electric force can be either attractive or repulsive. These two forces are important because one holds the solar system together (gravity) and the other holds atoms and molecules together (electricity).

COULOMB'S LAW OF ELECTROSTATICS

The force between two point charges q_1 and q_2 is directly proportional to the product of their magnitudes and inversely proportional to the square of the distance separating them, r.

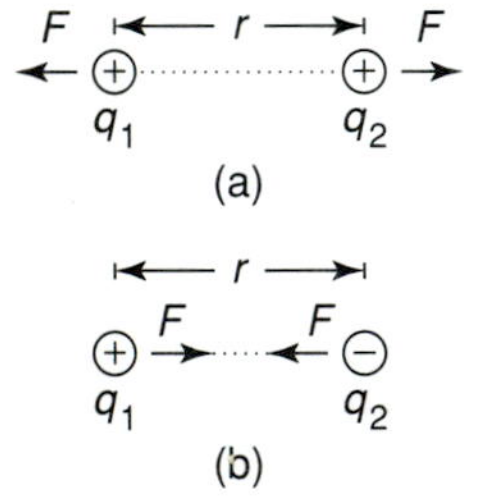

Figure 16.9
(a) Two like charges a distance r apart repel each other with a force F. (b) Two unlike charges attract.

Fig. 16.9(a) shows the repulsive force between two like (positive in this case) charges separated by a distance r. The attractive force between two unlike charges is shown in Fig. 16.9(b).

We must use a *proportionality constant k* in writing Coulomb's law as an equation to take into account the air or other medium between the charges. Written in equation form, Coulomb's law becomes

$$F = \frac{kq_1q_2}{r^2}$$

where F = force of attraction or repulsion (in newtons)
$k = 9.00 \times 10^9 \text{ N m}^2/\text{C}^2$ (k was found by experiment)
q_1, q_2 = size of charges (in coulombs)
r = distance between the charges (in metres)

The force between the charges is a vector quantity that acts on each charge.

EXAMPLE

Two charges, each with magnitude $+6.50 \times 10^{-5}$ C, are separated by a distance of 0.200 cm. Find the force of repulsion between them.

Data:

$$q_1 = q_2 = +6.50 \times 10^{-5} \text{ C}$$
$$r = 0.200 \text{ cm} = 0.00200 \text{ m} = 2.00 \times 10^{-3} \text{ m}$$

$k = 9.00 \times 10^9 \text{ N m}^2/\text{C}^2$
$F = ?$

Basic Equation:

$$F = \frac{kq_1q_2}{r^2}$$

Working Equation: Same

Substitution:

$$F = \frac{(9.00 \times 10^9 \text{ N m}^2/\text{C}^2)(6.50 \times 10^{-5} \text{ C})(6.50 \times 10^{-5} \text{ C})}{(2.00 \times 10^{-3} \text{ m})^2}$$

$$= 9.51 \times 10^6 \text{ N}$$

PROBLEMS 16.3

1. Two identical charges, each -8.00×10^{-5} C, are separated by a distance of 25.0 cm. What is the force of repulsion?
2. The force of repulsion between two identical positive charges is 0.800 N when the charges are 1.00 m apart. Find the value of each charge.
3. A charge of $+3.0 \times 10^{-6}$ C exerts a force of 940 N on a charge of $+6.0 \times 10^{-6}$ C. How far apart are the charges?
4. A charge of -3.0×10^{-8} C exerts a force of 0.045 N on a charge of $+5.0 \times 10^{-7}$ C. How far apart are the charges?
5. When a -9.0-μC charge is placed 0.12 cm from a charge q in a vacuum, the force between the two charges is 850 N. What is the value of q?
6. How far apart are two identical charges of $+6.00$ μC if the force between them is 25.0 N?
7. Three charges are located along the x-axis. Charge A $(+3.00 \times 10^{-6}$ C) is located at the origin. Charge B $(+5.50 \times 10^{-6}$ C) is located at $x = +0.400$ m. Charge C $(-4.60 \times 10^{-6}$ C) is located at $x = +0.750$ m. Find the total force (and direction) on charge B.
8. Find the total force (and direction) on charge A in Problem 7.
9. Find the total force (and direction) on charge C in Problem 7.

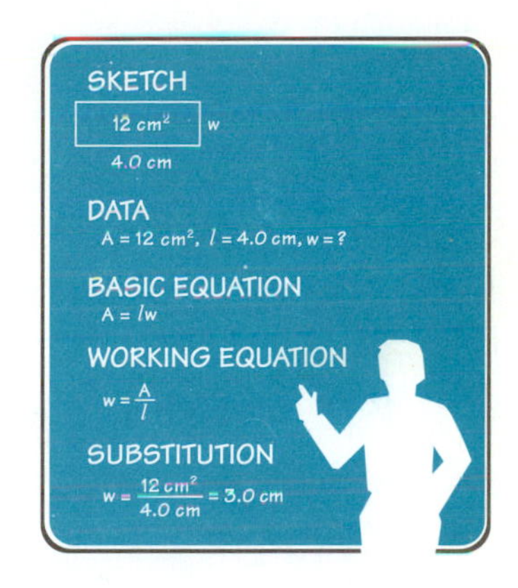

16.4 ELECTRIC FIELDS

So far, we have discussed electrification due to the brushing of electrons from a surface. The concept of the electric field is also an important part of the study of static electricity. Two magnets may either attract or repel each other even though they may not be touching each other. This illustrates the idea of the "field" in that even though they are not touching each other, there is an invisible region around each one that affects the other magnet if that magnet is placed in the region.

In terms of static electricity, an **electric field** exists where an electric force (of attraction or repulsion) acts on a charge brought into the area. A charged balloon put on a wall illustrates this principle (Fig. 16.10). Note that the balloon attracts the wall even without physical contact. The balloon has acquired a negative charge through friction, but the wall surface acquires a positive charge, produced by the electric field of the charged balloon. Such an invisible electric field is present around every charged object.

Figure 16.10
Common electric field. Balloon "sticks" to a vertical wall due to static electricity.

Static electricity can be a real hazard in industry as well as a curiosity and sometimes a nuisance in daily life. The electrical spark from static electricity, particularly in synthetic fiber textile mills, is extremely dangerous. Also, some workers in cosmetic factories in which aerosol (spray) products are made with hydrocarbon (petroleum type) propellants are required to wear cotton clothes rather than those made with synthetic fibers.

Lightning is simply a huge static electricity spark produced in the atmosphere by moving air masses; it is a tremendous discharge (Fig. 16.11).

Figure 16.11
Lightning: static electrical discharge.

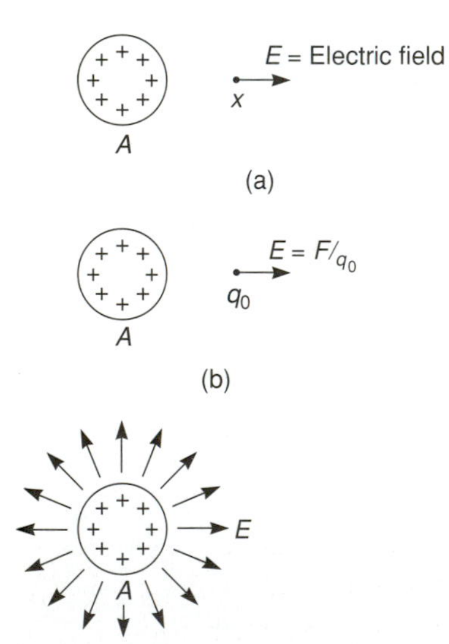

Figure 16.12
An electric field is created in the space around a charged object.

Gravitational fields are set up around objects such as the earth. These fields exert an attractive force on nearby objects. A ball of mass m, for example, released near the surface of the earth has a force $F = mg$ pulling it down to the earth's surface as studied earlier in Section 4.5. In a similar way, an electric field, E, is set up in the space surrounding a charged object. Fig. 16.12(a) shows a charged object A. At the point x, an electric field exists that will exert a force on any charged object placed there. In Fig. 16.12(b) a test charge q_0 is placed at point x. The electric field, E, at x is defined to be $E = F/q_0$ where F is the force exerted on charge q_0 by the charged object A. Fig. 16.12(c) shows the direction of the electric field surrounding object A.

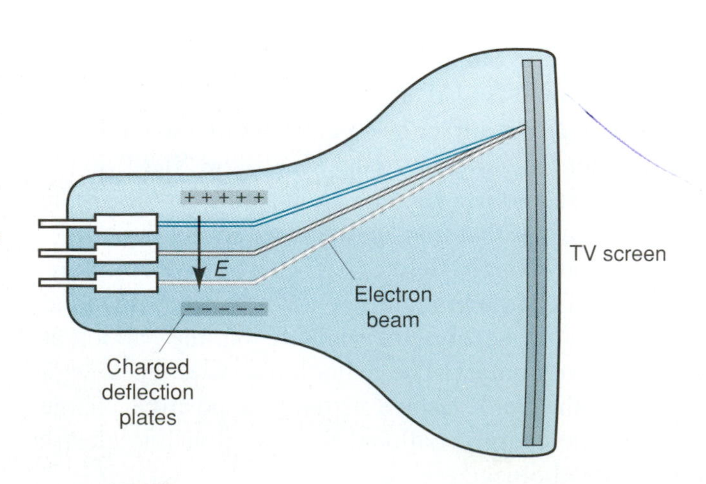

Figure 16.13

Electric fields are used for many applications in electronics and elsewhere. An ink-jet printer may use the deflection caused by an electric field on charged ink droplets to direct ink to the appropriate spot on the paper. In a similar way, the electron beam in a TV picture tube is deflected to the correct spots on the screen to produce a picture (Fig. 16.13).

GLOSSARY

Conduction A transfer of charge from one place to another. (p. 394)
Coulomb's Law States that the force between two point charges is directly proportional to the product of their magnitudes and inversely proportional to the square of the distance between them. (p. 396)
Electric Field An electric field is said to exist in an area where an electric force acts on a charge brought into the area. (p. 397)
Electron A negatively charged particle that is found in every atom. (p. 392)
Induction A method of charging a part of an electroscope or other object by bringing a charged object near to, but not touching, the electroscope. (p. 394)
Neutron A neutral particle that is found in the nucleus of most atoms. (p. 392)
Proton A positively charged particle that is found in the nucleus of every atom. (p. 392)

FORMULA

16.3 Coulomb's law: $F = \dfrac{kq_1q_2}{r^2}$

REVIEW QUESTIONS

1. The atomic particle that carries a positive charge is the
   (a) neutron. (b) proton.
   (c) electron. (d) none of the above.
2. The atomic particle that carries a negative charge is the
   (a) neutron. (b) proton.
   (c) electron. (d) none of the above.
3. The process by which an object becomes charged when it comes in contact with a charged object is called
   (a) induction. (b) electrification.
   (c) conduction. (d) none of the above.
4. The process by which an object becomes permanently charged when it comes near a charged object requires that the first object be
   (a) an insulator. (b) touched by another object.
   (c) a conductor. (d) none of the above.
5. In your own words, describe how materials can become charged by electrification.
6. What particles make up an atom?
7. What particles are located in the nucleus (center) of an atom?
8. Where are electrons located in an atom?
9. What are the two types of charge? What atomic particle carries each type of charge?
10. Describe the process of charging an electroscope by conduction.
11. Describe the process of charging an electroscope by induction.
12. In your own words, describe Coulomb's law of electrostatics.
13. Describe an electric field.
14. Describe lightning.

Chapter 17

Direct Current Electricity

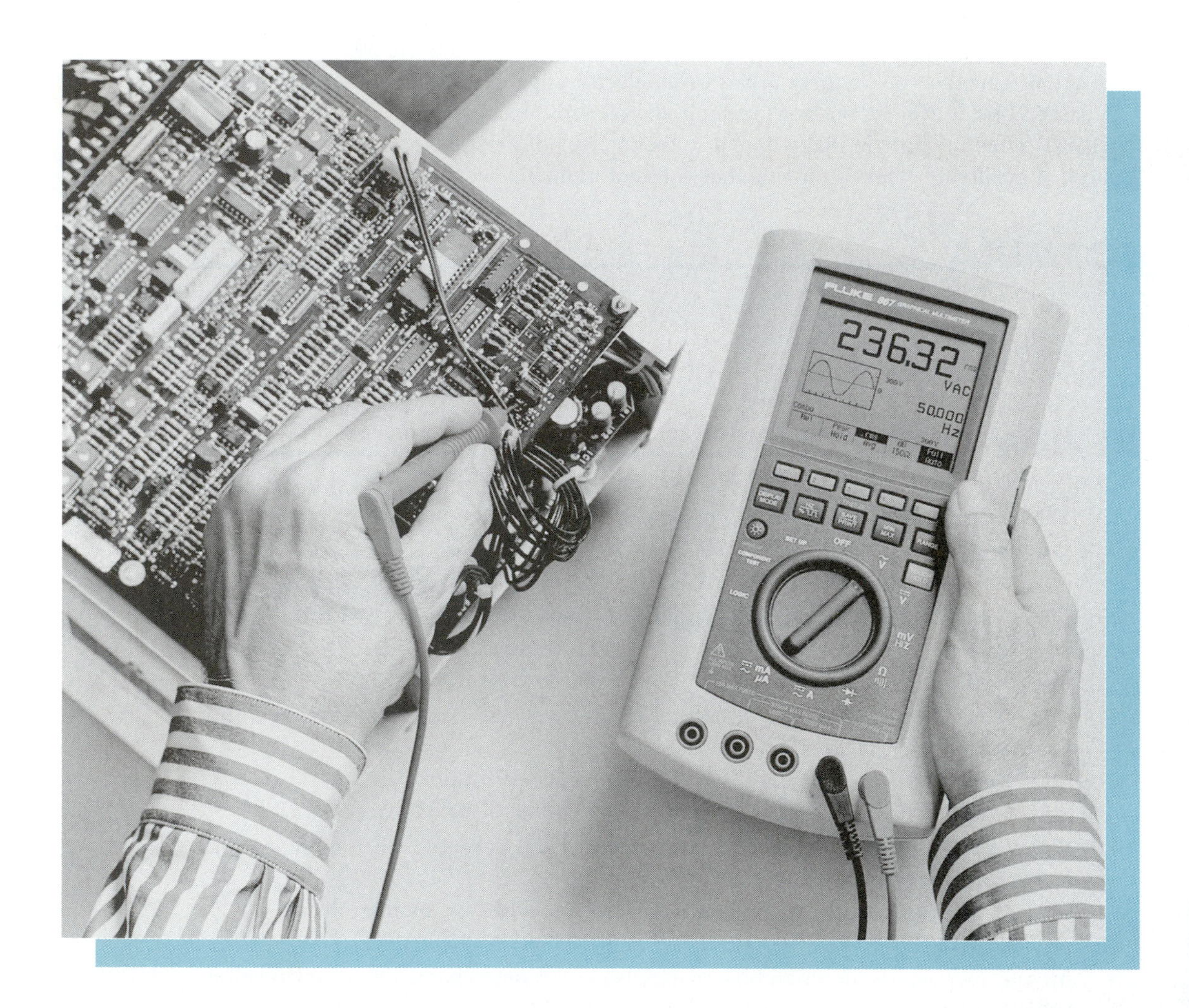

Electricity is the best means of transmitting energy for many purposes. We now consider the delivery of energy by electric currents and introduce common electrical terms, including circuit, volt, ampere, load, and resistance. Then we address series and parallel circuits and electrical instruments.

OBJECTIVES

The major goals of this chapter are to enable you to:

1. Describe the characteristics of electricity.
2. Use Ohm's law to solve electrical flow problems.
3. Use electrical symbols to describe circuits.
4. Determine current, voltage, and resistance in simple circuits.

17.1 SIMPLE CIRCUITS

Electrons moving in a wire make up a current in the wire. When the electron current flows in only one direction (Fig. 17.1), it is called **direct current** (dc). Current that changes direction is called **alternating current** (ac). Alternating current will be considered in Chapter 20.

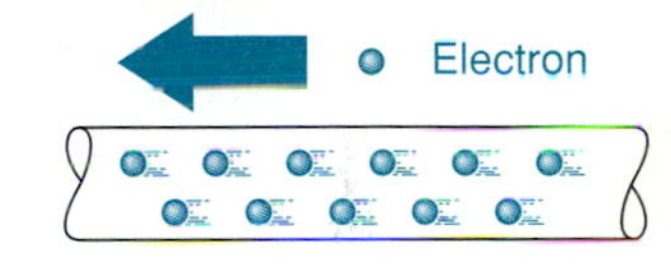

Figure 17.1
Current flowing in only one direction is direct current.

An electric current is a convenient and cost-effective means of transmitting energy. Technicians face many situations every day that require energy to do a particular task. To drill a hole in a metal block (Fig. 17.2), energy must be supplied and transformed into mechanical energy to turn the drill bit. The problem the technician faces is how to supply energy to the machine being used in a form that the machine can turn into useful work. Electricity is often the most satisfactory means of transmitting energy.

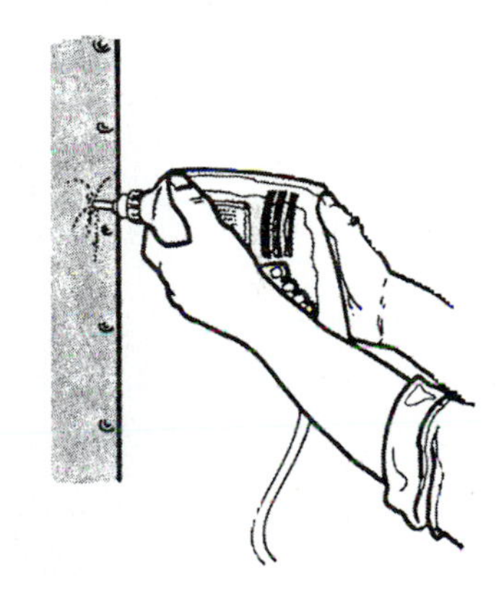

Figure 17.2
Changing electric energy to mechanical energy in drilling.

We begin our study of the use of electricity in transferring energy by looking at an example. Figure 17.3 shows a circuit of a simple flashlight. An **electric circuit** is a conducting loop in which electrons carrying electric energy may be transferred from a suitable source to do useful work and returned back to the source. Energy is stored in the battery. When the switch is closed, energy is transmitted to the light, and the light glows.

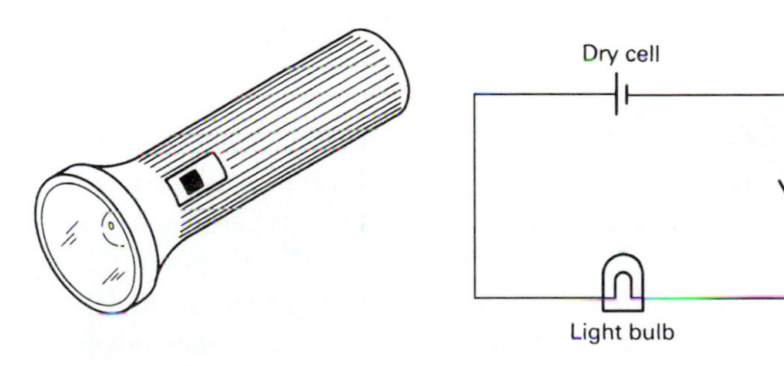

Figure 17.3
Simple electric circuit.

Compared with static electricity, current electricity is the flow of energized electrons through an electron carrier called a conductor (Fig. 17.4). The electrons move from the energy **source** (the battery, here) to the **load** (where the transmitted energy is turned into useful work). There they lose energy picked up in the source. Consider each part of the circuit and determine its function.

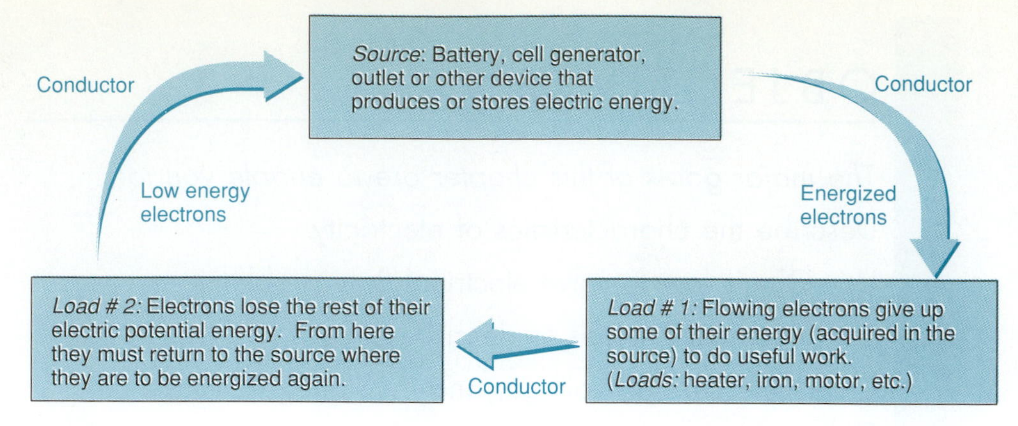

Figure 17.4
Flow of energized electrons through a conductor.

The Source of Energy

The dry cell (Fig. 17.5) is a device that converts chemical energy to electrical energy. How the cell does this will be studied in Chapter 18. Here, we simply state that, by chemical action, electrons are given energy in the cell. When energy is given to electrons in this manner, their electrical potential energy is raised.

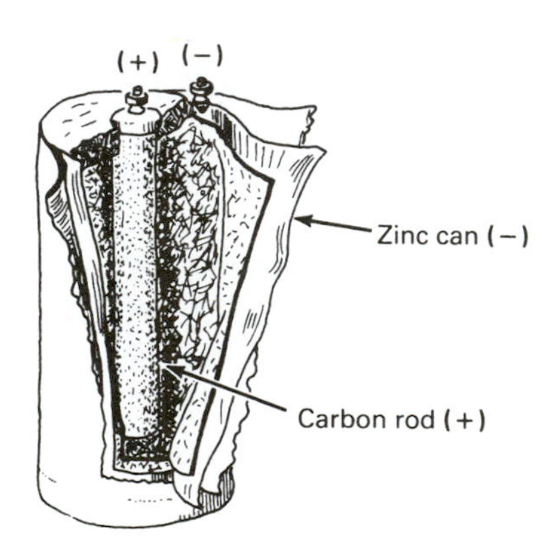

Figure 17.5
Chemical energy is changed to electric energy in a dry cell.

What does "electric potential energy" mean? The flow of charge in an electric circuit is often compared to the flow of water in a hydraulic system, as shown in Fig. 17.6(b). Water naturally flows from a position of high potential to a position of low potential and performs work in the process, such as turning a waterwheel or turbine. A pump is needed to return the water from its low-potential position to its high-potential position. There is a *difference in potential* due to its position. Work has been done in lifting it against gravity to the higher position. In a source of electric energy something similar happens. In the source (the battery), work is done on electrons that gives them potential energy. This potential difference between the energized electrons [at the negative (−) pole] and the low potential energy electrons [at the positive (+) pole] causes the electrons to flow from one point (−) to the other (+) when connected [Fig. 17.6(a)].

Think of this potential difference between the two points as an electric field set up between the poles that drives the electrons in the external circuit from the negative pole to the positive pole. The energized electrons collect at the source's

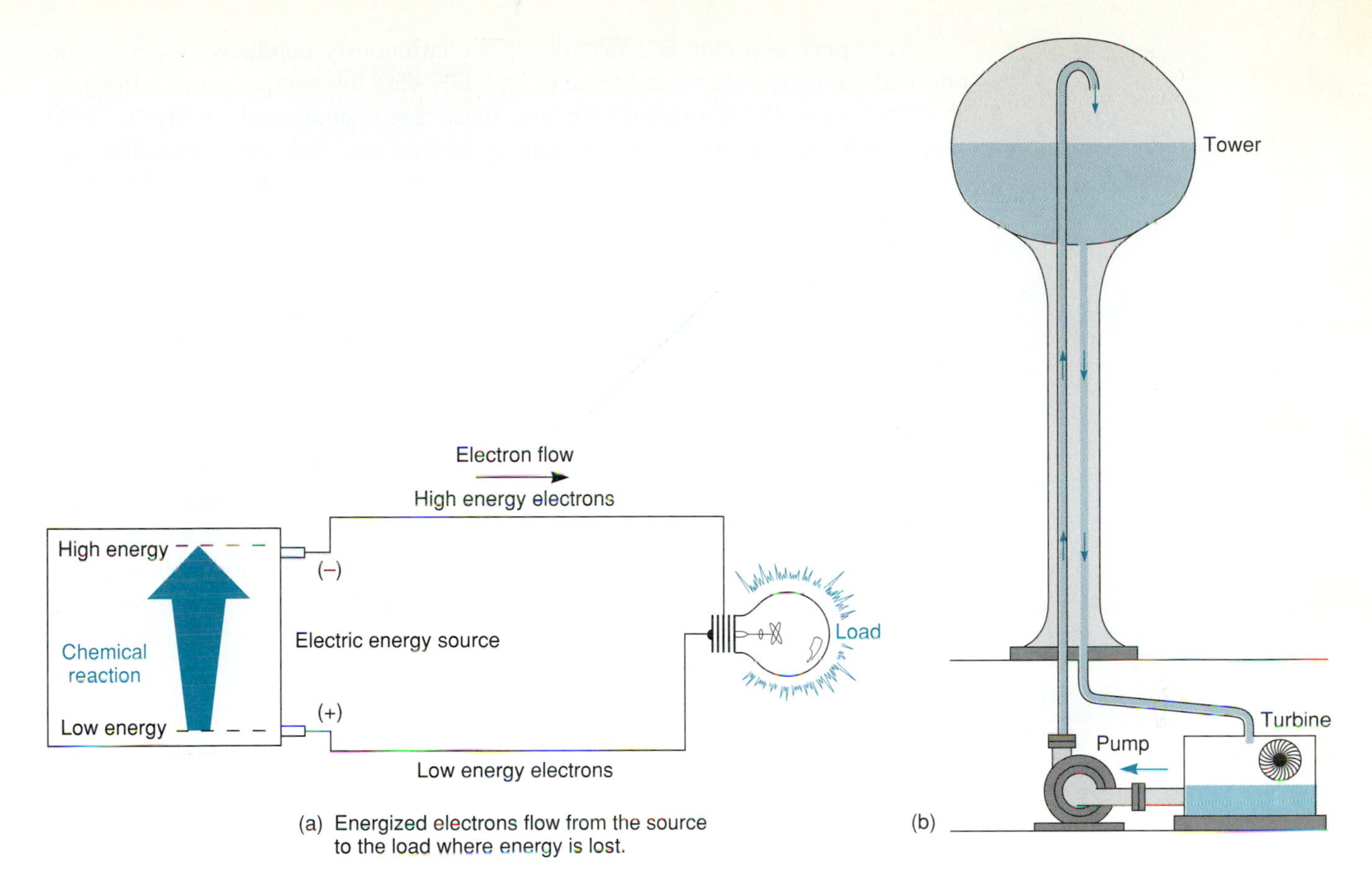

Figure 17.6
The flow of charge in an electric circuit is often compared to the flow of water in a hydraulic system.

negative pole, repel each other, and flow through the circuit to the positive pole. They lose their potential electric energy to the load.

The Conductor

A conductor carries or transfers the electric charge to the load [Fig. 17.6(a), the light bulb]. **Conductors** are substances (such as copper) that have large numbers of free electrons (electrons that are free to move throughout the conductor). As high-energy electrons from the dry cell pass through the conductor, they collide with other electrons in the conductor. These electrons then carry the energy farther along the wire until they collide again and transfer energy on through the wire.

Silver, copper, and aluminum are metals that allow electrons to pass freely through and thus are good conductors. Other metals offer more opposition to the flow of electrons and are poorer conductors (Fig. 17.7). Substances that do not allow electrons to pass readily are called **insulators.** Common insulators are rubber, wool, silk, glass, wood, distilled water, and dry air.

A small number of materials, called **semiconductors,** fall between conductors and insulators. Their importance is due to the fact that these materials under certain conditions allow current to flow in one direction only. Silicon is an example of a semiconductor and is used in transistors and integrated circuits (IC's).

(a) Good conductor

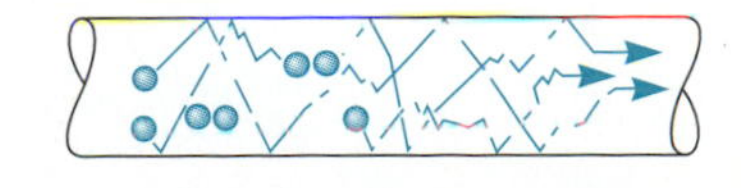

(b) Poor conductor

Figure 17.7

A **superconductor** is a material that continuously conducts electrical current without resistance when cooled to typically very low temperatures, often near absolute zero. H. Kamerlingh-Omnes discovered superconductivity in 1911 shortly after he discovered how to liquefy helium gas. He determined that mercury metal lost its resistance at temperatures just below 4.2 K, the boiling point of helium. Scientists are currently finding materials in which superconductivity exists at higher temperatures. The ultimate goal is finding a material under conditions at which superconductivity exists at room temperature.

The Load

Figure 17.8
Electric energy is changed to light and heat in the load.

In the load, electrons lose their energy. The load converts the electric energy to other useful forms. In a light bulb, electric energy is changed to light and heat (Fig. 17.8). An electric motor changes electrical energy to mechanical energy. The load may be a complex motor or only a simple resistor with heat the only new form of energy. The electrons do not collect and remain in the load, but continue back to the low-energy side of the battery (+). There they may be energized again for another trip through the circuit.

Current

The flow of electrons through a conductor is called **current.** We could count the electrons passing a point during a certain time to get the rate of flow. This is impractical because the flow of electrons is so large (about 10^{18}/s). To have a workable unit of electric charge, we define a charge of 6.25×10^{18} electrons as 1 *coulomb*. The *ampere* (A) is the rate of flow of 1 coulomb of charge passing a point in 1 second. We define a unit for the rate of flow of charge as follows:

$$1 \text{ ampere (A)} = \frac{1 \text{ coulomb (C)}}{1 \text{ second (s)}}$$

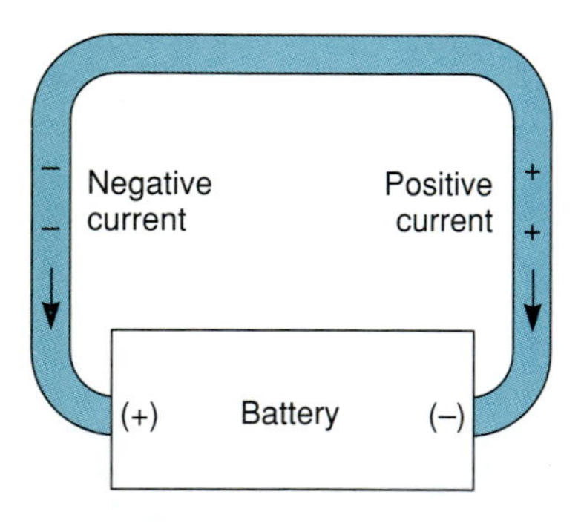

Figure 17.9
Current flow as positive current or negative current.

As mentioned earlier, the charge carriers in metals are electrons. In some other conductors, such as electrolytes (conducting liquid solutions), the charge carriers may be positive or negative or both. An agreement must be made to determine which charge carriers should be assumed in our following discussions.

Note that positive charges would flow in the opposite direction (toward the negative terminal) from that followed by negative charges (toward the positive terminal) when a battery is connected to a circuit (Fig. 17.9). A positive current moving in one direction is equivalent for almost all measurements to a current of negative charges flowing in the opposite direction.

In this book we assume that the charge carriers are positive, and we draw our current arrows in the direction that a positive charge would flow. This is the practice of the majority of engineers and technicians. If you encounter the negative-current convention at a later date, you should remember that a negative current flows in the opposite direction from that of a positive current. Regardless of the method used, the analysis of a situation by either method will give the correct result. Some of the rules discussed later, such as the right-hand rule for finding the direction of the magnetic field, will be different if the negative-current convention is used.

Voltage

We have seen that current flows in a circuit because of the difference in potential of the different points in the circuit. Work is done as a charge moves from one point to another in an electric field. Work is required to move a charge from one point to another. Such points differ in electric potential. The *potential difference* between two points in an electric field is the work done per unit of charge as the charge is moved between two points. That is,

$$\text{potential difference} = \frac{\text{work}}{\text{charge}}$$

In *sources,* the raising of the potential energy of electrons which results in a potential difference is called **emf** *(E).* In *circuits,* the lowering of the potential energy of electrons in a load is called **voltage drop** across the load.

The *volt* (V) is the unit of both emf and voltage drop. We define the volt as the potential difference between two points if 1 joule of work is produced or used in moving 1 coulomb of charge from one point to another.

$$1 \text{ volt (V)} = \frac{1 \text{ joule (J)}}{1 \text{ coulomb (C)}}$$

Resistance

Not all substances and not even all metals are good conductors of electricity. Those with few free electrons tend to have greater opposition to the flow of charge. This opposition is called **resistance.** The unit of resistance is the *ohm* (Ω). It is not a fundamental unit and is discussed in Section 17.2.

The resistance of a wire is determined by several factors. Among these are

1. *Temperature.* An increase in temperature results in an increase in resistance in a wire, for most metals. Other materials, such as semiconductors, show a decrease in resistance with increasing temperature.
2. *Length.* Resistance varies directly with length. If we double the length of a given wire, the resistance is doubled [Fig. 17.10(a)].
3. *Cross-sectional area.* Resistance varies inversely with cross-sectional area. If we double the cross-section of a wire, the resistance is *halved.* This is sim-

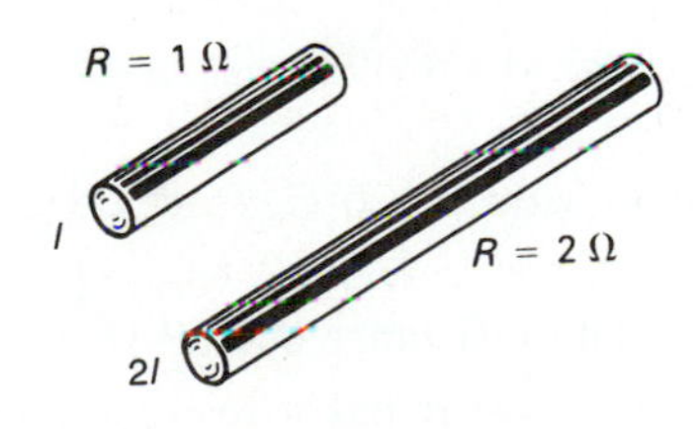

(a) Resistance varies directly with length.

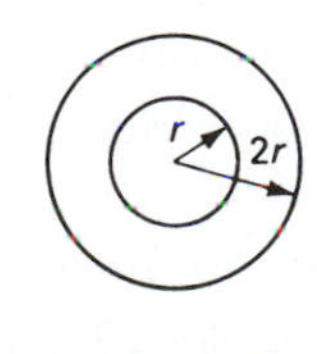

(b) Doubling the radius more than doubles the cross-sectional area.

Figure 17.10

ilar to water flowing through two pipes. It flows more easily through the larger pipe. (Note that doubling the radius of a wire [Fig. 17.10(b)] *more than* doubles the cross-sectional area: $A = \pi r^2$.)

4. *Material.* Resistance depends on the nature of the material. For example, copper is a better conductor than steel. The conducting characteristic of various materials is described by resistivity. **Resistivity** (ρ) is the resistance of a specified amount of the material.

These factors are related by the equation:

$$R = \frac{\rho l}{A}$$

where R = resistance (Ω)
ρ = resistivity (Ω cm)
l = length (cm)
A = cross-sectional area (cm^2)

EXAMPLE

Find the resistance of a copper wire 20.0 m long with cross-sectional area of 6.56×10^{-3} cm^2 at 20°C. The resistivity of copper at 20°C is 1.72×10^{-6} Ω cm.

Data:

$$l = 20.0 \text{ m} = 2.00 \times 10^3 \text{ cm}$$
$$A = 6.56 \times 10^{-3} \text{ cm}^2$$
$$\rho = 1.72 \times 10^{-6} \ \Omega \text{ cm}$$
$$R = ?$$

Basic Equation:

$$R = \frac{\rho l}{A}$$

Working Equation: Same

Substitution:

$$R = \frac{(1.72 \times 10^{-6} \ \Omega \ \cancel{\text{cm}})(2.00 \times 10^3 \ \cancel{\text{cm}})}{6.56 \times 10^{-3} \ \cancel{\text{cm}^2}}$$
$$= 0.524 \ \Omega$$

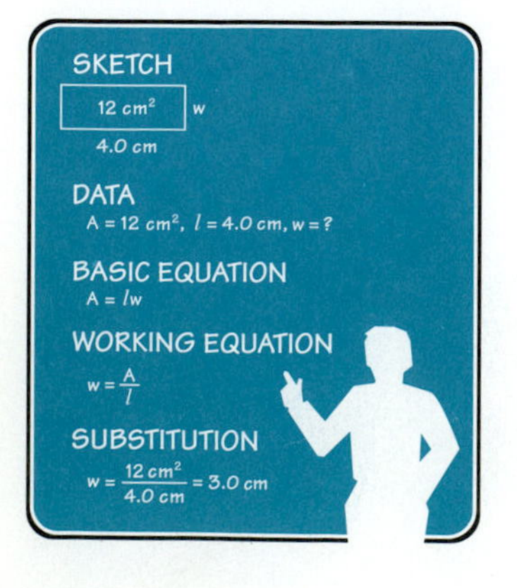

■ PROBLEMS 17.1

1. Find the resistance of 78.0 m of No. 20 aluminum wire at 20°C. ($\rho = 2.83 \times 10^{-6}$ Ω cm, $A = 2.07 \times 10^{-2}$ cm^2.)
2. Find the resistance of 315 ft of No. 24 copper wire with resistance 0.0262 Ω/ft.
3. Find the resistance per foot of No. 22 copper wire if $58\bar{0}$ ft has a resistance of 9.57 Ω.
4. At 77°F, $10\bar{0}$ ft of No. 18 copper wire has a resistance of 0.651 Ω. Find the resistance of $50\bar{0}$ ft of this wire.
5. Find the resistance of 475 m of No. 20 copper wire at 20°C. ($\rho = 1.72 \times 10^{-6}$ Ω cm, $A = 2.07 \times 10^{-2}$ cm^2.)

6. Find the resistance of $10\bar{0}$ m of No. 20 copper wire at 20°C. ($\rho = 1.72 \times 10^{-6}\ \Omega$ cm, $A = 2.07 \times 10^{-2}$ cm^2.)
7. Find the resistance of 50.0 m of No. 20 aluminum wire at 20°C, ($\rho = 2.83 \times 10^{-6}\ \Omega$ cm, $A = 2.07 \times 10^{-2}$ cm^2.)
8. Find the length of copper wire with resistance 0.0262 Ω/ft and total resistance 3.00 Ω.
9. Find the cross-sectional area of copper wire at 20°C that is 60.0 m long and has resistivity $\rho = 1.72 \times 10^{-6}\ \Omega$ cm and resistance 0.788 Ω.
10. Find the length of a copper wire with resistance 0.0262 Ω/ft and total resistance 5.62 Ω.

17.2 OHM'S LAW

When a voltage is applied *across* a material that conducts electrical current, the relationship between current *through* the material and voltage across it depends upon the type of material as shown in Fig. 17.11. The straight-line relationship shown in Fig. 17.11(a) is typical of many materials, including metal conductors. Other materials, such as semiconductors shown in Fig. 17.11(b), show a nonlinear relationship between I and V. For the materials with a straight-line relationship, the equation relating I and V was determined by a German physicist, Georg Simon Ohm (1787–1854). The relationship is called **Ohm's law** (see Fig. 17.12).

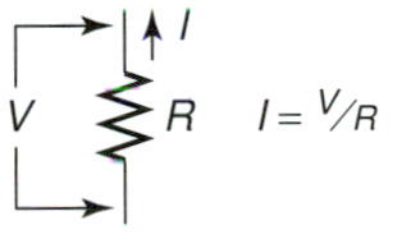

Figure 17.12
The relationship between the voltage *across* a resistance and the current *through* the resistance is described by Ohm's law.

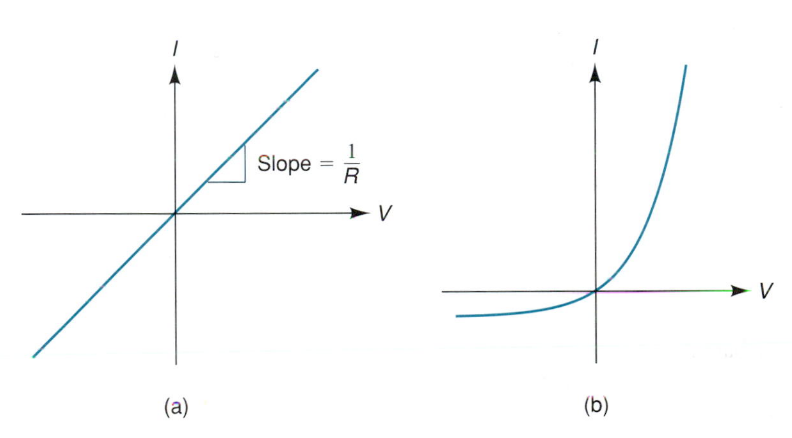

Figure 17.11
(a) Straight-line (linear) *I-V* characteristic (typical of resistors). (b) Nonlinear *I-V* characteristic (typical of semiconductor diodes).

Ohm's law:

$$I = \frac{V}{R}$$

where I = current *through* the resistance
V = voltage drop *across* the resistance
R = resistance

Ohm's law can also be written:

$$I = \frac{E}{R}$$

where E = emf of the source of electrical energy

Ohm's law applies to dc circuits containing linear resistors and those ac circuits containing only resistance. It may be applied to the whole circuit or to any part of it.

Ohm's law should aid us now in understanding resistance. As we mentioned earlier, the ohm (Ω) is a derived unit. From Ohm's law,

$$I = \frac{V}{R}$$

Solving for R,

$$R = \frac{V}{I}$$

Substituting units,

$$\boxed{\Omega = \frac{V}{A}}$$

EXAMPLE 1

A heating element on an electric range operating on 240 V has a resistance of 30.0 Ω. What current does it draw?

Data:

$$E = 240 \text{ V}$$
$$R = 30.0\ \Omega$$
$$I = ?$$

Basic Equation:

$$I = \frac{E}{R}$$

Working Equation: Same

Substitution:

$$I = \frac{240 \text{ V}}{30.0\ \Omega}$$
$$= 8.0 \text{ V}/\Omega$$
$$= 8.0 \text{ A}$$

$$\boxed{\text{A} = \frac{V}{\Omega}}$$

EXAMPLE 2

A flashlight bulb is connected to two dry cells with an equivalent voltage of 3.0 V. If it draws 15 mA, what is its resistance?

Sketch:

R = ?
E = 3.0 V
I = 15 mA

Data:

$$E = 3.0 \text{ V}$$
$$I = 15 \text{ mA} = 15 \times 10^{-3} \text{ A} = 0.015 \text{ A}$$
$$R = ?$$

Basic Equation:

$$I = \frac{E}{R}$$

Working Equation:

$$R = \frac{E}{I}$$

Substitution:

$$\begin{aligned} R &= \frac{3.0 \text{ V}}{0.015 \text{ A}} \\ &= 2\bar{0}0 \text{ V/A} \\ &= 2\bar{0}0\ \Omega \end{aligned}$$

■ PROBLEMS 17.2

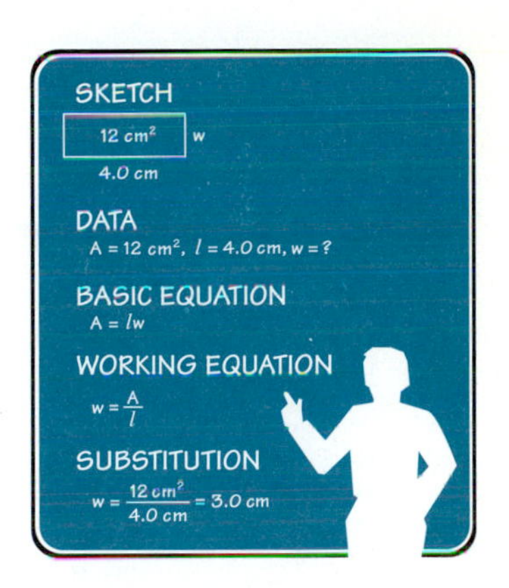

1. A heating element operates on 115 V. If it has a resistance of 24.0 Ω, what current does it draw?
2. A given coffeepot operates on 12.0 V. If it draws 2.50 A, find its resistance.
3. An electric heater draws a maximum of 14.0 A. If its resistance is 15.7 Ω, on what voltage is it operating?
4. A heating coil operates on $22\bar{0}$ V. If it draws 15.0 A, find its resistance.
5. If a resistance draws 0.750 A on 115 V, what is its resistance?
6. What current does a 75.0-Ω resistance draw on 115 V?
7. A heater operates on $22\bar{0}$ V. If it draws 12.5 A, what is its resistance?
8. What current does a 50.0-Ω resistance draw on 115 V?
9. What current does a 175-Ω resistance draw on $22\bar{0}$ V?
10. A heater draws 3.50 A on 115 V. What is its resistance?
11. (a) What current does a 150-Ω resistance draw on a $1\bar{0}$-V battery? (b) What voltage battery would produce 3 times the current in (a)? (c) What current would a 75-Ω resistor draw on the $1\bar{0}$-V battery?
12. A heater draws 4.25 A on 32.0 V. (a) What is the resistance of the heater? (b) What resistance heater would draw 8.50 A on 32.0 V? ■

17.3 SERIES CIRCUITS

In order to communicate about problems in electricity, technicians have developed a "language" of their own. It is a picture language using symbols and diagrams. The circuit diagram is the most common and useful way to show a circuit. Note how each component (part) of the picture in Fig. 17.13(a) is represented by its symbol in the symbol diagram in its relative position in Fig. 17.13(b). The light bulb may be represented as a resistance. Then the circuit diagram would appear as in Fig. 17.13(c). Some of the symbols used most often appear in Fig. 17.14.

There are two basic types of circuits: series and parallel. A fuse in a house is wired in series with the outlets. The outlets themselves are wired in parallel. A study of series and parallel circuits is basic to a study of electricity.

An electric circuit with only one path for the current to flow (Fig. 17.15) is called a **series circuit.** The current in a series circuit is the same throughout. That

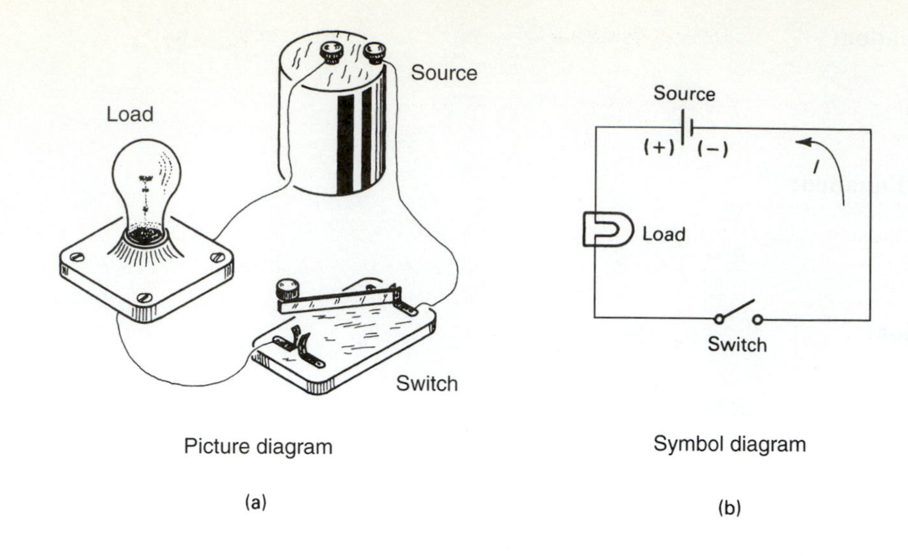

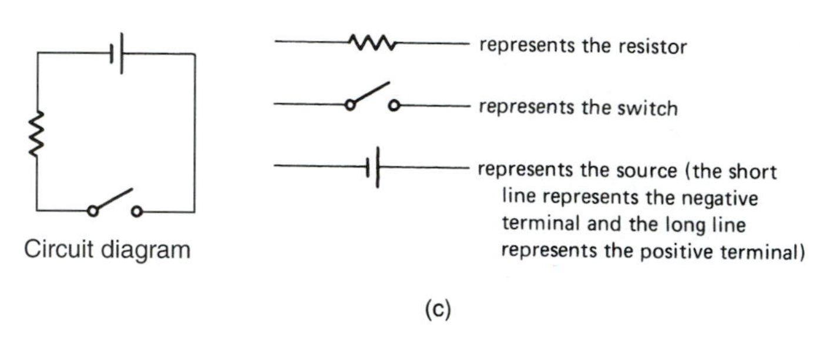

Figure 17.13
A series circuit.

is, the current flows out of one resistor and into the next resistor. Therefore, the total current is the same as the current flowing through each resistance in the circuit.

SERIES

$$I = I_1 = I_2 = I_3 = \cdots$$

where I = total current
I_1 = current through R_1
I_2 = current through R_2
I_3 = current through R_3

In a series circuit the emf of the source equals the sum of the separate voltage drops in the circuit (Fig. 17.16).

SERIES

$$E = V_1 + V_2 + V_3 + \cdots$$

where E = emf of the source
V_1 = voltage drop across R_1
V_2 = voltage drop across R_2
V_3 = voltage drop across R_3

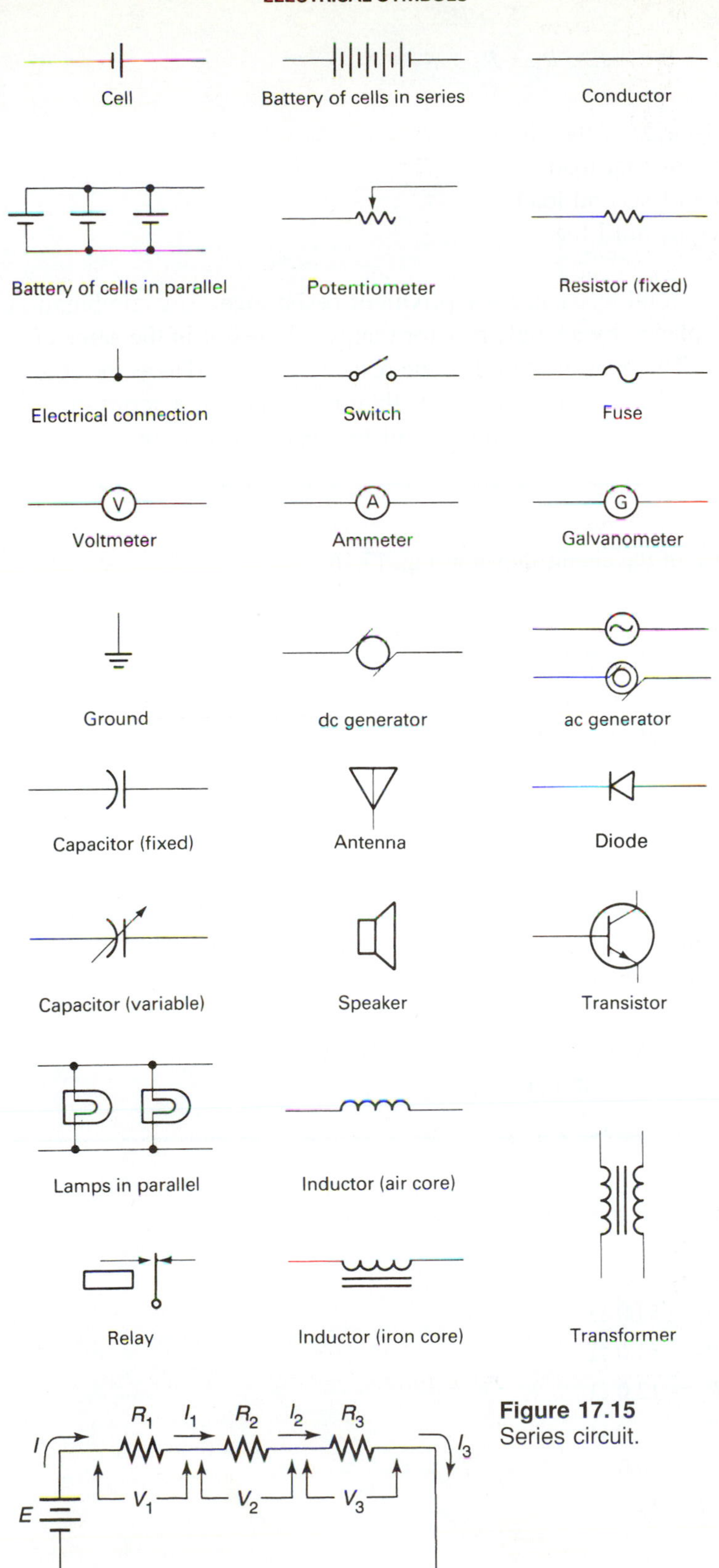

Figure 17.14

Figure 17.15
Series circuit.

The resistance of the conducting wires is very small and will be neglected here. The total resistance of a series circuit equals the sum of all the resistances in the circuit.

SERIES

$$R = R_1 + R_2 + R_3 + \cdots$$

where R = total resistance of the circuit
R_1 = resistance of first load
R_2 = resistance of second load
R_3 = resistance of third load

Another term for total resistance is **equivalent resistance.** Any combination of resistors can be replaced by a single resistor that would result in the same current flow and voltage. The resistance of this one resistor is referred to as the equivalent resistance of the combination. The equivalent resistance of a series combination is larger than the resistance of any one of the resistors in series.

EXAMPLE 1

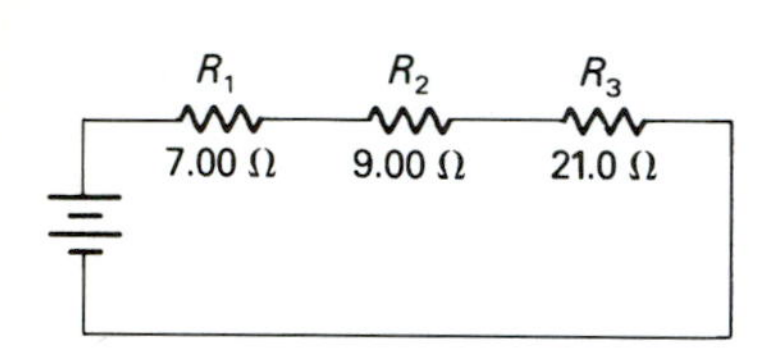

Figure 17.16

Find the total resistance of the circuit shown in Fig. 17.16.

Data:

$$R_1 = 7.00\ \Omega$$
$$R_2 = 9.00\ \Omega$$
$$R_3 = 21.0\ \Omega$$
$$R = ?$$

Basic Equation:

$$R = R_1 + R_2 + R_3$$

Working Equation: Same

Substitution:

$$R = 7.00\ \Omega + 9.00\ \Omega + 21.0\ \Omega$$
$$= 37.0\ \Omega$$

EXAMPLE 2

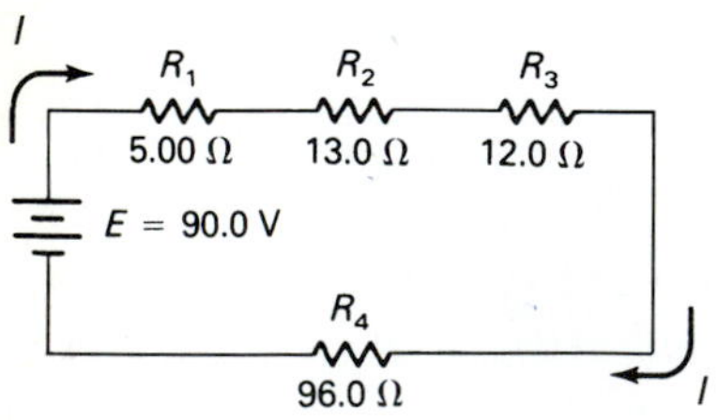

Figure 17.17

Find the current in the circuit shown in Fig. 17.17.

Data:

$$R_1 = 5.00\ \Omega$$
$$R_2 = 13.0\ \Omega$$
$$R_3 = 12.0\ \Omega$$
$$R_4 = 96.0\ \Omega$$
$$E = 90.0\ \text{V}$$
$$I = ?$$

Basic Equations:

$$R = R_1 + R_2 + R_3 + R_4 \quad \text{and} \quad I = \frac{E}{R}$$

Working Equations: Same

Substitutions:

$$R = 5.00\ \Omega + 13.0\ \Omega + 12.0\ \Omega + 96.0\ \Omega$$
$$= 126.0\ \Omega$$

$$I = \frac{90.0\ \text{V}}{126.0\ \Omega}$$
$$= 0.714\ \text{A}$$

EXAMPLE 3

Find the value of R_3 in the circuit shown in Fig. 17.18.

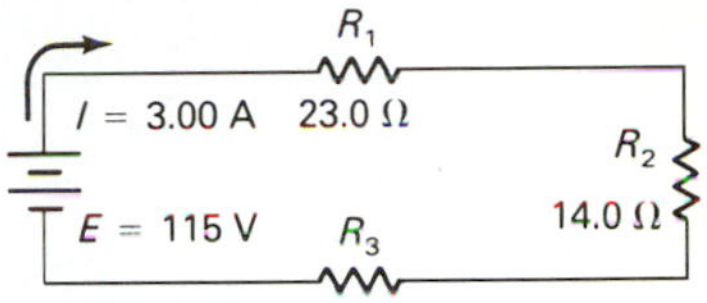

Data:

$$I = 3.00\ \text{A}$$
$$E = 115\ \text{V}$$
$$R_1 = 23.0\ \Omega$$
$$R_2 = 14.0\ \Omega$$
$$R_3 = ?$$

Basic Equations:

$$I = \frac{E}{R} \quad \text{and} \quad R = R_1 + R_2 + R_3$$

Working Equations:

$$R = \frac{E}{I} \quad \text{and} \quad R_3 = R - R_1 - R_2$$

Substitutions:

$$R = \frac{115\ \text{V}}{3.00\ \text{A}}$$
$$= 38.3\ \Omega$$

$$R_3 = 38.3\ \Omega - 23.0\ \Omega - 14.0\ \Omega$$
$$= 1.3\ \Omega$$

Figure 17.18

EXAMPLE 4

Find the voltage drop across R_3 in Example 3.

Data:

$$I = I_3 = 3.00\ \text{A}$$
$$R_3 = 1.3\ \Omega$$
$$V_3 = ?$$

Basic Equation:

$$I_3 = \frac{V_3}{R_3}$$

Working Equation:

$$V_3 = I_3 R_3$$

Substitution:

$$V_3 = (3.00 \text{ A})(1.3 \text{ }\Omega)$$
$$= 3.9 \text{ V}$$

A fuse is connected in series in an electrical circuit.

Table 17.1 summarizes the characteristics of series circuits.

Table 17.1 Characteristics of Series Circuits

	Series
Current	$I = I_1 = I_2 = I_3 = \cdots$
Equivalent Resistance	$R = R_1 + R_2 + R_3 + \cdots$
Voltage	$E = V_1 + V_2 + V_3 + \cdots$

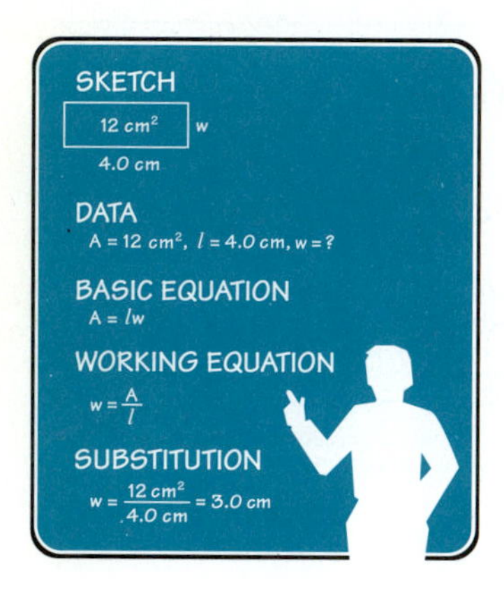

PROBLEMS 17.3

1. Three resistors of 2.00 Ω, 5.00 Ω, and 6.50 Ω are connected in series with a 24.0-V battery. Find the total resistance of the circuit.
2. Find the current in Problem 1.
3. Find the equivalent resistance in the circuit shown in Fig. 17.19.
4. Find the current through R_2 in Problem 3.
5. Find the current in the circuit shown in Fig. 17.20.
6. Find the voltage drop across R_1 in Problem 5.
7. What emf is needed for the circuit shown in Fig. 17.21?
8. Find the voltage drop across R_3 in Problem 7.
9. Find the equivalent resistance in the circuit shown in Fig. 17.22.
10. Find R_3 in the circuit in Problem 9.
11. Find the values of R_1, R_2, and R_3 in Fig. 17.23.
12. Find the values of V_1, R_2, and V_3 in Fig. 17.24.
13. Find the values of R_1, V_2, and R_3 in Fig. 17.25.

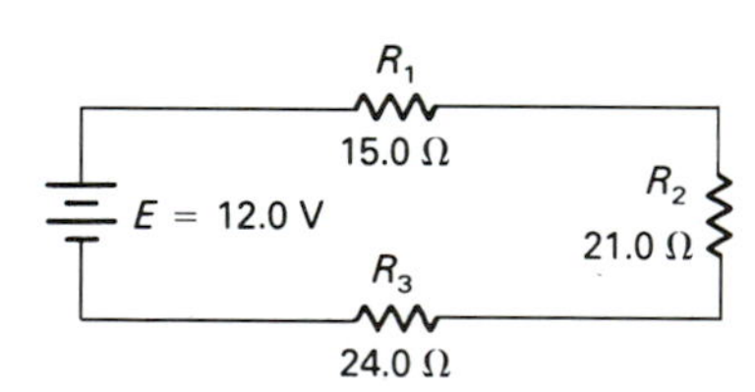

Figure 17.19

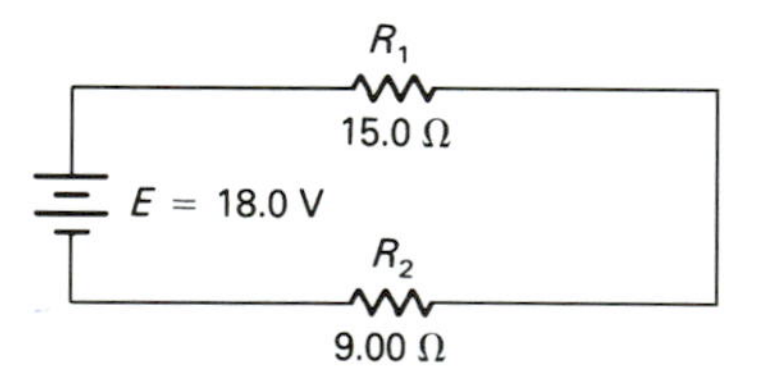

Figure 17.20

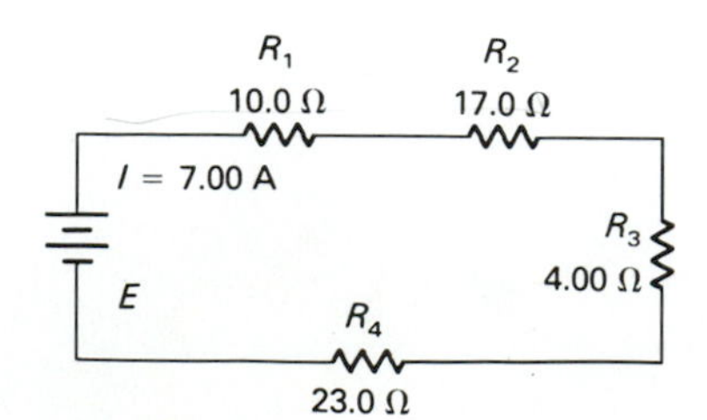

Figure 17.21

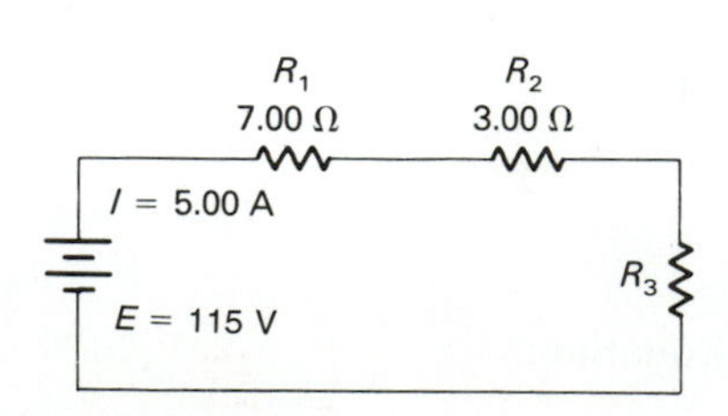

Figure 17.22

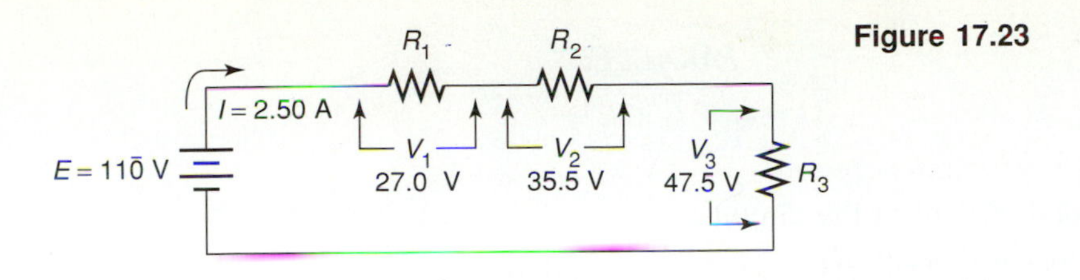

Figure 17.23

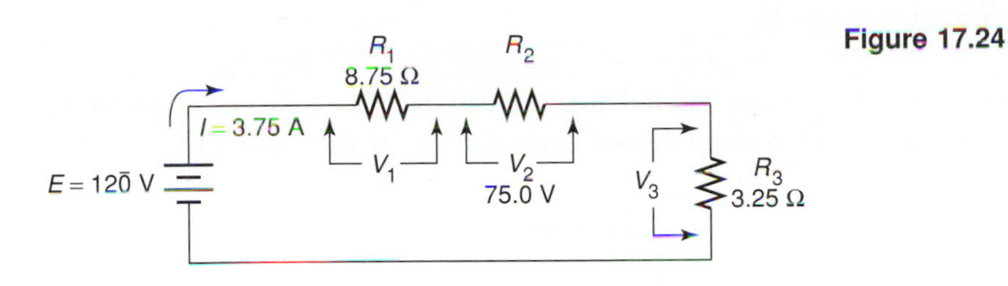

Figure 17.24

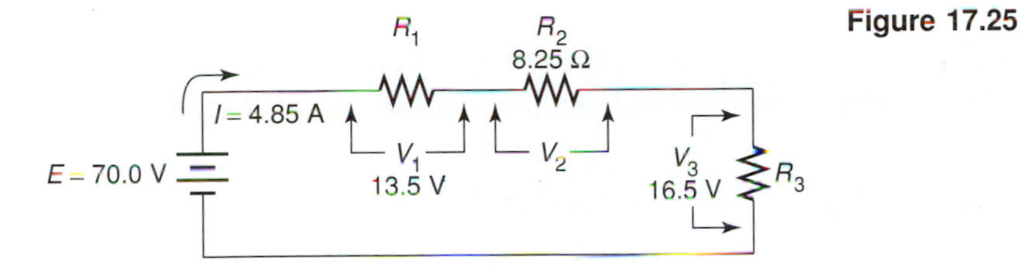

Figure 17.25

17.4 PARALLEL CIRCUITS

An electric circuit with more than one path for the current to flow (Fig. 17.26) is called a **parallel circuit.** All resistors connected in parallel have their ends connected to two common points (nodes) in the circuit (points A and B in Fig. 17.26). The current in a parallel circuit is divided among the branches of the circuit (Fig. 17.27). How it is divided depends on the resistance of each branch. The paths with the least resistance allow the largest currents to flow. Since the current divides, the current from the source equals the sum of the currents through each of the branches.

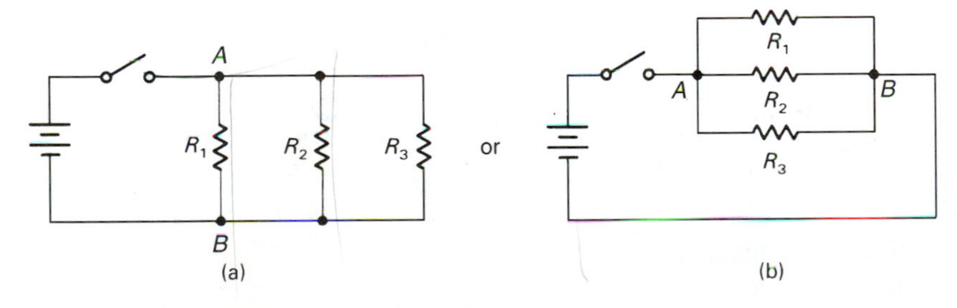

Figure 17.26
Different ways to represent a parallel circuit.

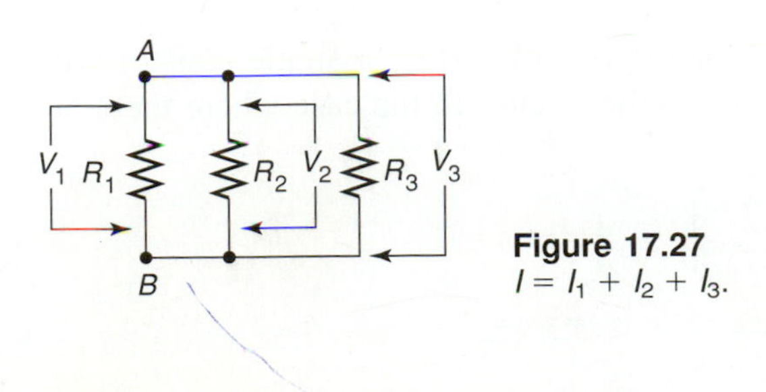

Figure 17.27
$I = I_1 + I_2 + I_3$.

> *PARALLEL*
>
> $$I = I_1 + I_2 + I_3 + \cdots$$

where I = total current in the circuit
I_1 = current through R_1
I_2 = current through R_2
I_3 = current through R_3

Since the ends of all resistors in parallel are connected to the same common points (nodes) in the circuit, the voltage across each resistor is the same (Fig. 17.27).

> *PARALLEL*
>
> $$V_1 = V_2 = V_3 = \cdots$$

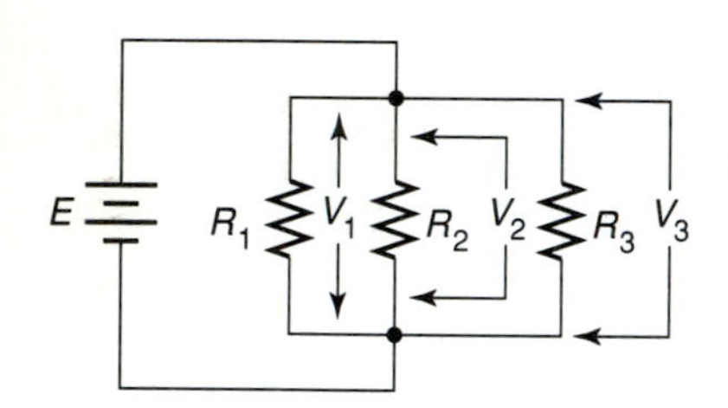

Figure 17.28
$E = V_1 = V_2 = V_3$.

The emf of the source is the same as the voltage drop across each resistance in the circuit if there are no other (series) elements in the circuit (Fig. 17.28).

> *PARALLEL WITH VOLTAGE SOURCE*
>
> $$E = V_1 = V_2 = V_3 = \cdots$$

where E = emf of the source
V_1 = voltage drop across R_1
V_2 = voltage drop across R_2
V_3 = voltage drop across R_3

Therefore, several different loads requiring the same voltage are connected in parallel.

Any combination of resistors can be replaced by a single resistor that would result in the same current flow and voltage. The resistance of this one resistor is referred to as the *equivalent resistance* of the combination. The equivalent resistance of a parallel circuit is less than the resistance of any single branch of the circuit. To find the equivalent resistance, use the formula

> *PARALLEL*
>
> $$\frac{1}{R} = \frac{1}{R_1} + \frac{1}{R_2} + \frac{1}{R_3} + \cdots$$

where R = equivalent resistance
R_1 = resistance of R_1
R_2 = resistance of R_2
R_3 = resistance of R_3

If the parallel combination of resistors is replaced by a single resistor with the resistance R, the same current flows in the circuit. In the case where there are only two resistors in parallel, then

$$\frac{1}{R} = \frac{1}{R_1} + \frac{1}{R_2}$$

or

$$R = \frac{R_1 R_2}{R_1 + R_2}$$

(See Fig. 17.29.)

For comparison to parallel circuits, consider the water system shown in Fig. 17.30(a).

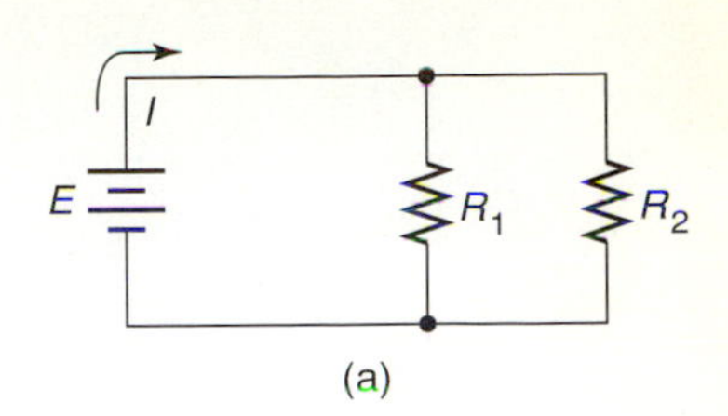

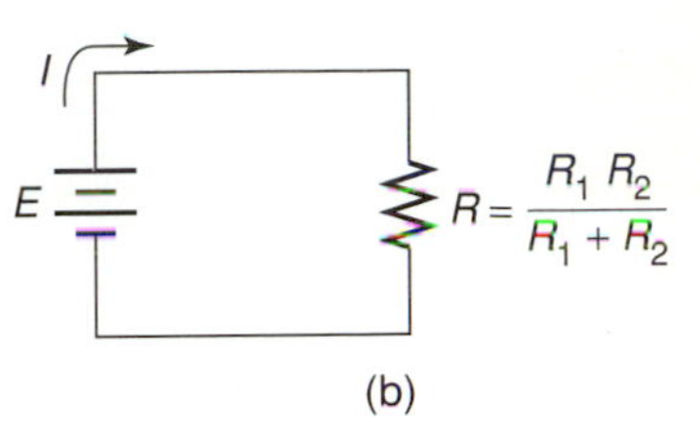

Figure 17.29
Resistor R in part (b) is equivalent to the pair of resistors R_1 and R_2 connected in parallel in part (a).

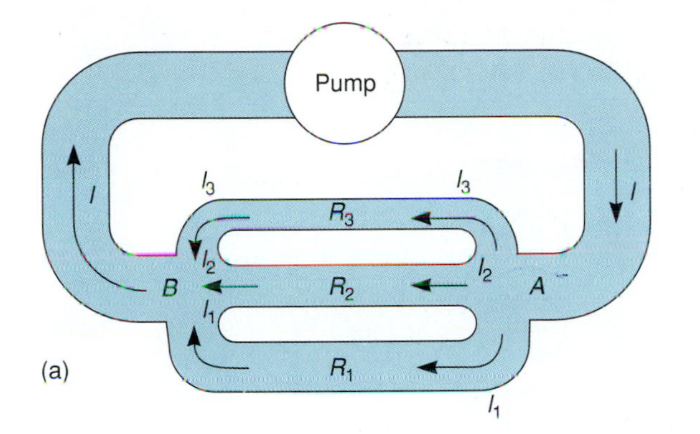

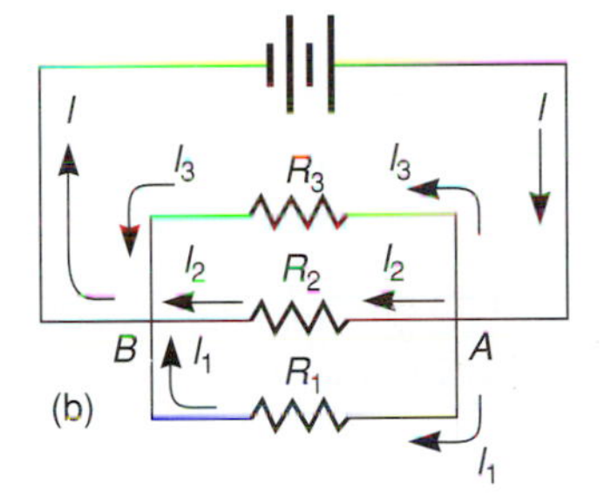

Figure 17.30
A water system may be compared to a parallel electrical circuit.

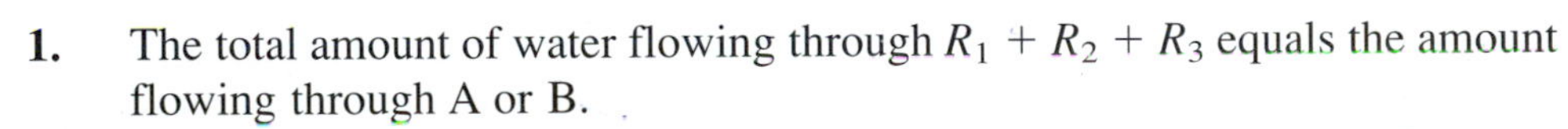

1. The total amount of water flowing through $R_1 + R_2 + R_3$ equals the amount flowing through A or B.
2. The water flowing past point A divides into the three branches R_1, R_2, and R_3.
3. The larger pipes have *less* opposition to water flow than do the smaller pipes. Because R_1 has a larger cross-sectional area than R_2 or R_3, it has less opposition to the flow of water and, therefore, carries more water than R_2 or R_3.

Similarly, in a parallel electrical circuit as in Fig. 17.30(b):

1. The total amount of current flowing through $R_1 + R_2 + R_3$ equals the amount flowing through A or B.
2. The current flowing past point A divides into the three branches R_1, R_2, and R_3.
3. The smaller resistors have *less* opposition to current flow and therefore carry larger currents.

EXAMPLE 1

Find the equivalent resistance of the circuit shown in Fig. 17.31.

Data:

$$R_1 = 7.00\ \Omega$$
$$R_2 = 9.00\ \Omega$$
$$R_3 = 12.0\ \Omega$$
$$R = ?$$

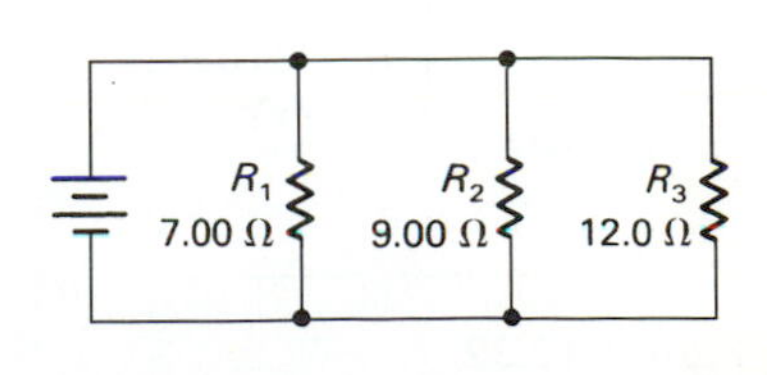

Figure 17.31

Basic Equation:

$$\frac{1}{R} = \frac{1}{R_1} + \frac{1}{R_2} + \frac{1}{R_3}$$

Working Equation:

When using this formula, you should solve for the reciprocal of the unknown, then substitute.

Substitution:

$$\frac{1}{R} = \frac{1}{7.00\ \Omega} + \frac{1}{9.00\ \Omega} + \frac{1}{12.0\ \Omega}$$

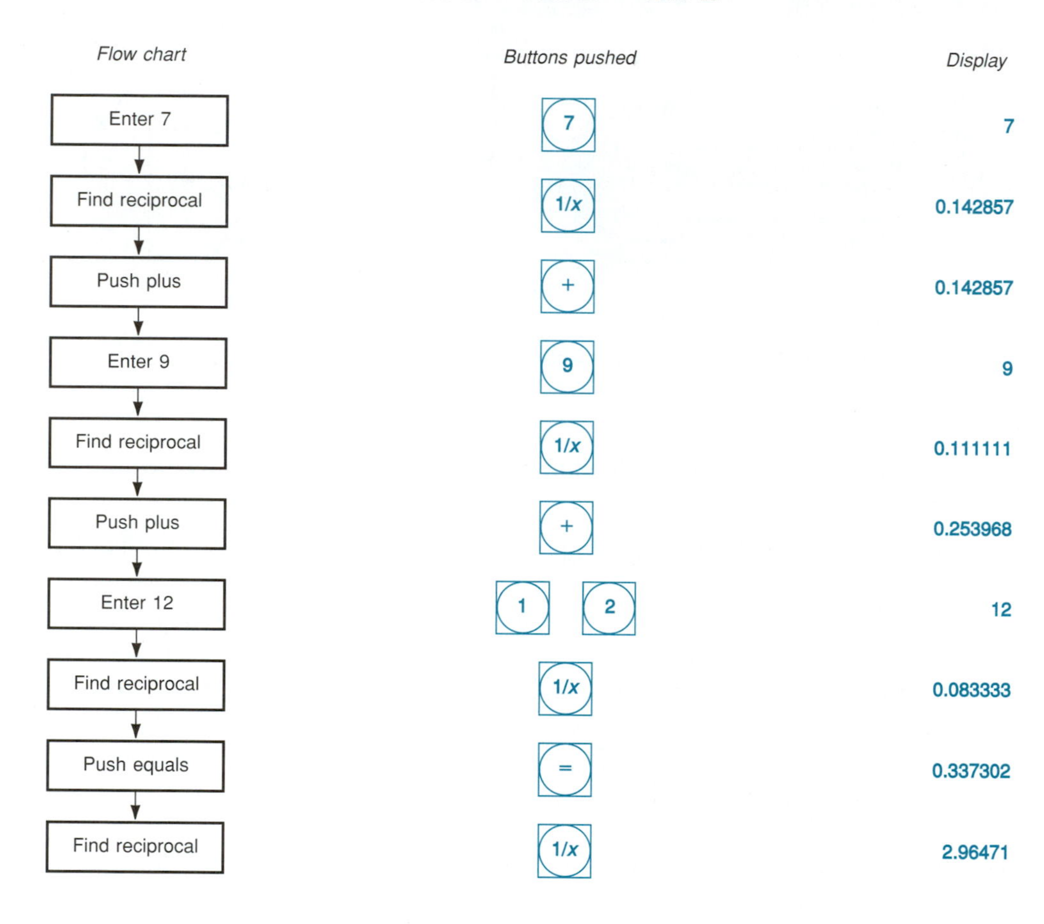

Thus, $R = 2.96\ \Omega$

EXAMPLE 2

Find the total current in the circuit shown in Fig. 17.32.

Data:

$$R_1 = 23.0\ \Omega$$
$$R_2 = 14.0\ \Omega$$
$$R_3 = 5.00\ \Omega$$
$$E = 90.0\ \text{V}$$
$$I = ?$$

$R_1 = 23.0\ \Omega$, $R_2 = 14.0\ \Omega$, $R_3 = 5.00\ \Omega$, $E = 90.0$ V

Figure 17.32

First, find the equivalent resistance, R. Second, find the total current, I. To find R:

Basic Equation:

$$\frac{1}{R} = \frac{1}{R_1} + \frac{1}{R_2} + \frac{1}{R_3}$$

Working Equation: Same

Substitution:

$$\frac{1}{R} = \frac{1}{23.0\ \Omega} + \frac{1}{14.0\ \Omega} + \frac{1}{5.00\ \Omega}$$

Using a calculator flowchart as in Example 1, we find

$$R = 3.18\ \Omega$$

To find I:

Basic Equation:

$$I = \frac{E}{R}$$

Working Equation: Same

Substitution:

$$I = \frac{90.0\ \text{V}}{3.18\ \Omega}$$
$$= 28.3\ \text{A}$$

EXAMPLE 3

Find the current through R_2 in Fig. 17.32 from Example 2.

Data:

$$R_2 = 14.0\ \Omega$$
$$E = 90.0\ \text{V} = V_2$$
$$I_2 = ?$$

Basic Equation:

$$I_2 = \frac{V_2}{R_2}$$

Working Equation: Same

Substitution:

$$I_2 = \frac{90.0\ \text{V}}{14.0\ \Omega}$$
$$= 6.43\ \text{A}$$

EXAMPLE 4

Find the equivalent resistance and the value of R_3 in the circuit shown in Fig. 17.33.

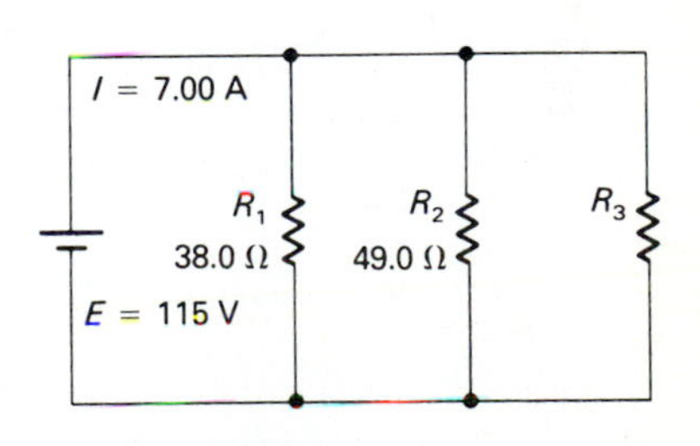

Figure 17.33

Data:

$$E = 115\ \text{V}$$
$$I = 7.00\ \text{A}$$
$$R_1 = 38.0\ \Omega$$
$$R_2 = 49.0\ \Omega$$
$$R_3 = ?$$

First find R:

Basic Equation:

$$I = \frac{E}{R}$$

Working Equation:

$$R = \frac{E}{I}$$

Substitution:

$$R = \frac{115 \text{ V}}{7.00 \text{ A}}$$
$$= 16.4\ \Omega$$

To find R_3:

Basic Equation:

$$\frac{1}{R} = \frac{1}{R_1} + \frac{1}{R_2} + \frac{1}{R_3}$$

Working Equation:

$$\frac{1}{R_3} = \frac{1}{R} - \frac{1}{R_1} - \frac{1}{R_2}$$

Substitution:

$$\frac{1}{R_3} = \frac{1}{16.4\ \Omega} - \frac{1}{38.0\ \Omega} - \frac{1}{49.0\ \Omega}$$

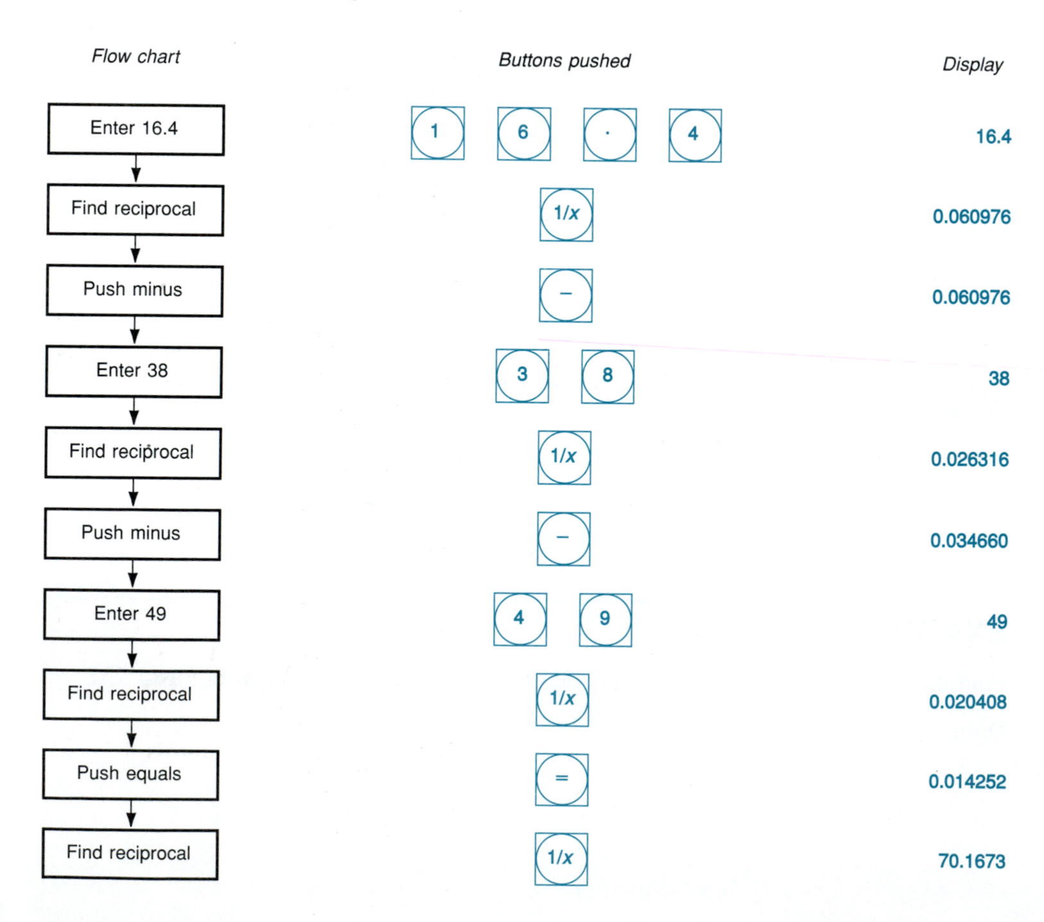

Thus, $R_3 = 70.2\ \Omega$.

The characteristics of parallel circuits are summarized in Table 17.2.

Table 17.2 Characteristics of Parallel Circuits

	Parallel
Current	$I = I_1 + I_2 + I_3 + \cdots$
Resistance	$\frac{1}{R} = \frac{1}{R_1} + \frac{1}{R_2} + \frac{1}{R_3} + \cdots$
Voltage	$E = V_1 = V_2 = V_3 = \cdots$

■ PROBLEMS 17.4

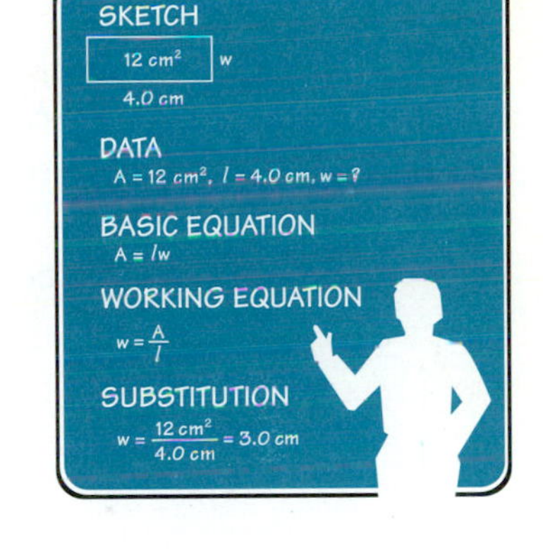

1. Find the equivalent resistance in the circuit shown in Fig. 17.34.
2. (a) What is the total current in the circuit in Problem 1?
   (b) What is the current through R_1?
   (c) What is the current through R_2?
3. (a) Find I_2 (current through R_2) in the circuit shown in Fig. 17.35.
   (b) Find I_3.
   (c) Find I_1.
4. Find the total current in the circuit in Problem 3.
5. Find the equivalent resistance in the circuit in Problem 3.
6. Find the resistance of R_3 in the circuit in Fig. 17.36.
7. (a) What is the current through R_1 in Problem 6?
   (b) What is the current through R_3?
8. What is the equivalent resistance in the circuit shown in Fig. 17.37?
9. (a) What emf is required for the circuit in Problem 8?
   (b) What is the voltage drop across each resistance?
10. What is the current through each resistance in Problem 8?

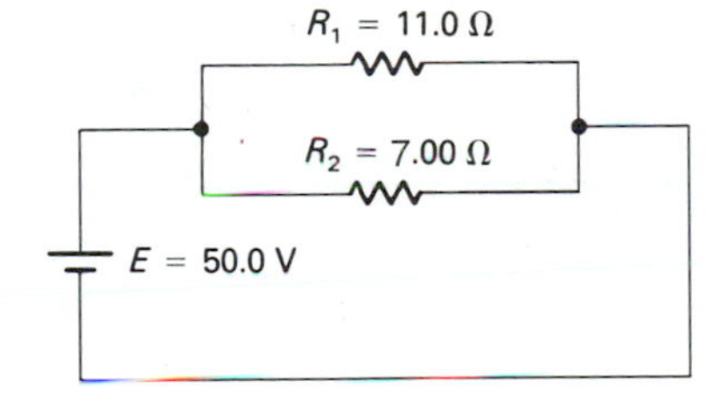

Figure 17.34

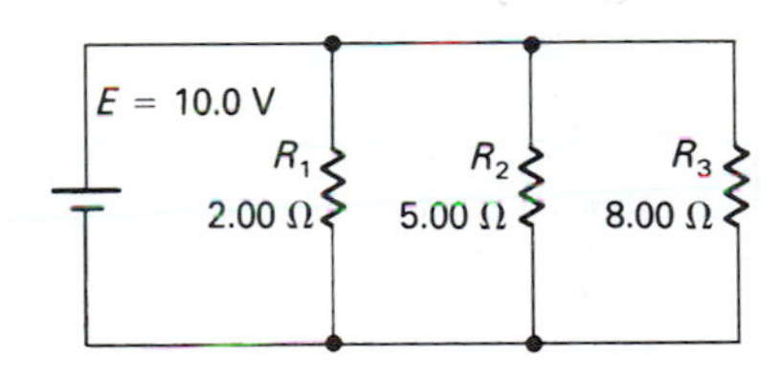

Figure 17.35

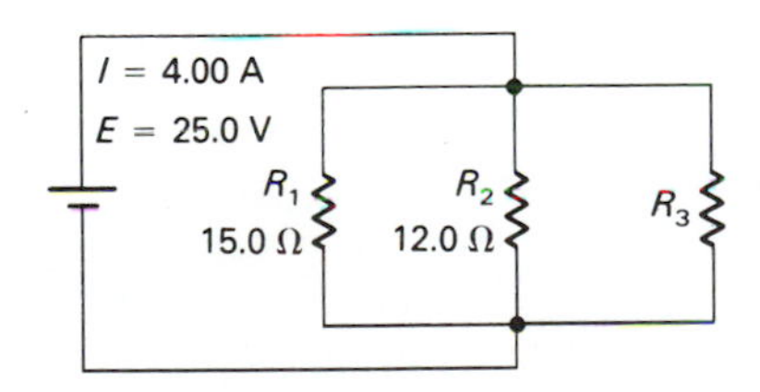

Figure 17.36

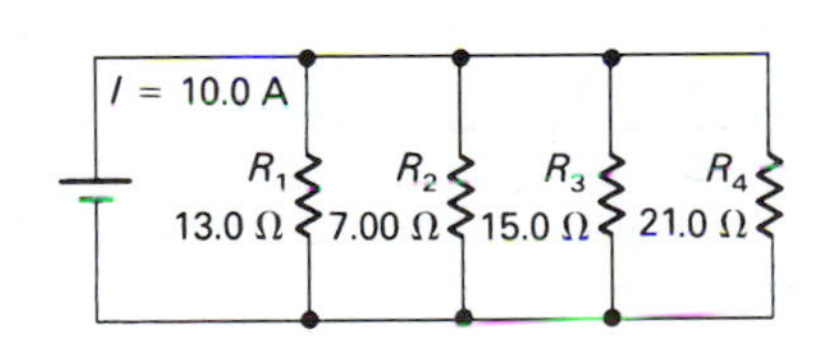

Figure 17.37 ■

17.5 SERIES–PARALLEL CIRCUITS

Many circuits cannot be solved directly because of the number and arrangement of the resistances. To simplify this kind of circuit, we apply the rules for series and parallel circuits to find an equivalent circuit that reduces to a circuit with one resistance.

EXAMPLE 1

Circuit B in Fig. 17.38 is equivalent to circuit A, where $R_4 = R_1 + R_2$. Then, circuit C is equivalent to circuit B, where $\frac{1}{R_5} = \frac{1}{R_3} + \frac{1}{R_4}$.

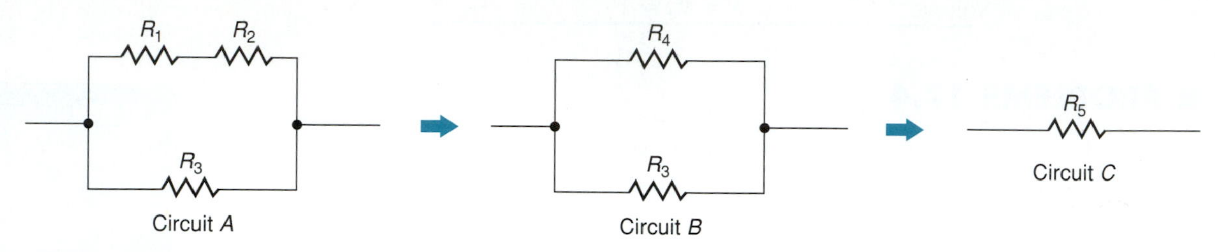

Figure 17.38
Circuit A can be replaced by circuit C, where R_5 is the equivalent resistance.

EXAMPLE 2

Circuit B in Fig. 17.39 is equivalent to circuit A, where $\frac{1}{R_4} = \frac{1}{R_2} + \frac{1}{R_3}$. Then, circuit C is equivalent to circuit B, where $R_5 = R_1 + R_4$.

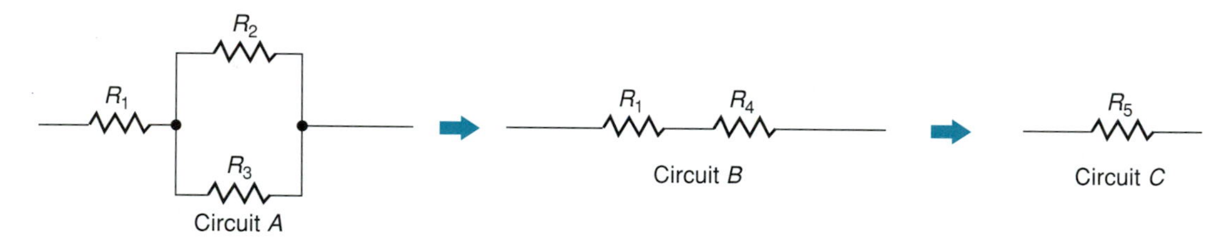

Figure 17.39
Circuit A can be replaced by circuit C, where R_5 is the equivalent resistance.

EXAMPLE 3

Find the total current in the circuit shown in Fig. 17.40.

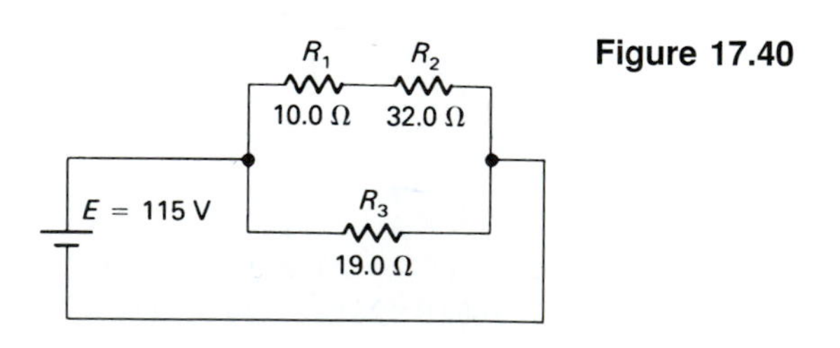

Figure 17.40

Solution:

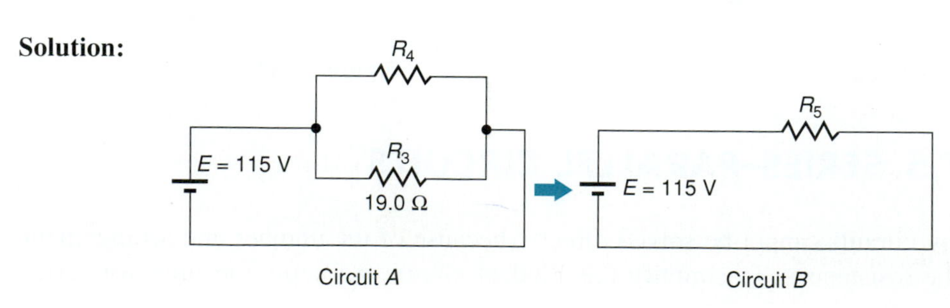

Circuit A is equivalent to the circuit in Fig. 17.40. Then, circuit B is equivalent to circuit A.

Data:

$$E = 115 \text{ V}$$
$$R_1 = 10.0\ \Omega$$
$$R_2 = 32.0\ \Omega$$
$$R_3 = 19.0\ \Omega$$
$$R_4 = R_1 + R_2 = 10.0\ \Omega + 32.0\ \Omega = 42.0\ \Omega$$
$$I = ?$$

First, find the equivalent resistance, R_5. Second, find the total current, I.

To find R_5:

Basic Equation:

$$\frac{1}{R_5} = \frac{1}{R_3} + \frac{1}{R_4}$$

Working Equation: Same

Substitution:

$$\frac{1}{R_5} = \frac{1}{19.0\ \Omega} + \frac{1}{42.0\ \Omega}$$

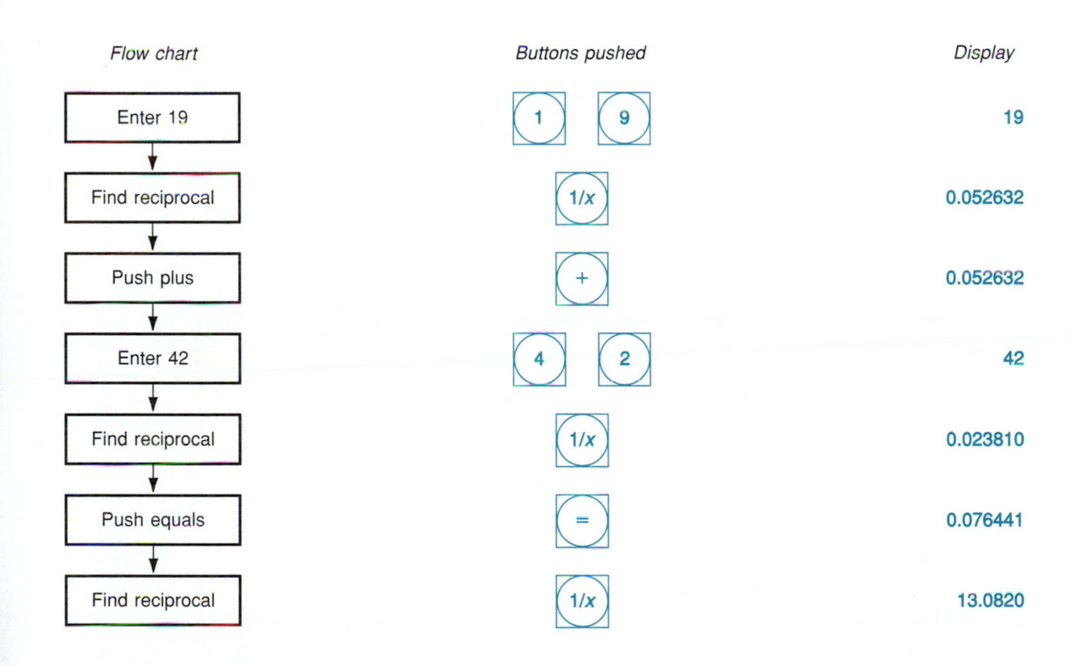

Thus, $R_5 = 13.1\ \Omega$.

To find I:

Basic Equation:

$$I = \frac{E}{R_5}$$

Working Equation: Same

Substitution:

$$I = \frac{115 \text{ V}}{13.1\ \Omega}$$
$$= 8.78 \text{ A}$$

EXAMPLE 4

Find the equivalent resistance in the circuit shown in Fig. 17.41.

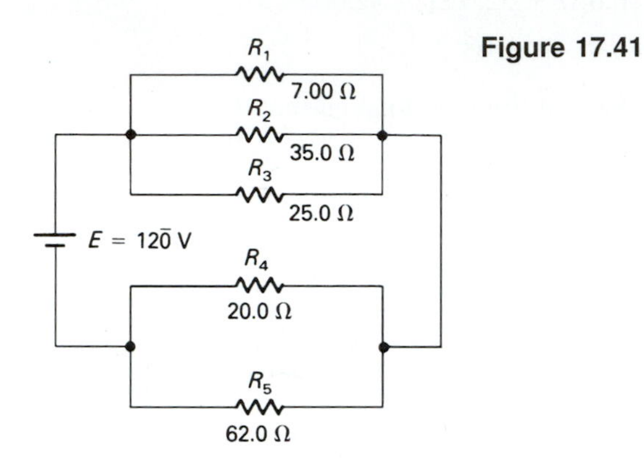

Figure 17.41

Solution:

Circuit *A* Circuit *B*

Circuit *A* is equivalent to the circuit in Fig. 17.41, and circuit *B* is equivalent to circuit *A*.

Data:

$$R_1 = 7.00\ \Omega$$
$$R_2 = 35.0\ \Omega$$
$$R_3 = 25.0\ \Omega$$
$$R_4 = 20.0\ \Omega$$
$$R_5 = 62.0\ \Omega$$
$$E = 12\bar{0} \text{ V}$$
$$R_8 = ?$$

First, find R_6. Second, find R_7. Third, find the equivalent resistance, R_8.
To find R_6:

Basic Equation:

$$\frac{1}{R_6} = \frac{1}{R_1} + \frac{1}{R_2} + \frac{1}{R_3}$$

Working Equation: Same

Substitution:

$$\frac{1}{R_6} = \frac{1}{7.00\ \Omega} + \frac{1}{35.0\ \Omega} + \frac{1}{25.0\ \Omega}$$

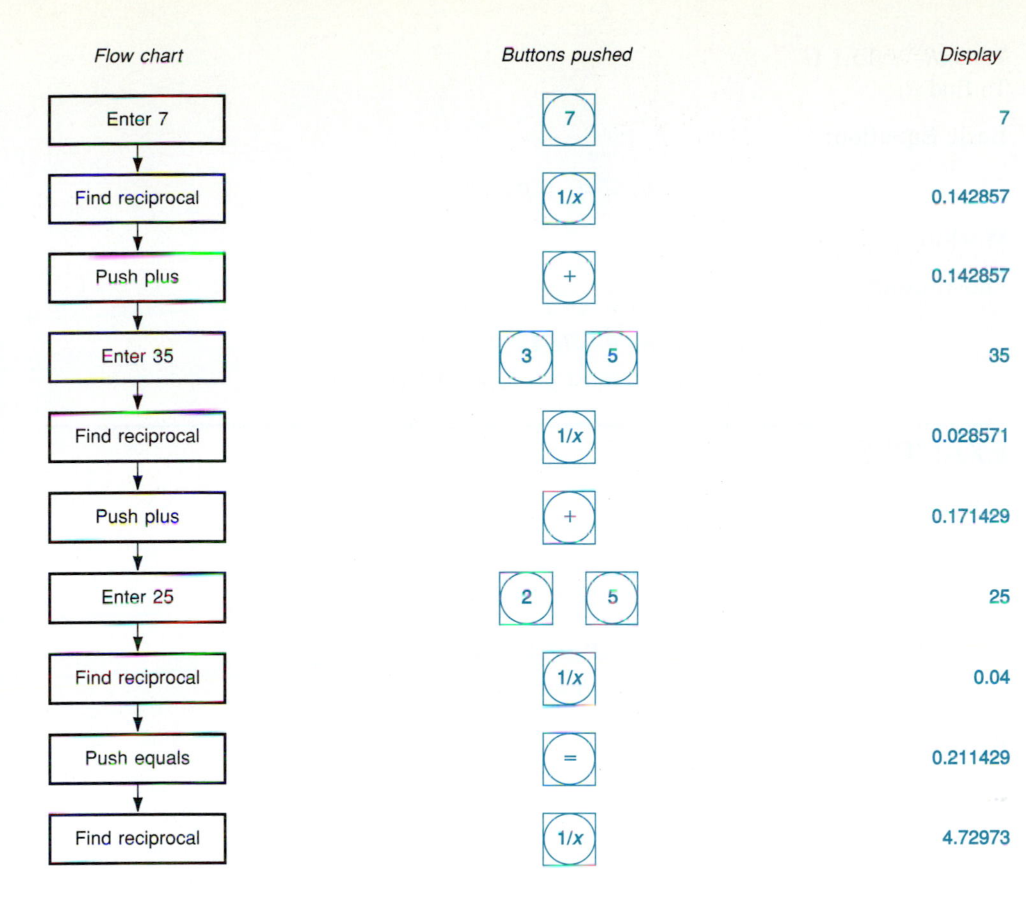

Thus, $R_6 = 4.73\ \Omega$.
To find R_7:

Basic Equation:

$$\frac{1}{R_7} = \frac{1}{R_4} + \frac{1}{R_5}$$

Working Equation: Same

Substitution:

$$\frac{1}{R_7} = \frac{1}{20.0\ \Omega} + \frac{1}{62.0\ \Omega}$$

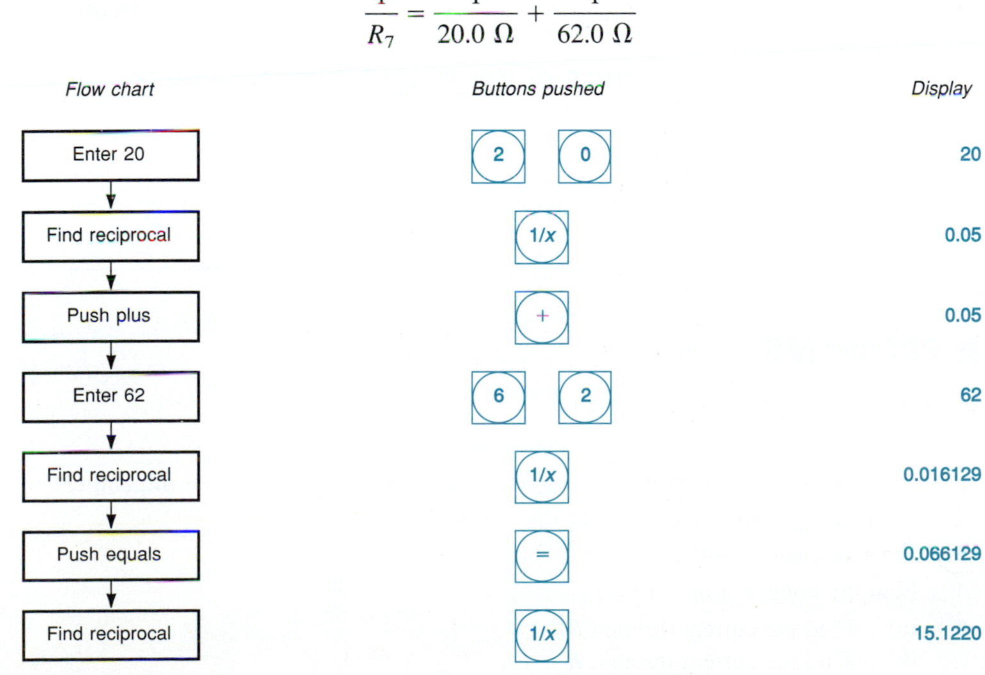

Thus, $R_7 = 15.1\ \Omega$.
To find R_8:

Basic Equation:

$$R_8 = R_6 + R_7$$

Working Equation: Same

Substitution:

$$\begin{aligned} R_8 &= 4.73\ \Omega + 15.1\ \Omega \\ &= 19.83\ \Omega \text{ or } 19.8\ \Omega \end{aligned}$$

EXAMPLE 5

Find the total current in Example 4.

Data:

$$\begin{aligned} E &= 12\bar{0}\text{ V} \\ R_8 &= 19.8\ \Omega \\ I &= ? \end{aligned}$$

Basic Equation:

$$I = \frac{E}{R_8}$$

Working Equation: Same

Substitution:

$$\begin{aligned} I &= \frac{12\bar{0}\text{ V}}{19.8\ \Omega} \\ &= 6.06\text{ A} \end{aligned}$$

Table 17.3 summarizes the characteristics of series and parallel circuits.

Table 17.3 Characteristics of Series and Parallel Circuits

	Series	*Parallel*
Current	$I = I_1 = I_2 = I_3 = \cdots$	$I = I_1 + I_2 + I_3 + \cdots$
Resistance	$R = R_1 + R_2 + R_3 + \cdots$	$\frac{1}{R} = \frac{1}{R_1} + \frac{1}{R_2} + \frac{1}{R_3} + \cdots$
Voltage	$E = V_1 + V_2 + V_3 + \cdots$	$E = V_1 = V_2 = V_3 = \cdots$

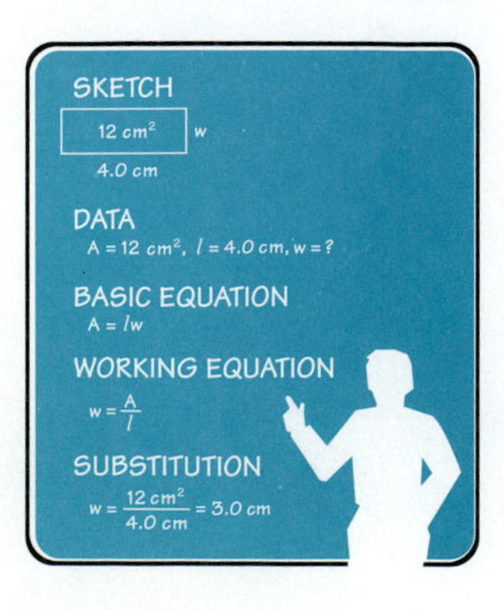

■ PROBLEMS 17.5

Use Fig. 17.42 in Problems 1 through 5.

1. (a) Which resistances are connected in parallel?
   (b) What is the equivalent resistance of the resistances connected in parallel?
2. Find the equivalent resistance of the entire circuit.
3. Find the current in R_1.
4. Find the voltage drop across R_1.
5. (a) Find the current through R_3.
   (b) Find the current through R_2.

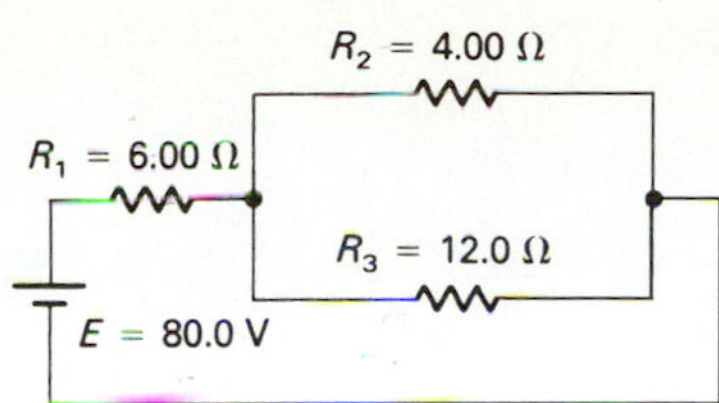

Figure 17.42

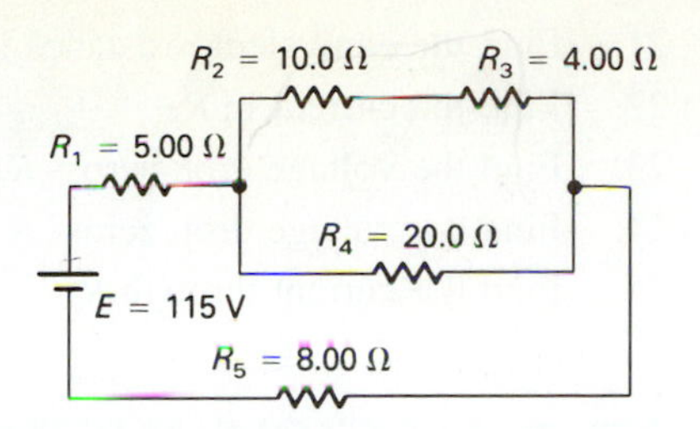

Figure 17.43

Use Fig. 17.43 in Problems 6 through 12.

6. What is the equivalent resistance of the resistances connected in parallel?
7. Find the equivalent resistance of the circuit.
8. Find the current in R_1.
9. What is the voltage drop across the parallel part of the circuit?
10. Find the current through R_3.
11. Find the current through R_5.
12. What is the voltage drop across R_3?

Use Fig. 17.44 in Problems 13 through 20.

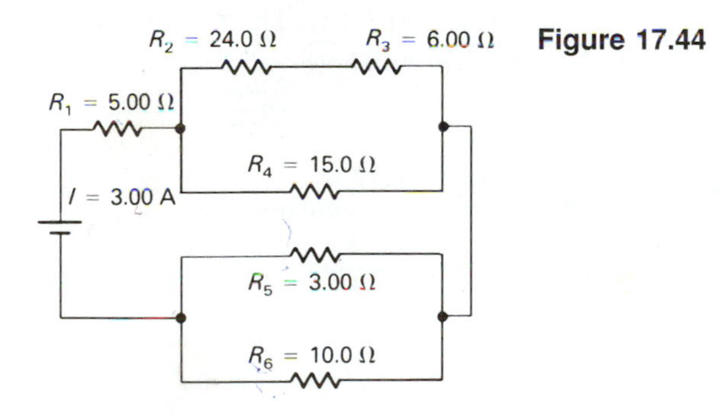

Figure 17.44

13. Find the equivalent resistance of the parallel arrangement in the upper branch.
14. Find the equivalent resistance of the parallel arrangement in the lower branch.
15. Find the equivalent resistance of the entire circuit.
16. What emf is required for the given current flow in the circuit?
17. Find the voltage drop across the parallel arrangement in the upper branch.
18. Find the voltage drop across R_4.
19. Find the voltage drop across R_6.
20. Find the current through R_6.

Use Fig. 17.45 in Problems 21 through 25.

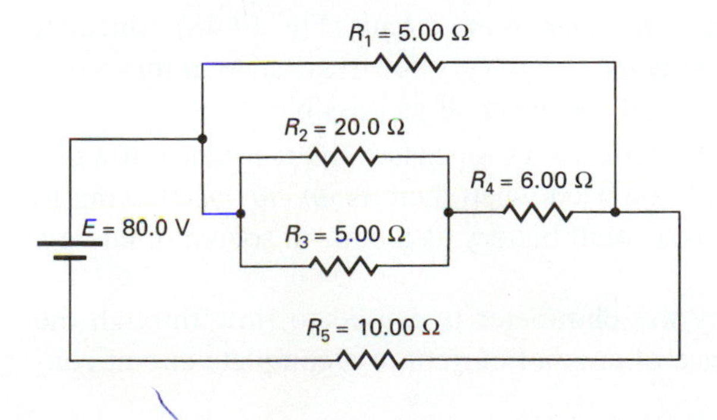

Figure 17.45

21. Find the equivalent resistance in the circuit.
22. Find the current in R_5.
23. Find the voltage drop across R_5.
24. Find the voltage drop across R_4.
25. Find the current through R_2. ■

17.6 ELECTRIC INSTRUMENTS

In the laboratory we use several kinds of electric meters for measurements. Great care must be taken to avoid passing a large current through the meters. Meters are fragile instruments, and abuse will ruin them. A large current will burn out the meter. Measurements in ac circuits and dc circuits require different kinds of meters or different settings on the same **multimeter.**

Digital instruments have more than one range on which readings are made (Fig. 17.46). Autorange meters adjust ranges automatically. The reading and use of the different modes of operation will be studied in the laboratory.

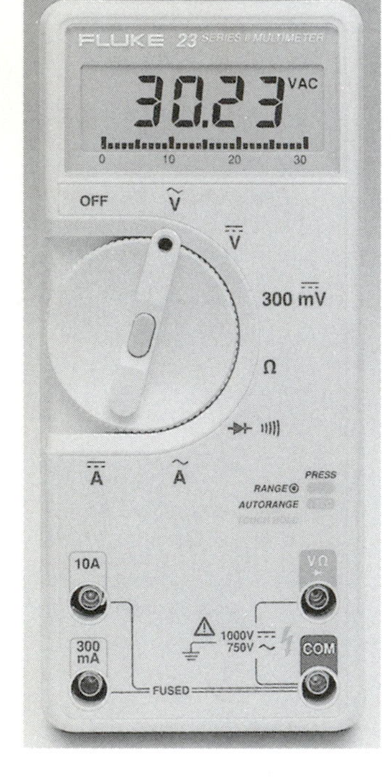

Figure 17.46
Digital multimeter.

1. *Voltmeter.* In this mode, the **voltmeter** instrument measures the voltage across (or the difference in potential between) two points in a circuit. It should *always* be connected in *parallel* with the part of the circuit across which we wish to measure the voltage drop (Fig. 17.47). The voltmeter is a high-resistance instrument and draws very little current.

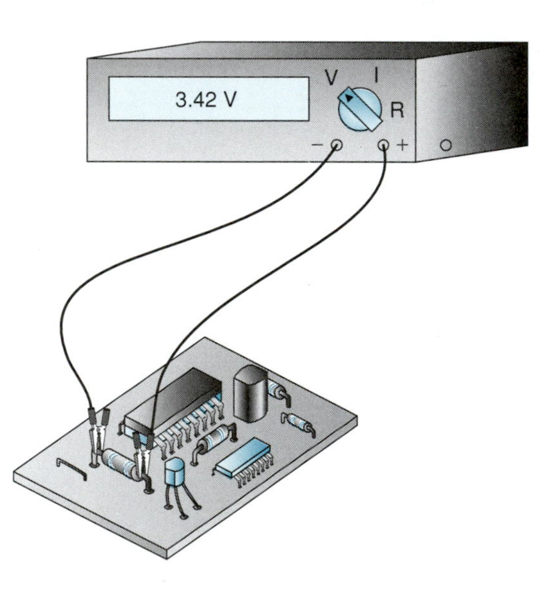

Figure 17.47
A voltmeter measures the voltage across two points in an electrical circuit.

2. *Ammeter.* The **ammeter** mode measures the current flowing through a circuit. Therefore, it is connected in *series* in the circuit (Fig. 17.48). Since all the current flows through the meter, it has very low resistance in this mode so that its effect on the circuit will be as small as possible.
3. *Ohmmeter.* The **ohmmeter** mode is used to measure the resistance of a circuit component. It should only be used when there is *no current* flowing in the circuit. The ohmmeter has a small battery as a built-in source of energy.

A small current provided by the ohmmeter is caused to flow through the component under test. The presence of another current or a complete circuit con-

Figure 17.48
An ammeter measures the current flowing through an electrical circuit.

nected to the component under test will distort the resistance reading since the rest of the circuit may allow some of this test current to flow "around" the component under test. To avoid this problem, the component should be tested in isolation from any complete circuit (Fig. 17.49).

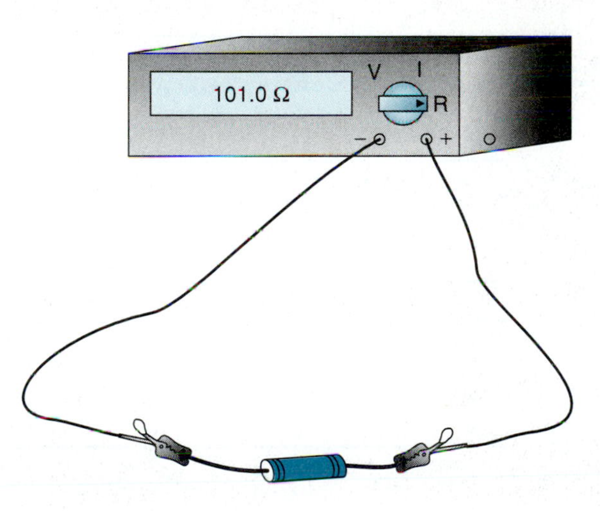

Figure 17.49
An ohmmeter measures the resistance of a component in an electrical circuit.

Another instrument is the galvanometer, a very sensitive instrument that is used to detect the presence and direction of *very small* currents.

■ PROBLEMS 17.6

Using the formulas for series and parallel circuits, fill in the blanks in the tables shown opposite each circuit. In the blanks across from Battery under

> *V*: Write the emf of the battery.
> *I*: Write the total current in the circuit.
> *R*: Write the equivalent or total resistance of the entire circuit.

In the blanks across from R_1 under

> *V*: Write the voltage drop across R_1.
> *I*: Write the current flowing through R_1.
> *R*: Write the resistance of R_1.

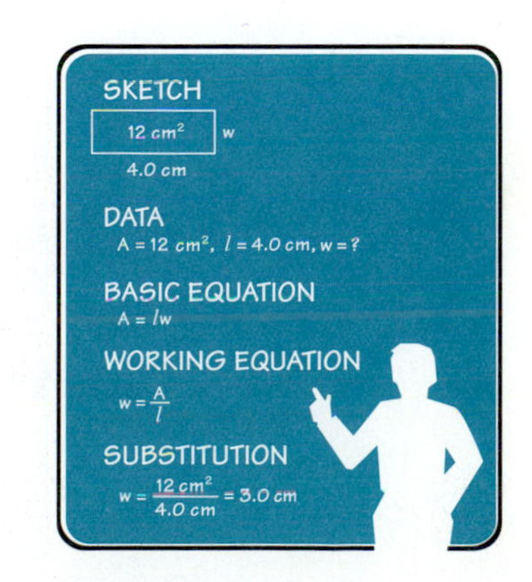

In the blanks across from R_2, R_3, . . . , fill in the appropriate numbers under *V*, *I*, and *R*. (Begin by looking for key information given in the table and work from there.)

1.

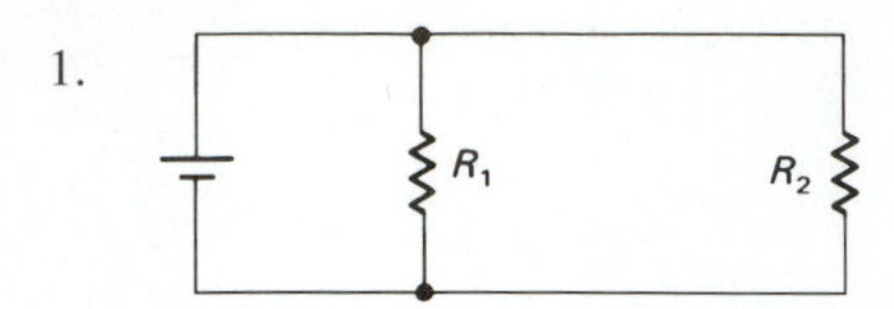

	V	*I*	*R*
Battery	12.0 V	A	Ω
R_1	V	A	2.00 Ω
R_2	V	A	4.00 Ω

2.

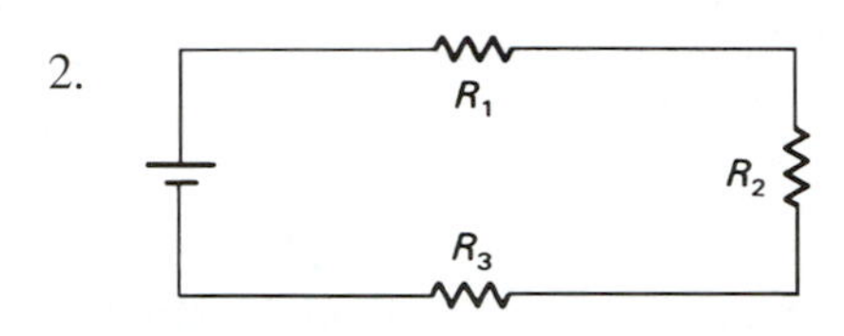

	V	*I*	*R*
Battery	V	A	Ω
R_1	V	2.00 A	4.00 Ω
R_2	V	A	6.00 Ω
R_3	V	A	8.00 Ω

3.

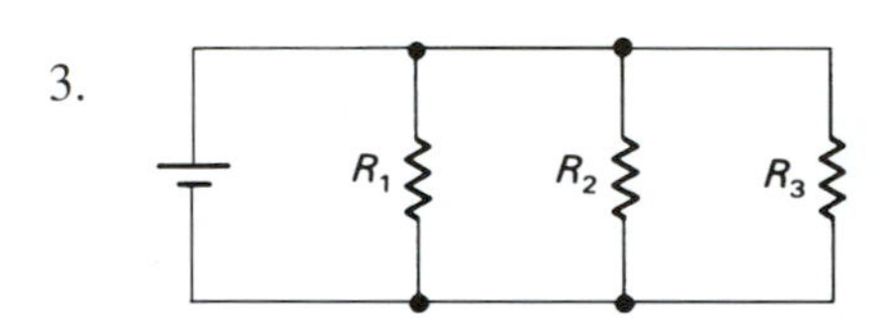

	V	*I*	*R*
Battery	V	A	Ω
R_1	V	2.00 A	Ω
R_2	V	3.00 A	12.0 Ω
R_3	V	1.00 A	Ω

4.

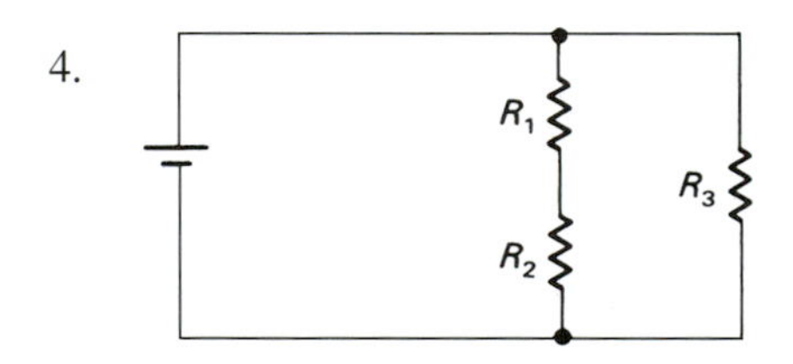

	V	*I*	*R*
Battery	12.0 V	2.00 A	Ω
R_1	V	A	6.00 Ω
R_2	V	A	4.00 Ω
R_3	V	A	15.0 Ω

5.

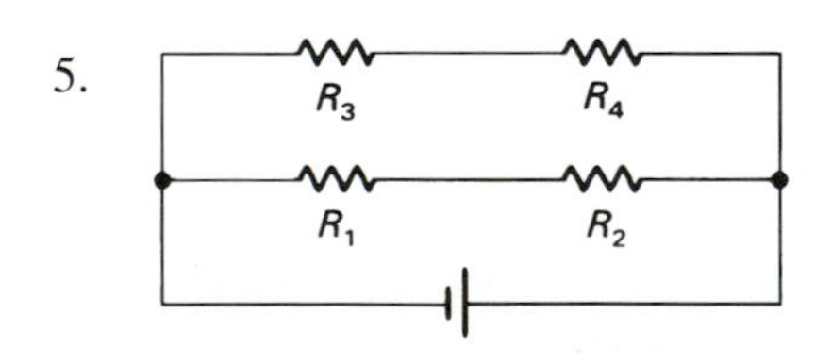

	V	*I*	*R*
Battery	50.0 V	5.00 A	Ω
R_1	V	2.00 A	Ω
R_2	25.0 V	A	Ω
R_3	10.0 V	A	Ω
R_4	V	3.00 A	Ω

6.

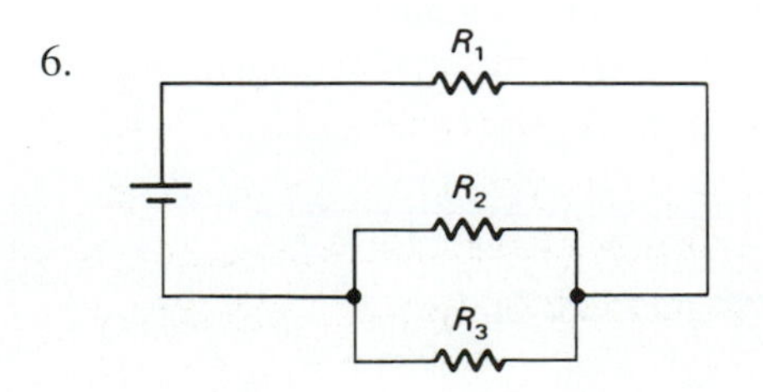

	V	*I*	*R*
Battery	24.0 V	A	Ω
R_1	8.00 V	A	Ω
R_2	V	4.00 A	Ω
R_3	V	2.00 A	Ω

7.

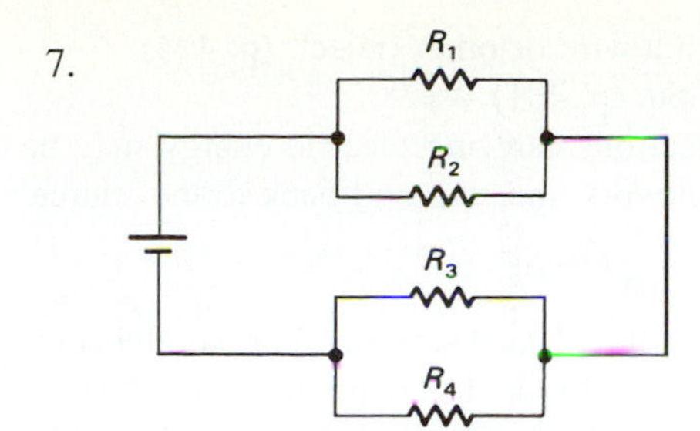

	V	I	R
Battery	V	A	Ω
R_1	12.0 V	A	2.00 Ω
R_2	V	A	4.00 Ω
R_3	24.0 V	A	4.00 Ω
R_4	V	A	8.00 Ω

8.

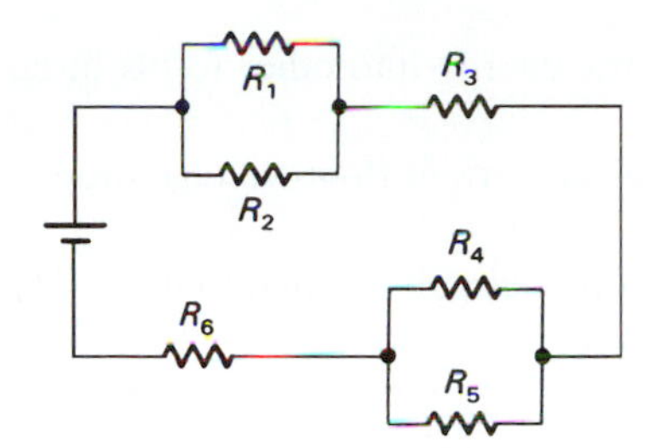

	V	I	R
Battery	30.0 V	A	Ω
R_1	6.00 V	3.00 A	Ω
R_2	V	2.00 A	Ω
R_3	V	A	3.00 Ω
R_4	V	1.00 A	Ω
R_5	8.00 V	A	Ω
R_6	V	A	Ω

9.

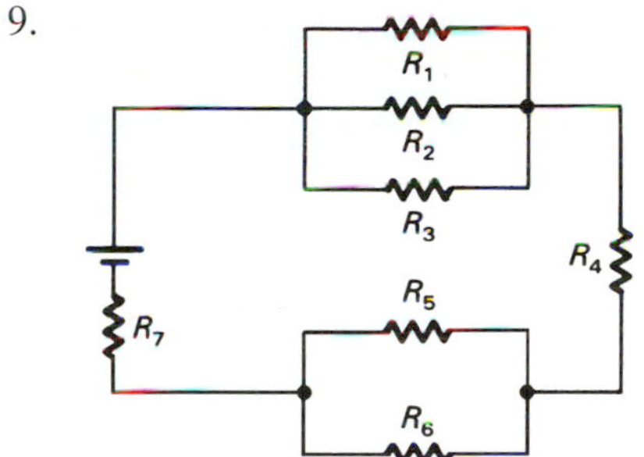

	V	I	R
Battery	V	12.0 A	Ω
R_1	V	A	Ω
R_2	18.0 V	2.00 A	Ω
R_3	V	A	3.00 Ω
R_4	V	A	4.00 Ω
R_5	V	A	2.00 Ω
R_6	V	8.00 A	Ω
R_7	6.00 V	A	Ω

10.

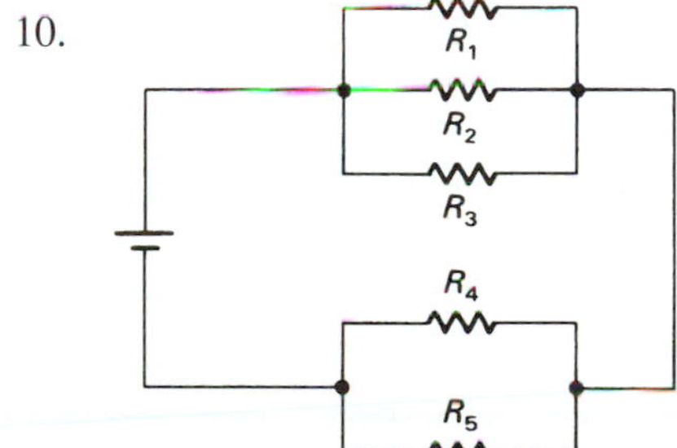

	V	I	R
Battery	46.0 V	A	Ω
R_1	V	3.00 A	Ω
R_2	V	4.00 A	Ω
R_3	V	A	6.00 Ω
R_4	V	3.00 A	Ω
R_5	V	7.00 A	Ω

11.

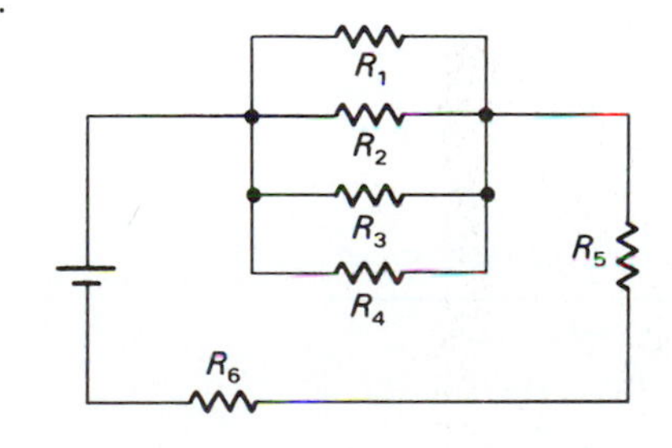

	V	I	R
Battery	V	A	Ω
R_1	V	A	20.0 Ω
R_2	10.0 V	A	Ω
R_3	V	A	4.00 Ω
R_4	V	1.00 A	Ω
R_5	V	5.00 A	5.00 Ω
R_6	V	A	6.00 Ω

GLOSSARY

Alternating Current Current that changes direction. (p. 401)

Ammeter An instrument that measures the current flowing in a circuit. (p. 428)

Conductor A substance through which electrons can flow to carry current. Typically, has a large number of electrons that are free to move about in the substance. (p. 403)

Current The rate at which charge passes through a wire or other object. (p. 404)

Direct Current Current that flows in one direction. (p. 401)

Electric Circuit A conducting loop in which electrons carrying electric energy may be transferred from a suitable source to do useful work and returned back to the source. (p. 401)

emf The potential difference across a source. (p. 405)

Equivalent Resistance The single resistance that can replace a series and/or parallel combination in a circuit and provide the same current flow and voltage drop. (p. 412)

Insulator A substance that does not allow electric current to flow through it readily. (p. 403)

Load The object in a circuit that changes the electric energy into other forms of energy or work. (p. 401)

Multimeter An instrument that can be used to measure current flow, voltage drop, or resistance. (p. 428)

Ohm's Law States that the voltage drop across part of a circuit is equal to the product of the current and the resistance, $V = I \times R$. (p. 407)

Ohmmeter An instrument that measures the resistance of a circuit component. (p. 428)

Parallel Circuit An electric circuit with more than one path for the current to flow. The current is divided among the branches of the circuit. (p. 415)

Resistance The opposition to current flow. Commonly measured in units of ohms. (p. 405)

Resistivity The resistance of a specified amount of a material. Measures the resistance of a substance to the flow of electric current. (p. 406)

Semiconductors A small number of materials that fall between conductors and insulators in their ability to conduct electric current. (p. 403)

Series Circuit An electric circuit with only one path for the current to flow. The current in a series circuit is the same throughout. (p. 409)

Source The object that supplies electric energy for the flow of electric charge (electrons) in a circuit. (p. 401)

Superconductor A material that continuously conducts electrical current without resistance when cooled to typically very low temperatures, often near absolute zero. (p. 404)

Voltage Drop The potential difference across a load in a circuit. (p. 405)

Voltmeter An instrument that measures the difference in potential between two points in a circuit. (p. 428)

FORMULAS

17.1 $R = \frac{\rho l}{A}$

17.2 Ohm's law: $I = \frac{V}{R}$

17.3

Characteristics of Series Circuits

Current	$I = I_1 = I_2 = I_3 = \cdots$
Resistance	$R = R_1 + R_2 + R_3 + \cdots$
Voltage	$E = V_1 + V_2 + V_3 + \cdots$

17.4

Characteristics of Parallel Circuits

Current	$I = I_1 + I_2 + I_3 + \cdots$
Resistance	$\frac{1}{R} = \frac{1}{R_1} + \frac{1}{R_2} + \frac{1}{R_3} + \cdots$
Voltage	$E = V_1 = V_2 = V_3 = \cdots$

REVIEW QUESTIONS

1. The resistance of a wire is dependent on all of the following except
   (a) temperature. (b) cross-sectional area.
   (c) length. (d) material.
   (e) voltage.
2. Which of the following are good electrical conductors?
   (a) Aluminum (b) Wood
   (c) Glass (d) Distilled water
   (e) Silver
3. The total resistance in a circuit containing resistors connected in series is given by
   (a) the sum of the individual resistances. (b) the sum of the inverse of the individual resistances.
   (c) the sum of the currents. (d) the sum of the voltages.
4. The current in a parallel circuit is given by
   (a) the sum of the inverse currents.
   (b) the sum of the voltages.
   (c) the sum of the currents in the branches.
   (d) none of the above.
5. The flow of electrons through a conductor is called _____.
6. The unit of current is the _____.
7. The unit of emf is the _____.
8. The unit of resistance is the _____.
9. What effect does doubling the diameter of a wire have on the wire's resistance?
10. In your own words, explain Ohm's law.
11. Differentiate between a series and a parallel circuit.
12. Differentiate between the equivalent resistance in a series circuit and a parallel circuit.
13. In using an electrical instrument, with what range should you start when making a measurement?
14. Explain how a water system compares to a parallel circuit.
15. How does the current change in a circuit if the resistance increases by a factor of 2?
16. How does the current change in a circuit if the voltage is increased by a factor of 2?
17. How would the resistance of a wire change if the length were to be increased by a factor of 2?
18. Explain the concept of electrical potential.
19. Explain the transfer of energy that occurs in a circuit that includes a dry cell and two lamps in series.

REVIEW PROBLEMS

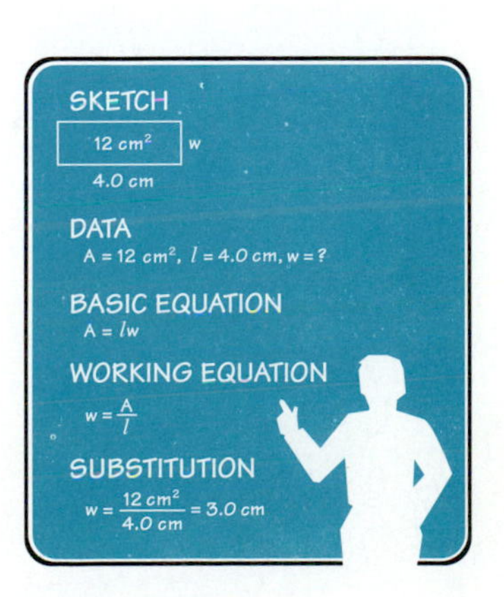

1. Find the resistance of 85.5 m of No. 20 aluminum wire. ($\rho = 2.83 \times 10^{-6}\ \Omega$ cm, $A = 2.07 \times 10^{-2}$ cm^2)
2. At 75°F, $12\overline{0}$ ft of wire has a resistance of 0.743 Ω. Find the resistance of $56\overline{0}$ ft of this wire.
3. Find the resistance of 134 m of No. 20 copper wire at 20°C. ($\rho = 1.72 \times 10^{-6}\ \Omega$ cm, $A = 2.07 \times 10^{-2}$ cm^2)
4. Find the length of a copper wire with resistance 0.0273 Ω/ft and total resistance 3.97 Ω.

5. Find the cross-sectional area of copper wire at 20°C that is 55.4 m long and has resistivity $\rho = 1.79 \times 10^{-6}$ Ω cm and resistance 0.943 Ω.
6. A heating element operates on 115 V. If it has a resistance of 15.4 Ω, what current does it draw?
7. A heating coil operates on $22\bar{0}$ V. If it draws 8.75 A, find its resistance.
8. What current does a 234-Ω resistance draw on 115 V?
9. Four resistors of 3.40 Ω, 6.54 Ω, 8.32 Ω, and 1.34 Ω are connected in series with a 12.0-V battery. Find the total resistance of the circuit.
10. Find the current in Problem 9.
11. Find the emf in the circuit shown in Fig. 17.50.
12. Find the equivalent resistance in the circuit shown in Fig. 17.51.

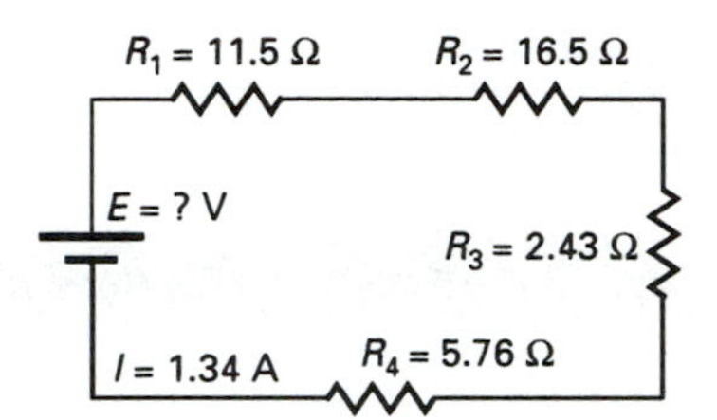

Figure 17.50

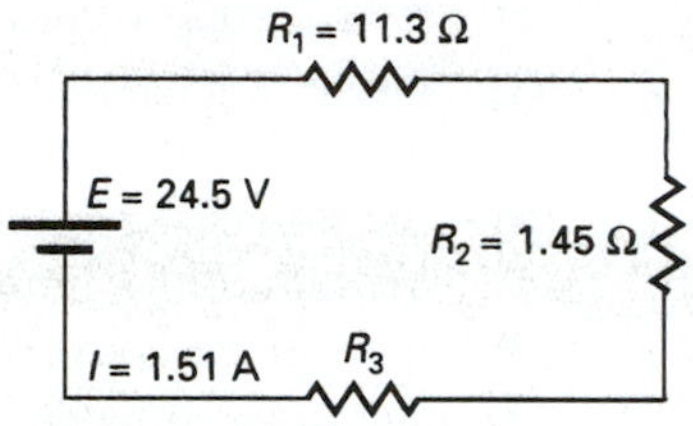

Figure 17.51

13. Find R_3 in the circuit in Problem 12.
14. Find the equivalent resistance in the circuit shown in Fig. 17.52.
15. Find the current in Fig. 17.52.
16. Find the current through R_1 in Fig. 17.52.
17. Find the current through R_2 in Fig. 17.52.
18. Find the equivalent resistance in the circuit of Fig. 17.53.

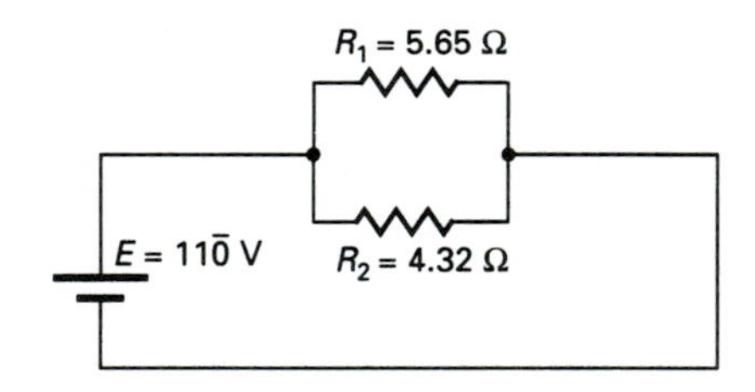

Figure 17.52

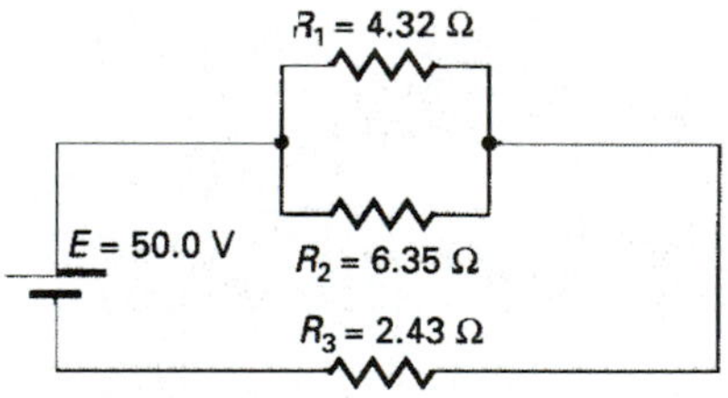

Figure 17.53

19. Find the current through R_3 in Fig. 17.53.
20. Find the current through R_1 in Fig. 17.53; through R_2.
21. Find the equivalent resistance in Fig. 17.54.

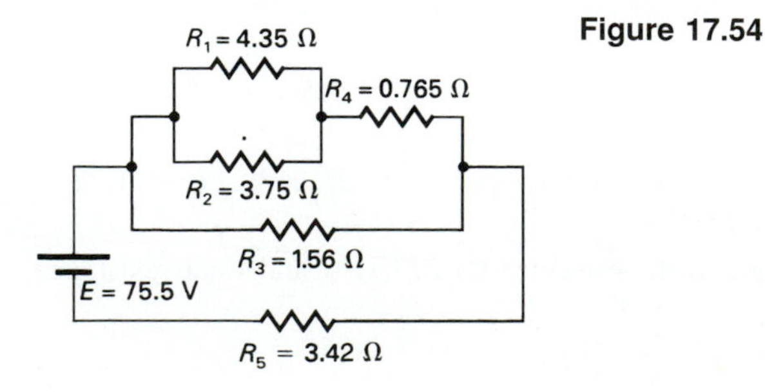

Figure 17.54

22. Find the current in R_5 in Fig. 17.54.
23. Find the voltage drop across R_5 in Fig. 17.54.
24. Find the current in R_1 in Fig. 17.54.
25. Find the voltage drop across R_1 in Fig. 17.54.
26. Using the formulas for series and parallel circuits, fill in the blanks in the table below for the circuit of Fig. 17.55.

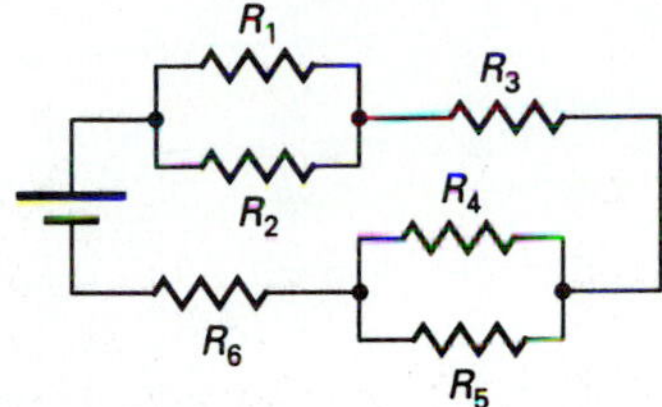

Figure 17.55

	V	I	R
Battery	35.0 V	A	Ω
R_1	5.00 V	2.75 A	Ω
R_2	V	1.95 A	Ω
R_3	V	A	2.80 Ω
R_4	V	0.97 A	Ω
R_5	7.50 V	A	Ω
R_6	V	A	Ω

Chapter 18

dc SOURCES

Batteries are becoming more important in our lives because of the growing popularity of portable communications and computing products. Cells in series will be considered along with electrical power.

OBJECTIVES

The major goals of this chapter are to enable you to:

1. Understand the nature of cells and batteries.
2. Analyze circuits with cells in series and parallel.
3. Find electrical power.

18.1 LEAD STORAGE CELL

Lead storage batteries are used in automobiles and in many other types of vehicles and machinery. A battery is a group of cells connected together. Each cell consists of a positive plate and a negative plate in a conducting solution. These lead cells are **secondary cells,** which means that they are rechargeable. The passing of an electric current through the cell to restore the original chemicals is called **recharging.** Cells, such as the dry cell, that cannot be efficiently recharged are called **primary cells.** ***Note:*** The cells are all connected together in series (Fig. 18.1). Six storage cells of 2.0 V each connected in series give 12.0 V for an automobile storage battery.

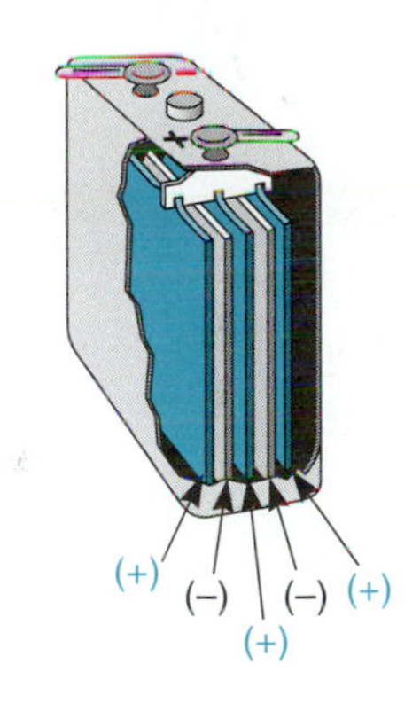

Figure 18.1
Lead storage battery.

Lead storage cells are made up of two kinds of lead plates (lead and lead oxide) submerged in a solution of distilled water and sulfuric acid (Fig. 18.2). This acid solution is called an **electrolyte.** The chemical action between the lead plates and the acid solution produces large numbers of free electrons at the negative (−) pole of the battery. These electrons have a large amount of electric potential energy, which is used in the load in the circuit (for instance, to operate headlights or to turn a starter motor).

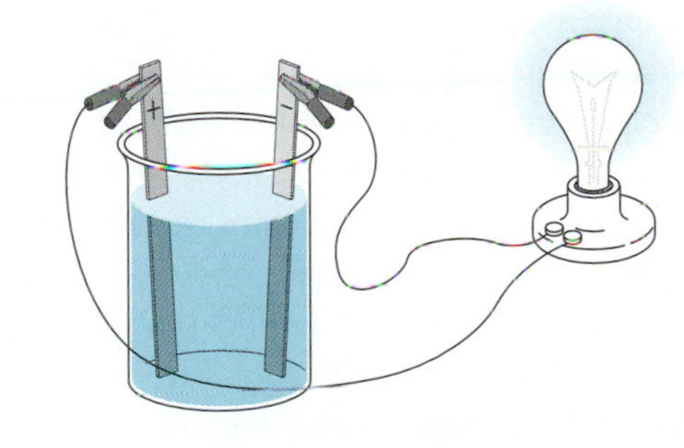

Figure 18.2
Simple storage cell.

As the electric energy is used in the load, the battery must be recharged. This is done by a generator or an alternator. Such devices provide an electric current to reverse the chemical reaction taking place in the battery. The recharging process extends the life of the battery, which would otherwise be very short.

18.2 DRY CELL

The dry cell is the most widely used primary cell. This kind of cell is used in flashlights and portable products such as cellular phones, notebook computers, drills, etc. A **dry cell** consists of an electrolyte and two electrodes of unlike materials, one of which reacts chemically with the electrolyte. The carbon–zinc dry cell is made of a carbon rod, which is the positive (+) terminal or pole, and a zinc can, which acts as the negative (−) terminal (Fig. 18.3). In between is a paste of chemicals and water that reacts with the terminals to provide energized electrons. These cells are available in a wide range of sizes. Common battery voltages range from 1.5 to 9 V. Nine volts require a series stack of six cells.

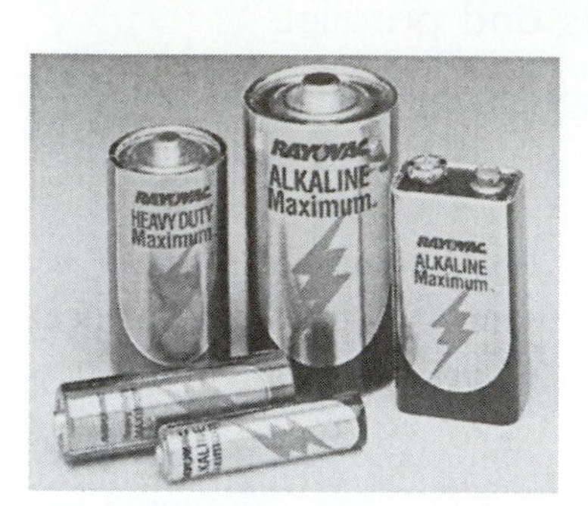

Figure 18.3
Dry cell batteries.

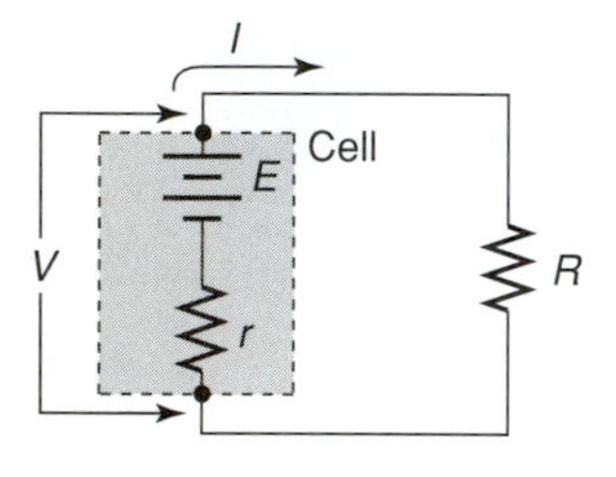

Figure 18.4
Effect of the internal resistance of a dry cell.

The dry cell, as well as the lead cell, has resistance within the cell itself which opposes the movement of the electrons. This is called the **internal resistance** (r) of the cell. Every cell has internal resistance. Because current flows in the cell, the emf of the cell is reduced by the voltage drop across the internal resistance (Fig. 18.4). The voltage applied to the external circuit is then

$$V = E - Ir$$

where V = voltage applied to circuit
E = emf of the cell
I = current through cell
r = internal resistance of cell

When the current or voltage available from a single cell is inadequate for a particular job, we usually connect two or more cells in a parallel or series arrangement.

Alkaline cells resemble carbon–zinc cells but are five to eight times longer lasting. An alkaline cell has a highly porous zinc anode that oxidizes more readily than does the carbon–zinc cell's anode. Its electrolyte is a strong alkali solution called *potassium hydroxide*. This compound conducts electricity inside the cell very well and enables the alkaline cell to deliver relatively high currents with greater efficiency than that of carbon–zinc cells.

Nickel–cadmium, metal hydride, and lithium batteries are now also in common use. Their advantage over other types of dry cells is that they are rechargeable. Lithium batteries have greater capacity than nickel–cadmium or hydride batteries.

Another type of dc power source is the solar cell. It is commonly made from a semiconductor material (typically amorphous silicon or gallium arsenide). Many small calculators operate with power supplied by solar cells.

18.3 CELLS IN SERIES AND PARALLEL

To connect cells in series, the positive terminal of one is connected to the negative terminal of the next cell. This procedure is continued until the desired number of cells are all connected (Fig. 18.5). The rules for cells connected in series and parallel are similar to those for simple resistances.

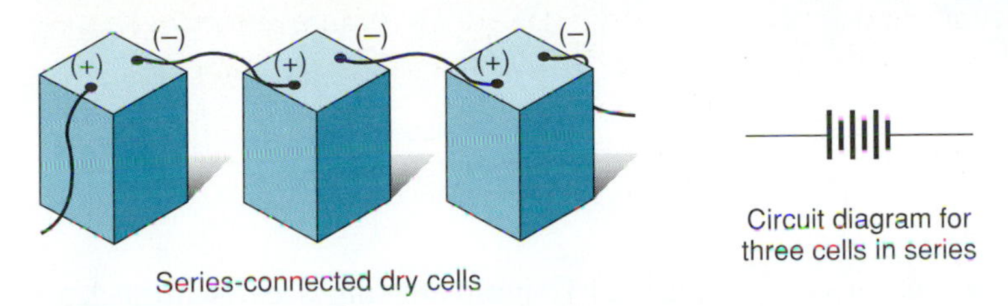

Circuit diagram for three cells in series

Figure 18.5

> *CELLS IN SERIES*
>
> **1.** The current in the circuit equals the current in any single cell:
>
> $$I = I_1 = I_2 = I_3 = \cdots$$
>
> **2.** The internal resistance of the battery equals the sum of the individual internal resistances of the cells:
>
> $$r = r_1 + r_2 + r_3 + \cdots$$
>
> **3.** The emf of the battery equals the sum of the emf's of the individual cells:
>
> $$E = E_1 + E_2 + E_3 + \cdots$$

EXAMPLE 1

Two 6.00-V cells with internal resistance of 0.100 Ω each are connected in series to form a battery with a current of 0.750 A in each cell (Fig. 18.6).

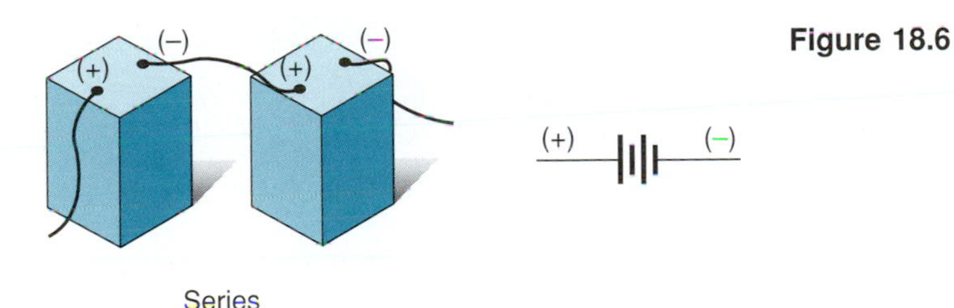

Figure 18.6

(a) What is the emf of the battery?
(b) Find the internal resistance of the battery.
(c) Find the current in the external circuit.

(a) $E = E_1 + E_2 = 6.00 \text{ V} + 6.00 \text{ V} = 12.00 \text{ V}$ (Rule 3)
(b) $r = r_1 + r_2 = 0.100\ \Omega + 0.100\ \Omega = 0.200\ \Omega$ (Rule 2)
(c) $I = 0.750 \text{ A}$ (Rule 1)

To connect cells in parallel, the positive terminals of all the cells are connected together and the negative terminals are all connected together (Fig. 18.7).

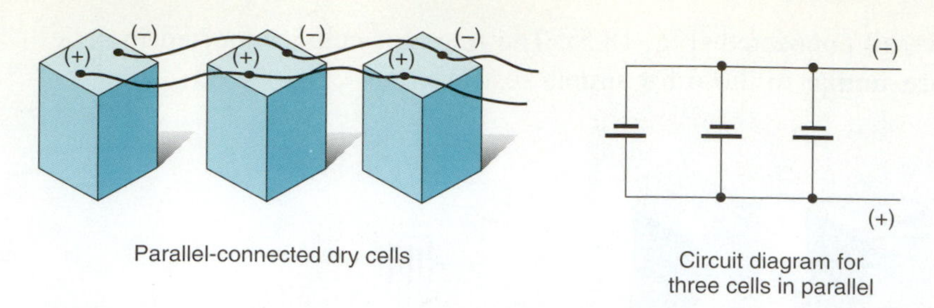

Figure 18.7

A common example of cells in parallel is jump-starting a car with a dead battery (Fig. 18.8). It is not common to find cells hooked in parallel because a mismatch of output voltages could cause problems. The leads from the external circuit may be connected to any positive and negative terminals. (The external circuit is all of the circuit *outside* the battery or cell.)

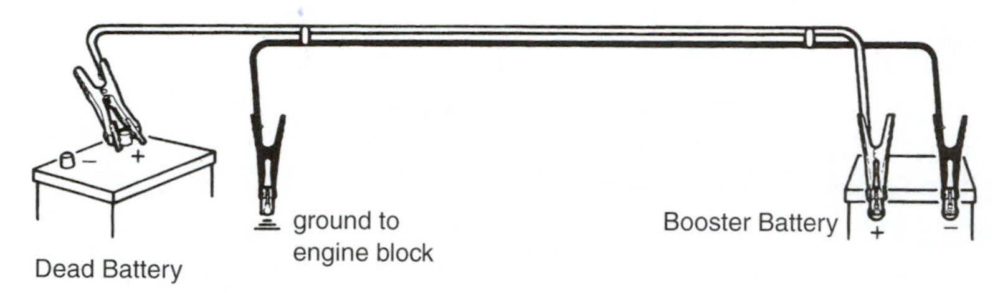

Figure 18.8
How to properly make connections to jump-start a car with a dead battery.

CELLS IN PARALLEL

1. The total current equals the sum of the individual currents in each cell:
$$I = I_1 + I_2 + I_3 + \cdots$$
2. The internal resistance equals the resistance of one cell divided by the number of cells:*
$$r = \frac{r \text{ of one cell}}{\text{number of cells}}$$
3. The emf of the battery equals the emf of any single cell:
$$E = E_1 = E_2 = E_3 = \cdots$$

EXAMPLE 2

Four cells, each 1.50 V and an internal resistance of 0.0500 Ω, are connected in parallel to form a battery with a current output of 0.250 A in each cell (Fig. 18.9).

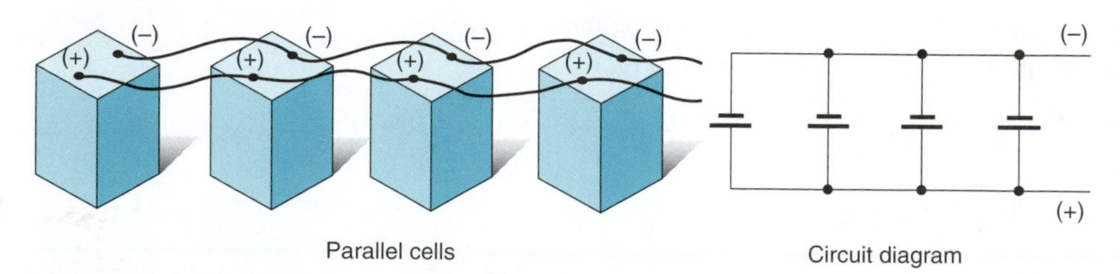

Figure 18.9

*This formula works only when all the cells have the same internal resistance. Otherwise, a formula similar to that for resistors in parallel must be used.

(a) What is the emf of the battery?
(b) Find the internal resistance of the battery.
(c) Find the current in the external circuit.

(a) $E = 1.50 \text{ V}$ (Rule 3)

(b) $r = \dfrac{r \text{ of one cell}}{\text{number of cells}} = \dfrac{0.0500\ \Omega}{4} = 0.0125\ \Omega$ (Rule 2)

(c) $I = I_1 + I_2 + I_3 + I_4 = 0.250 \text{ A} + 0.250 \text{ A} + 0.250 \text{ A} + 0.250 \text{ A} = 1.000 \text{ A}$ (Rule 1)

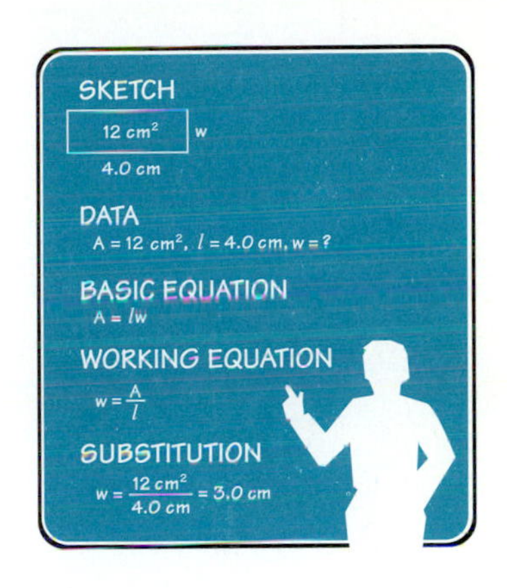

PROBLEMS 18.3

1. A cell has an emf of 1.50 V and an internal resistance of 0.0450 Ω. If there is 0.250 A in the cell, what voltage is applied to the external circuit?
2. The voltage applied to a circuit is 11.8 V when the current through the battery is 0.500 A. If the internal resistance of the battery is 0.150 Ω, what is the emf of the battery?
3. The emf of a battery is 12.0 V. If the internal resistance is 0.300 Ω and the voltage applied to the circuit is 11.6 V, what is the current through the battery?
4. Three 1.50-V cells, each with an internal resistance of 0.0500 Ω, are connected in series to form a battery with a current of 0.850 A in each cell.
   (a) Find the current in the external circuit.
   (b) What is the emf of the battery?
   (c) Find the internal resistance of the battery.
5. Five 9.00-V cells, each with internal resistance of 0.100 Ω and current output of 0.750 A, are connected in parallel to form a battery in a certain circuit.
   (a) Find the current in the external circuit.
   (b) What is the emf of the battery?
   (c) Find the internal resistance of the battery.
6. Find the current in the circuit shown in Fig. 18.10.
7. Find the current in the circuit shown in Fig. 18.11.

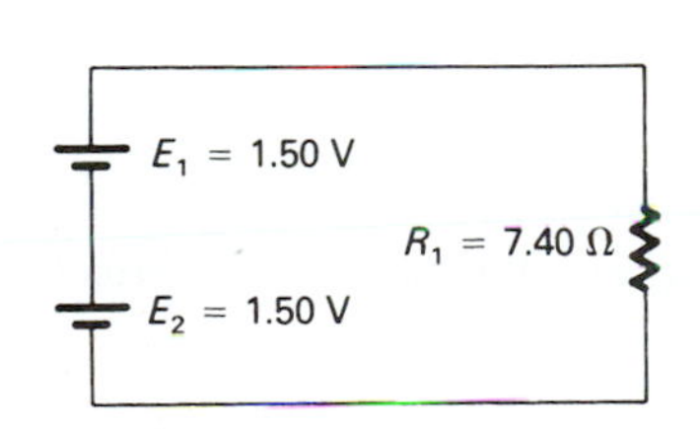

Figure 18.10

Figure 18.11

8. If the current in the circuit in Fig. 18.12 is 1.20 A, what is the value of R?
9. Find the current in the circuit shown in Fig. 18.13.
10. Find the total resistance in the circuit shown in Fig. 18.13.

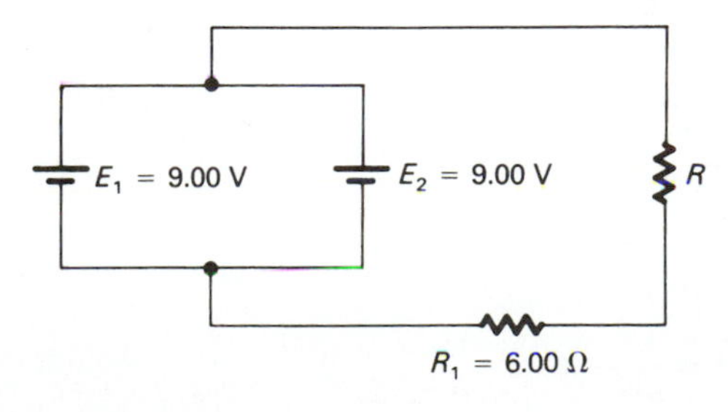

Figure 18.12

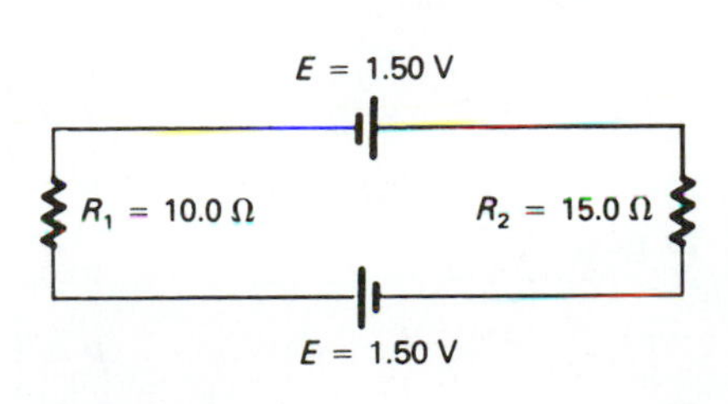

Figure 18.13

18.4 ELECTRICAL POWER

Tremendous quantities of energy are used by industry. This energy is, of course, not free but is sold by power companies. The rate of consuming energy is called **power.** The unit of power is the watt. One *watt* is the power generated by a current of 1 A flowing because of a potential difference of 1 V. A volt is a joule/coulomb (J/C); an ampere is a coulomb/second (C/s). Their product is

$$\mathrm{VA} = \frac{\mathrm{J}}{\mathrm{C}} \cdot \frac{\mathrm{C}}{\mathrm{s}} = \frac{\mathrm{J}}{\mathrm{s}}$$

Thus, 1 watt = 1 J/s.
Hence, power is

$$P = VI$$

where P = power (watts)
V = voltage drop
I = current

This equation applies to components of dc circuits and to whole dc circuits as well as to ac circuits with resistance only.

Recalling Ohm's law, $I = V/R$, we find two other equations for power: Given

$$P = VI$$

Substitute for V using $V = IR$ to obtain

$$P = (IR)I$$

Simplified, this yields

$$P = I^2R$$

Note from the following unit analysis that amps squared times ohms gives watts:

$$\mathrm{A}^2\,\Omega = \mathrm{A}^2 \cdot \frac{\mathrm{V}}{\mathrm{A}} = \mathrm{A\,V} = \frac{\mathrm{C}}{\mathrm{s}} \cdot \frac{\mathrm{J}}{\mathrm{C}} = \frac{\mathrm{J}}{\mathrm{s}} = \mathrm{W}$$

Also, given

$$P = I^2R$$

Substitute

$$I = \frac{V}{R}$$

to get

$$P = \left(\frac{V}{R}\right)^2 R = \frac{V^2}{R^2} \cdot R$$

or

$$P = \frac{V^2}{R}$$

EXAMPLE 1

A soldering iron draws 7.50 A in a 115-V circuit. What is its wattage rating?

Data:

$$I = 7.50 \text{ A}$$
$$V = 115 \text{ V}$$
$$P = ?$$

Basic Equation:

$$P = VI$$

Working Equation: Same

Substitution:

$$P = (115 \text{ V})(7.50 \text{ A})$$
$$= 863 \text{ W}$$

Therefore, a soldering iron drawing 7.50 A in a 115-V circuit has a rating of 863 W.

EXAMPLE 2

A hand drill draws 4.00 A and has a resistance of 14.6 Ω. What power does it use?

Data:

$$I = 4.00 \text{ A}$$
$$R = 14.6\ \Omega$$
$$P = ?$$

Basic Equation:

$$P = I^2R$$

Working Equation: Same

Substitution:

$$P = (4.00 \text{ A})^2(14.6\ \Omega)$$
$$= 234 \text{ W}$$

Thus, a drill that draws 4.00 A with a resistance of 14.6 Ω has a rating of 234 W.

Since the watt is a relatively small unit, the kilowatt (1 kW = 1000 watts) is commonly used in industry.

Although we speak of "paying our power bill," what power companies actually sell is **energy.** Energy is sold in kilowatt-hours (kWh). The amount of energy consumed is equal to the power used times the time it is used. Therefore,

$$\text{energy} = \text{power} \times \text{time}$$

or

$$\text{energy (in kWh)} = (VI)t$$

$$\text{number of kWh} = VIt$$

when V is in volts, I is in amperes, and t is time in hours. Note that electrical energy can be expressed in other units (joules), but kilowatt-hours are commonly used. This equation is useful in finding the cost of electric energy. Cost is measured in cents per kilowatt-hour. The cost of operating an electric device may be found as follows:

$$\text{cost} = \text{energy} \times \text{cost per unit energy}$$

$$\text{cost} = (\text{kWh})\left(\frac{\text{cents}}{\text{kWh}}\right)$$

$$\text{cost (in cents)} = \text{power (in W)} \times \text{hours} \times \frac{1 \text{ kW}}{1000 \text{ W}} \times \frac{\text{cents}}{\text{kWh}}$$

conversion factor

EXAMPLE 3

An iron is rated at 550 W. How much would it cost to operate it for 2.50 h at \$0.08/kWh?

Data:

$$P = 550 \text{ W}$$
$$t = 2.50 \text{ h}$$
$$\text{rate} = \$0.08/\text{kWh}$$
$$\text{cost} = ?$$

Basic Equation:

$$\text{cost} = Pt\left(\frac{\text{kW}}{1000 \text{ W}}\right)\left(\frac{\text{cents}}{\text{kWh}}\right)$$

Working Equation: Same

Substitution:

$$\text{cost} = (550 \text{ W})(2.50 \text{ h})\left(\frac{\text{kW}}{1000 \text{ W}}\right)\left(\frac{\$0.08}{\text{kWh}}\right)$$
$$= \$0.11$$

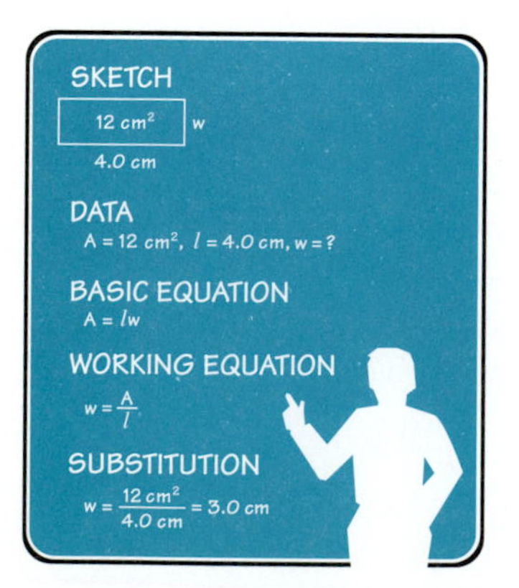

■ PROBLEMS 18.4

1. A heater draws 8.70 A on a $11\bar{0}$-V line. What is its wattage rating?
2. What power is needed for a sander that draws 3.50 A and has a resistance of 6.70 Ω?
3. How many amperes will a 75.0-W lamp draw on a $11\bar{0}$-V line?
4. Find the resistance of the lamp in Problem 3.
5. How many amperes will a $75\bar{0}$-W lamp draw on a $11\bar{0}$-V circuit?
6. Find the cost to operate the lamp in Problem 5 for 40.0 h if the cost of energy is \$0.07 per kWh.
7. Six 50.0-W bulbs are operated on a 115-V circuit. They are in use for 25.0 h in a certain month. If energy costs \$0.075 per kWh, what is the cost of operating them for the month?
8. A small furnace expends 3.00 kW of power. If the cost of operation of the furnace is \$3.84 for a 24.0-h period, what is the cost of energy per kWh?

9. Will a 20.0-A fuse blow if a $10\bar{0}0$-W hair dryer, a $12\bar{0}0$-W electric skillet, and a $11\bar{0}0$-W toaster are all used at once on a $11\bar{0}$-V line?
10. How long could you operate a $10\bar{0}0$-W soldering iron for $0.50 if the cost of energy is $0.075/kWh?
11. Find the cost of operating a 1.50-A motor on a $11\bar{0}$-V circuit for 2.00 h at $0.08/kWh.
12. Find the cost of operating a 2.50-A motor on a $11\bar{0}$-V circuit for 3.00 h at $0.07/kWh.
13. Find the cost of operating a 3.00-A motor on a $11\bar{0}$-V circuit for 2.00 h at $0.07/kWh.
14. How many amperes will a $6\bar{0}$-W lamp draw on a $11\bar{0}$-V line?
15. Using the following table, list two different combinations of bulbs and appliances that could be used on a $2\bar{0}$-A circuit breaker.

Appliance	*Power Rating*
Light bulb (60 W)	60 W
Fluorescent bulb (40 W)	40 W
12″ TV	55 W
Projection TV	1500 W
Personal computer	550 W
Hand drill	400 W
Microwave oven	1000 W

16. Using the preceding table, list two different combinations that could be used on a $3\bar{0}$-A circuit breaker. ■

GLOSSARY

Dry Cell A battery made of a carbon rod and a metal can with a chemical paste in between that reacts with the terminals to provide energized electrons. (p. 438)
Electrolyte An acid solution that produces large numbers of free electrons at the negative pole of a battery. (p. 437)
Energy Work delivered to an electrical component or appliance (power × time). (p. 443)
Internal Resistance The resistance within a cell itself, which opposes movement of the electrons. (p. 438)
Power Energy per unit time consumed in a circuit. (p. 442)
Primary Cell A cell or battery that cannot be recharged. (p. 437)
Recharging The passing of an electric current through a secondary cell to restore the original chemicals. (p. 437)
Secondary Cell A rechargeable type of battery. (p. 437)

FORMULAS

18.2 $V = E - Ir$

18.3 *Cells in series:*

$$I = I_1 = I_2 = I_3 = \cdots$$

$$r = r_1 + r_2 + r_3 + \cdots$$

$$E = E_1 + E_2 + E_3 + \cdots$$

Cells in parallel:

$$I = I_1 + I_2 + I_3 + \cdots$$

$$r = \frac{r \text{ of one cell}}{\text{number of cells}}$$

$$E = E_1 = E_2 = E_3 = \cdots$$

18.4 $P = VI$

$$P = I^2R$$

$$P = \frac{V^2}{R}$$

$$\text{energy} = \text{power} \times \text{time}$$

$$\text{cost (in cents)} = \text{power (in W)} \times \text{hours} \times \frac{1\text{ kW}}{1000\text{ W}} \times \frac{\text{cents}}{\text{kWh}}$$

REVIEW QUESTIONS

1. The emf of a battery with cells connected in series equals the sum of the
   (a) internal resistances of the cells. (b) emf of the individual cells.
   (c) current in the individual cells.
2. The current in a battery with cells connected in series equals the
   (a) internal resistances of the cells. (b) emf of the individual cells.
   (c) current in the individual cells.
3. The current in a battery with cells connected in parallel equals the
   (a) current in one cell. (b) internal resistance.
   (c) sum of the currents in each cell.
4. Examples of dry cells include
   (a) lead–zinc cells. (b) nickel–cadmium cells.
   (c) carbon–zinc cells. (d) fuel cells.
5. Distinguish between a primary and a secondary cell.
6. Explain recharging.
7. Describe the function of an electrolyte.
8. In your own words, describe the manner in which a secondary cell produces electrical energy.
9. What is the effect of the internal resistance of a cell?
10. The unit of electrical power is the ______________.
11. In your own words, explain the relationship between power, voltage, and current.
12. Do we pay the utility company for our power use or our energy use? Explain.
13. Explain the relationship between power, voltage, and resistance.
14. Explain the relationship between power, current, and resistance.
15. If the current in a circuit is increased by a factor of 2 and the voltage stays constant, how does the power change?
16. If the resistance in a circuit decreases by a factor of 2 and the voltage stays constant, how does the power change?
17. If the voltage and current in a circuit each decrease by a factor of 2, how does the power change?
18. If the current increases in a circuit by a factor of 2 and the voltage stays constant, how would the cost of operating the circuit change?

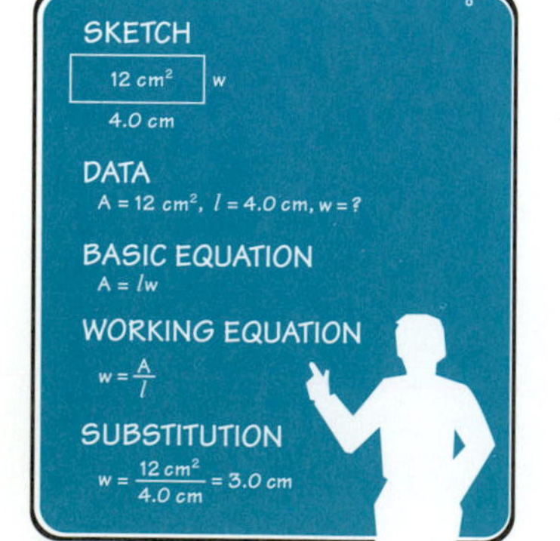

REVIEW PROBLEMS

1. A cell has an emf of 1.44 V and an internal resistance of 0.0550 Ω. If there is 0.135 A in the cell, what voltage is applied to the external circuit?

2. The voltage applied to a circuit is 12.0 V when the current through the battery is 0.858 A. If the internal resistance of the battery is 0.245 Ω, what is the emf?
3. Six 6.00-V cells, each with an internal resistance of 0.0987 Ω and current output of 0.658 A, are connected in parallel to form a battery in a certain circuit. What is the current in the external circuit?
4. What is the emf of the battery in Problem 3?
5. What is the internal resistance of the battery in Problem 3?
6. Find the current in the circuit shown in Fig. 18.14.

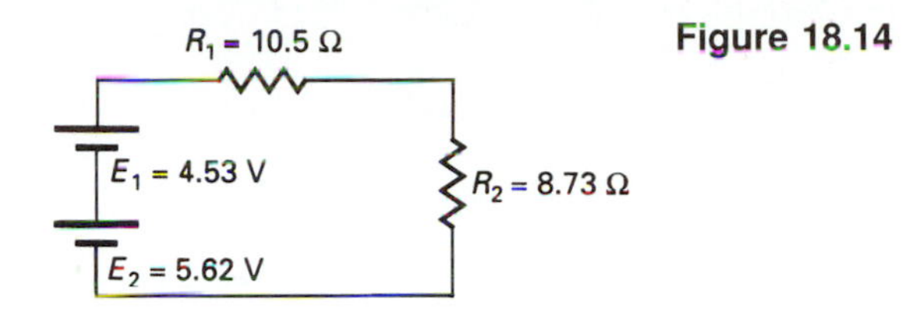

Figure 18.14

7. Find the total resistance in the circuit shown in Fig. 18.15.

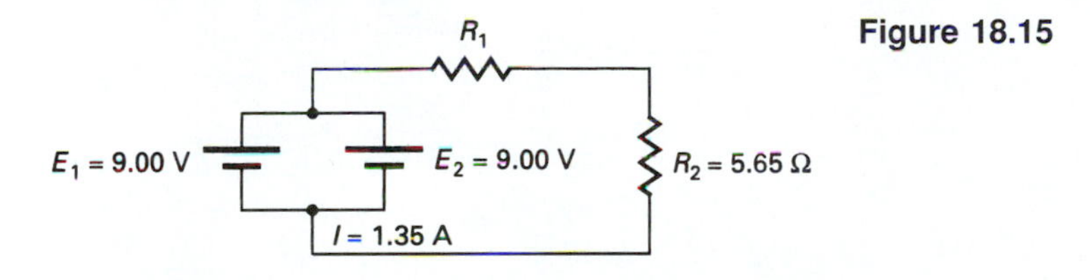

Figure 18.15

8. What power is needed for a drill that draws 2.45 A and has a resistance of 6.55 Ω on a $11\bar{0}$-V circuit?
9. How many amperes will a $15\bar{0}$-W light bulb draw on a $11\bar{0}$-V circuit?
10. What is the cost to operate the lamp in Problem 9 for 135 h if the cost of energy is $0.05/kWh?
11. If the cost of energy is $0.043/kWh, how long could you operate a motor that draws 0.40 A on a 110-V line for $0.45?
12. How many amperes will a $1\bar{0}0$-W lamp draw on a 110-V line?

Chapter 19

MAGNETISM

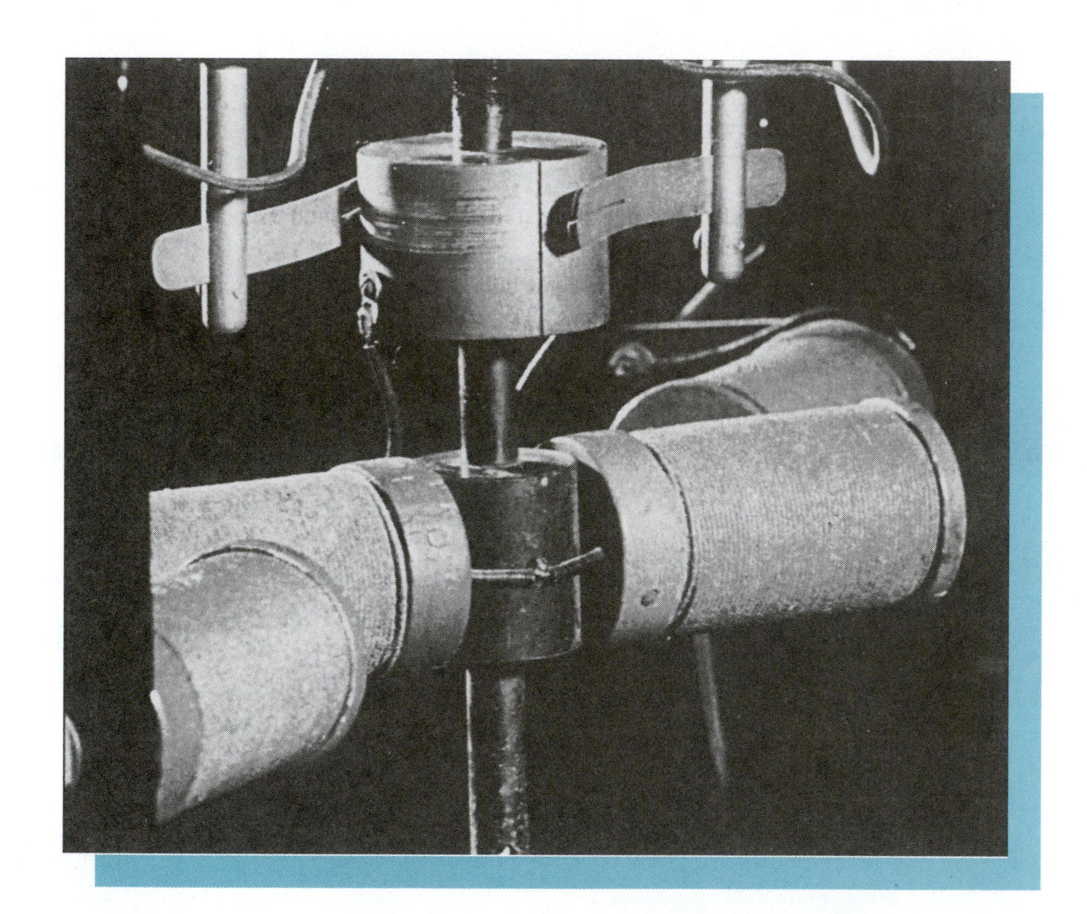

Magnetism is now being used to replace wheels on levitated high-speed trains. Electromagnetism is at the heart of generation of electrical power. We consider how generators and motors use these principles in the production and consumption of electricity.

OBJECTIVES

The major goals of this chapter are to enable you to:

1. Describe the nature of magnetism and magnetic effect of electric currents.
2. Understand how induced magnetism and current are related.
3. Distinguish between generators and motors, and describe the principles that apply to both.

19.1 INTRODUCTION TO MAGNETISM

Many devices that use or produce electric energy depend on the relation of magnetism and electric currents. Motors and meters are designed to use the fact that electric currents in wires behave like magnets. Generators produce electric current due to the movement of wires near very large magnets.

In this chapter we investigate the basic properties of magnets and the relation between currents and magnetism. In later sections on generators, motors, and transformers, we will use the basic principles of magnetism that are developed here.

Certain kinds of metals have been found to have the ability to attract pieces of iron, steel, and some other metals (Fig. 19.1). Metals that have this ability are said to be **magnetic.** Deposits of iron ore that are naturally magnetic have been found. This ore is called *lodestone*.

Artificial magnets can be made from iron, steel, and several special alloys such as Permalloy™ and alnico. We will discuss the process of creating artificial magnets later. Materials that can be made into magnets are called *magnetic materials*. Most materials are nonmagnetic (examples are wood, aluminum, copper, and zinc).

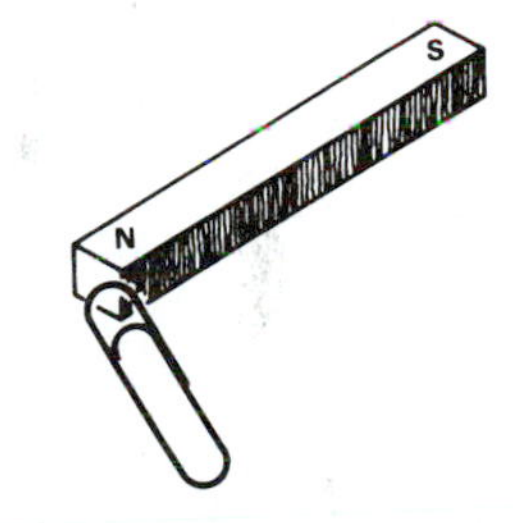

Figure 19.1
Magnetic materials attract iron and steel.

Forces between Magnets

Suppose that a bar magnet is suspended by a string so that it is free to rotate. It will rotate until one end points north and the other south (Fig. 19.2). The end that points north is called the north-seeking pole, or *north* (N) *pole*. The other end is the south-seeking pole, or *south* (S) *pole*.

If the north pole of another bar magnet is brought near the north pole of this magnet, the two like poles will repel [Fig. 19.3(a)]. The south pole of one magnet will attract the north pole of the other [Fig. 19.3(b)]. In summary:

> Like magnetic poles repel each other, and unlike magnetic poles attract each other.

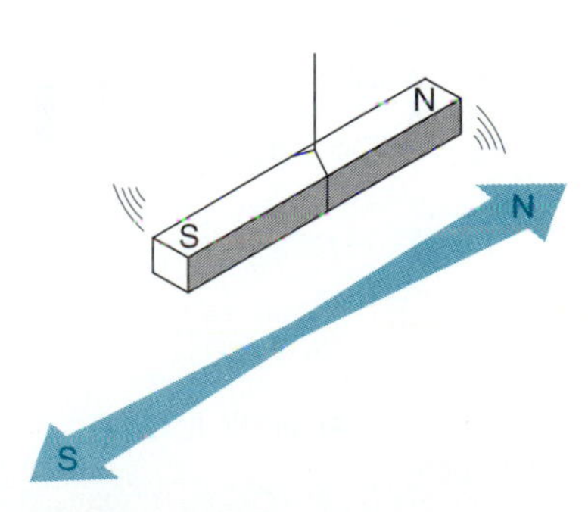

Figure 19.2
A suspended magnet will rotate to line up north and south.

A **compass** (Fig. 19.4) is simply a small magnetic needle that is free to rotate on a bearing.

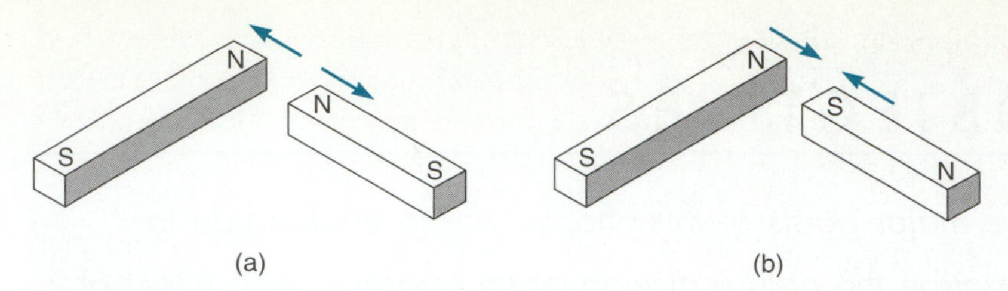

Figure 19.3
Unlike poles attract and like poles repel each other.

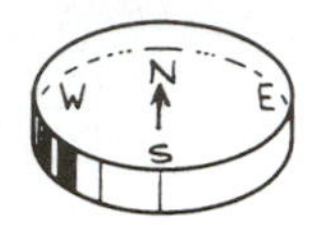

Figure 19.4
Simple compass.

Magnetic Fields of Force

There is a **magnetic field** near a magnetic pole. The existence of this field can be detected by using another magnet. We can represent this field of force by drawing lines that indicate the direction of the force exerted on a north pole placed in the field. The field of a bar magnet can be mapped by moving a small compass around the magnet as shown in Fig. 19.5. These resulting lines are called **flux lines** (lines of force).

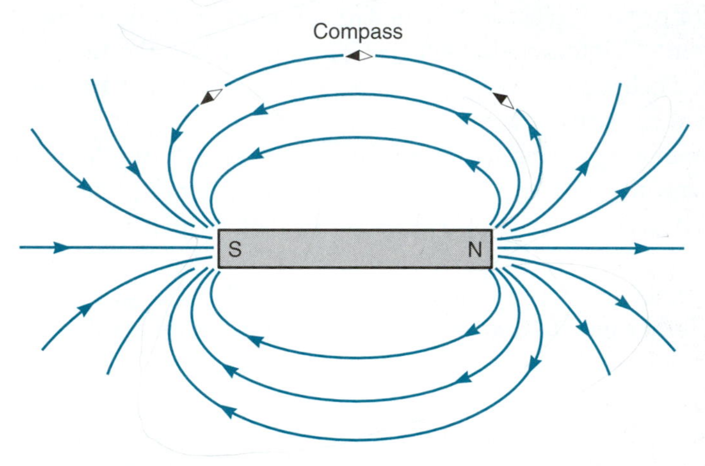

Figure 19.5
Mapping the field of a bar magnet.

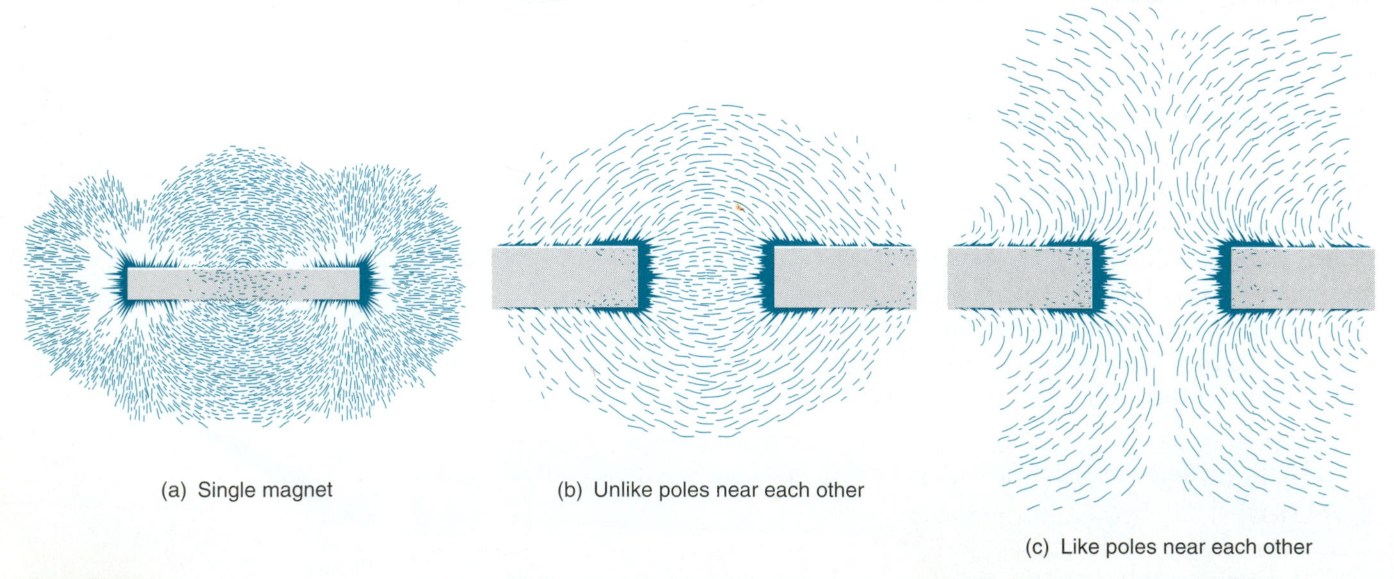

Figure 19.6
Flux lines shown by iron filings.

The flux lines can also be found by sprinkling iron filings on a sheet of paper laid over a magnet (Fig. 19.6). The fields of combinations of magnets can also be found in this way. Although the iron filing patterns are two-dimensional, the magnetic field around a magnet is actually a three-dimensional field, as shown in Fig. 19.7.

Figure 19.7
A magnetic field is three-dimensional.

Earth's three-dimensional magnetic field is shown in Fig. 19.8. Many puzzling aspects of earth's magnetic field have not been resolved. The north magnetic pole and the north geographic pole (sometimes called *true north*) are at different locations. The axis of rotation and the magnetic field axis are slightly different and change approximately 10 minutes of arc each year. Even more puzzling, scientific evidence indicates that earth's magnetic field reverses completely every few hundred thousand years without significantly affecting earth's rotational or orbital motions.

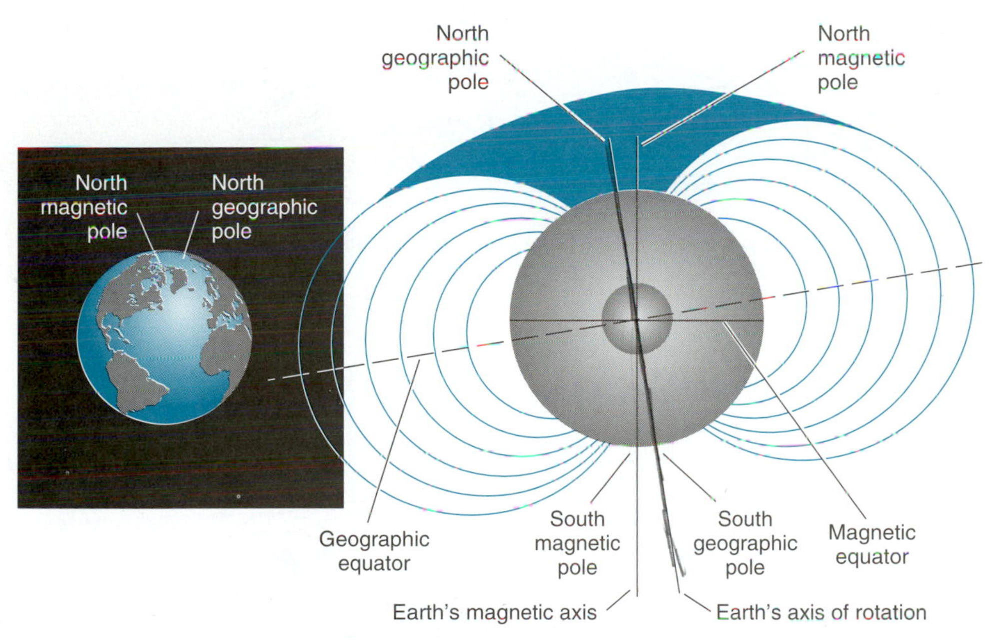

Figure 19.8
The earth's magnetic field is similar to that of other magnets. The geographic and magnetic poles do not coincide; similarly, the geographic and magnetic equators differ.

19.2 MAGNETIC EFFECTS OF CURRENTS

When a current passes through a conductor, it sets up a magnetic field. A compass placed near the current shows the direction of this magnetic field. We can show this by connecting a battery to a wire (refer to Fig. 19.9 for this discussion). A compass needle is placed under the wire as in Fig. 19.9(a). When the switch is closed, the needle deflects as in Fig. 19.9(b). If the terminals of the battery are reversed, the needle deflects in the opposite direction [Fig. 19.9(c)].

When the compass needle is placed on top of the conductor, the direction of deflection is reversed in each case [Fig. 19.9(d)–(f)]. When the current in a wire flows in a given direction, the flux lines point in one direction below the wire and in the opposite direction above the wire.

The field actually curves around the straight current-carrying wire [Fig. 19.10(a)]. Iron filings on a sheet of paper perpendicular to a current-carrying wire show the shape of the field. The magnetic field is stronger for large currents than

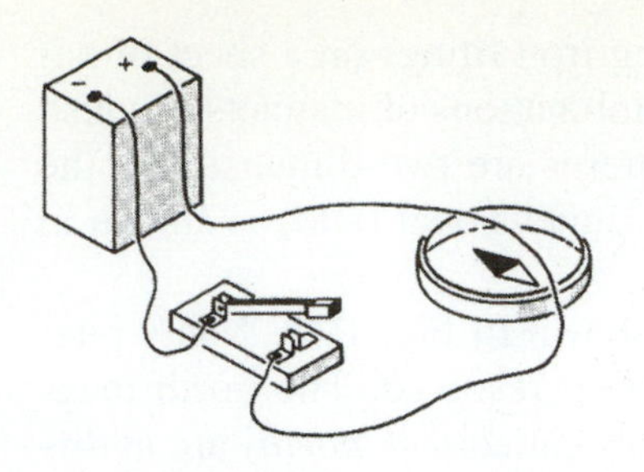

(a) Compass below conductor, switch open.

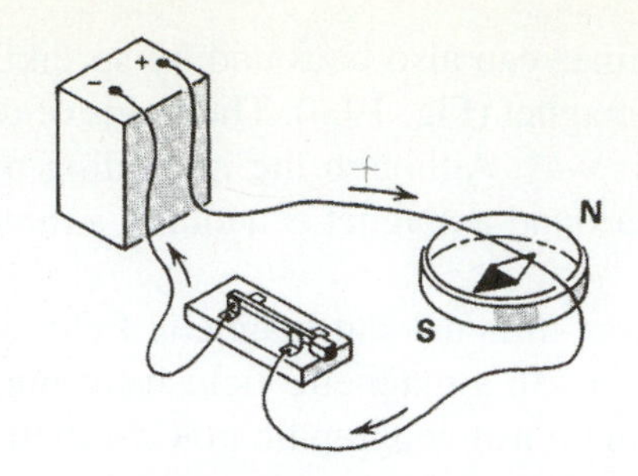

(b) Compass below conductor, switch closed.

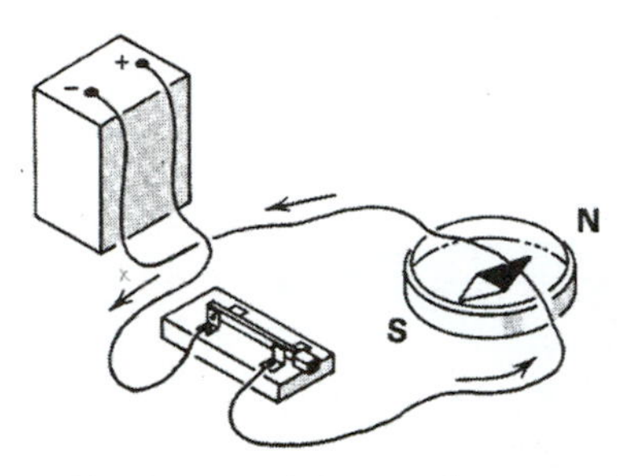

(c) Compass below conductor, switch closed. Battery terminals reversed.

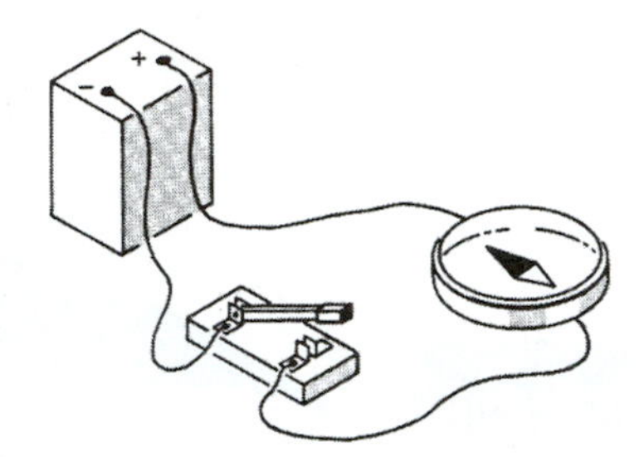

(d) Compass above conductor, switch open.

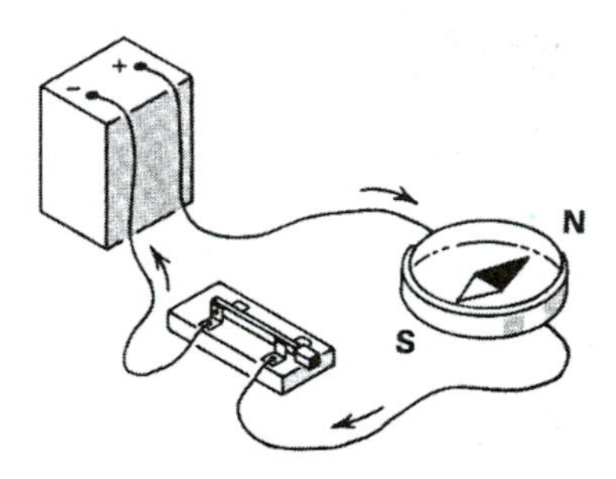

(e) Compass above conductor, switch closed.

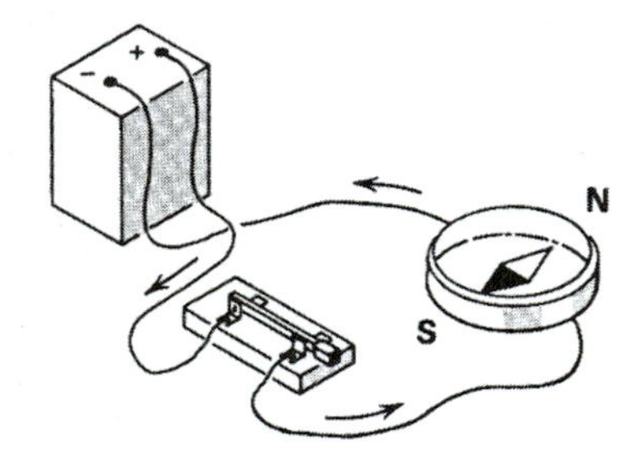

(f) Compass above conductor, switch closed. Battery terminals reversed.

Figure 19.9
Magnetic effects of currents.

for small currents. The direction of the field near a current in a straight wire is shown in Fig. 19.10(b) and given by the following rule:

> *AMPÈRE'S RULE*
>
> Hold the conductor in your right hand, with your thumb extended in the direction of the current. Your fingers circle the wire in the direction of the flux lines.

The magnetic field near a long current-carrying wire, measured in units of teslas, is circular about the wire and given by Ampère's law:

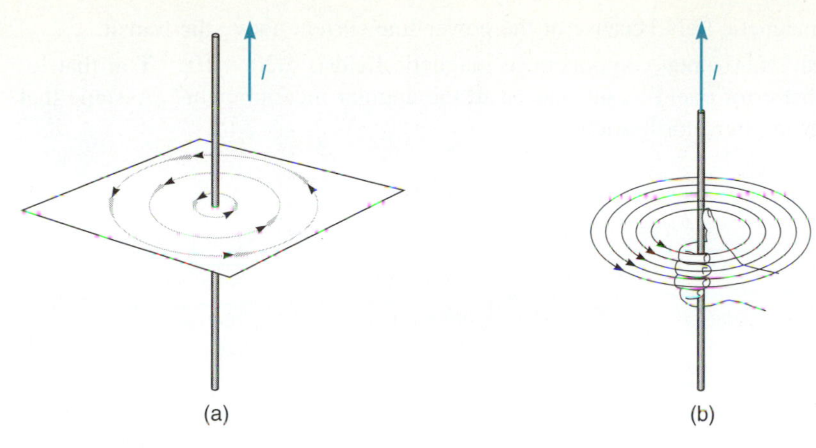

Figure 19.10
(a) Field around a wire. (b) Direction of the field around a current-carrying wire.

$$B = \frac{\mu_0 I}{2\pi R}$$

where B = magnetic field, in tesla
I = current through the wire
R = perpendicular distance from the center of the wire
$\mu_0 = 4\pi \times 10^{-7}$ T m/A

The magnetic field, B, has the unit tesla (T) and is defined in terms of electric current by the constant μ_0, the permeability constant. The value of μ_0 is not experimentally determined but is an assigned value that explicitly defines magnetic field in terms of electric current.

EXAMPLE 1

A power line carrying $40\overline{0}$ A is 9.00 m above a transit used by a surveying student in Fig. 19.11(a).

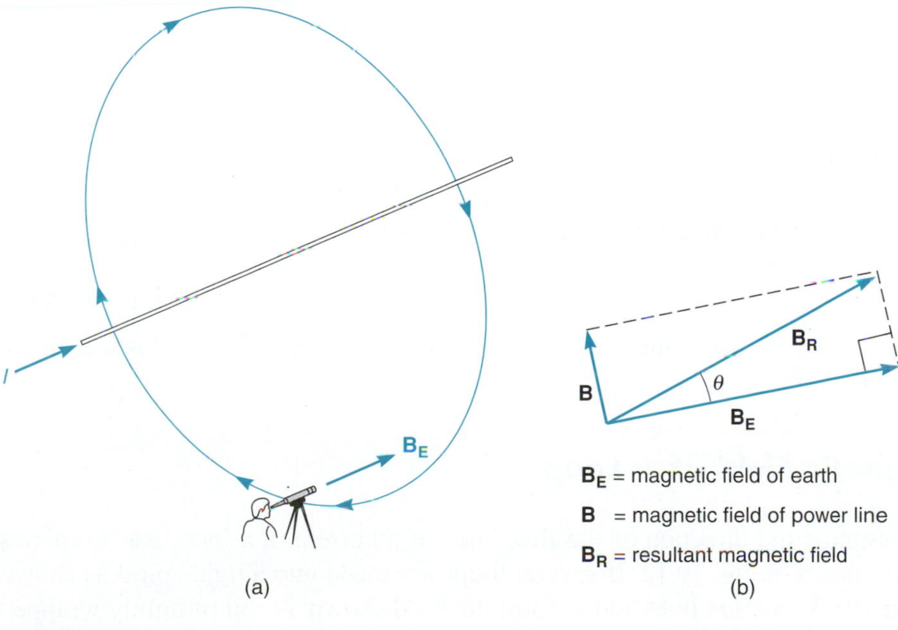

Figure 19.11

(a) Find the magnetic field because of the power-line current above the transit.

(b) If the earth's horizontal component of magnetic field is 5.20×10^{-5} T at that location, what error could be introduced in the angular measurement? (Assume that the power line runs north–south.)

(a) Data

$$I = 40\bar{0} \text{ A}$$
$$R = 9.00 \text{ m}$$
$$\mu_0 = 4\pi \times 10^{-7} \text{ T m/A}$$
$$B = ?$$

Basic Equation:

$$B = \frac{\mu_0 I}{2\pi R}$$

Working Equation: Same

Substitution:

$$B = \frac{(4\pi \times 10^{-7} \text{ T } \cancel{\text{m}}/\cancel{\text{A}})(40\bar{0} \text{ } \cancel{\text{A}})}{2\pi(9.00 \text{ } \cancel{\text{m}})}$$
$$= 8.89 \times 10^{-6} \text{ T}$$

Therefore, the magnetic field from the power line is 8.89×10^{-6} T. With the current from south to north, Ampère's rule shows the direction of B to be east to west.

(b) The angle that the resultant vector [the earth's field plus the wire's field $(B_E + B)$] makes with B_E would be the angular error θ.

Data:

$$B_E = 5.20 \times 10^{-5} \text{ T} \quad \text{earth's component}$$
$$B = 8.89 \times 10^{-6} \text{ T} \quad \text{[from part (a)]}$$
$$\theta = ?$$

Basic Equation:

$$\tan \theta = \frac{B}{B_E}$$

Working Equation: Same

Substitution:

$$\tan \theta = \frac{8.89 \times 10^{-6} \text{ } \cancel{\text{T}}}{5.20 \times 10^{-5} \text{ } \cancel{\text{T}}} = 0.171$$
$$\theta = 9.7°$$

The bearing on the surveying student's transit could be in error by 9.7° because of the power line.

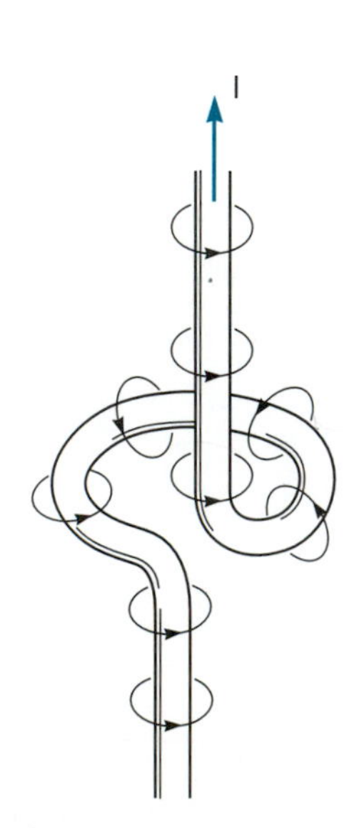

Figure 19.12
Magnetic field of a loop.

Magnetic Field of a Loop

To determine the direction of the flux lines of a current in a loop, use Ampère's rule as shown in Fig. 19.12. If several loops are made into a tight spiral as shown in Fig. 19.13, the flux lines add to form the field shown. A coil of tightly wrapped

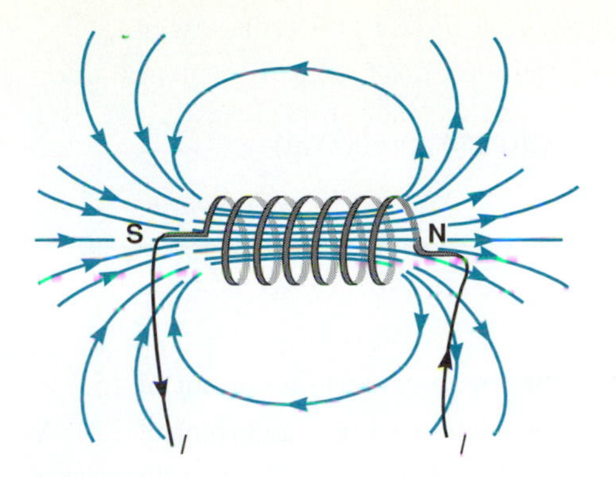

Figure 19.13
Magnetic field of a coil.

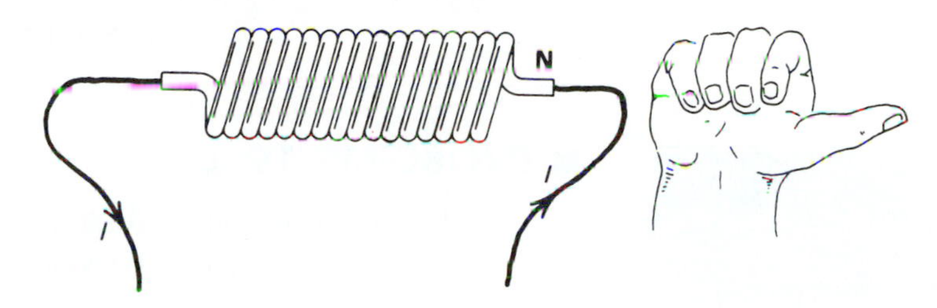

Figure 19.14
Polarity of a solenoid.

wire is called a **solenoid.** The left side of this solenoid acts like a south magnetic pole. The right side acts like a north magnetic pole. This polarity could be found by using a compass. The rule for finding the polarity of a solenoid is this:

> Hold the solenoid in your right hand so that your fingers circle it in the same direction as the current. Your thumb points to the north pole of the solenoid (Fig. 19.14).

For a long coil that is tightly turned, the field strength at its center is

$$B = \mu_0 I n$$

where B = magnetic field in the region at the center of the solenoid
μ_0 = permeability constant, $4\pi \times 10^{-7}$ T m/A
I = current through the solenoid
n = number of turns per unit length of solenoid

The longer the solenoid is with respect to its radius, the more uniform the magnetic field is inside the solenoid; and for an infinitely long solenoid, the value of B is uniform throughout.

EXAMPLE 2

Find the magnetic field at the center of a solenoid that is 0.425 m long, 0.075 m in diameter, and has three layers of $85\bar{0}$ turns each, when 0.250 A flows throughout.

Data:

$$I = 0.250 \text{ A}$$
$$n = \frac{3 \times 85\bar{0} \text{ turns}}{0.425 \text{ m}} = 60\bar{0}0 \text{ turns/m}$$
$$\mu_0 = 4\pi \times 10^{-7} \text{ T m/A}$$
$$B = ?$$

Note: The length, 42.5 cm, can be considered "long" as compared with the radius, 3.75 cm (about 11 times).

Basic Equation:

$$B = \mu_0 I n$$

Working Equation: Same

Substitution:

$$B = (4\pi \times 10^{-7}\ \text{T m/A})(0.250\ \text{A})(60\bar{0}0/\text{m})$$
$$= 1.88 \times 10^{-3}\ \text{T}$$

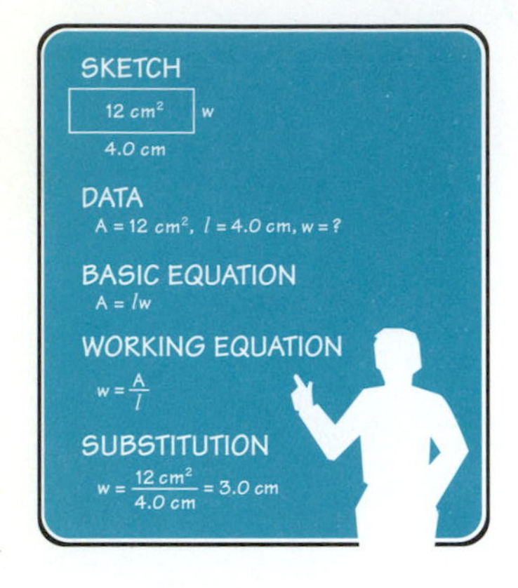

■ PROBLEMS 19.2

1. Find the magnetic field at 0.250 m from a long wire carrying a current of 15.0 A.
2. Find the magnetic field at 0.500 m from a long wire carrying a current of 7.50 A.
3. What is the current in a wire if the magnetic field is 5.75×10^{-6} T at a distance of 2.00 m from the wire?
4. A power line runs north–south carrying 675 A and is 5.00 m above a transit used by a surveyor.
   (a) What is the magnetic field at the transit because of the power-line current?
   (b) If the earth's horizontal component of magnetic field is 5.20×10^{-5} T, what error is introduced in the surveyor's angular measurement?
5. Find the magnetic field at 0.350 m from a long wire carrying a current of 3.00 A.
6. Find the current in a wire if the magnetic field is 3.50×10^{-6} T at a distance of 2.50 m from the wire.
7. A solenoid has $100\bar{0}$ turns of wire, is 0.320 m long, and carries a current of 5.00 A. What is the magnetic field at the center of the solenoid? Assume that its length is long in comparison with its diameter.
8. A solenoid has $300\bar{0}$ turns of wire and is 0.350 m long. What current is required to produce a magnetic field of 0.100 T at the center of the solenoid? Assume that its length is long in comparison with its diameter.
9. A small solenoid is 0.150 m in length, 0.0150 m in diameter, and has $60\bar{0}$ turns of wire. What current is required to produce a magnetic field of 1.25×10^{-3} T at the center of the solenoid?
10. A solenoid has $25\bar{0}0$ turns of wire and is 0.200 m long. What current is required to produce a magnetic field of 0.100 T at the center of the solenoid? Assume that its length is long in comparison with its diameter. ■

19.3 INDUCED MAGNETISM AND ELECTROMAGNETS

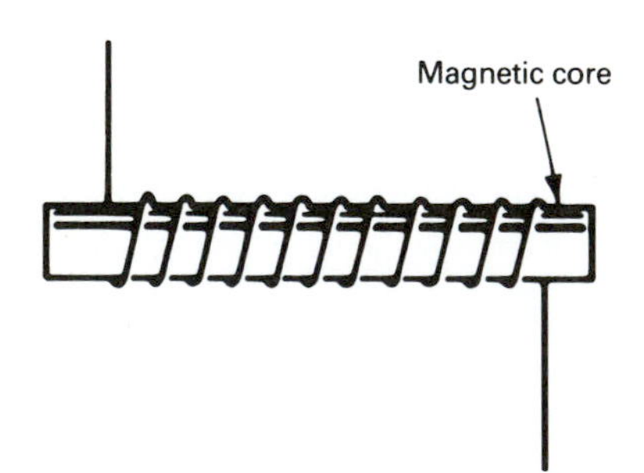

Figure 19.15
Simple electromagnet.

When a magnetic material such as iron is placed in the core of a current-carrying solenoid (Fig. 19.15), the material becomes very strongly magnetized. This is called **induced magnetism.** The solenoid and magnetic core are called an **electromagnet.** When the current through the coil is turned off, the strength of the induced magnet decreases, but some remains. When the core is removed, a magnetic field remains in the core. In materials such as soft iron, very little magnetic field remains in the core after the current flow stops. In other materials, such as steel, alnico, and Permalloy, a much stronger field remains. The latter materials are used for permanent magnets. However, they are undesirable for use as a core in an induction motor. Soft-iron cores are often used for this application because less energy is required to reverse the polarity of the induced magnetic field.

A magnet can be thought to consist of many atoms, each behaving like a small magnet. In each atom, the electrons orbit about the nucleus and each electron spins about its own axis, producing small current loops that generate magnetic fields. In most materials, these current loops are arranged so that their magnetic fields point in different directions. The result is that the magnetism of one loop (atom) is canceled out by its neighbors.

In magnetic materials the atomic magnetic fields line up with each other in regions called *magnetic domains* when no field is present [Fig. 19.16(a)]. Each domain has a magnetic field direction to which most atoms in the domain are aligned. When no magnetic field is present, the orientations of the domains are random. However, when an external field is present, the domains tend to line up with the field, causing the domain boundaries to shift [Fig. 19.16(b)]. In high electric fields, nearly all the material is aligned [Fig. 19.16(c)]. When the external magnetic field is removed, some materials such as the alloy alnico retain the aligned domains, creating a permanent magnet.

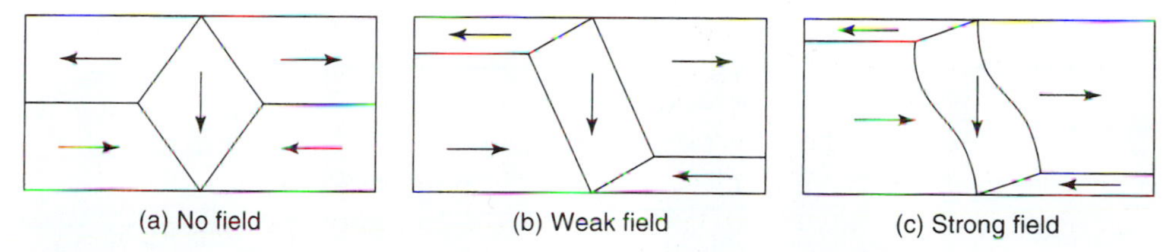

Figure 19.16

19.4 INDUCED CURRENT

When a magnet is moved so that its flux lines cut across a wire, an emf is induced in the wire, which is known as **induced current.** The strength of this induced emf depends on the strength of the magnetic field and on the rate at which the flux lines are cut by moving the magnet or wire. Increasing the strength of the field or increasing the rate at which the flux lines are cut also causes the current to increase.

While the magnet shown in Fig. 19.17(a) is moved downward, the galvanometer indicates that a current flows through the wire. If the magnet shown in Fig. 19.17(b) is moved upward, the induced current is in the opposite direction.

A current would also flow in the wire if the circuit is closed, the wire is moved, and the magnet is stationary [Fig. 19.17(c)]. The current is produced by the relative motion of the magnet and wire. In commercial generators, magnets are spun inside a set of coils of wire. The induced emf is increased by replacing the single wire with a coil of many turns. For example, tripling the number of turns triples the induced emf [Fig. 19.17(d)].

19.5 GENERATORS

The induction of an emf in a coil of wire can be used to produce electric power. A coil of wire rotating in a magnetic field produces a fluctuating (changing) emf in the wire. This is the simplest kind of **generator** to build in the laboratory, so we will study its operation here and compare it to commercial generators where the magnets (electromagnets) are rotated.

The current produced by rotating the wire through the magnetic field is called an alternating current (ac). As side A of the current loop in Fig. 19.18(a) passes downward by the north pole, the induced current is in one direction. As side A (same side of the rotating loop) passes upward by the south pole as in Fig. 19.18(b),

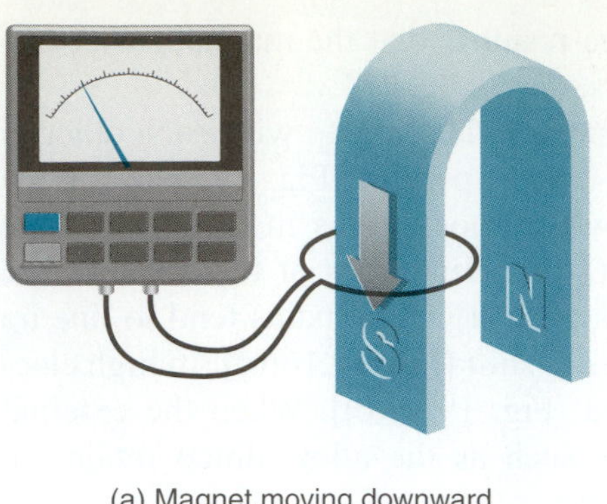

(a) Magnet moving downward

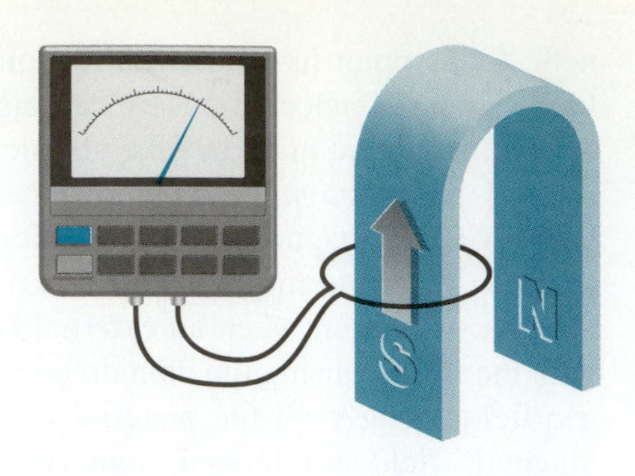

(b) Magnet moving upward; current flows in opposite direction

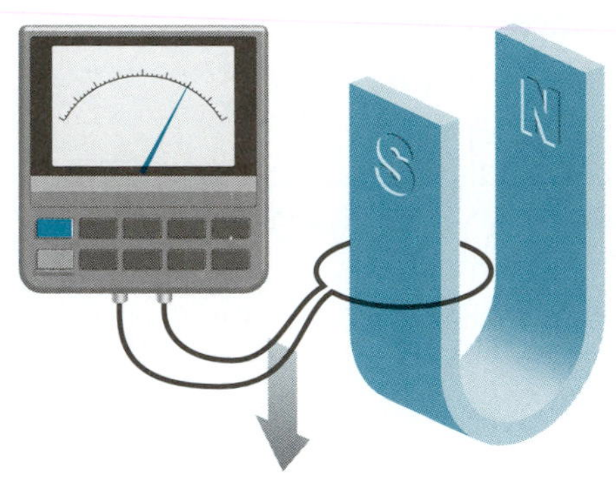

(c) Wire moving downward

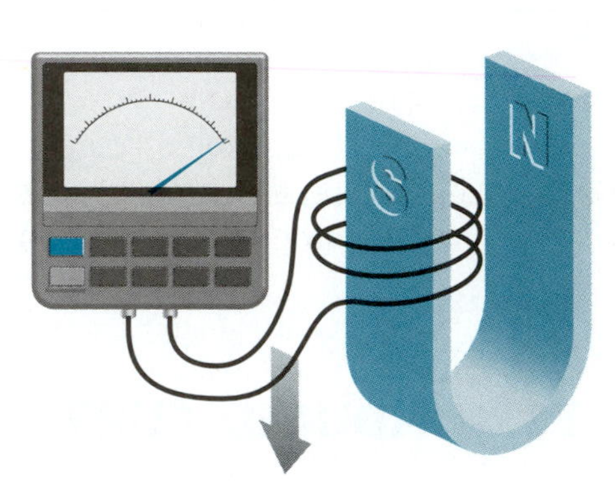

(d) Coil moving downward increases the current flow in same direction

Figure 19.17
Induced current in a wire.

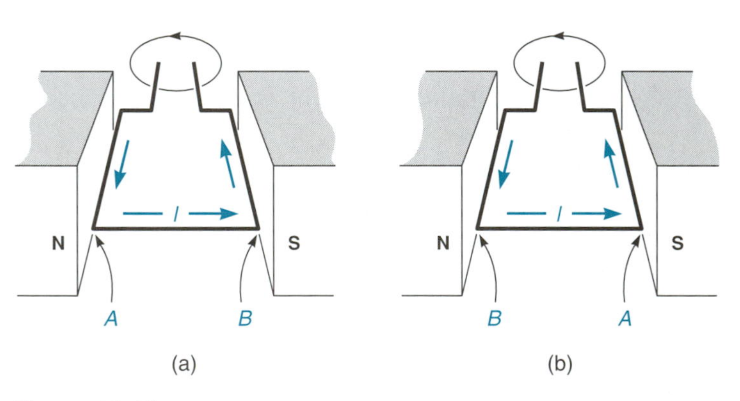

(a) (b)

Figure 19.18
Direction of current flow in a moving wire in a magnetic field.

the induced current is in the opposite direction. The result is an alternating current induced in the rotating wire. As side *B* (the other side of the rotating loop) passes upward and then downward, the current in it also alternates.

The direction of current flow in a wire as it moves between the north and south magnet poles is illustrated in Fig. 19.19. In beginning position (1) in Fig. 19.19, the current is zero. As the wire passes down through position (2), the current builds and reaches a maximum value in position (3). The current then becomes smaller through position (4) until it is zero again in position (5).

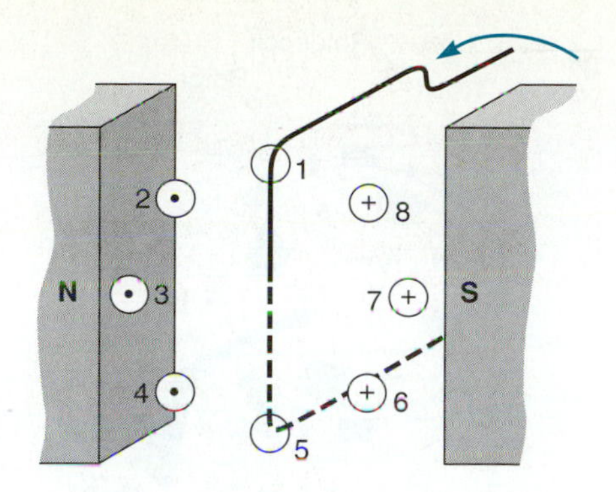

Figure 19.19
Induced current in a wire moving in a magnetic field.

As the wire begins to pass back up through the magnetic field in position (6), the current begins to build and reaches a maximum value in the opposite direction in position (7). It then becomes smaller in position (8) and falls to zero again as it reaches position (1) again. The cycle is then repeated for the next rotation of the wire through the magnetic field.

A graph of the induced current showing the relative magnitude of the current for the changing positions of the wire is shown in Fig. 19.20.

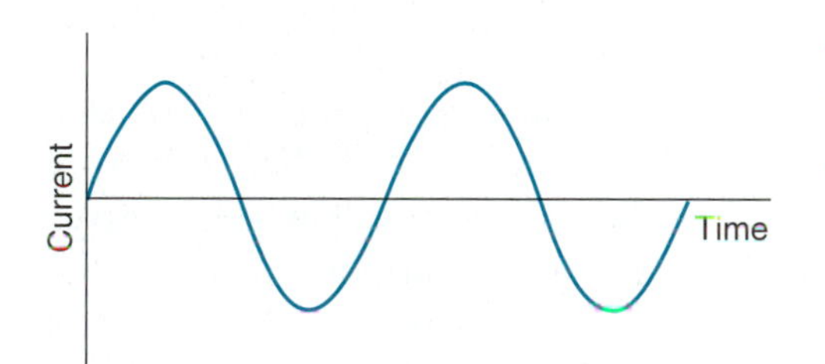

Figure 19.20
Graph of induced current as the wire changes positions in a magnetic field.

One cycle is produced by one revolution of the wire. The time required for one cycle depends on the rotational speed of the coil. If the coil rotates 60 times each second, an alternating current of frequency 60 hertz (cycles per second) is produced. The current produced in the coil is conducted by brushes on slip rings to the external circuit as shown in Fig. 19.21. The rotating coil is called the *rotor* or **armature,** and the *field magnets* are called the **stator.**

The generator does not actually create electric energy; it changes the mechanical energy of rotation into electric energy. The energy to turn the rotor may be supplied by water falling down a waterfall, a diesel engine, or a steam turbine.

Power companies use large commercial ac generators to produce the current they need to supply to their customers. These generators work similarly to the generator discussed here, but they have many coils and use electromagnets instead of permanent magnets. The large generators used by electric power companies can produce voltages as large as 13,000 V and currents up to 10 A. The alternator used in automobiles is an ac generator that produces about 13 V and up to 40 A.

dc Generators

By the use of a special device called a **commutator,** the ac generator can be used to produce direct current. The commutator is a split ring that replaces the slip

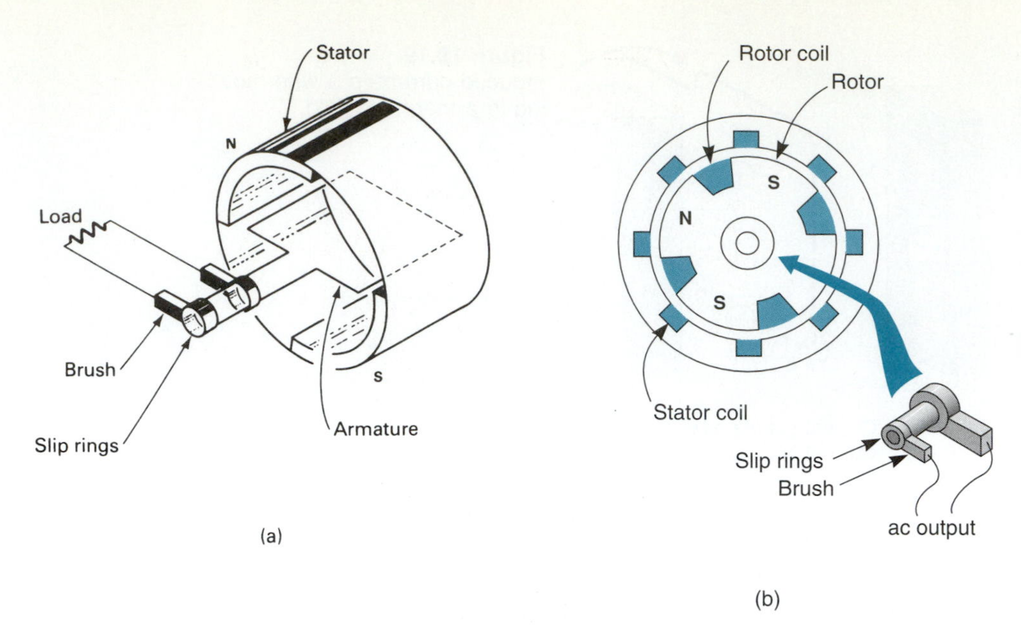

Figure 19.21
Current in a generator.

rings as shown in Fig. 19.22(a). When side *A* of the coil passes upward along the north pole, the induced current flows in the direction shown in Fig. 19.22(b) and is picked up by brush 1. The current in the external circuit is also shown.

When side *B* of the coil passes upward along the north pole, the induced current flows in the direction shown and is picked up by brush 1 [Fig. 19.22(c)]. The current in the external circuit is in the same direction as it was when *A* passed along the north pole [Fig. 19.22(d)]. Thus, this is a direct current.

The current produced by this dc generator does not have the same value at all times. A graph of the induced current is shown in Fig. 19.23(a). Commercial dc generators that are used for industrial purposes contain many coils. The output current has almost the same value at all times due to the large number of coils [Fig. 19.23(b)].

19.6 THE MOTOR PRINCIPLE

We have seen that like poles of magnets repel each other. A magnet that is pivoted will spin due to the repulsion of another magnet nearby as shown in Fig. 19.24(a). We can construct an electromagnet by wrapping wire around an iron core and running a current through the wire [Fig. 19.24(b)]. The north pole of the electromagnet will be repelled by a north pole of another magnet. The electromagnet will turn until its south pole is next to the north pole of the permanent magnet [Fig. 19.24(c)].

If we could suddenly change the polarity of the electromagnet (often called the *armature*), the magnet would repel the north pole and the electromagnet would continue to spin [Fig. 19.24(d)]. If a dc current supply is used [Fig. 19.24(e)], this change can be made by using a commutator (split ring) to change the direction of the current in the electromagnet. Changing the direction of the current flowing through the coil of the electromagnet changes the poles.

As the current changes direction, the electromagnet spins due to the repulsion of like poles. A shaft may be connected to the electromagnet so that the ro-

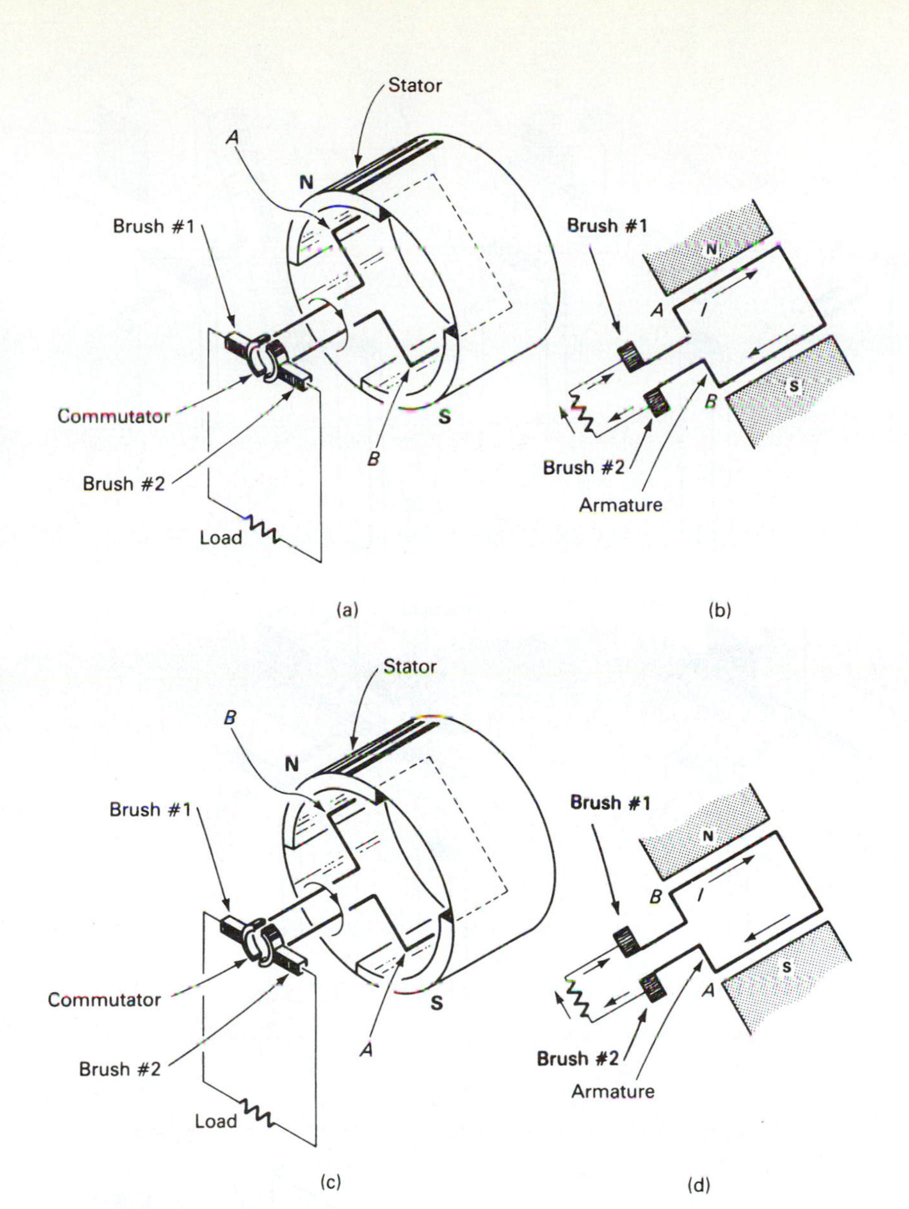

Figure 19.22
Current in a dc generator.

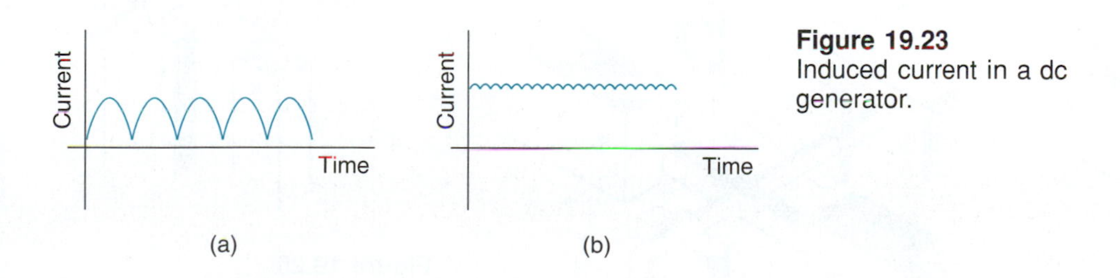

Figure 19.23
Induced current in a dc generator.

tational motion can be used to do work. This device is called a **motor.** A motor converts electric energy to mechanical energy and thus performs the reverse function of a generator (Fig. 19.25).

If ac current is supplied to the electromagnet, slip rings are used instead of a commutator. The use of alternating current makes the commutator unnecessary. The changes in direction are supplied by the ac current itself (Fig. 19.26).

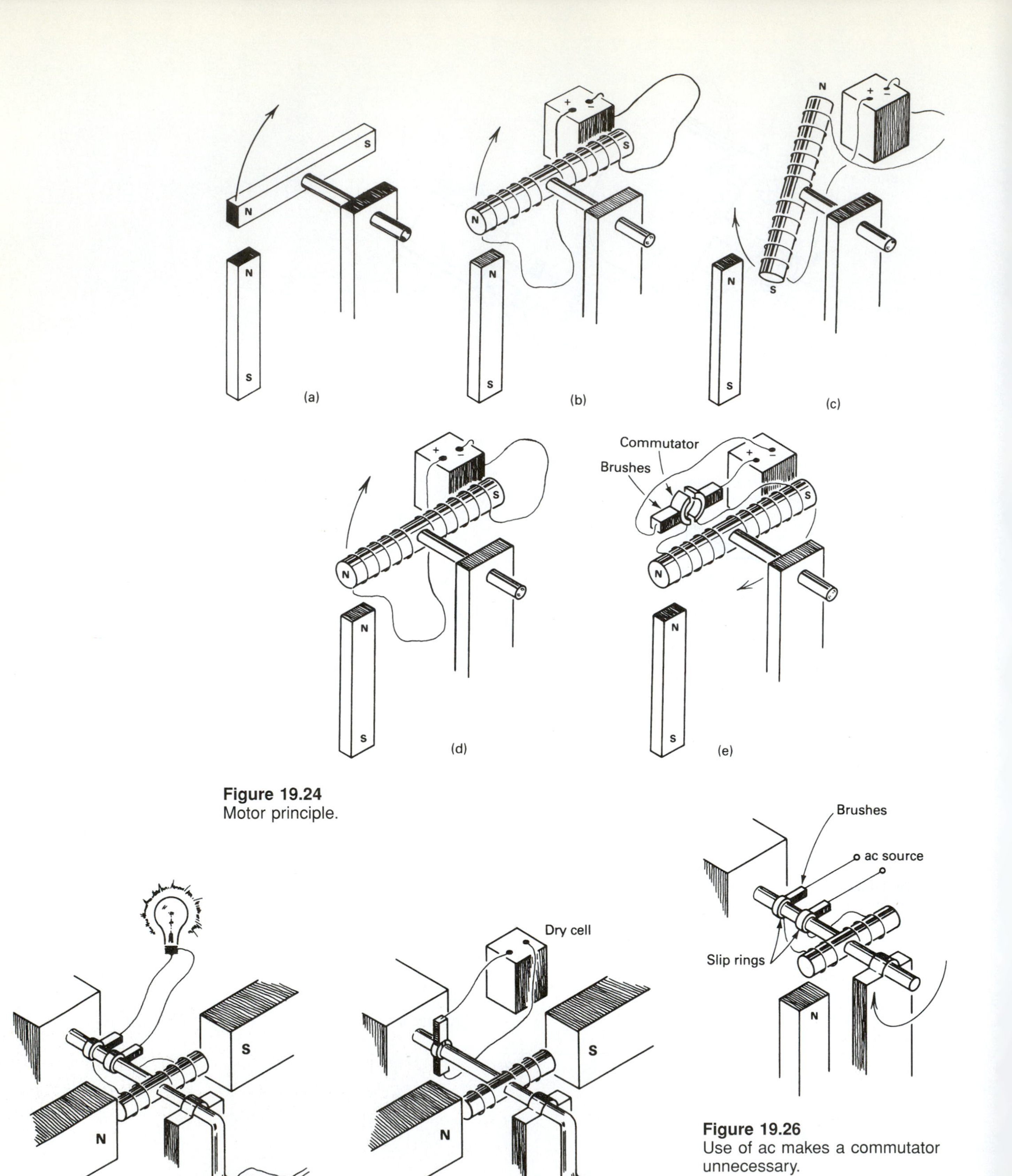

Figure 19.24
Motor principle.

Figure 19.26
Use of ac makes a commutator unnecessary.

Figure 19.25
A motor performs the reverse function of a generator.

Industrial Motors

Commercial motors operate in the same way as the motors just discussed. However, they usually use electromagnets in place of the permanent magnets and are much more complex. Slip rings are not necessary in ac motors. The current in the rotating electromagnet can be induced in the same way a current is induced in a generator.

Motors can be designed for many different purposes. Heavily loaded motors need certain types of starters. The torque and power outputs can be greatly varied by differences in design. Several types of ac motors are discussed here.

1. The **universal motor** (Fig. 19.27) can be run on either ac or dc power. Slip rings are used in this type of motor as in dc motors. This motor is often used in small hand tools and appliances.
2. The **induction motor** (Fig. 19.28) is the most widely used ac motor. The rotating electromagnet is not connected to a power source by slip rings. Instead, the current in the electromagnet is induced by a moving magnetic field caused by the ac current.

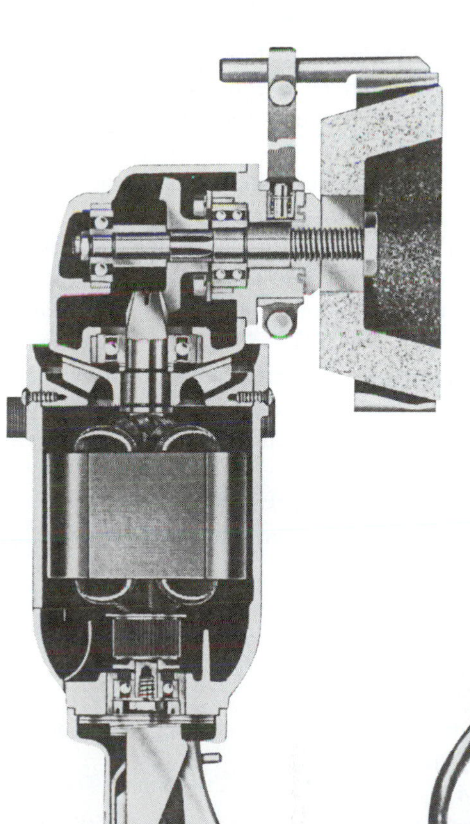

Figure 19.27
Universal motor used in a grinder. (Courtesy of Thor Power Tool Company.)

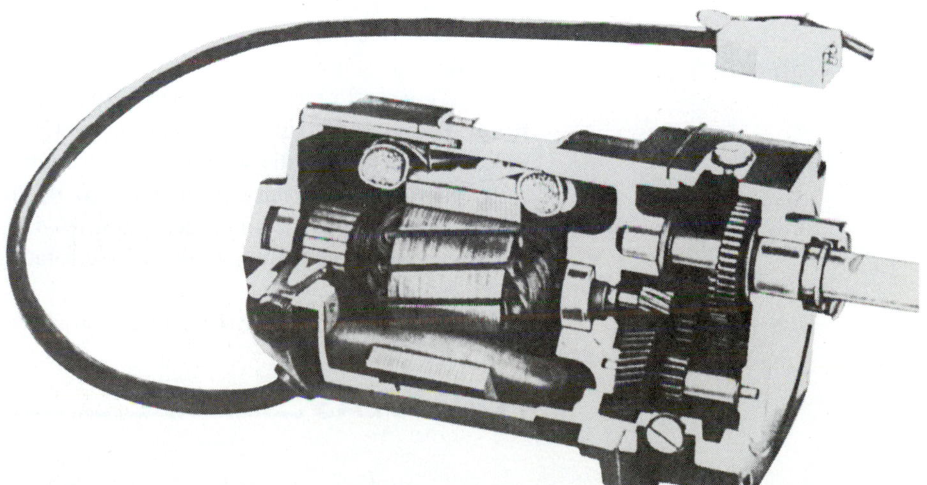

Figure 19.28
Induction motor. (Courtesy of Bodine Electric Company.)

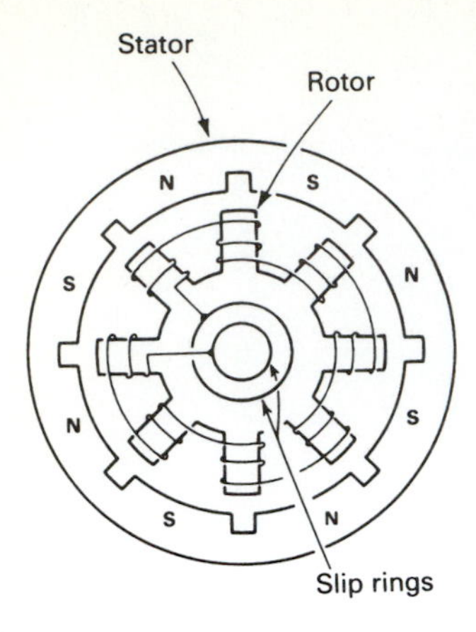

Figure 19.29
Synchronous motor.

3. The **synchronous motor** (Fig. 19.29) is very similar to the slip ring ac motor discussed earlier. The rotating electromagnet is supplied with current through slip rings. The speed of rotation of a synchronous motor is constant and depends on the number of coils and on the frequency of the power supply. The synchronous motor will work only when operated with an ac power source of the frequency for which it is designed. The word *synchronous* is derived from *syn,* meaning same, and *chrona,* or time. Synchronous motors are used to operate clocks and other devices needing accurate speed control.

GLOSSARY

Ampère's Rule States that the flux lines around a current-carrying wire circle the wire in the direction that your right-hand fingers would point if you held the conductor in your hand with your thumb extended in the direction of the current. (p. 452)
Armature The rotating coil or electromagnet in a generator. (p. 459)
Commutator A device in a generator that is used to produce a direct current. Composed of a split ring that replaces the slip rings in an ac generator and produces a direct current in the circuit connected to the split ring of the generator. (p. 459)
Compass A small magnetic needle that is free to rotate on a bearing. The needle's north pole points to the south magnetic pole of the earth. (p. 449)
Electromagnet A combination of a solenoid and a magnetic material, such as iron, in the core of the solenoid. When a current is passed through the solenoid, the magnetic fields of the atoms in the magnetic material line up to produce a strong magnetic field. (p. 456)
Flux Lines Lines indicating the direction of the magnetic field near a magnetic pole. (p. 450)
Generator An apparatus consisting of a coil of wire rotating in a magnetic field. A current is induced in the coil converting mechanical energy into electrical energy. (p. 457)
Induced Current A current produced in a circuit by the emf produced by motion of the circuit through the flux lines of a magnetic field. (p. 457)
Induced Magnetism Magnetism produced in a magnetic material such as iron when the material is placed in a magnetic field, such as that produced by a solenoid. (p. 456)
Induction Motor An ac motor with an electromagnetic current induced by the moving magnetic field of the ac current. (p. 463)
Magnetic Property of metals or other materials that can attract iron. (p. 449)
Magnetic Field A field of force near a magnetic pole or a current that can be detected using a magnet. (p. 450)
Motor A device that is composed of an armature and a stator. The armature rotates in the field of the stator when a current is passed through the armature. Used to convert electrical energy to mechanical energy. (p. 461)
Rotor The rotating coil in a generator. (p. 459)
Solenoid A coil of tightly wrapped wire. Commonly used to create a strong magnetic field by passing current through the wire. (p. 455)
Stator The field magnets in a generator. (p. 459)
Synchronous Motor An ac motor whose rotating electromagnet is supplied with current by slip rings. (p. 464)
Universal Motor A motor that can be run on either ac or dc power. (p. 463)

FORMULAS

19.2 $B = \dfrac{\mu_0 I}{2\pi R}$

$B = \mu_0 I n$

REVIEW QUESTIONS

1. The presence of a magnetic force field may be detected by using
   (a) a compass. (b) iron filings.
   (c) a magnet. (d) all of the above.
2. The deflection of a compass needle placed near a current-carrying wire shows
   (a) the magnetic field of the sun.
   (b) the magnetic field of the wire.
   (c) the electric field.
3. Ampère's rule relates
   (a) the strength of a magnetic field to the magnetic pole.
   (b) the direction of a magnetic field surrounding a current-carrying wire.
   (c) the direction of a magnetic field near a bar magnet.
   (d) none of the above.
4. The unit used to express the strength of a magnetic field is the _____.
5. Describe how a strong magnetic field can be produced in a solenoid.
6. Describe how to determine the direction of a magnetic field in a solenoid.
7. Describe how a magnetic field is induced by a current-carrying coil surrounding a core of magnetic material.
8. Describe how a generator produces current.
9. Describe the function of a commutator.
10. Describe how a motor works.
11. What is a synchronous motor, and how does it work?
12. Distinguish between a universal motor and an induction motor.
13. Distinguish between an armature and a stator.
14. Describe how an electromagnet works.
15. If the current in a solenoid is increased by a factor of 2, how does the magnetic field change?
16. If the radius of a solenoid decreases by a factor of 2, how does the magnetic field change?
17. If the number of turns per inch in a solenoid were increased by a factor of 4, how would the magnetic field change?
18. Describe how to find the flux lines near a bar magnet.
19. How is alternating current produced by a generator?

REVIEW PROBLEMS

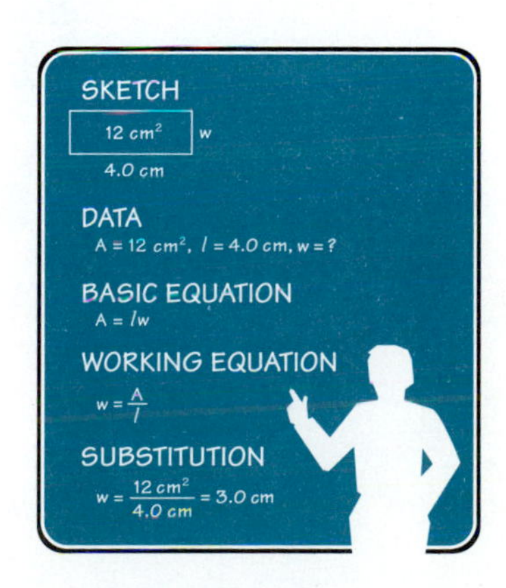

1. Find the magnetic field at 0.255 m from a long wire carrying a current of 1.38 A.
2. Find the magnetic field at 0.365 m from a long wire carrying a current of 8.95 A.
3. What is the current in a wire if the magnetic field is 4.75×10^{-6} T at a distance of 1.75 m from the wire?
4. A solenoid has $20\bar{0}0$ turns of wire, is 0.452 m long, and carries a current of 4.55 A. What is the magnetic field at the center of the solenoid?
5. A solenoid has 2750 turns of wire and is 0.182 m long. What current is required to produce a magnetic field of 0.235 T at the center of the solenoid?
6. A power line (running north–south) carrying $50\bar{0}$ A is 7.00 m above a transit used by a surveyor. What error is induced in the compass used by the surveyor? (Assume the earth's horizontal component of magnetic field is 5.20×10^{-5} T.)

Chapter 20

ALTERNATING CURRENT ELECTRICITY

Ordinary household current is alternating current. The current produced by rotating a loop through a magnetic field is alternating current and constantly changing. We now consider the nature and characteristics of alternating current, transformation of voltage, and various devices used in electrical circuits. The digital circuits in computers use a form of alternating current that produces two voltage levels (high and low, which represent the 0's and 1's of binary logic).

OBJECTIVES

The major goals of this chapter are to enable you to:

1. Express the nature and characteristics of alternating current.
2. Understand the use of transformers in changing voltage.
3. Apply inductance and inductive reactance in circuits.
4. Use capacitance and capacitive reactance in circuits.

20.1 WHAT IS ALTERNATING CURRENT?

In Chapter 19 we learned that the current produced by rotating a loop of wire through a magnetic field alternates in direction. This current is called an *alternating current.* Alternating current is used much more frequently in industry and everyday experience than is direct current. The main reason for this is that transmission of power over long distances is more efficient with alternating current than with direct current.

As its name implies, **alternating current** (ac) is current that flows in one direction in a conductor, then changes direction and flows in the other direction. The direction of flow changes many times in one second. Ordinary household current is 60-Hz current. This means that the voltage goes through 60 complete cycles, positive to negative and back again each second.

Every time the current repeats itself—flows, changes direction, flows, and changes direction—it goes through one *cycle* (Fig. 20.1). The reason for this alternation is that this is how current comes from electric generators. The emf and

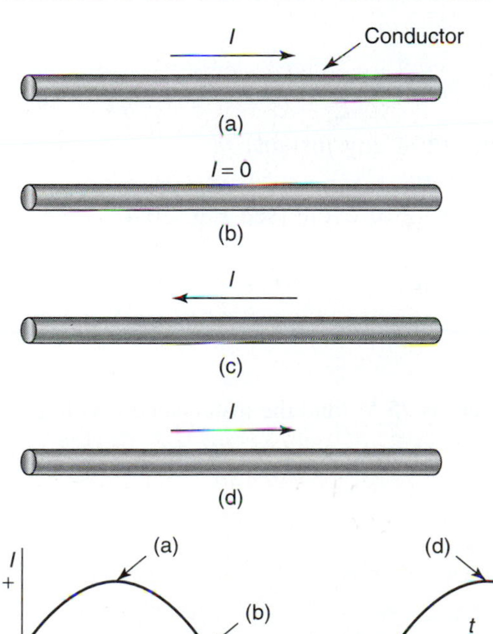

Figure 20.1
Alternating current changes direction as it flows in a conductor.

current produced by a generator do not alternate instantly between maximum values in each direction, but they build up to maximum values and then decrease, change direction, and build to maximum values in the other direction (Fig. 20.2).

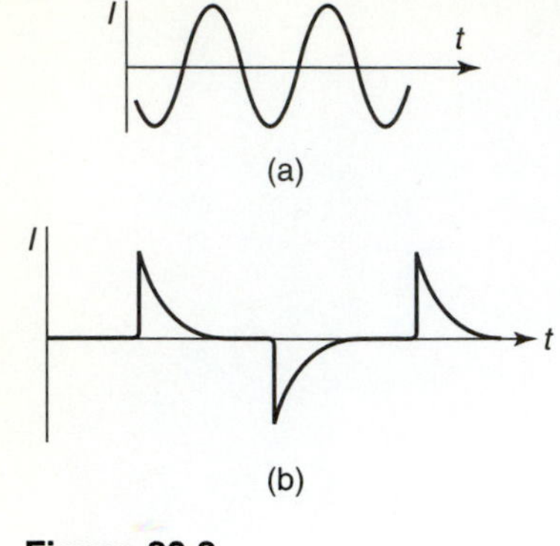

Figure 20.2
Alternating current does not alternate instantly between maximum values but builds to maximum values as in part (a). Current in a computer digital circuit varies as shown in part (b).

Direct current is usually a steady flow at a constant value. Graphically, it can be represented as shown in Fig. 20.3. Alternating current, however, is constantly changing. To graphically represent ac, we must show that it builds up and drops off. This can be demonstrated by the curve shown in Fig. 20.4, called a *sine curve*. We form the curve by rotating a vector **V** about a point and plotting the vertical components of **V.** Rotating **V** through 360° graphs one cycle.

The graph in Fig. 20.4 shows the ac-current curve. A graph of ac voltage is also a sine curve. The current and voltage of ac are constantly changing. **Instantaneous current** is the current at any instant of time; **instantaneous voltage** is the voltage at any instant of time. We can find the value of current or voltage at any instant by using the fact that each makes a sine curve (Fig. 20.5).

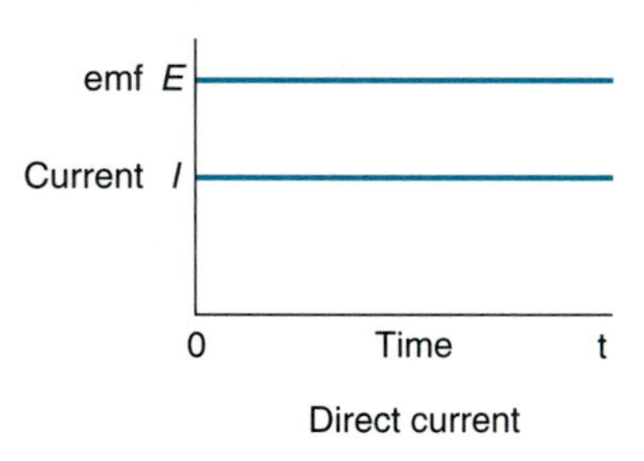

Figure 20.3
Direct current does not alternate but is steady.

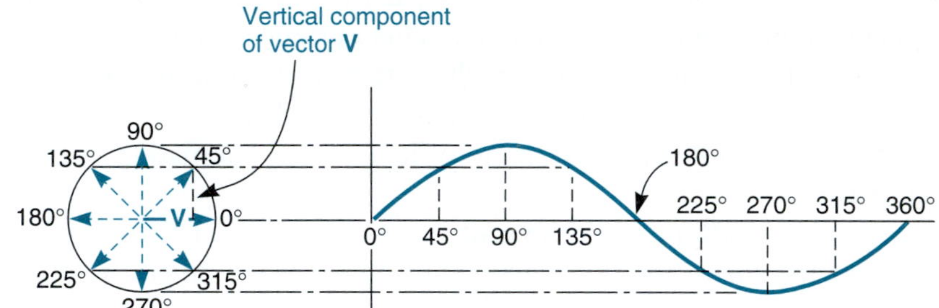

Figure 20.4
Graph of alternating current.

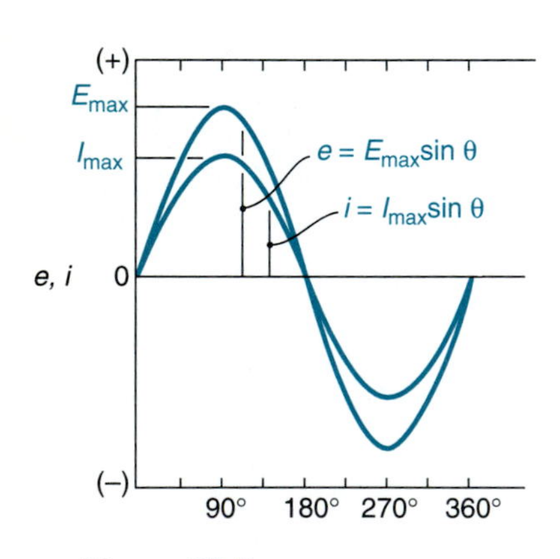

Figure 20.5
Instantaneous values of current and voltage.

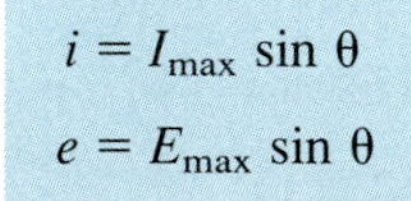

$$i = I_{max} \sin \theta$$

$$e = E_{max} \sin \theta$$

where i = instantaneous current (current at any instant)
I_{max} = maximum instantaneous current
θ = angle measured from beginning of cycle (see Fig. 20.4)
e = instantaneous voltage
E_{max} = maximum instantaneous voltage

EXAMPLE

The maximum voltage in an alternating current is 75 V. Find the instantaneous voltage at $\theta = 35°$.

Data:

$$E_{max} = 75 \text{ V}$$
$$\theta = 35°$$
$$e = ?$$

Basic Equation:

$$e = E_{max} \sin \theta$$

Working Equation: Same

Substitution:

$$e = (75 \text{ V})(\sin 35°)$$
$$= 43 \text{ V}$$

Both curves in Fig. 20.6(a) show e and i reaching a maximum at the same time and falling to zero at the same time. When this occurs, they are said to be "in phase." When there is only resistance in the circuit, e and i are in phase in electrical circuits. Later we will study some ac components that will cause e and i to be out of phase [see Fig. 20.6(b)].

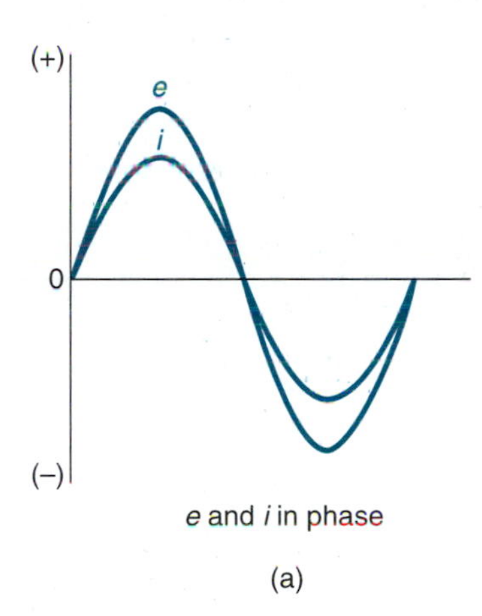

e and *i* in phase

(a)

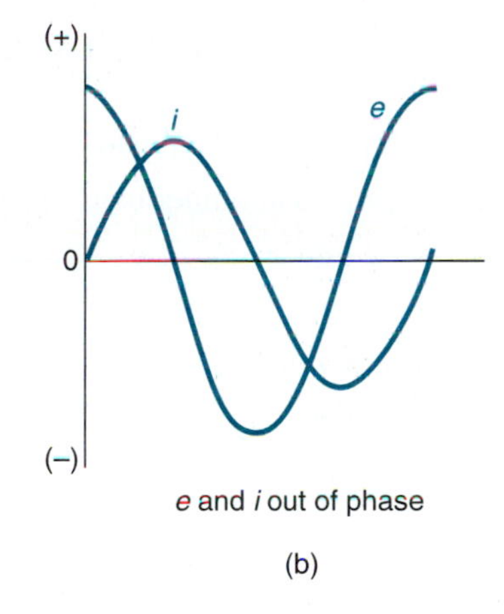

e and *i* out of phase

(b)

Figure 20.6

■ PROBLEMS 20.1

1. What is the maximum voltage in an ac circuit in which the instantaneous voltage at $\theta = 35.0°$ is 27.0 V?
2. The instantaneous voltage in an ac circuit at $\theta = 65.0°$ is 82.0 V. What is the maximum voltage?
3. If the maximum ac voltage on a line is 165 V, what is the instantaneous voltage at $\theta = 45.0°$?
4. The maximum current in an ac circuit is 8.00 A. Find the instantaneous current at $\theta = 60.0°$.
5. The instantaneous current in an ac circuit is 6.50 A at $\theta = 45.0°$. Find the maximum current.
6. What is the maximum voltage in an ac circuit where the instantaneous voltage at $\theta = 51.0°$ is 14.5 V?
7. If the maximum ac voltage on a line is 145 V, what is the instantaneous voltage at $\theta = 35.0°$?
8. The maximum current in an ac circuit is 5.75 A. What is the instantaneous current at $\theta = 80.0°$?
9. Find the maximum current in an ac circuit where the instantaneous current at $\theta = 45.0°$ is 4.00 A.
10. The instantaneous voltage in an ac circuit at $\theta = 55.0°$ is $45\bar{0}$ V. Find the maximum voltage.
11. If the ratio of I/I_{max} is 0.45 at a given time, what is θ?
12. If $I = 1.23$ A and $I_{max} = 3.41$ A, what is θ?
13. If $I_{max} = 4.59$ A and $I = 4.32$ A, what is θ? ■

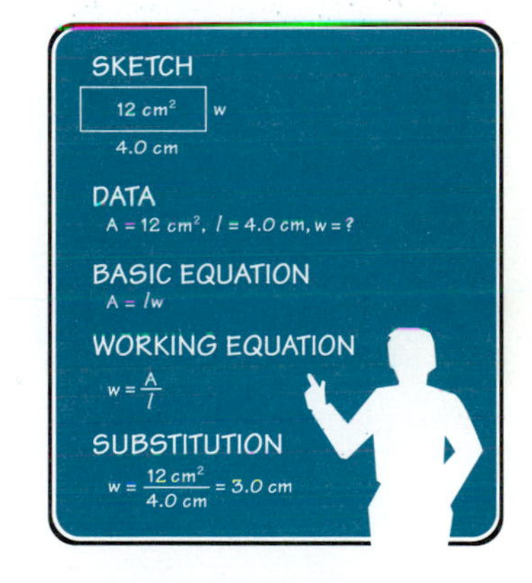

20.2 EFFECTIVE VALUES OF ac

A direct measurement of ac is difficult because it is constantly changing. The most useful value of ac is based on its heating effect and is called its **effective value.** The effective value of an alternating current is the number of amperes that produce the same amount of heat in a resistance as an equal number of amperes of a steady direct current.

The numerical factors in the following equations are derived from an average of the sine-wave time variation of the ac current. When this time average is taken, the factors $\sqrt{2} = 1.414$ and $1/\sqrt{2} = 0.707$ are found.

$$I = 0.707\ I_{\text{max}}$$
$$I_{\text{max}} = 1.41\ I$$

where I = effective value of current (sometimes called rms value)
I_{max} = maximum instantaneous current

EXAMPLE

The current supplied to a woodworking shop is rated at 10.0 A. What is the maximum value of the current supplied?

Data:

$$I = 10.0 \text{ A}$$
$$I_{\text{max}} = ?$$

Basic Equation:

$$I_{\text{max}} = 1.41\ I$$

Working Equation: Same

Substitution:

$$I_{\text{max}} = 1.41\ (10.0 \text{ A}) = 14.1 \text{ A}$$

The effective value for ac voltage may be expressed similarly:

$$E = 0.707\ E_{\text{max}}$$
$$E_{\text{max}} = 1.41\ E$$

where E = effective value of voltage
E_{max} = maximum instantaneous voltage

When we say a house is wired for 120 volts, we are using the effective value of the voltage. Actually, the voltage varies between +170 V and −170 V during each cycle.

Unless otherwise stated, ac voltage and current are *always* expressed in terms of effective or rms values.

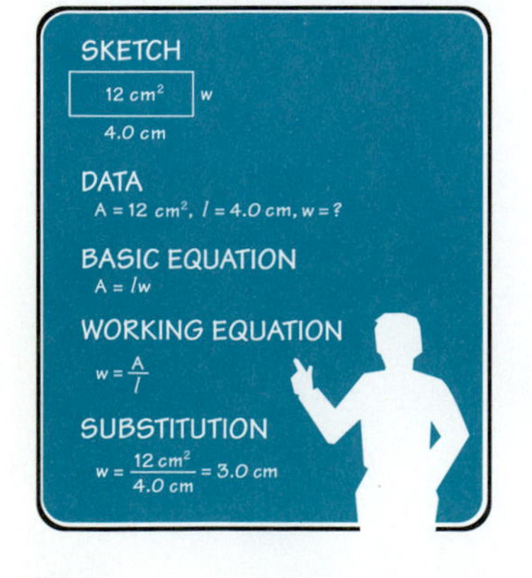

■ PROBLEMS 20.2

1. Find the effective value of an ac voltage whose maximum voltage is 2250 V.
2. Find the maximum current in an ac circuit with a current of 6.00 A.

3. Find the effective value of an ac voltage whose maximum voltage is 165 V.
4. Find the maximum current in an ac circuit with a current of 4.00 A.
5. Find the effective value of a current in an ac circuit that reaches a maximum of 17.0 A.
6. Find the effective value of an ac voltage whose maximum voltage is 1150 V.
7. Find the maximum current in an ac circuit with a current of 8.50 A.
8. Find the effective value of an ac voltage whose maximum voltage is 135 V.
9. Find the maximum current in an ac circuit with a current of 7.00 A.
10. Find the effective value of a current in an ac circuit that reaches a maximum of 125 A.

20.3 ac POWER

When the load has only resistance, power in ac circuits is found in the same way as in dc circuits.

$$P = I^2R$$

or

$$P = VI \qquad \text{(using } V = IR\text{)}$$

or

$$P = \frac{V^2}{R} \qquad \text{(using } I = V/R\text{)}$$

EXAMPLE 1

What power is expended in a resistance of 37.0 Ω if it has a current of 0.480 A flowing through it?

Data:

$$R = 37.0\ \Omega$$
$$I = 0.480\ \text{A}$$
$$P = ?$$

Basic Equation:

$$P = I^2R$$

Working Equation: Same

Substitution:

$$P = (0.480\ \text{A})^2(37.0\ \Omega)$$
$$= 8.52\ \text{W}$$

EXAMPLE 2

What power is expended in a load of 12.0 Ω resistance if the voltage drop across it is $11\bar{0}$ V?

Data:

$$R = 12.0\ \Omega$$
$$V = 11\bar{0}\ \text{V}$$
$$P = ?$$

Basic Equation:

$$P = \frac{V^2}{R}$$

Working Equation: Same

Substitution:

$$P = \frac{(11\bar{0}\text{ V})^2}{12.0\ \Omega}$$

$$= 1010\text{ W} \quad \text{or} \quad 1.01\text{ kW}$$

The preceding relationships are true only when e and i are in phase. Phase differences produced by capacitance and inductance in an ac circuit are due to reactance. Capacitance, inductance, and reactance will be studied later.

Note that, in the graphs comparing dc and ac power (Fig. 20.7), ac power varies but is always positive (+). The sign indicates only the direction of the current. Even so, p is positive in calculations because the product of $-e$ and $-i$ is positive: $p = (-e)(-i) = ei$.

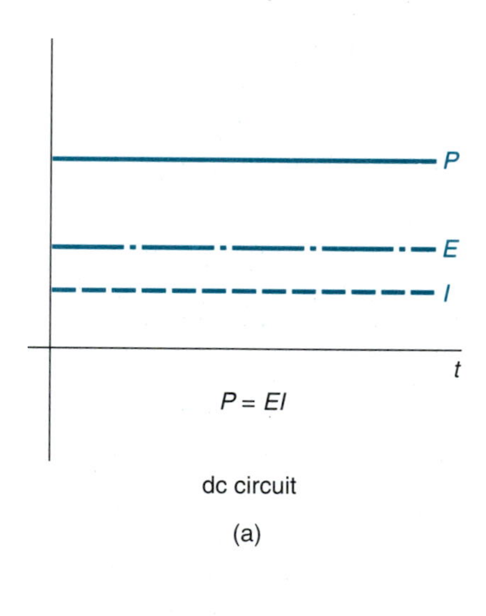

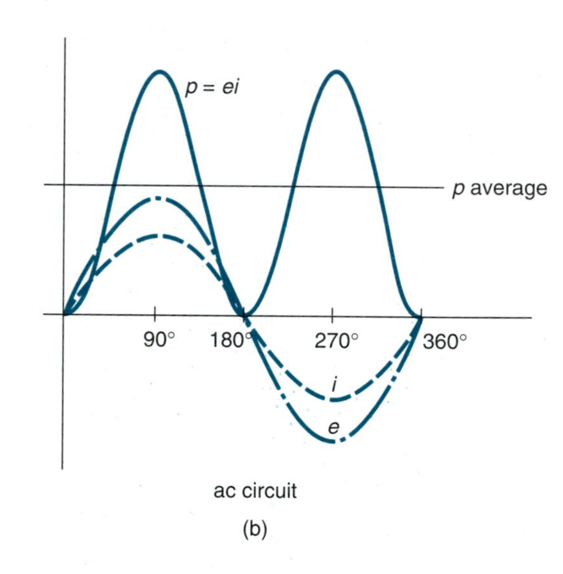

Figure 20.7
dc and ac power compared.

SKETCH
12 cm²
w
4.0 cm
DATA
A = 12 cm², l = 4.0 cm, w = ?
BASIC EQUATION
A = lw
WORKING EQUATION
w = A/l
SUBSTITUTION
w = 12 cm²/4.0 cm = 3.0 cm

■ PROBLEMS 20.3

1. A soldering iron is rated at 35$\bar{0}$ W. If the current in the iron is 4.00 A, what is the resistance of the iron?
2. What power is developed by a device that draws 6.00 A and has a resistance of 12.0 Ω?
3. Find the output power of a transformer that has an output voltage of 50$\bar{0}$ V and current of 7.00 A.
4. A heater operates on a 11$\bar{0}$-V line and is rated at 45$\bar{0}$ W. What is the resistance of the element?
5. A heating element draws 6.00 A on a 22$\bar{0}$-V line. What power is expended in the element?
6. A resistance coil has a resistance of 32.0 Ω. If it expends 375 W of power, what is the current in the coil?
7. What power is used by a heater that has a resistance of 12.0 Ω and draws a current of 7.00 A?

8. A heater operates on a $11\bar{0}$-V line and is rated at $75\bar{0}$ W. What is the resistance of the heater element?
9. A resistance coil has a resistance of $11\bar{0}$ Ω. If it draws a current of 5.00 A, what power is dissipated?
10. What power is used by a heater that has a resistance of 19.5 Ω and draws a current of 5.55 A?

20.4 TRANSFORMERS AND POWER

The uses of direct current in industry are somewhat limited. Primary applications are in charging storage batteries, electroplating, the generation of alternating current, electrolysis, electromagnets, and automobile ignition systems. In all cases, however, ac can be changed to dc by a simple device called a **rectifier.**

Much more can be done with alternating current. From the kitchen toaster to the largest industrial motors, ac finds wide application. There are very practical reasons for this. The voltage of ac can be easily and efficiently changed in transformers to give almost any desired values. Actually, ac can be used for most purposes just as efficiently as dc. One advantage of ac is that it can be transmitted over large distances with very little heat loss. Heat lost in any electrical device is found by the formula:

$$\text{heat loss (power) } P = I^2R$$

The energy wasted as heat can be reduced by making the current smaller. Transformers reduce the current by increasing the voltage, since

$$P = \frac{E^2}{R}$$

The major advantage of ac over dc is that ac voltage can easily be changed to meet our needs.

EXAMPLE 1

A plant generates 50.0 kW ($50,\bar{0}00$ W) of power to be sent to a substation on a line with a resistance of 3.00 Ω. We know that some power will be lost as heat during the transmission. The power lost is $P_{\text{lost}} = I^2R$.

(a) How much power is lost if the transmission is at 1150 V?
(b) What percent of the power generated is lost in transmission at 1150 V?
(c) How much power is lost if the transmission is at 11,500 V?
(d) What percent of the power generated is lost in transmission at 11,500 V?
(e) Compare the power losses at the two different transmission voltages.

At 1150 V:

(a)

$$P = VI$$

$$I = \frac{P}{V}$$

$$I = \frac{50,\bar{0}00 \text{ W}}{1150 \text{ V}}$$

$$= 43.5 \text{ A}$$

$$P_{\text{lost}} = I^2R$$
$$P_{\text{lost}} = (43.5\text{ A})^2(3.00\ \Omega)$$
$$= 5680\text{ W}$$
$$= 5.68\text{ kW}$$

(b)

$$\%_{\text{lost}} = \frac{\text{power lost}}{\text{power generated}} \times 100\%$$
$$\%_{\text{lost}} = \frac{5.68\text{ kW}}{50.0\text{ kW}} \times 100\% = 11.4\%$$

At 11,500 V:

(c)

$$P = VI$$
$$I = \frac{P}{V}$$
$$I = \frac{50,\bar{0}00\text{ W}}{11,500\text{ V}}$$
$$= 4.35\text{ A}$$
$$P_{\text{lost}} = I^2R$$
$$P_{\text{lost}} = (4.35\text{ A})^2(3.00\ \Omega)$$
$$= 56.8\text{ W}$$
$$= 0.0568\text{ kW}$$

(d)

$$\%_{\text{lost}} = \frac{\text{power lost}}{\text{power generated}} \times 100\%$$
$$\%_{\text{lost}} = \frac{0.0568\text{ kW}}{50.0\text{ kW}} \times 100\% = 0.114\%$$

(e) This example shows that whereas 11.4% of the power is lost during transmission at 1150 V, only 0.114% is lost at 11,500 V, so by increasing the voltage, the current is correspondingly lowered, and the power wasted in transmission is greatly reduced.

Changing Voltage with Transformers

The transformer is a device that is used to change the voltage to reduce the current and thereby lessen the power loss. A **transformer** consists of two coils of wire wrapped on an iron core (Fig. 20.8). It is used to change the voltage of electricity. When an alternating current passes through the **primary coil,** it induces an alternating magnetic field in the core (Fig. 20.9). This magnetic field in turn induces an alternating voltage in the **secondary coil.** The magnitude of the voltage induced in the secondary coil depends on

1. The voltage applied to the primary coil.
2. The number of turns in the primary coil.

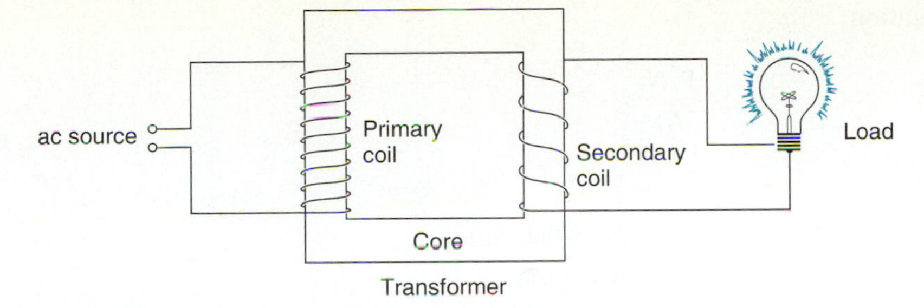

Figure 20.8
Basic components of a transformer.

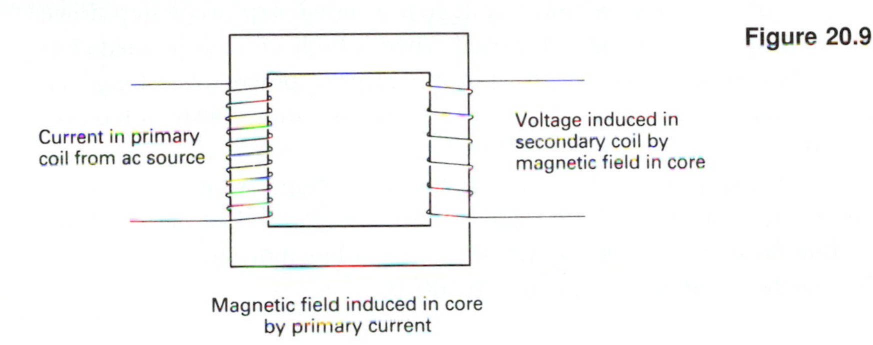

Figure 20.9

3. The number of turns in the secondary coil.
4. The power lost between primary and secondary coils.

If we assume no power loss between the primary and secondary coils, we have the following equation:

$$\frac{V_P}{V_S} = \frac{N_P}{N_S}$$

where V_P = primary voltage
V_S = secondary voltage
N_P = number of primary turns
N_S = number of secondary turns

EXAMPLE 2

A transformer on a neon sign has $10\bar{0}$ turns in its primary coil and $15,\bar{0}00$ turns in its secondary coil. If the voltage applied to the primary is $11\bar{0}$ V, what is the secondary voltage?

Data:

$$V_P = 11\bar{0}\text{ V}$$
$$N_P = 10\bar{0}\text{ turns}$$
$$N_S = 15,\bar{0}00\text{ turns}$$
$$V_S = ?$$

Basic Equation:

$$\frac{V_P}{V_S} = \frac{N_P}{N_S}$$

Working Equation:

$$V_S = \frac{V_P N_S}{N_P}$$

Substitution:

$$V_S = \frac{(11\bar{0}\text{ V})(15,\bar{0}00\text{ turns})}{10\bar{0}\text{ turns}}$$

$$= 16,500\text{ V} \quad \text{or} \quad 16.5\text{ kV}$$

Transformers used to raise or lower voltage are called step-up or step-down transformers. **Step-up transformers** are used when a high voltage is needed to operate X-ray tubes, neon signs, and to transmit electric power over long distances. A step-up transformer raises the voltage by having more turns in the secondary than in the primary coil [Fig. 20.10(a)].

Step-down transformers are used to lower the voltage from high-voltage transmission lines to regular 110 V and 220 V for home and industrial use. Voltage is lowered in the step-down transformer because it has more turns in the primary coil than in the secondary coil [Fig. 20.10(b)].

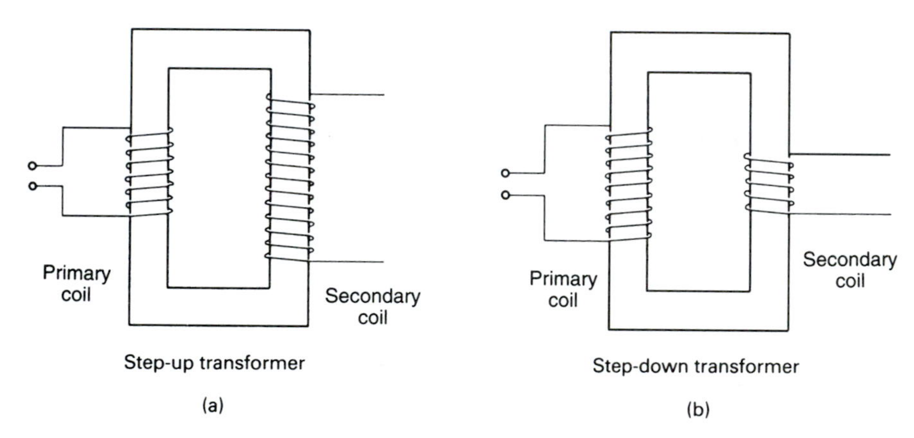

Figure 20.10

Auto transformers are used when a variable output voltage is needed. In this type of transformer, contact can be made across a variable number of the secondary coils using a brush contact. The output voltage is therefore variable from nearly zero to some maximum value. This type of transformer is often used to supply ac power to resistive heater elements to control the heating output.

Transformers do not create energy. In fact, some energy is lost during the change of voltage. Energy losses in transformers are of three types:

1. *Copper losses.* These result from the resistance of the copper wires in the coils and are unavoidable.
2. *Magnetic losses* (called **hysteresis losses**). Some energy is lost (turned into heat) by reversing the magnetism in the core.
3. *Eddy currents.* When a mass of metal (the core) is subjected to a changing magnetic field, **eddy currents** are set up in the metal that do no useful work, waste energy, and produce heat. These losses can be lessened by *laminating* the core. Instead of using a solid block of metal for the core [Fig. 20.11(a)],

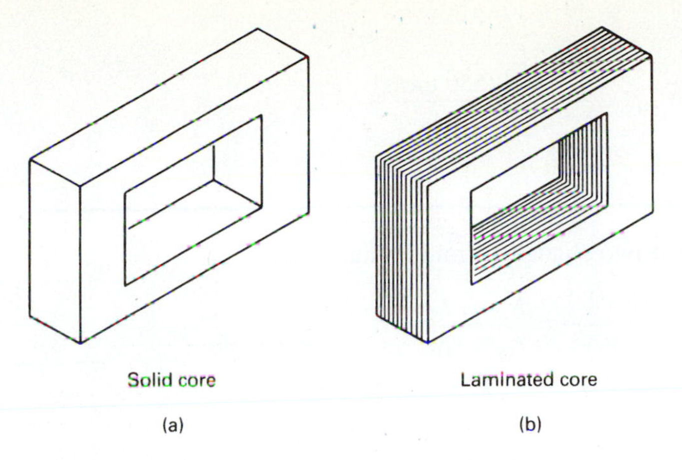

Figure 20.11

thin sheets of metal with insulated surfaces are used [Fig. 20.11(b)], reducing these induced currents. These ac eddy currents cause the laminations to vibrate, producing the characteristic transformer "hum."

When a transformer steps up the voltage applied to its primary, it reduces the current. Energy is conserved—we cannot get any more electrical energy out of a transformer than we put into it. The relationship between primary and secondary currents is

$$\frac{I_S}{I_P} = \frac{N_P}{N_S}$$

where I_S = current in secondary coil
I_P = current in primary coil
N_P = number of turns in primary
N_S = number of turns in secondary

EXAMPLE 3

The primary current in a transformer is 10.0 A. If the primary coil has $55\bar{0}$ turns and the secondary has $25\bar{0}0$ turns, what current flows in the secondary coil?

Data:

$$N_P = 55\bar{0} \text{ turns}$$
$$I_P = 10.0 \text{ A}$$
$$N_S = 25\bar{0}0 \text{ turns}$$
$$I_S = ?$$

Basic Equation:

$$\frac{I_S}{I_P} = \frac{N_P}{N_S}$$

Working Equation:

$$I_S = \frac{I_P N_P}{N_S}$$

Substitution:

$$I_S = \frac{(10.0\text{ A})(55\bar{0}\text{ turns})}{25\bar{0}0\text{ turns}}$$
$$= 2.20\text{ A}$$

It follows from the last two shaded formulas that

$$\frac{V_P}{V_S} = \frac{N_P}{N_S} = \frac{I_S}{I_P}$$

so

$$\frac{V_P}{V_S} = \frac{I_S}{I_P}$$

From this we obtain,

$$I_P V_P = I_S V_S$$

or, using $P = IV$,

$$P_P = P_S$$

which shows that power is conserved between the primary and secondary coils under these assumptions.

EXAMPLE 4

The power in the primary coil of a transformer is 375 W. If the current in the secondary is 11.4 A, what is the voltage in the secondary?

Data:

$$P_P = 375\text{ W}$$
$$I_S = 11.4\text{ A}$$
$$V_S = ?$$

Basic Equation:

$$P_P = P_S \qquad \text{and} \qquad P_S = V_S I_S$$

Working Equations:

$$P_P = V_S I_S \qquad (\textit{Note:}\ \text{Substitute for } P_S).$$
$$V_S = \frac{P_P}{I_S}$$

Substitution:

$$V_S = \frac{375\text{ W}}{11.4\text{ A}}$$
$$= 32.9\text{ V}$$

Good transformers are more than 98% efficient. This is very important in power transmission. It is impractical to generate electricity at high voltage, but high voltage is desirable for transmission. Therefore, transformers are used to step up the voltage for transmission. High voltage is unsuitable, though, for consumer use, so transformers are used to reduce the voltage. A simplified diagram of a power distribution system is shown in Fig. 20.12.

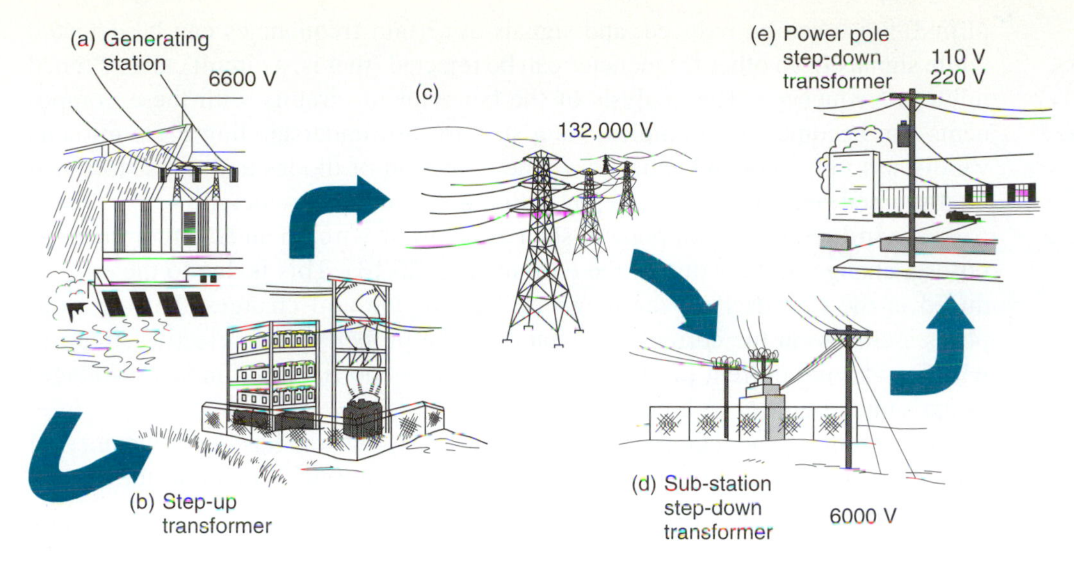

Figure 20.12
Power distribution system.

■ PROBLEMS 20.4

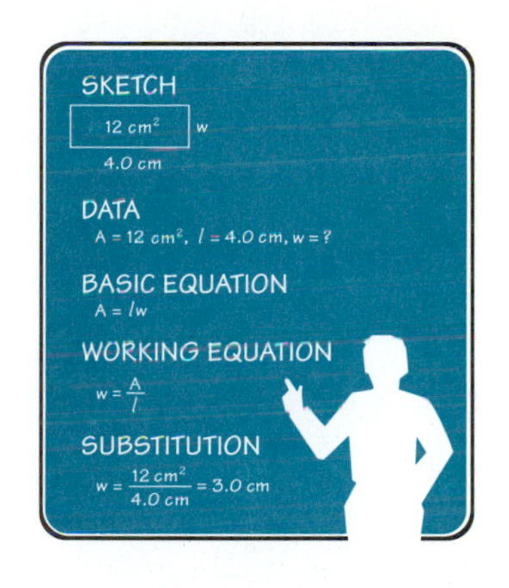

1.	$V_P = 30.0$ V $V_S = 45.0$ V $N_S = 15.0$ turns Find N_P.	2.	$V_P = 25\overline{0}$ V $N_P = 73\overline{0}$ turns $N_S = 275$ turns Find V_S.	3.	$I_P = 6.00$ A $I_S = 4.00$ A $V_P = 39.0$ V Find V_S.

4. A step-up transformer on a 115-V line provides a voltage of $23\overline{0}0$ V. If the primary coil has 65.0 turns, how many turns does the secondary have?
5. A step-down transformer on a 115-V line provides a voltage of 11.5 V. If the secondary coil has 30.0 turns, how many turns does the primary have?
6. A transformer has 20.0 turns in the primary coil and $22\overline{0}0$ turns in the secondary. If the primary voltage is 12.0 V, what is the secondary voltage?
7. If there is a current of 9.00 A in the primary in Problem 3, find the current in the secondary.
8. If the voltage in the secondary coil of a transformer is $11\overline{0}$ V and the current in it is 15.0 A, what power does it supply?
9. A neon sign has a transformer that changes electricity from $11\overline{0}$ V to $15,\overline{0}00$ V. If the primary current is 8.00 A, find the current in the secondary coil.
10. Find the power in the primary in Problem 9.
11. A transformer has an output power of 990 W. If the current in the secondary coil is 0.45 A, what is the voltage in the secondary?
12. The current in the secondary coil of a transformer is 5.00 A. Find the voltage in the secondary if the power is 775 W.
13. A transformer steps down $66\overline{0}0$ V to $12\overline{0}$ V. If the secondary current is 14.0 A, what is the primary current?
14. Find the power in the primary coil in Problem 13. ■

20.5 INDUCTANCE

Electronic circuitry in televisions, radios, computers, and electronic instruments has many components other than resistors. These components include capacitors, inductors, diodes, and transistors. With these components weak signals can be am-

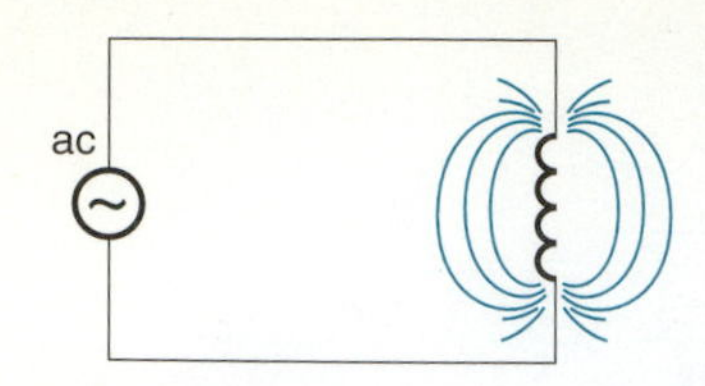

Figure 20.13
Induced emf opposing current change in an ac circuit.

plified, noise can be reduced, and signals at certain frequencies can be detected while signals from other frequencies can be rejected (that is, a circuit can be "tuned in" to a frequency). The analysis of the behavior of circuits with these components can become very complex. As a start toward understanding these circuits, we discuss inductors and capacitors. The operation of diodes and transistors will be discussed only briefly. We begin with a discussion of inductors.

An **inductor** is a component, such as a coil of wire, in an ac circuit that opposes a change in the value of the current (Fig. 20.13). This is due to the emf induced in the coil itself as the magnetic field of the coil changes. This emf opposes a change in the current. Inductance is the property of an electric circuit in which a varying current produces a varying magnetic field that induces voltages in the same circuit or in a nearby circuit.

The unit of inductance *(L)* is the henry (H). A coil has an inductance of 1 henry if an emf of 1 volt is induced when the current changes at the rate of 1 A/s. A henry can be expressed as Ω s.

$$1 \text{ henry} = 1\ \Omega\ \text{s}$$

The henry is a large unit. A more practical unit is the millihenry (mH), which is one one-thousandth of a henry.

Inductance can be illustrated by connecting a coil with a large number of turns and a lamp in series. When connected to a dc source, the lamp burns brightly [Fig. 20.14(a)]. However, when this circuit is connected to an ac power source of the same voltage, the lamp is dimmer because of the inductance of the coil [Fig. 20.14(b)]. The circuit symbol for inductance is shown in Fig. 20.15.

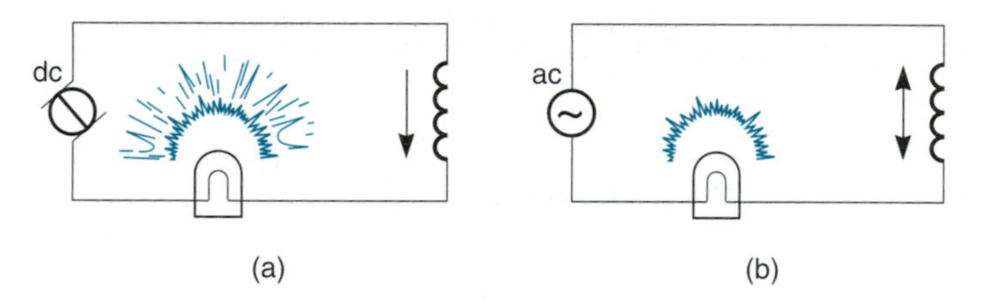

Figure 20.14
A coil in an ac circuit produces inductance.

L

Figure 20.15
Circuit diagram symbol for inductance.

20.6 INDUCTIVE REACTANCE

The opposition to ac current flow in an inductor is called **inductive reactance** and is measured in ohms. This is usually represented by X_L. The inductive reactance of a coil is directly proportional to frequency and is found by the following:

$$X_L = 2\pi f L$$

where X_L = inductive reactance
f = frequency of the ac voltage, expressed in hertz (cycles per second), such as 60 Hz or 60/s
L = inductance, in henries

If inductance is given in mH, a conversion must be made to H (henries).

The current in a circuit that has only an ac voltage source and an inductor is given by

$$I = \frac{E}{X_L}$$

where I = current
E = voltage
X_L = inductive reactance

EXAMPLE

A coil with inductance 1.00 mH is connected to a 60.0-kHz ac power source of $11\bar{0}$ V. What is the current in the circuit?

Sketch:

$E = 11\bar{0}$ V
$f = 60.0$ kHz
$L = 1.00$ mH

Data:

$E = 11\bar{0}$ V
$L = 1.00 \text{ mH} = 1.00 \times 10^{-3}$ H
$f = 60.0 \text{ kHz} = 60.0 \times 10^3 \text{ Hz} = 6.00 \times 10^4/\text{s}$
$I = ?$

Basic Equations:

$$X_L = 2\pi f L \quad \text{and} \quad I = \frac{E}{X_L}$$

Working Equations: Same

Substitutions:

$$\begin{aligned} X_L &= 2\pi(6.00 \times 10^4/\text{s})(1.00 \times 10^{-3}\ \text{H}) \\ &= 377\ \frac{\text{H}}{\text{s}} \\ &= 377\ \frac{\not{\text{H}}}{\not{\text{s}}}\left(\frac{1\ \Omega\ \not{\text{s}}}{1\ \not{\text{H}}}\right) \quad \text{(note conversion factor)} \\ &= 377\ \Omega \end{aligned}$$

$$\begin{aligned} I &= \frac{E}{X_L} \\ I &= \frac{11\bar{0}\ \text{V}}{377\ \Omega} \\ &= 0.292\ \frac{\not{\text{V}}}{\not{\Omega}}\left(\frac{\text{A}\ \not{\Omega}}{\not{\text{V}}}\right) \quad \text{(note conversion factor)} \\ &= 0.292\ \text{A} \end{aligned}$$

In an inductor the current lags behind the voltage by a quarter of a cycle (Fig. 20.16). The maximum voltage in a 60-Hz circuit thus occurs

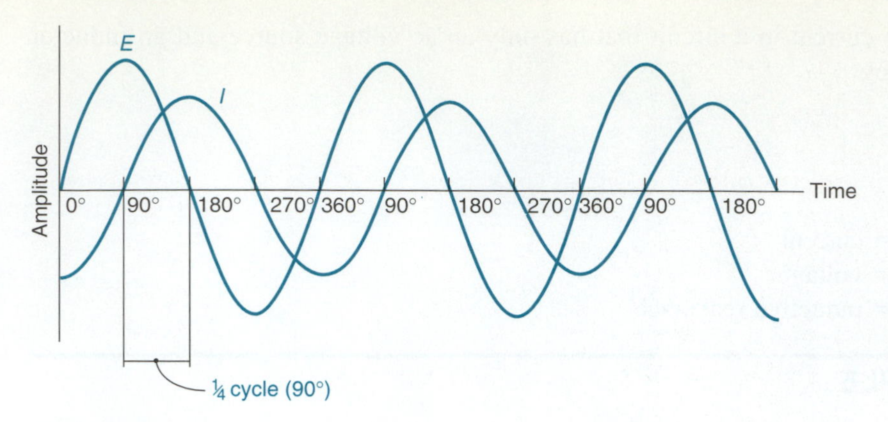

Figure 20.16
In an inductive circuit the current lags behind the voltage by one fourth of a cycle.

$$\frac{1}{4} \times \frac{1}{60}\text{ s} = \frac{1}{240}\text{ s}$$

before the maximum current. The current lag is usually measured in degrees. A quarter of a cycle is 90°.

The impedance of an inductor increases with frequency.

SKETCH
12 cm²
w
4.0 cm
DATA
A = 12 cm², l = 4.0 cm, w = ?
BASIC EQUATION
A = lw
WORKING EQUATION
$w = \frac{A}{l}$
SUBSTITUTION
$w = \frac{12\text{ cm}^2}{4.0\text{ cm}} = 3.0\text{ cm}$

■ PROBLEMS 20.6

Find the inductive reactance (in ohms) of each inductance at the given frequency.

1. $L = 3.00$ mH, $f = 60.0$ Hz
2. $L = 20.0$ mH, $f = 75.0$ Hz
3. $L = 70.0$ mH, $f = 10.0$ kHz
4. $L = 8.00$ mH, $f = 8.00$ kHz
5. $L = 425$ μH, $f = 15.0$ MHz
6. $L = 655$ μH, $f = 125$ MHz

Find the current (in amperes) in each inductive circuit.

7. $L = 30.0$ mH, $f = 125$ Hz, $E = 14.0$ V
8. $L = 1.00$ mH, $f = 125$ kHz, $E = 145$ V
9. $L = 5.00$ mH, $f = 2.00$ kHz, $E = 50.0$ V
10. $L = 30.0$ mH, $f = 7.00$ MHz, $E = 75.0$ V
11. $L = 72.0$ μH, $f = 2.00$ MHz, $E = 105$ V
12. $L = 525$ μH, $f = 25.0$ MHz, $E = 65.0$ V ■

20.7 INDUCTANCE AND RESISTANCE IN SERIES

Most ac circuits have resistance in the form of lights or resistors in addition to inductance (Fig. 20.17). The current lags behind the voltage by any amount of time greater than zero and as large as a quarter cycle.

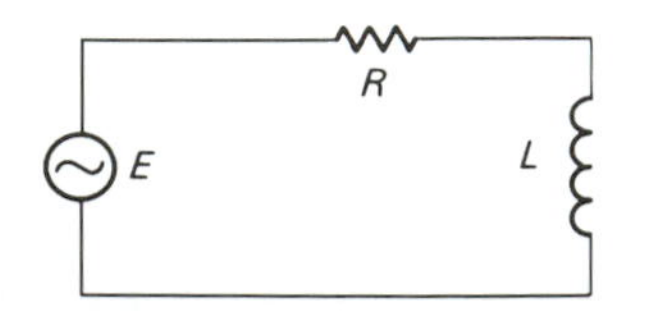

Figure 20.17
ac current with resistance and inductance.

The effect of both the resistance and the inductance on a circuit is called the **impedance.** Ohm's law in an ac circuit can be written as

$$I = \frac{E}{Z}$$

where I = current
E = voltage
Z = impedance

The impedance of a series circuit containing a resistance and an inductance is

$$Z = \sqrt{R^2 + X_L^2}$$
$$Z = \sqrt{R^2 + (2\pi f L)^2}$$

where Z = impedance
R = resistance
X_L = inductive reactance
f = frequency
L = inductance

The impedance can be represented as a vector as the hypotenuse of the right triangle shown in Fig. 20.18. The resistance is always drawn as a vector pointing in the positive x-direction. The inductive reactance is drawn as a vector pointing in the positive y-direction. The angle ϕ shown is the **phase angle** and equals the amount by which the current lags behind the voltage. The phase angle is given by

$$\tan \phi = \frac{X_L}{R}$$

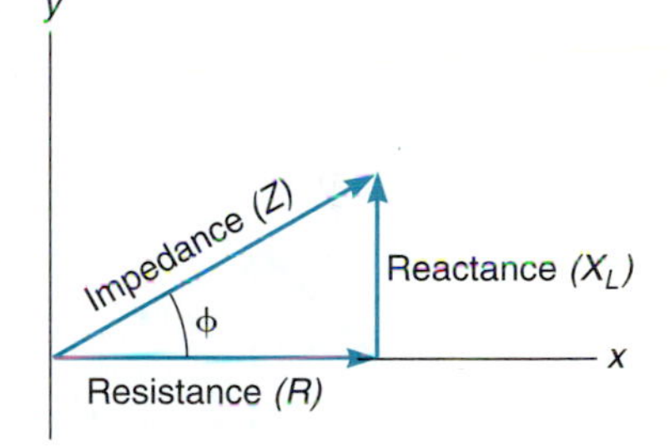

Figure 20.18
Graphic representation of impedance.

EXAMPLE

A lamp of resistance 40.0 Ω is connected in series with an inductance of 95.0 mH. This circuit is connected to a 115-V, 60.0-Hz power supply. (a) What is the current in the circuit? (b) What is the phase angle?

Sketch:

$R = 40.0\ \Omega$
$E = 115$ V
$f = 60.0$ Hz
$L = 95.0$ mH

Data:

(a)

$E = 115$ V
$f = 60.0$ Hz $= 60.0$/s
$R = 40.0\ \Omega$
$L = 95.0$ mH $= 95.0 \times 10^{-3}$ H $= 0.0950$ H
$Z = ?$
$I = ?$

Basic Equations:

$$Z = \sqrt{R^2 + (2\pi f L)^2} \quad \text{and} \quad I = \frac{E}{Z}$$

Working Equations: Same

Substitutions: First calculate the impedance.

$$Z = \sqrt{(40.0\ \Omega)^2 + [2\pi(60.0/\text{s})(0.0950\ \text{H})]^2}$$
$$= \sqrt{16\bar{0}0\ \Omega^2 + (35.8\ \text{H/s})^2}$$
$$= \sqrt{16\bar{0}0\ \Omega^2 + \left[\left(35.8\ \frac{\text{H}}{\text{s}}\right)\left(\frac{1\ \Omega\ \text{s}}{1\ \text{H}}\right)\right]^2} \quad \text{(note conversion factor)}$$
$$= \sqrt{16\bar{0}0\ \Omega^2 + 1280\ \Omega^2}$$
$$= \sqrt{2880\ \Omega^2}$$
$$= 53.7\ \Omega$$
$$I = \frac{E}{Z}$$
$$I = \frac{115\ \text{V}}{53.7\ \Omega}\left(\frac{1\ \text{A}\ \Omega}{1\ \text{V}}\right)$$
$$= 2.14\ \text{A}$$

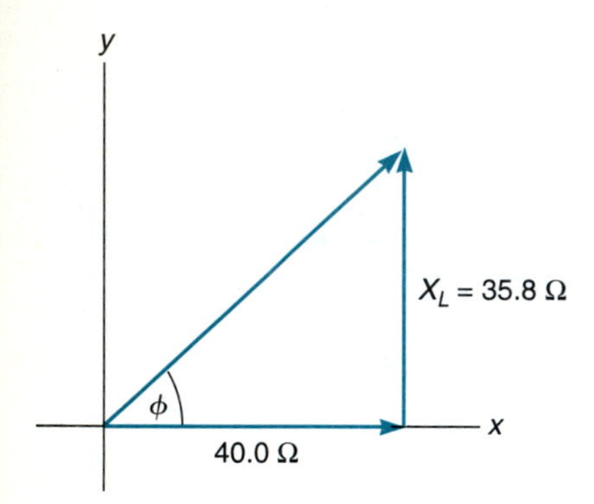

Figure 20.19

(b) To find the phase angle ϕ, first find X_L. Then, construct the vector right triangle as in Fig. 20.19 to find ϕ.

Data:

$$f = 60.0/\text{s}$$
$$L = 0.0950\ \text{H}$$
$$X_L = ?$$
$$\phi = ?$$

Basic Equations:

$$X_L = 2\pi f L \qquad \text{and} \qquad \tan\phi = \frac{X_L}{R}$$

Working Equations: Same

Substitutions:

$$X_L = 2\pi(60.0/\text{s})(0.0950\ \text{H})$$
$$= 35.8\ \frac{\text{H}}{\text{s}}$$
$$= 35.8\ \frac{\text{H}}{\text{s}}\left(\frac{1\ \Omega\ \text{s}}{1\ \text{H}}\right)$$
$$= 35.8\ \Omega$$

[***Note:*** This value can also be taken directly from the first working equation in part (a).]

Then, from Fig. 20.19 we have

$$\tan\phi = \frac{35.8\ \Omega}{40.0\ \Omega} = 0.895$$
$$\phi = 41.8^\circ$$

PROBLEMS 20.7

Find the impedance (in ohms) of each circuit.

1. $R = 20\bar{0}\ \Omega$, $L = 10.0$ mH, $f = 1.25$ kHz
2. $R = 12.0\ \Omega$, $L = 1.00$ mH, $f = 90\bar{0}$ Hz
3. $R = 1.00\ \text{k}\Omega$, $L = 50.0$ mH, $f = 10.0$ kHz
4. $R = 2.00\ \text{k}\Omega$, $L = 70.0$ mH, $f = 5.00$ kHz
5. $R = 30\bar{0}\ \Omega$, $L = 2.00$ mH, $f = 3.00$ kHz
6. Find the phase angle in Problem 1.
7. Find the phase angle in Problem 2.
8. Find the phase angle in Problem 3.
9. Find the phase angle in Problem 4.
10. Find the phase angle in Problem 5.
11. Find the current in Problem 1 if the voltage is 45.0 V.
12. Find the current in Problem 2 if the voltage is 10.0 V.
13. Find the current in Problem 3 if the voltage is 15.0 V.
14. Find the current in Problem 4 if the voltage is 12.0 V.
15. Find the current in Problem 5 if the voltage is 6.00 V.

20.8 CAPACITANCE

An important component of many ac circuits is the capacitor. A **capacitor** consists of two conductors that are usually parallel plates separated by a thin insulator. The plates are often made of a metal foil rolled to a convenient size. Capacitors are represented in circuit diagrams, as shown in Fig. 20.20.

Figure 20.20
Circuit diagram symbol for a capacitor.

The unique property of a capacitor is that it can build up and store charge. When a capacitor is connected to a battery, electrons flow from the negative terminal to one capacitor plate as shown in Fig. 20.21(b). When the capacitor is removed from the battery, the charges remain on the capacitor. If the capacitor is then connected to a resistor [Fig. 20.21(c)], electrons will flow through the circuit until the capacitor has lost its charge.

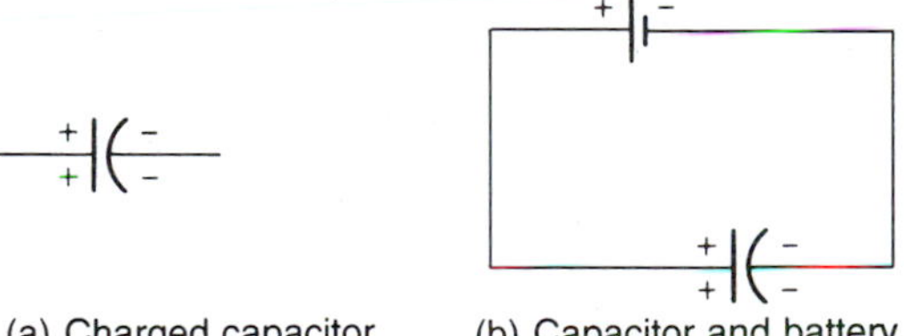

(a) Charged capacitor.

(b) Capacitor and battery in a circuit.

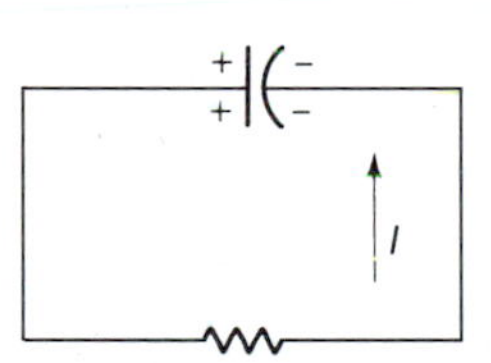

(c) An electron current flows when the capacitor is connected to a resistance.

Figure 20.21

A capacitor will block a direct current from flowing in a circuit once the capacitor is charged. A low-frequency ac voltage in a capacitive circuit (Fig. 20.22) will cause only a small current to flow because of this blocking nature of a capacitor. A high-frequency ac voltage source will cause a larger current to flow in the capacitive circuit of Fig. 20.22 because a current is required to quickly change the polarity of the capacitor voltage. Capacitors can therefore be used to tune the frequency response of circuits, allowing the blocking of low-frequency electrical

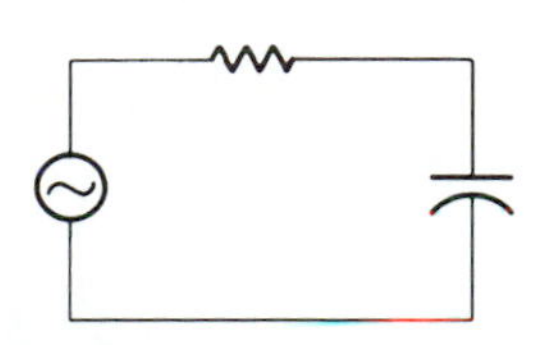

Figure 20.22
Capacitor in an ac circuit.

signals and tuning the resonance frequency of circuits as described in Section 20.11.

The unit of capacitance is the *farad* (F). A more practical unit is the microfarad (μF or 10^{-6} F). The effect of a capacitor on a circuit is inversely proportional to frequency and is measured as **capacitive reactance**. It is measured in ohms and given by

$$X_C = \frac{1}{2\pi f C}$$

where X_C = capacitive reactance
f = frequency
C = capacitance (farads)

$$1 \text{ F} = 1 \text{ s}/\Omega$$

In a circuit that contains only capacitors, the current *leads* the voltage by 90° (one-quarter cycle) as shown in Fig. 20.23.

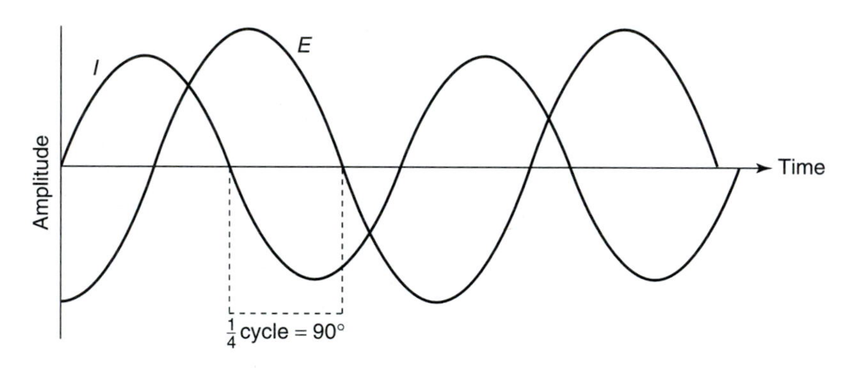

Figure 20.23
Current leads the voltage by one-quarter cycle.

EXAMPLE

Find the capacitive reactance of a 10.0-μF capacitor in a circuit of frequency 1.00 kHz.

Data:

$$C = 10.0\ \mu\text{F} = 10.0 \times 10^{-6}\ \text{F} = 1.00 \times 10^{-5}\ \text{F}$$
$$f = 1.00\ \text{kHz} = 1.00 \times 10^{3}/\text{s}$$
$$X_C = ?$$

Basic Equation:

$$X_C = \frac{1}{2\pi f C}$$

Working Equation: Same

Substitution:

$$X_C = \frac{1}{2\pi(1.00 \times 10^{3}/\text{s})(1.00 \times 10^{-5}\ \text{F})}$$

$$= \frac{1}{\left(6.28 \times 10^{-2} \frac{\text{F}}{\not{s}}\right)\left(\frac{1 \not{s}}{1 \not{F}\,\Omega}\right)} \quad \text{(note conversion factor)}$$

$$= 15.9\ \Omega$$

■ PROBLEMS 20.8

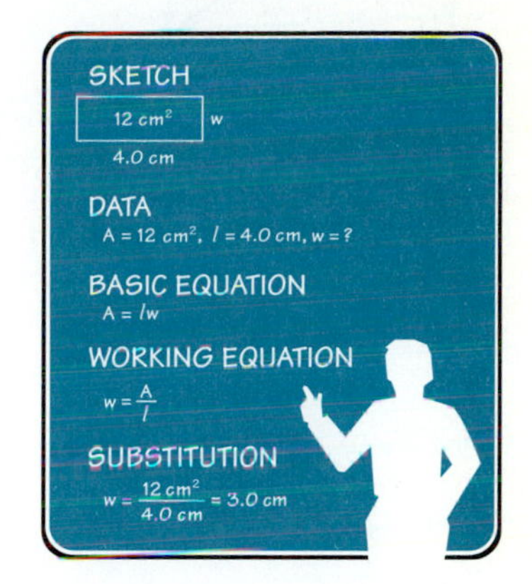

Find the capacitive reactance (in ohms) in each ac circuit.

1. $C = 20.0\ \mu\text{F}, f = 1.00$ kHz
2. $C = 7.00$ mF, $f = 10\bar{0}$ Hz
3. $C = 0.600\ \mu\text{F}, f = 0.100$ kHz
4. $C = 30.0$ mF, $f = 2.50$ MHz
5. $C = 0.800\ \mu\text{F}, f = 0.250$ MHz
6. Find the capacitive reactance of a 15.0-μF capacitor in a circuit of frequency 60.0 Hz.
7. Find the capacitive reactance of a 45.0-μF capacitor in a circuit of frequency 60.0 kHz.
8. Find the capacitive reactance of a 6.00-mF capacitor in a circuit with frequency $10\bar{0}$ Hz.
9. Find the capacitive reactance of a $33\bar{0}$-μF capacitor in a circuit with frequency $30\bar{0}$ Hz.
10. Find the capacitive reactance of a 222-μF capacitor in a circuit of frequency $12\bar{0}$ Hz. ■

20.9 CAPACITANCE AND RESISTANCE IN SERIES

The combined effect of capacitance and resistance in series is measured by the impedance, Z, of the circuit.

$$Z = \sqrt{R^2 + X_C^2}$$

$$Z = \sqrt{R^2 + \left(\frac{1}{2\pi fC}\right)^2}$$

where Z = impedance
R = resistance
X_C = capacitive reactance
f = frequency
C = capacitance

The current is given by Ohm's law:

$$I = \frac{E}{Z}$$

where I = current
E = voltage
Z = impedance

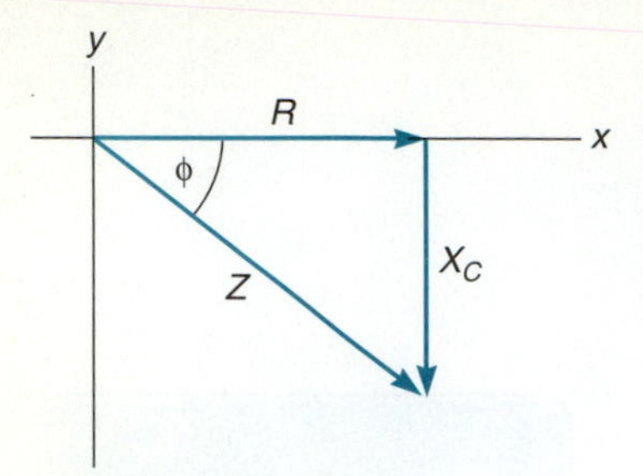

Figure 20.24
Determination of the phase angle in an ac circuit.

The phase angle can be found by drawing the resistance as a vector in the positive x-direction and the capacitive impedance as a vector in the negative y-direction as shown in Fig. 20.24. The phase angle gives the amount by which the voltage lags behind the current.

$$\tan \phi = \frac{X_C}{R}$$

EXAMPLE

What current will flow in a 60.0-Hz ac circuit that includes a $11\bar{0}$-V source, a capacitor of 90.0 μF, and a 16.0-Ω resistance in series? Also find the phase angle.

Sketch:

Sketch: $R = 16.0\ \Omega$; $C = 90.0\ \mu\text{F}$; $E = 11\bar{0}\ \text{V}$; $f = 60.0\ \text{Hz}$

Data:

$$E = 11\bar{0}\ \text{V}$$
$$f = 60.0\ \text{Hz} = 60.0/\text{s}$$
$$R = 16.0\ \Omega$$
$$C = 90.0\ \mu\text{F} = 90.0 \times 10^{-6}\ \text{F} = 9.00 \times 10^{-5}\ \text{F}$$
$$Z = ?$$
$$I = ?$$

Basic Equations:

$$Z = \sqrt{R^2 + \left(\frac{1}{2\pi fC}\right)^2} \quad \text{and} \quad I = \frac{E}{Z}$$

Working Equations: Same

Substitutions:

First, find Z:

$$\begin{aligned} Z &= \sqrt{(16.0\ \Omega)^2 + \left(\frac{1}{2\pi(60.0/\text{s})(9.00 \times 10^{-5}\ \text{F})}\right)^2} \\ &= \sqrt{256\ \Omega^2 + \left(\frac{1}{0.0339\ \text{F/s}}\right)^2} \\ &= \sqrt{256\ \Omega^2 + \left(29.5\ \frac{\text{s}}{\text{F}} \times \frac{1\ \Omega\ \text{F}}{1\ \text{s}}\right)^2} \\ &= \sqrt{256\ \Omega^2 + 87\bar{0}\ \Omega^2} \\ &= 33.6\ \Omega \end{aligned}$$

Then use Z to find I:

$$I = \frac{E}{Z}$$

$$I = \frac{11\bar{0}\text{ V}}{33.6\ \Omega}$$

$$= 3.27\ \frac{\text{V}}{\Omega}\left(\frac{1\ \Omega\text{ A}}{1\text{ V}}\right)$$

$$= 3.27\text{ A}$$

Then find the phase angle:

$$\tan\phi = \frac{X_C}{R}$$

$$\tan\phi = \frac{29.5\ \Omega}{16.0\ \Omega} = 1.84$$

$$\phi = 61.5°$$

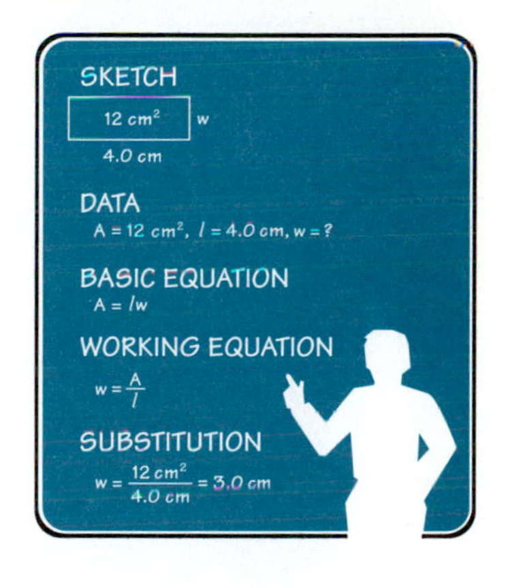

PROBLEMS 20.9

Find the impedance (in ohms) in each ac circuit.

1. $R = 1.00\text{ k}\Omega$, $C = 1.00\ \mu\text{F}$, $E = 10\bar{0}\text{ V}$, $f = 10\bar{0}\text{ Hz}$
2. $R = 375\ \Omega$, $C = 5.00\ \mu\text{F}$, $E = 20.0\text{ V}$, $f = 1.00\text{ kHz}$
3. $R = 4.80\text{ k}\Omega$, $C = 45.0\ \mu\text{F}$, $E = 15.0\text{ V}$, $f = 1.75\text{ kHz}$
4. $R = 145\text{ m}\Omega$, $C = 10.0\ \mu\text{F}$, $E = 7.00\text{ mV}$, $f = 72.5\text{ kHz}$
5. $R = 10.0\text{ m}\Omega$, $C = 5.00\ \mu\text{F}$, $E = 15.0\text{ mV}$, $f = 10.0\text{ kHz}$
6. Find the phase angle in Problem 1.
7. Find the phase angle in Problem 2.
8. Find the phase angle in Problem 3.
9. Find the phase angle in Problem 4.
10. Find the phase angle in Problem 5.
11. Find the current in Problem 1.
12. Find the current in Problem 2.
13. Find the current in Problem 3.
14. Find the current in Problem 4.
15. Find the current in Problem 5.

20.10 CAPACITANCE, INDUCTANCE, AND RESISTANCE IN SERIES

Many circuits that are important in the design of electronic equipment contain all three types of circuit elements discussed in this chapter. The impedance of a circuit containing resistance, capacitance, and inductance in series can be found from the equation

$$Z = \sqrt{R^2 + (X_L - X_C)^2}$$

where Z = impedance
R = resistance

X_L = inductive reactance
X_C = capacitive reactance

The vector diagram for this type of circuit is shown in Fig. 20.25. The phsae angle is given by

$$\tan \phi = \frac{X_L - X_C}{R}$$

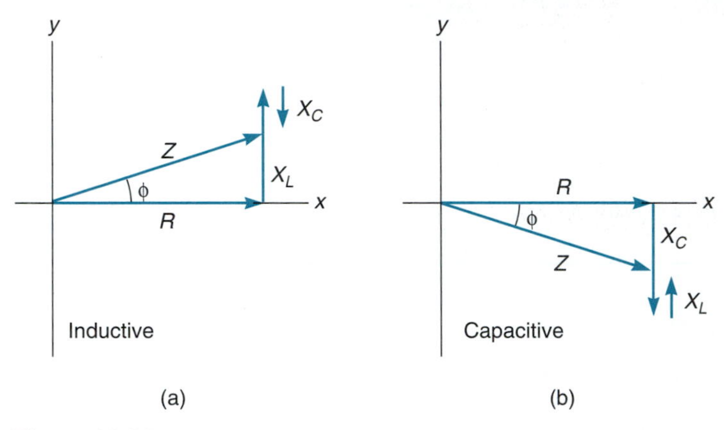

Figure 20.25
Vector diagrams: (a) inductive reactance; (b) capacitive reactance.

In a circuit containing R, L, and C components, the circuit is *inductive* if $X_L > X_C$ and the current lags behind the voltage. A circuit is *capacitive* if $X_C > X_L$, in which case the voltage lags the current. If $X_C = X_L$, the circuit is *resistive;* the voltage and current are *in phase*. If the circuit power is to be maximized, keep the voltage and current in phase. The current in this type of circuit is given by

$$I = \frac{E}{Z} = \frac{E}{\sqrt{R^2 + \left(2\pi f L - \frac{1}{2\pi f C}\right)^2}}$$

EXAMPLE

An ac circuit contains a $10\bar{0}$-Ω resistance, a 10.0-μF capacitor, and a 10.0-mH inductance in series with a 25.0-V, $20\bar{0}$-Hz source. Find the impedance and the current.

Sketch:

R = $10\bar{0}$ Ω
E = 25.0 V
f = $20\bar{0}$ Hz
L = 10.0 mH
C = 10.0 μF

Data:

$R = 10\bar{0}\ \Omega$
$C = 10.0\ \mu\text{F} = 10.0 \times 10^{-6}\ \text{F} = 1.00 \times 10^{-5}\ \text{F}$

$L = 10.0 \text{ mH} = 10.0 \times 10^{-3} \text{ H} = 1.00 \times 10^{-2} \text{ H}$

$E = 25.0 \text{ V}$

$f = 20\bar{0} \text{ Hz} = 20\bar{0}/\text{s}$

$Z = ?$

$I = ?$

Basic Equations:

$$X_L = 2\pi f L$$

$$X_C = \frac{1}{2\pi f C}$$

$$Z = \sqrt{R^2 + (X_L - X_C)^2}$$

$$I = \frac{E}{Z}$$

Working Equations: Same

Substitutions:

First find X_L:

$$X_L = 2\pi f L$$

$$X_L = 2\pi(20\bar{0}/\text{s})(1.00 \times 10^{-2} \text{ H})$$

$$= 12.6 \frac{\cancel{\text{H}}}{\cancel{\text{s}}}\left(\frac{1\ \Omega\ \cancel{\text{s}}}{1\ \cancel{\text{H}}}\right)$$

$$= 12.6\ \Omega$$

Then find X_C:

$$X_C = \frac{1}{2\pi f C}$$

$$X_C = \frac{1}{2\pi(20\bar{0}/\text{s})(1.00 \times 10^{-5} \text{ F})}$$

$$= 79.6\ \Omega$$

$$\frac{1}{\frac{\text{F}}{\text{s}}} = 1 \div \frac{\text{F}}{\text{s}} = 1 \times \frac{\text{s}}{\text{F}} = \frac{\text{s}}{\text{F}} \times \frac{1\ \Omega\ \text{F}}{1\ \text{s}} = \Omega$$

Then find the impedance:

$$Z = \sqrt{R^2 + (X_L - X_C)^2}$$

$$Z = \sqrt{(10\bar{0}\ \Omega)^2 + (12.6\ \Omega - 79.6\ \Omega)^2}$$

$$= \sqrt{(10\bar{0}\ \Omega)^2 + (-67.0\ \Omega)^2}$$

$$= \sqrt{1.00 \times 10^4\ \Omega^2 + 4490\ \Omega^2}$$

$$= 12\bar{0}\ \Omega$$

Then find the current:

$$I = \frac{E}{Z}$$

$$I = \frac{25.0 \text{ V}}{12\bar{0}\ \Omega}$$

$$= 0.208 \frac{\cancel{\text{V}}}{\cancel{\Omega}}\left(\frac{1\ \text{A}\ \cancel{\Omega}}{1\ \cancel{\text{V}}}\right)$$

$$= 0.208 \text{ A}$$

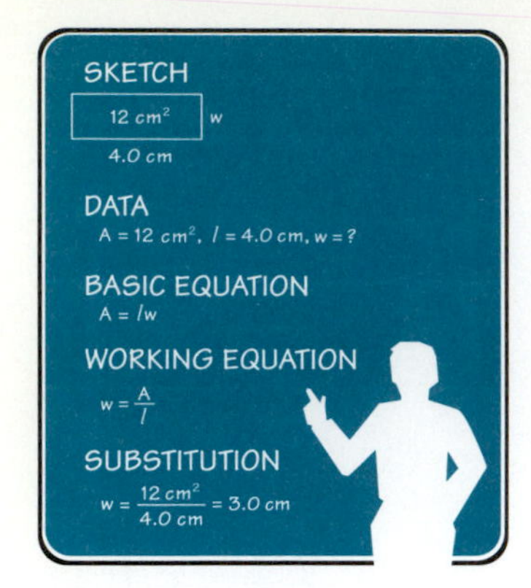

■ PROBLEMS 20.10

Find the impedance and current in each ac circuit.

1. $R = 25.0\ \Omega$, $L = 50.0$ mH, $C = 50.0\ \mu$F, $f = 60.0$ Hz, $E = 5.00$ V
2. $R = 225\ \Omega$, $L = 10.0$ mH, $C = 0.200\ \mu$F, $f = 1.00$ kHz, $E = 15.0$ V
3. $R = 1.00$ kΩ, $L = 10.0$ mH, $C = 30.0$ mF, $f = 10.0$ kHz, $E = 15.0$ V
4. $R = 1.00$ kΩ, $L = 0.700$ H, $C = 30.0\ \mu$F, $f = 60.0$ Hz, $E = 8.00$ V
5. A circuit contains a $15\bar{0}$-Ω resistance, a 35.0-μF capacitor, and a 0.600-H inductance in series with a 6.00-V, $12\bar{0}$-Hz source. Find the impedance and the current.
6. A circuit contains a 225-Ω resistance, a 5.00-μF capacitor, and a 0.550-H inductance in series with a 7.50-V, 60.0-Hz source. Find the impedance and the current.
7. A circuit contains a 175-Ω resistance, a 4.50-μF capacitor, and a 0.735-H inductance in series with a 5.00-V, $10\bar{0}$-Hz source. Find the impedance and the current.
8. A circuit contains a 575-Ω resistance, a $10\bar{0}$-μF capacitor, and a 0.400-H inductance in series with a $10\bar{0}$-V, $60\bar{0}$-Hz source. Find the impedance and the current.
9. A circuit contains a $45\bar{0}$-Ω resistance, a 35.0-μF capacitor, and a 45.0-mH inductance in series with a 25.0-V, 1.00-kHz source. Find the impedance and the current.
10. A circuit contains a 375-Ω resistance, a $50\bar{0}$-μF capacitor, and a 0.500-H inductance in series with a 55.0-V, $50\bar{0}$-Hz source. Find the impedance and the current. ■

20.11 RESONANCE

The current in a circuit containing resistance, capacitance, and inductance is given by the equation

$$I = \frac{E}{\sqrt{R^2 + (X_L - X_C)^2}}$$

When the inductive reactance equals the capacitive reactance, they nullify each other, and the current is given by

$$I = \frac{E}{R}$$

which is its maximum possible value. When this condition exists, the circuit is in **resonance** with the applied voltage. To have resonance, it is essential the circuit have both capacitance and inductance.

Resonant circuits are used in radios and televisions. The frequency of a certain station is tuned in when a resonant circuit (antenna circuit) is adjusted to that frequency (Fig. 20.26). This is accomplished by changing the capacitance until the capacitive reactance equals the inductive reactance. The applied voltage is the radio signal picked up by the antenna. A variable capacitor has one set of plates (usually aluminum) mounted on a rotating shaft. The plates have air between them, and the capacitance varies with the amount of overlap of the plates as they are rotated between each other.

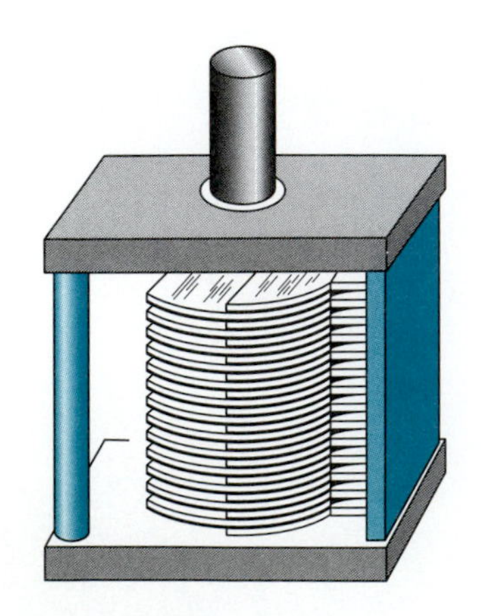
Figure 20.26
Variable capacitor.

The resonant frequency occurs when $X_L = X_C$; power transfer is maximized at resonance as discussed in the preceding section. We find this frequency as follows:

$$X_L = X_C$$

by substitution:

$$2\pi fL = \frac{1}{2\pi fC}$$

Solving for f^2 gives:

$$f^2 = \frac{1}{4\pi^2 LC}$$

then,

$$f = \frac{1}{\sqrt{4\pi^2 LC}}$$

or

$$f = \frac{1}{2\pi\sqrt{LC}}$$

The circuit can be adjusted to any frequency by varying the capacitance or the inductance.

EXAMPLE

Find the resonant frequency of a circuit containing a 5.00-nF capacitor in series with a 2.60-μH inductor.

Sketch:

$C = 5.00$ nF

$L = 2.60$ μH

Data:

$$C = 5.00 \text{ nF} = 5.00 \times 10^{-9} \text{ F}$$
$$L = 2.60 \text{ μH} = 2.60 \times 10^{-6} \text{ H}$$
$$f = ?$$

Basic Equation:

$$f = \frac{1}{2\pi\sqrt{LC}}$$

Working Equation: Same

Substitution:

$$f = \frac{1}{2\pi\sqrt{(2.60 \times 10^{-6} \text{ H})(5.00 \times 10^{-9} \text{ F})}}$$

$$= \frac{1}{2\pi\sqrt{1.30 \times 10^{-14} \cancel{\text{H}}\,\cancel{\text{F}} \left(\frac{1\,\cancel{\Omega}\,\text{s}}{1\,\cancel{\text{H}}}\right)\left(\frac{1\,\text{s}}{1\,\cancel{\text{F}}\,\cancel{\Omega}}\right)}}$$

$$= \frac{1}{7.16 \times 10^{-7} \text{ s}}$$

$$= 1.40 \times 10^6 \frac{\text{cycles}}{\text{s}} \quad \text{or} \quad 14\bar{0}0 \frac{\text{kilocycles}}{\text{s}} \quad \text{or} \quad 14\bar{0}0 \text{ kHz}$$

This frequency is in the AM radio band.

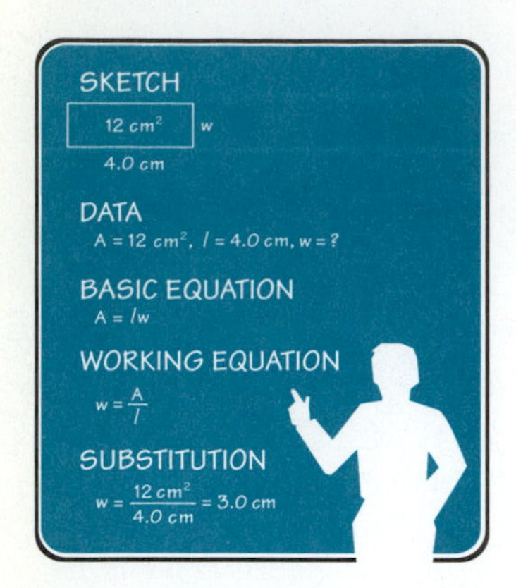

■ PROBLEMS 20.11

Find the resonant frequency in each ac circuit.

1. $L = 1.00\ \mu\text{H}$ and $C = 4.00\ \mu\text{F}$
2. $L = 2.00\ \mu\text{H}$ and $C = 35.0\ \mu\text{F}$
3. $L = 2.50\ \mu\text{H}$ and $C = 7.00\ \mu\text{F}$
4. $L = 2.65\ \mu\text{H}$ and $C = 35.0\ \mu\text{F}$
5. $L = 42.5\ \mu\text{H}$ and $C = 40.0\ \mu\text{F}$
6. Find the resonant frequency of a circuit containing a 25.0-μF capacitor in series with a 75.0-μH inductor.
7. Find the resonant frequency of a circuit containing a 33.0-μF capacitor in series with a 43.5-μH inductor.
8. Find the resonant frequency of a circuit containing a 10.0-μF capacitor in series with a 37.5-μH inductor.
9. Find the resonant frequency of a circuit containing a 4.00-μF capacitor in series with a $10\bar{0}$-μH inductor.
10. Find the resonant frequency of a circuit containing a 3.75-μF capacitor in series with a $30\bar{0}$-μH inductor. ■

20.12 RECTIFICATION AND AMPLIFICATION

It is often necessary to change ac into dc to provide dc for charging batteries or to power the integrated circuits (IC's) of computers and other electronic units. This process is called **rectification.** A device that accomplishes this is called a *diode*. Early diodes were constructed as vacuum tubes. Modern diodes are made of a semiconductor material. Diodes are usually less than $\frac{1}{8}$ in. long (Fig. 20.27).

A **diode** allows current to flow in only one direction. It is similar to a turn-

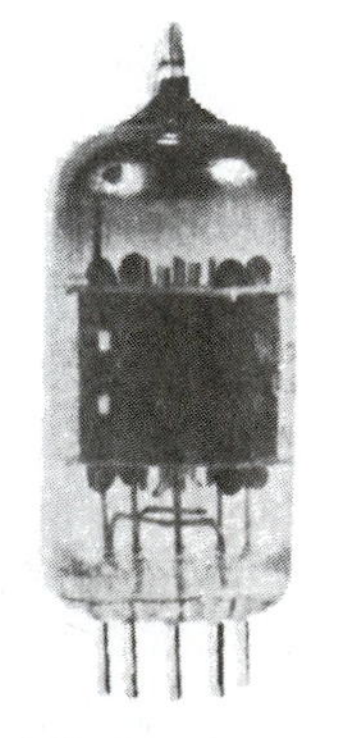

(a) Early vacuum tube.

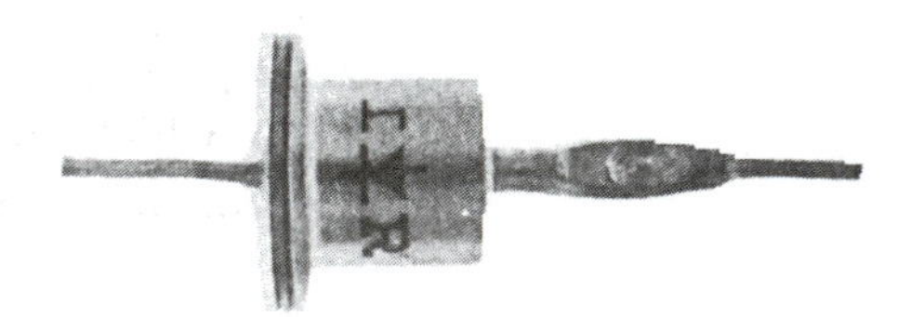

(b) Semiconductor diode (enlarged)

Figure 20.27

stile that revolves in only one direction. People may pass the turnstile in one direction but are blocked when they attempt to pass in the opposite direction (Fig.

20.28). A diode allows electrons to pass in only one direction and not in the other (Fig. 20.29).

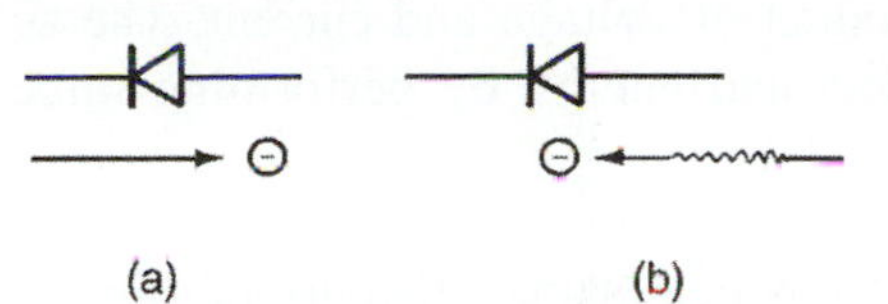

Figure 20.29
A diode allows electrons to flow in only one direction.

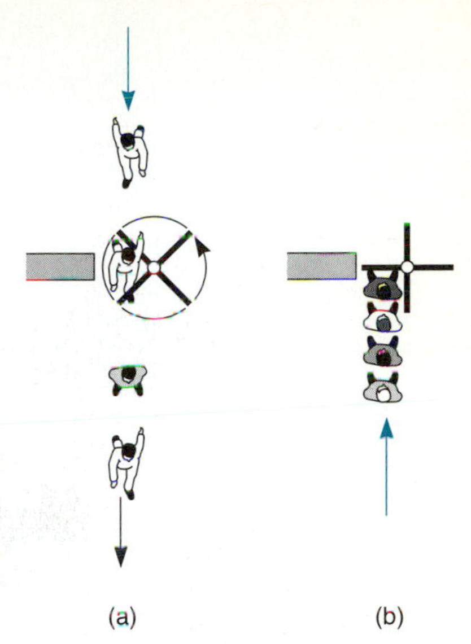

Figure 20.28
People moving through a one-way turnstile is like a diode.

Thomas Edison found that when a wire filament was heated near a metal plate, an electron charge would begin to flow from the filament across space to the plate. This is called the *Edison effect* and was the beginning of the electronics industry. The electron emitter filament is called the *cathode*. The plate is called the *anode*. The entire device is called a *diode*. The diode allows electrons to flow only in one direction, which is the process of rectification.

So a rectifier allows alternating current to pass only in one direction and is thus changed to a direct current (Fig. 20.30). Additional circuit devices can be added to the rectifier that will smooth out the direct current so that it appears as shown in Fig. 20.31.

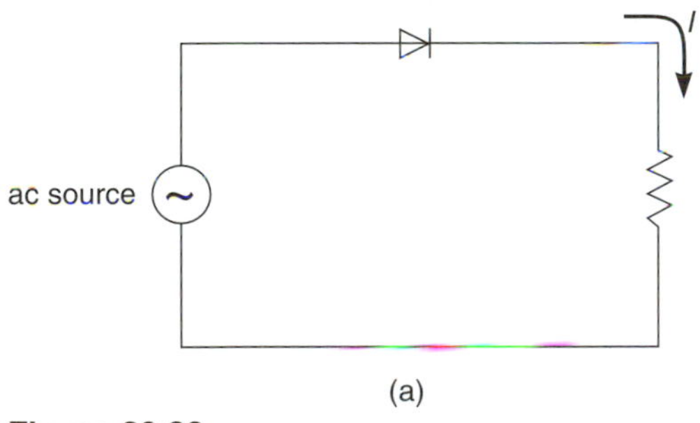

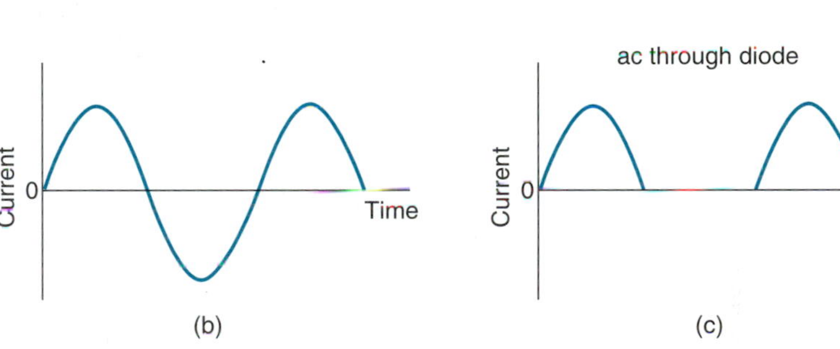

Figure 20.30
A diode can change ac to dc.

Rectifiers are used in automobiles to change the alternating current produced by the alternator into direct current.

It is often necessary to increase the strength of an electronic signal. This is referred to as **amplification.** Radios, stereos, and many other instruments contain one or more amplifier circuits. Early amplifiers utilized vacuum tubes together with other components. The transistor has replaced the vacuum tube in most circuitry because of its smaller size and power consumption. A *transistor* amplifier is typically composed of one or more transistors in addition to capacitors, resistors, and possibly inductors to provide the amplifier with the desired gain (amplification), frequency response, and power output. Amplifiers composed of individual transistors, resistors, and capacitors have been replaced for many applications by *integrated circuits* (IC's), which are only slightly larger in size than some transistors. An IC may contain millions of tiny transistors, diodes, resistors, and capacitors on a small chip of silicon less than 1 cm square. In addition to amplifying signals, IC's have been designed to serve as memory or logic units in computers or other applications. These IC's can be programmed to perform arithmetic operations, as in a calculator, or to perform control operations, as in most appliances.

Figure 20.31
Rectified ac current.

20.13 COMMERCIAL GENERATOR POWER OUTPUT

The power output of the generator is the product of voltage and current. The ac generator converts mechanical energy to electrical energy by performing three functions:

1. *Production of voltage:* electrical pressure, which pushes the current through the loads
2. *Production of power current:* current converted into heat, light, and mechanical power
3. *Production of magnetizing current:* current transferred back and forth for magnetizing purposes in the generation of electrical power, called *reactive kVA* (kilovolt-amperes)

Apparent Power and Reactive kVA

If the current and voltage are not in phase, the resultant product of current and voltage is **apparent power** instead of actual power. Apparent power is measured in kVA (kilovolt-amperes). **Actual power** is the product of apparent power and the **power factor:**

$$\text{power factor} = \frac{\text{actual power}}{\text{apparent power}}$$

where the actual power is measured in kW, the apparent power is measured in kVA and is called *reactive kVA,* and the power factor is a unitless ratio less than 1. Note that 1 VA = 1 W.

Mathematically, the power factor is equal to the cosine of the angle by which the current lags behind (or in rare cases leads) the voltage. The power factor is really a correction factor that must be applied to determine actual power produced. The situation is very similar to finding the amount of work done when a force and the motion are not in the same direction.

EXAMPLE

Find the actual power produced by a generating system that produces 13,600 kVA with a power factor of 0.900.

Data:

$$\text{apparent power} = 13{,}600 \text{ kVA}$$
$$\text{power factor} = 0.900$$
$$\text{actual power} = ?$$

Basic Equation:

$$\text{power factor} = \frac{\text{actual power}}{\text{apparent power}}$$

Working Equation:

$$\text{actual power} = (\text{apparent power})(\text{power factor})$$

Substitution:

$$\text{actual power} = (13{,}600 \text{ kVA})(0.900)$$
$$= 12{,}200 \text{ kVA}$$
$$= 12{,}200 \text{ kW}$$

■ PROBLEMS 20.13

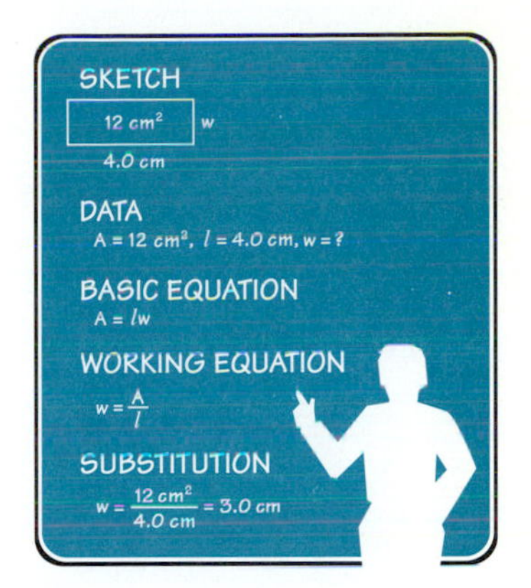

1. Find the actual power produced by a generating station that produces 12,600 kVA with a power factor of 0.85.
2. A generating station operates with a power factor of 0.91. What actual power is available on the transmission lines if the apparent power is 12,800 kVA?
3. Find the apparent power produced by a generating station whose actual power is 120,000 kW and whose power factor is 0.90.
4. Find the apparent power produced by a generating station whose actual power is 1,900,000 kW and whose power factor is 0.80.
5. A generating station operates with a power factor of 0.88. What actual power is available on the transmission lines if the apparent power is 11,500 kVA?
6. Find the apparent power produced by a generating station whose actual power is 2,350,000 kW and whose power factor is 0.85.
7. Find the actual power produced by a generating station that produces 23,800 kVA with a power factor of 0.81.
8. A generating station operates with a power factor of 0.84. What actual power is available on the transmission lines if the apparent power is 13,500 kVA?
9. Find the apparent power produced by a generating station whose actual power is 350,000 kW and whose power factor is 0.86.
10. Find the apparent power produced by a generating station whose actual power is 1,250,000 kW and whose power factor is 0.82.
11. Find the power factor of a generating station whose actual power is 55,800 kW and whose apparent power is 63,400 kVA.
12. Find the power factor of a generating station whose apparent power is 645,000 kVA and whose actual power is 587,000 kW. ■

GLOSSARY

Actual Power A measure of the actual power available to be converted into other forms of energy. The resultant product of current and voltage. When the current and voltage are not in phase, the actual power is less than the apparent power. (p. 496)

Alternating Current A current that flows in one direction in a conductor, changes direction, and then flows in the other direction. (p. 467)

Amplification The process of increasing the strength of an electronic signal. (p. 495)

Apparent Power The product of the maximum voltage and the maximum current. When the current and voltage are out of phase, the apparent power is greater than the actual power. (p. 496)

Capacitor A circuit component consisting of two parallel plates separated by a thin insulator. Can build up and store charge. (p. 485)

Capacitive Reactance A measure of the opposition to ac current flow by a capacitor. (p. 486)

Diode A device that allows current to flow through it in only one direction. (p. 494)

Eddy Current A current induced in the core of a transformer, producing heat. (p. 476)

Effective Value A value based on the heating effect of a current or voltage. For a sine-wave variation, the effective value is equal to 0.707 of the maximum value. (p. 470)

Hysteresis Loss Energy lost in a transformer due to heating effects in the core. (p. 476)

Impedance The combined effect of resistance, inductance, and capacitance in a circuit. (p. 482)

Inductive Reactance A measure of the opposition to ac current flow by an inductor. (p. 480)

Inductor A circuit component, such as a coil, in which an induced emf opposes any current change in a circuit. (p. 480)

Instantaneous Current The current at any instant of time. (p. 468)

Instantaneous Voltage The voltage at any instant of time. (p. 468)

Phase Angle The angle between the resistance and impedance vectors in a circuit. (p. 483)

Power Factor The ratio of the actual power to the apparent power. (p. 496)

Primary Coil The coil of a transformer that is connected to an ac source. (p. 474)

Rectification The process of changing ac into dc. (p. 494)

Rectifier A device that changes ac to dc. (p. 473)

Resonance A condition in a circuit when the inductive reactance equals the capacitive reactance and they nullify each other. The current that flows in the circuit is then at its maximum value. (p. 492)

Secondary Coil The coil of a transformer in which a voltage is induced by the current flowing in the primary coil. (p. 474)

Step-Down Transformer A transformer that is used to reduce the voltage available to drive the primary coil. (p. 476)

Step-Up Transformer A transformer that is used to produce a higher voltage than that available to drive the primary coil. (p. 476)

Transformer A device composed of two coils (primary and secondary) and a magnetic core. Used to step up or step down a voltage. (p. 474)

FORMULAS

20.1 $i = I_{\text{max}} \sin \theta$

$e = E_{\text{max}} \sin \theta$

20.2 $I = 0.707 I_{\text{max}}$

$E = 0.707 E_{\text{max}}$

20.3 $P = I^2 R = VI = \dfrac{V^2}{R}$

20.4 $\dfrac{V_P}{V_S} = \dfrac{N_P}{N_S}$

$\dfrac{I_S}{I_P} = \dfrac{N_P}{N_S}$

$I_P V_P = I_S V_S$

20.6 $X_L = 2\pi f L$

$I = \dfrac{E}{X_L}$

20.7 $I = \dfrac{E}{Z}$

$Z = \sqrt{R^2 + X_L^2}$

$Z = \sqrt{R^2 + (2\pi f L)^2}$

$$\tan \phi = \frac{X_L}{R}$$

20.8 $$X_C = \frac{1}{2\pi f C}$$

20.9 $$Z = \sqrt{R^2 + X_C^2}$$

$$Z = \sqrt{R^2 + \left(\frac{1}{2\pi f C}\right)^2}$$

$$I = \frac{E}{Z}$$

$$\tan \phi = \frac{X_C}{R}$$

20.10 $$Z = \sqrt{R^2 + (X_L - X_C)^2}$$

$$\tan \phi = \frac{X_L - X_C}{R}$$

$$I = \frac{E}{Z} = \frac{E}{\sqrt{R^2 + \left(2\pi f L - \frac{1}{2\pi f C}\right)^2}}$$

20.11 $$I = \frac{E}{\sqrt{R^2 + (X_L - X_C)^2}}$$

$$f = \frac{1}{2\pi\sqrt{LC}}$$

20.13 $$\text{power factor} = \frac{\text{actual power}}{\text{apparent power}}$$

REVIEW QUESTIONS

1. Which of the following describes alternating current electricity?
   (a) It can be produced by rotating a loop of wire through a magnetic field.
   (b) It flows in one direction for a period of time and then reverses direction.
   (c) It goes through one cycle when it flows in one direction and then reverses direction.
   (d) All of the above.
2. The voltage (e) and the current (i) in an alternating current circuit are in phase when
   (a) the peak values of both e and i occur at different times.
   (b) the peak values of e and i occur at the same time but their zero values do not.
   (c) the peak values and the zero values of e and i occur simultaneously.
   (d) none of the above.
3. Which of the following affect the voltage induced in the secondary coil of a transformer?
   (a) The current through the primary coil
   (b) The resistance of the primary coil
   (c) The number of turns on the primary coil
   (d) None of the above
4. Which of the following contribute to the energy loss in a transformer?
   (a) Resistance of the copper wires
   (b) Reversing the magnetic field in the core

(c) Induced currents in the core
(d) The emf in the outside circuit
(e) All of the above

5. Differentiate between maximum current and effective current.
6. Differentiate between maximum voltage and instantaneous voltage.
7. Explain how power in an ac circuit is related to voltage and current.
8. Explain how power in an ac circuit is related to voltage and resistance.
9. If the number of turns in the secondary coil of a transformer is doubled, how does the output voltage change?
10. The unit of inductance is the _____.
11. Discuss the importance of inductive reactance.
12. How does the inductive reactance depend on frequency?
13. Does the current lead or lag the voltage in an inductive circuit?
14. Describe how energy is stored in a capacitor. How can the stored energy be used?
15. Does the current lead or lag the voltage in a capacitive circuit?
16. How does the reactance of a capacitor depend on frequency?
17. Discuss the condition that leads to resonance.
18. What is the function of a diode in a circuit?
19. Distinguish between amplification and rectification.
20. Is the phase angle always constant in a circuit containing resistive, capacitive, and inductive elements?

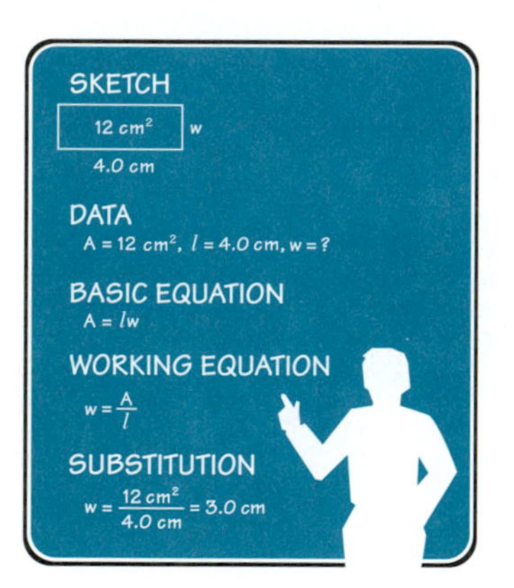

REVIEW PROBLEMS

1. What is the maximum voltage in a circuit when the instantaneous value of the voltage is 95.4 V at $\theta = 62°$?
2. If the maximum ac voltage on a line is 185 V, what is the instantaneous voltage at $\theta = 41°$?
3. If the maximum ac voltage on a line is 175 V, what is the instantaneous voltage at $\theta = 23°$?
4. What is the effective value of an ac voltage whose maximum voltage is 135 V?
5. What is the maximum current in a circuit with current 6.35 A?
6. What power is developed by a device that draws 6.87 A and has a resistance of 15.4 Ω?
7. A heating element draws 4.50 A on a $11\bar{0}$-V line. What power is expended in the element?
8. What power is used by a heater that has a resistance of 22.3 Ω and draws a current of 7.65 A?
9. A step-up transformer on a 115-V line provides a voltage of 2050 V. If the primary coil has 75.0 turns, how many turns does the secondary coil have?
10. If there is a current of 4.55 A in the primary coil in Problem 9, what is the current in the secondary?
11. Find the power in the primary coil in Problems 9 and 10.
12. An inductance of 48.0 mH is connected in series with a lamp of resistance 23.0 Ω. This circuit is connected to a 115-V, 60.0-Hz power supply. (a) What is the current in the circuit? (b) what is the phase angle? (c) What is the voltage drop across the inductance?
13. A lamp of resistance 47.5 Ω is connected in series with an inductance of 43.2 mH. This circuit is connected to a 115-V, 60.0-Hz power supply. (a) What is the current in the circuit? (b) What is the phase angle? (c) What is the voltage drop across the resistance?

14. What current will flow in a 60.0-Hz ac series circuit that includes a $11\bar{0}$-V source, a resistor of 19.5 Ω, and a capacitor of 57.4 μF? What is the phase angle? What is the voltage across the resistor? Across the capacitor?
15. A resistor of 21.6 Ω and a capacitor of 38.5 μF are connected in series with a 60.0-Hz ac source with a voltage of $11\bar{0}$ V. What is the current in the circuit? What is the voltage across the capacitor? What is the phase angle?
16. A circuit contains a 175-Ω resistance, a 25.0-μF capacitor, and a 62.0-mH inductance in series with a $11\bar{0}$-V, 60.0-Hz source. Find the impedance and the current.
17. A circuit contains a 115-Ω resistance, a 35.0-μF capacitor, and a 65.0-mH inductance in series with a $11\bar{0}$-V, 60.0-Hz source. Find the impedance and the current.
18. Find the resonant frequency of a circuit containing a 7.50-μF capacitor in series with a 3.70-μH inductance, a 633-Ω resistor, and a $11\bar{0}$-V, 60.0-Hz source. Find the impedance and the current.
19. Find the resonant frequency of a circuit containing a 4.70-μF capacitor in series with a 4.50-μH inductance, a 25.0-Ω resistor, and a $11\bar{0}$-V, 60.0-Hz source. Find the impedance and the current.
20. Find the apparent power produced by a generating station whose actual power is 2,900,000 kW and whose power factor is 0.85.

Chapter 21

LIGHT

For centuries scientists have sought to explain the nature of light—why it is reflected and why it is refracted. Even the speed of light has been the subject of study since the seventeenth century.

Light seems to exhibit some characteristics of both particles and waves. The study of the measurement of light is photometry. We begin by examining the nature of light.

OBJECTIVES

The major goals of this chapter are to enable you to:

1. Gain an understanding of the nature of light.
2. Determine solutions to speed of light problems.
3. Contrast the wave and particle characteristics of light.
4. Apply principles of photometry to technical problems.

21.1 NATURE OF LIGHT

The search for an explanation for the nature of light has been going on for many centuries. A number of famous scientists, including Isaac Newton, Christian Huygens, Albert Einstein, and Louis de Broglie, have made major contributions to the current theory of light and its interaction with matter.

In the seventeenth century many of the foundations for modern scientific theories were put forward. Two conflicting theories for the nature of light were proposed. The experimental observations that had to be explained were the following:

1. *Straight-line propagation of light*. An application of this property of light is found in survey work, in which sight lines are commonly used (Fig. 21.1).
2. **Reflection** *at the boundary between two different media*. Examples of this are the reflection of light at the boundary between air and water or air and glass.

Figure 21.1
Surveyor using the straight-line travel characteristic of light.

3. **Refraction,** *or bending, of light as it passes through the boundary between two media, such as air and water* (Fig. 21.2). This bending of light makes objects under water appear to be closer to the surface and farther away from the observer than they really are when viewed from above the surface. It also makes a straight object partially submerged in water appear bent at the surface.

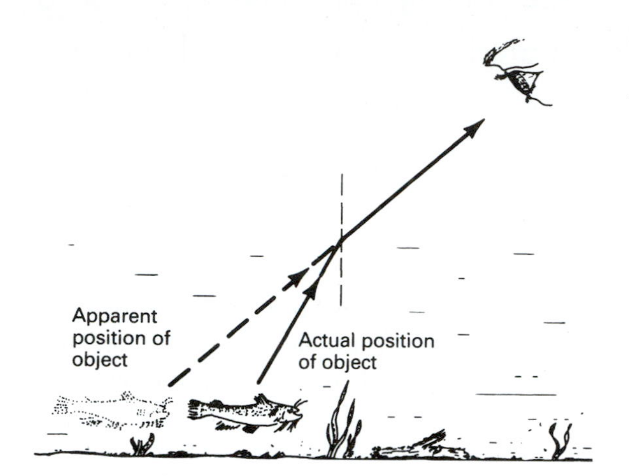

Figure 21.2
Refraction of light as it passes between media.

The two conflicting theories mentioned above are the wave theory and the particle theory of light. Christian Huygens proposed the **wave theory** of light. According to the wave theory, light consists of waves traveling out from sources of light in a way that is similar to water waves traveling out from the point at which a stone is dropped into the water. A similar type of wave behavior is sound propagation out from a source of sound. All three types of waves mentioned here—light, water, and sound—travel in straight lines. They also reflect off surfaces or boundaries between media. Refraction of these waves is also observed.

Isaac Newton proposed the **particle theory** of light as an alternative explanation of the experimental observations. Newton thought light was made up of streams of particles. These particles of light, which Newton referred to as *corpuscles,* he felt behaved in a manner similar to particles of matter, which travel in a straight line if the net force acting on them is zero. Particles of matter also rebound or reflect off surfaces. An example of this is a rubber ball bouncing off a wall. A change in the velocity of a particle could produce a change in direction. Newton's particle theory persisted until the early nineteenth century, when diffraction and interference of light were observed. Since these properties could only be explained by the wave theory, the particle theory fell out of favor.

Beginning in the late nineteenth century, experiments were developed that confirmed the electromagnetic theory and showed visible light to be a small portion of the electromagnetic spectrum (Fig. 21.3). Such waves are all similar in nature, travel at the same speed in a vacuum (3.00×10^8 m/s or 186,000 mi/s), but differ in their frequencies and wavelengths. Note from Fig. 21.3 that as the frequency increases to the right, the wavelength decreases.

The electromagnetic theory, though, is still not a complete explanation of the nature of light. Early in this century, the **photoelectric effect** was discovered: the emission (or giving off) of electrons by a substance when struck by electromagnetic radiation. Albert Einstein received a Nobel prize for his work in this

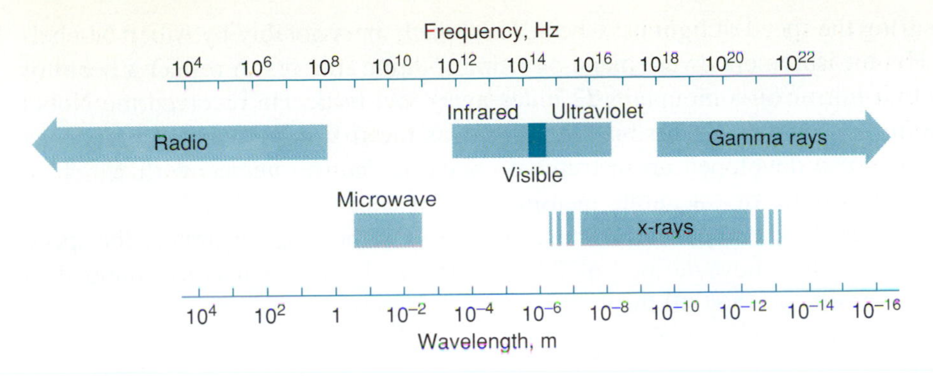

Figure 21.3
Electromagnetic spectrum.

area. The energy of the emitted electron is related to the energy (frequency) of the incoming light in the same way that the energy of a billiard ball is related to the energy of the cue ball that strikes it. The particle theory of light explains the photoelectric effect while the wave theory cannot. The electromagnetic wave theory fails to fully explain the nature of these emissions of electrons.

Later work by Max Planck and Albert Einstein developed the idea that light was energy radiated at the speed of light in the form of wave packets of energy, which were called **photons.** Further work showed that in some circumstances light acted like a stream of particles. Based on the work of Planck and Einstein, the **quantum theory** was developed. This theory states that energy, including electromagnetic radiation, is not absorbed or radiated continuously but is radiated in multiples of definite and discontinuous units.

Light therefore appears to have at least a dual character, having properties of both waves and particles (Fig. 21.4). This dual character may be shown by considering how energy may be transported from one point to another.

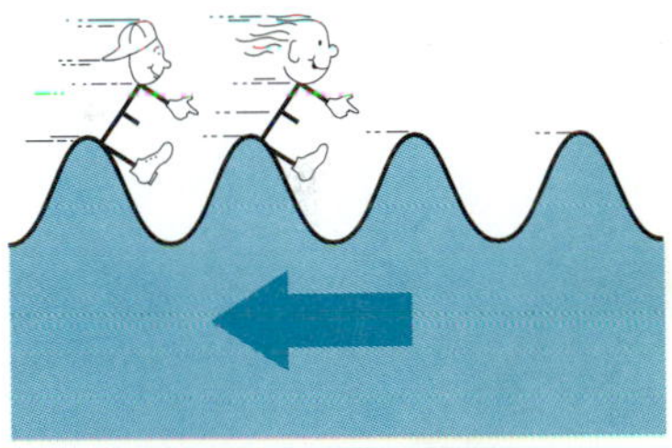

Figure 21.4
Light has properties of both particles and waves.

The first method is to transport particles of matter that carry energy with them. Examples of this method are electron conduction in a wire, shooting a bullet, natural gas pipelines, and gasoline transport.

The second method is the propagation of a wave disturbance through the medium between two points. Sound and water waves are examples of this method of energy transport.

Light is unusual in that it appears to combine characteristics of each method. When light is traveling through a medium, it appears to behave like a wave, with the following characteristics: (1) reflection at the surface of a medium, (2) refraction when passing from one medium to another, (3) interference (cancellation) when two waves are properly superimposed, and (4) diffraction (bending) when the waves pass the corners of an obstacle. When light interacts with matter, such as when it is absorbed or emitted, it behaves as if it were a massless particle.

21.2 THE SPEED OF LIGHT

One of the most important measured quantities in physics is the **speed of light.** At first, light was assumed to be instantaneous. Galileo probably first suggested that time was required for light to travel from place to place. A Danish astronomer, Olaus Roemer, made the first estimate of the speed of light from his study of the time of eclipse of one of the moons of the planet Jupiter as viewed from different places of the orbit of the earth in 1675. Since then, laboratory methods for measuring the speed of light have been developed, most notably by Albert Michelson. His measurements were made by using rotating mirrors to reflect a beam of light to a mirror on a mountain 22 miles away and back. He received the Nobel Prize in physics in 1907 for his work and was the first American to receive the prize. He also developed an instrument called an *interferometer,* with which in 1920 he made the first accurate measurement of a star's diameter.

Modern laboratory methods have been used to measure accurately the speed of light, which is now defined as 299,792,458 m/s. This is usually rounded to 3.00×10^8 m/s or 186,000 mi/s.

As stated earlier, light is one form of a class of radiation called *electromagnetic radiation,* which also includes radio and television waves, infrared, gamma rays, and x-rays. A chart of the entire electromagnetic spectrum is shown in Fig. 21.3.

The distance traveled by any form of electromagnetic radiation can be found by substituting the speed of light c into the equation $s = vt$ as follows:

$$s = ct$$

where s = distance
c = speed of light, 3.00×10^8 m/s or 186,000 mi/s
t = time

EXAMPLE

Find the distance (in mi) traveled by an x-ray in 0.100 s.

Data:

$$c = 186{,}000 \text{ mi/s}$$
$$t = 0.100 \text{ s}$$
$$s = ?$$

Basic Equation:

$$s = ct$$

Working Equation: Same

Substitution:

$$s = (186{,}000 \text{ mi/}\cancel{\text{s}})(0.100 \cancel{\text{s}})$$
$$= 18{,}600 \text{ mi}$$

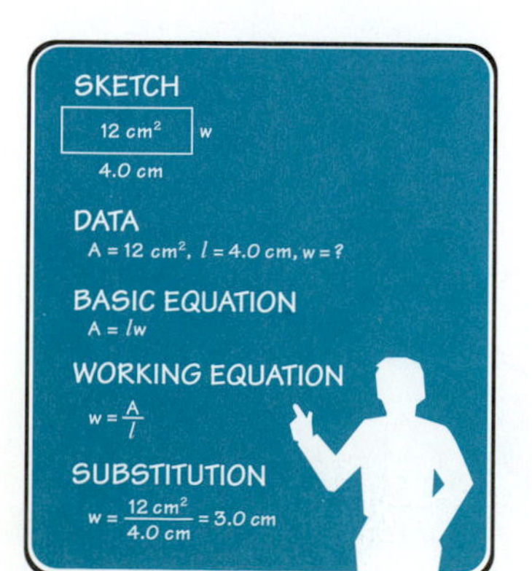

Very large distances, such as those between stars, cannot be conveniently expressed in common distance units. Astronomers therefore use the unit light-year to measure such distances. A **light-year** is the distance traveled by light in one earth year, so 1 light-year equals 9.5×10^{15} m or 5.9×10^{12} mi.

PROBLEMS 21.2

1. Find the distance (in metres) traveled by a radio wave in 5.00 s.
2. Find the distance (in metres) traveled by a light wave in 6.40 s.
3. A television signal is sent to a communications satellite that is 20,000 mi above a relay station. How long does it take for the signal to reach the satellite?

4. How long does it take for a radio signal from earth to reach an astronaut on the moon? The distance from the earth to the moon is 2.40×10^5 mi.
5. The sun is 9.30×10^7 mi from the earth. How long does it take light to travel from the sun to the earth?
6. A radar wave is bounced off an airplane and returns to the radar receiver in 2.50×10^{-5} s. How far (in km) is the airplane from the radar receiver?
7. How long does it take for a radio wave to travel $30\bar{0}0$ mi across the U.S.?
8. How long does it take for a flash of light to travel $10\bar{0}$ m?
9. How long does it take for a police radar beam to travel to a truck and back if the truck is 115 m from the radar unit?
10. How far away (in km) is an airplane if the radar wave returns to the scanning radar unit in 1.24×10^{-3} s?
11. How long does it take for light to reach the earth from Mars when the separation of the two planets is at its smallest? The earth's orbital radius is 143 million kilometres. The orbital radius of Mars is 218 million kilometres. How long does it take when the separation is at its maximum?
12. If it takes 4.5 years for light to reach the earth from Alpha Centauri, the closest star to the earth other than the sun, what is the distance (in miles) to the next nearest neighbor (Barnard's Star) which is 25.0% farther away?
13. How long does it take light to reach the earth from Jupiter when the separation of the two planets is at its smallest? The earth's orbital radius is 143 million kilometres. The orbital radius of Jupiter is 725 million kilometres. ■

21.3 LIGHT AS A WAVE

Light and the other forms of electromagnetic radiation are composed of oscillations in the electric and magnetic fields that exist in space. These oscillations are set up by rapid movement of charged particles such as electrons in radio antennas and electrons in a hot object such as a light bulb filament. All waves are characterized by the distance that separates two points on the wave which are at the same point of vibration (Fig. 21.5). This distance is called the **wavelength** and is denoted by the Greek lowercase letter lambda, λ. The wavelength of visible light ranges from about 4×10^{-7} m to 7.6×10^{-7} m. The human eye perceives light in the visible spectrum as one or more colors depending upon the frequency or wavelength of the light hitting the retina of the eye. The longest visible wavelengths ($\lambda = \sim 7.5 \times 10^{-7}$ m), which are also the smallest frequencies, are perceived as red. The shortest visible wavelengths ($\lambda = \sim 4 \times 10^{-7}$ m), which are

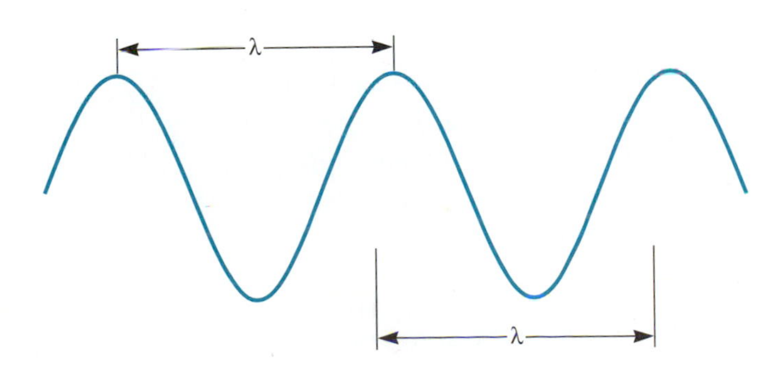

Figure 21.5
The wavelength of a repeating wave is the distance between two successive corresponding points.

the highest frequencies, are perceived as blue. The wavelengths of other electromagnetic radiations are found in Fig. 21.3.

Another characteristic of waves is the **frequency,** f. Frequency is the number of complete oscillations per second. The measurement unit of frequency (cycles/s) is named the hertz (Hz) after Heinrich Hertz, a leader in the study of electromagnetic theory. For example, 1 cycle per second = 1 Hz = 1/s. Since a "cycle" has no units, it does not appear in the hertz unit.

The following basic relationship exists for all electromagnetic waves, which relates the frequency, wavelength, and the velocity of the wave.

$$c = \lambda f$$

where c = speed of light, 3.00×10^8 m/s or 186,000 mi/s
f = frequency
λ = wavelength

EXAMPLE

Find the frequency of a light wave with a wavelength of 5.00×10^{-7} m.

Data:

$$\lambda = 5.00 \times 10^{-7} \text{ m}$$
$$c = 3.00 \times 10^8 \text{ m/s}$$
$$f = ?$$

Basic Equation:

$$c = \lambda f$$

Working Equation:

$$f = \frac{c}{\lambda}$$

Substitution:

$$f = \frac{3.00 \times 10^8 \text{ m/s}}{5.00 \times 10^{-7} \text{ m}}$$
$$= 6.00 \times 10^{14} \text{ Hz} \qquad \text{(or cycles/s)}$$

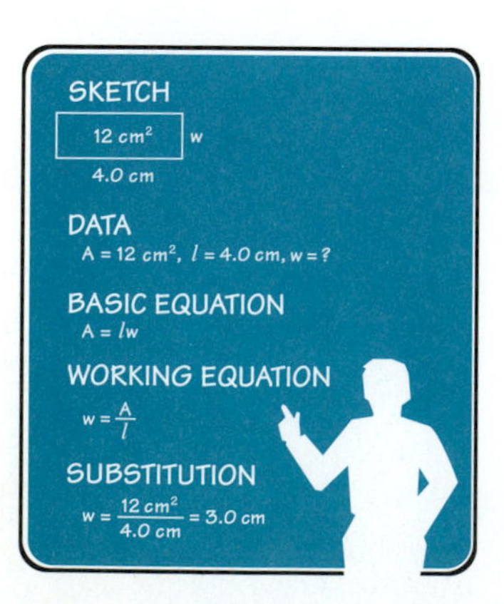

■ PROBLEMS 21.3

1. $c = 3.00 \times 10^8$ m/s
   $\lambda = 4.55 \times 10^{-5}$ m
   $f = ?$
2. $c = 3.00 \times 10^8$ m/s
   $\lambda = 9.70 \times 10^{-10}$ m
   $f = ?$
3. $c = 3.00 \times 10^8$ m/s
   $f = 9.70 \times 10^{11}$ Hz
   $\lambda = ?$
4. $c = 3.00 \times 10^8$ m/s
   $f = 24.2$ MHz
   $\lambda = ?$
5. $c = 3.00 \times 10^8$ m/s
   $f = 45.6$ MHz
   $\lambda = ?$
6. $c = 3.00 \times 10^8$ m/s
   $f = 415$ Hz
   $\lambda = ?$
7. $c = 3.00 \times 10^8$ m/s
   $\lambda = 6.59 \times 10^{12}$ m
   $f = ?$ Hz
8. $c = 3.00 \times 10^8$ m/s
   $\lambda = 9.23$ km
   $f = ?$ Hz

9. Find the wavelength of a radio wave from an AM station broadcasting at a frequency of $14\bar{0}0$ kHz.
10. Find the wavelength of a radio wave from an FM station broadcasting at a frequency of $10\bar{0}$ MHz.
11. Find the frequency of an electromagnetic wave if its wavelength is 85.5 m.
12. Find the frequency of an electromagnetic wave if its wavelength is 3.25×10^{-8} m.
13. Find the frequency of blue light.
14. Find the frequency of red light.
15. Find the frequency of yellow light if its wavelength is midway between those of red and blue photons. ■

21.4 LIGHT AS A PARTICLE

As mentioned earlier, light sometimes behaves as if it were a massless particle. These particles are called *photons* and each carries a portion of the energy of the wave. This energy is given by

$$E = hf$$

where f = frequency
$h = 6.62 \times 10^{-34}$ J s (Planck's constant)
E = energy

EXAMPLE

What is the energy of a photon of electromagnetic radiation with frequency 1.00×10^{12} Hz?

Data:

$$h = 6.62 \times 10^{-34} \text{ J s}$$
$$f = 1.00 \times 10^{12} \text{ Hz} = 1.00 \times 10^{12}/\text{s}$$
$$E = ?$$

Basic Equation:

$$E = hf$$

Working Equation: Same

Substitution:

$$E = (6.62 \times 10^{-34} \text{ J s})(1.00 \times 10^{12}/\text{s})$$
$$= 6.62 \times 10^{-22} \text{ J}$$

■ PROBLEMS 21.4

1. What is the energy of a photon of electromagnetic radiation with frequency 8.95×10^{10} Hz?
2. What is the frequency of a photon of electromagnetic radiation with energy 3.96×10^{-22} J?
3. What is the energy of a photon of electromagnetic radiation with frequency 4.55×10^{8} Hz?

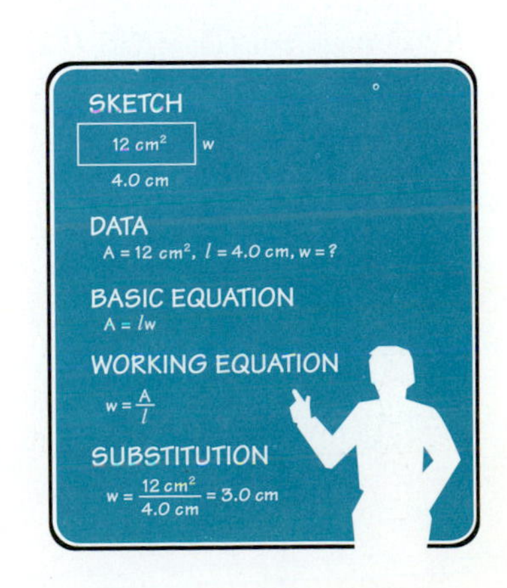

4. Find the frequency of electromagnetic radiation with energy 2.00×10^{-24} J.
5. Find the frequency of electromagnetic radiation with energy 5.50×10^{-26} J.
6. Find the energy of a photon of electromagnetic radiation with frequency 2.50×10^{12} Hz.
7. Find the frequency of electromagnetic radiation with energy 3.65×10^{-23} J.
8. Find the energy of a photon of electromagnetic radiation with frequency 9.20×10^{16} Hz.
9. Find the energy of a red photon.
10. Find the energy of a blue photon.
11. Find the energy of a yellow photon if its wavelength is midway between those of red and blue photons. ■

21.5 PHOTOMETRY

Recall that light is produced along with other forms of radiation when substances are heated like our greatest source of natural light, the sun. Light may be produced in other ways, however, such as electrons bombarding gas molecules in neon lights or chemically, like fireflies. Our most common source of artificial light, however, is the incandescent lamp.

Incandescent lamps (like light bulbs) produce light by the heating of a material (the filament) and the giving off of a wide range of radiation in addition to visible light. Such objects, which produce light, are called **luminous.** On the other hand, objects like the moon, which are not producers of light but only reflect light from another source, are called **illuminated.** When light strikes the surface of most objects, some light is reflected, some transmitted, and some absorbed.

The study of the measurement of light is called **photometry.** Two important measurable quantities in photometry are the *luminous intensity, I,* of a light source, and the *illumination* of a surface, *E.*

Luminous intensity measures the brightness of a light source. The unit for luminous intensity *(I)* is the candle or **candela** (cd). The early use of certain candles for standards of intensity led to the name of the unit. We now use a platinum source at a certain temperature as the standard for comparison. Another unit, the **lumen,** ℓm, is often used for the measurement of the intensity of a source. One candle produces 4π lumens (ℓm).

$$1 \text{ candle} = 4\pi \ \ell\text{m}$$

To determine the intensity rating in lumens of a 40-W bulb, we can use this conversion to find

$$35 \text{ cd} \times \frac{4 \ \pi \ \ell\text{m}}{1 \text{ cd}} = 440 \ \ell\text{m}$$

Thus, a 40-W light bulb rated at 35 cd has an intensity rating of 440 ℓm.

The illumination on a surface may be varied by either changing the luminous intensity of the source (use a brighter bulb) or changing the position of the source (moving it closer to or farther from the surface to be illuminated). Of course, the illumination is also less if the surface illuminated is slanted and not directly facing the light source.

The amount of illumination on a surface varies inversely with the square of the distance from the source. For example, if the distance of the illuminated surface from the source is doubled, the illumination is reduced to one fourth of its former intensity. This can be illustrated by considering a point source of light at the center of concentric (having the same center) spheres (Fig. 21.6). The solid angle is measured in units called *steradians*.

If the source radiates light uniformly in all directions, the light is uniformly distributed over a spherical surface centered at the source. Since the surface area of a sphere is $4\pi r^2$, the illumination, E, at the surface is given by

$$E = \frac{I}{4\pi r^2}$$

where E = illumination
I = intensity of the source (in lumens)
r = distance between the source and the illuminated surface

The unit of illumination, E, is the *lux:*

$$1\ \ell\text{m/m}^2 = 1\ \text{lux}$$

We assume the surface being illuminated is perpendicular to the source.

Note that the inverse square law behavior of light is similar to the two inverse square laws studied earlier, those of gravity and of the electric force between charges described by Coulomb's law.

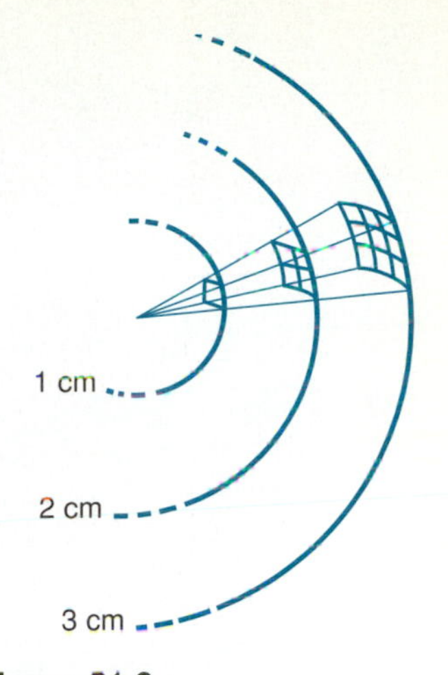

Figure 21.6
The amount of illumination on a surface varies inversely with the square of the distance from the source.

EXAMPLE 1

Find the illumination E on a surface located 2.00 m from a source with an intensity of $40\bar{0}\ \ell$m.

Data:

$$I = 40\bar{0}\ \ell\text{m}$$
$$r = 2.00\ \text{m}$$
$$E = ?$$

Basic Equation:

$$E = \frac{I}{4\pi r^2}$$

Working Equation: Same

Substitution:

$$E = \frac{40\bar{0}\ \ell\text{m}}{4\pi(2.00\ \text{m})^2}$$
$$= 7.96\ \ell\text{m/m}^2 \quad \text{or} \quad 7.96\ \text{lux}$$

The unit used for illumination is the lux (ℓm/m^2) in the metric system as shown above and the ℓm/ft^2 in the English system. Another unit often used is the foot-candle, which is equal to 1 ℓm/ft^2

$$1\ \text{ft-candle} = 1\ \ell\text{m/ft}^2$$

EXAMPLE 2

Find the illumination 4.00 ft from a source with an intensity of $60\bar{0}$ ℓm.

Data:

$$I = 60\bar{0}\ \ell\text{m}$$
$$r = 4.00\ \text{ft}$$
$$E = ?$$

Basic Equation:

$$E = \frac{I}{4\pi r^2}$$

Working Equation: Same

Substitution:

$$E = \frac{60\bar{0}\ \ell\text{m}}{4\pi(4.00\ \text{ft})^2}$$
$$E = 2.98\ \frac{\ell\text{m}}{\text{ft}^2} \times \frac{1\ \text{ft-candle}}{1\ \ell\text{m/ft}^2}$$
$$= 2.98\ \text{ft-candles}$$

$$\frac{\ell\text{m}}{\text{ft}^2} \times \frac{1\ \text{ft-candle}}{1\ \ell\text{m/ft}^2} = \frac{\ell\text{m}}{\text{ft}^2} \times 1\ \text{ft-candle} \times \frac{\text{ft}^2}{\ell\text{m}}$$
$$= \text{ft-candle}$$

In photography, photoelectric cells are used in light meters or exposure meters to measure illumination for taking photographs. The electricity produced is proportional to the illumination and is directly calibrated on the instrument scale. The units of measurement of such meters, however, are not standardized, and the scale may be arbitrarily selected.

A **laser** is a light source that produces a narrow beam with high intensity. Laser is an acronym for "light amplification by stimulated emission of radiation." The first laser was constructed in 1960 by T. H. Maiman.

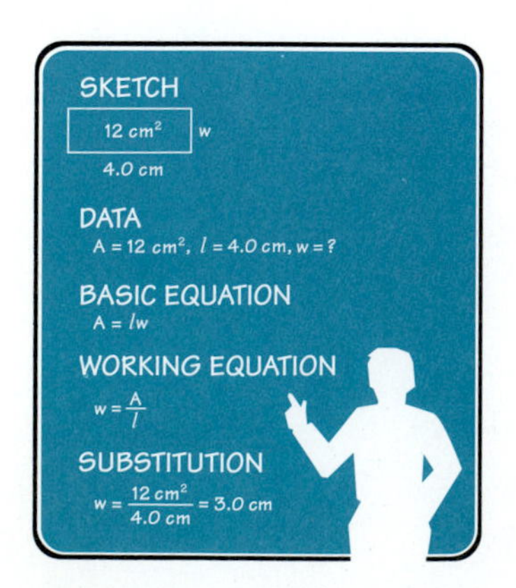

■ PROBLEMS 21.5

1. $I = 48.0$ cd
$I =$ _____ ℓm
2. $I = 342$ cd
$I =$ _____ ℓm
3. $I = 765$ ℓm
$I =$ _____ cd
4. $I = 432$ ℓm
$I =$ _____ cd
5. $I = 75.0$ cd
$I =$ _____ ℓm
6. $I = 65\bar{0}$ ℓm
$I =$ _____ cd
7. $I = 90\bar{0}$ ℓm
$r = 7.00$ ft
$E = ?$
8. $I = 741$ ℓm
$r = 6.50$ m
$E = ?$
9. $I = 893$ ℓm
$r = 3.25$ ft
$E = ?$
10. $E = 4.32$ lux
$r = 9.00$ m
$I = ?$
11. $E = 10.5$ ft-candles
$r = 6.00$ ft
$I = ?$
12. Find the intensity of a light source that produces an illumination of 5.50 ft-candles a distance of 9.85 ft from the source.

13. Find the intensity of a light source that produces an illumination of 2.39 lux a distance of 4.50 m from the source.
14. Find the intensity of a light source that produces an illumination of 5.28 lux a distance of 6.50 m from the source.
15. If an observer triples her distance from a light source:
    (a) Does the illumination at that point increase or decrease?
    (b) In what proportion does the illumination increase or decrease?
16. If the illuminated surface is slanted at an angle of 35.0°, what part of the full-front illumination is lost?
17. Find the illumination on a surface due to three light sources, each with intensity 150 ℓm located at distances of 2.00 m, 2.70 m, and 2.98 m from the surface.
18. Find the intensity of two identical light sources located 1.40 m and 1.96 m from a point where the illumination is 3.54 $\ell m/m^2$.
19. Find the intensity of two identical light sources located 0.88 m and 1.12 m from a point where the illumination is 5.86 $\ell m/m^2$. ■

GLOSSARY

Candela A unit of light source intensity. (p. 510)
Frequency The number of complete oscillations per second of a wave. (p. 508)
Illuminated Objects that are not producers of light, but only reflect it. (p. 510)
Laser A light source that produces a narrow beam with high intensity. An acronym for "light amplification by stimulated emission of radiation." (p. 512)
Light Radiant energy that can be seen by the human eye.
Light-Year The distance that light travels in one year: 9.5×10^{15} m or 5.9×10^{12} mi. (p. 506)
Lumen A unit of luminous intensity for a light source. (p. 510)
Luminous Objects that produce light. (p. 510)
Particle Theory Describes the behavior of light that can be explained in terms of light photons (or particles) such as the photoelectric effect. (p. 504)
Photoelectric Effect The emission of electrons by a surface when struck by electromagnetic radiation. (p. 504)
Photometry The study of the measurement of light. (p. 510)
Photons Wave packets of energy that carry light and other forms of electromagnetic radiation. (p. 505, 509)
Quantum Theory The theory developed by Planck and Einstein that states that energy, including electromagnetic radiation, is radiated or absorbed in multiples of certain units of energy. (p. 505)
Reflection The turning back of all or part of a beam of light at a surface. (p. 503)
Refraction The bending of light as it passes through the boundary between two media, such as air and water. (p. 504)
Speed of Light The speed at which light and other forms of electromagnetic radiation travel. 3.00×10^8 m/s in vacuum. (p. 505)
Wave Theory Describes the behavior of light that can be explained in terms of waves of light, such as refraction and diffraction. (p. 504)
Wavelength The distance between two successive corresponding points on a wave. (p. 507)

FORMULAS

21.2 $s = ct$

21.3 $c = \lambda f$

21.4 $E = hf$

21.5 $E = \frac{I}{4\pi r^2}$

REVIEW QUESTIONS

1. Which of the following are examples of electromagnetic radiation?
   (a) Gamma rays (b) Sound waves
   (c) Radio waves (d) Water waves
   (e) Visible light
2. The particle theory of light explains
   (a) diffraction of light around a sharp edge. (b) refraction of light at a boundary.
   (c) the photoelectric effect. (d) none of the above.
3. A light-year equals
   (a) the time it takes light to travel from the sun to earth.
   (b) the distance from the sun to earth.
   (c) the distance to the nearest star other than the sun.
   (d) the distance light travels in one year.
4. Light behaves
   (a) as a massive particle. (b) always as a wave.
   (c) sometimes as a wave. sometimes as a particle (d) none of the above.
5. Does the wavelength of light depend on its frequency? Explain.
6. How does the energy of a photon of light depend on its frequency?
7. How does the intensity of illumination depend on the distance from a source radiating uniformly in all directions?
8. In your own words, explain how the speed of light has been measured.
9. Does light always travel at the same velocity? Explain.
10. What name is given to the entire range of waves that are similar to visible light?
11. Who proposed the particle theory of light?
12. Who developed the wave packet theory of light?
13. Who made the first estimate of the speed of light?
14. How was the first estimate of the speed of light made?
15. What is the unit of luminous intensity?
16. In your own words, explain luminous intensity.

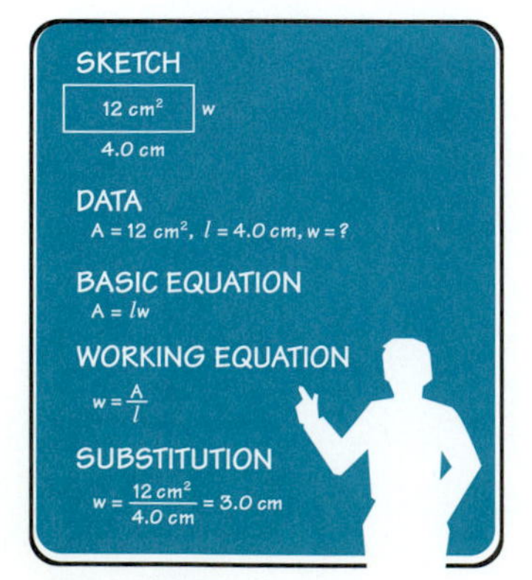

REVIEW PROBLEMS

1. Find the distance (in metres) traveled by a radio wave in 21.5 h.
2. A radar wave that is bounced off an airplane returns to the radar receiver in 3.78×10^{-5} s. How far (in miles) is the airplane from the radar receiver?
3. How long does it take for a police radar beam to travel to a car and back if the car is 0.245 mi from the radar unit?
4. How long does it take for a pulse of laser light to return to a police speed detector after bouncing off a speeding car 0.274 mi away?
5. How long does it take for a radio signal to travel from the earth to a communications satellite 22,500 mi above the surface of the earth?
6. Find the wavelength of a radio wave from an AM station broadcasting at a frequency of 1230 kHz.

7. Find the frequency of a radio wave if its wavelength is 46.5 m.
8. Find the frequency of a light wave if its wavelength is 5.415×10^{-8} m.
9. What is the energy of a photon with frequency 1.45×10^{11} Hz?
10. What is the frequency of a photon with energy of 4.75×10^{-23} J?
11. What is the energy of a photon with frequency 8.25×10^{15} Hz?
12. Find the intensity of the light source necessary to produce an illumination of 3.75 ft-candles at a distance of 6.75 ft from the source.
13. Find the intensity of the light source necessary to produce an illumination of 4.86 lux at a distance of 9.25 m from the source.
14. What is the intensity of the light source required to produce the illumination of Problem 13 if the distance to the light source is doubled?
15. What are the maximum and minimum transit times for light traveling from Jupiter to Mars? The orbital radii are 215 million kilometres for Mars and 725 million kilometres for Jupiter.
16. Find the intensity of two identical light sources located 0.454 m and 0.538 m from a point where the illumination is 8.46 $\ell m/m^2$.
17. Find the illumination on a surface due to three light sources, each with intensity 125 ℓm, located at distances of 1.85 m, 1.92 m, and 2.43 m from the surface.

Chapter 22

REFLECTION AND REFRACTION

The nature of light may still be somewhat of a mystery. However, its characteristics have been the subject of intensive study for hundreds of years. Light may be transmitted, reflected, or absorbed by a medium.

Anyone wearing glasses can appreciate the refraction of light as it bends passing from one medium to another. The index of refraction is a tool of the scientist to describe the ability of certain substances to bend light as it passes through them.

Our examination of the behavior of light begins with the study of images and reflection.

OBJECTIVES

The major goals of this chapter are to enable you to:

1. Express the laws of reflection.

2. Locate and describe images formed by plane, convex, and concave mirrors.

3. Apply the mirror formula to image formation.

4. Express the law of refraction.

5. Describe total internal reflection.

6. Locate and describe images formed by converging and diverging lenses.

22.1 MIRRORS AND IMAGES

Although the nature of light is complex, we know very well how light behaves. Every day we experience and unconsciously use our knowledge of what light does and depend on the ways it works. We are able to see objects because of the first characteristic we will study: reflection. **Reflection** is the name given to turning back a beam of light from a surface. Unlike sound, light does not require a medium (some kind of matter) to travel through and may be transmitted or passed through empty space. When light does strike a medium, the light may be reflected, absorbed, transmitted, or a combination of the three.

Mirrors show how light may be reflected. Any dark cloth shows how light may be absorbed. Window glass illustrates how light may be transmitted, or passed through a medium.

A medium may be classified according to how well light may be transmitted through it.

1. ***Transparent:*** almost all light passes through
   Examples: window glass, clear water
2. ***Translucent:*** some but not all light passes through
   Examples: murky water, light fog, skylight panels for farm buildings, stained glass
3. ***Opaque:*** almost all light reflected or absorbed
   Examples: wood, metal, plaster

These classifications are relative because some light is reflected from the surface of any medium, whereas some passes into or through it.

In studying reflection we look at what happens when light is turned back from a surface. The beam of a flashlight directed at a mirror shows several things about reflection. First, upon striking the surface of the glass, some of the light is reflected in all directions. This is called *scattering*. If there were no scattering, no light would reach our eye and we would be unable to observe the beam at all. However, only a very small part of the beam of light is scattered. Rough or uneven surfaces produce more scattering than do smooth ones. This scattering of light by uneven surfaces is called **diffusion.** Diffused lighting has many applications at home and in industry where bright glare is not desirable.

Nearly complete reflection (with very little scattering) is called **regular** (or specular) **reflection.** Regular reflection occurs when parallel or nearly parallel rays of light (such as sunlight and spotlight beams) are still parallel after being reflected from a surface (Fig. 22.1). Note that in Fig. 22.1, the incoming rays are referred to as *incident rays.*

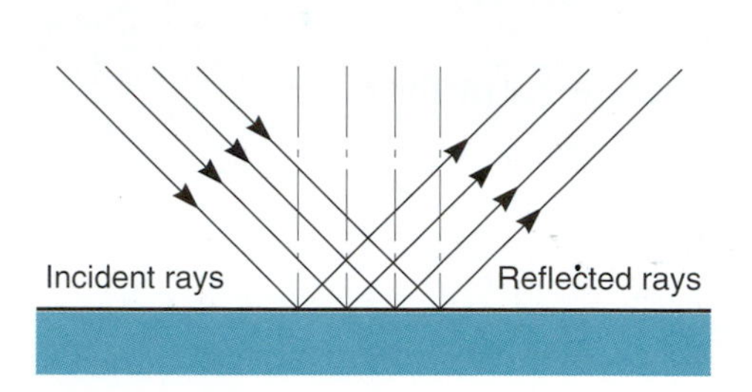

Figure 22.1
Regular reflection (reflected rays are parallel).

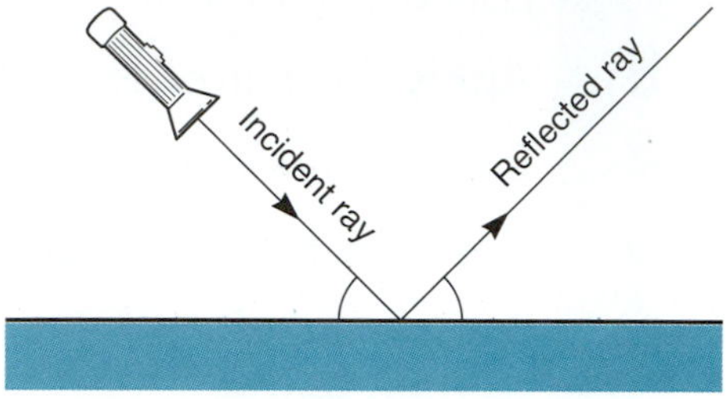

Figure 22.2
On a regular surface, the reflected rays leave at the same angle as the incident rays.

A flashlight beam on a mirror in a darkened room also shows something else about light striking a regular reflecting surface: The reflected rays of light leave the surface at the same angle at which the incident (coming-in) rays strike the surface (Fig. 22.2). Expressed another way, the angles measured from the normal or the perpendicular to the reflecting surface are equal. These angles are shown in Fig. 22.3.

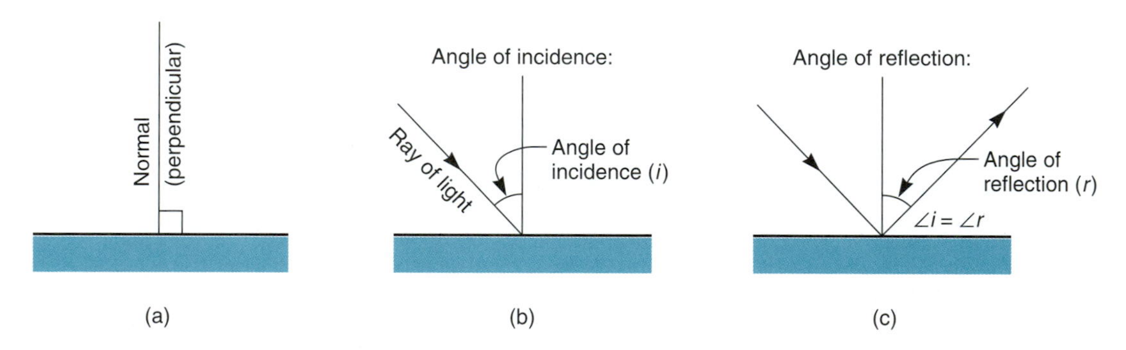

Figure 22.3
The angle of incidence is equal to the angle of reflection.

The same principle applies to curved surfaces (Fig. 22.4). This behavior of light rays is defined by the following law:

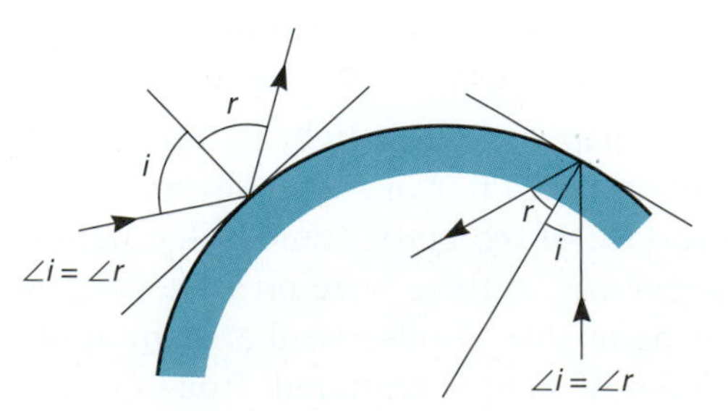

Figure 22.4
For curved surfaces as well $\angle i = \angle r$.

FIRST LAW OF REFLECTION

The angle of incidence *(i)* is equal to the angle of reflection *(r)*; that is,

$$\angle i = \angle r$$

Further observation of the light beam readily shows a second law:

SECOND LAW OF REFLECTION

The incident ray, the reflected ray, and the normal (perpendicular) to the surface all lie in the same plane.

These laws of reflection apply not only to light, but to all kinds of waves.

Mirrors of glass or any highly reflecting surface have countless practical applications, from rear-view mirrors in automobiles to watching for shoplifting in stores.

We look next at how images are formed by three widely used kinds of mirrors: plane, concave, and convex. **Plane mirrors** are flat. **Concave mirrors** are curved away from the observer [like the inside of a bowl; Fig. 22.5(a)], and **convex mirrors** are curved toward the observer [like Christmas tree ornaments; Fig. 22.5(b)].

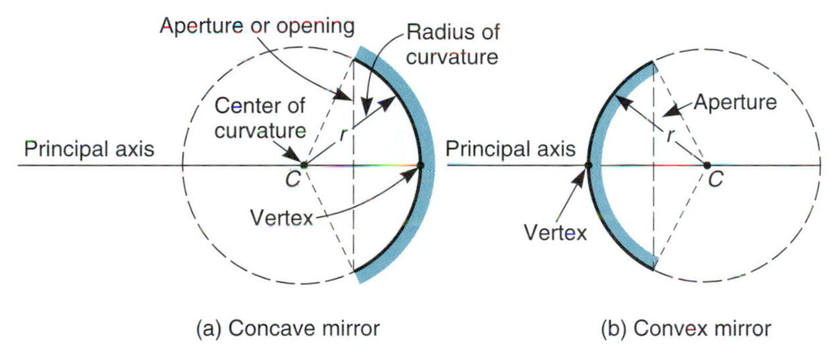

Figure 22.5
Spherical mirrors.

For convenience we will use spherical (sections of spheres) mirrors, although parabolic mirrors (reflecting surface in the shape of a parabola; Fig. 22.6) have wider practical use.

We consider next how images are formed by plane, concave, and convex mirrors. Images formed by mirrors may be **real images** (images actually formed by rays of light) or **virtual images** (images that only appear to the eye to be formed by rays of light).

Real images made by a single mirror are always inverted (upside down) and may be larger, smaller, or the same size as the object. They can be shown on a screen. Virtual images are always erect and may be larger, smaller, or the same size as the object. They cannot be shown on a screen.

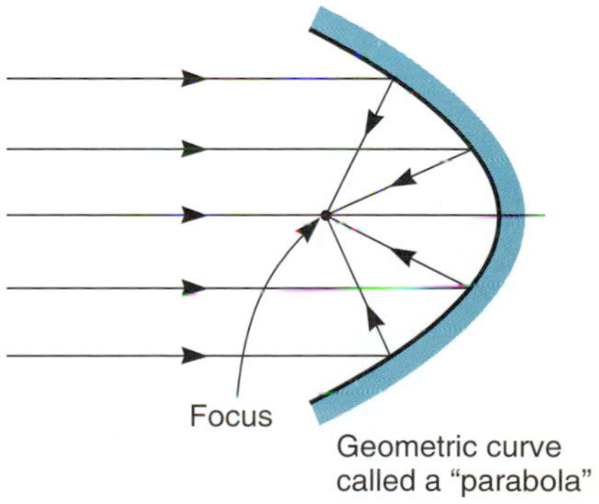

Focus

Geometric curve called a "parabola"

Figure 22.6
Parabolic mirror.

22.2 IMAGES FORMED BY PLANE MIRRORS

Plane mirror images are always erect, virtual, and appear as far behind the mirror as the distance the object is in front of the mirror. Note that plane mirrors also reverse right and left, so the right hand held in front of a plane mirror appears, in the mirror, to be a left hand.

We use light-ray diagrams as a method to illustrate how our eyes see images in the various kinds of mirrors. We do this by representing rays of light and lines of sight with straight lines (Figs. 22.7 and 22.8). This method can be used to construct diagrams and locate the images formed. Simply view the object from two or more separate places and construct the light-ray lines.

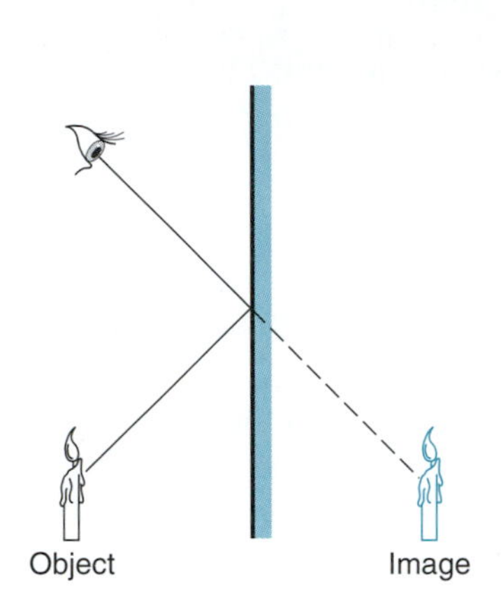

Figure 22.7
Plane mirror image.

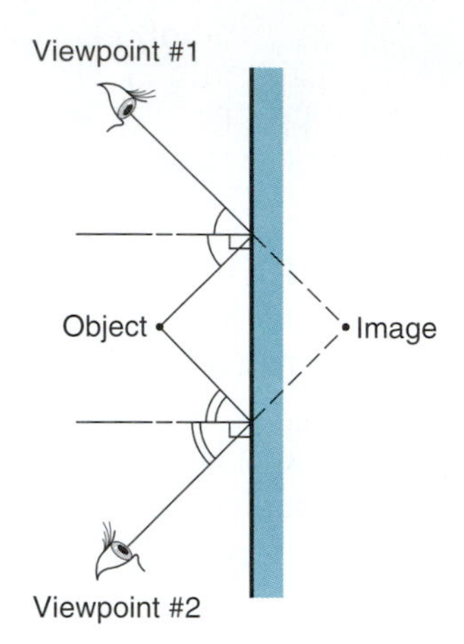

Figure 22.8
Location of an image in a plane mirror.

22.3 IMAGES FORMED BY CONCAVE MIRRORS

Figure 22.9
An inverted image is seen in a large spoon.

Find a shiny tablespoon and look at your image in it (Fig. 22.9). Now turn it over and look again. Of course, the images are very different, in that one is erect and the other inverted (upside down). We use ray diagrams to show just why this happens. As we shall see, the kind of image produced depends on the location of the object with respect to the mirror.

Becoming familiar with how light is reflected from spherical surfaces will enable us to construct these diagrams. Fig. 22.10(a) shows a spherical mirror with the key terms identified. The focal point (F) is the point through which all rays parallel to the principal axis will be reflected as shown in Fig. 22.10(b). Also note

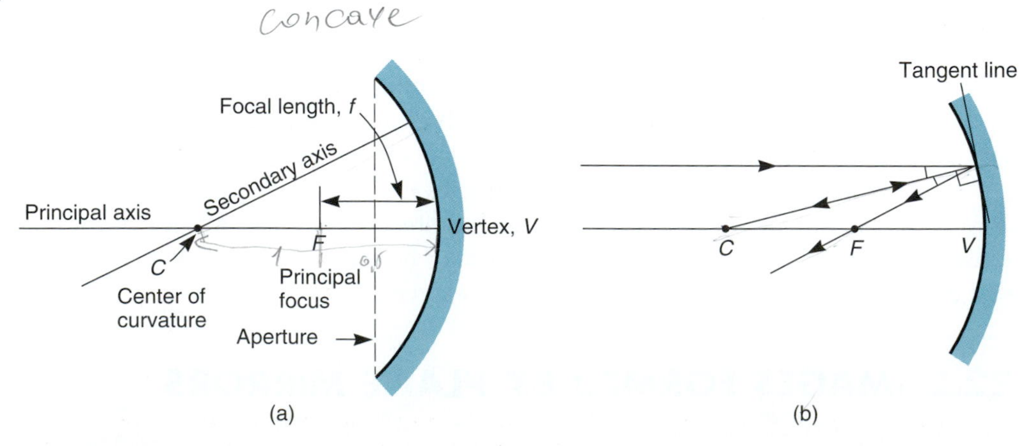

Figure 22.10
Spherical mirror.

that any ray through the center of curvature is reflected straight back because any such ray is perpendicular to the surface of the mirror at the point the ray strikes the mirror. For mirrors with small apertures [Fig. 22.10(a)], the focal length *(f)* is one-half the radius of curvature *(R)*. That is, $f = \frac{R}{2}$.

If the object is placed at the focal point, no image will be formed because the rays of light will be reflected back parallel to the principal axis (Fig. 22.11). The location of the reflected ray may be found by using the laws of reflection (Fig. 22.12). The angle of incidence is equal to the angle of reflection. The normal or perpendicular is a line made between the center of curvature and the point the light ray strikes the mirror.

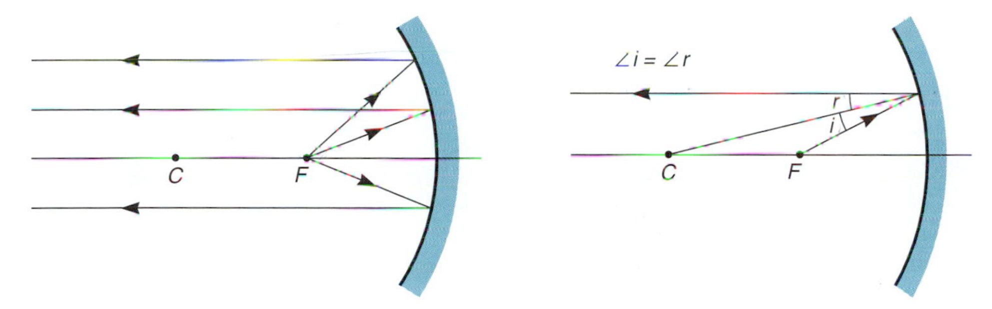

Figure 22.11
No image is formed if the object is located at the focal point.

Figure 22.12

Now consider the more common case, in which the object is beyond the center of curvature [for example, looking into a tablespoon: Fig. 22.13(a)]. Note that again we use the fact that a ray parallel to the principal axis is reflected through the focal point and a ray through the focal point is reflected parallel to the principal axis. Then, where the two rays intersect, a point on the image is formed. (In this case it is the flame of the candle. Can you see that the candle base image lies on the principal axis because the object base is also on that line?) The same method would be used for the case where the candle base extends below the principal axis [Fig. 22.13(b)].

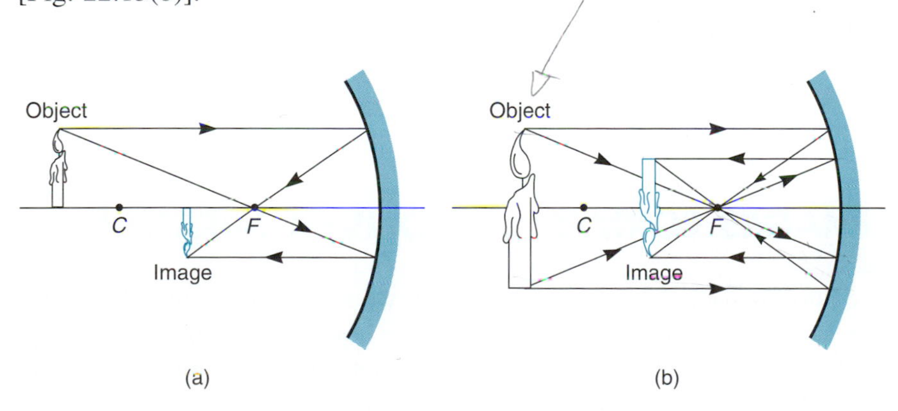

Figure 22.13
Images formed in spherical mirrors.

Now apply these principles to the diagrams of other images formed by concave mirrors (Fig. 22.14). Decide whether the image is real or virtual; erect or inverted; larger, smaller, or the same size; and where located. Note that the only

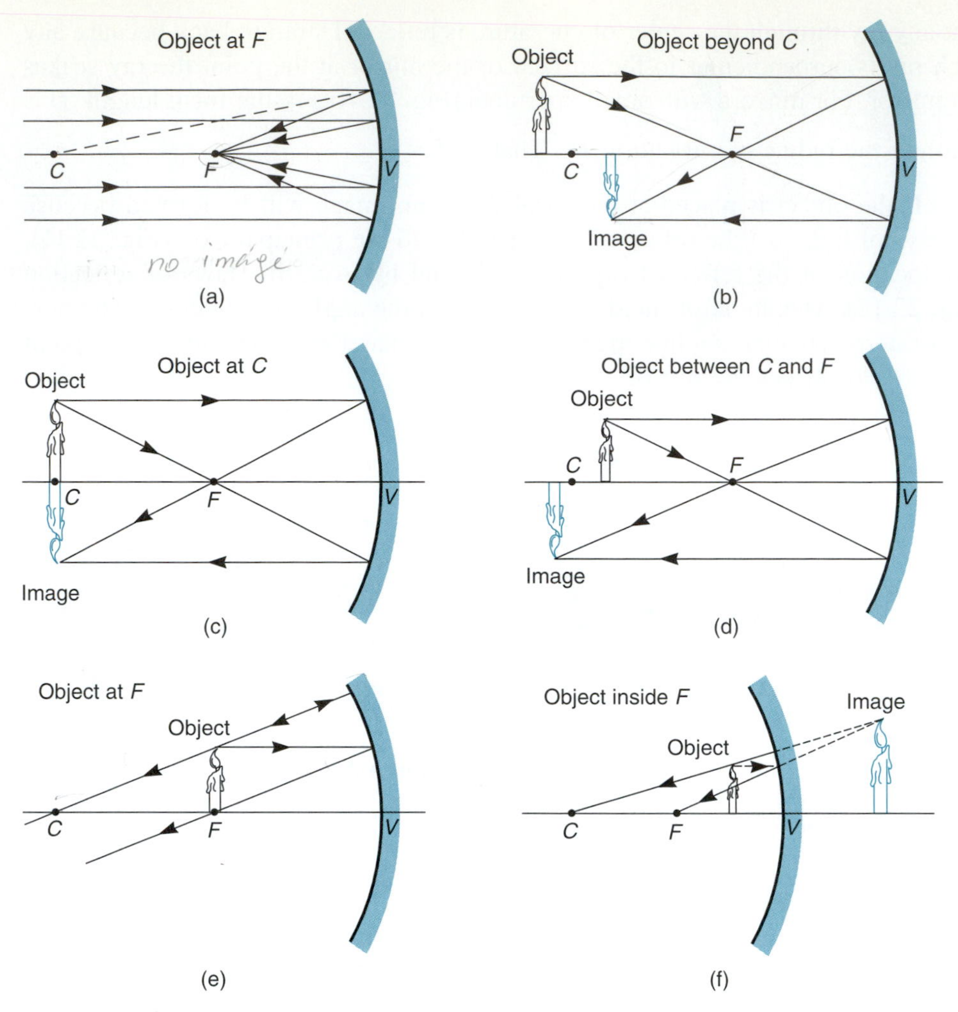

Figure 22.14
Formation of images in concave mirrors.

time a virtual image is produced is when the object is between the focal point and the mirror [Fig. 22.14(f)]. The construction of the diagram is the same as the other cases except that the light rays are converging (coming together) and must be extended behind the mirror (where the image appears to be) forming the virtual image.

22.4 IMAGES FORMED BY CONVEX MIRRORS

By looking into the back side of our tablespoon, we see an erect, virtual, smaller image. Use the diagram shown in Fig. 22.15 to see how such an image is formed. Curved surface mirrors are used in some telescopes, spotlights, and automobile headlights. But because spherical mirrors produce clear images over only a very small portion of their surfaces (small aperture), the surfaces used commercially are actually another geometric shape, that of a parabola or parabolic, as mentioned before. For apertures wider than about 10°, spherical mirrors produce fuzzy images because all parallel rays are not reflected through the focal point. This is called *spherical aberration.*

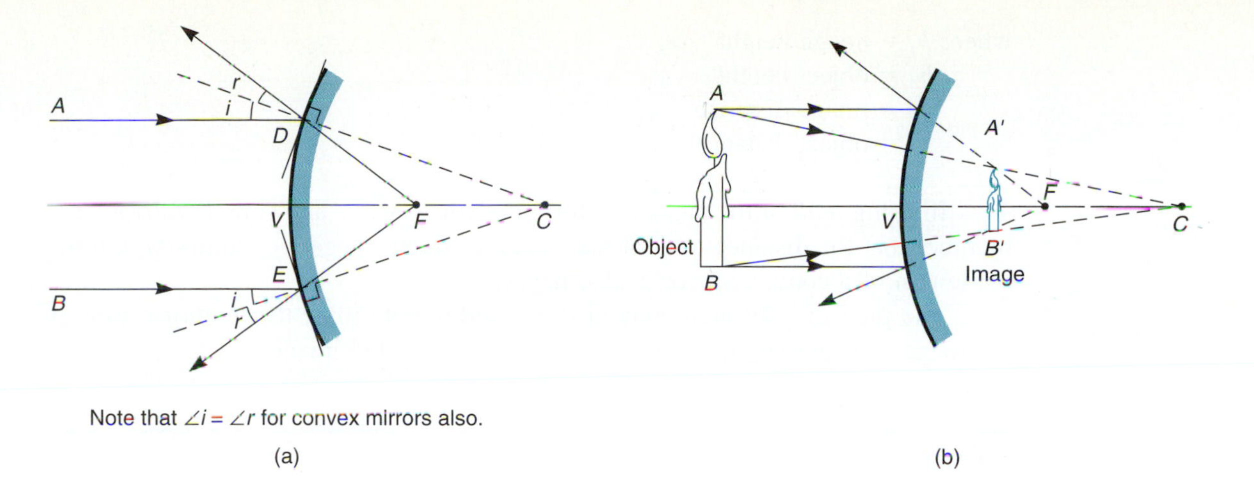

Figure 22.15
Formation of images in convex mirrors.

22.5 THE MIRROR FORMULA

As we might expect from the previous cases, the **focal point** (distance from the focus to the mirror), the distance from the object to the mirror, and the distance from the image to the mirror are all related (Fig. 22.16). This relationship can be expressed as the *mirror formula:*

$$\frac{1}{f} = \frac{1}{s_o} + \frac{1}{s_i}$$

where f = focal length of mirror
s_o = distance of object from mirror
s_i = distance of image from mirror

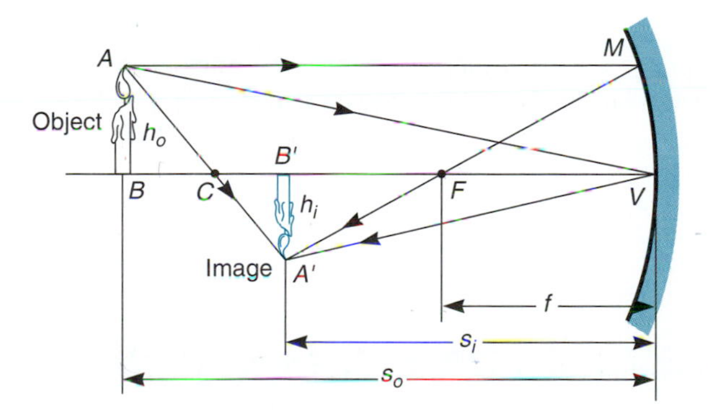

Figure 22.16
The mirror formula is determined by f, s_o, and s_i.

Therefore, if two of the three distances f, s_o, and s_i are known, the third can be calculated.

A second formula shows how the height of the object and the height of the image depend on the object distance and the image distance:

$$\frac{h_i}{h_o} = \frac{s_i}{s_o}$$

where h_i = image height
h_o = object height
s_i = image distance
s_o = object distance

In using *both* of the preceding formulas for concave and convex mirrors, remember that the distance to a virtual image is always negative; similarly, the focal length of a convex mirror is also negative.

The preceding formula may also be used to determine the magnification of an image. A negative magnification implies an inverted image.

EXAMPLE

An object 10.0 cm in front of a convex mirror forms an image 5.00 cm behind the mirror. What is the focal length of the mirror?

Sketch:

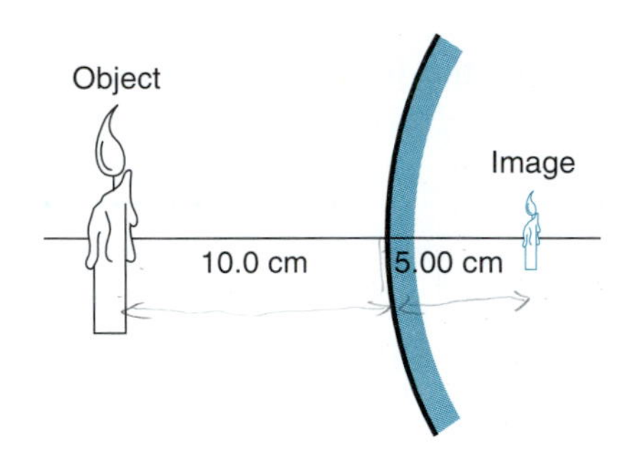

Data:

$s_o = 10.0$ cm

$s_i = -5.00$ cm   ***Note:*** The image is virtual (appears behind the mirror) so s_i is given a (−) sign to show this [won't f also be (−)?]

$f = ?$

Basic Equation:

$$\frac{1}{f} = \frac{1}{s_o} + \frac{1}{s_i}$$

Working Equation: Same

Substitution:

$$\frac{1}{f} = \frac{1}{10.0 \text{ cm}} + \frac{1}{-5.00 \text{ cm}}$$

Using a calculator as in Chapter 17, we find that

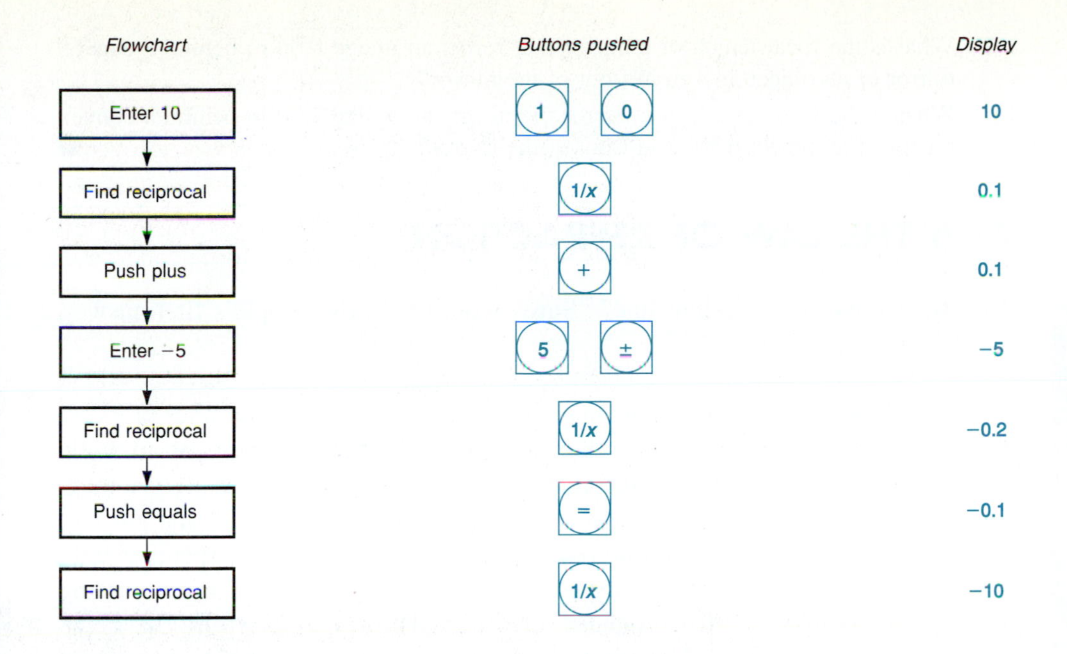

Thus,

$$f = -10.0 \text{ cm}.$$

Remember that f and s_i may be negative only when forming virtual images and/or using convex mirrors.

PROBLEMS 22.5

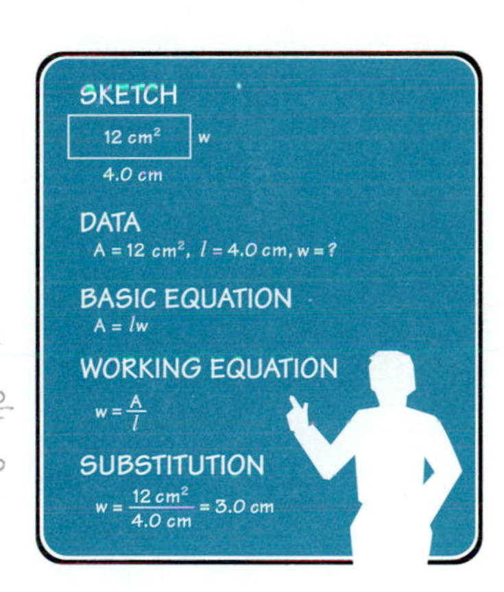

Use the formulas $\frac{1}{f} = \frac{1}{s_o} + \frac{1}{s_i}$ and $\frac{h_i}{h_o} = \frac{s_i}{s_o}$ for Problems 1–8.

1. Given $s_o = 1.65$ cm and $s_i = 6.00$ cm, find f.
2. Given $f = 15.0$ cm and $s_i = 3.00$ cm, find s_o.
3. Given $s_i = 14.5$ cm and $f = 10.0$ cm, find s_o.
4. Given $s_i = -10.0$ cm and $f = -5.00$ cm, find s_o.
5. Given $s_o = 7.35$ cm and $s_i = 17.0$ cm, find f.
6. Given $h_i = 2.75$ cm, $h_o = 4.50$ cm, and $s_i = 6.00$ cm, find s_o.
7. Given $h_o = 12.0$ cm, $s_i = 13.0$ cm, and $s_o = 25.0$ cm, find h_i.
8. Given $h_i = 3.50$ cm, $h_o = 2.50$ cm, and $s_i = 15.5$ cm, find s_o.
9. If an object is 2.50 m tall and 8.60 m from a large mirror with an image formed 3.75 m from the mirror, what is the height of the image formed?
10. An object 30.0 cm tall is located 10.5 cm from a concave mirror with focal length 16.0 cm.
    (a) Where is the image located?
    (b) How high is it?
11. An object and its image in a concave mirror are the same height when the object is 20.0 cm from the mirror. What is the focal length of the mirror?
12. An object 12.6 cm in front of a convex mirror forms an image 6.00 cm behind the mirror. What is the focal length of the mirror?
13. Find the focal length of a mirror that forms an image 3.55 cm behind the mirror of an object 24.5 cm in front of the convex mirror.

14. What is the focal length of a mirror that forms an image 5.66 m behind a convex mirror of an object 34.4 m in front of the mirror?
15. What is the focal length of a mirror that forms an image 2.30 m behind a convex mirror of an object 6.50 m in front of the mirror? ■

22.6 THE LAW OF REFRACTION

Figure 22.17
Refraction is the bending of light at an angle as it passes from one medium to another.

Does light travel in a straight line? "Sure" would be most people's first answer. But does it always? Fig. 22.17.

The answers to our questions may be found in the study of another property of light—refraction. **Refraction** is the bending of light as it passes at an angle from one medium to another of different optical density. Optical density is a measurable property of a material that is related to the speed of light in that particular material. For example, water is optically denser than air and the speed of light in water is less than the speed of light in air. This change of speed when passing from one medium to another produces refraction. The wave shown in Fig. 22.18 illustrates how this occurs. Note that when passing from one medium to another perpendicular to the surface (called the *interface*) the wave is not bent, although the speed of the wave is slowed.

When the wave passes obliquely (at an angle) through, the entire wave front does not all strike the surface at the same time. The first part of the wave to strike the glass is slowed before the part striking later—thus the bending of the wave (Fig. 22.19). Draw your own diagram to show whether a fish in a pond, as viewed from the bank, is actually nearer the surface or the bottom of the pond than it appears to be.

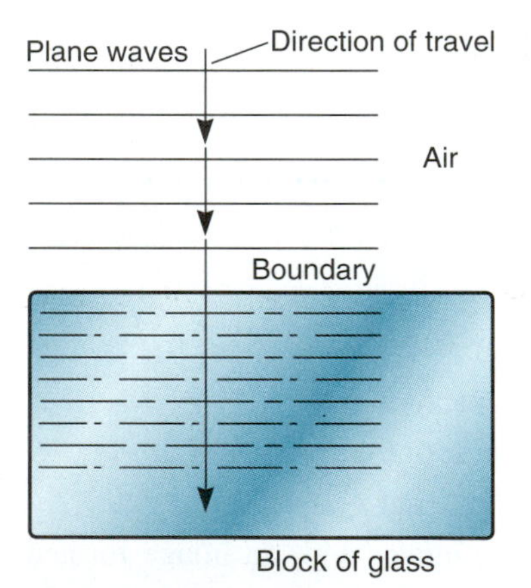

Figure 22.18
The speed of light is different in different mediums.

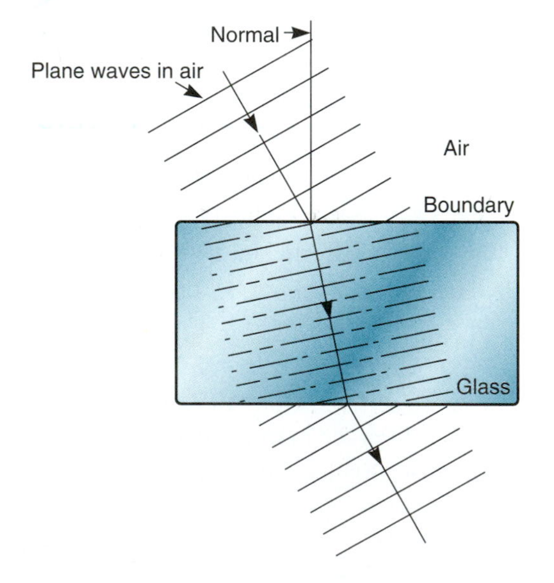

Figure 22.19
The wave bends when all parts of the wave don't strike the glass at the same time. It also bends when leaving the glass at an angle.

As we might expect, since the speed of light increases when it leaves a denser medium to enter the air, bending also occurs. In this case, however, instead of the light bending toward the perpendicular or normal, the light is bent away from the normal when passing from the denser medium to the less dense (Fig. 22.19). This illustrates the following law:

LAW OF REFRACTION

When a beam of light passes, at an angle, from a medium of less optical density to a denser medium, the light is bent *toward* the normal. When a beam of light passes, at an angle, from a medium of more optical density to one less dense, the light is bent *away from* the normal.

Note: The angles of incidence and refraction are measured from the *normal* as in Fig. 22.20.

A Dutch mathematician, Willebord Snell, found a formula to identify a property of substances that pass light called the **index of refraction.** This index is a constant for a particular material and is independent of the angle the light strikes. It may be expressed in two ways:

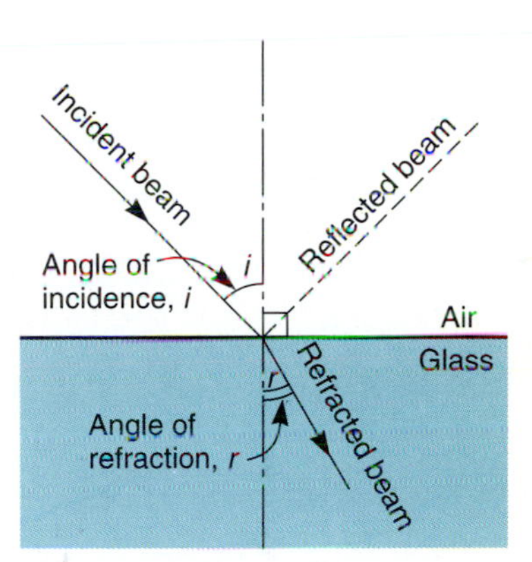

Figure 22.20
The angles of incidence and refraction are measured from the normal.

1. As **Snell's law,** the index of refraction, n, equals the sine of the angle of incidence divided by the sine of the angle of refraction, where the incident ray is passing through a vacuum or air:

$$n = \frac{\sin i}{\sin r}$$

2. As the ratio of the speed of light in a vacuum (nearly the same as in air) to the speed of light in the particular substance:

$$n = \frac{\text{speed of light in vacuum}}{\text{speed of light in substance}}$$

EXAMPLE 1

The angle of incidence of light passing from air to water is 59.0°. The angle of refraction is 41.0°. What is the index of refraction of the water?

Data:

$$i = 59.0°$$
$$r = 41.0°$$
$$n = ?$$

Basic Equation:

$$n = \frac{\sin i}{\sin r}$$

Working Equation: Same

Substitution:

$$n = \frac{\sin 59.0°}{\sin 41.0°}$$
$$= 1.31$$

EXAMPLE 2

The index of refraction of water is 1.33. What is the speed of light in water?

Data:

$$n = 1.33$$
$$c = 3.00 \times 10^8 \text{ m/s}$$
$$v_{\text{water}} = ?$$

Basic Equation:

$$n = \frac{\text{speed of light in vacuum}}{\text{speed of light in substance}}$$

Working Equation:

$$\text{speed of light in water} = \frac{\text{speed of light in vacuum}}{n}$$

Substitution:

$$\text{speed of light in water} = \frac{3.00 \times 10^8 \text{ m/s}}{1.33}$$
$$= 2.26 \times 10^8 \text{ m/s}$$

The index of refraction for some common substances is given in Table 22.1.

Table 22.1 Indices of Refraction for Various Substances

Substance	*Index of Refraction*
Air, dry (STP)	1.00029
Alcohol, ethyl	1.360
Benzene	1.501
Carbon dioxide (STP)	1.00045
Carbon disulfide	1.625
Carbon tetrachloride	1.459
Diamond	2.417
Glass, crown flint	1.575
Lucite	1.50
Quartz, fused	1.45845
Water, distilled	1.333
Water vapor (STP)	1.00025

22.7 TOTAL INTERNAL REFLECTION

If the angle of refraction is 90° or greater, a beam of light does not leave the medium but is reflected back inside the medium (Fig. 22.21). This is called **total internal reflection.** Total internal reflection occurs when the angle of incidence is greater than the **critical angle.** The critical angle is the smallest angle that produces an angle of refraction of 90° or more. It may be expressed by the formula

$$\sin i_c = \frac{1}{n}$$

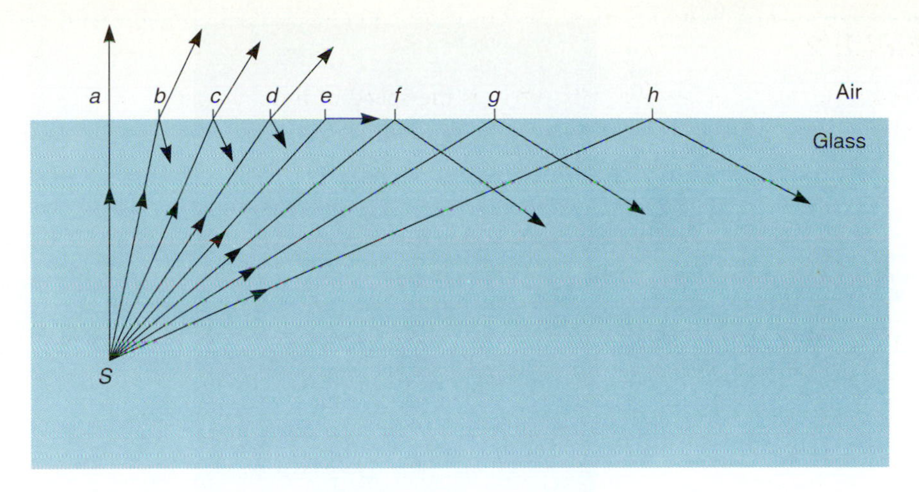

Figure 22.21
A point light source S is shown in glass. For all angles of incidence less than the critical angle θ_c, the light ray is refracted and leaves the glass as shown at a, b, c, and d. When the angle of incidence equals the critical angle θ_c, the refracted ray points along the glass–air surface as shown at e. For all angles of incidence greater than the critical angle θ_c, total internal reflection of the light ray occurs as shown at f, g, and h.

where i_c = critical angle of incidence
n = index of refraction of denser medium

Note: The incident ray is passing through a vacuum or air.

EXAMPLE

What is the critical angle of incidence for water that has an index refraction of 1.33?

Data:

$$n = 1.33$$
$$i_c = ?$$

Basic Equation:

$$\sin i_c = \frac{1}{n}$$

Working Equation: Same

Substitution:

$$\sin i_c = \frac{1}{1.33} = 0.752$$
$$i_c = 49°$$

Where there is total internal reflection, no light enters the air; it is totally reflected within the glass (Fig. 22.22). The property of having a very small critical angle gives a diamond its brilliance by multiple internal reflections before the light passes out through the top. An example of the practical application of this principle is fiber optics. Light may be transferred inside flexible glass or plastic fibers, which are transparent but keep nearly all the light inside because the light is reflected along the inside surface of the fiber.

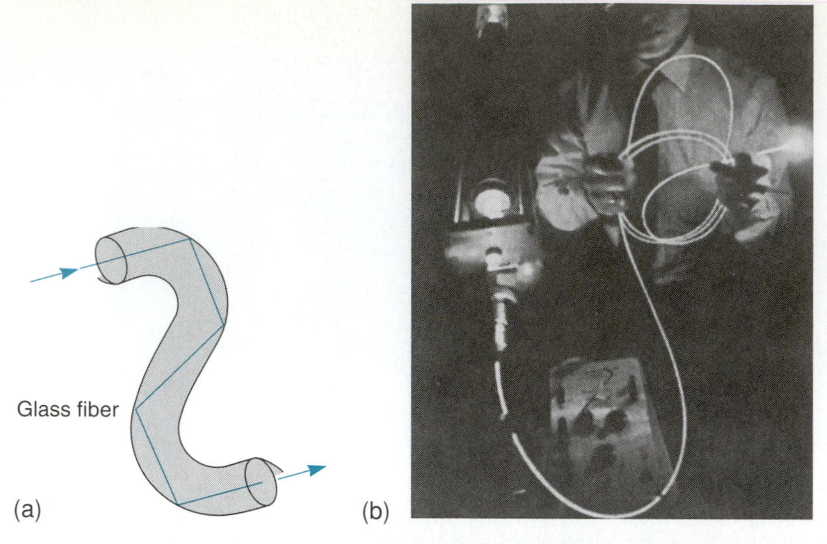

Figure 22.22
(a) Light travels inside an optical or glass fiber like a light pipe in which the light is always incident at an angle greater than the critical angle. Thus, no light escapes the optical fiber by refraction. Fiber optics is essential to light-wave communications systems. (b) Total internal reflection within the tiny fibers of this light pipe makes it possible to transmit light in complex paths with minimal loss.

22.8 TYPES OF LENSES

Many technical applications use the principles of refraction ranging from testing the nature of liquids to microscopes to eyeglasses. Attention is now directed to the use of refraction in applications using lenses. Lenses may be converging or diverging. **Converging lenses** bend the light passing through them to some point beyond the lens. Converging lenses are always thicker in the middle than on the edges [Fig. 22.23(a)]. **Diverging lenses** bend the light passing through them so as to spread the light. Diverging lenses are thicker on the edges than at the center [Fig. 22.23(b)].

Understanding how light is bent in lenses may be made easier by looking at the bending of light passing through a prism (Fig. 22.24). Recall and apply the law of refraction, noting that light is bent toward the normal when passing into the glass and away from the normal when passing from the glass back to the air.

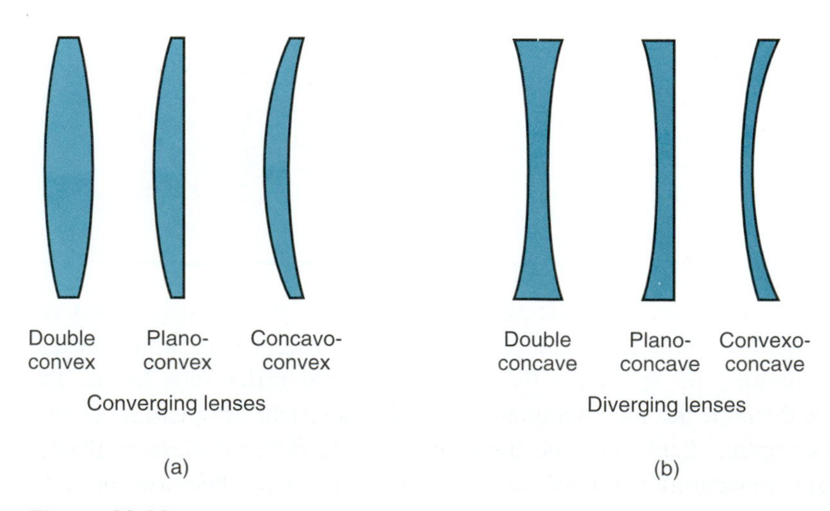

Figure 22.23

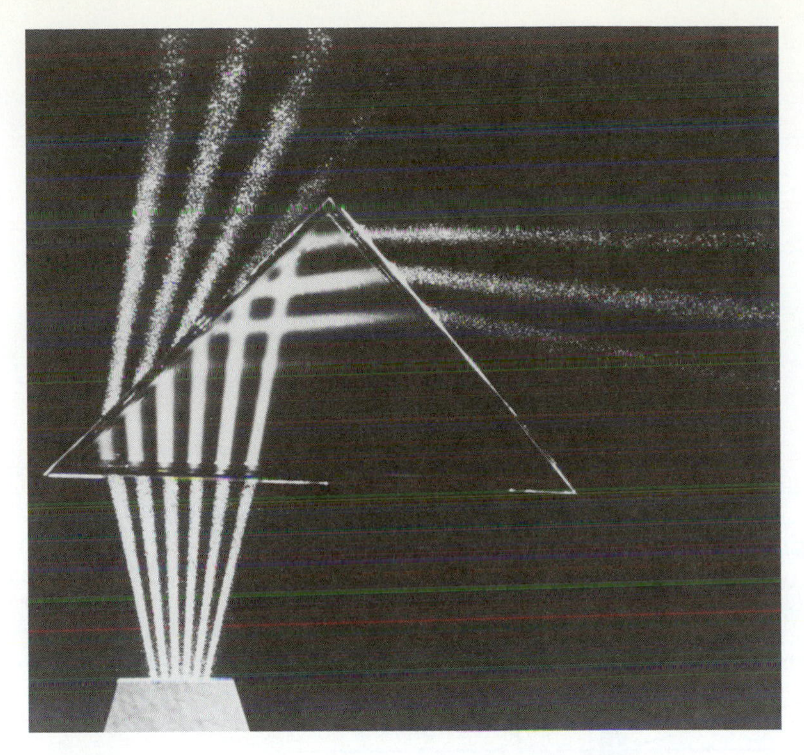

Figure 22.24
Bending of light passing through a prism. Note that some point light sources are refracted up through the prism and some are totally internally reflected back inside the prism before being refracted through the prism to the right.

22.9 IMAGES FORMED BY CONVERGING LENSES

As with mirrors, we use light-ray diagrams to help us to understand how light can be bent with lenses. Every lens has a focal length—that distance from the lens center to the point where parallel beams directed through the lens come together if converging or *appear* to come together if diverging (Fig. 22.25).

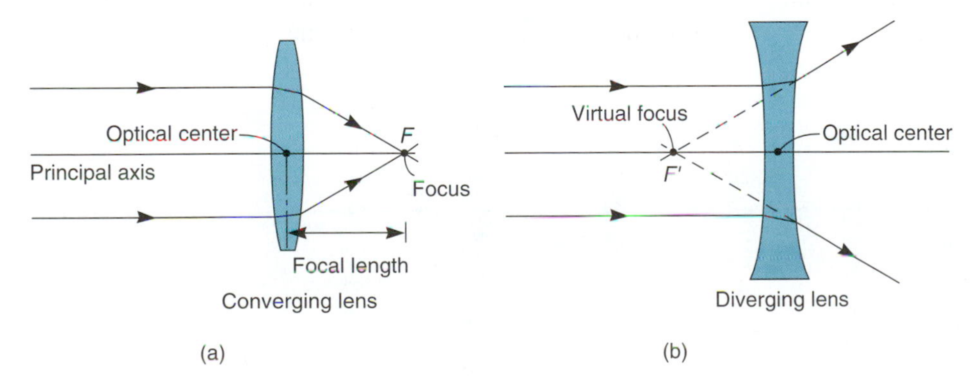

Figure 22.25

The location of the focus depends upon the curvature of the lens and the index of refraction of the glass or material of which the lens is made. Rays of light passing through the optical center of the lens are refracted so little that we may consider them as going straight through (Fig. 22.26).

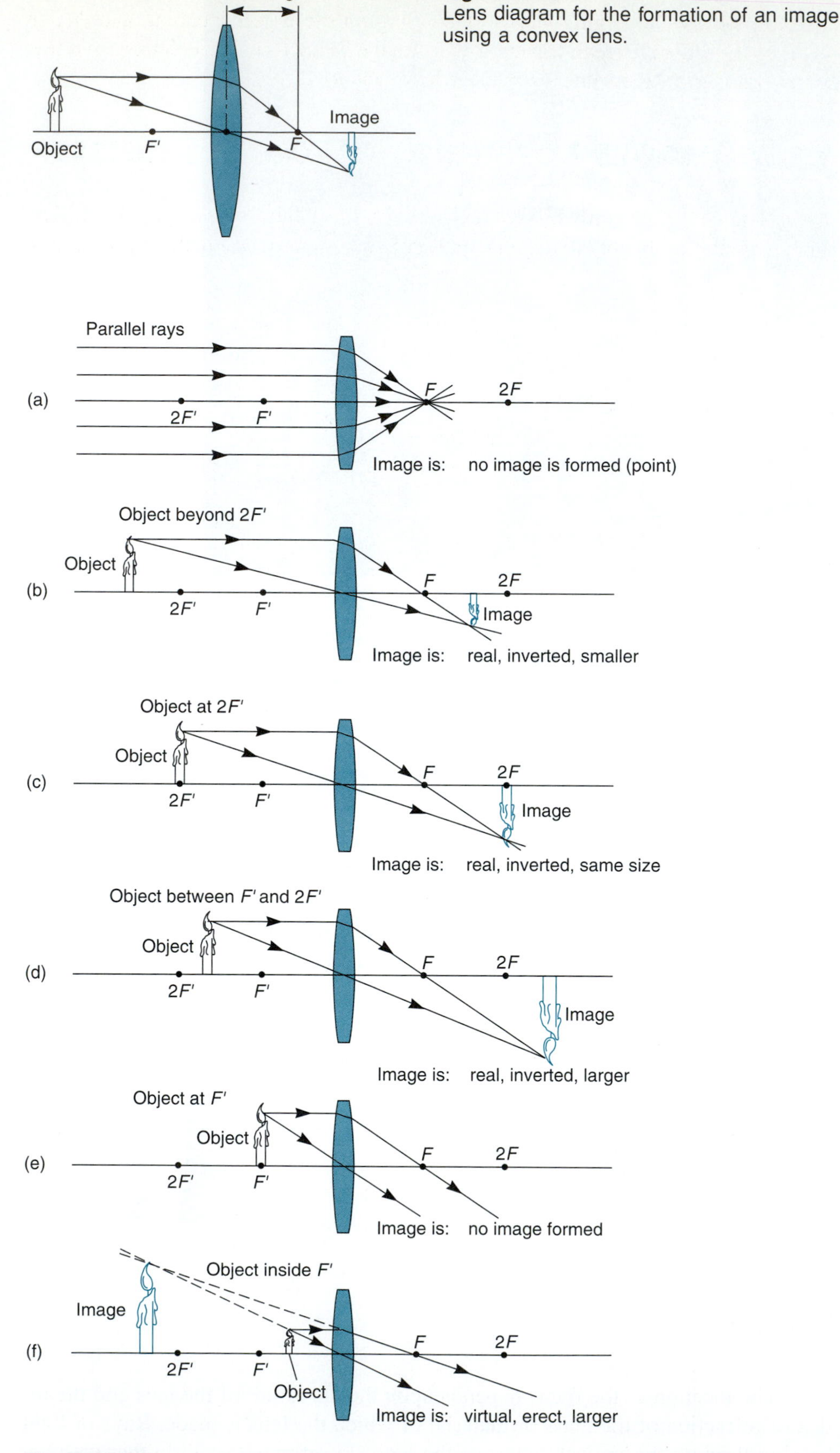

Figure 22.26
Lens diagram for the formation of an image using a convex lens.

Figure 22.27
Images formed by converging lenses.

Now apply these principles to the diagrams shown in Fig. 22.27 of other images formed by converging lenses, depending on the object location, and decide whether the image is real, virtual, or no image formed at all; erect or inverted; larger, smaller, or the same size; and where it is located.

22.10 IMAGES FORMED BY DIVERGING LENSES

Virtual images are the only images produced by diverging lenses (Fig. 22.28). The same formulas that apply to mirrors also apply to converging and diverging lenses.

$$\frac{1}{f} = \frac{1}{s_o} + \frac{1}{s_i}$$

where f = focal length
s_o = object distance from lens center
s_i = image distance from lens center

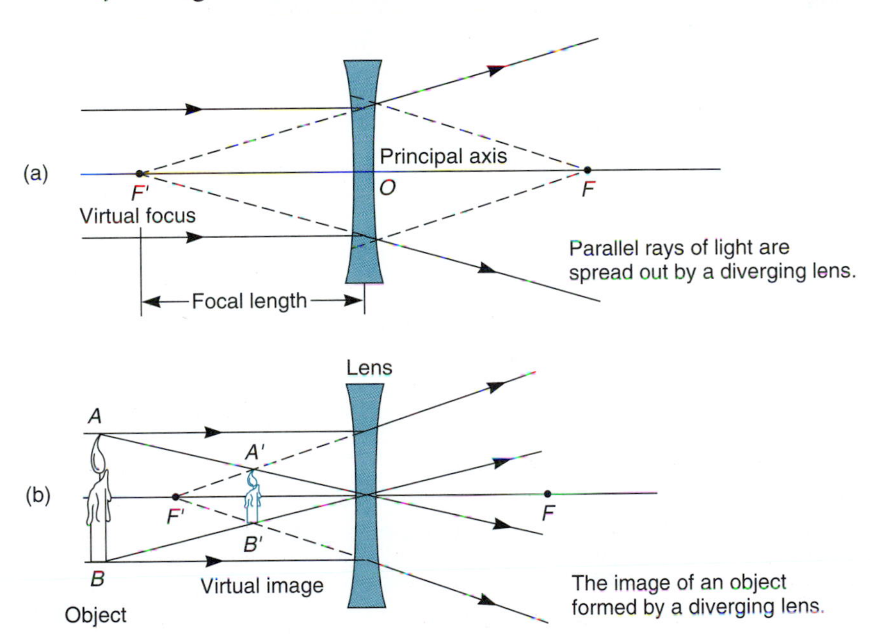

Figure 22.28
Images formed by diverging lenses.

Therefore, if two of the three distances, f, s_o, and s_i are known, the third can be calculated. Also,

$$\frac{h_i}{h_o} = \frac{s_i}{s_o}$$

where h_i = height of image
h_o = height of object
s_i = image distance from lens center
s_o = object distance from lens center

So if three of the four quantities h_o, h_i, s_o, and s_i are know, the fourth can be calculated.

Remember, when the image is virtual, s_i is negative ($-$), and for diverging lenses both s_i and f are negative.

EXAMPLE

An object 3.00 cm tall is placed 24.0 cm from a converging lens. A real image is formed 8.00 cm from the lens.

(a) What is the focal length of the lens?

(b) What is the size of the image?

(a) Data:

$$s_o = 24.0 \text{ cm}$$
$$s_i = 8.00 \text{ cm}$$
$$f = ?$$

Basic Equation:

$$\frac{1}{f} = \frac{1}{s_o} + \frac{1}{s_i}$$

Working Equation: Same

Substitution:

$$\frac{1}{f} = \frac{1}{24.0 \text{ cm}} + \frac{1}{8.00 \text{ cm}}$$

Using a calculator as before, we find that

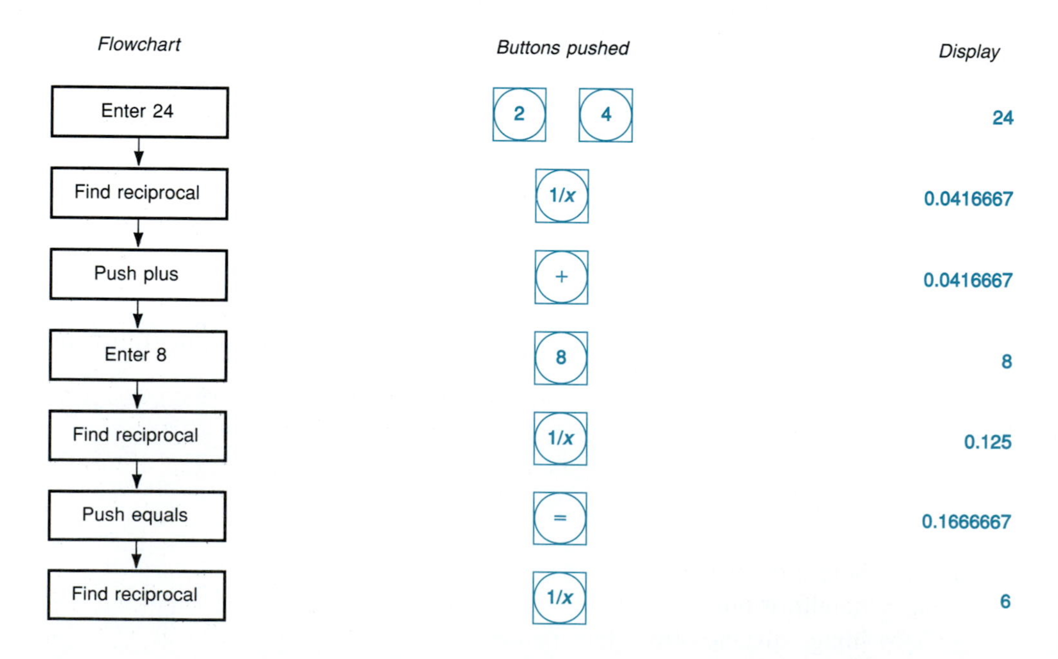

Thus,

$$f = 6.00 \text{ cm}$$

(b) Data:

$$s_o = 24.0 \text{ cm}$$
$$s_i = 8.00 \text{ cm}$$
$$h_o = 3.00 \text{ cm}$$
$$h_i = ?$$

Basic Equation:

$$\frac{h_i}{h_o} = \frac{s_i}{s_o}$$

Working Equation:

$$h_i = \frac{s_i h_o}{s_o}$$

Substitution:

$$h_i = \frac{(8.00 \text{ cm})(3.00 \text{ cm})}{24.0 \text{ cm}}$$
$$= 1.00 \text{ cm}$$

PROBLEMS 22.10

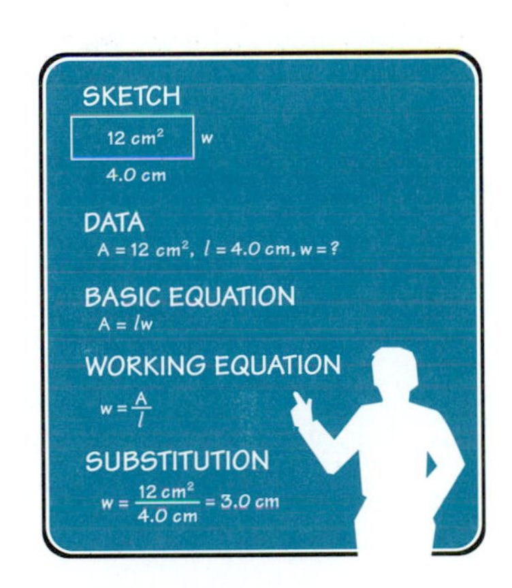

1. Find the index of refraction of a medium for which the angle of incidence of a light beam is 31.5° and angle of refraction is 25.6°.
2. If the index of refraction of a medium is 2.40 and the angle of incidence is 14.6°, what is the angle of refraction?
3. If the index of refraction of a liquid is 1.50, find the speed of light in that liquid.
4. The angle of incidence of light passing from air to a liquid is 38.0°. The angle of refraction is 24.5°. What is the index of refraction of the liquid?
5. If the critical angle of a liquid is 42.4°, find the index of refraction for that liquid.
6. If the index of refraction of a substance is 2.45, find its critical angle of incidence.
7. A converging lens has a focal length of 15.0 cm. If it is placed 48.0 cm from an object, how far from the lens will the image be formed?
8. An object 2.50 cm tall is placed 20.0 cm from a converging lens. A real image is formed 9.00 cm from the lens.
   (a) What is the focal length of the lens?
   (b) What is the size of the image?
9. The focal length of a lens is 5.00 cm. How far from the lens must the object be to produce an image 1.50 cm from the lens?
10. If the distance from the lens in your eye to the retina is 19.0 mm, what is the focal length of the lens when reading a sign 40.0 cm from the lens?
11. An object 5.00 cm tall is placed 15.0 cm from a converging lens, and a real image is formed 7.50 cm from the lens.
    (a) What is the focal length of the lens?
    (b) What is the size of the image?
12. An object 4.50 cm tall is placed 18.0 cm from a converging lens with a focal length of 26.0 cm.
    (a) What is the location of the image?
    (b) What is its size?
13. What are the size and location of an image produced by a converging lens with a focal length of 19.5 cm of an object 5.76 cm from the lens and 1.45 cm high?

14. What are the size and location of an image produced by a convex lens with a focal length of 14.5 cm of an object 10.5 cm from the lens and 2.35 cm high?
15. What is the focal length of a convex lens that produces an image twice as large as the object at a distance of 13.3 cm from the object? ■

GLOSSARY

Concave Mirror A mirror with a surface that curves away at the center of the mirror from an observer. (p. 519)
Converging Lens A lens that bends the light passing through it to some point beyond the lens. Converging lenses are thicker in the middle. (p. 530)
Convex Mirror A mirror with a surface that curves out at the center of the mirror toward an observer. (p. 519)
Critical Angle The angle of incidence at or above which all light striking a surface is totally internally reflected. (p. 528)
Diffusion Scattering of light by an uneven surface. (p. 518)
Diverging Lens A lens that bends the light passing through it so as to spread the light. Diverging lenses are thicker at the edges than at the center. (p. 530)
First Law of Reflection States that the angle of incidence is equal to the angle of reflection. (p. 519)
Focal Point The distance of the focus from the mirror. (p. 523)
Index of Refraction A measure of the optical density of a material. Equal to the ratio of the speed of light in vacuum to the speed of light in the material. (p. 527)
Law of Refraction States that when a beam of light passes, at an angle, from a medium of less optical density to a denser media, the light is bent toward the normal. When a beam passes from a medium of more optical density to one less dense, the light is bent away from the normal. (p. 527)
Opaque A medium that absorbs or reflects almost all light. (p. 517)
Plane Mirror A mirror with a flat surface. (p. 519)
Real Image An image actually formed by rays of light. (p. 519)
Reflection The turning back of all or part of a beam of light at a surface. (p. 517)
Refraction The bending of light as it passes at an angle from one medium to another of different optical density. (p. 526)
Regular Reflection Reflection of light with very little scattering. (p. 518)
Second Law of Reflection States that the incident ray, the reflected ray, and the normal (perpendicular) to the reflecting surface all lie in the same plane. (p. 519)
Snell's Law States that the index of refraction equals the sine of the angle of incidence divided by the sine of the angle of refraction. (p. 527)
Total Internal Reflection A condition such that the light striking a surface does not pass through the surface, but is completely reflected back. (p. 528)
Translucent A medium through which not all light passes. (p. 517)
Transparent A medium through which almost all light passes. (p. 517)
Virtual Image An image that only appears to the eye to be formed by rays of light. (p. 519)

FORMULAS

22.5 $$\frac{1}{f} = \frac{1}{s_o} + \frac{1}{s_i}$$

$$\frac{h_i}{h_o} = \frac{s_i}{s_o}$$

22.6 $n = \dfrac{\sin i}{\sin r}$

$$n = \frac{\text{speed of light in vacuum}}{\text{speed of light in substance}}$$

22.7 $\sin i_c = \dfrac{1}{n}$

22.10 $\dfrac{1}{f} = \dfrac{1}{s_o} + \dfrac{1}{s_i}$

$$\frac{h_i}{h_o} = \frac{s_i}{s_o}$$

REVIEW QUESTIONS

1. Stained glass is an example of
   (a) a transparent material.
   (b) a translucent material.
   (c) an opaque material.
   (d) none of the above.
2. A virtual image may be
   (a) larger than the object.
   (b) smaller than the object.
   (c) erect.
   (d) all of the above.
   (e) none of the above.
3. A real image may be
   (a) erect.
   (b) shown on a screen.
   (c) formed by a flat mirror.
   (d) none of the above.
4. Explain the difference between diffusion and regular reflection.
5. In your own words, explain the first law of reflection.
6. In your own words, explain the second law of reflection.
7. Describe the type of images formed by plane mirrors.
8. Differentiate between real and virtual images.
9. Differentiate between a concave and a convex mirror.
10. Explain the effect of spherical aberration.
11. For a given focal length mirror, how does the image distance change if the object distance is decreased?
12. For a given object distance from a mirror, how does the image distance change if the focal length is increased?
13. The index of refraction depends on
    (a) the focal length.
    (b) the speed of light.
    (c) the image distance.
    (d) none of the above.
14. Snell's law involves
    (a) the lens equation.
    (b) the index of refraction.
    (c) the focal length.
    (d) none of the above.
15. Differentiate between converging and diverging lenses.
16. Give several examples of total internal reflection.
17. In your own words, explain the law of refraction.
18. How does the speed of light in a high index of refraction material compare to the speed of light in a vacuum?
19. In your own words, explain why light waves are refracted at a boundary between two materials.
20. What type of images are formed by diverging lenses?
21. What type of images are formed by converging lenses?

22. How do water waves affect the escape of light from below the surface of the water?
23. Explain why a fish under water appears to be at a different depth below the surface than it actually is. Does it appear deeper or shallower?
24. Does light always travel in a straight line? Explain.
25. Explain how total internal reflection allows light in a glass fiber to be guided along the fiber.
26. Under what conditions will a converging lens form a virtual image?
27. Under what conditions will a converging lens form a real image that is the same size as the object?
28. Under what conditions will a diverging lens form a virtual image that is smaller than the object?

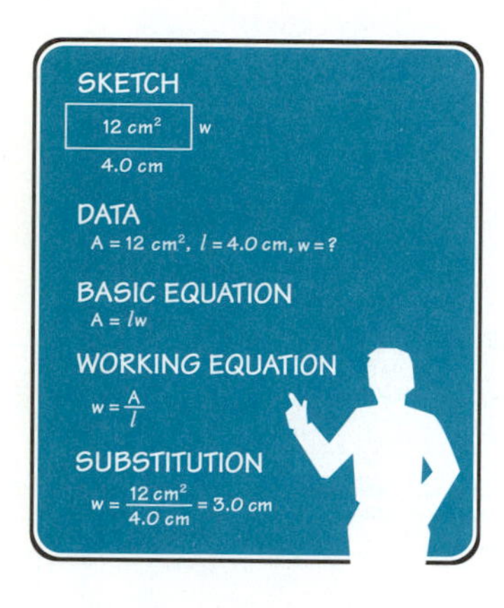

REVIEW PROBLEMS

1. Using $\frac{1}{f} = \frac{1}{s_o} + \frac{1}{s_i}$, $s_o = 3.50$ cm, and $s_i = 7.25$ cm, find f.
2. Using $\frac{1}{f} = \frac{1}{s_o} + \frac{1}{s_i}$, $s_o = 8.50$ cm, and $f = 25.0$ cm, find s_i.
3. Using $\frac{h_i}{h_o} = \frac{s_i}{s_o}$, $h_o = 6.50$ cm, $s_i = 7.50$ cm, and $s_o = 14.0$ cm, find h_i.
4. If an object is 3.75 m tall and is 7.35 m from a large mirror with an image formed 4.35 m from the mirror, what is the height of the image formed?
5. An object 43.0 cm tall is located 23.4 cm from a concave mirror with focal length 21.4 cm.
   (a) Where is the image located?
   (b) How high is the image?
6. An object and its image in a concave mirror are the same height when the object is 45.3 cm from the mirror. What is the focal length of the mirror?
7. The angle of incidence of light passing from air to a liquid is 41.0°. The angle of incidence is 29.0°. Find the index of refraction of the liquid.
8. If the index of refraction of a liquid is 1.44, find the speed of light in that liquid.
9. If the critical angle of a liquid is 45.6°, find the index of refraction for that liquid.
10. If the index of refraction of a substance is 1.50, find its critical angle of incidence.
11. A converging lens has a focal length of 12.0 cm. If it is placed 36.0 cm from an object, how far from the lens will the image be formed?
12. An object 4.50 cm tall is placed 20.0 cm from a converging lens. A real image is formed 12.0 cm from the lens.
    (a) What is the focal length of the lens?
    (b) What is the size of the image?
13. The focal length of a lens is 4.00 cm. How far from the lens must the object be to produce an image 7.20 cm from the lens?
14. What is the focal length of a convex lens that produces an image three times as large as the object at a distance of 25.0 cm from the object?
15. What is the focal length of a mirror that forms an image 3.44 m behind a convex mirror of an object 5.33 m in front of the mirror?
16. What are the size and location of an image produced by a convex lens with a focal length of 21.0 cm of an object 11.5 cm from the lens and 3.25 cm high?

Chapter 23

MODERN PHYSICS

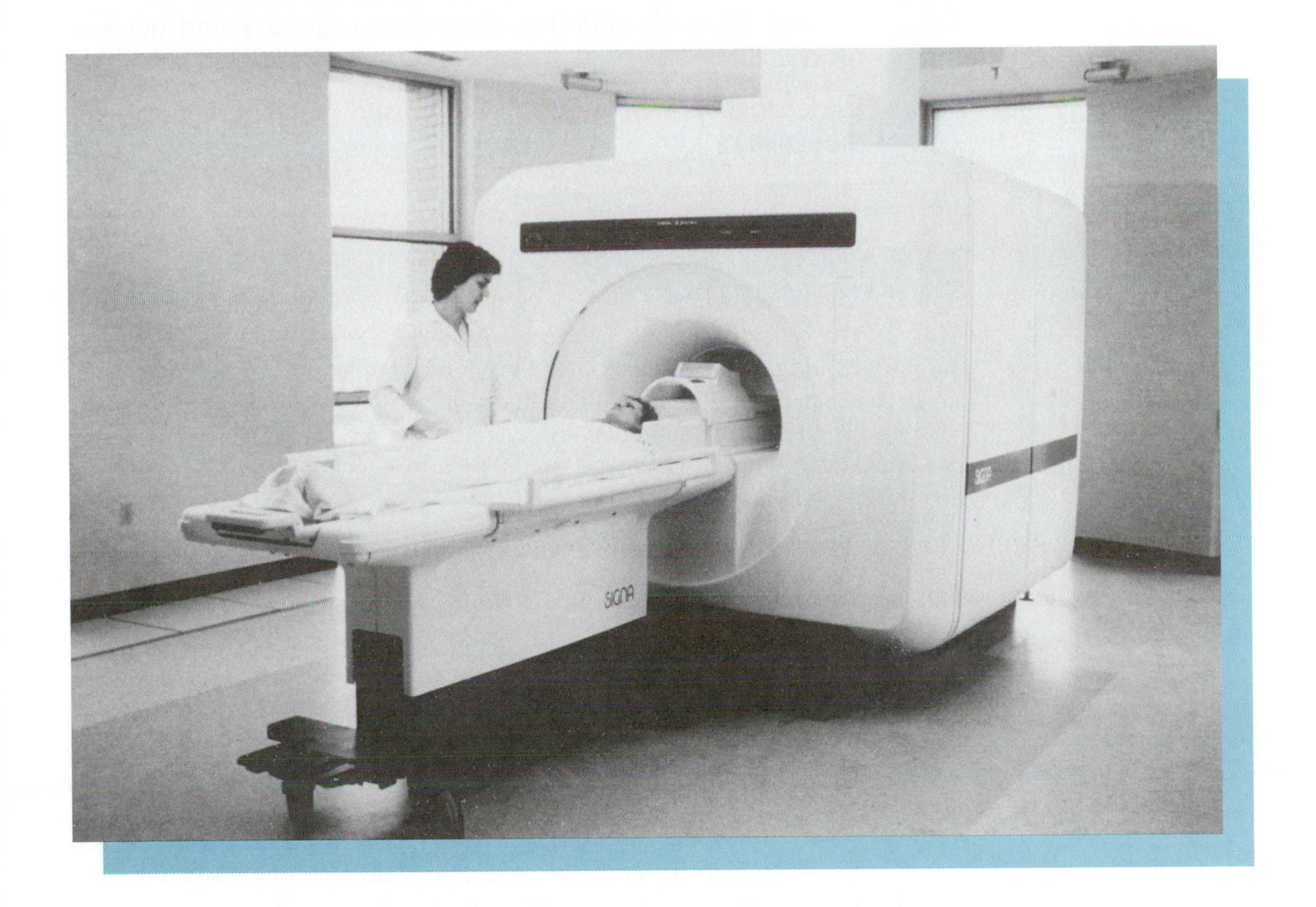

To this point we have studied classical or Newtonian physics. Physics is also concerned with the building blocks of all matter—atoms—and the subatomic particles that make up atoms and the forces between them.

Ancient Greek philosophers thought that all matter could be reduced to four basic elements: air, earth, fire, and water. Modern physics probes deeply into the nature of matter, establishes that even the atom is not the ultimate particle of matter and continues the search describing the mass, size, and energy relationships among the ever expanding list of known subatomic particles.

OBJECTIVES

The major goals of this chapter are to enable you to:

1. Understand the structure and properties of the atomic nucleus.
2. Analyze problems of radioactive decay.
3. Develop an understanding of nuclear reactions and nuclear fusion.
4. Understand principles of detection and measurement of radioactivity.

23.1 STRUCTURE AND PROPERTIES OF THE ATOMIC NUCLEUS

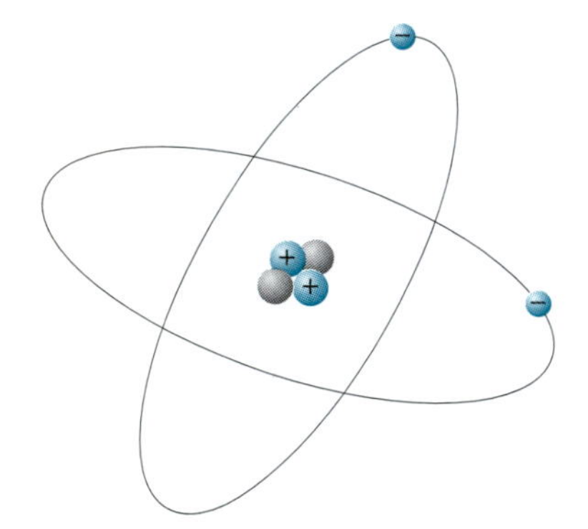

Figure 23.1
The nucleus is composed of neutrons and protons.

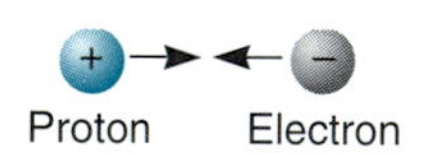

Figure 23.2
The attractive strong force operates at short distances.

The nucleus of an atom contains over 99.9% of the mass of an atom, yet it occupies less than one-trillionth of 1% of the volume of an atom. The particles that make up the nucleus are called *nucleons*. There are two types of nucleons: **protons,** which carry a positive charge, and **neutrons,** which are electrically neutral (Fig. 23.1). Protons have a mass of

$$m_p = 1.6726 \times 10^{-27} \text{ kg}$$

while neutrons have a mass of

$$m_n = 1.6749 \times 10^{-27} \text{ kg}$$

They are both much more massive than the negatively charged **electron,** which has a mass of

$$m_e = 9.1094 \times 10^{-31} \text{ kg}$$

Negatively charged electrons are attracted to the nucleus by the positively charged protons (Fig. 23.2). The size of the positive charge on a proton is equal but opposite in sign to the charge on an electron. An electrically neutral atom contains an equal number of electrons and protons.

The simplest atom, hydrogen, contains one proton in its nucleus. Other atoms contain two or more protons together with some number of neutrons in their nuclei (Fig. 23.3). The role of neutrons in the nucleus of these more complex atoms is to bind the positively charged protons together and prevent the nucleus from flying apart under the repulsive force between the protons. In addition to the repulsive **Coulomb force** (electric force) between protons, there is a nuclear force, referred to as the **strong force.** This is a very short-range attractive force between all nucleons (neutrons and protons) independent of their charge. The neutrons provide additional strong force to overcome Coulomb repulsion. Two neutrons in a nucleus with these two protons provide the additional attractive strong force that holds this nucleus together. For heavier atoms it is necessary to have more neutrons than protons to hold the nucleus together (Fig. 23.4). For example, there are

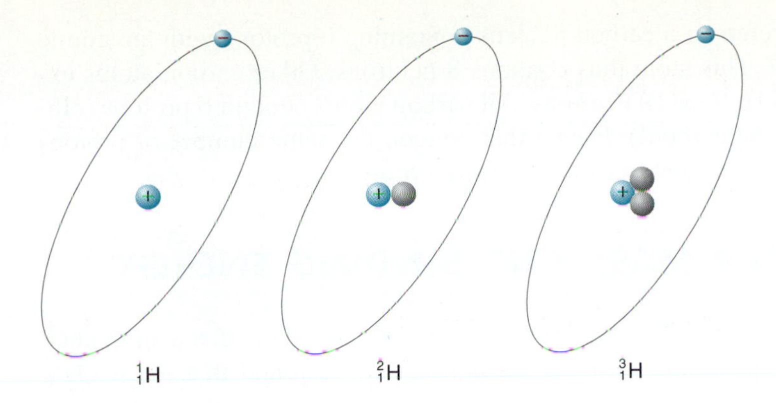

Figure 23.3
Three isotopes of hydrogen. Each has a single electron and a single proton, but different numbers of neutrons.

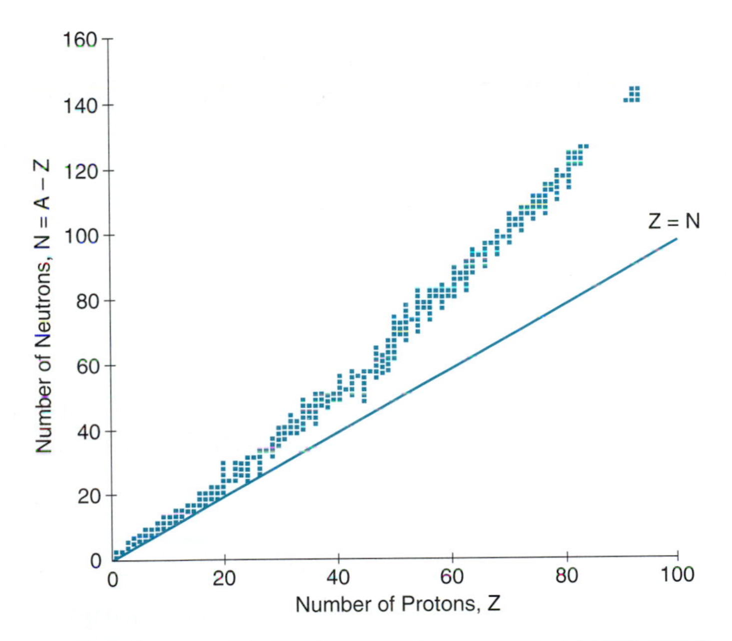

Figure 23.4
Number of neutrons versus number of protons for stable nuclei (dots).

82 protons and 126 neutrons in the common form of lead. Atoms heavier than lead cannot be completely stabilized even by the addition of extra neutrons.

The number of protons in a nucleus is referred to as the **atomic number,** Z. The total number of nucleons (protons and neutrons) in a nucleus is called the **atomic mass number,** A. The neutron number, N, which is given by $N = A - Z$, can be different for nuclei with the same atomic number. The different types of nuclei are referred to as **nuclides.** Each type of nuclide is specified using a symbol of the form

$$^A_Z X$$

where X = the chemical symbol for the element
A = the atomic mass number
Z = the atomic number.

For example, $^{14}_{6}C$ refers to a carbon nucleus containing 6 protons with an atomic mass number of 14. This atom thus contains 8 neutrons. Other carbon atoms exist that contain 5, 6, 7, 9, or 10 neutrons. All carbon atoms contain 6 protons (otherwise, it wouldn't be carbon!). Nuclei that contain the same number of protons but a different number of neutrons are called **isotopes.**

23.2 NUCLEAR MASS AND BINDING ENERGY

When the total mass of a nucleus is compared with the sum of the individual masses of the protons and neutrons when unbound, it is found that the nuclear mass is smaller. How can this be? It would be like combining a dozen 1-kg objects together and ending up with less than 12 kg of mass! Early in the twentieth century, Einstein stated a physical law that mass and energy are equivalent forms of "matter." According to this **principle of equivalence,** mass can be converted into energy according to the relation

$$E = \Delta mc^2$$

where Δm = the change in mass
c = the speed of light, 3.00×10^8 m/s

When the total mass in a reaction is decreased by some amount, energy is given off in the form of radiation or kinetic energy. The energy is given by the equation above. The difference in mass, converted using the equation above, is referred to as the **binding energy of the nucleus.** This energy represents the total energy that must be put into a nucleus to break it apart into separate nucleons. Note that in this chapter we often use more than three significant digits because of the precision of atomic and nuclear measurements.

EXAMPLE 1

Find the binding energy of a $^{4}_{2}He$ nucleus. The mass of the neutral $^{4}_{2}He$ atom is 6.6463×10^{-27} kg.

Data: The mass of two individual neutrons, two individual protons, and two individual electrons (remember, the mass was given for the neutral He atom including the orbital electrons) is given by

$$\begin{aligned} m &= 2m_p + 2m_n + 2m_e \\ &= 2(1.6726 \times 10^{-27} \text{ kg}) + 2(1.6749 \times 10^{-27} \text{ kg}) + 2(9.1094 \times 10^{-31} \text{ kg}) \\ &= 6.6968 \times 10^{-27} \text{ kg} \end{aligned}$$

Then since the mass of neutral He atom = 6.6463×10^{-27} kg,

$$\begin{aligned} \Delta m &= 6.6968 \times 10^{-27} \text{ kg} - 6.6463 \times 10^{-27} \text{ kg} \\ &= 0.0505 \times 10^{-27} \text{ kg} = 5.05 \times 10^{-29} \text{ kg} \end{aligned}$$

Now we find E.

Basic Equation:

$$E = \Delta mc^2$$

Working Equation: Same

Substitution:

$$\begin{aligned} E &= (5.05 \times 10^{-29}\ \text{kg})(3.00 \times 10^{8}\ \text{m/s})^2 \\ &= 4.55 \times 10^{-12}\ \text{kg m}^2/\text{s}^2 \qquad (1\ \text{J} = 1\ \text{kg m}^2/\text{s}^2) \\ &= 4.55 \times 10^{-12}\ \text{J} \end{aligned}$$

Nucleon masses are often given in unified **atomic mass units** (u). A neutral $^{12}_{6}C$ atom is defined to be 12.000000 u. A neutron is 1.008665 u, a proton 1.007276 u, and a neutral hydrogen atom 1.007825 u. The conversion factor relating atomic mass units and mass is

$$1\ \text{u} = 1.6605 \times 10^{-27}\ \text{kg}$$

Another way to express the mass of atomic and nuclear particles is in million electron volts (MeV). An electron volt is the energy that one electron would gain in passing through an electrical potential difference of 1 V. One MeV is given by

$$1\ \text{MeV} = 1.602 \times 10^{-13}\ \text{J}$$

EXAMPLE 2

The MeV unit can also be expressed in terms of its equivalent mass by using the equation $E = \Delta mc^2$ as follows:

Data:

$$\begin{aligned} E &= 1\ \text{MeV} = 1.602 \times 10^{-13}\ \text{J} \\ c &= 2.998 \times 10^{8}\ \text{m/s} \\ \Delta m &= ? \end{aligned}$$

(*Note:* we need to use 4 significant digits here.)

Basic Equation:

$$E = \Delta mc^2$$

Working Equation:

$$\Delta m = \frac{E}{c^2}$$

Substitution:

$$\begin{aligned} \Delta m &= \frac{1.602 \times 10^{-13}\ \text{J}}{(2.998 \times 10^{8}\ \text{m/s})^2} \times \frac{1\ \text{kg m}^2/\text{s}^2}{1\ \text{J}} \\ &= 1.782 \times 10^{-30}\ \text{kg} \end{aligned}$$

Next, convert kilograms to atomic mass units.

$$1.782 \times 10^{-30}\ \text{kg} \times \frac{1\ \text{u}}{1.6605 \times 10^{-27}\ \text{kg}} = 1.073 \times 10^{-3}\ \text{u}$$

Thus,

$$1\ \text{MeV} = 1.602 \times 10^{-13}\ \text{J} = 1.782 \times 10^{-30}\ \text{kg} = 1.073 \times 10^{-3}\ \text{u}$$

The mass of several atomic particles is listed in the following table in the three units described here.

	Mass		
Particle	*kg*	*u*	*MeV/c²*
Electron	9.1094×10^{-31}	0.00054858	0.51100
Proton	1.67262×10^{-27}	1.007276	938.27
Neutron	1.67493×10^{-27}	1.008665	939.57
^{1}H atom	1.67353×10^{-27}	1.007825	938.78

The **average binding energy per nucleon** in a nucleus is the total binding energy of the nucleus divided by the total number of nucleons A. Fig. 23.5 gives the average binding energy per nucleon in MeV for stable nuclei. For $^{4}_{2}$He the average binding energy is 7.1 MeV. Binding energies are largest for A between about 20 and 60. This allows nuclei below 20 and above 60 to undergo reactions or decay that produce substantial amounts of energy, as we will see in later sections.

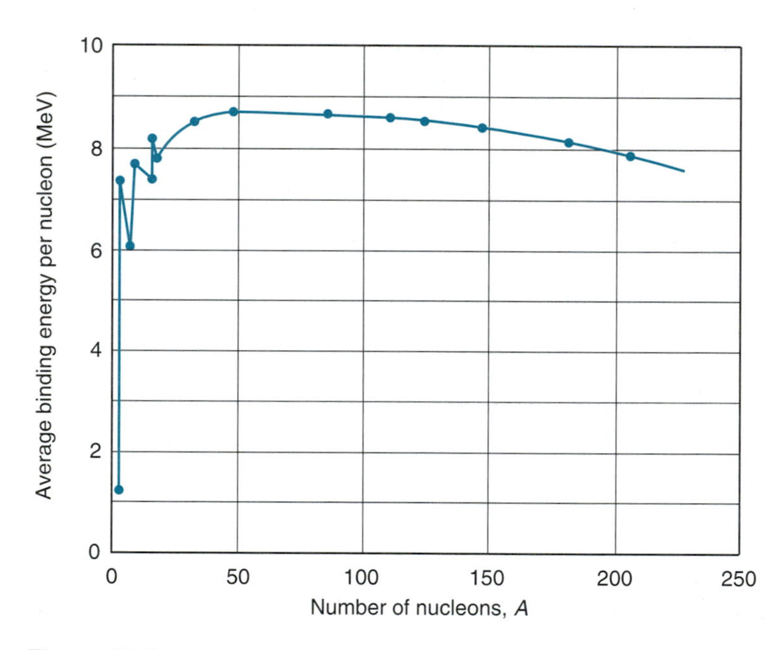

Figure 23.5
Average binding energy per nucleon versus mass number A for stable nuclei.

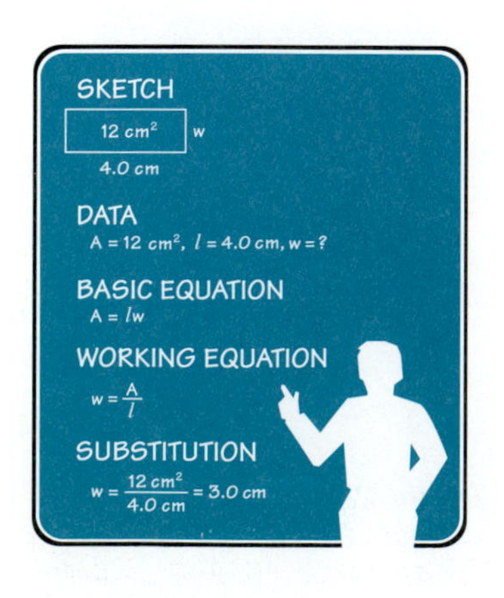

■ PROBLEMS 23.2

1. Find the mass in kilograms of the $^{232}_{92}$U atom if its mass in atomic mass units is 232.037131 u.
2. Find the mass in kilograms of the $^{228}_{90}$Th atom if its mass in atomic mass units is 228.028716 u.
3. Find the binding energy of a $^{4}_{2}$He nucleus in MeV.
4. Find the binding energy of a $^{232}_{92}$U nuclei in MeV if the mass of the neutral U atom is 232.037131 u.
5. Find the binding energy of a $^{228}_{90}$Th nuclei in MeV if the mass of the neutral Th atom is 228.028716 u.
6. Estimate the total binding energy for $^{48}_{22}$Ti from Fig. 23.5.
7. Estimate the total binding energy for $^{90}_{40}$Zr from Fig. 23.5. ■

23.3 RADIOACTIVE DECAY

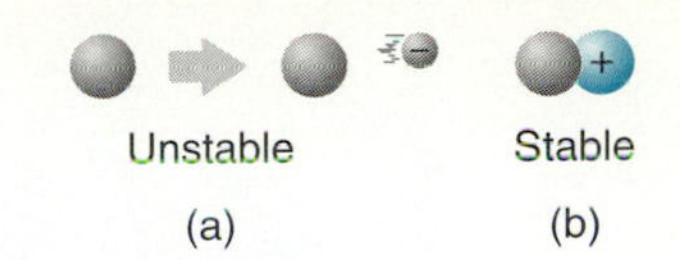

Figure 23.6
A neutron by itself is unstable, but is stable with a proton. The lone neutron decays into a proton and an electron.

The size of an atomic nucleus is limited by the fact that neutrons are unstable. An isolated neutron will decay into a proton and an electron in about 12 minutes on average. In the presence of protons, the neutron is more stable (Fig. 23.6). In many atomic nuclei, neutrons are stable for many billions of years. For heavy atoms, the large number of neutrons required to hold the protons together leads to a situation where there are not enough protons around to prevent one or more of the neutrons from decaying into a proton and an electron. This type of nuclear decay is one of several types that is called **radioactive decay.**

All the elements heavier than bismuth exhibit radioactive decay. The radioactive elements exhibit three types of decay. During the decay, one or more radioactive "rays" are emitted from the nucleus: These are alpha (α), beta (β), and gamma (γ) rays. **Alpha rays** are found to have a positive charge, since they can be observed to curve in the direction that known positive charges curve in a magnetic field. **Beta rays** are found to have a negative charge, since they curve in the opposite direction in a magnetic field. **Gamma rays** are found not to be affected by a magnetic field and therefore must be uncharged (see Fig. 23.7).

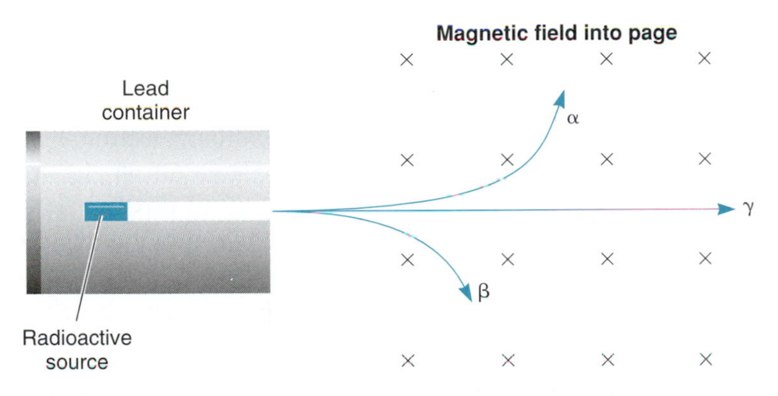

Figure 23.7.
Alpha and beta rays curve in opposite directions in a magnetic field; gamma rays are undeflected.

An α ray is further found to be made of particles having two neutrons and two protons. This combination of particles is called an **alpha particle.** The helium nucleus is identical to an α particle. As we discuss later, the α particle (helium nucleus) is the most stable nuclear combination and can therefore be easily formed in a nuclear reaction.

A β ray is found to consist of a number of electrons. These electrons are emitted from neutrons in a nucleus that decay into protons. A γ ray is found to have no mass and is composed of photons of electromagnetic radiation. Gamma rays are similar to light but have much higher energy.

23.4 ALPHA DECAY

A nucleus that emits an α particle becomes a different nucleus since it has lost two protons and two neutrons (Fig. 23.8).

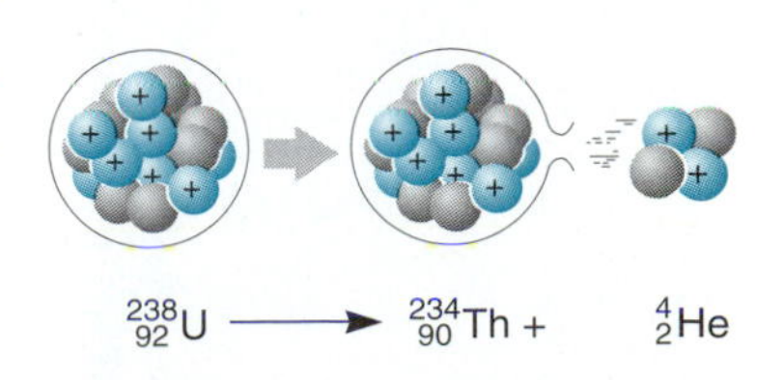

$^{238}_{92}U \longrightarrow\ ^{234}_{90}Th +\ ^{4}_{2}He$

Figure 23.8
Uranium 238 is unstable and decays by giving off an α particle.

When radium 226 ($^{226}_{88}$Ra) gives off an α particle, it becomes a nucleus with $A = 226 - 4 = 222$ and $Z = 88 - 2 = 86$. This nucleus is that of a new element, radon ($^{222}_{86}$Rn). This nuclear reaction is written in the form

$$^{226}_{88}\text{Ra} \longrightarrow {}^{222}_{86}\text{Rn} + {}^{4}_{2}\text{He}$$

Alpha decay occurs because there are not enough neutrons in the nucleus to keep it stable. The total energy released in radioactive decay is called the **disintegration energy, Q,** and is given by

$$Q = (M_p - M_d - m_\alpha)c^2$$

where M_p = the mass of the parent nucleus
M_d = the mass of the daughter nucleus
m_α = the mass of the α particle
c = the speed of light

This disintegration energy is in the form of kinetic energy of the α particle as it moves away from the nucleus and the kinetic energy of the recoiling nucleus that moves away in the opposite direction.

23.5 BETA DECAY

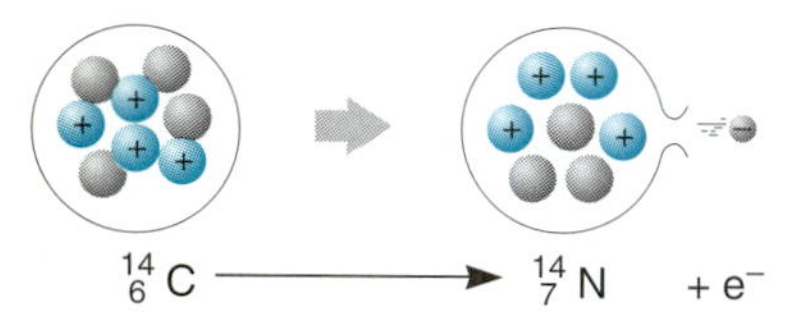

Figure 23.9
Carbon 14 can decay into a nitrogen 14 atom by giving off a β particle (an electron).

A nucleus can also decay by the emission of a β particle (an electron). How can a nucleus, which is made up of only protons and neutrons, emit an electron? One of the neutrons in the nucleus can change into a proton and give off an electron in a manner similar to the decay of free neutrons. In the process of β decay, a nucleus thus changes its charge, and the atom will have to pick up an additional electron in its charge clouds to remain neutral. The element involved in this process is therefore changed into the element with one more electron. An example of this process is the β decay that changes carbon into nitrogen (Fig. 23.9):

$$^{14}_{6}\text{C} \longrightarrow {}^{14}_{7}\text{N} + e^-$$

23.6 NUCLEAR REACTIONS

Nuclear reactions take place when a nucleus is struck by a neutron, a γ ray, or another nucleus giving rise to an interaction. Many nuclear reactions have been observed since Ernest Rutherford in 1919 observed the reaction between α particles and nitrogen that formed oxygen and a proton. One of the most significant findings occurred in the 1930s when attempting to use neutron bombardment of existing nuclei to produce new elements. The heaviest known element at the time was uranium. While attempting to form heavier elements using neutron bombardment of uranium, a process was observed that has had a large impact on the world. It was found that the bombardment of uranium with neutrons can produce lighter nuclei, each nearly half the size of uranium along with a large number of neutrons (Fig. 23.10). This process became known as **nuclear fission.** The reaction can be written

$$\text{n} + {}^{235}_{92}\text{U} \longrightarrow {}^{236}_{92}\text{U} \longrightarrow {}^{141}_{58}\text{Ba} + {}^{92}_{36}\text{Kr} + 3\text{n}$$

A substantial amount of energy is released in this process as the mass of the uranium nucleus and the bombarding neutron is larger than the fission fragments

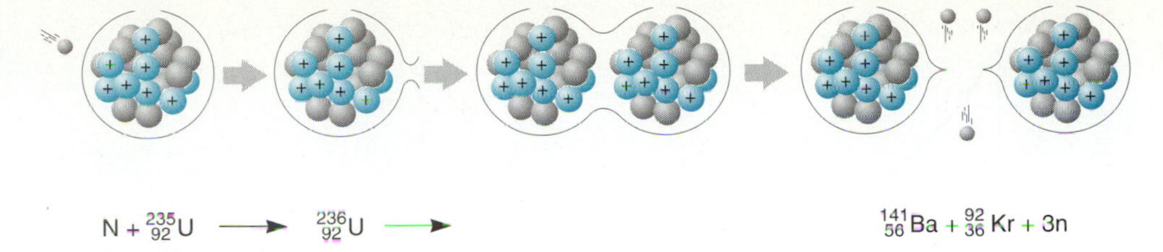

Figure 23.10
Fission of 235 U after neutron capture.

produced by the reaction. The total energy release per fission is approximately 200 MeV. This is an extremely large amount of energy per atom!

The large number of neutrons released in this reaction can be used to create further reactions. The process is called a **chain reaction.** Enrico Fermi and his associates succeeded in producing the first self-sustaining chain reaction at the University of Chicago in 1942. Chain reactions are used in nuclear reactors to produce electric power, to produce intense neutron sources for research and medical use, and in atomic bombs. Many of the end products of fission reactions are radioactive for long periods of time. These radioactive wastes must be disposed of with extreme care.

23.7 NUCLEAR FUSION

Light nuclei can interact to form heavier nuclei with the release of energy because the total mass of the reaction products can be less than that of the initial nuclei and particles. This process is called **nuclear fusion.** An example of a sequence of fusion reactions is

$$ {}^{1}_{1}H + {}^{1}_{1}H \longrightarrow {}^{2}_{1}H + e^{+} \qquad (0.42 \text{ MeV}) $$

$$ {}^{1}_{1}H + {}^{2}_{1}H \longrightarrow {}^{3}_{2}He + \gamma \qquad (5.49 \text{ MeV}) $$

$$ {}^{3}_{2}He + {}^{3}_{2}He \longrightarrow {}^{4}_{2}He + {}^{1}_{1}H + {}^{1}_{1}H \qquad (12.86 \text{ MeV}) $$

This reaction sequence is constantly going on in many stars, including our sun. Fusion is the source of energy in stars. The sun's energy is believed to be produced largely by this sequence of reactions, which starts with protons and produces helium nuclei (α particles) with the release of substantial energy. Nuclear fusion has been used in hydrogen bombs. Fusion is also a subject of considerable research as a "clean" energy source because the end products of the reaction are not radioactive.

23.8 RADIOACTIVE HALF-LIFE

The radioactivity of an isotope is commonly measured by the decay rate or half-life (Fig. 23.11). The decay of radioactive isotopes is a completely random process. That is, it cannot be predicted exactly when a given nucleus will decay. It is possible, however, to say what the probability is in any given time interval. For a short (compared to the half-life) time interval, this probability can be expressed by the formula

$$ P = \lambda \Delta t $$

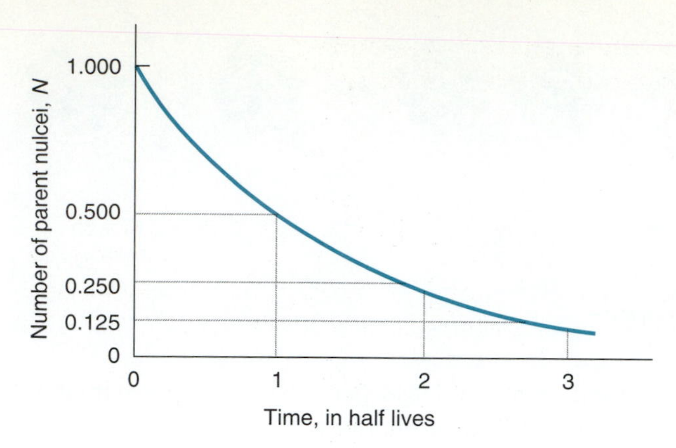

Figure 23.11
An exponential decrease of parent nuclei is observed.

where P = the probability of decay
λ = the decay rate
Δt = the time interval

The **decay rate** is the probability per unit time that a decay will occur.

EXAMPLE 1

Find the probability of decay of a single atom of ${}^{14}_{6}C$ during the next 10.0 years. The decay rate for this isotope is 1.21×10^{-4} per year.

Data:

$$\lambda = 1.21 \times 10^{-4}/\text{yr}$$
$$\Delta t = 10.0 \text{ years}$$
$$P = ?$$

Basic Equation:

$$P = \lambda \Delta t$$

Working Equation: Same

Substitution:

$$P = (1.21 \times 10^{-4}/\text{yr})(10.0 \text{ yr})$$
$$= 1.21 \times 10^{-3}$$

Thus the probability that a single atom of ${}^{14}_{6}C$ will decay in the next 10 years is 1.21×10^{-3}. This means that there is one chance in $1/P = 826$ that the atom will decay.

For a large number of nuclei in a sample, it can be stated that the **half-life** is the amount of time that is required for half of the radioactive atoms in a sample to decay. The half-life of radium 226 is 1620 years. In the first 1620 years after measurement begins, half of the radium 226 atoms will decay. In the next 1620 years, half of the remaining radium 226 atoms will decay, leaving one-fourth of the initial sample. The decay can be expressed by the equation

$$N = N_0 e^{-\lambda t}$$

where N = the number of remaining radioactive isotopes
N_0 = the original number of radioactive isotopes
e = the number 2.718
λ = the decay constant
t = the elapsed time

(Note that calculations involving the number $e = 2.718$ can easily be performed on most calculators using the e^x function key. Simply enter the exponent x and then push the e^x key. The number represented by e^x is then displayed.) The half-life $T_{1/2}$ is related to the decay constant by the equation

$$T_{1/2} = \frac{0.693}{\lambda}$$

EXAMPLE 2

Find the amount of radioactive radium remaining after 2450 years if the original amount of radon was 3.54×10^{23} atoms. The half-life of radium is 1620 years.

Data:

$$N_0 = 3.54 \times 10^{23}$$
$$\lambda = \frac{0.693}{T_{1/2}} = \frac{0.693}{1620 \text{ yr}} = 4.28 \times 10^{-4}/\text{yr}$$
$$t = 2450 \text{ yr}$$
$$N = ?$$

Basic Equation:

$$N = N_0 e^{-\lambda t}$$

Working Equation: Same

Substitution:

$$\begin{aligned} N &= (3.54 \times 10^{23})e^{-(4.28 \times 10^{-4}/\text{yr})(2450 \text{ yr})} \\ &= (3.54 \times 10^{23})e^{-1.0486} \\ &= 1.24 \times 10^{23} \text{ atoms} \end{aligned}$$

Half-lives of atoms range from a small fraction of a second to billions of years. For short times, radioactive half-lives can be measured by waiting until half the atoms decay. For longer times the rate at which a sample decays can be measured by using one of several types of radiation detectors.

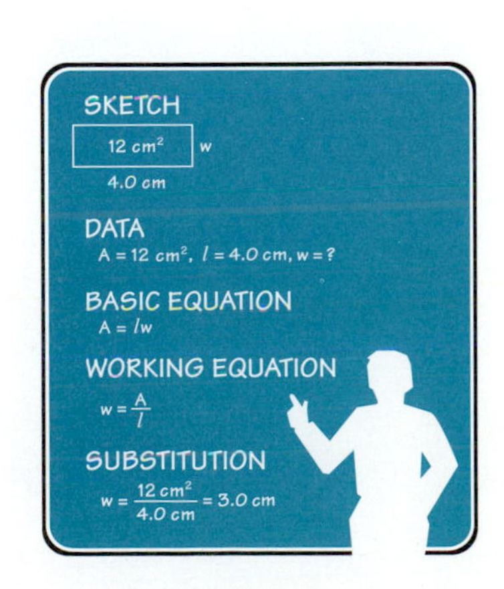

PROBLEMS 23.8

1. Find the quantity of uranium 238 atoms remaining from an original sample of 5.50×10^{20} atoms after 2.45 billion years if the half-life of uranium 238 is 4.50 billion years.
2. Find the quantity of ${}^{14}_{6}\text{C}$ remaining from an original sample of 3.75×10^{21} atoms after 1000 years. ${}^{14}_{6}\text{C}$ has a half-life of 5730 years.
3. Find the probability that a single ${}^{14}_{6}\text{C}$ atom will decay in the next 20.0 years. ${}^{14}_{6}\text{C}$ has a half-life of 5730 years.
4. Find the probability that a single ${}^{238}_{92}\text{U}$ atom will decay in the next year. The half-life of uranium 238 is 4.50 billion years.

5. Find the percent of a $^{14}_{6}C$ sample that will decay in the next $300\bar{0}$ years. The half-life of carbon 14 is 5730 years.
6. Find the percent of a sample that will decay in the next second if the half-life is 2.35 s.
7. Find the probability that a radioactive radium atom will decay in the next $10\bar{0}$ years if the half-life is 1620 years.
8. Find the probability that a uranium 138 atom will decay in the next $10\bar{0}0$ years if the half-life of uranium is 4.50 billion years. ■

23.9 RADIATION PENETRATING POWER

The three types of radiation (α, β, and γ) rays can be stopped by matter. Some are stopped more easily than others (Fig. 23.12). Alpha rays, which are massive and carry a double positive charge, are stopped quite easily during "collisions" with atoms in the material. These collisions occur when the positively charged α particle comes close enough to an atom to feel a Coulomb force from some of the electrons or protons. The α particle gives up its energy to the material through these "collisions." A few pieces of paper or a few centimetres of air are sufficient to stop them. As soon as an α particle slows up enough to pick up two electrons while passing through matter, it becomes a harmless helium atom.

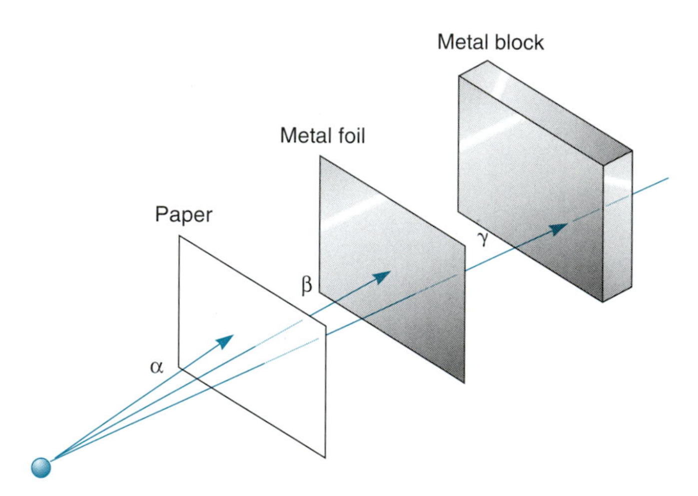

Figure 23.12
Penetrating power of different forms of nuclear radiation.

Beta rays, which are electrons, can easily be stopped by several thousandths of an inch of a metal, such as aluminum foil. They lose their energy through many collisions with electrons in atoms in the material they are passing through. They become electrons in the material, indistinguishable from others.

Gamma rays have the greatest penetrating power because they have no charge and therefore cannot be electrically attracted or repelled by electrons or protons. The γ ray, which is an energetic photon, is slowed down only through a direct hit of an electron or a nucleus. A dense material, such as a heavy metal, is the best absorber of γ radiation.

23.10 RADIATION DAMAGE

As radiation passes through matter and loses its energy, it can do much damage. Many materials will become weakened and brittle upon exposure to substantial levels of radiation. This is a major engineering problem that has had to be solved for nuclear reactors.

Radiation damage to living organisms can be very severe. As radiation is absorbed in material, atoms become ionized as electrons are captured or emitted. If these atoms are basic in bonding molecules together, the molecule may break apart. The functioning of the living cell may be altered substantially if a large number of key molecules are damaged. If the damage occurs to the DNA molecule that controls the growth and replication of the cell, the cell may be severely damaged and die. Obviously, if a large number of cells die, the organism will be sick (radiation sickness) or die. If the damaged cell replicates itself rapidly with faulty DNA, cancer may be the result. To prevent this type of damage during normal medical x-ray procedures, it is common for patients to wear lead aprons around tissue not being imaged by the x-ray. This is also why x-ray technicians step behind a lead brick wall when x-rays are being taken.

Radiation can be useful for medical diagnostics (x-rays and radioactive tracers) and radiation therapy. In radiation therapy, cancer cells can be destroyed by the localized application to a tumor.

23.11 DETECTION AND MEASUREMENT OF RADIATION

Since we cannot see, touch, or feel nuclear radiation, it is necessary to have instruments available to detect the presence of radiation. The most common detector of radiation is the Geiger counter (Fig. 23.13). This instrument contains a cylindrical metal tube filled with gas. A wire runs along the center of the metal tube. A large quantity of voltage (approximately 1000 volts) is maintained across the wire and cylinder. When a charged particle enters the tube and ionizes some of the gas, the voltage on the tube causes the gas to "break down" and conduct an electrical pulse between the wire and the cylinder. This electrical pulse is detected by electronics that count the number of pulses. A loudspeaker is often hooked up to the electronics, allowing a "click" to be heard each time a charged particle enters the tube.

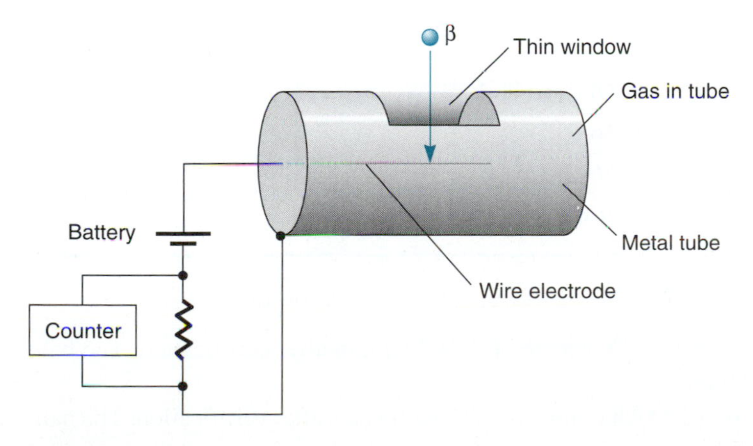

Figure 23.13
Diagram of a Geiger counter.

Since radiation can cause severe damage to living organisms and materials, it is necessary to be able to measure the amount, or *dose,* of radiation. There are two common units of *source activity:* the **curie** (Ci), which is defined as

$$1 \text{ Ci} = 3.70 \times 10^{10} \text{ disintegrations/s}$$

and the **becquerel** (Bq), which is defined as

$$1 \text{ Bq} = 1 \text{ disintegration/s}$$

The activity of a source may be specified at a given time. At a later time, the activity is decreased according to the equation

$$A = \lambda N = \lambda N_0 e^{-\lambda t} = A_0 e^{-\lambda t}$$

where A = the activity
N = the remaining quantity of radioactive isotopes
N_0 = the original quantity
λ = the decay rate
t = the time
A_0 = the original activity
e = 2.718

EXAMPLE

Find the activity of a $^{222}_{88}\text{Ra}$ source 6.54 days after it was originally certified to have an activity of 0.356 Ci. The half-life of $^{222}_{88}\text{Ra}$ is 3.82 days.

Data:

$$t = 6.54 \text{ days}$$
$$A_0 = 0.356 \text{ Ci}$$
$$T_{1/2} = 3.82 \text{ days}$$
$$\lambda = \frac{0.693}{T_{1/2}} = \frac{0.693}{3.82 \text{ days}} = 0.181/\text{day}$$
$$A = ?$$

Basic Equation:

$$A = A_0 e^{-\lambda t}$$

Working Equation: Same

Substitution:

$$\begin{aligned} A &= (0.356 \text{ Ci})e^{-(0.181/\text{day})(6.54 \text{ day})} \\ &= (0.356 \text{ Ci})e^{-1.184} \\ &= (0.356 \text{ Ci})(0.306) \\ &= 0.109 \text{ Ci} \end{aligned}$$

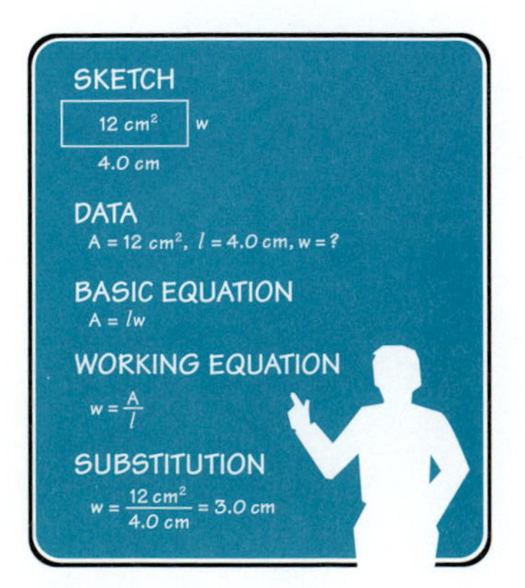

PROBLEMS 23.11

1. Find the activity of a 1.24-Ci sample of $^{13}_{7}\text{N}$ 20.0 min after certification. The half-life of $^{13}_{7}\text{N}$ is 10.0 min.
2. Find the activity of a 2.64-Ci sample of $^{14}_{6}\text{C}$ $40\bar{0}0$ years after certification. The half-life of $^{14}_{6}\text{C}$ is 5370 years.

3. Find the activity of a 0.476-Ci sample of $^{24}_{11}Na$ 36.5 h after certification. The half-life of $^{24}_{11}Na$ is 14.95 h.
4. Find the activity of a 3.98-Ci sample of $^{11}_{6}C$ 10.3 h after certification. The half-life of $^{11}_{6}C$ is 20.4 min. ■

23.12 ATOMIC STRUCTURE AND ATOMIC SPECTRA

In the first section of this chapter we studied the structure and properties of the atomic nucleus. In the remainder of the chapter we will see how the properties of the nucleus determine the structure of the atom and then how that in turn determines the chemical properties of the atom.

The model of the atom presented earlier in this chapter has a very small nucleus surrounded by electrons. A serious problem for this model is that because of the attractive force between the electrons and the protons in the nucleus, the electrons would be expected to eventually fall into the nucleus. It might be assumed that the electrons could exist in stable orbits around the nucleus in a manner similar to that of planets orbiting a star. However, because the electrons are charged, they should radiate away their energy in the form of electromagnetic radiation and then fall into the nucleus. Niels Bohr, a Danish scientist, proposed a solution to this problem by suggesting a theory for the behavior of matter. According to this theory, the energy of the electron would be quantized. That is, the energy is restricted only to certain values. Each of these energy values was shown by Professor Bohr to be given by the equation

$$E = \frac{-kZ^2}{n^2}$$

where E = the energy of electron
k = a constant = 2.179×10^{-18} J or 13.595 eV
Z = the atomic number (the number of positive charges in the nucleus)
n = the integer that characterizes the orbit ($n = 1, 2, 3, 4, \ldots$), also called the quantum number.

The distance of the electron from the nucleus increases as the integer n increases. In the lowest energy "state" or lowest orbit, the energy is minimum. This lowest-energy orbit is called the **ground state** of the electron. The higher-energy orbits are called **excited states** (Fig. 23.14).

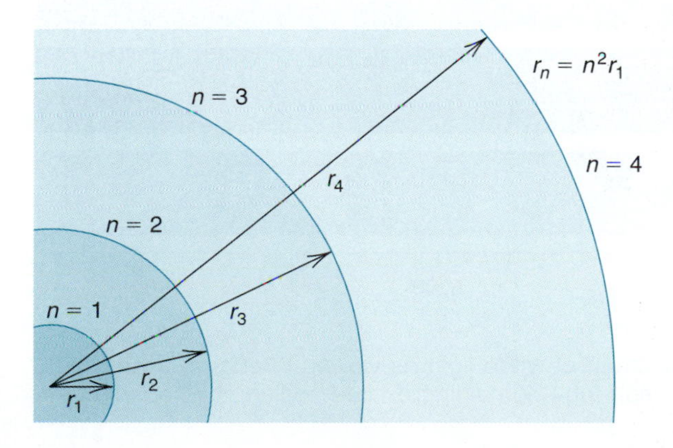

Figure 23.14
The circular orbits of the Bohr model of the hydrogen atom.

EXAMPLE

An electron on a hydrogen atom is in the $n = 2$ orbit. What is its energy?

Data:

$$n = 2$$
$$Z = 1$$
$$E = ?$$

Basic Equation:

$$E = -\frac{kZ^2}{n^2}$$

Working Equation: Same

Substitution:

$$E = \frac{-(13.595 \text{ eV})(1)^2}{2^2}$$
$$= -3.40 \text{ eV}$$

Neon lights work on the principle that certain colors of light are given off by different atoms when the electrons in the atoms are moved from the ground state to an excited state by an electrical current passing through the neon tube containing the gas at low pressure. As the electrons return to their ground state, they give off their excess energy in the form of electromagnetic radiation. For a neon tube containing hydrogen, the color given off is blue. Other atoms give off other colors, leading to the rich colors available through this type of light tube. The spectrum of light given off by an atom in a neon tube is composed of a number of discrete lines as shown in Fig. 23.15.

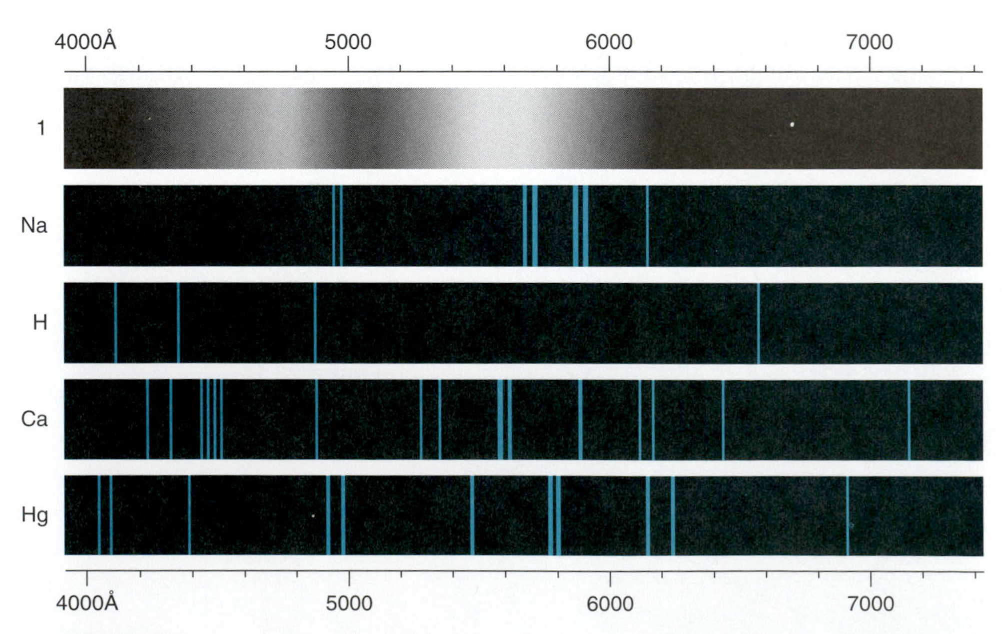

Figure 23.15
A comparison of the continuous spectrum of white light as viewed through a prism and the line spectra of the light from a neon tube containing excited sodium, hydrogen, calcium, and mercury atoms.

■ PROBLEM 23.12

1. Find the energy of the ground state orbit in the hydrogen atom.
2. Find the energy of the $n = 3$ state of the hydrogen atom.
3. Find the energy given off by a hydrogen atom undergoing a transition from the $n = 3$ to the $n = 1$ state.
4. Find the wavelength of the photon given off in the transition in Problem 3.
5. Find the wavelength of a photon given off in a hydrogen filled neon tube undergoing a transition from the $n = 3$ to the $n = 2$ orbit. ■

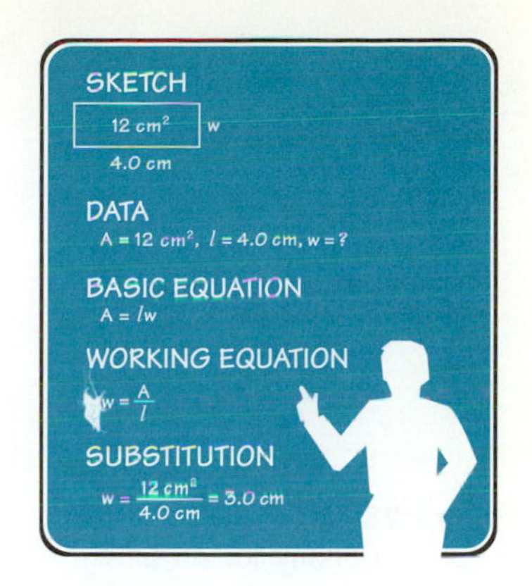

23.13 THE QUANTUM MECHANICAL MODEL OF THE ATOM AND ATOMIC PROPERTIES

The agreement between the Bohr model and the energy levels of the hydrogen atom are excellent. However, his theory does not account for the properties of more complex atoms with more than one electron. Now we will learn about the quantum mechanical model of the atom, which gives a good description of more complex atoms.

The **quantum mechanical model** of atoms is based on the theory, that matter, just like light, sometimes behaves as a particle and sometimes as a wave. This wave theory of the atom can be used in a simple way to understand the Bohr theory of the atom. In the Bohr theory, discussed in the previous section, the electron can only have certain discrete energy levels. Other energy levels are not "allowed." When an electron is behaving like a wave, its orbital path must consist of an integral number of wavelengths, otherwise destructive interference would occur (Fig. 23.16).

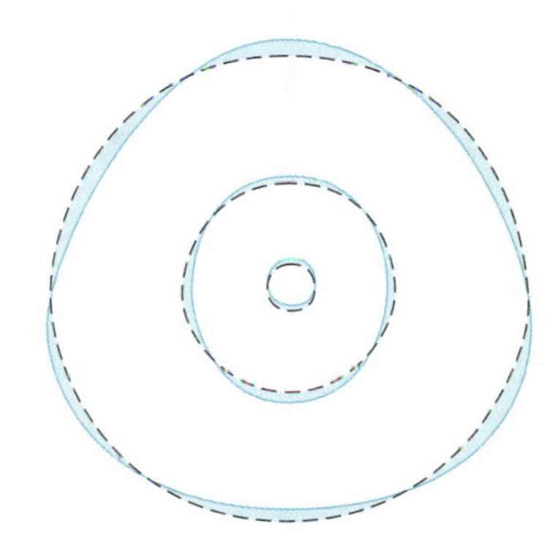

Figure 23.16
An orbit consisting of a fractional number of wavelengths of the electron cannot persist because destructive interference would occur.

When an electron is behaving like a wave, its exact location cannot be determined; instead, we can only calculate using wave equations for the electron where the electron is likely to be. We can find a probability from these equations that describes for any position in space how likely the electron is to be at that point at any given time (Fig. 23.17). Chemists describe the region of space in

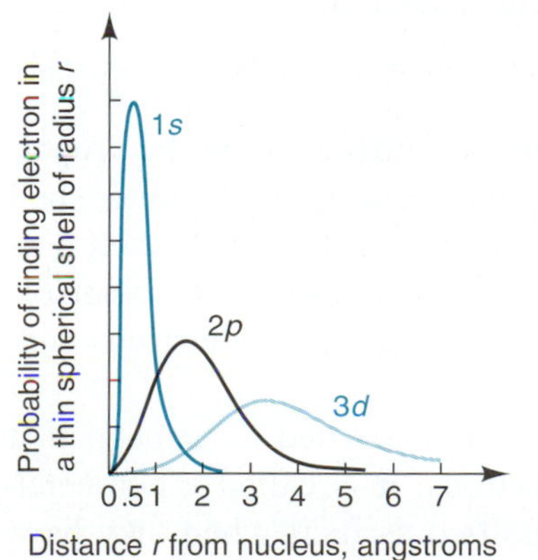

Figure 23.17
The probability of finding the electron at a distance *r* from the nucleus for the three lowest orbitals of a hydrogen atom.

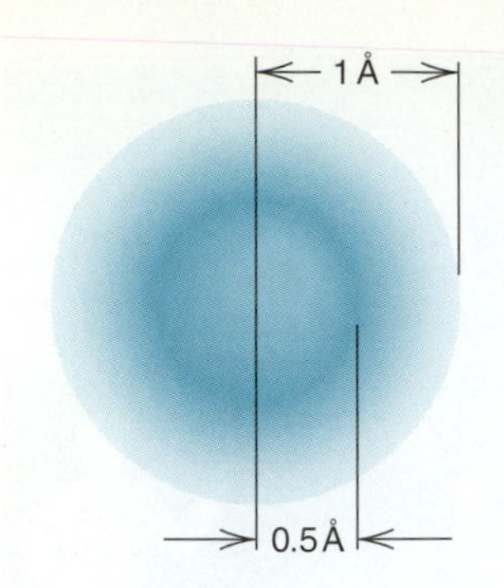

Figure 23.18
Electron density for an electron in the ground state of a hydrogen atom.

which the electron is likely to be as the atomic **orbital**. In the ground state of the hydrogen atom, the electron spends most of its time in a spherical region centered about 0.5 Å from the nucleus as shown in Fig. 23.18.

Orbitals are characterized by a number, n, which is called the principal quantum number. It can take on integer values, $n = 1, 2, 3, 4, \ldots$. Just as in the Bohr model, the orbitals extend further from the nucleus as n increases. As the orbitals get further from the nucleus, there are different orbital "shapes" that become possible for each principal quantum number. Each of these shapes is described by a secondary quantum number or letter ($s, p, d, f, \ldots$) (Fig. 23.19).

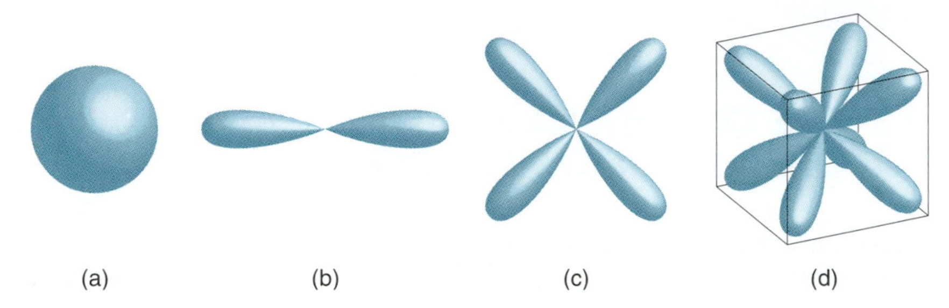

Figure 23.19
Orbital shapes for (a) an *s* subshell, (b) a *p* subshell, (c) a *d* subshell, and (d) an *f* subshell.

Each possible shape represents an orbital that can contain up to two electrons. In the ground state (lowest energy state) of an atom, the electrons fill up the lowest energy orbitals. The types of orbitals for each principal quantum number are shown in Table 23.1.

Table 23.1 Maximum Number of Electrons in Atomic Orbitals

n Value	*Types of Orbitals*	*Maximum Number of Electrons*
1	1*s*	2
2	2*s*, 2*p*	8
3	3*s*, 3*p*, 3*d*	18
4	4*s*, 4*p*, 4*d*, 4*f*	32
5	5*s*, 5*p*, 5*d*, 5*f*, 5*g*	50

Electrons fill up the lowest energy orbitals in any given atom. Table 23.2 summarizes the number of electrons in each orbit for the atomic elements with up to 21 protons in the nucleus and therefore up to 21 electrons in orbitals. Note in Table 23.2 that the $n = 4$ orbital starts filling up (see potassium) before the $n = 3$ orbital is completely filled. This is because the 4*s* orbital has lower energy than the 3*d* orbitals.

The theory of electron orbits provides a basis for understanding the properties of different atoms, depending upon how many electrons are in the outer shell and how much energy it takes to add or remove an electron from that shell during chemical reactions between atoms. Each atom attempts to have its outer orbit complete and does this by attempting to borrow, lend, or share electrons with other atoms surrounding it in all matter.

On the basis of an atoms atomic structure, it is classified as a metal if it tends to lend electrons. If it tends to borrow electrons, it is called a nonmetal. Copper is an example of a metal. Atoms are classified as inert when they have complete outer shells and therefore tend not to borrow or share electrons. Helium and neon are examples of inert atoms. They both have completely filled outer shells (see Table 23.2). The fewer electrons an atom tends to borrow, lend, or share, the more reactive it tends to be in chemical reactions.

Table 23.2 Table of the First 21 Elements

Element	*Atomic No.*	*Number of Protons*	*Number of Electrons*	*Electrons in $n = 1$ Orbit*	*Electrons in $n = 2$ Orbit*	*Electrons in $n = 3$ Orbit*	*Electrons in $n = 4$ Orbit*
Hydrogen	1	1	1	1			
Helium	2	2	2	2			
Lithium	3	3	3	2	1		
Beryllium	4	4	4	2	2		
Boron	5	5	5	2	3		
Carbon	6	6	6	2	4		
Nitrogen	7	7	7	2	5		
Oxygen	8	8	8	2	6		
Fluorine	9	9	9	2	7		
Neon	10	10	10	2	8		
Sodium	11	11	11	2	8	1	
Magnesium	12	12	12	2	8	2	
Aluminum	13	13	13	2	8	3	
Silicon	14	14	14	2	8	4	
Phosphorus	15	15	15	2	8	5	
Sulfur	16	16	16	2	8	6	
Chlorine	17	17	17	2	8	7	
Argon	18	18	18	2	8	8	
Potassium	19	19	19	2	8	8	1
Calcium	20	20	20	2	8	8	2
Scandium	21	21	21	2	8	9	2

■ PROBLEMS 23.13

1. What atom has the $n = 2$ orbital filled?
2. What atom has the $n = 1$ orbital filled?
3. What atom is one electron short of having the $n = 2$ orbital filled?
4. What atom has just one electron in the $n = 3$ orbital? ■

23.14 PERIODIC TABLE OF THE ELEMENTS

Well before the atomic theory of the atom was developed, Dmitri Mendeleev in 1889 proposed a table containing the elements arranged in a periodic manner that showed various trends in the properties of atoms. The existence of eight columns in his table and the trends in the properties of the atoms was later understood in terms of the atomic structure described in the last section.

The horizontal rows of the **Periodic Table** (see Table 23.3) are called periods or rows. There are seven periods, each of which starts with an atom having one valence electron and ends with a complete outer shell structure (an inert element). The first three rows consist of 2, 8, and 8 elements. Rows 4 and 5 are longer rows consisting of 18 atoms, and row 6 has 32 elements, most of period 7 elements are radioactive and do not occur in nature. The number of elements in each row can be understood in terms of the atomic theory described earlier and the number of electrons in each orbital (see Table 23.2).

Table 23.3 Periodic Table of the Elements and Trends in Chemical Properties

Key: Symbol — Cl 17 — Atomic number; Atomic mass* — 35.453; $3p^5$ — Electron configuration

Group I	*Group II*	*Transition Elements*										*Group III*	*Group IV*	*Group V*	*Group VI*	*Group VII*	*Group 0*
H 1 1.0079 $1s^1$																	He 2 4.00260 $1s^2$
Li 3 6.94 $2s^1$	Be 4 9.01218 $2s^2$											B 5 10.81 $2p^1$	C 6 12.011 $2p^2$	N 7 14.0067 $2p^3$	O 8 15.9994 $2p^4$	F 9 18.9984 $2p^5$	Ne 10 20.18 $2p^6$
Na 11 22.9898 $3s^1$	Mg 12 24.305 $3s^2$											Al 13 26.9815 $3p^1$	Si 14 28.0855 $3p^2$	P 15 30.974 $3p^3$	S 16 32.06 $3p^4$	Cl 17 35.453 $3p^5$	Ar 18 39.948 $3p^6$
K 19 39.0983 $4s^1$	Ca 20 40.08 $4s^2$	Sc 21 44.9559 $3d^14s^2$	Ti 22 47.9 $3d^24s^2$	V 23 50.9415 $3d^34s^2$	Cr 24 51.996 $3d^54s^1$	Mn 25 54.938 $3d^54s^2$	Fe 26 55.847 $3d^64s^2$	Co 27 58.9332 $3d^74s^2$	Ni 28 58.7 $3d^84s^2$	Cu 29 63.546 $3d^{10}4s^1$	Zn 30 65.39 $3d^{10}4s^2$	Ga 31 69.73 $4p^1$	Ge 32 72.6 $4p^2$	As 33 74.9216 $4p^3$	Se 34 78.96 $4p^4$	Br 35 79.904 $4p^5$	Kr 36 83.80 $4p^6$
Rb 37 85.47 $5s^1$	Sr 38 87.62 $5s^2$	Y 39 88.9059 $4d^15s^2$	Zr 40 91.22 $4d^25s^2$	Nb 41 92.9064 $4d^45s^1$	Mo 42 95.94 $4d^55s^1$	Tc 43 (98) $4d^55s^2$	Ru 44 101.07 $4d^75s^1$	Rh 45 102.906 $4d^85s^1$	Pd 46 106.4 $4d^{10}5s^0$	Ag 47 107.868 $4d^{10}5s^1$	Cd 48 112.41 $4d^{10}5s^2$	In 49 114.82 $5p^1$	Sn 50 118.7 $5p^2$	Sb 51 121.75 $5p^3$	Te 52 127.60 $5p^4$	I 53 126.90 $5p^5$	Xe 54 131.3 $5p^6$
Cs 55 132.905 $6s^1$	Ba 56 137.33 $6s^2$	57-71†	Hf 72 178.49 $5d^26s^2$	Ta 73 180.95 $5d^36s^2$	W 74 183.85 $5d^46s^2$	Re 75 186.207 $5d^56s^2$	Os 76 190.2 $5d^66s^2$	Ir 77 192.22 $5d^76s^2$	Pt 78 195.08 $5d^96s^1$	Au 79 196.97 $5d^{10}6s^1$	Hg 80 200.59 $5d^{10}6s^2$	Tl 81 204.38 $6p^1$	Pb 82 207.2 $6p^2$	Bi 83 208.980 $6p^3$	Po 84 (209) $6p^4$	At 85 (210) $6p^5$	Rn 86 (222) $6p^6$
Fr 87 (223) $7s^1$	Ra 88 226.025 $7s^2$	89-103‡	Rf 104 (261) $6d^27s^2$	Ha 105 (262) $6d^37s^2$	106 (263)	107 (262)	108 (265)	109 (266)									

†Lanthanide series	La 57 138.906 $5d^16s^2$	Ce 58 140.12 $5d^14f^16s^2$	Pr 59 140.908 $4f^36s^2$	Nd 60 144.24 $4f^46s^2$	Pm 61 (145) $4f^56s^2$	Sm 62 150.4 $4f^66s^2$	Eu 63 151.96 $4f^76s^2$	Gd 64 157.25 $5d^14f^76s^2$	Tb 65 158.925 $5d^14f^86s^2$	Dy 66 162.50 $4f^{10}6s^2$	Ho 67 164.930 $4f^{11}6s^2$	Er 68 167.26 $4f^{12}6s^2$	Tm 69 168.934 $4f^{13}6s^2$	Yb 70 173.04 $4f^{14}6s^2$	Lu 71 174.967 $5d^14f^{14}6s^2$
‡Actinide series	Ac 89 (227) $6d^17s^2$	Th 90 232.038 $6d^27s^2$	Pa 91 231.036 $5f^26d^17s^2$	U 92 238.029 $5f^36d^17s^2$	Np 93 237.048 $5f^46d^17s^2$	Pu 94 (244) $5f^66d^07s^2$	Am 95 (243) $5f^76d^07s^2$	Cm 96 (247) $5f^76d^17s^2$	Bk 97 (247) $5f^96d^07s^2$	Cf 98 (251) $5f^{10}6d^07s^2$	Es 99 (252) $5f^{11}6d^07s^2$	Fm 100 (257) $5f^{12}6d^07s^2$	Md 101 (258) $5f^{13}6d^07s^2$	No 102 (259) $6d^07s^2$	Lr 103 (260) $6d^17s^2$

*Atomic mass values averaged over isotopes in percentages they occur on earth's surface. For many unstable elements, mass number of the most stable known isotope is given in parentheses.

Metals are found on the left of the table, with the most active metals in the lower left corner. Nonmetals are found on the right side. The noble or inert gases are on the far right. The acid-forming properties increase toward the right. Base-forming properties increase toward the left. Other properties of atoms, including the atomic weight and the atomic size, are given in common forms of the Periodic Table.

PROBLEMS 23.14

1. Which of the two elements, Br and Ca, would be expected to be more acidic?
2. Which of the two elements, Br and Ca, would be expected to be a stronger base?
3. Which of the elements Cl, Ar, I, Ca, K is inert?
4. Which of the elements C, N, O, Fe, is a metal?

GLOSSARY

Alpha Particle A particle having two neutrons and two protons; the most stable nuclear combination. (p. 545)
Alpha Ray A kind of radioactive ray emitted from a nucleus during decay; has a positive charge. (p. 545)
Atomic Mass Number The total number of nucleons in a nucleus. (p. 541)
Atomic Mass Unit A unit of measure of atomic nucleon mass in which $1 \text{ u} = 1.6605 \times 10^{-27}$ kg. (p. 543)
Atomic Number The number of protons in a nucleus. (p. 541)
Average Binding Energy per Nucleon The total binding energy of the nucleus divided by the total number of nucleons. (p. 544)
Becquerel (Bq) Unit of source activity; 1 Bq = 1 disintegration/s. (p. 552)
Beta Ray A kind of radioactive ray emitted from a nucleus during decay; consists of a number of electrons; has a negative charge. (p. 545)
Binding Energy of the Nucleus The total energy that must be put into a nucleus to break it apart into separate nucleons. (p. 542)
Bohr Model A model of the atom that works well for the simple hydrogen atom. (p. 553)
Chain Reaction The process of using the large number of neutrons released in nuclear fission to create further reactions. (p. 547)
Coulomb Force An electric, repulsive force between protons. (p. 540)
Curie (Ci) Unit of source activity; $1 \text{ Ci} = 3.70 \times 10^{10}$ disintegrations/s. (p. 552)
Decay Rate The probability per unit time that a decay of radioactive isotopes will occur. (p. 548)
Disintegration Energy The total energy released in radioactive decay; in the form of kinetic energy. (p. 546)
Electron A negative charge. (p. 540)
Excited State A high-energy orbit for an electron in an atom. (p. 553)
Gamma Ray A kind of radioactive ray emitted from a nucleus during decay; has no mass; made up of photons; is uncharged. (p. 545)
Ground State The lowest energy level for an electron in an atom. (p. 553)
Half-Life The amount of time required for half of the radioactive atoms in a sample to decay. (p. 548)
Isotopes Nuclei that contain the same number of protons but a different number of neutrons. (p. 542)
Neutron A fundamental particle that makes up the nucleus of an atom; neutrally charged. (p. 540)
Nuclear Fission The bombardment of uranium with neutrons, which produces lighter nuclei along with a large number of neutrons. (p. 546)

Nuclear Fusion The process of light nuclei interacting to form heavier nuclei with the release of energy. (p. 547)

Nuclide A specific type of atom characterized by its nuclear properties, such as the number of neutrons and protons and the energy state of its nucleus. (p. 541)

Orbital A region of space surrounding an atomic nucleus in which an electron has a high probability of being found. (p. 556)

Periodic Table A table containing all atomic elements arranged according to their atomic numbers, which can be used to predict their chemical properties. (p. 557)

Principle of Equivalence A physical law stating that mass and energy are equivalent forms of matter; stated by Einstein. (p. 542)

Proton A fundamental particle that makes up the nucleus of the atom; positively charged. (p. 540)

Quantum Mechanical Model A model of the atom that is based on electrons behaving like waves. (p. 555)

Radioactive Decay A type of nuclear decay which occurs in heavy atoms when there are not enough protons to prevent one or more neutrons from decaying into a proton and an electron. (p. 545)

Strong Force An attractive force between all nucleons (neutrons and protons) independent of their charge. (p. 540)

FORMULAS

23.2 $E = \Delta mc^2$

23.4 $Q = (M_p - M_d - m_\alpha)c^2$

23.8 $P = \lambda \Delta t$

$$N = N_0 e^{-\lambda t}$$

$$T_{1/2} = \frac{0.693}{\lambda}$$

23.11 $A = \lambda N = \lambda N_0 e^{-\lambda t} = A_0 e^{-\lambda t}$

23.12 $E = \dfrac{-kZ^2}{n^2}$

REVIEW QUESTIONS

1. Which of the following are nuclear particles?
   (a) neutrons
   (b) protons
   (c) nucleons
   (d) atoms
   (e) all of the above
2. The amount of radioactive material remaining after a period of time is related to the
   (a) atomic mass.
   (b) pressure.
   (c) half-life.
   (d) volume
   (e) none of the above.
3. The Einstein's equivalence principle relates to
   (a) weight and time.
   (b) space and gravity.
   (c) mass and energy.
   (d) all of the above.
   (e) none of the above.
4. Describe the similarities of protons and neutrons. Describe the differences.
5. Describe the differences between the electrical force and the strong force.
6. If the strong force suddenly became much weaker in a nucleus while the electrical force remained unchanged, what might happen to the nucleus?

7. What is the difference among the following atoms: ${}^{11}_{6}C$, ${}^{12}_{6}C$, and ${}^{13}_{6}C$? What are the similarities?
8. Explain the principle of equivalence of matter and energy in your own words.
9. Explain the term *electron volt* in your own words.
10. Describe the importance of the neutron in atomic nuclei.
11. Describe an α ray in your own words.
12. Describe a γ ray in your own words.
13. Describe a β ray in your own words.
14. What was the first nuclear reaction observed in a laboratory? Who observed it?
15. What important discovery was made by Enrico Fermi?
16. Explain a self-sustaining chain reaction.
17. Describe nuclear fusion.
18. Describe nuclear fission.
19. Describe the term *decay rate*.
20. What fraction of a radioactive sample has not decayed after 4 half-lives have elapsed?
21. What damage can be caused to living organisms by radiation?
22. What medical uses does radiation have?
23. The energy from nuclear fusion appears in the form of what kind of energy?
24. Describe the Bohr model of the atom.
25. Explain the difference between the ground state and the excited states of electrons in atoms.

REVIEW PROBLEMS

1. Find the mass in kilograms of the ${}^{15}_{8}O$ atom if its mass in atomic mass units is 15.003065 u.
2. Find the mass in kilograms of the ${}^{19}_{9}F$ atom if its mass in atomic mass units is 18.998404 u.
3. Find the mass in kilograms of the ${}^{166}_{68}Er$ atom if its mass in atomic mass units is 165.930292 u.
4. Find the binding energy in MeV of a ${}^{86}_{38}Sr$ atom if the mass of the neutral Sr atom is 85.909266 u.
5. Estimate the total binding energy for ${}^{102}_{44}Ru$ from Fig. 23.5.
6. Estimate the total binding energy for ${}^{153}_{63}Eu$ from Fig. 23.5.
7. Estimate the total binding energy for ${}^{187}_{75}Re$ from Fig. 23.5.
8. Find the quantity of osmium 191 atoms remaining from an original sample of 8.25×10^{13} atoms after 54 days. The half-life of osmium 191 is 15.4 days.
9. Find the quantity of iodine 131 atoms remaining from an original sample of 8.33×10^{18} atoms after 34.4 days. The half-life of iodine 131 is 8.04 days.
10. Find the probability that a single iodine 131 atom will decay in the next 30.0 min. The half-life of iodine 131 is 8.04 days.
11. Find the probability that a single osmium 191 atom will decay in the next 10.0 min. The half-life of osmium 191 is 15.4 days.
12. Find the probability that a single strontium 88 atom will decay in the next year. The half-life of strontium 88 is 29.1 yr.
13. Find the percentage of a strontium 88 sample that will decay in the next 2.40 yr. The half-life of strontium 88 is 29.1 yr.
14. Find the percentage of an osmium 191 sample that will decay in the next 2.00 h. The half-life of osmium 191 is 15.4 days.

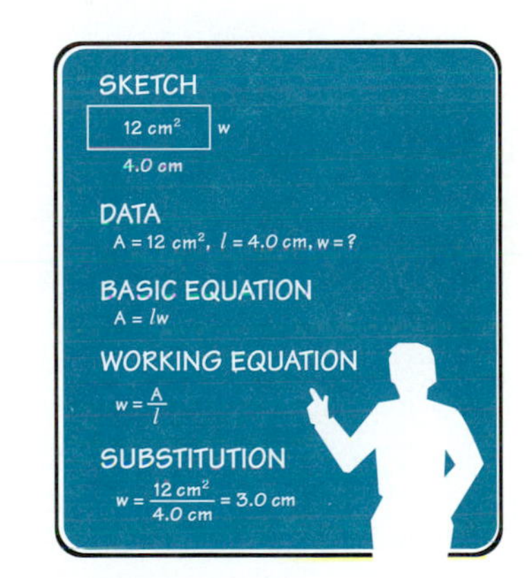

15. Find the activity of a 2.43-Ci sample of osmium 191 43.3 days after certification. The half-life of osmium 191 is 15.4 days.
16. Find the activity of a 3.79-Ci sample of nitrogen 13 43.0 min after certification. The half-life of nitrogen 13 is 10.0 min.
17. Find the activity of a 9.41-Ci sample of carbon 11 95.4 min after certification. The half-life of carbon 11 is 20.4 min.
18. Find the probability that a radioactive radium atom will decay in the next $1\bar{0}0$ yr if its half-life is $3\bar{0}0$ yr.
19. Find the probability that a uranium 138 atom will decay in the next $10,\bar{0}00$ years if its half-life is 4.5 billion years.
20. Find the energy of the electron in the $n = 5$ orbital of a hydrogen atom.
21. Find the energy of the photon given off when the electron in problem 20 transitions down to the $n = 3$ orbit.
22. List four inert atoms from the Periodic Table.
23. List four metals from the Periodic Table.

Appendix A

ALGEBRA REVIEW

A.1 SIGNED NUMBERS

Signed numbers have many applications to the study of physics. The rules for working with signed numbers follow.

Adding Signed Numbers

To add two positive numbers, add their absolute values.* A positive sign may or may not be placed before the result. It is usually omitted.

EXAMPLE 1

Add:

(a) $\begin{array}{r} +4 \\ \underline{+7} \\ +11 \end{array}$ or 11

(b) $(+3) + (+5) = +8$ or 8

To add two negative numbers, add their absolute values and place a negative sign before the result.

EXAMPLE 2

Add:

(a) $\begin{array}{r} -2 \\ \underline{-5} \\ -7 \end{array}$

(b) $(-6) + (-7) = -13$

(c) $(-8) + (-4) = -12$

To add a negative number and a positive number, find the difference of their absolute values. The sign of the number having the larger absolute value is placed before the result.

EXAMPLE 3

Add:

(a) $\begin{array}{r} +4 \\ \underline{-6} \\ -2 \end{array}$ (b) $\begin{array}{r} -2 \\ \underline{+8} \\ +6 \end{array}$ (c) $\begin{array}{r} -8 \\ \underline{+3} \\ -5 \end{array}$ (d) $\begin{array}{r} +9 \\ \underline{-4} \\ +5 \end{array}$

---

*The absolute value of a number is its nonnegative value. For example, the absolute value of −6 is 6; the absolute value of +10 is 10; and the absolute value of 0 is 0.

(e) $(+7) + (-2) = +5$ (f) $(-9) + (+6) = -3$
(g) $(-3) + (+10) = +7$ (h) $(+4) + (-12) = -8$

To add three or more signed numbers:

1. Add the positive numbers.
2. Add the negative numbers.
3. Add the sums from steps 1 and 2 according to the rules for addition of signed numbers.

EXAMPLE 4

Add: $(-2) + 4 + (-6) + 10 + (-7)$.

Step 1: $\begin{array}{r} +4 \\ \underline{+10} \\ +14 \end{array}$ Step 2: $\begin{array}{r} -2 \\ -6 \\ \underline{-7} \\ -15 \end{array}$ Step 3: $\begin{array}{r} -15 \\ \underline{+14} \\ -1 \end{array}$

Therefore, $(-2) + 4 + (-6) + 10 + (-7) = -1$.

Subtracting Signed Numbers

To subtract two signed numbers, change the sign of the *number being subtracted* and *add* according to the rules for addition.

EXAMPLE 5

Subtract:

(a) Subtract: $\begin{array}{r} +3 \\ \underline{+7} \\ -4 \end{array}$ $\leftrightarrow$ Add: $\begin{array}{r} +3 \\ \underline{-7} \\ -4 \end{array}$ To subtract, change the sign of the number being subtracted, +7, and add.

(b) Subtract: $\begin{array}{r} -9 \\ \underline{-6} \\ -3 \end{array}$ $\leftrightarrow$ Add: $\begin{array}{r} -9 \\ \underline{+6} \\ -3 \end{array}$ To subtract, change the sign of the number being subtracted, −6, and add.

(c) Subtract: $\begin{array}{r} +8 \\ \underline{-4} \\ +12 \end{array}$ $\leftrightarrow$ Add: $\begin{array}{r} +8 \\ \underline{+4} \\ +12 \end{array}$

(d) Subtract: $\begin{array}{r} -6 \\ \underline{+8} \\ -14 \end{array}$ $\leftrightarrow$ Add: $\begin{array}{r} -6 \\ \underline{-8} \\ -14 \end{array}$

(e) $(+6) - (+8) = (+6) + (-8) = -2$ To subtract, change the sign of the number being subtracted, +8, and add.

(f) $(-3) - (-5) = (-3) + (+5) = +2$

(g) $(+10) - (-3) = (+10) + (+3) = +13$

(h) $(-5) - (+2) = (-5) + (-2) = -7$

When more than two signed numbers are involved in subtraction, change the sign of *each* number being subtracted and add the resulting signed numbers.

EXAMPLE 6

Subtract: $(-2) - (+4) - (-1) - (-3) - (+5)$
$= (-2) + (-4) + (+1) + (+3) + (-5)$.

Step 1:		Step 2:		Step 3:	
	+1		−2		+4
	+3		−4		−11
	+4		−5		−7
			−11		

Therefore, $(-2) - (+4) - (-1) - (-3) - (+5) = -7$.

When combinations of addition and subtraction of signed numbers occur in the same problem, change *only* the sign of each number being subtracted. Then add the resulting signed numbers.

EXAMPLE 7

Find the result:

$$(-2) + (-4) - (-3) - (+6) + (+1) - (+2) + (-7) - (-5)$$
$$= (-2) + (-4) + (+3) + (-6) + (+1) + (-2) + (-7) + (+5)$$

Step 1:		Step 2:		Step 3:	
	+3		−2		+9
	+1		−4		−21
	+5		−6		−12
	+9		−2		
			−7		
			−21		

Therefore, $(-2) + (-4) - (-3) - (+6) + (+1) - (+2) + (-7) - (-5) = -12$

Multiplying Signed Numbers

To multiply two signed numbers:

1. If the signs of the numbers are both positive or both negative, find the product of their absolute values. This product is always positive.
2. If the signs of the numbers are unlike, find the product of their absolute values and place a negative sign before the result.

EXAMPLE 8

Multiply:

(a)		(b)		(c)		(d)	
	+3		−5		−6		+2
	+4		−8		+7		−3
	+12		+40		−42		−6

(e) $(+3)(+5) = +15$ (f) $(-7)(-8) = +56$
(g) $(-1)(+6) = -6$ (h) $(+4)(-2) = -8$

To multiply more than two signed numbers, first multiply the absolute values of the numbers. If there is an odd number of negative factors, place a negative sign before the result. If there is an even number of negative factors, the product is positive. ***Note:*** An *even* number is a number divisible by 2.

EXAMPLE 9

Multiply:
(a) $(+5)(-6)(+2)(-1) = +60$
(b) $(-3)(-3)(+4)(-5) = -180$

Dividing Signed Numbers

The rules for dividing signed numbers are similar to those for multiplying signed numbers. To divide two signed numbers:

1. If the signs of the numbers are both positive or both negative, divide their absolute values. This quotient is always positive.
2. If the two numbers have different signs, divide their absolute values and place a negative sign before the quotient.

Note: Division by 0 is undefined.

EXAMPLE 10

Divide:

(a) $\dfrac{+10}{+2} = +5$ (b) $\dfrac{-18}{-3} = +6$ (c) $\dfrac{+20}{-4} = -5$ (d) $\dfrac{-24}{+2} = -12$

■ PROBLEMS A.1

Perform the indicated operations.

1. $(-5) + (-6)$
2. $(+1) + (-10)$
3. $(-3) + (+8)$
4. $(+5) + (+7)$
5. $(-5) + (+3)$
6. $0 + (-3)$
7. $(-7) - (-3)$
8. $(+2) - (-9)$
9. $(-4) - (+2)$
10. $(+4) - (+7)$
11. $0 - (+3)$
12. $0 - (-2)$
13. $(-9)(-2)$
14. $(+4)(+6)$
15. $(-7)(+3)$
16. $(+5)(-8)$
17. $(+6)(0)$
18. $(0)(-4)$
19. $\dfrac{+36}{+12}$
20. $\dfrac{-9}{-3}$
21. $\dfrac{+16}{-2}$
22. $\dfrac{-15}{+3}$
23. $\dfrac{0}{+6}$
24. $\dfrac{4}{0}$
25. $(+2) + (-1) + (+10)$
26. $(-7) + (+2) + (+9) + (-8)$
27. $(-9) + (-3) + (+3) + (-8) + (+4)$
28. $(+8) + (-2) + (-6) + (+7) + (-6) + (+9)$
29. $(-4) - (+5) - (-4)$
30. $(+3) - (-5) - (-6) - (+5)$
31. $(-7) - (-4) - (+6) - (+4) - (-5)$
32. $(-8) - (+7) - (+3) - (-7) - (-8) - (-2)$
33. $(+5) + (-2) - (+7)$
34. $(-3) - (-8) - (+3) + (-9)$
35. $(-2) - (+1) - (-10) + (+12) + (-9)$
36. $(-1) - (-11) + (+2) - (-10) + (+8)$
37. $(+3)(-5)(+3)$
38. $(-1)(+2)(+2)(-1)$
39. $(+2)(-4)(-6)(-3)(+2)$
40. $(-1)(+3)(-2)(-4)(+5)(-1)$ ■

A.2 POWERS OF 10

The ability to work quickly and accurately with powers of 10 is important in scientific and technical fields.

> When multiplying two powers of 10, add the exponents. That is,
>
> $$10^a \times 10^b = 10^{a+b}$$

EXAMPLE 1

Multiply:

(a) $(10^6)(10^3) = 10^{6+3} = 10^9$
(b) $(10^4)(10^2) = 10^{4+2} = 10^6$
(c) $(10^1)(10^{-3}) = 10^{1+(-3)} = 10^{-2}$
(d) $(10^{-2})(10^{-5}) = 10^{[-2+(-5)]} = 10^{-7}$

When dividing two powers of 10, subtract the exponents as follows:

$$10^a \div 10^b = 10^{a-b}$$

EXAMPLE 2

Divide:

(a) $\frac{10^7}{10^4} = 10^{7-4} = 10^3$
(b) $\frac{10^3}{10^5} = 10^{3-5} = 10^{-2}$
(c) $\frac{10^{-2}}{10^{+3}} = 10^{(-2)-(+3)} = 10^{-5}$
(d) $\frac{10^4}{10^{-2}} = 10^{4-(-2)} = 10^6$

To raise a power of 10 to a power, multiply the exponents as follows:

$$(10^a)^b = 10^{ab}$$

EXAMPLE 3

Find each power:

(a) $(10^2)^3 = 10^{(2)(3)} = 10^6$
(b) $(10^{-3})^2 = 10^{(-3)(2)} = 10^{-6}$
(c) $(10^4)^{-5} = 10^{(4)(-5)} = 10^{-20}$
(d) $(10^{-3})^{-4} = 10^{(-3)(-4)} = 10^{12}$

Next, we will show that $10^0 = 1$. To do this, we need to use the substitution principle, which states that

$$\text{if } a = b \text{ and } a = c, \text{ then } b = c$$

First,

$$\frac{10^n}{10^n} = 10^{n-n} \quad \text{To divide powers, subtract the exponents.}$$
$$= 10^0$$

Second,

$$\frac{10^n}{10^n} = 1 \quad \text{Any number other than zero divided by itself equals 1.}$$

That is, since

$$\frac{10^n}{10^n} = 10^0 \quad \text{and} \quad \frac{10^n}{10^n} = 1$$

then $10^0 = 1$.

We also will use the fact that $\frac{1}{10^a} = 10^{-a}$. To show this,

$$\frac{1}{10^a} = \frac{10^0}{10^a} \quad (1 = 10^0)$$

$$= 10^{0-a} \quad \text{To divide powers, subtract the exponents.}$$

$$= 10^{-a}$$

We also need to show that $\frac{1}{10^{-a}} = 10^a$.

$$\frac{1}{10^{-a}} = \frac{10^0}{10^{-a}}$$

$$= 10^{0-(-a)}$$

$$= 10^a$$

In summary,

$$10^0 = 1 \qquad \frac{1}{10^a} = 10^{-a} \qquad \frac{1}{10^{-a}} = 10^a$$

Actually, any number (except zero) raised to the zero power equals 1.

■ PROBLEMS A.2

Do as indicated. Express the results using positive exponents.

1. $(10^5)(10^3)$
2. $10^6 \div 10^2$
3. $(10^2)^4$
4. $(10^{-2})(10^{-3})$
5. $\frac{10^3}{10^5}$
6. $(10^{-3})^3$
7. $10^5 \div 10^{-2}$
8. $(10^{-2})^{-3}$
9. $(10^4)(10^{-1})$
10. $\frac{10^0}{10^{-4}}$
11. $(10^0)(10^{-4})$
12. $\frac{10^{-4}}{10^{-3}}$
13. $(10^0)^{-2}$
14. 10^{-3}
15. $\frac{1}{10^{-5}}$
16. $\frac{(10^4)(10^{-2})}{(10^6)(10^3)}$
17. $\frac{(10^{-2})(10^{-3})}{(10^3)^2}$
18. $\frac{(10^2)^4}{(10^{-3})^2}$
19. $\left(\frac{1}{10^3}\right)^2$
20. $\left(\frac{10^2}{10^{-3}}\right)^2$
21. $\left(\frac{10 \cdot 10^2}{10^{-1}}\right)^2$
22. $\left(\frac{1}{10^{-3}}\right)^2$
23. $\frac{(10^4)(10^{-2})}{10^{-8}}$
24. $\frac{(10^4)(10^6)}{(10^0)(10^{-2})(10^3)}$ ■

A.3 SOLVING LINEAR EQUATIONS

An equation is a mathematical sentence stating that two quantities are equal. To solve an equation means to find the number or numbers that can replace the variable in the equation to make the equation a true statement. The value we find that makes the equation a true statement is called the *root* of the equation. When the root of an equation is found, we say we have *solved* the equation.

If $a = b$, then $a + c = b + c$ or $a - c = b - c$. (If the same quantity is added to or subtracted from both sides of an equation, the resulting equation is equivalent to the original equation.)

To solve an equation using this rule, think first of undoing what has been done to the variable.

EXAMPLE 1

Solve: $x - 5 = -9$.

$$x - 5 = -9$$

$$x - 5 + 5 = -9 + 5 \qquad \text{Undo the subtraction by adding 5 to both sides.}$$

$$x = -4$$

EXAMPLE 2

Solve: $x + 4 = 29$.

$$x + 4 = 29$$

$$x + 4 - 4 = 29 - 4 \qquad \text{Undo the addition by subtracting 4 from both sides.}$$

$$x = 25$$

If $a = b$, then $ac = bc$ or $a/c = b/c$ with $c \neq 0$. (If both sides of an equation are multiplied or divided by the same nonzero quantity, the resulting equation is equivalent to the original equation.)

EXAMPLE 3

Solve: $3x = 18$.

$$3x = 18$$

$$\frac{3x}{3} = \frac{18}{3} \qquad \text{Undo the multiplication by dividing both sides by 3.}$$

$$x = 6$$

EXAMPLE 4

Solve: $x/4 = 9$.

$$\frac{x}{4} = 9$$

$$4\left(\frac{x}{4}\right) = 4 \cdot 9 \qquad \text{Undo the division by multiplying both sides by 4.}$$

$$x = 36$$

When more than one operation is indicated on the variable in an equation, undo the additions and subtractions first, then undo the multiplications and divisions.

EXAMPLE 5

Solve: $3x + 5 = 17$.

$$3x + 5 = 17$$

$$3x + 5 - 5 = 17 - 5 \qquad \text{Subtract 5 from both sides.}$$

$$3x = 12$$

$$\frac{3x}{3} = \frac{12}{3} \qquad \text{Divide both sides by 3.}$$

$$x = 4$$

EXAMPLE 6

Solve: $2x - 7 = 10$.

$$
\begin{aligned}
2x - 7 &= 10 \\
2x - 7 + 7 &= 10 + 7 && \text{Add 7 to both sides.} \\
2x &= 17 \\
\frac{2x}{2} &= \frac{17}{2} && \text{Divide both sides by 2.} \\
x &= \frac{17}{2} = 8.5
\end{aligned}
$$

EXAMPLE 7

Solve: $x/5 - 10 = 22$.

$$
\begin{aligned}
\frac{x}{5} - 10 &= 22 \\
\frac{x}{5} - 10 + 10 &= 22 + 10 && \text{Add 10 to both sides.} \\
\frac{x}{5} &= 32 \\
5\left(\frac{x}{5}\right) &= 5(32) && \text{Multiply both sides by 5.} \\
x &= 160
\end{aligned}
$$

To solve an equation with variables on both sides:

1. Add or subtract either variable term from both sides of the equation.
2. Add or subtract from both sides of the equation the constant term that now appears on the same side of the equation with the variable. Then solve.

EXAMPLE 8

Solve: $3x + 6 = 7x - 2$.

$$
\begin{aligned}
3x + 6 &= 7x - 2 \\
3x + 6 - 3x &= 7x - 2 - 3x && \text{Subtract } 3x \text{ from both sides.} \\
6 &= 4x - 2 \\
6 + 2 &= 4x - 2 + 2 && \text{Add 2 to both sides.} \\
8 &= 4x \\
\frac{8}{4} &= \frac{4x}{4} && \text{Divide both sides by 4.} \\
2 &= x
\end{aligned}
$$

EXAMPLE 9

Solve: $4x - 2 = -5x + 10$.

$$
\begin{aligned}
4x - 2 &= -5x + 10 \\
4x - 2 + 5x &= -5x + 10 + 5x && \text{Add } 5x \text{ to both sides.} \\
9x - 2 &= 10
\end{aligned}
$$

$$9x - 2 + 2 = 10 + 2 \qquad \text{Add 2 to both sides.}$$
$$9x = 12$$
$$\frac{9x}{9} = \frac{12}{9} \qquad \text{Divide both sides by 9.}$$
$$x = \frac{4}{3}$$

To solve equations containing parentheses, first remove the parentheses and then proceed as before. The rules for removing parentheses are the following:

1. If the parentheses are preceded by a plus (+) sign, they may be removed without changing any signs.

   Examples:
   $$2 + (3 - 5) = 2 + 3 - 5$$
   $$3 + (x + 4) = 3 + x + 4$$
   $$5x + (-6x + 9) = 5x - 6x + 9$$

2. If the parentheses are preceded by a minus (−) sign, the parentheses may be removed if *all* the signs of the numbers (or letters) within the parentheses are changed.

   Examples:
   $$2 - (3 - 5) = 2 - 3 + 5$$
   $$5 - (x - 7) = 5 - x + 7$$
   $$7x - (-4x - 11) = 7x + 4x + 11$$

3. If the parentheses are preceded by a number, the parentheses may be removed if each of the terms inside the parentheses is multiplied by that (signed) number.

   Examples:
   $$2(x + 4) = 2x + 8$$
   $$-3(x - 5) = -3x + 15$$
   $$2 - 4(3x - 5) = 2 - 12x + 20$$

EXAMPLE 10

Solve: $3(x - 4) = 15$.

$$3(x - 4) = 15$$
$$3x - 12 = 15 \qquad \text{Remove parentheses.}$$
$$3x - 12 + 12 = 15 + 12 \qquad \text{Add 12 to both sides.}$$
$$3x = 27$$
$$\frac{3x}{3} = \frac{27}{3} \qquad \text{Divide both sides by 3.}$$
$$x = 9$$

EXAMPLE 11

Solve: $2x - (3x + 15) = 4x - 1$.

$$2x - (3x + 15) = 4x - 1$$
$$2x - 3x - 15 = 4x - 1 \qquad \text{Remove parentheses.}$$
$$-x - 15 = 4x - 1 \qquad \text{Combine like terms.}$$

$$
\begin{aligned}
-x - 15 + x &= 4x - 1 + x && \text{Add } x \text{ to both sides.}\\
-15 &= 5x - 1\\
-15 + 1 &= 5x - 1 + 1 && \text{Add 1 to both sides.}\\
-14 &= 5x\\
\frac{-14}{5} &= \frac{5x}{5} && \text{Divide both sides by 5.}\\
-2.8 &= x
\end{aligned}
$$

■ PROBLEMS A.3

Solve each equation.

1. $3x = 4$
2. $\frac{y}{2} = 10$
3. $x - 5 = 12$
4. $x + 1 = 9$
5. $2x + 10 = 10$
6. $4x = 28$
7. $2x - 2 = 33$
8. $4 = \frac{x}{10}$
9. $172 - 43x = 43$
10. $9x + 7 = 4$
11. $6y - 24 = 0$
12. $3y + 15 = 75$
13. $15 = \frac{105}{y}$
14. $6x = x - 15$
15. $2 = \frac{50}{2y}$
16. $9y = 67.5$
17. $8x - 4 = 36$
18. $10 = \frac{136}{4x}$
19. $2x + 22 = 75$
20. $9x + 10 = x - 26$
21. $4x + 9 = 7x - 18$
22. $2x - 4 = 3x + 7$
23. $-2x + 5 = 3x - 10$
24. $5x + 3 = 2x - 18$
25. $3x + 5 = 5x - 11$
26. $-5x + 12 = 12x - 5$
27. $13x + 2 = 20x - 5$
28. $5x + 3 = -9x - 39$
29. $-4x + 2 = -10x - 20$
30. $9x + 3 = 6x + 8$
31. $3x + (2x - 7) = 8$
32. $11 - (x + 12) = 100$
33. $7x - (13 - 2x) = 5$
34. $20(7x - 2) = 240$
35. $-3x + 5(x - 6) = 12$
36. $3(x + 117) = 201$
37. $5(2x - 1) = 8(x + 3)$
38. $3(x + 4) = 8 - 3(x - 2)$
39. $-2(3x - 2) = 3x - 2(5x + 1)$
40. $\frac{x}{5} - 2\left(\frac{2x}{5} + 1\right) = 28$ ■

A.4 SOLVING QUADRATIC EQUATIONS

A quadratic equation in one variable is one in which the largest exponent of the variable is 2. The most general quadratic equation in variable x is written as

$$ax^2 + bx + c = 0 \qquad (\text{where } a \neq 0)$$

EXAMPLE 1

Solve: $x^2 = 16$.

To solve a quadratic equation of this type, take the square root of both sides of the equation.

$$
\begin{aligned}
x^2 &= 16\\
x &= \pm 4 && \text{Take the square root of both sides.}
\end{aligned}
$$

In general, solve equations of the form $ax^2 = b$, where $a \neq 0$, as follows:

$$ax^2 = b$$

$$x^2 = \frac{b}{a} \qquad \text{Divide both sides by } a.$$

$$x = \pm\sqrt{\frac{b}{a}} \qquad \text{Take the square root of both sides.}$$

EXAMPLE 2

Solve: $2x^2 - 18 = 0$.

$$2x^2 - 18 = 0$$

$$2x^2 = 18 \qquad \text{Add 18 to both sides.}$$

$$x^2 = 9 \qquad \text{Divide both sides by 2.}$$

$$x = \pm 3 \qquad \text{Take the square root of both sides.}$$

EXAMPLE 3

Solve: $5y^2 = 100$.

$$5y^2 = 100$$

$$y^2 = 20 \qquad \text{Divide both sides by 5.}$$

$$y = \pm\sqrt{20} \qquad \text{Take the square root of both sides.}$$

$$y = \pm 4.47$$

The solutions of the general quadratic equation

$$ax^2 + bx + c = 0 \qquad (\text{where } a \neq 0)$$

are given by the formula (called the *quadratic formula*)

$$x = \frac{-b \pm \sqrt{b^2 - 4ac}}{2a}$$

where a = coefficient of the x^2 term
b = coefficient of the x term
c = constant term

The symbol ($\pm$) is used to combine two expressions or equations into one. For example, $a \pm 2$ means $a + 2$ or $a - 2$. Similarly,

$$x = \frac{-b \pm \sqrt{b^2 - 4ac}}{2a}$$

means

$$x = \frac{-b + \sqrt{b^2 - 4ac}}{2a} \qquad \text{or} \qquad x = \frac{-b - \sqrt{b^2 - 4ac}}{2a}$$

EXAMPLE 4

In the equation, $4x^2 - 3x - 7 = 0$, identify a, b, and c.

$$a = 4,\ b = -3,\ \text{and } c = -7$$

EXAMPLE 5

Solve $x^2 + 2x - 8 = 0$ using the quadratic formula.

First, $a = 1$, $b = 2$, and $c = -8$. Then

$$
\begin{aligned}
x &= \frac{-b \pm \sqrt{b^2 - 4ac}}{2a} \\
x &= \frac{-2 \pm \sqrt{(2)^2 - 4(1)(-8)}}{2(1)} \\
&= \frac{-2 \pm \sqrt{4 - (-32)}}{2} \\
&= \frac{-2 \pm \sqrt{36}}{2} \\
&= \frac{-2 \pm 6}{2} \\
&= \frac{-2 + 6}{2} \quad \text{or} \quad \frac{-2 - 6}{2} \\
&= \frac{4}{2} \quad \text{or} \quad \frac{-8}{2} \\
&= 2 \quad \text{or} \quad -4
\end{aligned}
$$

The solutions are 2 and -4.

If the number under the radical sign is not a perfect square, find the square root of the number by using a calculator and proceed as before.

EXAMPLE 6

Solve $4x^2 - 7x = 32$ using the quadratic formula.

Before identifying a, b, and c, the equation must be set equal to zero. That is,

$$4x^2 - 7x - 32 = 0$$

First, $a = 4$, $b = -7$, and $c = -32$. Then

$$
\begin{aligned}
x &= \frac{-b \pm \sqrt{b^2 - 4ac}}{2a} \\
&= \frac{-(-7) \pm \sqrt{(-7)^2 - 4(4)(-32)}}{2(4)} \\
&= \frac{7 \pm \sqrt{49 - (-512)}}{8} \\
&= \frac{7 \pm \sqrt{561}}{8} \\
&= \frac{7 \pm 23.7}{8} \qquad (\sqrt{561} = 23.7) \\
&= \frac{7 + 23.7}{8} \quad \text{or} \quad \frac{7 - 23.7}{8} \\
&= 3.84 \quad \text{or} \quad -2.09
\end{aligned}
$$

The approximate solutions are 3.84 and -2.09.

PROBLEMS A.4

Solve each equation.

1. $x^2 = 36$
2. $y^2 = 100$
3. $2x^2 = 98$
4. $5x^2 = 0.05$
5. $3x^2 - 27 = 0$
6. $2y^2 - 15 = 17$
7. $10x^2 + 4.9 = 11.3$
8. $2(32)(48 - 15) = v^2 - 27^2$
9. $2(107) = 9.8t^2$
10. $65 = \pi r^2$
11. $2.50 = \pi r^2$
12. $24^2 = a^2 + 16^2$

Find the values of a, b, and c, in each quadratic equation.

13. $3x^2 + x - 5 = 0$
14. $-2x^2 + 7x + 4 = 0$
15. $6x^2 + 8x + 2 = 0$
16. $5x^2 - 2x - 15 = 0$
17. $9x^2 + 6x = 4$
18. $6x^2 = x + 9$
19. $5x^2 + 6x = 0$
20. $7x^2 - 45 = 0$
21. $9x^2 = 64$
22. $16x^2 = 49$

Solve each quadratic equation using the quadratic formula.

23. $x^2 - 10x + 21 = 0$
24. $2x^2 + 13x + 15 = 0$
25. $6x^2 + 7x = 20$
26. $15x^2 = 4x + 4$
27. $6x^2 - 2x = 19$
28. $4x^2 = 28x - 49$
29. $18x^2 - 15x = 26$
30. $48x^2 + 9 = 50x$
31. $16.5x^2 + 8.3x - 14.7 = 0$
32. $125x^2 - 167x + 36 = 0$ ■

Appendix B

Scientific Calculators—Brief Instructions on Use

B.1 INTRODUCTION

There are several kinds and brands of calculators. Some are very simple to use; others do more difficult calculations. We demonstrate various operations on calculators that use algebraic logic, which follows the steps commonly used in mathematics.

To demonstrate how to use a calculator, we (1) use a flow chart, (2) show what buttons are pushed and the order in which they are pushed, and (3) show the display at each step. We assume that you know how to add, subtract, multiply, and divide on the calculator.

We have chosen to illustrate the most common types of calculators. Yours may differ in the number of digits displayed, or in the order of buttons pushed if yours has a function (F) or inverse button. If so, consult your manual.

B.2 SCIENTIFIC NOTATION

Numbers expressed in scientific notation can be entered into many calculators. The results may then also be given in scientific notation.

EXAMPLE 1

Multiply $(6.5 \times 10^{8})(1.4 \times 10^{-15})$ and write the result in scientific notation.

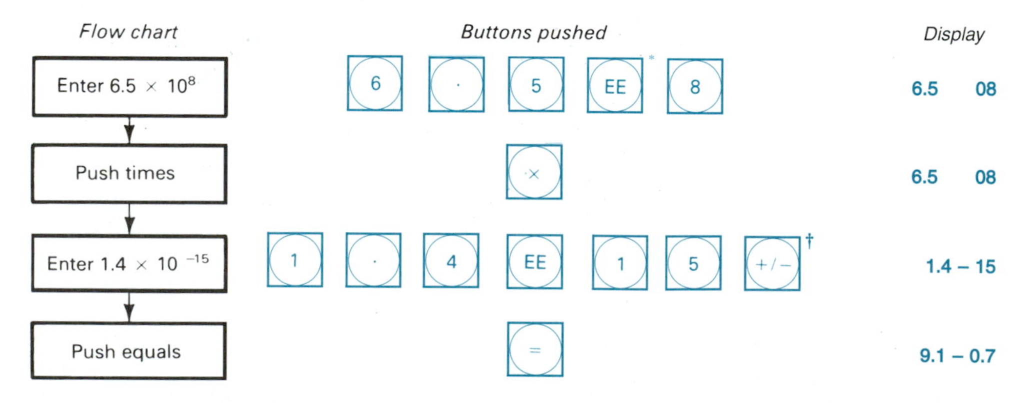

The result is 9.1×10^{-7}.

*Some calculators have a button marked EXP.

†This button is used to change the sign of the last number that has been entered. You should use this sign button before entering the number on some calculators.

EXAMPLE 2

Divide $\dfrac{3.24 \times 10^{-5}}{7.2 \times 10^{-12}}$ and write the result in scientific notation.

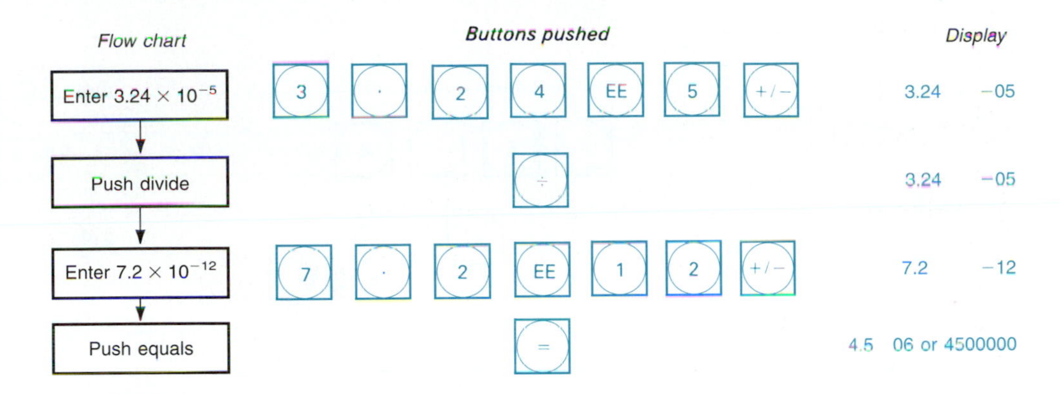

The result is 4.5×10^6.

EXAMPLE 3

Find the value of $\dfrac{(-6.3 \times 10^4)(-5.07 \times 10^{-9})(8.11 \times 10^{-6})}{(5.63 \times 10^{12})(-1.84 \times 10^7)}$ and write the result in scientific notation rounded to three significant digits.

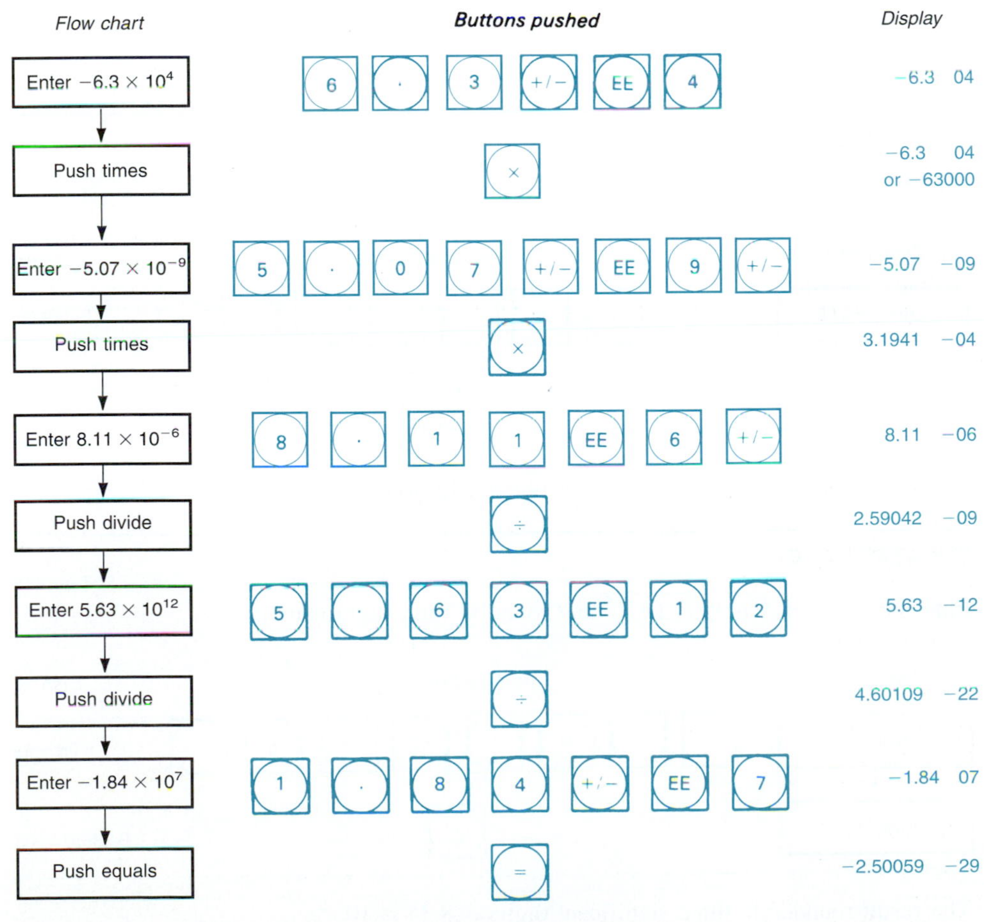

The result rounded to three significant digits is -2.50×10^{-29}.

B.3 SQUARES AND SQUARE ROOTS

EXAMPLE 1

Find the value of $(46.8)^2$.

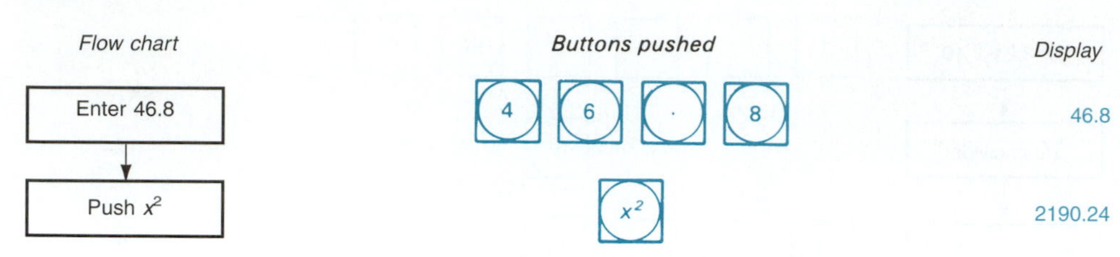

The result is 2190.24.

EXAMPLE 2

Find the value of $(6.3 \times 10^{-18})^2$.

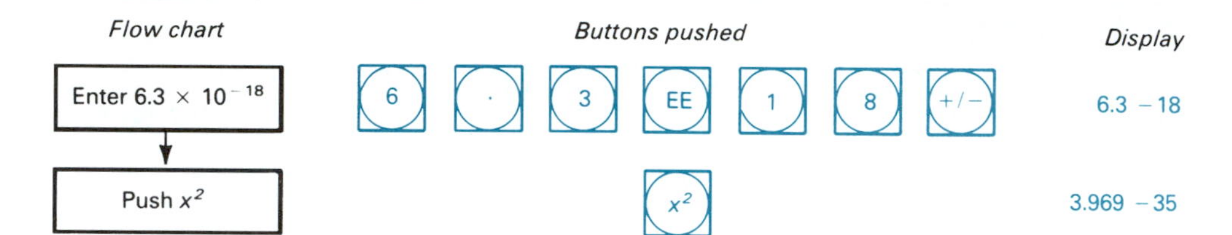

The result is 3.969×10^{-35}.

EXAMPLE 3

Find the value of $\sqrt{158.65}$ and round to four significant digits.

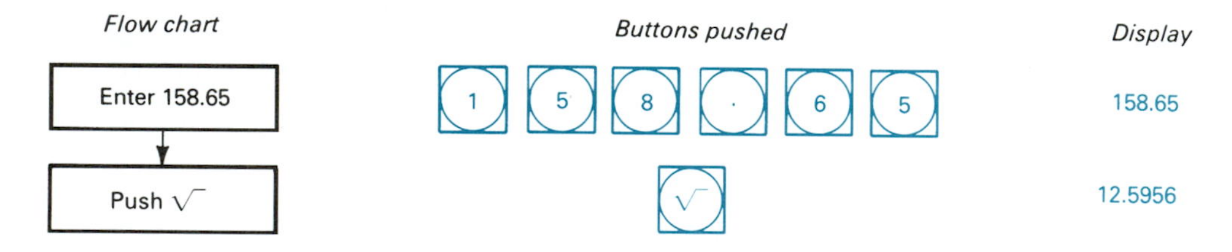

The result rounded to four significant digits is 12.60.

EXAMPLE 4

Find the value of $\sqrt{6.95 \times 10^{-15}}$ and round to three significant digits.

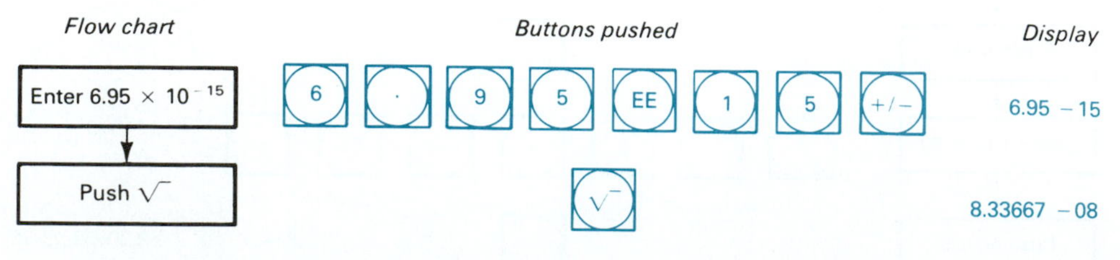

The result rounded to three significant digits is 8.34×10^{-8}.

EXAMPLE 5

Find the value of $\sqrt{15.7^2 + 27.6^2}$ and round to three significant digits.

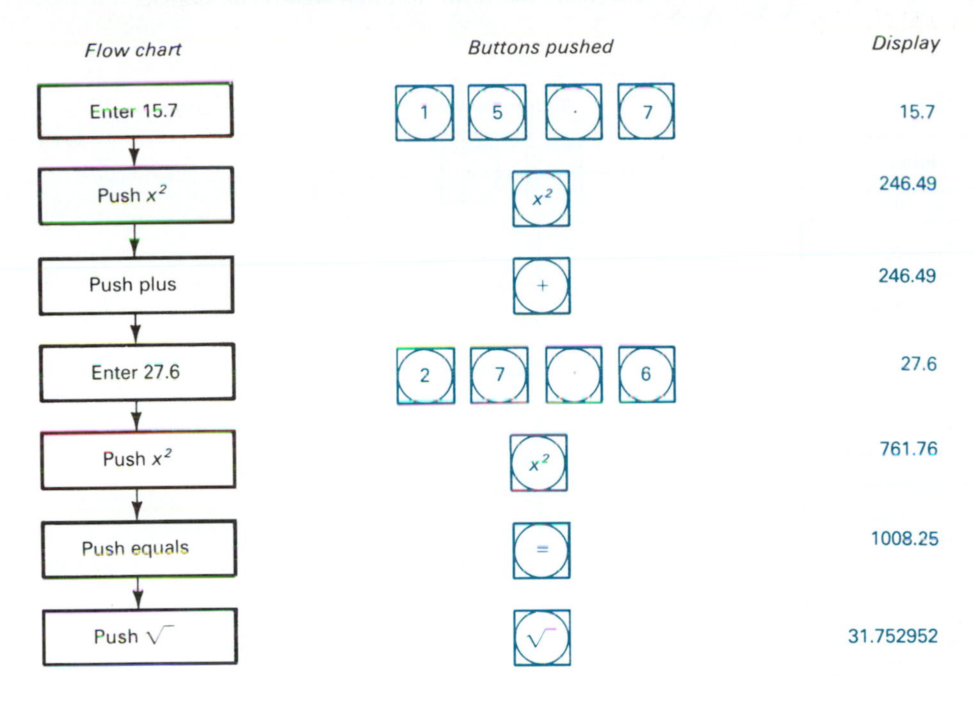

The result is 31.8 rounded to three significant digits.

B.4 USING A CALCULATOR WITH MEMORY

For some combinations of operations, a calculator with a memory is very helpful. A memory stores numbers for future use. Entries can then be added to or subtracted from what is already in the memory. ***Note:*** Consult the instruction manual if your calculator has more than one memory.

EXAMPLE 1

Find the value of $\frac{14}{\sqrt{5}} - \sqrt{\frac{15}{8}}$ and round the result to three significant digits.

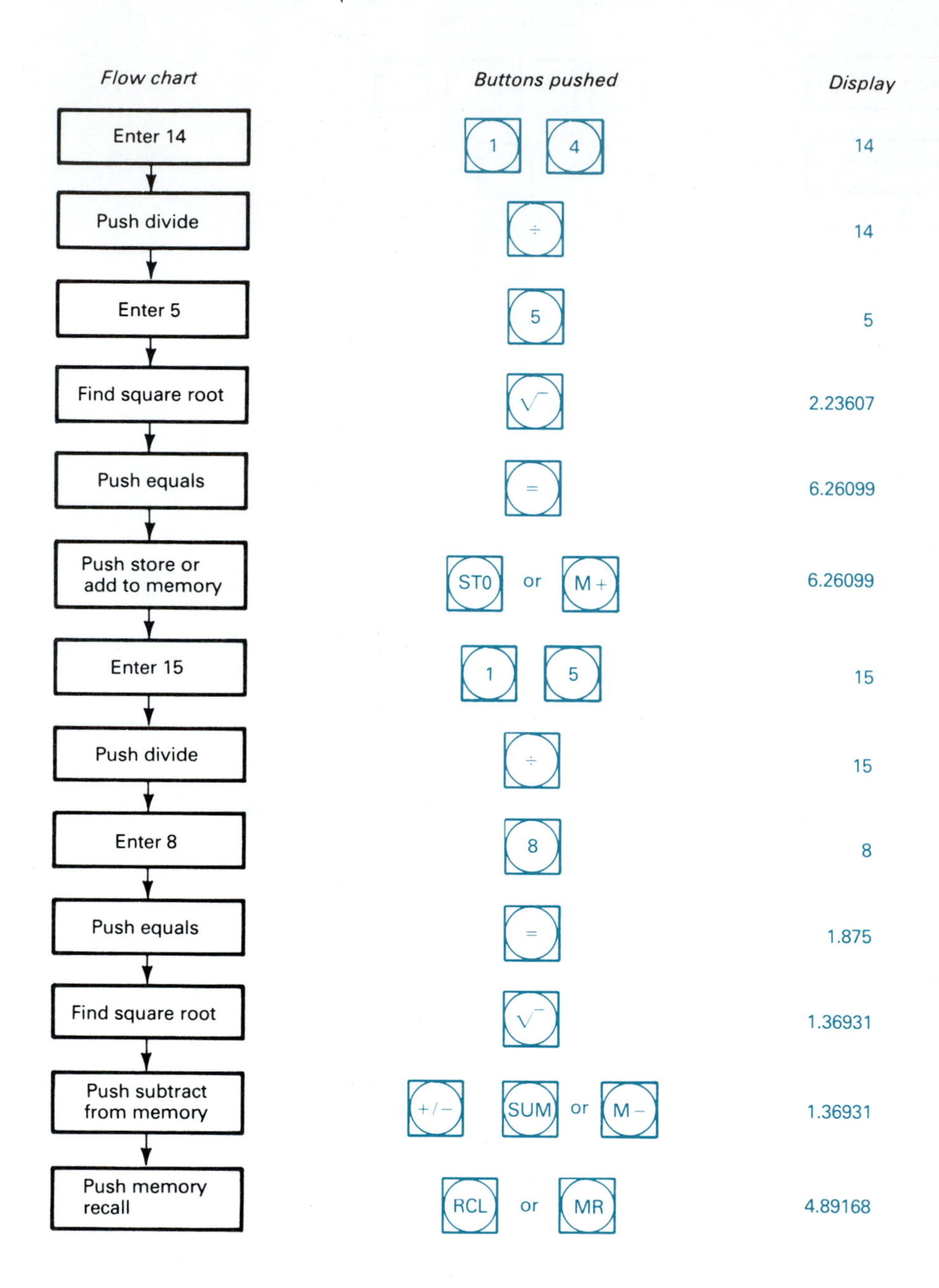

The result is 4.89 rounded to three significant digits.

EXAMPLE 2

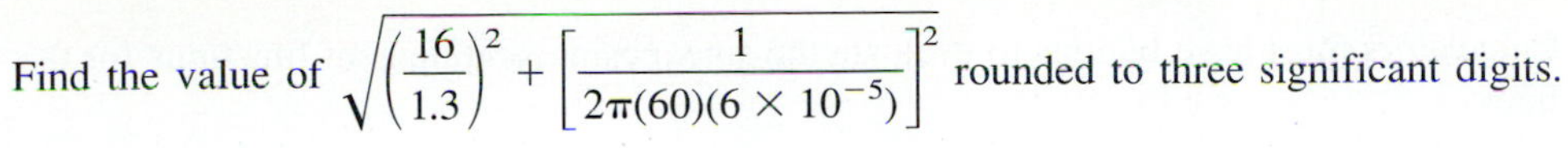

Find the value of $\sqrt{\left(\dfrac{16}{1.3}\right)^2 + \left[\dfrac{1}{2\pi(60)(6 \times 10^{-5})}\right]^2}$ rounded to three significant digits.

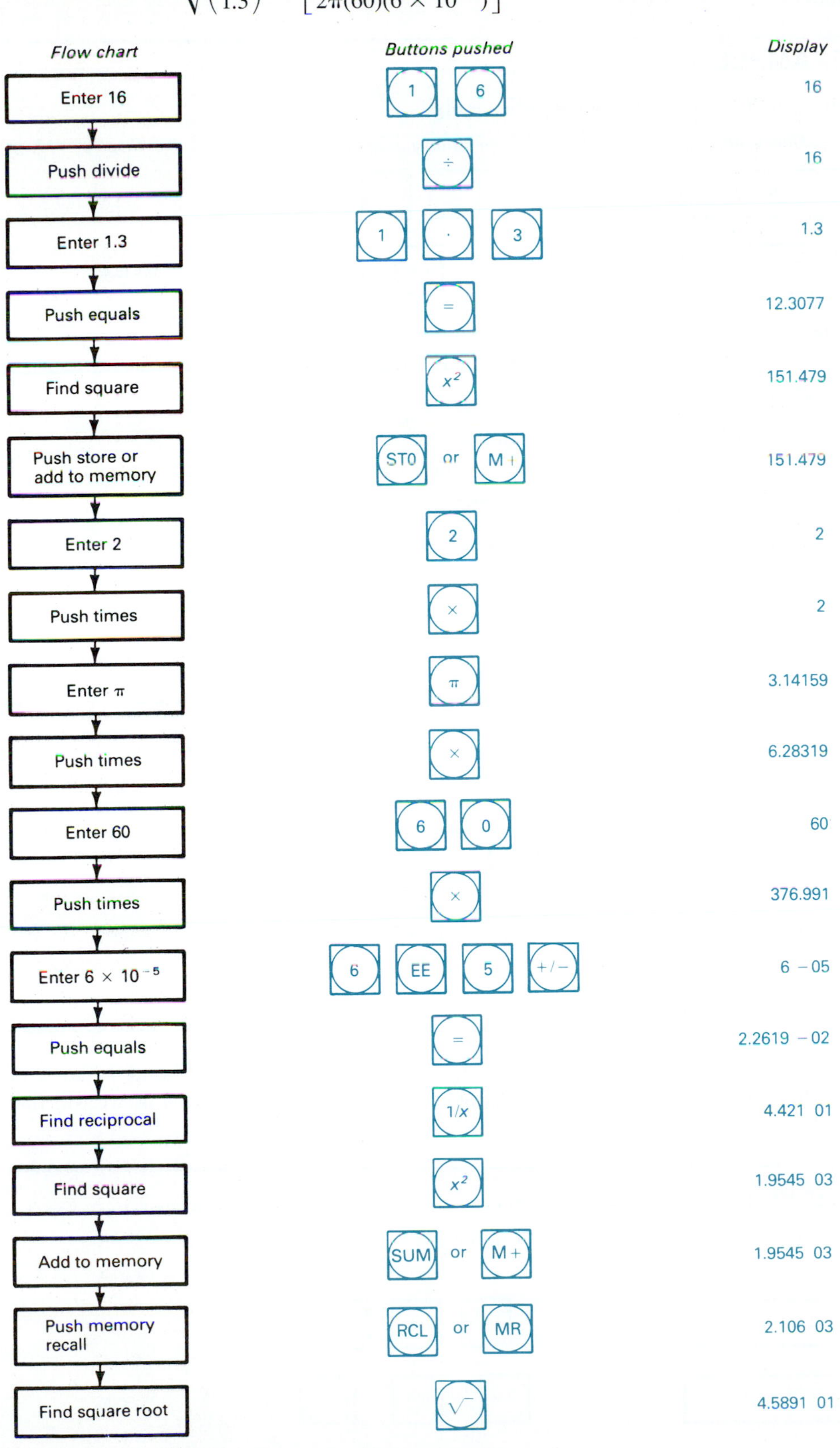

The result is 45.9 rounded to three significant digits.

B.5 TRIGONOMETRIC OPERATIONS

Calculators must have buttons to evaluate the sine, cosine, and tangent functions for this book.

EXAMPLE 1

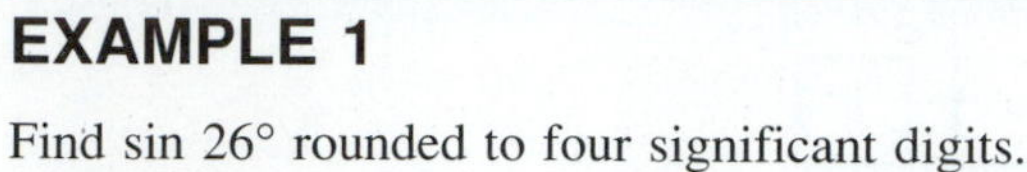

Find sin 26° rounded to four significant digits.

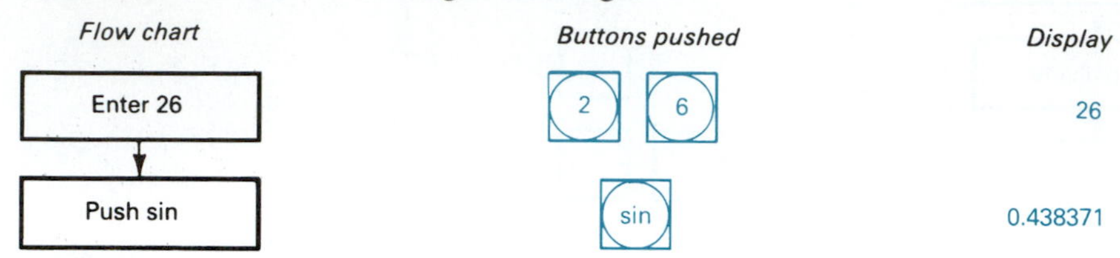

That is, sin 26° = 0.4384 rounded to four significant digits.

EXAMPLE 2

Find cos 36.75° rounded to four significant digits.

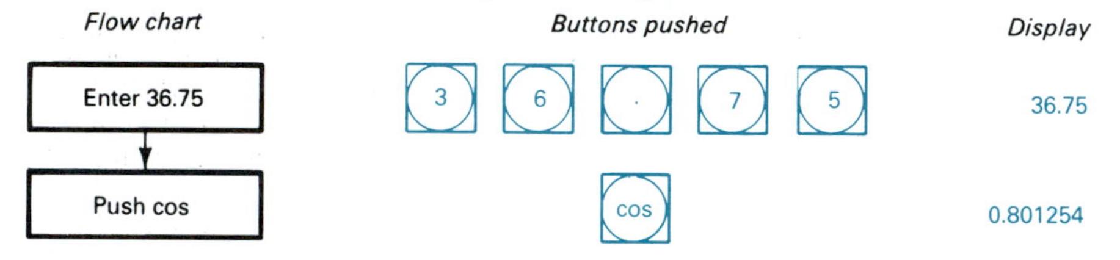

That is, cos 36.75° = 0.8013 rounded to four significant digits.

EXAMPLE 3

Find tan 70.6° rounded to four significant digits.

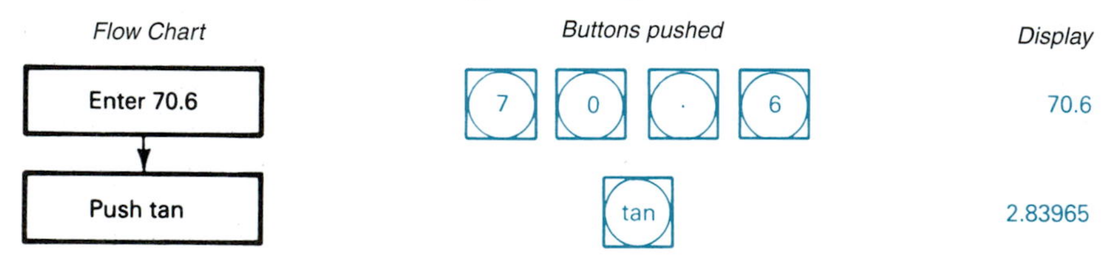

That is, tan 70.6° = 2.840 rounded to four significant digits.

Since we use right triangles almost exclusively, we show how to find the angle of a right triangle when the value of the trigonometric ratio is known. That is, we will find angle A when $0° \leq A \leq 90°$.

EXAMPLE 4

Given $\sin A = 0.4321$, find angle A to the nearest tenth of a degree.

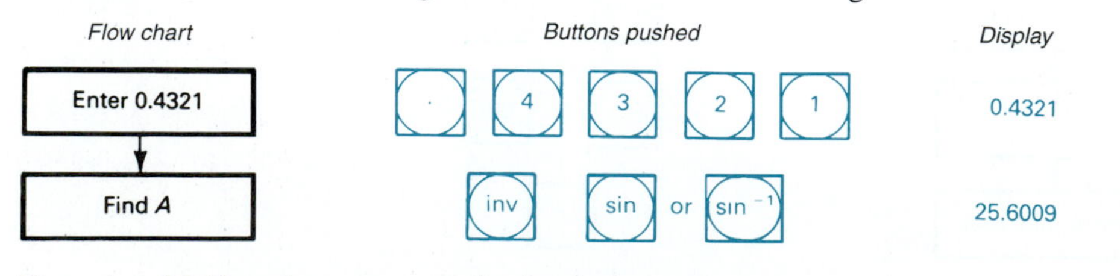

Thus, $A = 25.6°$ to the nearest tenth of a degree.

EXAMPLE 5

Given $\cos B = 0.6046$, find angle B to the nearest tenth of a degree.

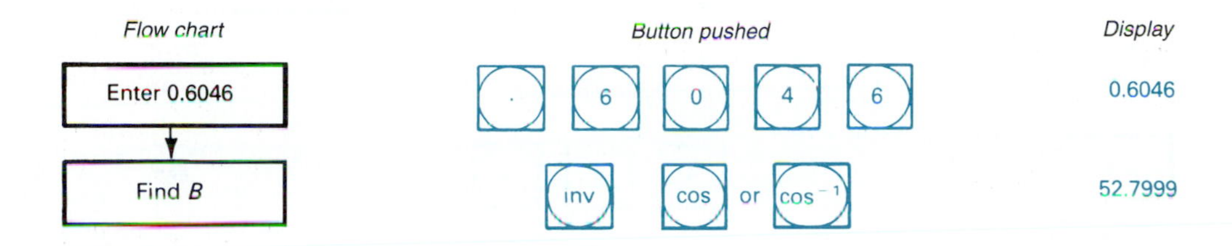

Thus, $B = 52.8°$ to the nearest tenth of a degree.

EXAMPLE 6

Given $\tan A = 2.584$, find angle A to the nearest tenth of a degree.

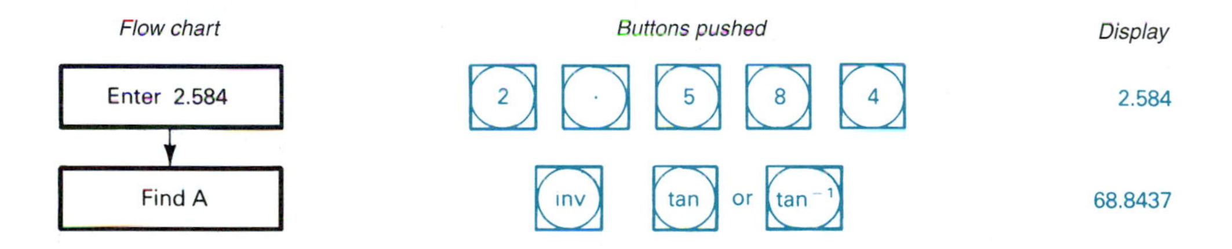

Thus, $A = 68.8°$ to the nearest tenth of a degree.

A trigonometric function often occurs in an expression that must be evaluated.

EXAMPLE 7

Given $a = (\tan 54°)(25.6 \text{ m})$, find a rounded to three significant digits.

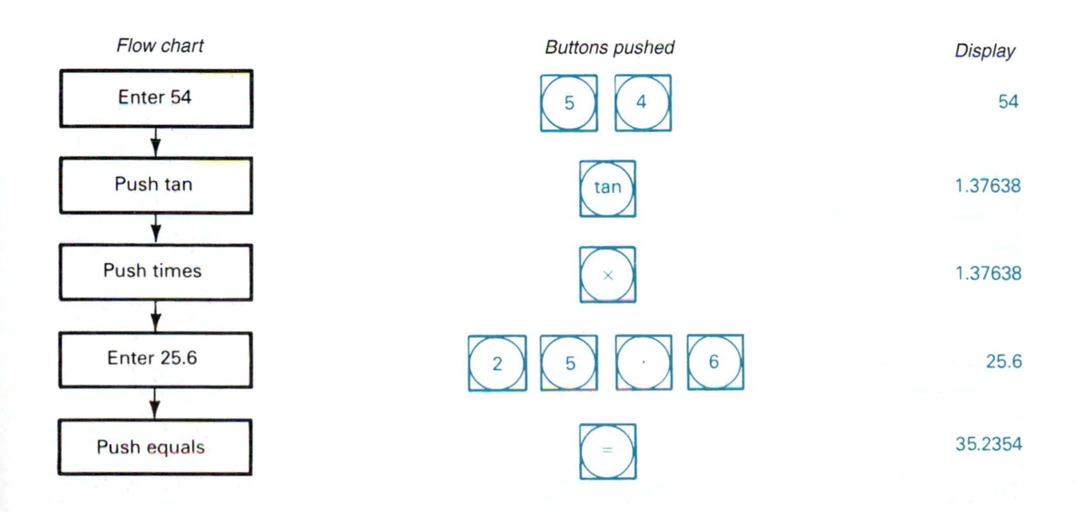

Thus, $a = 35.2$ m rounded to three significant digits.

EXAMPLE 8

Given $b = \dfrac{452 \text{ m}}{\cos 37.5°}$, find b rounded to three significant digits.

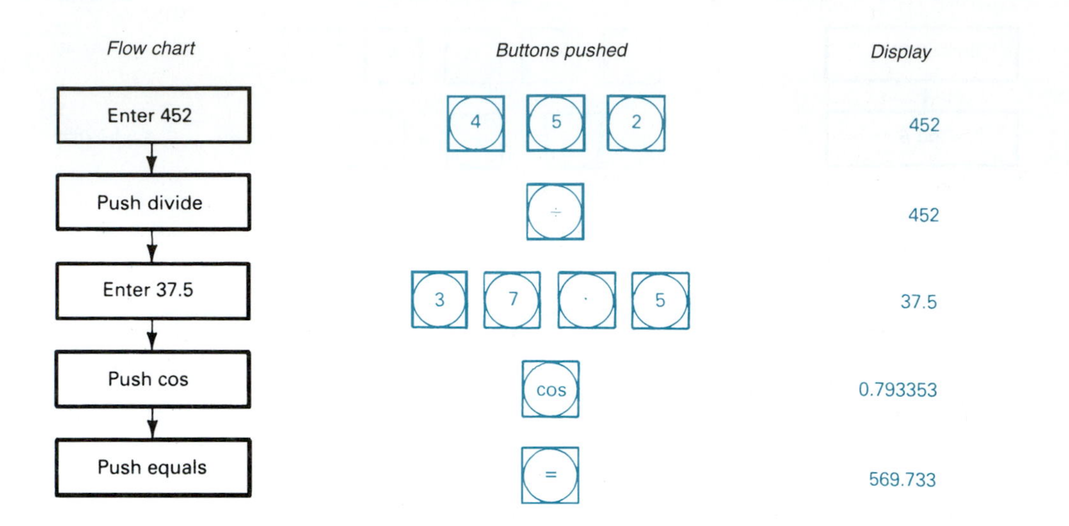

Thus, $b = 57\overline{0}$ m rounded to three significant digits.

B.6 USING THE y^x BUTTON

To raise a number to a power, use the y^x button as follows.

EXAMPLE 1

Find the value of 4^5.

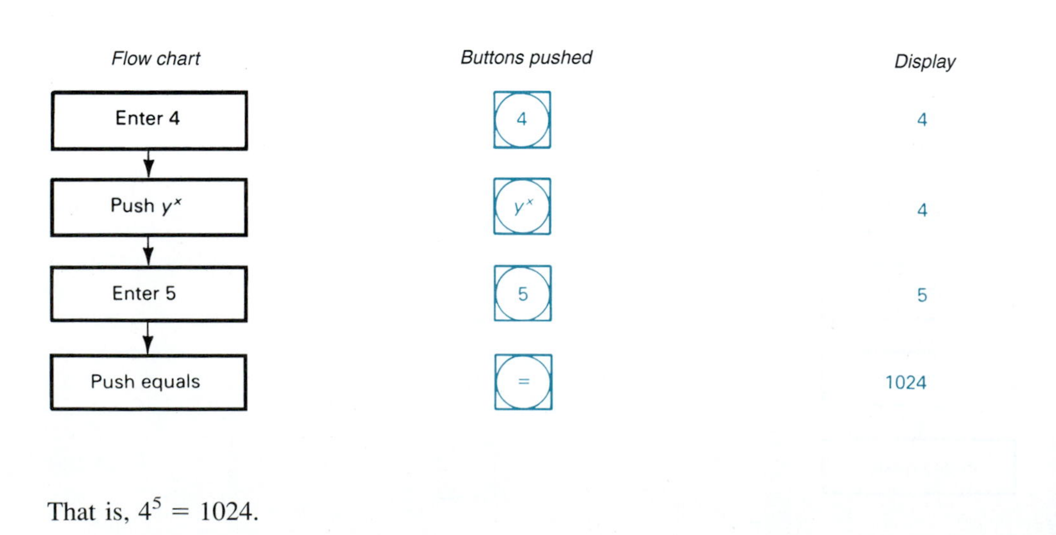

That is, $4^5 = 1024$.

EXAMPLE 2

Find the value of 1.5^{-4} rounded to three significant digits.

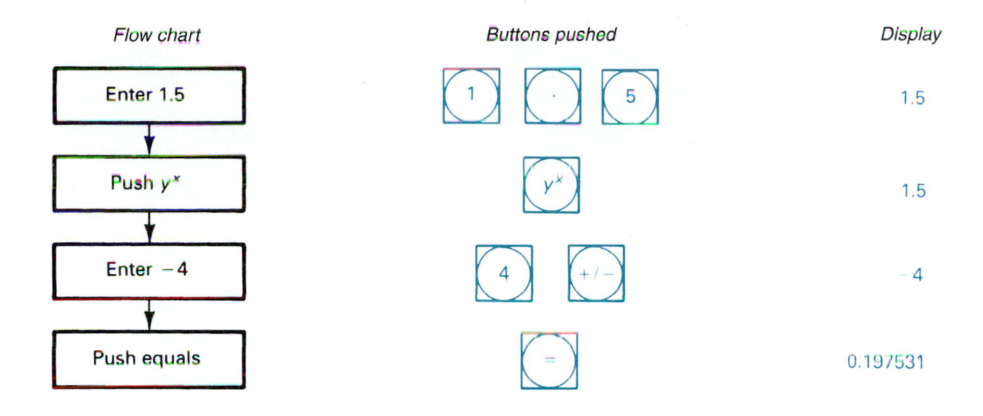

That is, $1.5^{-4} = 0.198$ rounded to three significant digits.

■ PROBLEMS B.6

Do as indicated and round each result to three significant digits.

1. $(6.43 \times 10^8)(5.16 \times 10^{10})$
2. $(4.16 \times 10^{-5})(3.45 \times 10^{-7})$
3. $(1.456 \times 10^{12})(-4.69 \times 10^{-18})$
4. $(-5.93 \times 10^9)(7.055 \times 10^{-12})$
5. $(7.45 \times 10^8) \div (8.92 \times 10^{18})$
6. $(1.38 \times 10^{-6}) \div (4.324 \times 10^6)$
7. $\dfrac{-6.19 \times 10^{12}}{7.755 \times 10^{-8}}$
8. $\dfrac{1.685 \times 10^{10}}{1.42 \times 10^{24}}$
9. $\dfrac{(5.26 \times 10^{-8})(8.45 \times 10^6)}{(-6.142 \times 10^9)(1.056 \times 10^{-12})}$
10. $\dfrac{(-2.35 \times 10^{-9})(1.25 \times 10^{11})(4.65 \times 10^{17})}{(8.75 \times 10^{23})(-5.95 \times 10^{-6})}$
11. $(68.4)^2$
12. $(3180)^2$
13. $\sqrt{46{,}500}$
14. $\sqrt{0.000634}$
15. $(1.45 \times 10^5)^2$
16. $(1.095 \times 10^{-18})^2$
17. $\sqrt{4.63 \times 10^{18}}$
18. $\sqrt{9.49 \times 10^{-15}}$
19. $\sqrt{(4.68)^2 + (9.63)^2}$
20. $\sqrt{(18.4)^2 - (6.5)^2}$
21. $18\sqrt{3} + \left(\dfrac{28.1}{19}\right)^2$
22. $\dfrac{8}{\sqrt{2}} + \sqrt{\dfrac{58}{14.5}}$
23. $25^2 - \sqrt{\dfrac{29.8}{0.0256}}$
24. $\dfrac{18.3}{6\sqrt{5}} - \left(\dfrac{225}{147}\right)^2$
25. $(12.6^2 + 21.5^2)^2 + (34.2^2 - 26.4^2)^2$
26. $\sqrt{21.4^2 + 18.7^2} + \sqrt{31.5^2 - 16.3^2}$
27. $\dfrac{91.4 - 48.6}{91.4 - 15.9}$
28. $\dfrac{14.7 + 9.6}{45.7 + 68.2}$
29. $\sqrt{\left(\dfrac{80.5}{25.6}\right)^2 + \left[\dfrac{1}{2\pi(60)(1.5 \times 10^{-7})}\right]^2}$

30. $\sqrt{\left(\frac{175}{36.5}\right)^2 + \left[\frac{1}{2\pi(60)(8.5 \times 10^{-10})}\right]^2}$

31. $\dfrac{(17.2)(11.6) + (8)(17.6) - (6)(16)}{(5)(15) + (8.5)(15) + (10)(26.5)}$

32. $\dfrac{(18.8)(5.5) + (7.75)(16.5) - (9.25)(13.85)}{(6.25)(12.5) + (4.75)(16.5) + (11.5)(14.1)}$

33. sin 13°
34. cos 22°
35. tan 52.3°
36. tan 31.25°
37. cos 59.36°
38. sin 84.55°
39. sin 48°
40. cos 48°
41. tan 75°
42. sin 8°
43. sin 8.7°
44. cos 35°

Find each angle rounded to the nearest tenth of a degree.

45. $\sin A = 0.6527$
46. $\cos B = 0.2577$
47. $\tan A = 0.4568$
48. $\sin B = 0.4658$
49. $\cos A = 0.5563$
50. $\tan B = 1.496$
51. $\sin B = 0.1465$
52. $\cos A = 0.4968$
53. $\tan B = 1.987$
54. $\sin A = 0.2965$
55. $\cos B = 0.3974$
56. $\tan A = 0.8885$

Find each angle to the nearest tenth of a degree between 0° and 90° and each side to three significant digits.

57. $b = (\sin 58.2°)(296 \text{ m})$
58. $a = (\cos 25.2°)(54.5 \text{ m})$
59. $c = \dfrac{37.5 \text{ m}}{\cos 65.2°}$
60. $b = \dfrac{59.7 \text{ m}}{\tan 41.2°}$
61. $\tan A = \dfrac{512 \text{ km}}{376 \text{ km}}$
62. $\cos B = \dfrac{75.2 \text{ m}}{89.5 \text{ m}}$
63. $a = (\cos 19.5°)(15.7 \text{ cm})$
64. $c = \dfrac{235 \text{ km}}{\sin 65.2°}$
65. $b = \dfrac{36.7 \text{ m}}{\tan 59.2°}$
66. $a = (\tan 5.7°)(135 \text{ m})$

Find the value of each power and round each result to three significant digits.

67. 12^4
68. 1.8^3
69. 0.46^5
70. 9^{-3}
71. 14^{-5}
72. 0.65^{-4}

Appendix C

TABLES

Table 1 English Weights and Measures

Units of length	Units of volume
Standard unit—inch (in. or ″)	*Liquid*
12 inches = 1 foot (foot or ′)	16 ounces (fl oz) = 1 pint (pt)
3 feet = 1 yard (yd)	2 pints = 1 quart (qt)
$5\frac{1}{2}$ yards or $16\frac{1}{2}$ feet = 1 rod (rd)	4 quarts = 1 gallon (gal)
5280 feet = 1 mile (mi)	*Dry*
Units of weight	2 pints (pt) = 1 quart (qt)
Standard unit—pound (lb)	8 quarts = 1 peck (pk)
16 ounces (oz) = 1 pound	4 pecks = 1 bushel (bu)
2000 pounds = 1 ton (T)	

Table 2 Conversion Table for Length

	cm	*m*	*km*	*in.*	*ft*	*mile*
1 centimetre =	1	10^{-2}	10^{-5}	0.394	3.28×10^{-2}	6.21×10^{-6}
1 metre =	100	1	10^{-3}	39.4	3.28	6.21×10^{-4}
1 kilometre =	10^5	1000	1	3.94×10^4	3280	0.621
1 inch =	2.54	2.54×10^{-2}	2.54×10^{-5}	1	8.33×10^{-2}	1.58×10^{-5}
1 foot =	30.5	0.305	3.05×10^{-4}	12	1	1.89×10^{-4}
1 mile =	1.61×10^5	1610	1.61	6.34×10^4	5280	1

Table 3 Conversion Table for Area

Metric	*English*
$1\ m^2 = 10{,}000\ cm^2$	$1\ ft^2 = 144\ in^2$
$= 1{,}000{,}000\ mm^2$	$1\ yd^2 = 9\ ft^2$
$1\ cm^2 = 100\ mm^2$	$1\ rd^2 = 30.25\ yd^2$
$= 0.0001\ m^2$	$1\ acre = 160\ rd^2$
$1\ km^2 = 1{,}000{,}000\ m^2$	$= 4840\ yd^2$
	$= 43{,}560\ ft^2$
	$1\ mi^2 = 640\ acres$

	m^2	cm^2	ft^2	in^2
1 square metre =	1	10^4	10.8	1550
1 square centimetre =	10^{-4}	1	1.08×10^{-3}	0.155
1 square foot =	9.29×10^{-2}	929	1	144
1 square inch =	6.45×10^{-4}	6.45	6.94×10^{-3}	1

1 circular mil = $5.07 \times 10^{-6}\ cm^2 = 7.85 \times 10^{-7}\ in^2$

1 hectare = $10{,}000\ m^2$ = 2.47 acres

Table 4 Conversion Table for Volume

Metric	*English*
$1\ m^3 = 10^6\ cm^3$	$1\ ft^3 = 1728\ in^3$
$1\ cm^3 = 10^{-6}\ m^3$	$1\ yd^3 = 27\ ft^3$
$= 10^3\ mm^3$	

	m^3	cm^3	L	ft^3	in^3
$1\ m^3$	1	10^6	1000	35.3	6.10×10^4
$1\ cm^3 =$	10^{-6}	1	1.00×10^{-3}	3.53×10^{-5}	6.10×10^{-2}
1 litre =	1.00×10^{-3}	1000	1	3.53×10^{-2}	61.0
$1\ ft^3 =$	2.83×10^{-2}	2.83×10^4	28.3	1	1728
$1\ in^3 =$	1.64×10^{-5}	16.4	1.64×10^{-2}	5.79×10^{-4}	1

1 U.S. fluid gallon = 4 U.S. fluid quarts = 8 U.S. pints = 128 U.S. fluid ounces = 231 in^3 = 0.134 ft^3
1 L = 1000 cm^3 = 1.06 qt   1 fl oz = 29.5 cm^3
1 ft^3 = 7.47 gal = 28.3 L

Table 5 Conversion Table for Mass

	g	*kg*	*slug*	*oz*	*lb*	*ton*
1 gram =	1	0.001	6.85×10^{-5}	3.53×10^{-2}	2.21×10^{-3}	1.10×10^{-6}
1 kilogram =	1000	1	6.85×10^{-2}	35.3	2.21	1.10×10^{-3}
1 slug =	1.46×10^4	14.6	1	515	32.2	1.61×10^{-2}
1 ounce =	28.4	2.84×10^{-2}	1.94×10^{-3}	1	6.25×10^{-2}	3.13×10^{-5}
1 pound =	454	0.454	3.11×10^{-2}	16	1	5.00×10^{-4}
1 ton =	9.07×10^5	907	62.2	3.2×10^4	2000	1

1 metric ton = 1000 kg = 2205 lb

Quantities in the shaded areas are not mass units. When we write, for example, 1 kg "=" 2.21 lb, this means that a kilogram is a mass that weighs 2.21 pounds under standard conditions of gravity ($g = 9.80\ m/s^2 = 32.2\ ft/s^2$).

Table 6 Conversion Table for Density

	$slug/ft^3$	kg/m^3	g/cm^3	lb/ft^3	lb/in^3
1 slug per ft^3 =	1	515	0.515	32.2	1.86×10^{-2}
1 kilogram per m^3 =	1.94×10^{-3}	1	0.001	6.24×10^{-2}	3.61×10^{-5}
1 gram per cm^3 =	1.94	1000	1	62.4	3.61×10^{-2}
1 pound per ft^3 =	3.11×10^{-2}	16.0	1.60×10^{-2}	1	5.79×10^{-4}
1 pound per in^3 =	53.7	2.77×10^4	27.7	1728	1

Quantities in the shaded areas are weight densities and, as such, are dimensionally different from mass densities.

Note that $D_w = D_m g$ where D_w = weight density
D_m = mass density
$g = 9.80\ m/s^2 = 32.2\ ft/s^2$

Table 7 Conversion Table for Time

	yr	*day*	*h*	*min*	*s*
1 year =	1	365	8.77×10^3	5.26×10^5	3.16×10^7
1 day =	2.74×10^{-3}	1	24	1440	8.64×10^4
1 hour =	1.14×10^{-4}	4.17×10^{-2}	1	60	3600
1 minute =	1.90×10^{-6}	6.94×10^{-4}	1.67×10^{-2}	1	60
1 second =	3.17×10^{-8}	1.16×10^{-5}	2.78×10^{-4}	1.67×10^{-2}	1

Table 8 Conversion Table for Speed

	ft/s	*km/h*	*m/s*	*mi/h*	*cm/s*
1 foot per second =	1	1.10	0.305	0.682	30.5
1 kilometre per hour =	0.911	1	0.278	0.621	27.8
1 metre per second =	3.28	3.60	1	2.24	100
1 mile per hour =	1.47	1.61	0.447	1	44.7
1 centimetre per second =	3.28×10^{-2}	3.60×10^{-2}	0.01	2.24×10^{-2}	1
1 mi/min = 88.0 ft/s = 60.0 mi/h					

Table 9 Conversion Table for Force

	N	*lb*	*oz*
1 newton =	1	0.225	3.60
1 pound =	4.45	1	16
1 ounce =	0.278	0.0625	1

Table 10 Conversion Table for Power

	Btu/h	*ft lb/s*	*hp*	*kW*	*W*
1 British thermal unit per hour =	1	0.216	3.93×10^{-4}	2.93×10^{-4}	0.293
1 foot pound per second =	4.63	1	1.82×10^{-3}	1.36×10^{-3}	1.36
1 horsepower =	2550	550	1	0.746	746
1 kilowatt =	3410	738	1.34	1	1000
1 watt =	3.41	0.738	1.34×10^{-3}	0.001	1

Table 11 Conversion Table for Pressure

	atm	*Inch of Water*	*mm Hg*	*N/m²* *(Pa)*	*lb/in²*	*lb/ft²*
1 atmosphere =	1	407	760	1.01×10^{5}	14.7	2120
1 inch of water[a] at 4°C =	2.46×10^{-3}	1	1.87	249	3.61×10^{-2}	5.20
1 millimetre of mercury[a] at 0°C =	1.32×10^{-3}	0.535	1	133	1.93×10^{-2}	2.79
1 newton per metre^2 (pascal) =	9.87×10^{-6}	4.02×10^{-3}	7.50×10^{-3}	1	1.45×10^{-4}	2.09×10^{-2}
1 pound per in^2 =	6.81×10^{-2}	27.7	51.7	6.90×10^{3}	1	144
1 pound per ft^2 =	4.73×10^{-4}	0.192	0.359	47.9	6.94×10^{-3}	1

[a]Where the acceleration of gravity has the standard value, $9.80 \text{ m/s}^2 = 32.2 \text{ ft/s}^2$.

Table 12 Mass and Weight Density[a]

Substance	*Mass Density (kg/m³)*	*Weight Density (lb/ft³)*
Solids		
Copper	8,890	555
Iron	7,800	490
Lead	11,300	708
Aluminum	2,700	169
Ice	917	57
Wood, white pine	420	26
Concrete	2,300	145
Cork	240	15
Liquids		
Water	$1{,}\bar{0}00$[b]	62.4
Seawater	1,025	64.0
Oil	870	54.2
Mercury	13,600	846
Alcohol	790	49.4
Gasoline	680	42.0
	A 0°C and 1 atm Pressure	*At 32°F and 1 atm Pressure*
Gases[a]		
Air	1.29	0.081
Carbon dioxide	1.96	0.123
Carbon monoxide	1.25	0.078
Helium	0.178	0.011
Hydrogen	0.0899	0.0056
Oxygen	1.43	0.089
Nitrogen	1.25	0.078
Ammonia	0.760	0.047
Propane	2.02	0.126

[a]The density of a gas is found by pumping the gas into a container, by measuring its volume and mass or weight, and then by using the appropriate density formula.
[b]Metric weight density of water = $98\bar{0}0$ N/m^3.

Table 13 Specific Gravity of Certain Liquids at Room Temperature (20°C or 68°F)

Liquid	*Specific Gravity*
Benzene	0.90
Ethyl alcohol	0.79
Gasoline	0.68
Kerosene	0.82
Mercury	13.6
Seawater	1.025
Sulfuric acid	1.84
Turpentine	0.87
Water	1.000

Table 14 Conversion Table for Energy, Work, and Heat

	Btu	*ft lb*	*J*	*cal*	*kWh*
1 British thermal unit =	1	778	1060	252	2.93×10^{-4}
1 foot pound =	1.29×10^{-3}	1	1.36	0.324	3.77×10^{-7}
1 joule =	9.48×10^{-4}	0.738	1	0.239	2.78×10^{-7}
1 calorie =	3.97×10^{-3}	3.09	4.19	1	1.16×10^{-6}
1 kilowatt-hour =	3410	2.66×10^{6}	3.60×10^{6}	8.60×10^{5}	1

Table 15 Heat Constants

			Specific Heat		*Heat of Fusion*		*Heat of Vaporization*	
	Melting Point (°C)	*Boiling Point (°C)*	*cal/g °C or kcal/kg °C or Btu/lb °F*	*J/kg °C*	*cal/g or kcal/kg*	*J/kg*	*cal/g or kcal/kg*	*J/kg*
Alcohol, ethyl	−117	78.5	0.58	2400	24.9	1.04×10^{5}	204	8.54×10^{5}
Aluminum	660	2057	0.22	920	76.8	3.21×10^{5}		
Brass	840		0.092	390				
Copper	1083	2330	0.092	390	49.0	2.05×10^{5}		
Glass			0.21	880				
Ice	0		0.51	2100	$8\bar{0}$	3.35×10^{5}		
Iron (steel)	1540	3000	0.115	481	7.89	3.30×10^{4}		
Lead	327	1620	0.031	130	5.86	2.45×10^{4}		
Mercury	−38.9	357	0.033	140	2.82	1.18×10^{4}	65.0	2.72×10^{5}
Silver	961	1950	0.056	230	26.0	1.09×10^{5}		
Steam			0.48	$2\bar{0}00$				
Water (liquid)	0	$10\bar{0}$	1.00	4190	$8\bar{0}$	3.35×10^{5}	$54\bar{0}$	2.26×10^{6}
Zinc	419	907	0.092	390	23.0	9.63×10^{4}		

Table 16 Coefficient of Linear Expansion

Material	*α (metric)*	*α (English)*
Aluminum	2.3×10^{-5}/C°	1.3×10^{-5}/F°
Brass	1.9×10^{-5}/C°	1.0×10^{-5}/F°
Concrete	1.1×10^{-5}/C°	6.0×10^{-6}/F°
Copper	1.7×10^{-5}/C°	9.5×10^{-6}/F°
Glass	9.0×10^{-6}/C°	5.1×10^{-6}/F°
Pyrex	3.0×10^{-6}/C°	1.7×10^{-6}/F°
Steel	1.3×10^{-5}/C°	6.5×10^{-6}/F°
Zinc	2.6×10^{-5}/C°	1.5×10^{-5}/F°

Table 17 Coefficient of Volume Expansion

Liquid	*β (metric)*	*β (English)*
Acetone	1.49×10^{-3}/C°	8.28×10^{-4}/F°
Alcohol, ethyl	1.12×10^{-3}/C°	6.62×10^{-4}/F°
Carbon tetrachloride	1.24×10^{-3}/C°	6.89×10^{-4}/F°
Mercury	1.8×10^{-4}/C°	1.0×10^{-4}/F°
Petroleum	9.6×10^{-4}/C°	5.33×10^{-4}/F°
Turpentine	9.7×10^{-4}/C°	5.39×10^{-4}/F°
Water	2.1×10^{-4}/C°	1.17×10^{-4}/F°

Table 18 Conversion Table for Charge

Charge on one electron = 1.60×10^{-19} coulomb
1 coulomb = 6.25×10^{18} electrons of charge
1 ampere-hour = 3600 C

Table 19 Copper Wire Table

Gauge No.	Diameter (mils)	Diameter (mm)	Cross Section cir mils	Cross Section in^2	Ohms per 1000 ft 25°C (77°F)	Ohms per 1000 ft 65°C (149°F)	Weight per 1000 ft (lb)
0000	460.0		212,000	0.166	0.0500	0.0577	641.0
000	410.0		168,000	0.132	0.0630	0.0727	508.0
00	365.0		133,000	0.105	0.0795	0.0917	403.0
0	325.0		106,000	0.0829	0.100	0.116	319.0
1	289.0	7.35	83,700	0.0657	0.126	0.146	253.0
2	258.0	6.54	66,400	0.0521	0.159	0.184	201.0
3	229.0	5.83	52,600	0.0413	0.201	0.232	159.0
4	204.0	5.19	41,700	0.0328	0.253	0.292	126.0
5	182.0	4.62	33,100	0.0260	0.319	0.369	100.0
6	162.0	4.12	26,300	0.0206	0.403	0.465	79.5
7	144.0	3.67	20,800	0.0164	0.508	0.586	63.0
8	128.0	3.26	16,500	0.0130	0.641	0.739	50.0
9	114.0	2.91	13,100	0.0103	0.808	0.932	39.6
10	102.0	2.59	10,400	0.00815	1.02	1.18	31.4
11	91.0	2.31	8,230	0.00647	1.28	1.48	24.9
12	81.0	2.05	6,530	0.00513	1.62	1.87	19.8
13	72.0	1.83	5,180	0.00407	2.04	2.36	15.7
14	64.0	1.63	4,110	0.00323	2.58	2.97	12.4
15	57.0	1.45	3,260	0.00256	3.25	3.75	9.86
16	51.0	1.29	2,580	0.00203	4.09	4.73	7.82
17	45.0	1.15	2,050	0.00161	5.16	5.96	6.20
18	40.0	1.02	1,620	0.00128	6.51	7.51	4.92
19	36.0	0.91	1,290	0.00101	8.21	9.48	3.90
20	32.0	0.81	1,020	0.000802	10.4	11.9	3.09
21	28.5	0.72	810	0.000636	13.1	15.1	2.45
22	25.3	0.64	642	0.000505	16.5	19.0	1.94
23	22.6	0.57	509	0.000400	20.8	24.0	1.54
24	20.1	0.51	404	0.000317	26.2	30.2	1.22
25	17.9	0.46	320	0.000252	33.0	38.1	0.970
26	15.9	0.41	254	0.000200	41.6	48.0	0.769
27	14.2	0.36	202	0.000158	52.5	60.6	0.610
28	12.6	0.32	160	0.000126	66.2	76.4	0.484
29	11.3	0.29	127	0.0000995	83.4	96.3	0.384
30	10.0	0.26	101	0.0000789	105	121	0.304
31	8.9	0.23	79.7	0.0000626	133	153	0.241
32	8.0	0.20	63.2	0.0000496	167	193	0.191
33	7.1	0.18	50.1	0.0000394	211	243	0.152
34	6.3	0.16	39.8	0.0000312	266	307	0.120
35	5.6	0.14	31.5	0.0000248	335	387	0.0954
36	5.0	0.13	25.0	0.0000196	423	488	0.0757
37	4.5	0.11	19.8	0.0000156	533	616	0.0600
38	4.0	0.10	15.7	0.0000123	673	776	0.0476
39	3.5	0.09	12.5	0.0000098	848	979	0.0377
40	3.1	0.08	9.9	0.0000078	1070	1230	0.0200

Table 20 Conversion Table for Plane Angles

	°	′	″	*rad*	*rev*
1 degree =	1	60	3600	1.75×10^{-2}	2.78×10^{-3}
1 minute =	1.67×10^{-2}	1	60	2.91×10^{-4}	4.63×10^{-5}
1 second =	2.78×10^{-4}	1.67×10^{-2}	1	4.85×10^{-6}	7.72×10^{-7}
1 radian =	57.3	3440	2.06×10^{5}	1	0.159
1 revolution =	360	2.16×10^{4}	1.30×10^{6}	6.28 or 2π	1

Table 21 The Greek Alphabet

Capital	*Lowercase*	*Name*
Α	α	alpha
Β	β	beta
Γ	γ	gamma
Δ	δ	delta
Ε	ϵ	epsilon
Ζ	ζ	zeta
Η	η	eta
Θ	θ	theta
Ι	ι	iota
Κ	κ	kappa
Λ	λ	lambda
Μ	μ	mu
Ν	ν	nu
Ξ	ξ	xi
Ο	ο	omicron
Π	π	pi
Ρ	ρ	rho
Σ	σ	sigma
Τ	τ	tau
Υ	υ	upsilon
Φ	φ	phi
Χ	χ	chi
Ψ	ψ	psi
Ω	ω	omega

Appendix D

Answers to Odd-Numbered Problems and to Chapter Review Questions and Problems

CHAPTER 1

1.2 Page 6

1. kilo **3.** hecto **5.** milli **7.** mega **9.** h **11.** m **13.** M **15.** c
17. 135 mm **19.** 28 kL **21.** 49 cg **23.** 75 hm **25.** 24 metres
27. 59 grams **29.** 27 millimetres **31.** 45 dekametres **33.** 26 megametres
35. metre **37.** litre and cubic metre **39.** second

1.3 Page 9

1. 3.26×10^2 **3.** 2.65×10^3 **5.** 8.264×10^2 **7.** 4.13×10^{-3} **9.** 6.43×10^0
11. 6.5×10^{-5} **13.** 5.4×10^5 **15.** 7.5×10^{-6} **17.** 5×10^{-8}
19. 7.32×10^{17} **21.** 86,200 **23.** 0.000631 **25.** 0.768 **27.** 777,000,000
29. 69.3 **31.** 96,100 **33.** 1.4 **35.** 0.0000084 **37.** 700,000,000,000
39. 0.00000072 **41.** 4,500,000,000,000 **43.** 0.000000000055

1.4 Pages 14–15

1. 1 metre **3.** 1 kilometre **5.** 1 kilometre **7.** cm **9.** m **11.** mm
13. km **15.** mm **17.** m **19.** km **21.** cm **23.** km **25.** cm **27.** km
29. mm **31.** cm **33.** 1000 **35.** 100 **37.** 0.1 **39.** 1000 **41.** 100
43. 0.001 **45.** 10 **47.** 0.25 km **49.** 178,000 m **51.** 8.3 m
53. 3750 mm **55.** 4,000,000 μm **57.** Answers vary
59. (a) 9 ft (b) 3 yd **61.** (a) 11,000 yd (b) 33,000 ft
63. 412.16 km **65.** 2.80 in. **67.** 6.1 m **69.** 9

1.5 Pages 23–26

1. 40 cm^2 **3.** 39 in^2 **5.** 22 in^2 **7.** 72 in^3 **9.** 40 cm^3 **11.** 1 litre
13. 1 cubic centimetre **15.** 1 square kilometre **17.** L **19.** m^2 **21.** m^3
23. ha **25.** mL **27.** m^3 **29.** L **31.** mL **33.** L **35.** m^2 **37.** L
39. ha **41.** m^3 **43.** m^2 **45.** L **47.** 1000 **49.** 0.1 **51.** 0.01 **53.** 10
55. 1 **57.** 1,000,000 **59.** 0.001 **61.** 10,000 **63.** 100 **65.** 0.01

67. 100 **69.** 7.5 L **71.** 1600 mL **73.** 275,000 mm^3 **75.** 4×10^9 mm^3
77. 275 mL **79.** 1000 L **81.** 7500 cm^3 **83.** 50 cm^2 **85.** 50,000 cm^2
87. 400 km^2 **89.** 45 ft^2 **91.** 13,935 cm^2 **93.** 0.75 ft^2 **95.** 156,816 in^2
97. 513 ft^3 **99.** 30.1 yd^3 **101.** 13,824 in^3 **103.** 623.2 cm^3 **105.** 36 cm^2
107. 76 cm^2 **109.** 40 mL

1.6 Pages 28–29

1. 1 gram **3.** 1 kilogram **5.** 1 kilogram **7.** kg **9.** kg **11.** metric ton
13. g **15.** mg **17.** kg **19.** g **21.** kg **23.** g **25.** g **27.** kg
29. kg **31.** metric ton **33.** kg **35.** 1000 **37.** 100 **39.** 0.1 **41.** 1000
43. 100 **45.** 0.001 **47.** 1,000,000 **49.** 575,000 mg **51.** 0.65 g **53.** 5 g
55. 30,000,000 mg **57.** 2.5 kg **59.** 0.4 mg **61.** 750 g **63.** 15,575 N
65. 890 N **67.** 450 lb **69.** 7.5 lb **71.** 36 oz **73.** 418.3 N

1.7 Page 32

1. second, s **3.** watt, W **5.** 1 milliampere **7.** 1 MW **9.** 1 A
11. 4.7 μA **13.** 7.5 kW **15.** 45 ns **17.** 125 MW **19.** 10^3 **21.** 10^9
23. 10^6 **25.** 10^3 **27.** 10^{12} **29.** 4000 mA **31.** 52,000 μW **33.** 6.8 A
35. 15,915 s **37.** 4×10^9 ns **39.** 0.03 MW

1.8 Pages 34–35

1. 3 **3.** 4 **5.** 2 **7.** 3 **9.** 5 **11.** 3 **13.** 5 **15.** 4 **17.** 3 **19.** 2
21. 1 **23.** 4 **25.** 2 **27.** 3 **29.** 4

1.9 Page 36

1. 1 V **3.** 1 m **5.** 0.0001 in. **7.** 10 km **9.** 0.01 m **11.** 0.00001 in.
13. 0.01 m **15.** 1 kg **17.** 0.0001 in. **19.** 1000 Ω **21.** 1 A **23.** 0.01 m^2
25. 100 kg **27.** 0.000001 A or 1×10^{-6} A **29.** 10,000 V or 1×10^4 V
31. (a) 15.7 in. (b) 0.018 in. **33.** (a) 16.01 cm (b) 0.734 cm
35. (a) 0.0350 A (b) 0.00040 A **37.** (a) $27,0\bar{0}0$ L (b) 4.75 L
39. (a) All have one significant digit (b) 50 Ω **41.** (a) 0.05 in. (b) 16.4 in.
43. (a) 0.65 m (b) 27.5 m **45.** (a) 0.00005 g (b) 0.75 g
47. (a) 3 V (b) 45,000 V **49.** (a) 20 Ω (b) $40\bar{0},000$ Ω

1.10 Pages 40–41

1. 14,200 ft **3.** 83.3 cm **5.** $7\bar{0},000$ V **7.** 802 m or 80,200 cm **9.** 18 A
11. 500 kg **13.** 41.0 g **15.** 3200 km **17.** 900,000 V **19.** 0.40 m or $4\bar{0}$ cm
21. 4900 m^2 **23.** $1,4\bar{0}0,000$ km^2 or 1.40×10^6 km^2 **25.** 737.7 m^2
27. 5560 cm^3 **29.** 2.91×10^7 in^3 **31.** $3\bar{0}$ ft **33.** 3.06 cm **35.** 75 km/h
37. $1\bar{0}00$ V/A **39.** 1100 ft lb/s **41.** 370 V/A **43.** 43.2 m **45.** 530 V^2/Ω
47. 4530 kg m/s^2 **49.** 10,300 m^3 **51.** 6100 m^2 **53.** 28,800 ft

Chapter 1 Review Questions Pages 42–43

1. c **2.** b **3.** b **4.** c **5.** a
6. (1) Pieces made separately may not fit together.
(2) Workers could not communicate directions to each other.
(3) Workers could not tell each other how much material to buy.
7. It is based on the decimal system.
8. (1) The distance to the moon.
(2) The thickness of aluminum foil.
9. Negative exponents are used to write numbers between 0 and 1 in scientific notation.

10. Any number with exponent 0 equals 1.
11. Yes
12. The surface that would be seen by cutting a geometric solid with a thin plate.
13. Yes **14.** Hectare **15.** Litre
16. (1) Medicines (2) Perfumes (3) Wine
17. Mass measures the quantity of matter. Weight measures gravitational pull on an object.
18. Newton (N) **19.** Millionth
20. Because nearly all measurements are approximate numbers rather than exact numbers.
21. No **22.** Yes

Chapter 1 Review Problems Pages 43–44

1. k **2.** m **3.** μ **4.** M **5.** 45 mg **6.** 138 cm **7.** 2.14×10^8 **8.** 3.36×10^0 **9.** 4.5×10^{-3} **10.** 17,200 **11.** 0.0066 **12.** 9.03 **13.** 1 L **14.** 1 MW **15.** 1 m^3 **16.** 0.25 **17.** 0.85 **18.** 5400 **19.** 550,000 **20.** 25,000,000 **21.** 75,000 **22.** 27,500 **23.** 0.035 **24.** 150,000 **25.** 500 **26.** 68.2 **27.** 11.0 **28.** 98.4 **29.** 968 **30.** 216 **31.** 212 **32.** 71.2 **33.** 4 h 20 min **34.** 3 **35.** 4 **36.** 2 **37.** 3 **38.** 0.1 ft **39.** 0.0001 s **40.** 1000 mi **41.** 100,000 V **42.** (a) 12.00 m (b) 0.008 m (c) 0.150 m (d) 2600 m **43.** (a) 18,050 L (b) 0.75 L (c) 0.75 L (d) 18,050 L **44.** 0.125 A **45.** 63,000 V **46.** 1,800,000 cm^3 **47.** 150 m^2 **48.** 9.73 kg m/s^2 **49.** 9.90 m^2 **50.** $7\bar{0}0$ cm^3

CHAPTER 2

2.1 Pages 49–50

1. $s = vt$ **3.** $m = \frac{w}{g}$ **5.** $R = \frac{E}{I}$ **7.** $g = \frac{\text{PE}}{mh}$ **9.** $h = \frac{v^2}{2g}$ **11.** $w = Pt$

13. $t = \frac{W}{P}$ **15.** $m = \frac{2(\text{KE})}{v^2}$ **17.** $s = \frac{W}{F}$ **19.** $I = \frac{V - E}{-r}$ or $I = \frac{E - V}{r}$

21. $P = \frac{\pi}{2R}$ **23.** $C = \frac{5F - 160}{9}$ or $C = \frac{5}{9}(F - 32)$ **25.** $f = \frac{1}{2\pi C X_C}$

27. $R_3 = R_T - R_1 - R_2 - R_4$ **29.** $I_P = \frac{I_S N_S}{N_p}$ **31.** $v_i = 2v_{\text{avg}} - v_f$

33. $s = \frac{v^2 - v_i^2 + 2as_i}{2a}$ **35.** $R = \frac{QJ}{I^2 t}$ **37.** $r = \sqrt{\frac{A}{\pi}}$ **39.** $d = \sqrt{\frac{kL}{R}}$

41. $I = \pm\sqrt{\frac{QJ}{Rt}}$

2.2 Page 52

1. (a) $A = bh$ (b) 162 cm^2 **3.** (a) $b = \frac{A}{h}$ (b) 7.50 cm **5.** (a) $c = P - a - b$ (b) 6.0 cm **7.** (a) $r = \frac{C}{2\pi}$ (b) 10.9 yd **9.** (a) $b = \frac{P - 2a}{2}$ or $b = \frac{P}{2} - a$ (b) 33.2 km **11.** (a) $h = \frac{V}{\pi r^2}$ (b) 6.11 m **13.** (a) $B = \frac{V}{h}$ (b) 154 m^2 **15.** (a) $b = \sqrt{A}$ (b) 21.6 in. **17.** (a) $C = 2\pi r$ (b) 121.6 m **19.** (a) $B = \frac{3V}{h}$ (b) 122.4 ft^2

2.3 Pages 57–59

1. 25,900 cm^3 **3.** 284 cm^3 **5.** 102.1 cm^2 **7.** 10,100 ft^3 **9.** 864 ft^3
11. 12.0 cm^2 **13.** 1.58 cm^2 **15.** 36.0 m **17.** 9.39 m **19.** 137 m
21. 65.5 ft **23.** $6\bar{0}$ panels **25.** 24.1 m^3 **27.** 4.44 yd^3 **29.** 266 in^3

Chapter 2 Review Questions Pages 59–60

1. c **2.** b **3.** a
4. (1) To find the volume of liquid storage tanks,
(2) To determine the amount of concrete needed for a driveway
5. As a shorthand way to designate different measured quantities of the same type
6. Most mistakes are made in problem solving by missing needed information or misinterpreting the information given.
7. Making a sketch helps to visualize what is happening in the problem.
8. The basic equation
9. The working equation is found by solving the basic equation for the unknown quantity.
10. Carrying the units through a problem shows whether the answer is the kind expected.
11. Making an estimate of the correct answer shows whether the solution is reasonable.

Chapter 2 Review Problems Page 60

1. (a) $m = \frac{F}{a}$ (b) $a = \frac{F}{m}$ **2.** $h = \frac{v^2}{2g}$ **3.** $v_f = \frac{2s}{t} - v_i$ **4.** $v = \sqrt{\frac{2\text{ KE}}{m}}$

5. 18 **6.** 12.0 m **7.** 2.19 m **8.** 122 cm^2 **9.** 12.0 cm **10.** 42.3 mm
11. 606 cm^2 **12.** 6.0 cm **13.** 6.27 m **14.** 14.4 m **15.** 430 cm^3
16. 4680 m^2

CHAPTER 3

3.2 Pages 65–66

1. 3.0 **3.** 1.4 **5.** 3.6 **7.**

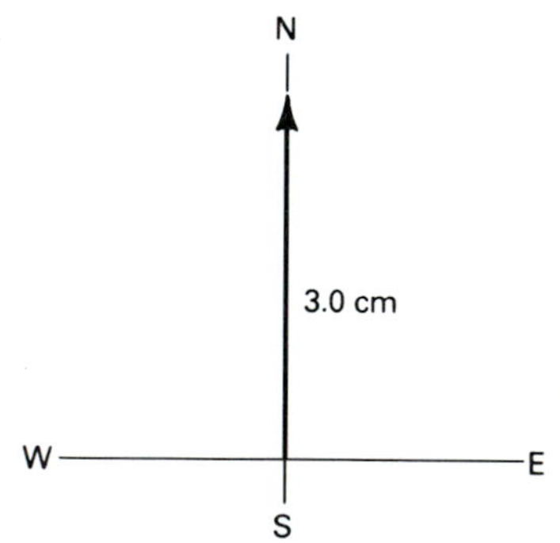

9. **11.** **13.** 2.0

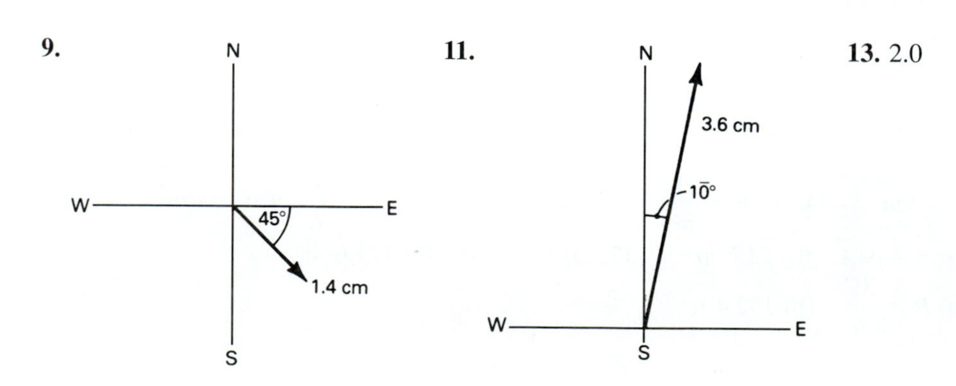

15. 2.8 **17.** 6.3 **19.**

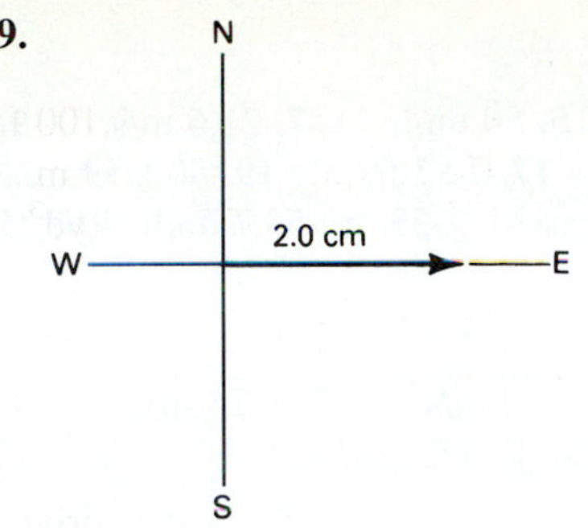

21.

N
W
E
45°
2.8 cm
S

23.

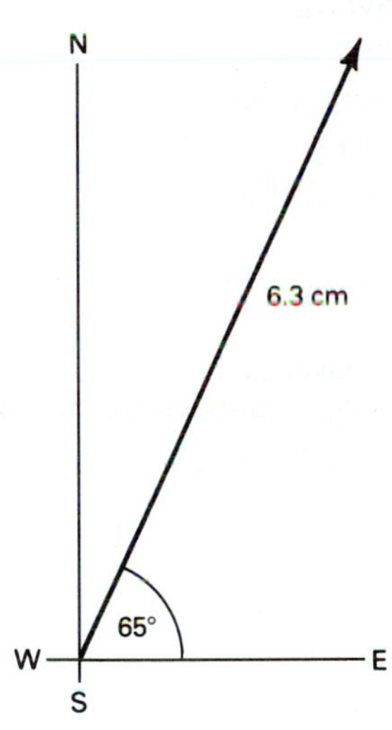

25. $1\frac{1}{4}$ **27.** $2\frac{5}{8}$ **29.** $\frac{15}{16}$

31.

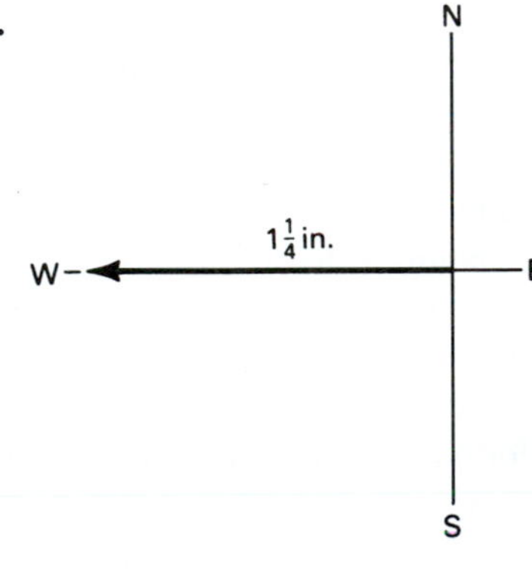

33.

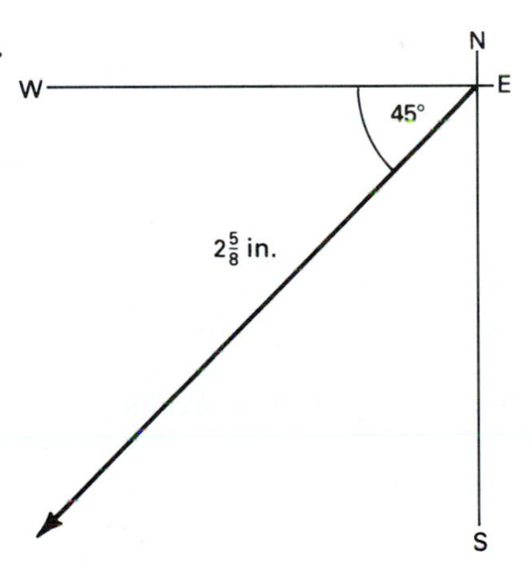

35.

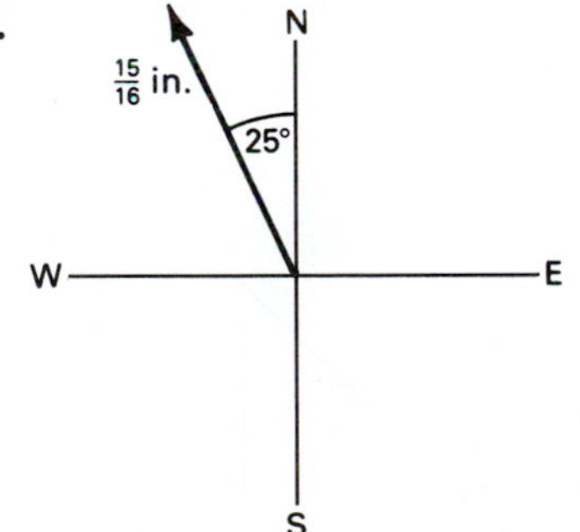

3.3 Pages 69–70

1. 61 km at 55° north of east **3.** 1300 mi at 1° west of south **5.** 36 km at 5° east of north **7.** 38 km at 25° north of west **9.** 1500 mi at 71° north of east **11.** 120 km at 72° south of east **13.** 47 mi at 49° north of east

3.4 Pages 72–73

1. 65 km/h **3.** $9\bar{0}$ km/h **5.** $5\bar{0}$ mi/h **7.** 21.6 m/s **9.** $12\bar{0}$ mi/h **11.** 25 m **13.** 25 m/s **15.** 79.2 km/h **17.** 72.2 ft/s **19.** $8\bar{0}$ km/h, east **21.** 125 mi/h, south **23.** 61.1 km/h at $3\bar{0}°$ south of east **25.** (a) 53.7 mi/h (b) 25.5 mi/h, west

3.5 Page 76

1. 15 m/s^2 **3.** $1\bar{0}$ ft/s^2 **5.** 3.2 m/s^2 **7.** 6.25 m/s^2 **9.** 0.206 m/s^2 **11.** 67.5 mi/h **13.** $72\bar{0}$ km/h **15.** 16.7 s

3.6 Pages 82–83

1. 5.05 m/s **3.** 127.8 m **5.** 2.13 ft/s^2 **7.** 9.00 mi/h **9.** $55\bar{0}$ ft **11.** 1.1 m/s^2 **13.** 43.1 m/s **15.** -1.6 m/s^2 **17.** (a) 23.5 m/s (b) 28.2 m **19.** (a) 3190 m (b) 25.5 s (c) 51.0 s **21.** (a) 1.02 s (b) 5.10 m (c) 24.3 m/s (d) 3.50 s **23.** (a) 44.0 ft/s (b) 4.21 s (c) 135 ft/s

3.8 Page 86

1. 24.8 cm **3.** 2.24 s **5.** 4.13 in. **7.** 1.36 s **9.** It is $\sqrt{2}$ times the original period.

Chapter 3 Review Questions Pages 87–88

1. d **2.** d **3.** c **4.** a
5. No; it is the study of the effect of forces on bodies.
6. Not necessarily; displacement has elements of both magnitude and direction.
7. (a) 12 oranges (b) 50 people (c) 7 inches
8. A vector is the representation of a quantity that has both magnitude and direction.
9. No
10. No; it is 8 only if both vectors are in the same direction.
11. Velocity has magnitude and direction whereas speed has only magnitude.
12. No; anything speeding up or slowing down has a changing velocity.
13. (a) An automobile speeding up or slowing down; (b) anything being dropped to the ground; (c) a bullet being fired
14. Acceleration is change in velocity; deceleration is a special case of acceleration where the object is slowing; acceleration need not be uniform and so may be subject to averaging.
15. 9.80 m/s^2 (metric) and 32.2 ft/s^2 (English)
16. Amplitude is maximum displacement.
17. Period is the time required for one full vibration. Frequency is the number of complete vibrations per unit of time.
18. No; the period of a pendulum is independent of its mass.

Chapter 3 Review Problems Page 88

1.

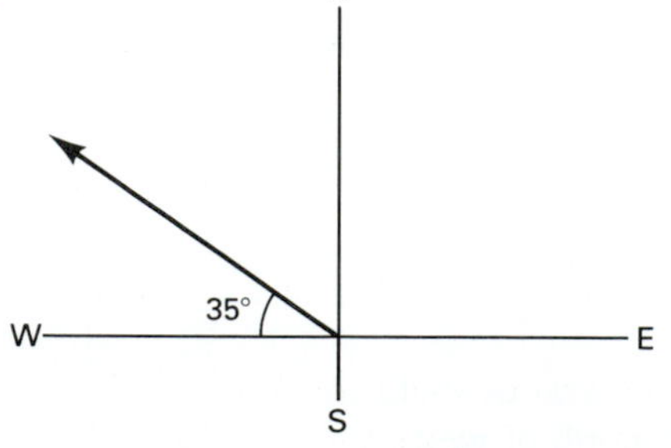

2.

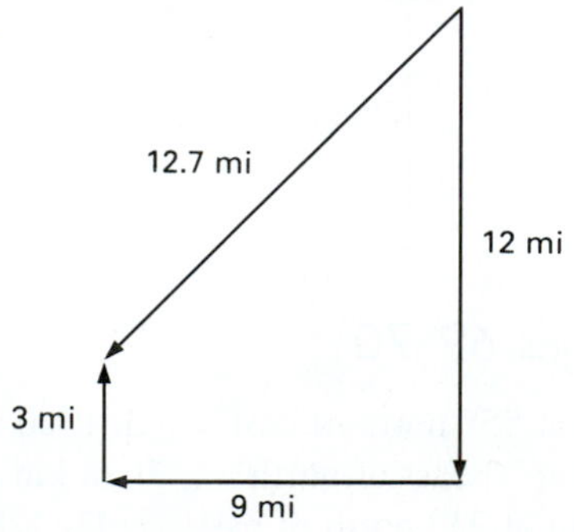

Scale: 1 cm = 3 mi

3. 25.5 mi **4.** 2.73 h **5.** 4.00 ft/s^2 **6.** 5.1 s **7.** 15.0 km/h **8.** 3.90 m/s
9. 3$\bar{0}$ h **10.** 13 m/s **11.** 4.00 m/s **12.** 2.6 m/s^2 **13.** (a) 205 m/s
(b) 20.9 s (c) 41.8 s **14.** (a) 37.0 m/s (b) 64.6 m **15.** 1.35 s **16.** 4.80 in.

CHAPTER 4

4.2 Page 95

1. 30.0 N **3.** 744 lb **5.** 252 lb **7.** 23$\bar{0}$ N **9.** 40.0 m/s^2 **11.** 11.7 m/s^2
13. 0.518 m/s^2 **15.** 1.39 ft/s^2 **17.** 5250 N **19.** 1320 lb **21.** 5740 N
23. 6.00 kg **25.** 12.8 m/s^2

4.3 Page 98

1. 380 N **3.** 1100 N **5.** 0.080 **7.** 25,000 lb **9.** 1600 N

4.4 Pages 100–101

1. 3.0 N; right **3.** 15.0 N; left **5.** 4 N; left **7.** 4.00 ft/s^2 **9.** 0.509 m/s^2

4.5 Page 103

1. 294 N **3.** 322 lb **5.** 1.73 kg **7.** 1220 kg **9.** 6.8×10^{11} kg
11. 11,300 N **13.** 85.4 slugs **15.** 13,200 N; 22$\bar{0}$0 N **17.** (a) 65 kg; (b) 110 N
19. Answers vary **21.** Answers vary **23.** (a) 3.57 slugs; (b) 303 lb
25. Answers vary

4.7 Pages 110–111

1. 80.0 kg m/s **3.** 765 slug ft/s **5.** 9.5×10^8 kg m/s **7.** 6.89×10^8 kg m/s
9. (a) 12,600 slug ft/s (b) 158 ft/s (c) 58$\bar{0}$0 lb; 2580 lb
11. (a) 55,200 kg m/s (b) 17$\bar{0}$ km/h **13.** 7.67 m/s **15.** (a) 14.0 m/s
(b) 1750 kg m/s **17.** (a) 0.00287 s (b) 12,$\bar{0}$00 N (c) 34.4 kg m/s
(d) 34.5 kg m/s

Chapter 4 Review Questions Pages 112–113

1. d **2.** b **3.** b **4.** c **5.** a **6.** b
7. (a) A bridge being supported over a river; (b) isometric exercises; (c) a magnet on a refrigerator
8. No **9.** No
10. A car hit from behind is forced into the car ahead of it. **11.** No
12. A body in motion continues in motion and a body at rest continues at rest unless a force acts on it.
13. Acceleration is change of velocity. **14.** 11.0 ft/s^2
15. Only if they have the same mass **16.** Yes; 3.00 lb = 13.35 N
17. More difficult; everything would slide.
18. No; a rock on the ground doesn't move.
19. Weight is a measure of gravitational pull. The moon, having less mass than the earth, exerts less gravitational pull.
20. For every force applied by object *A* to *B*, there is an equal force applied by *B* to *A* in the opposite direction.
21. For every action, there is an equal and opposite reaction. **22.** No
23. The longer the bat (applied force) is on the ball, the greater the impulse.
24. Total momentum in a system remains constant.
25. Momentum of the escaping gas molecules is equal to the momentum of the rocket.

Chapter 4 Review Problems Page 113

1. 22.5 N **2.** 3.00 m/s^2 **3.** 75.0 kg **4.** 19 ft/s^2 **5.** 19,000 N **6.** 64.0 N
7. 0.17 **8.** 3800 N **9.** 11 N **10.** 127 N **11.** 12 lb
12. 1.23×10^5 slug ft/s **13.** 0.204 m/s **14.** $15{,}\bar{0}00$ N s **15.** 8.5 kg m/s
16. 0.56 m/s **17.** (a) 12 m/s (b) 1800 kg m/s **18.** (a) 4.62×10^{-4} s
(b) 106,000 N (c) 49.0 kg m/s (d) 48.8 kg m/s

CHAPTER 5

5.1 Pages 119–122

1. *a* **3.** *c* **5.** *a* **7.** *B* **9.** *B* **11.** 0.9455 **13.** 1.804 **15.** 0.9799
17. 0.6477 **19.** 0.3065 **21.** 0.4617 **23.** 16° **25.** 43° **27.** 48°
29. 36.6° **31.** 46.5° **33.** 30.0° **35.** 22.28° **37.** 16.75° **39.** 35.50°
41. $B = 65.0°$; $a = 8.45$ m; $b = 18.1$ m **43.** $A = 47.7°$; $B = 42.3°$; $a = 12.4$ km
45. $A = 24.4°$; $B = 65.6°$; $c = 24.2$ mi **47.** $B = 7\bar{0}°$; $b = 24$ m; $c = 25$ m
49. $A = 49.35°$; $a = 17.98$ cm; $b = 15.44$ cm **51.** $b = 8.49$ cm
53. $c = 21.6$ mi **55.** $a = 10.2$ ft **57.** $c = 24.8$ cm **59.** $a = 8.60$ m
61. (a) 10.0° (b) 2.12 cm (c) 8.24 cm **63.** $C = 2.72$ in.; $D = 2.28$ in.
65. $b = 8.00$ cm; $c = 16.1$ cm

5.2 Pages 128–130

1.

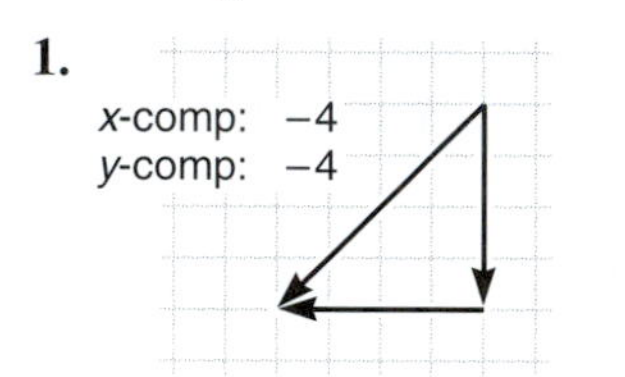

3.

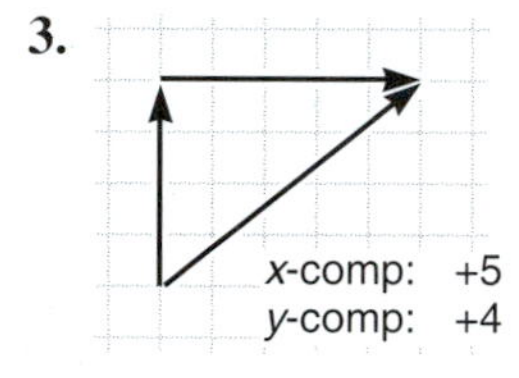

5.

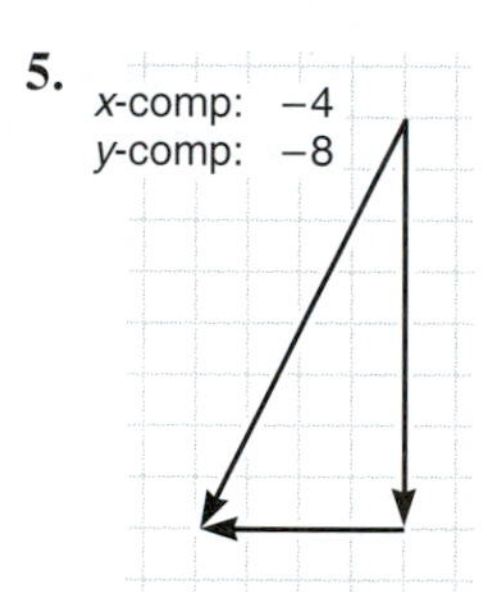

7.

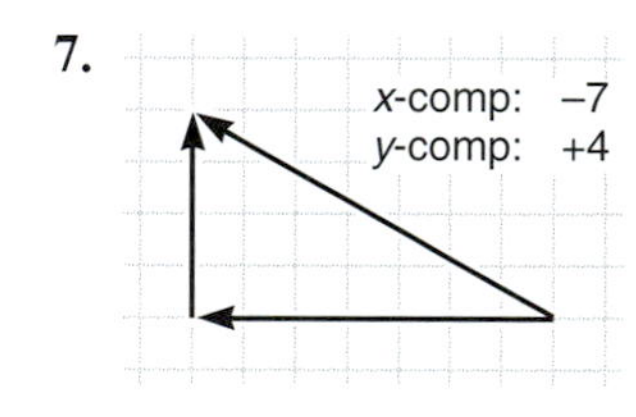

9.

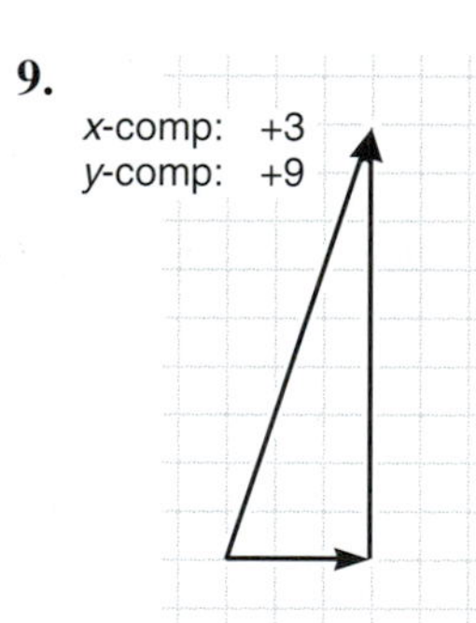

11.

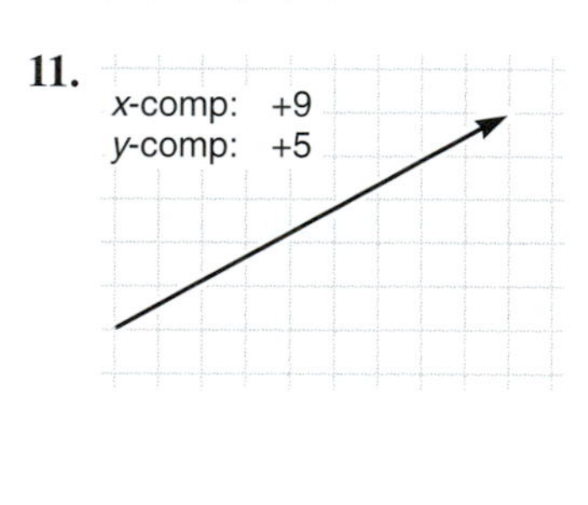

13.

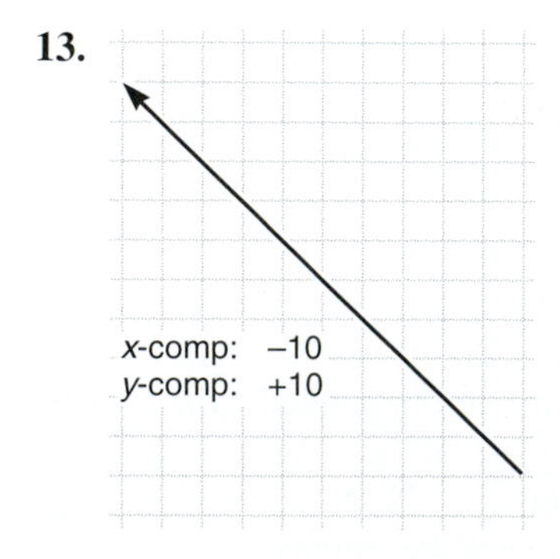

15.

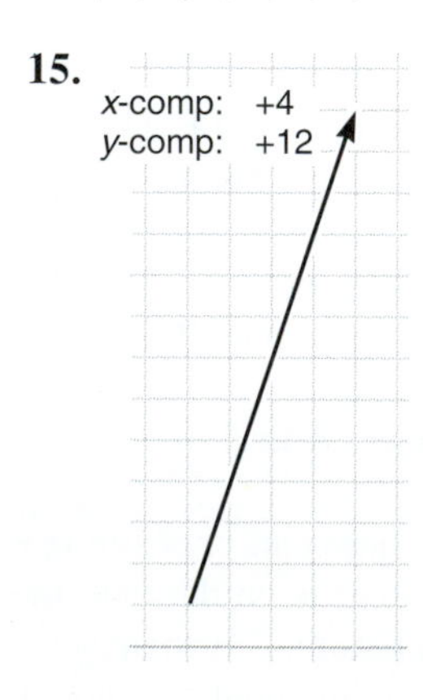

17.

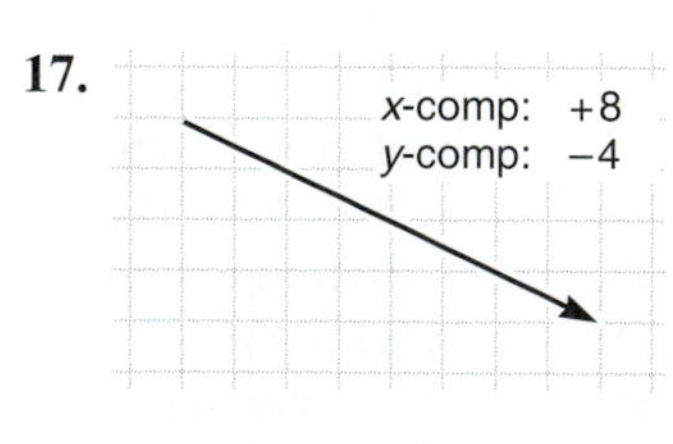

19.

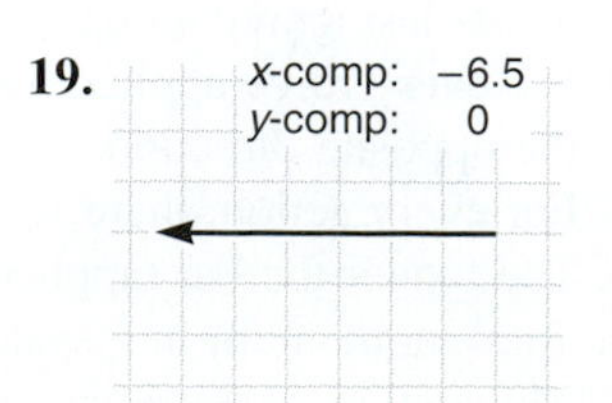

21.

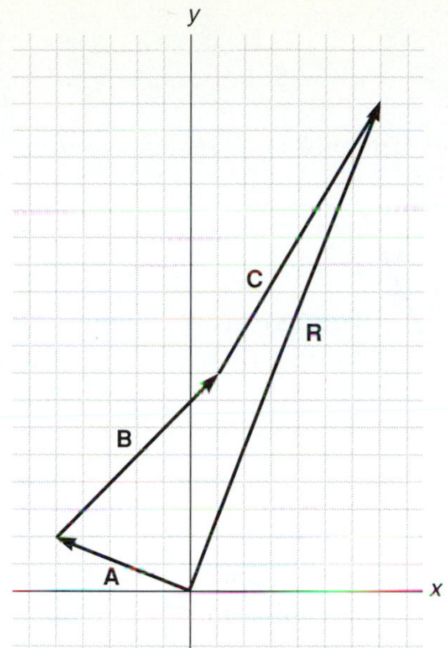

A + B + C = R

A + B + C = R

Vector	x-*component*	y-*component*
A	−5	2
B	6	6
C	6	10
R	7	18

23.

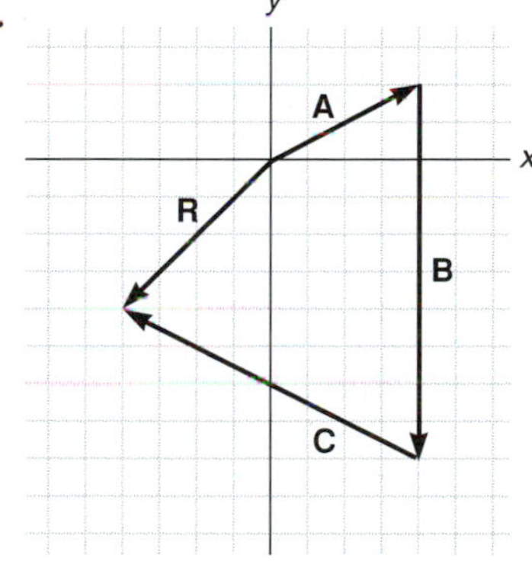

A + B + C = R

A + B + C = R

Vector	x-*component*	y-*component*
A	4	2
B	0	−10
C	−8	4
R	−4	−4

25.

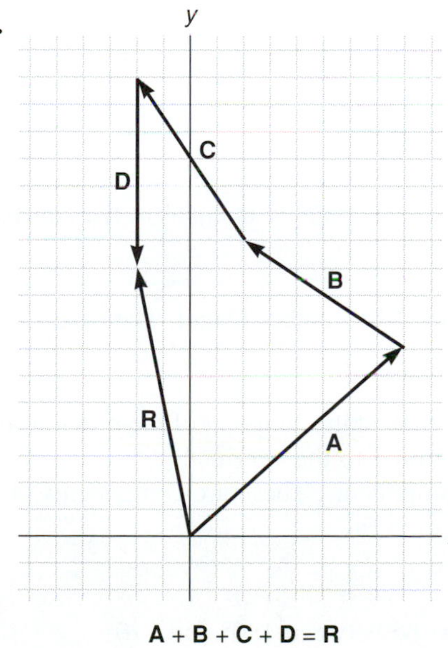

A + B + C + D = R

A + B + C + D = R

Vector	x-*component*	y-*component*
A	8	7
B	−6	4
C	−4	6
D	0	−7
R	−2	10

5.3 Pages 139–141

1.

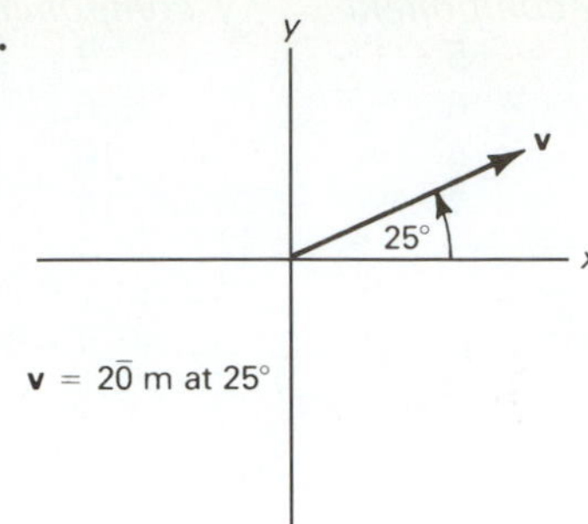

3.

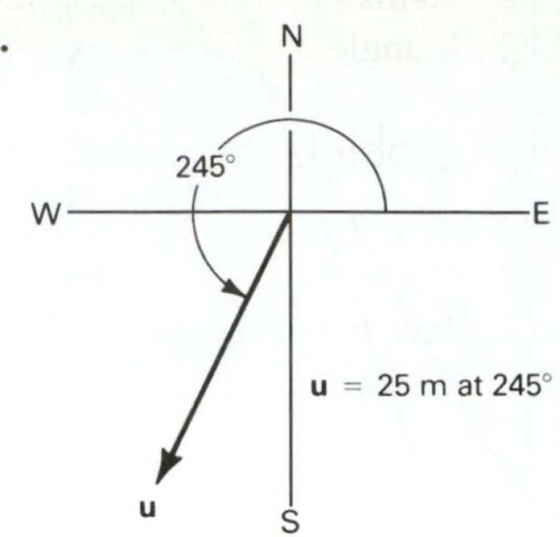

5.

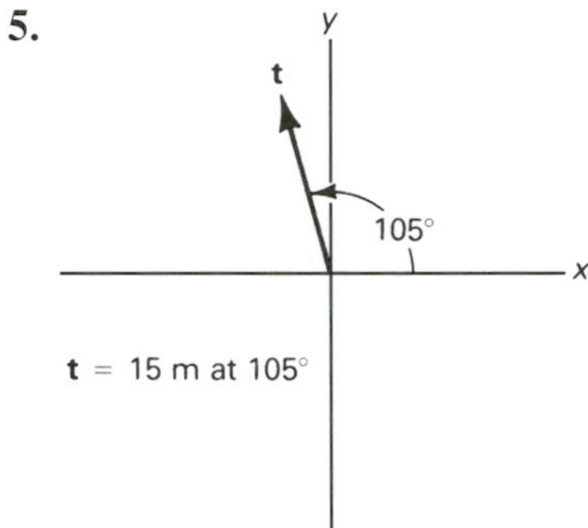

7.

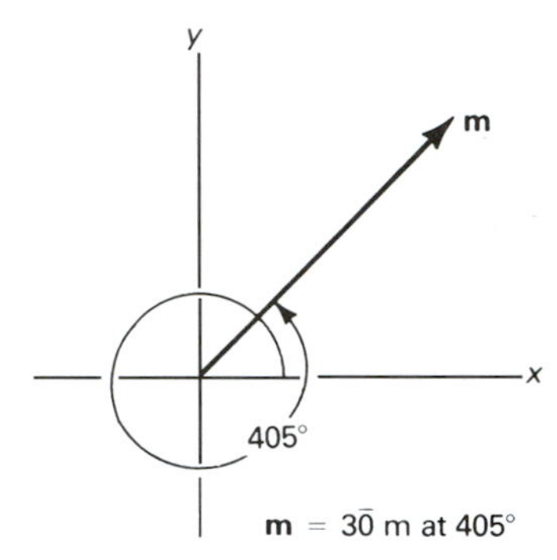

	x-*component*	y-*component*
9.	9.96 m	8.97 m
11.	−18.2 km	−45.1 km
13.	97.4 km/h	−14.4 km/h
15.	38.2 m	7.09 m
17.	6.17 km	−7.35 km
19.	−5.88 m/s	28.9 m/s

21. 10.0 m at 36.9° **23.** 26.2 mi at 315.0° **25.** 9.70 m/s at 98.3°
27. 53.3 m at 291.5° **29.** 22.3 mi at 48.8° **31.** 10.6 m/s at 155.7°
33. $37\bar{0}$ km/h at 90.0° **35.** 239 km/h at 190.8° **37.** 128 mi/h at 239.4°
39. 227 km/h at 162.4°

Chapter 5 Review Questions Page 143

1. d **2.** b **3.** a **4.** a
5. The hypotenuse is the side opposite the 90° angle.
6. sin, cos, tan
7. The sum of the squares of the two sides of a right triangle is equal to the square of the hypotenuse.
8. Yes
9. Two intersecting lines (axes) determine a plane (the number plane). The origin is the point where the axes intersect.
10. Yes **11.** Yes
12. Graph each vector placing the initial point of each at the endpoint of the previous one; the resultant is the vector from the very beginning to the very end.
13. Add the x-components and then add the y-components. These are components of the resultant.
14. No
15. Counterclockwise
16. In the third quadrant, the angle measure must be between 180° and 270°.

17. Complete the right triangle with the legs being the x- and y-components of the vector. Then find the lengths of x and y and determine their signs.
18. Complete the right triangle with the legs being the x- and y-axes.

Chapter 5 Review Problems Pages 143–144

1. 0.8290 **2.** 0.3963 **3.** 3.765 **4.** 69.9° **5.** 50.1° **6.** 17.4°
7. $A = 30.5°$; $B = 59.5°$; $c = 39.4$ cm **8.** $A = 61.1°$; $B = 28.9°$; $a = 53.4$ m
9. $A = 48.0°$; $a = 37.2$ km; $b = 33.5$ km **10.** $B = 50.6°$; $a = 32.9$ m; $c = 51.8$ m
11. 16.6 cm **12.** 0.663 km **13.** $c = 26.1$ ft; $A = 57.5°$; $B = 32.5°$ **14.** 14.3
15. $x = 6$; $y = -2$
16.

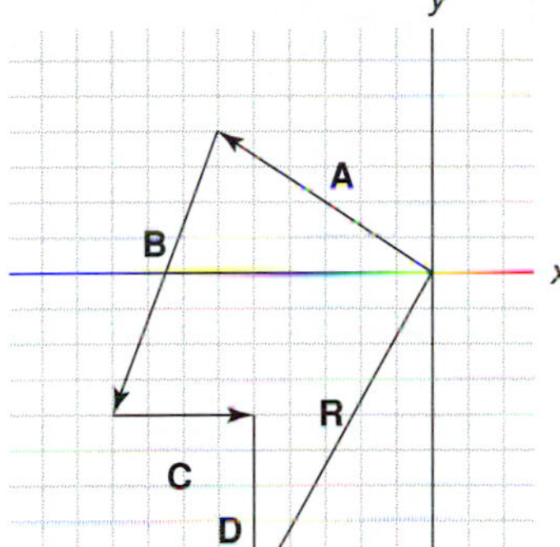

$\mathbf{A} + \mathbf{B} + \mathbf{C} + \mathbf{D} = \mathbf{R}$

Vector	*x-component*	*y-component*
A	−6	4
B	−3	−8
C	4	0
D	0	−5
R	−5	−9

17. $x = 11.3$ mm; $y = 6.50$ mm **18.** $x = -4.50$ cm; $y = -7.79$ cm
19. 226 km/h at 107.8° **20.** 33.4 km/h at 337.1°

CHAPTER 6

6.1 Pages 149–150

1. $3\bar{0}$ N (right) **3.** (a) $40\bar{0}$ N (b) $5\bar{0}$ N **5.** 1640 N at 55.6°
7. 2730 lb at 140.4° **9.** 4620 N at 123.2° **11.** 190 N at 126.3° from $\mathbf{F}_1$

6.2 Pages 158–160

1. $10\bar{0}$ N **3.** $26\bar{0}$ N **5.** $690\bar{0}$ N **7.** $57\bar{0}$ N **9.** Yes **11.** $\mathbf{F}_1 = 70.7$ N; $\mathbf{F}_2 = 70.7$ N **13.** $\mathbf{F}_1 = 823$ N; $\mathbf{F}_2 = 475$ N **15.** $\mathbf{F}_1 = 577$ lb; $\mathbf{F}_2 = 289$ lb
17. $\mathbf{F}_1 = 433$ lb; $\mathbf{F}_2 = 50\bar{0}$ lb **19.** $\mathbf{T}_1 = 1440$ lb; $\mathbf{T}_2 = 1440$ lb
21. $\mathbf{C} = 30\bar{0}0$ lb; $\mathbf{T} = 26\bar{0}0$ lb **23.** 5540 N **25.** $\mathbf{T} = 2330$ lb; $\mathbf{C} = 3690$ lb

Chapter 6 Review Questions Pages 161–162

1. b **2.** b **3.** a
4. c **5.** No (e.g., bridge) **6.** They are equal.
7. Supports exert forces up equal to the weight of the bridge down.
8. Equilibrium is the condition of a body where the net force acting on it is zero.
9. Toward the center of the earth **10.** They are in equilibrium. **11.** Compression
12. Tension **13.** It is a diagram showing how forces act on a body.
14. To describe them fully, each has magnitude and direction.
15. Yes; if equal opposing forces don't act on the same point, the tendency to rotate is produced.

Chapter 6 Review Problems Pages 162–164

1. 569 N (left) **2.** (a) $50\bar{0}$ lb (b) $5\bar{0}$ lb **3.** 3700 N at 156°
4. 8450 lb at 334.5° **5.** 94,600 N at 80.3° **6.** 1610 N at 114.2° from $\mathbf{F}_1$

7. 2150 N **8.** 470 N **9.** 43 lb **10.** 6.0 tons **11.** −525 **12.** −650 **13.** 54.0 lb **14.** 29.1 N **15.** 1080 N at 33.7° **16.** $\mathbf{F}_1$ = 645 N; $\mathbf{F}_2$ = 1520 N **17.** $\mathbf{F}_1$ = 5240 lb; $\mathbf{F}_2$ = 2210 lb **18.** **T** = 1790 N; **C** = 1370 N **19.** $\mathbf{T}_1$ = 348 lb; $\mathbf{T}_2$ = 426 lb; $\mathbf{T}_3$ = 475 lb **20.** $\mathbf{T}_1$ = 2900 N; $\mathbf{T}_2$ = 1900 N; $\mathbf{T}_3$ = 2200 N **21.** $\mathbf{T}_1$ = 6250 N; $\mathbf{T}_2$ = 6250 N; **C** = 6250 N

CHAPTER 7

7.1 Pages 171–172

1. 34.3 N m **3.** 917 kJ **5.** 24.4° **7.** 12,600 J **9.** 4410 J **11.** 34,000 ft lb **13.** 24.1 kJ **15.** 163,000 ft lb **17.** 6.2 MJ

7.2 Pages 177–178

1. 18.9 N m/s or 18.9 W **3.** 0.533 s **5.** 12.4 W **7.** (a) 1.68 hp (b) 219 N **9.** 59.4 s **11.** 1.49 kW **13.** 6.17 kW **15.** (a) 6.9 kW (b) 9.2 hp (c) 9.3 kW **17.** Four times **19.** (a) 110 passengers (b) 22 kW

7.3 Page 182

1. 2460 N m **3.** 217 J **5.** (a) 80.7 ft/s (b) 3.09×10^6 ft lb **7.** 14.2 ft/s **9.** (a) 294 kJ (b) 392 kJ **11.** 342 m/s **13.** 220 kJ **15.** (a) 2650 kW (b) 3550 hp (c) 1.59×10^8 J or 159 MJ **17.** 4

7.4 Page 186

1. 7.00 m/s **3.** 72.7 m/s **5.** 87.3 ft **7.** 49.5 m/s **9.** $19\overline{0}$ ft/s

11.

t	s	v	KE	h	PE	Total
0.000	0.00	0.00	0	300.0	11,760	11,760
1.000	4.90	9.80	190	295.1	11,570	11,760
2.000	19.6	19.60	770	280.4	10,990	11,760
3.000	44.1	29.40	1,730	255.9	10,030	11,760
4.000	78.4	39.20	3,070	221.6	8690	11,760
5.000	122.5	49.00	$4,8\overline{0}0$	177.5	6960	11,760
6.000	176.4	58.80	6,910	123.6	4850	11,760
7.000	240.1	68.60	9,410	59.9	2350	11,760
7.800	300.0	76.68	11,760	0.00	0.00	11,760

Chapter 7 Review Questions Pages 187–188

1. c **2.** a **3.** a **4.** d **5.** c **6.** No **7.** No **8.** $\text{J} = \left(\frac{\text{kg m}}{\text{s}^2}\right)(\text{m})$

9. No **10.** No **11.** Yes
12. By measuring the force applied, distance traveled, and time taken
13. It possesses the ability to do work (e.g., turn a wheel) in falling to the lower level.
14. (a) Elevator counterweights (b) roller coasters
15. Yes; $\text{KE} = \frac{1}{2}mv^2$ **16.** At its lowest point. **17.** At its highest point **18.** No
19. Yes **20.** The bolt has accelerated to a higher velocity.

Chapter 7 Review Problems Page 188

1. 3.6×10^6 J **2.** 0 **3.** 10.2 m **4.** 9.80×10^5 J **5.** (a) 27.2 W (b) 0.0365 hp **6.** 0.102 m **7.** 2.00 ft **8.** 1.41 m/s **9.** 1.40 m/s **10.** 303,000 ft lb **11.** 2350 J **12.** $3\overline{0}00$ J **13.** 1670 J **14.** $20\overline{0}0$ J **15.** 1880 J **16.** 31.3 m/s

CHAPTER 8

8.3 Pages 196–197

1. 14.8 **3.** 36.3 **5.** 52.4 **7.** 2.39 **9.** 48.8 **11.** 1.55 **13.** 2.27
15. 41.1 **17.** 4.00 **19.** 2.50 **21.** 4.00 **23.** 0.500

8.4 Page 199

1. 14.0 **3.** 271 **5.** 524 **7.** 48.8 **9.** 20.4 **11.** 438 lb **13.** 429 N
15. 2010 N **17.** 3.34

8.5 Pages 203–204

1. 1 **3.** 3 **5.** 6 **7.** 2
15. 2
17. 30.0 ft
19. 82.0 m
21. 260 N; 3.25 m
23. 325 N

9.

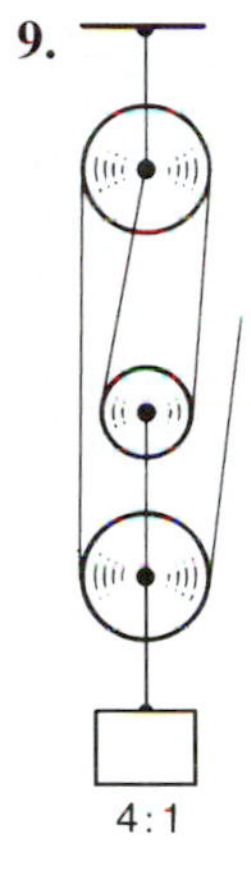

11.

13.

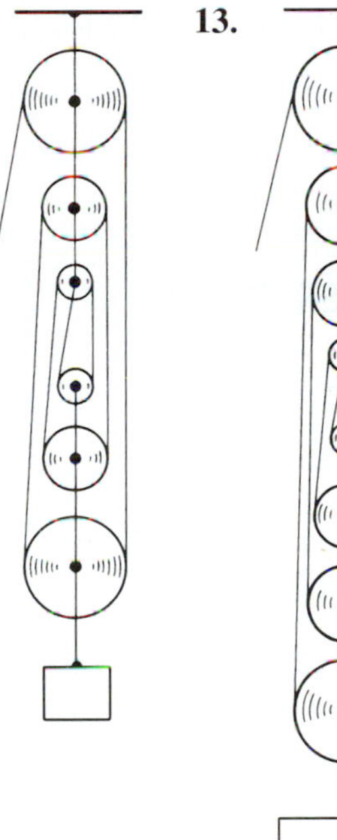

25. Mechanical Advantage (MA)

Number of Pulleys	1	2	3	4	5	6	7	8
Fixed	1	1	2	2	3	3	4	4
Movable	0	1	1	2	2	3	3	4
Fixed		0	1	1	2	2	3	3
Movable		1	1	2	2	3	3	4

8.6 Page 207

1. 12.8 **3.** 21.2 **5.** 36.3 **7.** 4.62 **9.** 5.61 **11.** 4.00 **13.** No
15. 3.84 **17.** 1.33 m

8.7 Pages 209–210

1. 2.30 **3.** 14.2 **5.** 2.28 **7.** 29.3 **9.** 35.2 **11.** 4.84 lb **13.** 2.23 cm
15. 37.7 **17.** 188 lb **19.** 33,500 N

8.9 Pages 212–213

1. 15.0 **3.** 40.0 **5.** 75.0 lb **7.** 125 N **9.** 1.44×10^5 N

Chapter 8 Review Questions Pages 215–216

1. d **2.** a **3.** b **4.** d **5.** b **6.** b
7. (a) Bicycles; (b) auto transmission; (c) high-speed drill **8.** Resistance force
9. Effort force × effort distance = resistance force × resistance distance
10. Mechanical advantage **11.** Efficiency **12.** No **13.** The fulcrum

14. Effort arm length divided by resistance arm length = MA of a lever
15. First class **16.** $F_R \times d_R = F_E \times d_E$
17. The opposite end of the resistance force with the effort force in between
18. Resistance force × resistance radius = effort force × effort radius
19. No; it depends on the radii.
20. A fixed pulley does not move. It is suspended by its center axle. A movable pulley is free to move and is suspended by the strand around the groove.
21. No **22.** MA = length of plane/height of plane
23. It is the distance a screw advances into the wood in one revolution.
24. It is greater because the handle of the jackscrew can be longer than the radius of the screwdriver.
25. Total MA is found by multiplying the MA's of each simple machine.

Chapter 8 Review Problems Pages 216–217

1. 2.00 **2.** 73.0% **3.** 2.13 **4.** (a) 540 N (b) 360 N **5.** 6.54 cm **6.** 2.60
7. 12 **8.** 675 N **9.** 292 N **10.** 3.00 **11.** 2.30 ft **12.** 16,400 N
13. 0.108 cm **14.** 151 **15.** 5 **16.** 4 **17.** (a) 64.0 (b) 4800 N **18.** 6.0

CHAPTER 9

9.1 Pages 223–224

1. (a) 40.8 rad (b) 2340° **3.** (a) 12.5 rev (b) 4500°
5. 154 rpm **7.** 8.38 rad/s **9.** 354 rev/s **11.** 6450 rpm
13. 14.2 rad/s **15.** $66\bar{0}$ rpm **17.** (a) 6.67 rev/s (b) $40\bar{0}$ rpm
(c) 41.9 rad/s **19.** (a) 0.0571 s (b) 87.5 rev **21.** 8.40 rad
23. 0.131 m **25.** 1.33 m/s **27.** (a) 68.6 rad/s (b) 12,300 rad
(c) 157 cm (d) 3080 m (e) 17.2 m/s **29.** (a) 50.5 rad/s
(b) 1520 rad (c) 502 m (d) 502 m **31.** 188 in./s or 15.7 ft/s
33. (a) 1680 km/h (b) 1450 km/h (c) 1190 km/h (d) 838 km/h

9.2 Pages 226–227

1. 96.0 lb ft **3.** 2.00 m **5.** 187 N **7.** 1.60×10^3 N **9.** 0.357 m
11. 159 lb **13.** $40\bar{0}$ N **15.** (a) 25 lb; (b) It is halved.

9.3 Page 229

1. 4350 N **3.** 79.4 slugs **5.** 5.53 m/s **7.** 1.92 m **9.** 3.60×10^3 lb
11. 28.8 m **13.** 5420 N **15.** 19 N **17.** $2\bar{0}00$ N

9.4 Page 233

1. 7260 **3.** 18.7 **5.** 1860 N m **7.** 343 hp **9.** 4.95/s
11. 4600 W **13.** (a) 299 hp (b) 599 hp **15.** 2.65 N m **17.** 37 kJ/min
19. 6.67/s **21.** 2800 W

9.6 Pages 241–243

1. 52 rpm **3.** 24 teeth **5.** 207 rpm **7.** 42.5 rpm **9.** 75 teeth
11. 63 teeth **13.** 144 rpm **15.** 60 teeth **17.** 9 teeth
19. Counterclockwise **21.** Clockwise **23.** Clockwise **25.** Clockwise
27. Clockwise **29.** 1050 rpm **31.** 576 rpm **33.** 1480 rpm **35.** 40 teeth
37. 20 teeth **39.** Gear *B* is reversed in all problems.

9.7 Pages 245–246

1. 2250 **3.** $36\bar{0}0$ **5.** 147 **7.** 62.5 rpm **9.** 22.5 in. **11.** Clockwise
13. Counterclockwise **15.** Counterclockwise **17.** The small pulley

Chapter 9 Review Questions Pages 247–248

1. d **2.** c **3.** a **4.** b
5. Curvilinear motion is motion along a curved path; rotational motion occurs when the body itself is spinning.
6. Radians and revolutions
7. A radian is an angle with its vertex at the center of a circle whose sides cut off an arc on the circle equal to its radius.
8. Angular displacement is the angle through which any point on a rotating body moves. It can be measured in radians or revolutions.
9. Linear velocity = angular velocity × radius
10. No; only when the applied force is perpendicular to the pedal shaft
11. Yes
12. No; centripetal force is toward the center and centrifugal force is away from the center.
13. No; it tends to cause a body to continue in a straight line (tangent to the curve).
14. Number of teeth of driver times number of revolutions of driver = number of teeth of driven gear times number of revolutions of driven gear.
15. An idler changes direction of rotation of the driver gear.
16. Opposite
17. Since the gears are connected, they both rotate together.
18. Diameters are used in similar equations rather than teeth.

Chapter 9 Review Problems Pages 248–249

1. (a) 26π or 81.7 rad (b) 4680° **2.** 1.7 rad/s **3.** 0.885 m/s **4.** 26.5 N m
5. 27.0 lb **6.** 7.50 N **7.** 2.75 m/s **8.** 5.67 ft lb/s **9.** 6.00 rad/s
10. 6.00 rpm **11.** 20 teeth **12.** 57.5 rpm **13.** Yes **14.** 105 rpm
15. 2.00 cm **16.** Counterclockwise **17.** $72\bar{0}$ rpm **18.** 32 teeth

CHAPTER 10

10.1 Pages 255–256

1. $10\bar{0}$ lb **3.** $50\bar{0}$ N **5.** $45\bar{0}$ N **7.** $40\bar{0}$ N **9.** 551 N; 331 N
11. 2.77×10^5 N; 1.07×10^4 N **13.** 2.67 m

10.2 Pages 259–261

1. 22.6 **3.** 42.0 **5.** 1.52×10^4 N; 9.85×10^3 N
7. 9.90×10^4 N; 8.82×10^4 N **9.** 22.5 lb; 20.5 lb
11. 8.99×10^4 N; 1.01×10^5 N **13.** 197 N; 118 N
15. 2390 N; 943 N **17.** 1020 N; 775 N
19. 1525 N (up) 4.54 m from *A*

Chapter 10 Review Questions Pages 261–262

1. a **2.** b **3.** b **4.** ΣF **5.** Yes
6. Even if the vector sum of opposing forces is zero, they must also be positioned so there is no rotation in the system.
7. Yes **8.** Choose a point through which a force acts to eliminate a variable.

9. (a) Stacking bricks; (b) riding a bicycle; (c) lifting any object; (d) hitting a baseball; (e) leaning into the wind
10. Because the weight of an object must be located at its center of gravity
11. No; only if the object is of uniform composition and shape

Chapter 10 Review Problems Page 262

1. 880 N **2.** 160 lb **3.** $12\bar{0}$ N **4.** 32.0 N **5.** $10\bar{0}$ N m **6.** 100.0 kg **7.** 16.6 lb **8.** 51.0 cm **9.** $56\bar{0}0$ kg **10.** End closest to truck, 62,300 N; other end, 47,700 N **11.** 343 N; 483 N **12.** 2580 N

CHAPTER 11

11.2 Pages 274–275

1. 46.2 in. **3.** 49.0 N **5.** $20\bar{0}$ N/m **7.** (a) 0.0250 cm (b) 1.56×10^6 N **9.** 3.57×10^5 lb **11.** 7.50 N **13.** 1.09×10^8 Pa or 109 MPa **15.** 15.6 cm

11.5 Pages 285–286

1. 2870 kg/m^3 **3.** 1750 lb **5.** 5600 cm^3 **7.** 1210 lb/ft^3 **9.** 2710 kg/m^3 **11.** 684 kg/m^3 **13.** 3.5 m^3 **15.** 58.8 lb/ft^3 **17.** 1.49 m^3 **19.** 2820 kg/m^3 **21.** (a) $10\bar{0}0$ L (b) 1500 L (c) 73.5 L **23.** 0.917 **25.** 7.8 **27.** 0.68 **29.** Floats **31.** Floats **33.** 5.8×10^{15}

Chapter 11 Review Questions Pages 287–288

1. a, c, and e **2.** all **3.** d **4.** d **5.** d **6.** b **7.** d **8.** e
9. Mass density refers to the mass per unit volume; weight density refers to weight per unit volume.
10. Yes; no.
11. Capillary action refers to the effect of surface tension of liquids that causes the level of liquid in small-diameter tubes to be higher or lower than that of the liquid in a large-diameter tube due to the adhesive force between the liquid and the tube.
12. Adhesion refers to the attractive force between different molecules; cohesion refers to the attractive force between similar molecules.
13. Surface tension of water allows the base of many insects' legs to be supported by the surface of water, allowing them to "walk" on the surface of a pond.
14. 1800 **15.** Pressure
16. Stress is directly proportional to strain as long as the elastic limit is not exceeded.
17. kPa
18. The specific gravity of an object can be found by dividing the density of the object by the density of water. The density can be found by determining the mass of the object on a scale and dividing by its volume.
19. Mass density **20.** Viscosity **21.** Elastic limit **22.** Solid, liquid, gas
23. An atom consists of one nucleus and its surrounding electrons; a molecule consists of two or more atoms.
24. A proton has a positive charge; a neutron is neutral.
25. Tension, compression, shear, bending, and twisting
26. The hydrometer measures the density of the battery electrolyte. The density is related to the amount of sulfuric acid in the electrolyte and therefore the charge on the battery. Temperature does affect the measurement.

Chapter 11 Review Problems Page 289

1. 165 N **2.** 24.6 cm **3.** 1.63×10^6 lb/in.; 0.0317 in. **4.** 311 MPa **5.** 2020 lb/ft^3 **6.** 1.3 N **7.** 8950 kg/m^3

8. 83.4 lb/ft^3 **9.** 55.8 cm^3 **10.** 21.9 m^3 **11.** 2160 kg
12. 59,300 lb **13.** 18.9 lb/ft^3 **14.** 54.8 lb/ft^3 **15.** 0.117 ft^3
16. 0.694 **17.** 77.8 lb **18.** 9.26 kg **19.** 2960 kg/m^3 **20.** sink

CHAPTER 12

12.1 Pages 298–299

1. 2.50 m × 0.80 m: 71 kPa; 2.50 m × 0.45 m: 130 kPa; 0.80 m × 0.45 m: 390 kPa
3. 21.7 lb/in^2 **5.** 16.5 lb/ft^3 **7.** 22,500 N **9.** 7.48 in.
11. 245 kPa **13.** 757 kg/m^3 **15.** 1.1×10^8 N **17.** 403 kPa
19. 34,800 kPa **21.** 9.24×10^8 N **23.** 3680 kPa **25.** 0.373 m

12.2 Pages 303–304

1. 60.0 lb **3.** 22.0 **5.** 24.0 **7.** 48 N **9.** 2.3×10^4 N
11. (a) 146 N (b) 11.6 N/cm^2 (c) 36.0
13. (a) 4800 N (b) 9.55 kPa (c) 9.55 kPa (d) 64.0
(e) The lift will exert twice the force on the large piston; the MA and the pressure will double.
(f) The lift will exert four times the force on the large piston; the MA and the pressure will be four times.

12.3 Pages 306–307

1. 118 lb/in^2 **3.** 293 kPa **5.** (a) 303.96 kPa (b) 202.64 kPa (c) 88.2 lb/in^2
(d) 506.6 kPa (e) 33.77 kPa (f) 25.33 kPa **7.** 586 kPa **9.** 30.3 lb/in^2
11. 306 kPa **13.** 1174 kPa **15.** 401 kPa

12.4 Page 309

1. 13.0 lb **3.** 3.9 N **5.** 75.5 lb **7.** 3.20 N **9.** 105 lb **11.** 431 N
13. 3.06×10^4 ft^3; 795 tons **15.** 7.65 m^3

12.5 Page 313

1. 358 L/min **3.** (a) 9.6 cm (b) 0.34 m/s **5.** (a) 34.1 ft/min (b) 136 ft/min
7. 16.7 cm

Chapter 12 Review Questions Pages 314–315

1. c **2.** b **3.** b **4.** c **5.** a **6.** kPa
7. Pressure is the force applied per unit area.
8. $F_s = \frac{1}{2}Ah\,D_w$
9. External pressure applied to a fluid is transmitted to all inner surfaces of the liquid's container.
10. A ship floats because it is lighter than an equal volume of water.
11. The spinning baseball creates a different air velocity on one side than on the other. This creates a higher pressure on one side that causes the ball to "curve."
12. The top side of the wing is curved more than the bottom, so that the velocity of air rushing past the top must be larger than that going past the bottom. This creates a low-pressure area at the top of the wing.
13. In streamline flow, all particles of the fluid passing a given point follow the same path. In turbulent flow, the particles passing a given point may follow a different path from that point.
14. A balloon filled with a gas with lower density than air, such as helium, rises.
15. Gauge pressure is the pressure measured relative to atmospheric pressure. Gauge pressure is used in an automobile tire gauge.

16. The pressures are identical. The forces are different.
17. This force depends on the horizontal surface area and the average pressure on that surface. The average pressure depends on the density and height of the liquid.
18. The pressure at the bottom is greater than at the top.
19. Smaller. The total force exerted on the brake pads must be larger than the force applied to the brake pedal.
20. Yes, but only if the fluid to be drawn through the straw is in an airtight sealed container.

Chapter 12 Review Problems Pages 315–316

1. 31.8 kPa **2.** 455 ft **3.** 30.3 m **4.** 9,3$\bar{0}$0 lb **5.** 196 kPa
6. 9.60×10^7 N **7.** 7.01×10^8 N **8.** 2.80×10^4 N **9.** 2610 kPa
10. 192 lb/in^2 **11.** 9410 kPa **12.** 77.2 lb **13.** 8.42 **14.** 35.5 N
15. 103 N **16.** 111 N/cm^2 **17.** 42.2 **18.** 303 kPa **19.** 554 kPa
20. 299 lb/in^2 **21.** 3.3 N **22.** 6.30 N **23.** 2.4 N **24.** 26,$\bar{0}$00 ft^3; 654 tons
25. 188 L/min **26.** 3610 L/min; 90,300 L **27.** (a) 49.3 N/m^3 (b) Alcohol

CHAPTER 13

13.1 Page 322

1. 25°C **3.** 125°C **5.** 293°F **7.** −12.2°C **9.** 203°F **11.** 298 K
13. 133°C **15.** 223 K **17.** 5727°C **19.** −38.9°C **21.** 29°C **23.** −321°F
25. 1204 **27.** 941 **29.** 1175 **31.** 590

13.2 Pages 324–325

1. 23 cal **3.** 1.21×10^6 ft lb **5.** 3.21×10^6 J or 3.21 MJ **7.** 4450 Btu
9. 2.62 MJ **11.** 42.2 MJ **13.** (a) 2.77×10^5 J/s (b) 277 kW

13.3 Pages 329–330

1. 0.021 ft^2 °F h/Btu **3.** 0.45 ft^2 °F h/Btu **5.** 1.7 ft^2 °F h/Btu
7. 8.1×10^6 Btu **9.** 11 J **11.** 1400 Btu **13.** 14 MJ **15.** 6800 J

13.4 Pages 331–332

1. 173 Btu **3.** 38$\bar{0}$0 Btu **5.** 2.1×10^5 J or 210 kJ **7.** 280 cal
9. 3.01×10^4 J or 30.1 kJ **11.** 64.0 kcal **13.** 39,000 Btu **15.** 29,000 kcal
17. 1.89×10^8 J or 189 MJ **19.** 8.07×10^7 J or 80.7 MJ **21.** 3310°C

13.5 Pages 334–335

1. 428°F **3.** 0.051 cal/g °C **5.** 95°F **7.** 81°C **9.** 0.104 cal/g °C
11. 286°C

13.6 Pages 341–342

1. 0.30 ft **3.** 0.10 m **5.** 200.10 m **7.** 0.54 ft **9.** 0.752 in.
11. 1.8 cm^2 **13.** 88.97 cm^2 **15.** 60.10 cm^3 **17.** 4.9 mm
19. 1.08×10^{-5}/°C **21.** 15 cm^3 **23.** 0.3397 m^3

13.7 Pages 343–344

1. 653 L **3.** 12.2 m^3 **5.** 3754 ft^3 **7.** 0.58 cm^3 **9.** $215 **11.** 23°C

13.8 Pages 354–355

1. 1100 cal **3.** 10,700 Btu **5.** 26,000 cal **7.** 6.70×10^6 J or 6.70 MJ
9. 3.39×10^6 J or 3.39 MJ **11.** 6070 Btu **13.** 11,650 Btu **15.** 5$\bar{0}$00 kcal

17. 3.12×10^6 J or 3.12 MJ **19.** 467,000 cal **21.** 3090 kcal

Chapter 13 Review Questions Pages 357–358

1. a, b, and d **2.** b and d **3.** a, b, and d **4.** a, b, and d **5.** d **6.** c

7. The total heat lost by warm objects is gained by cold objects.

8. 778 ft lb of work is equivalent to 1 Btu.

9. The Rankine scale. **10.** The Kelvin scale.

11. Each Celsius degree is 1.8 times the Fahrenheit degree. The freezing point of water is 0°C and 32°F. The boiling point of water is 100°C and 212°F.

12. Heat is an amount of energy. Temperature is a measure of the average kinetic energy of atoms and molecules in matter.

13. Burning fuel in a furnace. Burning fuel in an engine. Conversion of heat into electricity in a steam-driven generator.

14. Heat generated in a solid by a drill bit. Heat generated by rubbing two objects together. Heat generated in automobile brakes. All are due to friction.

15. Light clothing should be worn. It absorbs less heat.

16. It increases. The metal block increases in size, as does the hole.

17. 4 degrees Celsius. Most other liquids have their highest density at the temperature at which the change of state to a solid occurs.

18. Water below 4 degrees Celsius is less dense than water at 4 degrees Celsius and rises. Freezing therefore occurs at the top of a body of water.

19. Ten kilograms of ice at 0°C. The difference is the heat of fusion, which must be added to the ice to turn it to water.

20. The heat released when steam changes to water can cause severe burns.

21. Because water expands as it solidifies.

22. Conduction is the transfer of kinetic energy from atoms or molecules in a warmer location through nearby atoms or molecules to atoms or molecules in a colder section of the material. Convection is the transfer of kinetic energy from one region to another via the motion of warmer atoms or molecules from one region to another. Radiation is the transfer of energy through the emission and absorption of electromagnetic radiation.

23. Coolants boil at a higher temperature at high pressure. Therefore, the engine coolant can be at a higher temperature and transfer more heat from the engine to the atmosphere because of the greater temperature difference.

24. Heat is extracted from the outside air in winter to vaporize the refrigerant. The heat is given up by the condensing fluid to warm the inside air. In summer, heat is extracted from the inside air to vaporize the refrigerant. The condensing fluid then gives up this heat to the outside air.

Chapter 13 Review Problems Pages 358–359

1. 71°C **2.** 297 K **3.** 9230°F **4.** 335°C **5.** 10.3 cal
6. 1.56 kcal **7.** 3.38×10^5 ft lb **8.** 1.4×10^5 Btu
9. 2.4×10^4 J or 24 kJ **10.** 4.18×10^4 Btu **11.** 1.5×10^4 kcal
12. 1.22×10^4 kcal **13.** 1300°F **14.** 0.353 cal/g °C **15.** 85°C
16. 12.524 m **17.** 7.51 mm; 44.3 mm^2 **18.** 0.0295 cm^3 **19.** 2.57 L
20. 123 ft^3 **21.** 11,600 kcal **22.** $12\bar{0}0$ Btu **23.** 3440 kcal
24. 4.80×10^5 J

CHAPTER 14

14.1 Pages 362–363

1. 288 K **3.** 44°C **5.** 532°R **7.** $9\bar{0}$°F **9.** 258 cm^3 **11.** −16°F
13. $38\bar{0}$ m^3 **15.** 1430 L **17.** −59°C **19.** 30.5 L **21.** 107 K

14.2 Page 366

1. 265 cm^3 **3.** 75.0 kPa **5.** 2.08 kg/m^3 **7.** 0.180 lb/ft^3 **9.** 90.4 kPa
11. 3.34 kg/m^3 **13.** 801 kPa **15.** 19.0 ft^3 **17.** 162 psi
19. (a) 37.5 in^3 (b) 25.0 in^3 (c) $15\bar{0}$ in^3 **21.** 4.37 kg/m^3

14.3 Page 368

1. 1270 in^3 **3.** 506 m^3 **5.** −39°C **7.** 143°C **9.** $22\bar{0}0$ psi **11.** 399 kPa
13. $70\bar{0}$ kPa **15.** 3.59 m^3 **17.** 3770 mL

Chapter 14 Review Questions Page 369

1. c **2.** a **3.** c **4.** c **5.** b **6.** b

7. Standard pressure is 101.32 kPa or 14.7 lb/in^2 and standard temperature is 0°C or 32°F.

8. A temperature increase tends to cause a volume increase. A pressure increase tends to cause a volume decrease. If both temperature and pressure are increased, the volume change is given by the combined Charles' and Boyle's laws.

9. The temperature will increase.

10. Heating a gas increases the kinetic energy of the gas molecules. This causes an increase in pressure and volume.

11. When a gas is compressed, work must be done on it to decrease its volume. This work is transferred into increased kinetic energy of the molecules of the gas, causing a temperature increase.

12. When the volume is increased, the number of gas molecules striking the surface of the container per unit area decreases. Thus the pressure exerted by the gas molecules on the container decreases.

Chapter 14 Review Problems Page 370

1. 14.9 ft^3 **2.** 4.40 m^3 **3.** 131°F **4.** 30.9°C **5.** 34.1°C **6.** 138 kPa
7. 23.9 ft^3 **8.** 3.68 kg/m^3 **9.** 478 in^3 **10.** 19.9°C **11.** 2180 psi
12. 18°F **13.** 304 kPa **14.** −42.2 kPa **15.** 822 L **16.** 3.07 kg/m^3

CHAPTER 15

15.2 Pages 380–381

1. 2.00×10^{-3} s **3.** $80\bar{0}$ m/s **5.** 1.67×10^{-15} s **7.** 2.31 m
9. (a) 0.175 s (b) 5.71 Hz **11.** 6.00 m **13.** 3.35×10^4 m

15.4 Page 386

1. 337 m/s **3.** 317 m/s **5.** 5.00 s **7.** 2640 m **9.** 622 Hz **11.** 507 Hz
13. 653 Hz

Chapter 15 Review Questions Pages 389–390

1. a, b, and c **2.** c **3.** a **4.** b **5.** b **6.** a **7.** c and d

8. Interference is the result of two or more waves traveling through the same region at the same time. Diffraction is the bending of a single wave passing near an obstacle with an opening nearly the same size as the wavelength.

9. Two or more waves added together to form a larger wave is constructive interference. The formation of a smaller wave is destructive interference.

10. Sound would not be heard if some obstacle came between you and the stereo speakers and there were no nearby reflecting surfaces.

11. Waves passing through a break in a seawall.

12. It increases.

13. A wave is a periodic disturbance. A pulse is a one-time disturbance.
14. A sharp explosion creates a sound pulse.
15. The blue and violet light from the sun is diffracted by small molecules and dust particles in the air. The red light passes through without diffraction.
16. The blue light from the sun is diffracted by small molecules and dust particles in the atmosphere. This diffracted blue light gives the sky a blue color.
17. The speed of sound increases. The higher kinetic energy of air molecules at high temperature leads to a higher velocity of sound through the air.
18. A seismograph detects slight vibrations of the earth's surface by detecting the relative motion of the earth's surface and a massive object.
19. The velocity of sound is higher in water than in air. The higher density of water provides a higher velocity of sound.
20. Traffic radar guns depend on the Doppler effect to determine the speed of a car.
21. Motion toward an oncoming sound wave increases the frequency at which the maximum pressure regions in the sound wave strike an observer, therefore producing a higher-frequency sound. Motion away from an oncoming sound wave decreases the frequency at which the maximum pressure regions strike an observer, therefore producing a lower-frequency sound.
22. Sympathetic vibrations occur when an object vibrates at its natural resonance frequency in response to the vibration of a nearby object at some other frequency. Forced vibration is when an object vibrates at the same frequency as another nearby vibrating object.
23. Resonance occurs when an object vibrates at its natural resonance frequency in response to the vibration at the same frequency of a nearby object.
24. The light from these stars is shifted toward the red as a result of their motion away from the earth.

Chapter 15 Review Problems Page 390

1. 2.82×10^{-6} s or 2.82 μs **2.** 3.13 Hz **3.** 6.67×10^{14} Hz **4.** 1.50×10^{-15} s
5. 5.87 m/s **6.** 1.24 m **7.** 65 Hz **8.** 4.58 m 9. 383 m/s **10.** $31\bar{0}$ m/s
11. 0.967 s **12.** 29 s **13.** 602 Hz **14.** 5360 Hz **15.** $46\bar{0}0$ Hz
16. 368 Hz **17.** 5.59×10^{14} Hz

CHAPTER 16

16.3 Page 397

1. 922 N **3.** 1.3 cm **5.** 1.5×10^{-8} C **7.** −0.93 N **9.** −2.08 N

Chapter 16 Review Questions Page 399

1. (b) **2.** (c) **3.** (c) **4.** (b)
5. Materials can become charged when a charged object is brought nearby, inducing a polarization (separation) of charge on the material. If one side of the material is touched by another object, the charge at one side of the material can then be "drained" off, leaving the charge at the other side of the material.
6. Protons, electrons, and neutrons **7.** Protons and neutrons
8. Electrons are located in charge clouds surrounding the nucleus.
9. Positive and negative. Protons carry a positive charge; electrons carry a negative charge.
10. A charged object is brought into contact with the electroscope, thereby providing some of that charge to the electroscope.
11. A charged object is brought near the conducting ball on an electroscope causing a polarization (separation) of charge on the electroscope. The conducting ball is touched by a "ground," allowing one type of charge to leave the electroscope and go to ground, leaving the other charge behind.
12. Coulomb's law states that the force between two charges is directly proportional to the product of the magnitude of the charges and inversely proportional to the square of the separation between the two charges.

13. An electric field at a point represents the magnitude and direction of the force that would be exerted on a single unit of charge if placed at that point.
14. Lightning is the discharge of built-up static charge on a portion of a cloud.

CHAPTER 17

17.1 Pages 406–407

1. 1.07 Ω **3.** 0.0165 Ω/ft **5.** 3.95 Ω **7.** 6.84 Ω **9.** 0.131 cm^2

17.2 Page 409

1. 4.79 A **3.** $22\bar{0}$ V **5.** 153 Ω **7.** 17.6 Ω **9.** 1.26 A
11. (a) 0.067 A; (b) $3\bar{0}$ V; (c) 0.13 A

17.3 Pages 414–415

1. 13.50 Ω **3.** 60.0 Ω **5.** 0.750 A **7.** 378 V **9.** 23.0 Ω **11.** 10.8 Ω, 14.2 Ω, 19.0 Ω **13.** 2.78 Ω, 40.0 V, 3.40 Ω

17.4 Page 421

1. 4.28 Ω **3.** (a) 2.00 A; (b) 1.25 A; (c) 5.00 A **5.** 1.21 Ω
7. (a) 1.67 A; (b) 0.250 A **9.** (a) 29.9 V; (b) 29.9 V

17.5 Pages 426–428

1. (a) R_2, R_3; (b) 3.00 Ω **3.** 8.89 A **5.** (a) 2.23 A; (b) 6.68 A **7.** 21.24 Ω
9. 45 V **11.** 5.41 A **13.** 10.0 Ω **15.** 17.3 Ω **17.** 30.0 V **19.** 6.93 V
21. 13.3 Ω **23.** 60.2 V **25.** 0.370 A

17.6 Pages 429–431

1.

	V	I	R
Batt.	12.0 V	9.00 A	1.33 Ω
R_1	12.0 V	6.00 A	2.00 Ω
R_2	12.0 V	3.00 A	4.00 Ω

3.

	V	I	R
Batt.	36.0 V	6.00 A	6.00 Ω
R_1	36.0 V	2.00 A	18.0 Ω
R_2	36.0 V	3.00 A	12.0 Ω
R_3	36.0 V	1.00 A	36.0 Ω

5.

	V	I	R
Batt.	50.0 V	5.00 A	10.0 Ω
R_1	25.0 V	2.00 A	12.5 Ω
R_2	25.0 V	2.00 A	12.5 Ω
R_3	10.0 V	3.00 A	3.33 Ω
R_4	40.0 V	3.00 A	13.3 Ω

7.

	V	I	R
Batt.	36.0 V	9.00 A	4.00 Ω
R_1	12.0 V	6.00 A	2.00 Ω
R_2	12.0 V	3.00 A	4.00 Ω
R_3	24.0 V	6.00 A	4.00 Ω
R_4	24.0 V	3.00 A	8.00 Ω

9.

	V	I	R
Batt.	80.0 V	12.0 A	6.67 Ω
R_1	18.0 V	4.00 A	4.50 Ω
R_2	18.0 V	2.00 A	9.00 Ω
R_3	18.0 V	6.00 A	3.00 Ω
R_4	48.0 V	12.0 A	4.00 Ω
R_5	8.00 V	4.00 A	2.00 Ω
R_6	8.00 V	8.00 A	1.00 Ω
R_7	6.00 V	12.0 A	0.500 Ω

11.

	V	I	R
Batt.	65.0 V	5.00 A	13.0 Ω
R_1	10.0 V	0.500 A	20.0 Ω
R_2	10.0 V	1.00 A	10.0 Ω
R_3	10.0 V	2.50 A	4.00 Ω
R_4	10.0 V	1.00 A	10.0 Ω
R_5	25.0 V	5.00 A	5.00 Ω
R_6	30.0 V	5.00 A	6.00 Ω

Chapter 17 Review Questions Page 433

1. e **2.** a and e **3.** a **4.** c **5.** Current **6.** Ampere **7.** Volt
8. Ohm **9.** It decreases the resistance by a factor of 4.
10. The voltage drop across a segment of a circuit is equal to the product of the current through that segment and the resistance of that segment of the circuit.
11. Current flows sequentially through each portion of a series circuit. Current is divided among different segments of a parallel circuit.
12. The equivalent resistance in a series circuit is the sum of the resistance in the circuit. In a parallel circuit, the equivalent resistance is given by the reciprocal of the sum of the reciprocals of all resistances.
13. The highest range
14. Water flow is split and flows in parallel through different segments of a water distribution system. In a similar manner, current is divided and flows in parallel through resistors connected in parallel.
15. The current decreases by a factor of 2.
16. The current increases by a factor of 2.
17. The resistance would increase by a factor of 2.
18. Electrical charges move from regions of higher potential to regions of lower potential. Chemical reactions in batteries raise charges to higher potential energy. These charges can flow through a circuit and do work on circuit elements (create heat, light, or motion).
19. Chemical energy is transferred into electrical potential energy in the dry cell. Charges flow through the two lamps, giving up their energy in the form of light and heat.

Chapter 17 Review Problems Pages 433–435

1. 1.17 Ω **2.** 3.47 Ω **3.** 1.11 Ω **4.** 145 ft **5.** 0.0105 cm^2 **6.** 7.47 A
7. 25.1 Ω **8.** 0.491 A **9.** 19.60 Ω **10.** 0.612 A **11.** 48.5 V **12.** 16.2 Ω
13. 3.5 Ω **14.** 2.45 Ω **15.** 44.9 A **16.** 19.5 A **17.** 25.5 A **18.** 5.00 Ω
19. 10.0 A **20.** 5.95 A; 4.05 A **21.** 4.42 Ω **22.** 17.1 A **23.** 58.5 V **24.** 2.83 A
25. 12.3 V

26.

	V	*I*	*R*
Batt.	35.0 V	4.70 A	7.45 Ω
R_1	5.00 V	2.75 A	1.82 Ω
R_2	5.00 V	1.95 A	2.56 Ω
R_3	13.2 V	4.70 A	2.80 Ω
R_4	7.50 V	0.97 A	7.73 Ω
R_5	7.50 V	3.73 A	2.01 Ω
R_6	9.30 V	4.70 A	1.98 Ω

CHAPTER 18

18.3 Page 441

1. 1.49 V **3.** 1.33 A **5.** (a) 3.75 A (b) 9.00 V (c) 0.0200 Ω
7. 0.160 A **9.** 0.120 A

18.4 Pages 444–445

1. 957 W **3.** 0.682 A **5.** 6.82 A **7.** $0.56 **9.** Yes **11.** $0.026
13. $0.046 **15.** Oven, 2 fluorescent bulbs, 2 light bulbs; projection TV, personal computer, 2 lightbulbs.

Chapter 18 Review Questions Page 446

1. b **2.** c **3.** c **4.** b and c
5. Primary cells are not rechargeable, whereas secondary cells are.

6. An electric current flows in the reverse direction through a secondary cell, causing the normal chemical reaction to proceed in reverse, thus storing charge in the battery.
7. The electrolyte causes a chemical reaction at the plates and conducts current between the plates.
8. An electrolyte causes a chemical reaction at the plates that releases energy to force electrical charge to move through the battery and the outside circuit.
9. It decreases the voltage through the outside circuit. **10.** Watt
11. Power is given by the product of the voltage and current.
12. We pay for our energy use. Energy consumed is the total work done. Power is the instantaneous use of electricity; energy is the power times the time the power is used.
13. Power is given by the voltage squared divided by the resistance.
14. Power is given by the current squared times the resistance.
15. The power is increased by a factor of 2.
16. The power increases by a factor of 4.
17. The power decreases by a factor of 4. **18.** The cost increases by a factor of 2.

Chapter 18 Review Problems Pages 446–447

1. 1.43 V **2.** 12.2 V **3.** 3.95 A **4.** 6.00 V **5.** 0.0165 Ω **6.** 0.528 A
7. 6.67 Ω **8.** 39.3 W **9.** 1.36 A **10.** $1.01 **11.** 240 h **12.** 0.91 A

CHAPTER 19

19.2 Page 456

1. 1.20×10^{-5} T **3.** 57.5 A **5.** 1.71×10^{-6} T **7.** 0.0196 T **9.** 0.249 A

Chapter 19 Review Questions Page 465

1. d **2.** b **3.** b **4.** Tesla
5. A tightly wound solenoid with many turns per unit length carrying a large current will produce a high magnetic field.
6. Use Ampère's rule to find the direction of the flux line from any single turn in the solenoid. The magnetic field direction of the solenoid is in this direction.
7. The current-carrying coil produces a magnetic field that causes the magnetic domains in the magnetic material to align in the direction of the coil's field. This produces a stronger induced field in the core.
8. A moving magnet induces a current in the generator's coil.
9. The commutator is a split ring that allows the current produced by a generator always to flow in the same direction.
10. An induced magnetic field in the motor's electromagnet is repelled by the permanent magnet, causing the rotor to spin. The commutator in a dc motor allows the current through the electromagnet to change polarity, causing the rotor to continue spinning.
11. A synchronous motor rotates at a frequency that depends on the number of coils and the frequency of the ac power source. The motor has a number of poles along the stator that cause the rotor to spin at a fixed frequency.
12. A universal motor can operate on either dc or ac. The induction motor can operate only on ac.
13. The stator is a static magnet. The armature is an electromagnet that is free to rotate.
14. An electromagnet produces a strong magnetic field when a current is run through a solenoid. This magnetic field in turn induces a stronger magnetic field in a magnetic core.
15. The magnetic field increases by a factor of 2.
16. The field does not change as long as the length of the solenoid is much greater than the original diameter.
17. The magnetic field increases by a factor of 4.
18. The flux lines can be found by placing a small compass or magnetic filings near the magnet.

19. A spinning armature in the field of a stator crosses the magnetic flux lines of the stator, thereby inducing a current to flow in the armature coil. As the armature rotates and is reversed in the field, the current direction in the coil is reversed.

Chapter 19 Review Problems Page 465

1. 1.08×10^{-6} T **2.** 4.90×10^{-6} T **3.** 41.6 A **4.** 0.0253 T **5.** 12.4 A
6. 15.4°

CHAPTER 20

20.1 Page 469

1. 47.1 V **3.** 117 V **5.** 9.19 A **7.** 83.2 V **9.** 5.66 A **11.** 26.7°
13. 70.3°

20.2 Pages 470–471

1. 1590 V **3.** 117 V **5.** 12.0 A **7.** 12.0 A **9.** 9.87 A

20.3 Pages 472–473

1. 21.9 Ω **3.** $350\bar{0}$ W or 3.50 kW **5.** 1320 W or 1.32 kW **7.** 588 W
9. 2750 W or 2.75 kW

20.4 Page 479

1. 10.0 turns **3.** 58.5 V **5.** $30\bar{0}$ turns **7.** 6.00 A
9. 0.0587 A or 58.7 mA **11.** 2200 V **13.** 0.255 A

20.6 Page 482

1. 1.13 Ω **3.** $440\bar{0}$ Ω **5.** 40.1 kΩ **7.** 0.594 A **9.** 0.796 A
11. 0.116 A

20.7 Page 485

1. 215 Ω **3.** $330\bar{0}$ Ω **5.** 302 Ω **7.** 25.2° **9.** 72.3° **11.** 0.209 A
13. 4.55 mA **15.** 19.9 mA

20.8 Page 487

1. 7.96 Ω **3.** 2650 Ω or 2.65 kΩ **5.** 0.796 Ω **7.** 58.9 Ω
9. 1.61 Ω

20.9 Page 489

1. 1880 Ω **3.** $480\bar{0}$ Ω **5.** 3.18 Ω **7.** 4.85° **9.** 56.6° **11.** 53.2 mA
13. 3.13 mA **15.** 4.72 mA

20.10 Page 492

1. 42.4 Ω; 0.118 A **3.** 1180 Ω; 12.7 mA **5.** $44\bar{0}$ Ω; 13.6 mA
7. 206 Ω; 24.3 mA **9.** 529 Ω; 47.3 mA

20.11 Page 494

1. 79.6 kHz **3.** 38.0 kHz **5.** 3.86 kHz **7.** 4.20 kHz **9.** 7.96 kHz

20.13 Page 497

1. 10,700 kW **3.** 130,000 kVA **5.** 10,100 kW **7.** 19,000 kW
9. 410,000 kVA **11.** 0.880

Chapter 20 Review Questions Pages 499–500

1. d **2.** c **3.** a and e **4.** a, b, and c
5. The maximum current is the maximum instantaneous current. The effective value of an alternating current is the number of amperes that produce the same amount of heat in a resistance as an equal number of amperes of a steady direct current.
6. The maximum voltage is the maximum instantaneous voltage. The instantaneous voltage is the voltage at any instant.
7. Power is given by the product of the effective values of the voltage and current.
8. Power is given by the square of the effective value of the voltage divided by the resistance.
9. The output voltage doubles. **10.** Henry
11. Inductive reactance allows the analysis of circuits containing inductors.
12. They are directly proportional. **13.** It lags the voltage.
14. Energy is stored in the form of potential energy associated with a sheet of positive charge on one side of the capacitor and a sheet of negative charge on the other side.
15. It leads the voltage. **16.** They are inversely proportional.
17. Resonance occurs when the inductive reactance equals the capacitive reactance. The current is then given by its maximum value.
18. A diode allows current to flow in one direction but not in the reverse direction.
19. Amplification produces an increase in the value of a voltage or current in a circuit. Rectification produces a current or voltage in only one direction.
20. No, it depends on the frequency.

Chapter 20 Review Problems Pages 500–501

1. 110 V **2.** 120 V **3.** 68 V **4.** 95.4 V **5.** 8.95 A **6.** 727 W
7. 495 W **8.** 1310 W **9.** 1340 turns **10.** 0.255 A **11.** 523 W
12. (a) 3.92 A (b) 38.2° (c) 71.0 V **13.** (a) 2.29 A (b) 18.9° (c) 109 V
14. (a) 2.20 A (b) 67.1° (c) 42.9 V (d) 102 V **15.** (a) 1.52 A (b) 32.8 V (c) 72.6°
16. 194 Ω; 0.567 A **17.** 126 Ω; 0.873 A **18.** 30,200 Hz, 633 Ω, 0.174 A
19. 3.46×10^5 Hz, 565 Ω, 0.195 A **20.** 3.4×10^6 kVA

CHAPTER 21

21.2 Pages 506–507

1. 1.50×10^9 m **3.** 0.108 s **5.** $50\bar{0}$ s **7.** 16.1 ms **9.** 7.67×10^{-7} s
11. $25\bar{0}$ s; $12\bar{0}0$ s **13.** 1940 s

21.3 Pages 508–509

1. 6.59×10^{12} Hz **3.** 3.09×10^{-4} m **5.** 6.58 m **7.** 4.55×10^{-5} Hz
9. 214 m **11.** 3.51 MHz **13.** 7.5×10^{14} Hz **15.** 5.2×10^{14} Hz

21.4 Pages 509–510

1. 5.92×10^{-23} J **3.** 3.01×10^{-25} J **5.** 83.1 MHz **7.** 5.51×10^{10} Hz
9. 2.6×10^{-19} J **11.** 3.4×10^{-19} J

21.5 Pages 512–513

1. 603 ℓm **3.** 60.9 cd **5.** 942 ℓm **7.** 1.46 ft-cd **9.** 6.73 ft-cd
11. 4750 ℓm **13.** 608 ℓm **15.** (a) Decrease (b) $\frac{1}{9}$ **17.** 5.96 lux
19. 35.3 ℓm

Chapter 21 Review Questions Page 514

1. a, c, and e **2.** c **3.** d **4.** c
5. Yes; the wavelength varies inversely with the frequency.
6. The energy is directly proportional to the frequency.
7. The intensity falls off as 1 over the square of the distance.
8. The speed of light is measured by determining the time light takes to travel a measured distance and using the relationship $v = s/t$.
9. It always travels at the same velocity in a vacuum. It has a different velocity (slower) in any media.
10. Electromagnetic radiation **11.** Max Planck
12. Albert Einstein and Max Planck **13.** Olaus Roemer
14. By measuring the time difference for the start of the eclipse of the moons of Jupiter as viewed from different parts of earth's orbit **15.** Candela
16. Luminous intensity is the brightness of a light source.

Chapter 21 Review Problems Pages 514–515

1. 2.32×10^{13} m **2.** 3.52 mi **3.** 2.63×10^{-6} s or 2.63 μs **4.** 2.95×10^{-6} s or 2.95 μs **5.** 0.121 s **6.** 244 m **7.** 6.45×10^{6} Hz or 6.45 MHz
8. 5.54×10^{15} Hz **9.** 9.60×10^{-23} J **10.** 7.18×10^{10} Hz **11.** 5.46×10^{-18} J
12. 2150 ℓm **13.** 5230 ℓm **14.** 20,900 ℓm **15.** 3130 s; $17\bar{0}0$ s
16. 12.8 ℓm **17.** 7.29 lux

CHAPTER 22

22.5 Pages 525–526

1. 1.29 cm **3.** 32.2 cm **5.** 5.13 cm **7.** 6.24 cm **9.** 1.09 m **11.** 10.0 cm
13. −4.15 cm **15.** −3.56 m

22.10 Pages 535–536

1. 1.21 **3.** 2.00×10^{8} m/s **5.** 1.48 **7.** 21.8 cm **9.** −2.14 cm
11. (a) 5.00 cm; (b) 2.50 cm **13.** $s_i = -8.17$ cm, $h_i = 2.06$ cm **15.** 4.43 cm

Chapter 22 Review Questions Pages 537–538

1. a **2.** d **3.** a and b
4. Parallel light rays reflected off a rough surface may scatter in many different directions (diffusion) whereas parallel light rays reflected off a smooth surface will go off the surface as parallel rays.
5. At a smooth surface, the incident angle of a light ray is the same as the angle of the reflected ray.
6. At a smooth surface, the normal to the surface and both the incident and reflected rays lie in the same plane.
7. They are virtual, erect, and lie as far behind the mirror as the object is in front of the mirror.
8. A real image can be shown on a screen. A virtual image cannot.
9. Concave mirrors curve away from the observer; convex mirrors curve out toward the observer.

10. For large apertures, not all parallel rays are reflected through the focal point. This produces a fuzzy or aberrant image.
11. The image distance is increased. 12. The image distance is increased.
13. b 14. b
15. Converging lenses convert parallel rays into converging rays. Diverging lenses convert parallel rays into diverging rays.
16. Light propagating in an optical fiber, light reflected back into a swimming pool from an underwater light.
17. Light passing at an angle through an interface from a medium of low optical density to a medium of higher optical density is bent toward the normal to the interface.
18. The speed of light is lower in the high-index material.
19. A wave passing at an angle from one medium to another with a different wave velocity will be bent toward or away from the normal, depending on whether the speed is higher or lower in the new material.
20. Virtual 21. Real or virtual
22. They roughen the surface and scatter the light.
23. It appears shallower because of the refraction of light at the surface.
24. Light travels in a straight line unless it is reflected or refracted.
25. Light traveling at an angle greater than the critical angle is reflected from one side of the fiber to another as it propagates down the length of the fiber.
26. When the object is closer to the lens than the focal length.
27. When the object is located a distance from the lens equal to the focal length.
28. When the object is located outside the focal point.

Chapter 22 Review Problems Page 538

1. 2.36 cm **2.** −12.9 cm **3.** 3.48 cm **4.** 2.22 m **5.** (a) $25\bar{0}$ cm from mirror (b) 459 cm **6.** 22.7 cm **7.** 1.35 **8.** 2.08×10^8 m/s **9.** 1.40 **10.** 41.8° **11.** 18.0 cm **12.** (a) 7.50 cm (b) 2.70 cm **13.** 9.00 cm **14.** 6.25 cm **15.** −9.70 m **16.** −25.4 cm, 7.18 cm high

CHAPTER 23

23.2 Page 544

1. 3.8530×10^{-25} kg **3.** 28.4 MeV **5.** 1750 MeV **7.** 870 MeV

23.8 Pages 549–550

1. 3.77×10^{20} atoms **3.** 2.42×10^{-3} **5.** 30.4% **7.** 0.0428

23.11 Pages 552–553

1. 0.310 Ci **3.** 0.0875 Ci

23.12 Page 555

1. −13.595 eV **3.** 12.084 eV **5.** 6.57×10^{-7} m

23.13 Page 557

1. Neon **3.** Fluorine

23.14 Page 559

1. Bromine **3.** Argon

Chapter 23 Review Questions Pages 560–561

1. a, b, and c **2.** c **3.** c

4. Protons and neutrons are both nucleons that exert the attractive strong force on other nucleons. They both have similar masses, although the neutron is slightly more massive. The proton is a stable particle as an individual nucleon. The neutron, however, is unstable by itself. The proton has a positive charge, equal in magnitude to that of the electron. The neutron is uncharged.
5. The electrical force can be either attractive or repulsive and is exerted between charged particles. The strong force is always attractive and is exerted between nucleons, whether charged or not. The strong force is a very short-range force. The electrical force is exerted over larger distances.
6. The nucleus would expand in size due to the repulsive electrical force between protons. The nucleus might even break apart.
7. All three atoms have 6 protons in the nucleus and 6 electrons in the orbital shells. They all exhibit the chemical properties of carbon. They each have a different number of neutrons in the nucleus and therefore have a different mass.
8. Mass and energy are equivalent forms. Mass can be changed into energy under the proper conditions and vice versa.
9. An electron volt is the energy that is gained or lost by an electron in passing through a potential difference of 1 volt.
10. The neutron is the "glue" that binds the positively charged protons together in the nucleus. Without neutrons, the positively charged protons would be repelled by their similar electrical charge.
11. An alpha ray is composed of particles that have a double positive charge and are composed of two protons and two neutrons. The α particles are identical to the nucleus of a helium atom.
12. A γ ray is composed of very energetic photons of electromagnetic radiation.
13. A β ray is composed of β particles (electrons).
14. The formation of oxygen and a proton from the bombardment of nitrogen with α particles. It was observed by Ernest Rutherford in 1919.
15. He discovered that the neutron bombardment of uranium can result in the formation of lighter nuclei that are approximately one-half the mass of uranium.
16. A self-sustaining chain reaction is a nuclear reaction in which a sufficient number of neutrons are produced that will cause subsequent nuclear reactions to continue at a fixed rate.
17. Nuclear fusion reactions are reactions in which nuclei bombard each other, producing heavier nuclei. The original nuclei "fuse" together.
18. Nuclear fission reactions are reactions in which nuclei are bombarded by nuclear particles such as neutrons or α particles, causing the original nuclei to split apart.
19. The decay rate is the rate at which nuclei decay per time interval. **20.** 6.25%
21. Molecules in plant or animal cells can be damaged as a result of changes in the chemicals brought about by nuclear reactions produced by the radiation. Some of these damaged molecules may be in the genetic material, which might cause cells produced by the division of this cell to be defective.
22. It is used for diagnostic purposes by the injection or ingestion of radioactive tracer materials that may allow the identification of cancerous or defective cells. It can also be used to treat cancer by bombardment of the cancer cells, therefore destroying the cancerous material.
23. It appears in the form of radiation (γ rays) and the kinetic energy of α rays, β rays, and nuclei formed in the reaction. This kinetic energy can quickly be changed into thermal energy.
24. In the Bohr theory of the atom, the energy levels of the electrons are restricted to certain values; that is, the energy is quantized. The energy levels in a hydrogen atom are given by the equation $E = -kZ^2/n^2$.
25. The ground state is the lowest energy level for the electron in the atom. The excited states are the higher energy levels, which are unstable. An electron in an excited state will in time decay through lower-energy-level excited states to the ground state. The ground state has the smallest "orbital distance" from the nucleus.

Chapter 23 Review Problems Pages 561–562

1. 2.4913×10^{-26} kg **2.** 3.1547×10^{-26} kg **3.** 2.7553×10^{-25} kg **4.** 749 MeV **5.** 870 MeV **6.** 830 MeV **7.** 810 MeV **8.** 7.26×10^{12} atoms **9.** 4.29×10^{17} atoms **10.** 1.80×10^{-3} **11.** 3.13×10^{-4} **12.** 0.0238 **13.** 5.6% **14.** 0.37% **15.** 0.346 Ci **16.** 0.193 Ci **17.** 0.367 Ci **18.** 0.231 **19.** 1.54×10^{-6} **20.** − 0.5438 eV **21.** 0.967 eV **22.** He, Ne, Ar, Kr, Xe, Rn **23.** Cu, Ag, Au, Fe, Ti, Cr, Mn, etc.

APPENDIX A

A.1 Page 566

1. −11 **3.** 5 **5.** −2 **7.** −4 **9.** −6 **11.** −3 **13.** 18 **15.** −21 **17.** 0 **19.** 3 **21.** −8 **23.** 0 **25.** 11 **27.** −13 **29.** −5 **31.** −8 **33.** −4 **35.** 10 **37.** −45 **39.** −288

A.2 Page 568

1. 10^8 **3.** 10^8 **5.** $\frac{1}{10^2}$ **7.** 10^7 **9.** 10^3 **11.** $\frac{1}{10^4}$ **13.** 1 **15.** 10^5 **17.** $\frac{1}{10^{11}}$ **19.** $\frac{1}{10^6}$ **21.** 10^8 **23.** 10^{10}

A.3 Page 572

1. $\frac{4}{3}$ **3.** 17 **5.** 0 **7.** 17.5 **9.** 3 **11.** 4 **13.** 7 **15.** 12.5 **17.** 5 **19.** 26.5 **21.** 9 **23.** 3 **25.** 8 **27.** 1 **29.** $-\frac{11}{3}$ or $-3\frac{2}{3}$ **31.** 3 **33.** 2 **25.** 21 **37.** $\frac{29}{2}$ or $14\frac{1}{2}$ **39.** −6

A.4 Page 575

1. ±6 **3.** ±7 **5.** ±3 **7.** ±0.8 **9.** ±4.67 **11.** ±0.892 **13.** $a = 3$; $b = 1$; $c = -5$ **15.** $a = 6$; $b = 8$; $c = 2$ **17.** $a = 9$; $b = 6$; $c = -4$ **19.** $a = 5$; $b = 6$; $c = 0$ **21.** $a = 9$; $b = 0$; $c = -64$ **23.** 3; 7 **25.** $\frac{4}{3}$; $-\frac{5}{2}$ **27.** 1.95; −1.62 **29.** 1.69; −0.855 **31.** 0.725; −1.23

APPENDIX B

B.6 Pages 585–586

1. 3.32×10^{19} **3.** -6.83×10^{-6} **5.** 8.35×10^{-11} **7.** -7.98×10^{19} **9.** −68.5 **11.** 4680 **13.** 216 **15.** 2.10×10^{10} **17.** 2.15×10^9 **19.** 10.7 **21.** 33.4 **23.** 591 **25.** 609,000 **27.** 0.567 **29.** 1.77×10^4 **31.** 0.523 **33.** 0.225 **35.** 1.29 **37.** 0.510 **39.** 0.743 **41.** 3.73 **43.** 0.151 **45.** 40.7° **47.** 24.6° **49.** 56.2° **51.** 8.4° **53.** 63.3° **55.** 66.6° **57.** 252 m **59.** 89.4 m **61.** 53.7° **63.** 14.8 cm **65.** 21.9 m **67.** 20,700 **69.** 0.0206 **71.** 1.86×10^{-6}

INDEX

FORMULAS FROM GEOMETRY

Plane figures

In the following, a, b, c, d, and h are lengths of sides and altitudes, respectively.

		Perimeter	*Area*
Rectangle		$P = 2(a + b)$	$A = ab$
Square	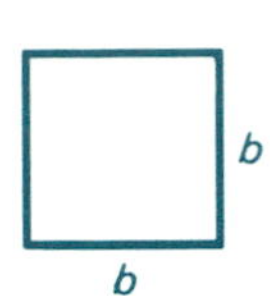	$P = 4b$	$A = b^2$
Parallelogram	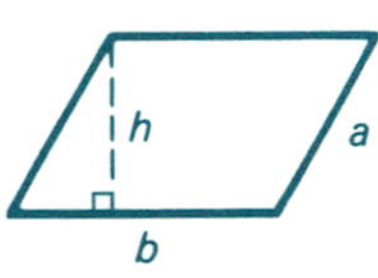	$P = 2(a + b)$	$A = bh$
Rhombus	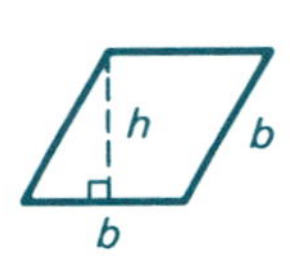	$P = 4b$	$A = bh$
Trapezoid	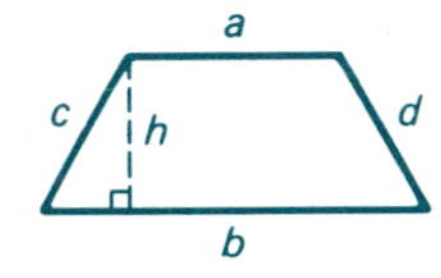	$P = a + b + c + d$	$A = \left(\frac{a + b}{2}\right)h$
Triangle	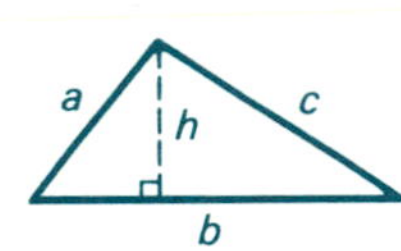	$P = a + b + c$	$A = \frac{1}{2}bh$

The sum of the measures of the angles of a triangle = 180°.

In a right triangle:

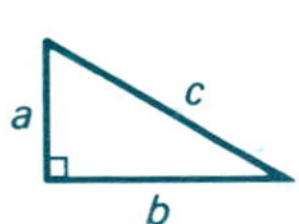

$$c^2 = a^2 + b^2 \text{ or } c = \sqrt{a^2 + b^2}$$

		Circumference	*Area*
Circle		$C = \pi d$ $C = 2\pi r$	$A = \pi r^2 \quad d = 2r$ $A = \frac{\pi d^2}{4}$

The sum of the measures of the central angles of a circle = 360°.